W9-CMK-470

LTE – The UMTS Long Term Evolution

LTE – The UMTS Long Term Evolution

From Theory to Practice

Second Edition

Stefania Sesia

ST-Ericsson, France

Issam Toufik

ETSI, France

Matthew Baker

Alcatel-Lucent, UK

A John Wiley & Sons, Ltd., Publication

This edition first published 2011

Registered office
John Wiley & Sons Ltd, The Atrium, Southern Gate, Chichester, West Sussex, PO19 8SQ, United Kingdom

For details of our global editorial offices, for customer services and for information about how to apply for permission to reuse the copyright material in this book please see our website at www.wiley.com.

Library of Congress Cataloging-in-Publication Data

Sesia, Stefania.
LTE–the UMTS long term evolution : from theory to practice / Stefania Sesia, Issam Toufik, Matthew Baker.– 2nd ed.
p. cm.
Includes bibliographical references and index.
ISBN 978-0-470-66025-6 (hardback)
1. Universal Mobile Telecommunications System. 2. Long-Term Evolution (Telecommunications) I. Toufik, Issam. II. Baker, Matthew (Matthew P.J.) III. Title.
TK5103.4883.S47 2011
621.3845′6–dc22
2010039466

A catalogue record for this book is available from the British Library.

Print ISBN: 9780470660256 (H/B)
ePDF ISBN: 9780470978511
oBook ISBN: 9780470978504
epub ISBN: 9780470978641

Printed in Great Britain by CPI Antony Rowe, Chippenham, Wiltshire.

Dedication

To my family.
Stefania Sesia

To my parents for their sacrifices and unconditional love. To my brother and sisters for their love and continual support. To my friends for being what they are.
Issam Toufik

To the glory of God, who 'so loved the world that He gave His only Son, that whoever believes in Him shall not perish but have eternal life'. — The Bible.
Matthew Baker

Chih-Pang Ko

Contents

Part I Network Architecture and Protocols 23

Editors' Biographies

Matthew Baker holds degrees in Engineering and Electrical and Information Sciences from the University of Cambridge. From 1996 to 2009 he worked at Philips Research where he conducted leading-edge research into a variety of wireless communication systems and techniques, including propagation modelling, DECT, Hiperlan and UMTS, as well as leading the Philips RAN standardization team. He has been actively participating in the standardization of both UMTS WCDMA and LTE in 3GPP since 1999, where he has been active in 3GPP TSG RAN Working Groups 1, 2, 4 and 5, contributing several hundred proposals. He now works for Alcatel-Lucent, which he joined in 2009, and he has been Chairman of 3GPP TSG RAN Working Group 1 since being elected to the post in August of that year. He is the author of several international conference papers and inventor of numerous patents. He is a Chartered Engineer, a Member of the Institution of Engineering and Technology and a Visiting Lecturer at the University of Reading, UK.

Stefania Sesia received her Ph.D. degree in Communication Systems and Coding Theory from both Eurecom (Sophia Antipolis, France) and ENST-Paris (Paris, France) in 2005. From 2002 to 2005 she worked at Motorola Research Labs, Paris, towards her Ph.D. thesis. In June 2005 she joined Philips/NXP Semiconductors (now ST-Ericsson) Research and Development Centre in Sophia Antipolis, France where she was technical leader and responsible for the High Speed Downlink Packet Access algorithm development. She has been participating in 3GPP TSG RAN Working Groups 1 and 4 standardization meetings. From 2007 to 2009 she was on secondment from NXP Semiconductors to the European Telecommunications Standard Institute (ETSI) acting as 3GPP TSG RAN and 3GPP TSG RAN Working Group 4 Technical Officer. She is currently back in ST-Ericsson as senior research and development engineer, actively participating in 3GPP TSG RAN Working Group 4 as a delegate. She is the author of several international IEEE conference and journal papers and many contributions to 3GPP, and inventor of numerous US and European patents.

Issam Toufik graduated in Telecommunications Engineering (majoring in Mobile Communication Systems) in 2002 from both ENST-Bretagne (Brest, France) and Eurecom (Sophia Antipolis, France). In 2006, he received his Ph.D. degree in Communication Systems from Eurecom/ENST-Paris, France. From June to August 2005 he worked for Samsung Advanced Institute of Technology (SAIT), South Korea, as a Research Engineer on LTE. In January 2007, he joined NXP Semiconductors/ST-Ericsson, Sophia Antipolis, France, as a Research and Development Engineer for UMTS and LTE algorithm development. In November 2009, he joined the European Telecommunications Standard Institute (ETSI) acting as 3GPP TSG

RAN and 3GPP TSG RAN Working Group 4 Technical Officer. He is the author of several international IEEE conference and journal papers and contributions to 3GPP, and inventor of numerous patents.

List of Contributors

Abe, Tetsushi, NTT DOCOMO
e-mail: abetet@nttdocomo.com

Ancora, Andrea, ST-Ericsson
e-mail: andrea.ancora@stericsson.com

Anderson, Nicholas, Research In Motion
e-mail: nianderson@rim.com

Baker, Matthew, Alcatel-Lucent
e-mail: matthew.baker@alcatel-lucent.com, m.p.j.baker.92@cantab.net

Bertrand, Pierre, Texas Instruments
e-mail: p-bertrand@ti.com

Bhattad, Kapil, Qualcomm
e-mail: kbhattad@qualcomm.com

Bury, Andreas, Blue Wonder Communications
e-mail: andreas.bury@bluwo.com

Chun, SungDuck,LG Electronics
e-mail: duckychun@lge.com

Classon, Brian, Huawei
e-mail: brian.classon@huawei.com

Damnjanovic, Jelena, Qualcomm
e-mail: jelenad@qualcomm.com

Farajidana, Amir, Qualcomm
e-mail: amirf@qualcomm.com

Fischer, Patrick, Bouygues Telecom
e-mail: pfischer@bouyguestelecom.fr

Geirhofer, Stefan, Qualcomm
e-mail: sgeirhofer@qualcomm.com

Gerstenberger, Dirk, Ericsson
e-mail: dirk.gerstenberger@ericsson.com

Gesbert, David, Eurecom
e-mail: david.gesbert@eurecom.fr

Godin, Philippe, Alcatel-Lucent
e-mail: philippe.godin@alcatel-lucent.com

Golitschek, Alexander, Panasonic
e-mail: alexander.golitschek@eu.panasonic.com

Gonsa, Osvaldo, Panasonic
e-mail: osvaldo.gonsa@eu.panasonic.com

Gorokhov, Alex, Qualcomm
e-mail: gorokhov@qualcomm.com

Hardouin, Eric, Orange Labs
e-mail: eric.hardouin@orange-ftgroup.com

Hu, Teck, Alcatel-Lucent
e-mail: teck.hu@alcatel-lucent.com

Hus, Olivier,
e-mail: olivierjhus@gmail.com

Jämsä, Tommi, Elektrobit
e-mail: tommi.jamsa@elektrobit.com

Jiang, Jing, Texas Instruments
e-mail: jing.jiang@ti.com

Kaltenberger, Florian, Eurecom
e-mail: florian.kaltenberger@eurecom.fr

Kazmi, Muhammad, Ericsson
e-mail: muhammad.kazmi@ericsson.com

Knopp, Raymond, Eurecom
e-mail: raymond.knopp@eurecom.fr

Laneman, J. Nicholas, University of Notre Dame
e-mail: jnl@nd.edu

Lee, YoungDae, LG Electronics
e-mail: leego@lge.com

Love, Robert, Motorola Mobility
e-mail: robert.love@motorola.com

Luo, Xiliang, Qualcomm
e-mail: xluo@qualcomm.com

Montojo, Juan, Qualcomm
e-mail: juanm@qualcomm.com

Moulsley, Tim, Fujitsu
e-mail: t.moulsley@btopenworld.com

Nakamura, Takaharu, Fujitsu
e-mail: n.takaharu@jp.fujitsu.com

Nakamura, Takehiro, NTT DOCOMO
e-mail: nakamurata@nttdocomo.co.jp

Nangia, Vijay, Motorola Mobility
e-mail: vijay.nangia@motorola.com

Nimbalker, Ajit, Motorola Mobility
e-mail: aijt.nimbalker@motorola.com

Palat, K. Sudeep, Alcatel-Lucent
e-mail: spalat@alcatel-lucent.com

Payne, Adrian, ERA Technology
e-mail: adrian.w.payne@gmail.com

Ranta-aho, Karri, Nokia Siemens Networks
e-mail: karri.ranta-aho@nsn.com

Rumney, Moray, Agilent
e-mail: moray_rumney@agilent.com

Sälzer, Thomas, Huawei
e-mail: thomas.salzer@huawei.com, thomas.salzer@gmx.de

Sayers, Tony, Ultra Electronics
e-mail: tony.sayers@talktalk.net

Sesia, Stefania, ST-Ericsson
e-mail: stefania.sesia@stericsson.com

Shen, Zukang, CATT
e-mail: shenzukang@catt.cn

Slock, Dirk, Eurecom
e-mail: dirk.slock@eurecom.fr

Suzuki, Hidetoshi, Panasonic
e-mail: Suzuki.Hidetoshi@jp.panasonic.com

Tomatis, Fabrizio, ST-Ericsson
e-mail: fabrizio.tomatis@stericsson.com

Tosato, Filippo, Toshiba
e-mail: filippo.tosato@toshiba-trel.com

Toufik, Issam, ETSI
e-mail: issam.toufik@etsi.org, issam.toufik@eurecom.fr

van der Velde, Himke, Samsung
e-mail: himke.vandervelde@samsung.com

van Rensburg, Cornelius, Huawei
e-mail: cdvanren@ieee.org

Whinnett, Nick, Picochip
e-mail: nickw@picochip.com
Yi, SeungJune, LG Electronics
e-mail: seungjune@lge.com

Ylitalo, Juha, Elektrobit
e-mail: juha.ylitalo@elektrobit.com

Foreword

GSM, and its evolution through GPRS, EDGE, WCDMA and HSPA, is the technology stream of choice for the vast majority of the world's mobile operators. Users have experienced increasing data rates, together with a dramatic reduction in telecommunications charges; they now expect to pay less but receive more. Therefore, in deciding the next steps, there must be a dual approach: seeking considerable performance improvement but at reduced cost. Improved performance must be delivered through systems which are cheaper to install and maintain. LTE and LTE-Advanced represent these next steps and will be the basis on which future mobile telecommunications systems will be built.

Many articles have already been published on the subject of LTE, varying from doctoral theses to network operator analyses and manufacturers' product literature. By their very nature, those publications have viewed the subject from one particular perspective, be it academic, operational or promotional. A very different approach is taken with this book. The authors come from a number of different spheres within the mobile telecommunications ecosystem and collectively bring a refreshing variety of perspectives. What binds the authors together is a thorough knowledge of the subject material which they have derived from their long experience within the standards-setting environment, the 3rd Generation Partnership Project (3GPP). LTE discussions started within 3GPP in 2004, so it is not a particularly new subject. In order to fully appreciate the thinking that conceived this technology, however, it is necessary to have followed the subject from the very beginning and to have witnessed the discussions that took place from the outset. Moreover, it is important to understand the thread that links academia, through research to standardization since it is widely acknowledged that by this route impossible dreams become market realities. Considerable research work has taken place to prove the viability of the technical basis on which LTE is founded and it is essential to draw on that research if any attempt is made to explain LTE to a wider audience. The authors of this book have not only followed the LTE story from the beginning but many have also been active players in WCDMA and its predecessors, in which LTE has its roots.

This book provides a thorough, authoritative and complete tutorial of the LTE system, now fully updated and extended to include LTE-Advanced. It gives a detailed explanation of the advances made in our theoretical understanding and the practical techniques that will ensure the success of this ground-breaking new radio access technology. Where this book is exceptional is that the reader will learn not just how LTE works but why it works.

I am confident that this book will earn its rightful place on the desk of anyone who needs a thorough understanding of the LTE and LTE-Advanced technology, the basis of the world's mobile telecommunications systems for the next decade.

Adrian Scrase, ETSI Vice-President,
International Partnership Projects

Preface to the Second Edition

> Research workers and engineers toil unceasingly on the development of wireless telegraphy. Where this development can lead, we know not. However, with the results already achieved, telegraphy over wires has been extended by this invention in the most fortunate way. Independent of fixed conductor routes and independent of space, we can produce connections between far-distant places, over far-reaching waters and deserts. This is the magnificent practical invention which has flowered upon one of the most brilliant scientific discoveries of our time!

These words accompanied the presentation of the Nobel Prize for Physics to Guglielmo Marconi in December 1909.

Marconi's success was the practical and commercial realization of wireless telegraphy – the art of sending messages without wires – thus exploiting for the first time the amazing capability for wireless communication built into our universe. While others worked on wireless telephony – the transmission of audio signals for voice communication – Marconi interestingly saw no need for this. He believed that the transmission of short text messages was entirely sufficient for keeping in touch.

One could be forgiven for thinking that the explosion of wireless voice communication in the intervening years has proved Marconi wrong; but the resurgence of wireless data transmission at the close of the twentieth century, beginning with the mobile text messaging phenomenon, or 'SMS', reveals in part the depth of insight Marconi possessed.

Nearly 100 years after Marconi received his Nobel prize, the involvement of thousands of engineers around the world in major standardization initiatives such as the 3rd Generation Partnership Project (3GPP) is evidence that the same unceasing toil of research workers and engineers continues apace.

While the first mobile communications standards focused primarily on voice communication, the emphasis now has returned to the provision of systems optimized for data. This trend began with the 3rd Generation Wideband Code Division Multiple Access (WCDMA) system designed in the 3GPP, and is now reaching fulfilment in its successor, the Long-Term Evolution (LTE). LTE was the first cellular communication system optimized from the outset to support packet-switched data services, within which packetized voice communications are just one part. Thus LTE can truly be said to be the heir to Marconi's heritage – the system, unknown indeed to the luminaries of his day, to which his developments have led.

LTE is an enabler. It is not technology for technology's sake, but technology with a purpose, connecting people and information to enable greater things to be achieved. It is already providing higher data rates than ever previously achieved in mobile communications, combined with wide-area coverage and seamless support for mobility without regard for the type of data being transmitted. To provide this level of functionality and flexibility, it is inevitable that the complexities of the LTE system have far surpassed anything Marconi could have imagined.

One aim of this book, therefore, is to chart an explanatory course through the LTE specifications, to support those who design LTE equipment.

The LTE specification documents themselves do not tell the whole story. Essentially they are a record of decisions taken – decisions which are often compromises between performance and cost, theoretical possibility and practical constraints. We aim therefore to give the reader a detailed insight into the evaluations and trade-offs which lie behind the technology choices inherent in LTE. The specifications also continue to develop, as new releases are produced, and this Second Edition is therefore fully updated to cover Release 9 and the first release of LTE-Advanced, Release 10.

Since the first version of LTE was developed, the theoretical understanding which gave rise to LTE has continued to advance, as the 'unceasing toil' of thousands of engineers continues with the aim of keeping pace with the explosive growth of mobile data traffic. Where the first version of LTE exploited Multiple-Input Multiple-Output (MIMO) antenna techniques to deliver high data rates, the evolution of LTE towards LTE-Advanced extends such techniques further for both downlink and uplink communication, together with support for yet wider bandwidths; meanwhile, heterogeneous (or hierarchical) networks, relaying and Coordinated MultiPoint (CoMP) transmission and reception start to become relevant in LTE-Advanced.

It is particularly these advances in underlying scientific understanding which this book seeks to highlight.

In selecting the technologies to include in LTE and LTE-Advanced, an important consideration is the trade-off between practical benefit and cost of implementation. Fundamental to this assessment is ongoing enhancement in understanding of the radio propagation environment and scenarios of relevance to deployments of LTE and LTE-Advanced. This has been built on significant advances in radio-channel modelling.

The advances in techniques and theoretical understanding continue to be supported by developments in integrated circuit technology and signal processing power which render them feasible where they would have been unthinkable only a few years ago.

Changes in spectrum availability and regulation also influence the development path of LTE towards LTE-Advanced, reinforcing the need for the new technology to be adaptable, capable of being scaled and enhanced to meet new global requirements and deployed in a wide range of different configurations.

With this breadth and depth in mind, the authorship of the chapters of the second edition of this book is even wider than that of the first edition, and again is drawn from all fields of the ecosystem of research and development that has underpinned the design of LTE. They work in the 3GPP standardization itself, in the R&D departments of companies active in LTE, for network operators as well as equipment manufacturers, in universities and in other collaborative research projects. They are uniquely placed to share their insights from the full range of perspectives.

To borrow Marconi's words, where LTE and LTE-Advanced will lead, we know not; but we can be sure that these will not be the last developments in wireless telegraphy.

Matthew Baker, *Stefania Sesia* and *Issam Toufik*

Acknowledgements

Like the first edition, the fully updated and expanded second edition of this book is first and foremost the fruit of a significant team effort, which would not have been successful without the expertise and professionalism displayed by all the contributors, as well as the support of their companies. The dedication of all the co-authors to their task, their patience and flexibility in allowing us to modify and move certain parts of their material for harmonization purposes, are hereby gratefully acknowledged. Particular thanks are due to ST-Ericsson, Alcatel-Lucent and ETSI for giving us the encouragement and working environment to facilitate such a time-consuming project. The help provided by ETSI, 3GPP and others in authorizing us to reproduce certain copyrighted material is also gratefully acknowledged. We would like to express our gratitude to the many experts who kindly provided advice, feedback, reviews and other valuable assistance. We believe their input in all its forms has made this book a more accurate, valuable and even enjoyable resource. These experts include Jacques Achard, Kevin Baum, Martin Beale, Keith Blankenship, Yufei Blankenship, Federico Boccardi, Kevin Boyle, Sarah Boumendil, Alec Brusilovsky, Paul Bucknell, Richard Burbidge, Aaron Byman, Emilio Calvanese Strinati, Choo Chiap Chiau, Anand Dabak, Peter Darwood, Merouane Debbah, Vip Desai, Marko Falck, Antonella Faniuolo, Jeremy Gosteau, Lajos Hanzo, Lassi Hentilä, Shin Horng Wong, Paul Howard, Howard Huang, Alan Jones, Yoshihisa Kishiyama, Achilles Kogiantis, Pekka Kyösti, Daniel Larsson, Jung-Ah Lee, Thierry Lestable, Gert-Jan van Lieshout, Andrew Lillie, Matti Limingoja, Huiheng Mai, Caroline Mathieson, Darren McNamara, Juha Meinilä, Tarik Muharemovic, Gunnar Nitsche, Jukka-Pekka Nuutinen, SungJun Park, Roope Parviainen, Paul Piggin, Claudio Rey, Safouane Sfar, Ken Stewart, Miloš Tesanovic, Paolo Toccacelli, Ludo Tolhuizen, Li Wang, Tim Wilkinson and Steve Zhang.

We would also like to acknowledge the efforts of all participants in 3GPP who, through innumerable contributions and intense discussions often late into the night, facilitated the completion of the LTE specifications for Releases 8, 9 and 10 in such a short space of time.

We would especially like to thank the publishing team at John Wiley & Sons, especially Tiina Ruonamaa, Susan Barclay, Jasmine Chang, Mariam Cheok, Sheena Deuchars, Caitlin Flint, Sarah Hinton, Anna Smart and Sarah Tilley for their professionalism and extensive support and encouragement throughout the preparation of both the first and second editions of this book.

Finally, it should be noted that this book is intended only as a guide to LTE and LTE-Advanced, and the reader should refer to the specifications published by 3GPP for definitive information. Any views expressed in this book are those of the authors and do not necessarily reflect the views of their companies. The editors welcome any suggestions to improve future editions of this book.

The Editors

Acknowledgements

List of Acronyms

* An asterisk indicates that the acronym can have different meanings depending on the context. The meaning is clearly indicated in the text when used.

3GPP 3rd Generation Partnership Project

3GPP2 3rd Generation Partnership Project 2

ABS Almost Blank Subframe

AC Access Class

ACI Adjacent Channel Interference

ACIR Adjacent Channel Interference Ratio

ACK Acknowledgement

ACLR Adjacent Channel Leakage Ratio

ACS Adjacent Channel Selectivity

ADC Analogue to Digital Converter

ADSL Asymmetric Digital Subscriber Line

AGI Antenna Gain Imbalance

A-GNSS Assisted Global Navigation Satellite System

AM Acknowledged Mode

AMC Adaptive Modulation and Coding

AMPS Analogue Mobile Phone System

AMR Adaptive MultiRate

ANR Automatic Neighbour Relation

ANRF Automatic Neighbour Relation Function

AoA Angle-of-Arrival

AoD Angle-of-Departure

APN Access Point Name

APP A-Posteriori Probability

ARFCN Absolute Radio Frequency Channel Number

ARIB Association of Radio Industries and Businesses

ARP Almost Regular Permutation*

ARP Allocation and Retention Priority*

ARQ Automatic Repeat reQuest

AS Access Stratum*

AS Angular Spread*

A-SEM Additional SEM

ATDMA Advanced TDMA

ATIS Alliance for Telecommunications Industry Solutions

AuC Authentication Centre

AWGN Additive White Gaussian Noise

BCC Base station Colour Code

BCH Broadcast CHannel

BCCH Broadcast Control CHannel

BCJR Algorithm named after its inventors, Bahl, Cocke, Jelinek and Raviv

BER Bit Error Rate

BLER BLock Error Rate

BM-SC Broadcast-Multicast Service Centre

BP Belief Propagation

BPRE Bits Per Resource Element

bps bits per second

BPSK Binary Phase Shift Keying

BSIC Base Station Identification Code

BSR Buffer Status Reports

CAPEX CAPital EXpenditure

CAZAC Constant Amplitude Zero AutoCorrelation

CB Circular Buffer

CBF Coordinated Beamforming

CC Component Carrier

CCCH Common Control CHannel

CCE Control Channel Element

CCI Co-Channel Interference

CCO Cell Change Order

CCSA China Communications Standards Association

CDD Cyclic Delay Diversity

CDF Cumulative Distribution Function

CDL Clustered Delay Line

CDM Code Division Multiplex(ed/ing)

CDMA Code Division Multiple Access

C/I Carrier-to-Interference ratio

CID Cell ID

CIF Carrier Indicator Field

CF Contention-Free

CFI Control Format Indicator

CFO Carrier Frequency Offset

CINR Carrier-to-Interference-and-Noise Ratio

CIR Channel Impulse Response

CM Cubic Metric

CMAS Commercial Mobile Alert Service

CMHH Constant Modulus HouseHolder

CN Core Network

CoMP Coordinated MultiPoint

CODIT UMTS Code DIvision Testbed

COFDM Coded OFDM

CP Cyclic Prefix

CPICH Common PIlot CHannel

CPR Common Phase Rotation

CPT Control PDU Type

CQI Channel Quality Indicator

CRC Cyclic Redundancy Check

CRE Cell Range Expansion

C-RNTI Cell Radio Network Temporary Identifier

CRS Common Reference Signal

CS Circuit-Switched*

CS Cyclic Shift*

CSA Common Subframe Allocation

CSG Closed Subscriber Group

CSI Channel State Information

CSI-RS Channel State Information RS

CSIT Channel State Information at the Transmitter

CTF Channel Transfer Function

CVA Circular Viterbi Algorithm

CVQ Channel Vector Quantization

CW Continuous-Wave

DAB Digital Audio Broadcasting

DAC Digital to Analogue Converter

DAI Downlink Assignment Index

dB deci-Bel

d.c. direct current

DCCH Dedicated Control CHannel

DCFB Direct Channel FeedBack

DCI Downlink Control Information

DFT Discrete Fourier Transform

DFT-S-OFDM DFT-Spread OFDM

Diffserv Differentiated Services

DL DownLink

DL-SCH DownLink Shared CHannel

DMB Digital Mobile Broadcasting

DM-RS DeModulation-RS

DOA Direction Of Arrival

DPC Dirty-Paper Coding

DRB Data Radio Bearer

DRX Discontinuous Reception

DS-CDMA Direct-Sequence Code Division Multiple Access

DSP Digital Signal Processor

DTCH Dedicated Traffic CHannel

DTX Discontinuous Transmission

DVB-H Digital Video Broadcasting – Handheld

DVB-T Digital Video Broadcasting – Terrestrial

DwPTS Downlink Pilot TimeSlot

ECGI E-UTRAN Cell Global Identifier

ECM EPS Connection Management

EDGE Enhanced Data rates for GSM Evolution

EESM Exponential Effective SINR Mapping

eICIC enhanced Inter-Cell Interference Coordination

EMEA Europe, Middle East and Africa

EMM EPS Mobility Management

eNodeB evolved NodeB

EPA Extended Pedestrian A

EPC Evolved Packet Core

EPG Electronic Programme Guide

ePHR extended Power Headroom Report

EPS Evolved Packet System

E-RAB E-UTRAN Radio Access Bearer

E-SMLC Evolved Serving Mobile Location Centre

ESP Encapsulating Security Payload

ETSI European Telecommunications Standards Institute

ETU Extended Typical Urban

ETWS Earthquake and Tsunami Warning System

E-UTRA Evolved-UTRA

E-UTRAN Evolved-UTRAN

EVA Extended Vehicular A

EVM Error Vector Magnitude

FACH Forward Access CHannel

FB Frequency Burst

FCC Federal Communications Commission

FCCH Frequency Control CHannel

FDD Frequency Division Duplex

FDE Frequency-Domain Equalizer

FDM Frequency Division Multiplexing

FDMA Frequency Division Multiple Access

FDSS Frequency-Domain Spectral Shaping

FFT Fast Fourier Transform

FI Framing Info

FIR Finite Impulse Response

FMS First Missing SDU

FSTD Frequency Switched Transmit Diversity

FTP File Transfer Protocol

FTTH Fibre-To-The-Home

GBR Guaranteed Bit Rate

GCL Generalized Chirp-Like

GERAN GSM EDGE Radio Access Network

GGSN Gateway GPRS Support Node

GMSK Gaussian Minimum-Shift Keying

GNSS Global Navigation Satellite System

GPRS General Packet Radio Service

GPS Global Positioning System

GSM Global System for Mobile communications

GT Guard Time

GTP GPRS Tunnelling Protocol

GTP-U GTP-User plane

HARQ Hybrid Automatic Repeat reQuest

HD-FDD Half-Duplex FDD

HeNB Home eNodeB

HFN Hyper Frame Number

HII High Interference Indicator

HLR Home Location Register

HRPD High Rate Packet Data

HSDPA High Speed Downlink Packet Access

HSPA High Speed Packet Access

HSPA+ High Speed Packet Access Evolution

HSS Home Subscriber Server

HSUPA High Speed Uplink Packet Access

HTTP HyperText Transfer Protocol

ICI Inter-Carrier Interference

ICIC Inter-Cell Interference Coordination

IDFT Inverse Discrete Fourier Transform

IETF Internet Engineering Task Force

IFDMA Interleaved Frequency Division Multiple Access

IFFT Inverse Fast Fourier Transform

i.i.d. Independent identically distributed

IM Implementation Margin

IMD Inter-Modulation Distortion

IMS IP Multimedia Subsystem

IMSI International Mobile Subscriber Identity

IMT International Mobile Telecommunications

InH Indoor Hotspot

IP Internet Protocol

IR Incremental Redundancy

IRC Interference Rejection Combining

ISD Inter-Site Distance

ISI Inter-Symbol Interference

IST-WINNER Information Society Technologies - Wireless world INitiative NEw Radio

ITU International Telecommunication Union

ITU-R ITU Radiocommunication sector

J-TACS Japanese Total Access Communication System

JT Joint Transmission

LA Local Area

LAC Local Area Code

LB Long Block

LBP Layered Belief Propagation

LBRM Limited Buffer Rate Matching

LCID Logical Channel ID

LDPC Low-Density Parity Check

L-GW LIPA GateWay

LI Length Indicator

LIPA Local IP Access

LLR Log-Likelihood Ratio

LMMSE Linear MMSE

LNA Low Noise Amplifier

LO Local Oscillator

LOS Line-Of-Sight

LPP LTE Positioning Protocol

LS Least Squares

LSF Last Segment Flag

LTE Long-Term Evolution

MA Metropolitan Area

MAC Medium Access Control

MAC-I Message Authentication Code for Integrity

MAN Metropolitan Area Network

MAP Maximum A posteriori Probability

MBL Mobility Load Balancing

MBMS Multimedia Broadcast/Multicast Service

MBMS GW MBMS GateWay

MBR Maximum Bit Rate

MBSFN Multimedia Broadcast Single Frequency Network

MCCH Multicast Control CHannel

MCE Multicell Coordination Entity

MCH Multicast CHannel

MCL Minimum Coupling Loss

MCS Modulation and Coding Scheme

Mcps Megachips per second

MDS Minimum Discernible Signal

MDT Minimization of Drive Tests

MeNB Macro eNodeB

MIB Master Information Block

MIMO Multiple-Input Multiple-Output

MIP Mobile Internet Protocol

MISO Multiple-Input Single-Output

ML Maximum Likelihood

MLD Maximum Likelihood Detector

MME Mobility Management Entity

MMSE Minimum MSE

MO Mobile Originated

MOP Maximum Output Power

MPS Multimedia Priority Service

M-PSK M-ary Phase-Shift Keying

MQE Minimum Quantization Error

MRB Multicast Radio Bearer

MRC Maximum Ratio Combining

M-RNTI MBMS Radio Network Temporary Identifier

MRO Mobility Robustness Optimization

MSA MCH Subframe Allocation

MSAP MCH Subframe Allocation Pattern

MSB Most Significant Bit

MSD Maximum Sensitivity Degradation

MSE Mean Squared Error

MSI MCH Scheduling Information

MSISDN Mobile Station International Subscriber Directory Number

MSP MCH Scheduling Period

MSR Maximum Sensitivity Reduction

MTC Machine-Type Communications

MTCH Multicast Traffic CHannel

MU-MIMO Multi-User MIMO

MUE Macro User Equipment

NACC Network Assisted Cell Change

NACK Negative ACKnowledgement

NACS NonAdjacent Channel Selectivity

NAS Non Access Stratum

NCC Network Colour Code

NCL Neighbour Cell List

NDI New Data Indicator

NF Noise Figure

NGMN Next Generation Mobile Networks

NLM Network Listen Mode

NLMS Normalized Least-Mean-Square

NLOS Non-Line-Of-Sight

NMT Nordic Mobile Telephone

NNSF NAS Node Selection Function

NodeB The base station in WCDMA systems

NR Neighbour cell Relation

NRT Neighbour Relation Table

O&M Operation and Maintenance

OBPD Occupied Bandwidth Power De-rating

OBW Occupied BandWidth

OCC Orthogonal Cover Code

OFDM Orthogonal Frequency Division Multiplexing

OFDMA Orthogonal Frequency Division Multiple Access

OPEX OPerational Expenditure

OSG Open Subscriber Group

OTDOA Observed Time Difference Of Arrival

OI Overload Indicator

OMA Open Mobile Alliance

OOB Out-Of-Band

P/S Parallel-to-Serial

PA Power Amplifier

PAN Personal Area Network

PAPR Peak-to-Average Power Ratio

PBCH Physical Broadcast CHannel

PBR Prioritized Bit Rate

PCC Policy Control and Charging*

PCC Primary Component Carrier*

PCCH Paging Control CHannel

P-CCPCH Primary Common Control Physical CHannel

PCEF Policy Control Enforcement Function

PCell Primary serving Cell

PCFICH Physical Control Format Indicator CHannel

PCG Project Coordination Group

PCH Paging CHannel

PCI Physical Cell Identity

P-CPICH Primary Common PIlot CHannel

PCRF Policy Control and charging Rules Function

PDCCH Physical Downlink Control CHannel

PDCP Packet Data Convergence Protocol

PDN Packet Data Network

PDP Power Delay Profile

PDSCH Physical Downlink Shared CHannel

PDU Protocol Data Unit

PF Paging Frame

PFS Proportional Fair Scheduling

P-GW PDN GateWay

PHICH Physical Hybrid ARQ Indicator CHannel

PHR Power Headroom Report

PLL Phase-Locked Loop

PLMN Public Land Mobile Network

P-MCCH Primary MCCH

PMCH Physical Multicast CHannel

PMI Precoding Matrix Indicators

PMIP Proxy MIP

PN Pseudo-Noise
PO Paging Occasion
PRACH Physical Random Access CHannel
PRB Physical Resource Block
P-RNTI Paging RNTI
PRG Precoder Resource block Group
PRS Positioning Reference Signal
PS Packet-Switched
P-SCH Primary Synchronization CHannel
PSD Power Spectral Density
PSS Primary Synchronization Signal
PTI Precoder Type Indication
PUCCH Physical Uplink Control CHannel
PUSCH Physical Uplink Shared CHannel
PVI Precoding Vector Indicator
PWS Public Warning System
QAM Quadrature Amplitude Modulation
QCI QoS Class Identifier
QoS Quality-of-Service
QPP Quadratic Permutation Polynomial
QPSK Quadrature Phase Shift Keying
RA Random Access
RAC Routing Area Code
RACH Random Access CHannel
RAN Radio Access Network
RAR Random Access Response
RA-RNTI Random Access Radio Network Temporary Identifier
RAT Radio Access Technology
RB Resource Block
RE Resource Element
REG Resource Element Group
RF Radio Frequency*
RF Resegmentation Flag*
RFC Request For Comments
RI Rank Indicator
RIM RAN Information Management
RIT Radio Interface Technology
RLC Radio Link Control
RLF Radio Link Failure
RLS Recursive Least Squares
RM Rate Matching*
RM Reed-Muller*
RMa Rural Macrocell
RN Relay Node
RNC Radio Network Controller
RNTI Radio Network Temporary Identifier
RNTP Relative Narrowband Transmit Power
ROHC RObust Header Compression
RoT Rise over Thermal
R-PDCCH Relay Physical Downlink Control Channel
RPRE Received Power per Resource Element
RPF RePetition Factor
R-PLMN Registered PLMN
RRC Radio Resource Control*
RRC Root-Raised-Cosine*
RRH Remote Radio Head
RRM Radio Resource Management
RS Reference Signal
RSCP Received Signal Code Power
RSRP Reference Signal Received Power
RSRQ Reference Signal Received Quality
RSSI Received Signal Strength Indicator
RSTD Reference Signal Time Difference
RTCP Real-time Transport Control Protocol
RTD Round-Trip Delay
RTP Real-time Transport Protocol
RTT Round-Trip Time
RV Redundancy Version
S/P Serial-to-Parallel
S1AP S1 Application Protocol
SAE System Architecture Evolution
SAP Service Access Point
SAW Stop-And-Wait
SB Short Block*
SB Synchronization Burst*
SBP Systematic Bit Puncturing

SCC Secondary Component Carrier

SC-FDMA Single-Carrier Frequency Division Multiple Access

SCH Synchronization CHannel

SCM Spatial Channel Model

SCME Spatial Channel Model Extension

SCTP Stream Control Transmission Protocol

SDMA Spatial Division Multiple Access

SDO Standards Development Organization

SDU Service Data Unit

SeGW Security GateWay

SEM Spectrum Emission Mask

SFBC Space-Frequency Block Code

SFDR Spurious-Free Dynamic Range

SFN System Frame Number

SGSN Serving GPRS Support Node

S-GW Serving GateWay

SI System Information

SIB System Information Block

SIC Successive Interference Cancellation

SIMO Single-Input Multiple-Output

SINR Signal-to-Interference plus Noise Ratio

SIP Session Initiation Protocol

SIR Signal-to-Interference Ratio

SI-RNTI System Information Radio Network Temporary Identifier

SISO Single-Input Single-Output*

SISO Soft-Input Soft-Output*

SLP SUPL Location Platform

S-MCCH Secondary MCCH

SMS Short Message Service

SN Sequence Number

SNR Signal-to-Noise Ratio

SO Segmentation Offset

SON Self-Optimizing Networks

SORTD Space Orthogonal-Resource Transmit Diversity

SPA Sum-Product Algorithm

SPS Semi-Persistent Scheduling

SPS-C-RNTI Semi-Persistent Scheduling C-RNTI

SR Scheduling Request

SRB Signalling Radio Bearer

SRIT Set of Radio Interface Technology

SRNS Serving Radio Network Subsystem

SRS Sounding Reference Signal

S-SCH Secondary Syncronization CHannel

SSS Secondary Synchronization Signal

STBC Space-Time Block Code

S-TMSI SAE-Temporary Mobile Subscriber Identity

STTD Space-Time Transmit Diversity

SU-MIMO Single-User MIMO

SUPL Secure User Plane Location

SVD Singular-Value Decomposition

TA Tracking Area

TAC Tracking Area Code

TACS Total Access Communication System

TAI Tracking Area Identity

TB Transport Block

TCP Transmission Control Protocol

TDC Time-Domain Coordination

TDD Time Division Duplex

TDL Tapped Delay Line

TDMA Time Division Multiple Access

TD-SCDMA Time Division Synchronous Code Division Multiple Access

TEID Tunnelling End ID

TF Transport Format

TFT Traffic Flow Template

TM Transparent Mode

TMD Transparent Mode Data

TNL Transport Network Layer

TNMSE Truncated Normalized Mean-Squared Error

TPC Transmitter Power Control

TPD Total Power De-rating

TPMI Transmitted Precoding Matrix Indicator

TR Tone Reservation

TSC Training Sequence Code

TSG Technical Specification Group

TTA Telecommunications Technology Association

TTC Telecommunications Technology Committee

TTFF Time To First Fix

TTI Transmission Time Interval

TU Typical Urban

UCI Uplink Control Information

UDP User Datagram Protocol

UE User Equipment

UL UpLink

ULA Uniform Linear Array

UL-SCH UpLink Shared CHannel

UM Unacknowledged Mode

UMa Urban Macrocell

UMB Ultra-Mobile Broadband

UMi Urban Microcell

UMTS Universal Mobile Telecommunications System

UP Unitary Precoding

UpPTS Uplink Pilot TimeSlot

US Uncorrelated-Scattered

USIM Universal Subscriber Identity Module

UTRA Universal Terrestrial Radio Access

UTRAN Universal Terrestrial Radio Access Network

VA Viterbi Algorithm

VCB Virtual Circular Buffer

VCO Voltage-Controlled Oscillator

VoIP Voice-over-IP

VRB Virtual Resource Block

WA Wide Area

WAN Wide Area Network

WCDMA Wideband Code Division Multiple Access

WFT Winograd Fourier Transform

WG Working Group

WiMAX Worldwide interoperability for Microwave Access

WINNER Wireless world INitiative NEw Radio

WLAN Wireless Local Area Network

WPD Waveform Power De-rating

WRC World Radiocommunication Conference

WSS Wide-Sense Stationary

WSSUS Wide-Sense Stationary Uncorrelated Scattering

ZC Zadoff–Chu

ZCZ Zero Correlation Zone

ZF Zero-Forcing

ZFEP Zero-Forcing Equal Power

1

Introduction and Background

Thomas Sälzer and Matthew Baker

1.1 The Context for the Long Term Evolution of UMTS

1.1.1 Historical Context

The Long Term Evolution of UMTS is one of the latest steps in an advancing series of mobile telecommunications systems. Arguably, at least for land-based systems, the series began in 1947 with the development of the concept of *cells* by Bell Labs, USA. The use of cells enabled the capacity of a mobile communications network to be increased substantially, by dividing the coverage area up into small cells each with its own base station operating on a different frequency.

The early systems were confined within national boundaries. They attracted only a small number of users, as the equipment on which they relied was expensive, cumbersome and power-hungry, and therefore was only really practical in a car.

The first mobile communication systems to see large-scale commercial growth arrived in the 1980s and became known as the 'First Generation' systems. The First Generation used analogue technology and comprised a number of independently developed systems worldwide (e.g. AMPS (Analogue Mobile Phone System, used in America), TACS (Total Access Communication System, used in parts of Europe), NMT (Nordic Mobile Telephone, used in parts of Europe) and J-TACS (Japanese Total Access Communication System, used in Japan and Hong Kong)).

Global roaming first became a possibility with the development of the 'Second Generation' system known as GSM (Global System for Mobile communications), which was based on digital technology. The success of GSM was due in part to the collaborative spirit in which it was developed. By harnessing the creative expertise of a number of companies working

LTE – The UMTS Long Term Evolution: From Theory to Practice, Second Edition.
Stefania Sesia, Issam Toufik and Matthew Baker.

together under the auspices of the European Telecommunications Standards Institute (ETSI), GSM became a robust, interoperable and widely accepted standard.

Fuelled by advances in mobile handset technology, which resulted in small, fashionable terminals with a long battery life, the widespread acceptance of the GSM standard exceeded initial expectations and helped to create a vast new market. The resulting near-universal penetration of GSM phones in the developed world provided an ease of communication never previously possible, first by voice and text message, and later also by more advanced data services. Meanwhile in the developing world, GSM technology had begun to connect communities and individuals in remote regions where fixed-line connectivity was non-existent and would be prohibitively expensive to deploy.

This ubiquitous availability of user-friendly mobile communications, together with increasing consumer familiarity with such technology and practical reliance on it, thus provides the context for new systems with more advanced capabilities. In the following section, the series of progressions which have succeeded GSM is outlined, culminating in the development of the system known as LTE – the Long Term Evolution of UMTS (Universal Mobile Telecommunications System).

1.1.2 LTE in the Mobile Radio Landscape

In contrast to transmission technologies using media such as copper lines and optical fibres, the radio spectrum is a medium shared between different, and potentially interfering, technologies.

As a consequence, regulatory bodies – in particular, ITU-R (International Telecommunication Union – Radiocommunication Sector) [1], but also regional and national regulators – play a key role in the evolution of radio technologies since they decide which parts of the spectrum and how much bandwidth may be used by particular types of service and technology. This role is facilitated by the *standardization* of families of radio technologies – a process which not only provides specified interfaces to ensure interoperability between equipment from a multiplicity of vendors, but also aims to ensure that the allocated spectrum is used as efficiently as possible, so as to provide an attractive user experience and innovative services.

The complementary functions of the regulatory authorities and the standardization organizations can be summarized broadly by the following relationship:

$$\text{Aggregated data rate} \quad = \quad \underbrace{\text{bandwidth}}_{\substack{\text{regulation and licences}\\ \text{(ITU-R, regional regulators)}}} \quad \times \quad \underbrace{\text{spectral efficiency}}_{\substack{\text{technology}\\ \text{and standards}}}$$

On a worldwide basis, ITU-R defines technology families and associates specific parts of the spectrum with these families. Facilitated by ITU-R, spectrum for mobile radio technologies is identified for the radio technologies which meet ITU-R's requirements to be designated as members of the *International Mobile Telecommunications* (IMT) family. Effectively, the IMT family comprises systems known as 'Third Generation' (for the first time providing data rates up to 2 Mbps) and beyond.

From the technology and standards angle, three main organizations have recently been developing standards relevant to IMT requirements, and these organisations continue to shape the landscape of mobile radio systems as shown in Figure 1.1.

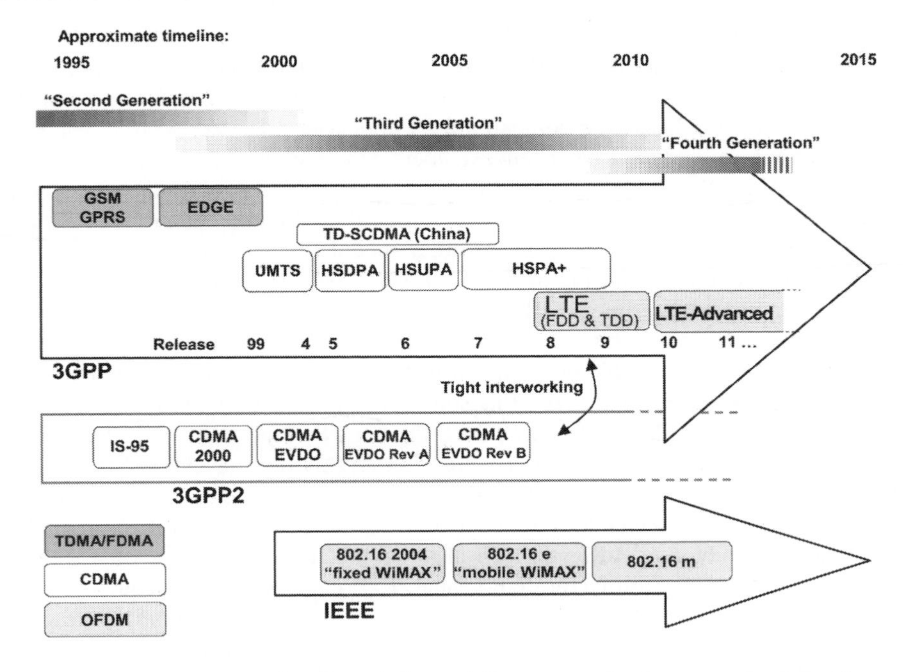

Figure 1.1: Approximate timeline of the mobile communications standards landscape.

The uppermost evolution track shown in Figure 1.1 is that developed in the 3rd Generation Partnership Project (3GPP), which is currently the dominant standards development group for mobile radio systems and is described in more detail below.

Within the 3GPP evolution track, three multiple access technologies are evident: the 'Second Generation' GSM/GPRS/EDGE family[1] was based on Time- and Frequency-Division Multiple Access (TDMA/FDMA); the 'Third Generation' UMTS family marked the entry of Code Division Multiple Access (CDMA) into the 3GPP evolution track, becoming known as *Wideband* CDMA (owing to its 5 MHz carrier bandwidth) or simply WCDMA; finally LTE has adopted Orthogonal Frequency-Division Multiplexing (OFDM), which is the access technology dominating the latest evolutions of all mobile radio standards.

In continuing the technology progression from the GSM and UMTS technology families within 3GPP, the LTE system can be seen as completing the trend of expansion of service provision beyond voice calls towards a multiservice air interface. This was already a key aim of UMTS and GPRS/EDGE, but LTE was designed from the start with the goal of evolving the radio access technology under the assumption that all services would be packet-switched, rather than following the circuit-switched model of earlier systems. Furthermore, LTE is accompanied by an evolution of the non-radio aspects of the complete system, under the term 'System Architecture Evolution' (SAE) which includes the Evolved Packet Core (EPC) network. Together, LTE and SAE comprise the Evolved Packet System (EPS), where both the core network and the radio access are fully packet-switched.

[1]The maintenance and development of specifications for the GSM family was passed to 3GPP from ETSI.

The standardization of LTE and SAE does not mean that further development of the other radio access technologies in 3GPP has ceased. In particular, the enhancement of UMTS with new releases of the specifications continues in 3GPP, to the greatest extent possible while ensuring backward compatibility with earlier releases: the original 'Release 99' specifications of UMTS have been extended with high-speed downlink and uplink enhancements (HSDPA[2] and HSUPA[3] in Releases 5 and 6 respectively), known collectively as 'HSPA' (High-Speed Packet Access). HSPA has been further enhanced in Release 7 (becoming known as HSPA+) with higher-order modulation and, for the first time in a cellular communication system, multistream 'MIMO'[4] operation, while Releases 8, 9 and 10 introduce support for multiple 5 MHz carriers operating together in downlink and uplink. These backward-compatible enhancements enable network operators who have invested heavily in the WCDMA technology of UMTS to generate new revenues from new features while still providing service to their existing subscribers using legacy terminals.

The first version of LTE was made available in Release 8 of the 3GPP specification series. It was able to benefit from the latest understanding and technology developments from HSPA and HSPA+, especially in relation to optimizations of the protocol stack, while also being free to adopt radical new technology without the constraints of backward compatibility or a 5 MHz carrier bandwidth. However, LTE also has to satisfy new demands, for example in relation to spectrum flexibility for deployment. LTE can operate in Frequency-Division Duplex (FDD) and Time-Division Duplex (TDD) modes in a harmonized framework designed also to support the evolution of TD-SCDMA (Time-Division Synchronous Code Division Multiple Access), which was developed in 3GPP as an additional branch of the UMTS technology path, essentially for the Chinese market.

A second version of LTE was developed in Release 9, and Release 10 continues the progression with the beginning of the next significant step known as LTE-Advanced.

A second evolution track shown in Figure 1.1 is led by a partnership organization similar to 3GPP and known as 3GPP2. CDMA2000 was developed based on the American 'IS-95' standard, which was the first mobile cellular communication system to use CDMA technology; it was deployed mainly in the USA, Korea and Japan. Standardization in 3GPP2 has continued with parallel evolution tracks towards data-oriented systems (EV-DO), to a certain extent taking a similar path to the evolutions in 3GPP. It is important to note that LTE will provide tight interworking with systems developed by 3GPP2, which allows a smooth migration to LTE for operators who previously followed the 3GPP2 track.

The third path of evolution has emerged from the IEEE 802 LAN/MAN[5] standards committee, which created the '802.16' family as a broadband wireless access standard. This family is also fully packet-oriented. It is often referred to as *WiMAX*, on the basis of a so-called 'System Profile' assembled from the 802.16 standard and promoted by the WiMAX Forum. The WiMAX Forum also ensures the corresponding product certification. While the first version, known as 802.16-2004, was restricted to fixed access, the following version 802.16e includes basic support of mobility and is therefore often referred to as 'mobile WiMAX'. However, it can be noted that in general the WiMAX family has not been designed with the same emphasis on mobility and compatibility with operators' core networks as the

[2]High-Speed Downlink Packet Access.

[3]High-Speed Uplink Packet Access.

[4]Multiple-Input Multiple-Output antenna system.

[5]Local Area Network / Metropolitan Area Network.

3GPP technology family, which includes core network evolutions in addition to the radio access network evolution. Nevertheless, the latest generation developed by the IEEE, known as 802.16m, has similar targets to LTE-Advanced which are outlined in Chapter 27.

The overall pattern is of an evolution of mobile radio towards flexible, packet-oriented, multiservice systems. The aim of all these systems is towards offering a mobile broadband user experience that can approach that of current fixed access networks such as Asymmetric Digital Subscriber Line (ADSL) and Fibre-To-The-Home (FTTH).

1.1.3 The Standardization Process in 3GPP

The collaborative standardization model which so successfully produced the GSM system became the basis for the development of UMTS. In the interests of producing truly global standards, the collaboration for both GSM and UMTS was expanded beyond ETSI to encompass regional Standards Development Organizations (SDOs) from Japan (ARIB and TTC), Korea (TTA), North America (ATIS) and China (CCSA), as shown in Figure 1.2.

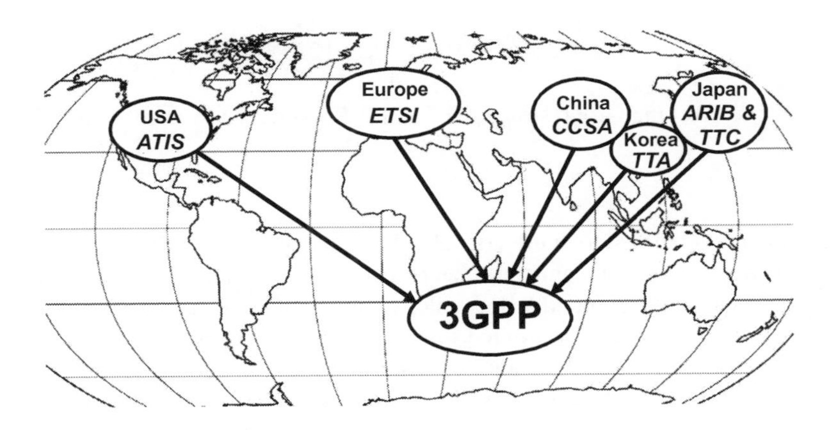

Figure 1.2: 3GPP is a global partnership of six regional SDOs.

So the 3GPP was born and by 2011 boasted 380 individual member companies.

The successful creation of such a large and complex system specification as that for UMTS or LTE requires a well-structured organization with pragmatic working procedures. 3GPP is divided into four Technical Specification Groups (TSGs), each of which is comprised of a number of Working Groups (WGs) with responsibility for a specific aspect of the specifications as shown in Figure 1.3.

A distinctive feature of the working methods of these groups is the consensus-driven approach to decision-making.

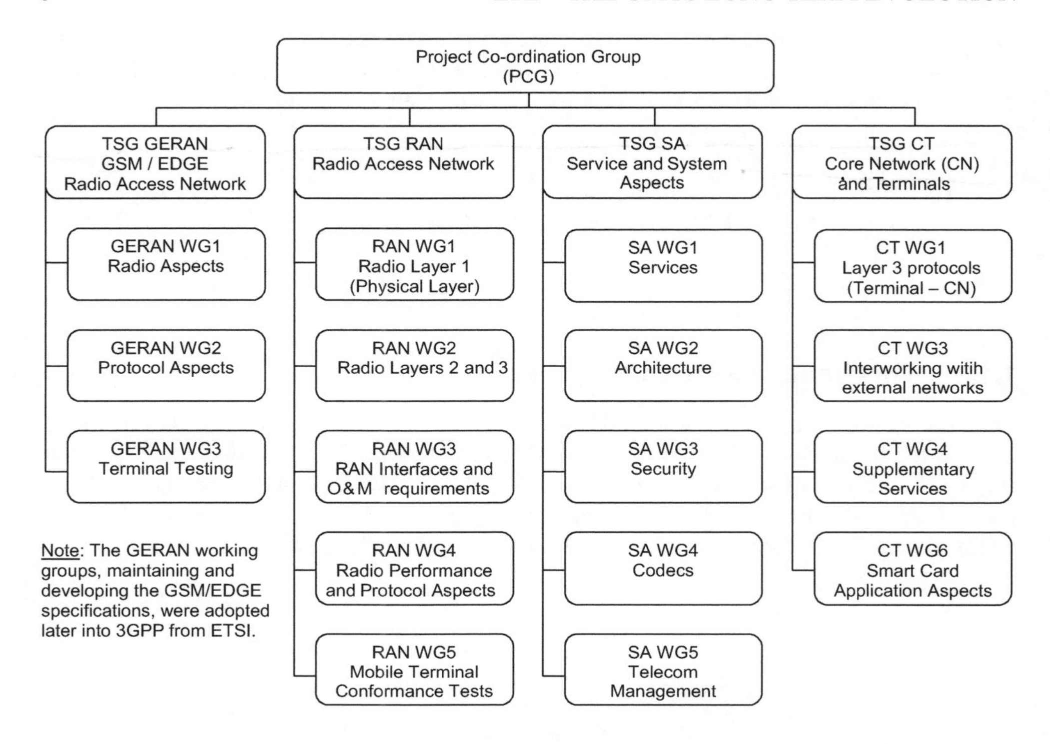

Figure 1.3: The Working Group structure of 3GPP. Reproduced by permission of © 3GPP.

All documents submitted to 3GPP are publicly available on the 3GPP website,[6] including contributions from individual companies, technical reports and technical specifications.

In reaching consensus around a technology, the WGs take into account a variety of considerations, including but not limited to performance, implementation cost, complexity and compatibility with earlier versions or deployments. Simulations are frequently used to compare performance of different techniques, especially in the WGs focusing on the physical layer of the air interface and on performance requirements. This requires consensus first to be reached around the simulation assumptions to be used for the comparison, including, in particular, understanding and defining the scenarios of interest to network operators.

The LTE standardization process was inaugurated at a workshop in Toronto in November 2004, when a broad range of companies involved in the mobile communications business presented their visions for the future evolution of the specifications to be developed in 3GPP. These visions included both initial perceptions of the *requirements* which needed to be satisfied, and proposals for *suitable technologies* to meet those requirements.

The requirements are reviewed in detail in Section 1.2, while the key technologies are introduced in Section 1.3.

[6] www.3gpp.org.

1.2 Requirements and Targets for the Long Term Evolution

Discussion of the key requirements for the new LTE system led to the creation of a formal 'Study Item' in 3GPP with the specific aim of 'evolving' the 3GPP radio access technology to ensure competitiveness over a ten-year time-frame. Under the auspices of this Study Item, the requirements for LTE Release 8 were refined and crystallized, being finalized in June 2005.

They can be summarized as follows:

- reduced delays, in terms of both connection establishment and transmission latency;
- increased user data rates;
- increased cell-edge bit-rate, for uniformity of service provision;
- reduced cost per bit, implying improved spectral efficiency;
- greater flexibility of spectrum usage, in both new and pre-existing bands;
- simplified network architecture;
- seamless mobility, including between different radio-access technologies;
- reasonable power consumption for the mobile terminal.

It can also be noted that network operator requirements for next generation mobile systems were formulated by the Next Generation Mobile Networks (NGMN) alliance of network operators [2], which served as an additional reference for the development and assessment of the LTE design. Such operator-driven requirements have also guided the development of LTE-Advanced (see Chapters 27 to 31).

To address these objectives, the LTE system design covers both the radio interface and the radio network architecture.

1.2.1 System Performance Requirements

Improved system performance compared to existing systems is one of the main requirements from network operators, to ensure the competitiveness of LTE and hence to arouse market interest. In this section, we highlight the main performance metrics used in the definition of the LTE requirements and its performance assessment.

Table 1.1 summarizes the main performance requirements to which the first release of LTE was designed. Many of the figures are given relative to the performance of the most advanced available version of UMTS, which at the time of the definition of the LTE requirements was HSDPA/HSUPA Release 6 – referred to here as the *reference baseline*. It can be seen that the target requirements for LTE represent a significant step from the capacity and user experience offered by the third generation mobile communications systems which were being deployed at the time when the first version of LTE was being developed.

As mentioned above, HSPA technologies are also continuing to be developed to offer higher spectral efficiencies than were assumed for the reference baseline. However, LTE has been able to benefit from avoiding the constraints of backward compatibility, enabling the inclusion of advanced MIMO schemes in the system design from the beginning, and highly flexible spectrum usage built around new multiple access schemes.

Table 1.1: Summary of key performance requirement targets for LTE Release 8.

		Absolute requirement	Release 6 (for comparison)	Comments
Downlink	Peak transmission rate	> 100 Mbps	14.4 Mbps	LTE in 20 MHz FDD, 2 × 2 spatial multiplexing. Reference: HSDPA in 5 MHz FDD, single antenna transmission
	Peak spectral efficiency	> 5 bps/Hz	3 bps/Hz	
	Average cell spectral efficiency	> 1.6–2.1 bps/Hz/cell	0.53 bps/Hz/cell	LTE: 2 × 2 spatial multiplexing, Interference Rejection Combining (IRC) receiver [3]. Reference: HSDPA, Rake receiver [4], 2 receive antennas
	Cell edge spectral efficiency	> 0.04–0.06 bps/Hz/user	0.02 bps/Hz/user	As above, 10 users assumed per cell
	Broadcast spectral efficiency	> 1 bps/Hz	N/A	Dedicated carrier for broadcast mode
Uplink	Peak transmission rate	> 50 Mbps	11 Mbps	LTE in 20 MHz FDD, single antenna transmission. Reference: HSUPA in 5 MHz FDD, single antenna transmission
	Peak spectral efficiency	> 2.5 bps/Hz	2 bps/Hz	
	Average cell spectral efficiency	> 0.66–1.0 bps/Hz/cell	0.33 bps/Hz/cell	LTE: single antenna transmission, IRC receiver [3]. Reference: HSUPA, Rake receiver [4], 2 receive antennas
	Cell edge spectral efficiency	> 0.02–0.03 bps/Hz/user	0.01 bps/Hz/user	As above, 10 users assumed per cell
System	User plane latency (two way radio delay)	< 10 ms		LTE target approximately one fifth of Reference.
	Connection set-up latency	< 100 ms		Idle state → active state
	Operating bandwidth	1.4–20 MHz	5 MHz	(initial requirement started at 1.25 MHz)
	VoIP capacity	NGMN preferred target expressed in [2] is > 60 sessions/MHz/cell		

The requirements shown in Table 1.1 are discussed and explained in more detail below. Chapter 26 shows how the overall performance of the LTE system meets these requirements.

1.2.1.1 Peak Rates and Peak Spectral Efficiency

For marketing purposes, the first parameter by which different radio access technologies are usually compared is the peak per-user data rate which can be achieved. This peak data rate generally scales according to the amount of spectrum used, and, for MIMO systems, according to the minimum of the number of transmit and receive antennas (see Section 11.1).

The peak data rate can be defined as the maximum throughput per user assuming the whole bandwidth being allocated to a single user with the highest modulation and coding scheme and the maximum number of antennas supported. Typical radio interface overhead (control channels, pilot signals, guard intervals, etc.) is estimated and taken into account for a given operating point. For TDD systems, the peak data rate is generally calculated for the downlink and uplink periods separately. This makes it possible to obtain a single value independent of the uplink/downlink ratio and a fair system comparison that is agnostic of the duplex mode. The maximum spectral efficiency is then obtained simply by dividing the peak rate by the used spectrum allocation.

The target peak data rates for downlink and uplink in LTE Release 8 were set at 100 Mbps and 50 Mbps respectively within a 20 MHz bandwidth,[7] corresponding to respective peak spectral efficiencies of 5 and 2.5 bps/Hz. The underlying assumption here is that the terminal has two receive antennas and one transmit antenna. The number of antennas used at the base station is more easily upgradeable by the network operator, and the first version of the LTE specifications was therefore designed to support downlink MIMO operation with up to four transmit and receive antennas. The MIMO techniques enabling high peak data rates are described in detail in Chapter 11.

When comparing the capabilities of different radio communication technologies, great emphasis is often placed on the peak data rate capabilities. While this is one indicator of how technologically advanced a system is and can be obtained by simple calculations, it may not be a key differentiator in the usage scenarios for a mobile communication system in practical deployment. Moreover, it is relatively easy to design a system that can provide very high peak data rates for users close to the base station, where interference from other cells is low and techniques such as MIMO can be used to their greatest extent. It is much more challenging to provide high data rates with good coverage and mobility, but it is exactly these latter aspects which contribute most strongly to user satisfaction.

In typical deployments, individual users are located at varying distances from the base stations, the propagation conditions for radio signals to individual users are rarely ideal, and the available resources must be shared between many users. Consequently, although the claimed peak data rates of a system are genuinely achievable in the right conditions, it is rare for a single user to be able to experience the peak data rates for a sustained period, and the envisaged applications do not usually require this level of performance.

A differentiator of the LTE system design compared to some other systems has been the recognition of these 'typical deployment constraints' from the beginning. During the design process, emphasis was therefore placed not only on providing a competitive peak data rate for use when conditions allow, but also importantly on *system level performance*, which was evaluated during several performance verification steps.

System-level evaluations are based on simulations of multicell configurations where data transmission from/to a population of mobiles is considered in a typical deployment scenario. The sections below describe the main metrics used as requirements for system level performance. In order to make these metrics meaningful, parameters such as the deployment scenario, traffic models, channel models and system configuration need to be defined.

The key definitions used for the system evaluations of LTE Release 8 can be found in an input document from network operators addressing the performance verification milestone in the LTE development process [5]. This document takes into account deployment scenarios and channel models agreed during the LTE Study Item [6], and is based on an evaluation methodology elaborated by NGMN operators in [7]. The reference deployment scenarios which were given special consideration for the LTE performance evaluation covered macrocells with base station separations of between 500 m and 1.7 km, as well as microcells using MIMO with base station separations of 130 m. A range of mobile terminal

[7] Four times the bandwidth of a WCDMA carrier.

speeds were studied, focusing particularly on the range 3–30 km/h, although higher mobile speeds were also considered important.

1.2.1.2 Cell Throughput and Spectral Efficiency

Performance at the cell level is an important criterion, as it relates directly to the number of cell sites that a network operator requires, and hence to the capital cost of deploying the system. For LTE Release 8, it was chosen to assess the cell level performance with full-queue traffic models (i.e. assuming that there is never a shortage of data to transmit if a user is given the opportunity) and a relatively high system load, typically 10 users per cell.

The requirements at the cell level were defined in terms of the following metrics:

- Average cell throughput [bps/cell] and spectral efficiency [bps/Hz/cell];
- Aaverage user throughput [bps/user] and spectral efficiency [bps/Hz/user];
- Cell-edge user throughput [bps/user] and spectral efficiency [bps/Hz/user] (the metric used for this assessment is the 5-percentile user throughput, obtained from the cumulative distribution function of the user throughput).

For the UMTS Release 6 reference baseline, it was assumed that both the terminal and the base station use a single transmit antenna and two receive antennas; for the terminal receiver the assumed performance corresponds to a two-branch Rake receiver [4] with linear combining of the signals from the two antennas.

For the LTE system, the use of two transmit and receive antennas was assumed at the base station. At the terminal, two receive antennas were assumed, but still only a single transmit antenna. The receiver for both downlink and uplink is assumed to be a linear receiver with optimum combining of the signals from the antenna branches [3].

The original requirements for the cell level metrics were only expressed as relative gains compared to the Release 6 reference baseline. The absolute values provided in Table 1.1 are based on evaluations of the reference system performance that can be found in [8] and [9] for downlink and uplink respectively.

1.2.1.3 Voice Capacity

Unlike full queue traffic (such as file download) which is typically delay-tolerant and does not require a guaranteed bit-rate, real-time traffic such as Voice over IP (VoIP) has tight delay constraints. It is important to set system capacity requirements for such services – a particular challenge in fully packet-based systems like LTE which rely on adaptive scheduling.

The system capacity requirement is defined as the number of satisfied VoIP users, given a particular traffic model and delay constraints. The details of the traffic model used for evaluating LTE can be found in [5]. Here, a VoIP user is considered to be in outage (i.e. not satisfied) if more than 2% of the VoIP packets do not arrive successfully at the radio receiver within 50 ms and are therefore discarded. This assumes an overall end-to-end delay (from mobile terminal to mobile terminal) below 200 ms. The system capacity for VoIP can then be defined as the number of users present per cell when more than 95% of the users are satisfied.

The NGMN group of network operators expressed a preference for the ability to support 60 satisfied VoIP sessions per MHz – an increase of two to four times what can typically be achieved in the Release 6 reference case.

1.2.1.4 Mobility and Cell Ranges

LTE is required to support communication with terminals moving at speeds of up to 350 km/h, or even up to 500 km/h depending on the frequency band. The primary scenario for operation at such high speeds is usage on high-speed trains – a scenario which is increasing in importance across the world as the number of high-speed rail lines increases and train operators aim to offer an attractive working environment to their passengers. These requirements mean that handover between cells has to be possible without interruption – in other words, with imperceptible delay and packet loss for voice calls, and with reliable transmission for data services.

These targets are to be achieved by the LTE system in typical cells of radius up to 5 km, while operation should continue to be possible for cell ranges of 100 km and more, to enable wide-area deployments.

1.2.1.5 Broadcast Mode Performance

The requirements for LTE included the integration of an efficient broadcast mode for high rate Multimedia Broadcast/Multicast Services (MBMS) such as mobile TV, based on a Single Frequency Network mode of operation as explained in detail in Chapter 13. The spectral efficiency requirement is given in terms of a carrier dedicated to broadcast transmissions – i.e. not shared with unicast transmissions.

In broadcast systems, the system throughput is limited to what is achievable for the users in the worst conditions. Consequently, the broadcast performance requirement was defined in terms of an achievable system throughput (bps) and spectral efficiency (bps/Hz) assuming a coverage of 98% of the nominal coverage area of the system. This means that only 2% of the locations in the nominal coverage area are in outage – where outage for broadcast services is defined as experiencing a packet error rate higher than 1%. This broadcast spectral efficiency requirement was set to 1 bps/Hz [10].

While the broadcast mode was not available in Release 8 due to higher prioritization of other service modes, Release 9 incorporates a broadcast mode employing Single Frequency Network operation on a mixed unicast-broadcast carrier.

1.2.1.6 User Plane Latency

User plane latency is an important performance metric for real-time and interactive services. On the radio interface, the minimum user plane latency can be calculated based on signalling analysis for the case of an unloaded system. It is defined as the average time between the first transmission of a data packet and the reception of a physical layer acknowledgement. The calculation should include typical HARQ[8] retransmission rates (e.g. 0–30%). This definition therefore considers the capability of the system design, without being distorted by the scheduling delays that would appear in the case of a loaded system. The round-trip latency is obtained simply by multiplying the one-way user plane latency by a factor of two.

LTE is also required to be able to operate with an IP-layer one-way data-packet latency across the radio access network as low as 5 ms in optimal conditions. However, it is recognized that the actual delay experienced in a practical system will be dependent on system loading and radio propagation conditions. For example, HARQ plays a key role in

[8]Hybrid Automatic Repeat reQuest – see Section 10.3.2.5.

maximizing spectral efficiency at the expense of increased delay while retransmissions take place, whereas maximal spectral efficiency may not be essential in situations when minimum latency is required.

1.2.1.7 Control Plane Latency and Capacity

In addition to the user plane latency requirement, call setup delay was required to be significantly reduced compared to previous cellular systems. This not only enables a good user experience but also affects the battery life of terminals, since a system design which allows a fast transition from an idle state to an active state enables terminals to spend more time in the low-power idle state.

Control plane latency is measured as the time required for performing the transitions between different LTE states. LTE is based on only two main states, 'RRC_IDLE' and 'RRC_CONNECTED' (i.e. 'active') (see Section 3.1).

LTE is required to support transition from idle to active in less than 100 ms (excluding paging delay and Non-Access Stratum (NAS) signalling delay).

The LTE system capacity is dependent not only on the supportable throughput but also on the number of users simultaneously located within a cell which can be supported by the control signalling. For the latter aspect, LTE is required to support at least 200 active-state users per cell for spectrum allocations up to 5 MHz, and at least 400 users per cell for wider spectrum allocations; only a small subset of these users would be actively receiving or transmitting data at any given time instant, depending, for example, on the availability of data to transmit and the prevailing radio channel conditions. An even larger number of non-active users may also be present in each cell, and therefore able to be paged or to start transmitting data with low latency.

1.2.2 Deployment Cost and Interoperability

Besides the system performance aspects, a number of other considerations are important for network operators. These include reduced deployment cost, spectrum flexibility and enhanced interoperability with legacy systems – essential requirements to enable deployment of LTE networks in a variety of scenarios and to facilitate migration to LTE.

1.2.2.1 Spectrum Allocations and Duplex Modes

As demand for suitable radio spectrum for mobile communications increases, LTE is required to be able to operate in a wide range of frequency bands and sizes of spectrum allocations in both uplink and downlink. LTE can use spectrum allocations ranging from 1.4 to 20 MHz with a single carrier and addresses all frequency bands currently identified for IMT systems by ITU-R [1] including those below 1 GHz.

This will include deploying LTE in spectrum currently occupied by older radio access technologies – a practice often known as 'spectrum refarming'.

New frequency bands are continually being introduced for LTE in a release-independent way, meaning that any of the LTE Releases can be deployed in a new frequency band once the Radio-Frequency (RF) requirements have been specified [11].

The ability to operate in both paired and unpaired spectrum is required, depending on spectrum availability (see Chapter 23). LTE provides support for FDD, TDD and half-duplex

FDD operation in a unified design, ensuring a high degree of commonality which facilitates implementation of multimode terminals and allows worldwide roaming.

Starting from Release 10, LTE also provides means for flexible spectrum use via aggregation of contiguous and non-contiguous spectrum assets for high data rate services using a total bandwidth of up to 100 MHz (see Chapter 28).

1.2.2.2 Inter-Working with Other Radio Access Technologies

Flexible interoperation with other radio access technologies is essential for service continuity, especially during the migration phase in early deployments of LTE with partial coverage, where handover to legacy systems will often occur.

LTE relies on an evolved packet core network which allows interoperation with various access technologies, in particular earlier 3GPP technologies (GSM/EDGE and UTRAN[9]) as well as non-3GPP technologies (e.g. WiFi, CDMA2000 and WiMAX).

However, service continuity and short interruption times can only be guaranteed if measurements of the signals from other systems and fast handover mechanisms are integrated in the LTE radio access design. LTE therefore supports tight inter-working with all legacy 3GPP technologies and some non-3GPP technologies such as CDMA2000.

1.2.2.3 Terminal Complexity and Cost

A key consideration for competitive deployment of LTE is the availability of low-cost terminals with long battery life, both in stand-by and during activity. Therefore, low terminal complexity has been taken into account where relevant throughout the LTE system, as well as designing the system wherever possible to support low terminal power consumption.

1.2.2.4 Network Architecture Requirements

LTE is required to allow a cost-effective deployment by an improved radio access network architecture design including:

- Flat architecture consisting of just one type of node, the base station, known in LTE as the *eNodeB* (see Chapter 2);
- Effective protocols for the support of packet-switched services (see Chapters 3 to 4);
- Open interfaces and support of multivendor equipment interoperability;
- efficient mechanisms for operation and maintenance, including self-optimization functionalities (see Chapter 25);
- Support of easy deployment and configuration, for example for so-called home base stations (otherwise known as femto-cells) (see Chapter 24).

[9]Universal Terrestrial Radio Access Network.

1.3 Technologies for the Long Term Evolution

The fulfilment of the extensive range of requirements outlined above is only possible thanks to advances in the underlying mobile radio technology. As an overview, we outline here three fundamental technologies that have shaped the LTE radio interface design: *multicarrier* technology, *multiple-antenna* technology, and the application of *packet-switching* to the radio interface. Finally, we summarize the combinations of capabilities that are supported by different categories of LTE mobile terminal in Releases 8 and 9.

1.3.1 Multicarrier Technology

Adopting a multicarrier approach for multiple access in LTE was the first major design choice. After initial consolidation of proposals, the candidate schemes for the downlink were Orthogonal Frequency-Division Multiple Access (OFDMA)[10] and Multiple WCDMA, while the candidate schemes for the uplink were Single-Carrier Frequency-Division Multiple Access (SC-FDMA), OFDMA and Multiple WCDMA. The choice of multiple-access schemes was made in December 2005, with OFDMA being selected for the downlink, and SC-FDMA for the uplink. Both of these schemes open up the frequency domain as a new dimension of flexibility in the system, as illustrated schematically in Figure 1.4.

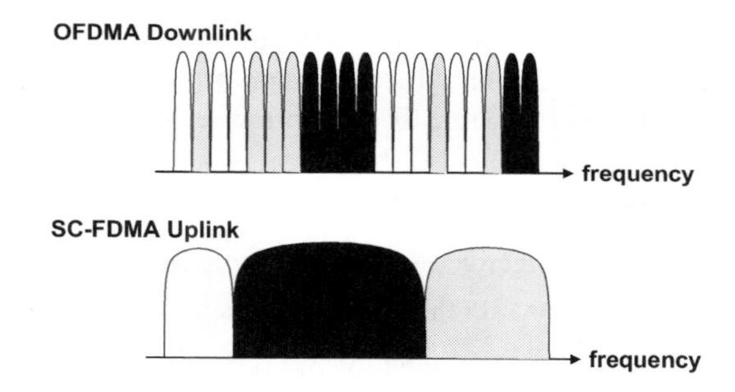

Figure 1.4: Frequency-domain view of the LTE multiple-access technologies.

OFDMA extends the multicarrier technology of OFDM to provide a very flexible multiple-access scheme. OFDM subdivides the bandwidth available for signal transmission into a multitude of narrowband subcarriers, arranged to be mutually orthogonal, which either individually or in groups can carry independent information streams; in OFDMA, this subdivision of the available bandwidth is exploited in sharing the subcarriers among multiple users.[11]

This resulting flexibility can be used in various ways:

[10]OFDM technology was already well understood in 3GPP as a result of an earlier study of the technology in 2003–4.

[11]The use of the frequency domain comes in addition to the well-known time-division multiplexing which continues to play an important role in LTE.

- Different spectrum bandwidths can be utilized without changing the fundamental system parameters or equipment design;
- Transmission resources of variable bandwidth can be allocated to different users and scheduled freely in the frequency domain;
- Fractional frequency re-use and interference coordination between cells are facilitated.

Extensive experience with OFDM has been gained in recent years from deployment of digital audio and video broadcasting systems such as DAB, DVB and DMB.[12] This experience has highlighted some of the key advantages of OFDM, which include:

- Robustness to time-dispersive radio channels, thanks to the subdivision of the wide-band transmitted signal into multiple narrowband subcarriers, enabling inter-symbol interference to be largely constrained within a guard interval at the beginning of each symbol;
- Low-complexity receivers, by exploiting frequency-domain equalization;
- Simple combining of signals from multiple transmitters in broadcast networks.

These advantages, and how they arise from the OFDM signal design, are explained in detail in Chapter 5.

By contrast, the transmitter design for OFDM is more costly, as the Peak-to-Average Power Ratio (PAPR) of an OFDM signal is relatively high, resulting in a need for a highly-linear RF power amplifier. However, this limitation is not inconsistent with the use of OFDM for *downlink* transmissions, as low-cost implementation has a lower priority for the base station than for the mobile terminal.

In the uplink, however, the high PAPR of OFDM is difficult to tolerate for the transmitter of the mobile terminal, since it is necessary to compromise between the output power required for good outdoor coverage, the power consumption, and the cost of the power amplifier. SC-FDMA, which is explained in detail in Chapter 14, provides a multiple-access technology which has much in common with OFDMA – in particular the flexibility in the frequency domain, and the incorporation of a guard interval at the start of each transmitted symbol to facilitate low-complexity frequency-domain equalization at the receiver. At the same time, SC-FDMA has a significantly lower PAPR. It therefore resolves to some extent the dilemma of how the uplink can benefit from the advantages of multicarrier technology while avoiding excessive cost for the mobile terminal transmitter and retaining a reasonable degree of commonality between uplink and downlink technologies.

In Release 10, the uplink multiple access scheme is extended to allow multiple clusters of subcarriers in the frequency domain, as explained in Section 28.3.6.

1.3.2 Multiple Antenna Technology

The use of multiple antenna technology allows the exploitation of the spatial-domain as another new dimension. This becomes essential in the quest for higher spectral efficiencies. As will be detailed in Chapter 11, with the use of multiple antennas the theoretically achievable spectral efficiency scales linearly with the minimum of the number of transmit and receive antennas employed, at least in suitable radio propagation environments.

[12] Digital Audio Broadcasting, Digital Video Broadcasting and Digital Mobile Broadcasting.

Multiple antenna technology opens the door to a large variety of features, but not all of them easily deliver their theoretical promises when it comes to implementation in practical systems. Multiple antennas can be used in a variety of ways, mainly based on three fundamental principles, schematically illustrated in Figure 1.5:

- **Diversity gain.** Use of the spatial diversity provided by the multiple antennas to improve the robustness of the transmission against multipath fading.
- **Array gain.** Concentration of energy in one or more given directions via precoding or beamforming. This also allows multiple users located in different directions to be served simultaneously (so-called multi-user MIMO).
- **Spatial multiplexing gain.** Transmission of multiple signal streams to a single user on multiple spatial layers created by combinations of the available antennas.

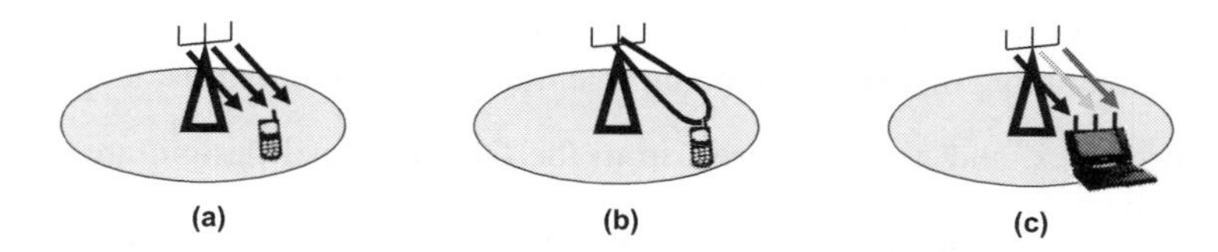

Figure 1.5: Three fundamental benefits of multiple antennas: (a) diversity gain; (b) array gain; (c) spatial multiplexing gain.

A large part of the LTE Study Item phase was therefore dedicated to the selection and design of the various multiple antenna features to be included in the first release of LTE. The final system includes several complementary options which allow for adaptability according to the network deployment and the propagation conditions of the different users.

1.3.3 Packet-Switched Radio Interface

As has already been noted, LTE has been designed as a completely packet-oriented multi-service system, without the reliance on circuit-switched connection-oriented protocols prevalent in its predecessors. In LTE, this philosophy is applied across all the layers of the protocol stack.

The route towards fast packet scheduling over the radio interface was already opened by HSDPA, which allowed the transmission of short packets having a duration of the same order of magnitude as the coherence time of the fast fading channel, as shown in Figure 1.6. This calls for a joint optimization of the physical layer configuration and the resource management carried out by the link layer protocols according to the prevailing propagation conditions. This aspect of HSDPA involves tight coupling between the lower two layers of the protocol stack – the MAC (Medium Access Control layer – see Chapter 4) and the physical layer.

In HSDPA, this coupling already included features such as fast channel state feedback, dynamic link adaptation, scheduling exploiting multi-user diversity, and fast retransmission protocols. In LTE, in order to improve the system latency, the packet duration was further reduced from the 2 ms used in HSDPA down to just 1 ms. This short transmission interval,

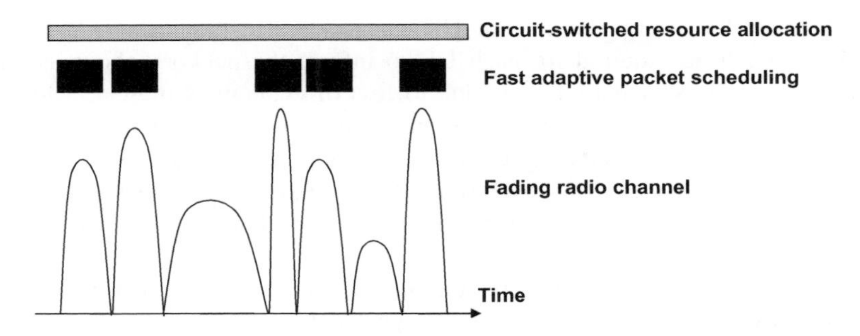

Figure 1.6: Fast scheduling and link adaptation.

together with the new dimensions of frequency and space, has further extended the field of cross-layer techniques between the MAC and physical layers to include the following techniques in LTE:

- Adaptive scheduling in both the frequency and spatial dimensions;
- Adaptation of the MIMO configuration including the selection of the number of spatial layers transmitted simultaneously;
- Link adaptation of modulation and code-rate, including the number of transmitted codewords;
- Several modes of fast channel state reporting.

These different levels of optimization are combined with very sophisticated control signalling.

1.3.4 User Equipment Categories

In practice it is important to recognize that the market for UEs is large and diverse, and there is therefore a need for LTE to support a range of categories of UE with different capabilities to satisfy different market segments. In general, each market segment attaches different priorities to aspects such as peak data rate, UE size, cost and battery life. Some typical trade-offs include the following:

- Support for the highest data rates is key to the success of some applications, but generally requires large amounts of memory for data processing, which increases the cost of the UE.
- UEs which may be embedded in large devices such as laptop computers are often not significantly constrained in terms of acceptable power consumption or the number of antennas which may be used; on the other hand, other market segments require ultra-slim hand-held terminals which have little space for multiple antennas or large batteries.

The wider the range of UE categories supported, the closer the match which may be made between a UE's supported functionality and the requirements of a particular market segment.

However, support for a large number of UE categories also has drawbacks in terms of the signalling overhead required for each UE to inform the network about its supported functionality, as well as increased costs due to loss of economies of scale and increased complexity for testing the interoperability of many different configurations.

The first release of LTE was therefore designed to support a compact set of five categories of UE, ranging from relatively low-cost terminals with similar capabilities to UMTS HSPA, up to very high-capability terminals which exploit the LTE technology to the maximum extent possible.

The five Release 8 UE categories are summarized in Table 1.2. It can be seen that the highest category of Release 8 LTE UE possesses peak data rate capabilities far exceeding the LTE Release 8 targets. Full details are specified in [12].

Table 1.2: Categories of LTE user equipment in Releases 8 and 9.

	UE category				
	1	2	3	4	5
Supported downlink data rate (Mbps)	10	50	100	150	300
Supported uplink data rate (Mbps)	5	25	50	50	75
Number of receive antennas required	2	2	2	2	4
Number of downlink MIMO layers supported	1	2	2	2	4
Support for 64QAM modulation in downlink	✔	✔	✔	✔	✔
Support for 64QAM modulation in uplink	✖	✖	✖	✖	✔
Relative memory requirement for physical layer processing (normalized to category 1 level)	1	4.9	4.9	7.3	14.6

Additional UE categories are introduced in Release 10, and these are explained in Section 27.5.

The LTE specifications deliberately avoid large numbers of optional features for the UEs, preferring to take the approach that if a feature is sufficiently useful to be worth including in the specifications then support of it should be mandatory. Nevertheless, a very small number of optional Release 8 features, whose support is indicated by each UE by specific signalling, are listed in [12]; such features are known as 'UE capabilities'. Some additional UE capabilities are added in later releases.

In addition, it is recognized that it is not always possible to complete conformance testing and Inter-Operability Testing (IOT) of every mandatory feature simultaneously for early deployments of LTE. Therefore, the development of conformance test cases for LTE was prioritized according to the likelihood of early deployment of each feature. Correspondingly, *Feature Group Indicators* (FGIs) are used for certain groups of lower priority mandatory features, to enable a UE to indicate whether IOT has been successfully completed for those features; the grouping of features corresponding to each FGI can be found in Annex B.1 of [13]. For UEs of Release 9 and later, it becomes mandatory for certain of these FGIs to be set to indicate that the corresponding feature(s) have been implemented and successfully tested.

1.3.5 From the First LTE Release to LTE-Advanced

As a result of intense activity by a larger number of contributing companies than ever before in 3GPP, the specifications for the first LTE release (Release 8) had reached a sufficient level of completeness by December 2007 to enable LTE to be submitted to ITU-R as a member of the IMT family of radio access technologies. It is therefore able to be deployed in IMT-designated spectrum, and the first commercial deployments were launched towards the end of 2009 in northern Europe.

Meanwhile, 3GPP has continued to improve the LTE system and to develop it to address new markets. In this section, we outline the new features introduced in the second LTE release, Release 9, and those provided by LTE Release 10, which begins the next significant step known as LTE-Advanced.

Increasing LTE's suitability for different markets and deployments was the first goal of Release 9. One important market with specific regulatory requirements is North America. LTE Release 9 therefore provides improved support for Public Warning Systems (PWS) and some accurate positioning methods (see Chapter 19). One positioning method uses the Observed Time Difference of Arrival (OTDOA) principle, supported by specially designed new reference signals inserted in the LTE downlink transmissions. Measurements of these positioning reference signals received from different base stations allow a UE to calculate its position very accurately, even in locations where other positioning means such as GPS fail (e.g. indoors). Enhanced Cell-ID-based techniques are also supported.

Release 9 also introduces support for a broadcast mode based on Single Frequency Network type transmissions (see Chapter 13).

The MIMO transmission modes are further developed in Release 9, with an extension of the Release 8 beamforming mode to support two orthogonal spatial layers that can be transmitted to a single user or multiple users, as described in Section 11.2.2.3. The design of this mode is forward-compatible for extension to more than two spatial layers in Release 10.

Release 9 also addresses specific deployments and, in particular, low power nodes (see Chapter 24). It defines new requirements for pico base stations and home base stations, in addition to improving support for Closed Subscriber Groups (CSG). Support for self-optimization of the networks is also enhanced in Release 9, as described in Chapter 25.

1.3.5.1 LTE-Advanced

The next version of LTE, Release 10, develops LTE to LTE-Advanced. While LTE Releases 8 and 9 already satisfy to a large extent the requirements set by ITU-R for the IMT-Advanced designation [14] (see Section 27.1), Release 10 will fully satisfy them and even exceed them in several aspects where 3GPP has set more demanding performance targets than those of ITU-R. The requirements for LTE-Advanced are discussed in detail in Chapter 27.

The main Release 10 features that are directly related to fulfilment of the IMT-Advanced requirements are:

- Carrier aggregation, allowing the total transmission bandwidth to be increased up to 100 MHz (see Chapter 28);
- Uplink MIMO transmission for peak spectral efficiencies greater than 7.5 bps/Hz and targeting up to 15 bps/Hz (see Chapter 29);

- Downlink MIMO enhancements, targeting peak spectral efficiencies up to 30 bps/Hz (see Chapter 29).

Besides addressing the IMT-Advanced requirements, Release 10 also provides some new features to enhance LTE deployment, such as support for relaying (see Chapter 30), enhanced inter-cell interference coordination (see Chapter 31) and mechanisms to minimize the need for drive tests by supporting extended measurement reports from the terminals (see Chapters 25 and 31).

1.4 From Theory to Practice

With commercial deployment of LTE now a reality, the advances in theoretical understanding and technology which underpin the LTE specifications are being exploited practically. This book is written with the primary aim of illuminating the transition from the underlying academic progress to the realization of useful advances in the provision of mobile communication services. Particular focus is given to the physical layer of the Radio Access Network (RAN), as it is here that many of the most dramatic technical advances are manifested. This should enable the reader to develop an understanding of the background to the technology choices in the LTE system, and hence to understand better the LTE specifications and how they may be implemented.

Parts I to IV of the book describe the features of LTE Releases 8 and 9, including indications of the aspects that are further enhanced in Release 10, while the details of the major new features of Release 10 are explained in Part V.

Part I sets the radio interface in the context of the network architecture and protocols, including radio resource management aspects, as well as explaining the new developments in these areas which distinguish LTE from previous systems.

In Part II, the physical layer of the RAN downlink is covered in detail, beginning with an explanation of the theory of the new downlink multiple access technology, OFDMA, in Chapter 5. This sets the context for the details of the LTE downlink design in Chapters 6 to 9. As coding, link adaptation and multiple antenna operation are of fundamental importance in fulfilling the LTE requirements, two chapters are then devoted to these topics, covering both the theory and the practical implementation in LTE.

Chapter 12 shows how these techniques can be applied to the system-level operation of the LTE system, focusing on applying the new degrees of freedom to multi-user scheduling and interference coordination.

Finally for the downlink, Chapter 13 covers broadcast operation – a mode which has its own unique challenges in a cellular system but which is nonetheless important in enabling a range of services to be provided to the end user.

Part III addresses the physical layer of the RAN uplink, beginning in Chapter 14 with an introduction to the theory behind the new uplink multiple access technology, SC-FDMA. This is followed in Chapters 15 to 18 with an analysis of the detailed uplink structure and operation, including the design of the associated procedures for random access, timing control and power control which are essential to the efficient operation of the uplink.

This leads on to Part IV, which examines a number of aspects of LTE related to its deployment as a mobile cellular system. Chapter 19 explains the UE positioning techniques introduced in Release 9. Chapter 20 provides a thorough analysis of the characteristics of the radio propagation environments in which LTE systems will be deployed, since an understanding of the propagation environment underpins much of the technology adopted for the LTE specifications. The new technologies and bandwidths adopted in LTE also have implications for the radio-frequency implementation of the mobile terminals in particular, and some of these are analysed in Chapter 21. The LTE system is designed to operate not just in wide bandwidths but also in a diverse range of spectrum allocation scenarios, and Chapter 23 therefore addresses the different duplex modes applicable to LTE and the effects that these may have on system design and operation. Chapter 24 addresses aspects of special relevance to deployment of low-power base stations such as Home eNodeBs and picocells, while Chapter 25 explains the advanced techniques for self-optimization of the network. Part IV concludes with a dedicated chapter examining a wide range of aspects of the overall system performance achievable with the first release of LTE.

Finally, Part V explains in detail the major new features included in Release 10 for LTE-Advanced, as 3GPP continues to respond to the ever-higher expectations of end-users. Chapters 28 to 30 address the technologies of carrier aggregation, enhanced MIMO and relaying respectively, and Chapter 31 covers enhanced Inter-Cell Interference Coordination, Minimization of Drive Tests and Machine-Type Communications. Chapter 32 provides an evaluation of the system performance achievable with LTE-Advanced Release 10, and concludes with a further look into the future.

References[13]

[1] ITU, International Telecommunications Union, www.itu.int/itu-r.

[2] NGMN, 'Next Generation Mobile Networks Beyond HSPA & EVDO – A white paper', www.ngmn.org, December 2006.

[3] J. H. Winters, 'Optimum Combining in Digital Mobile Radio with Cochannel Interference'. *IEEE Journal on Selected Areas in Communications*, Vol. 2, July 1984.

[4] R. Price and P. E. Green, 'A Communication Technique for Multipath Channels' in *Proceedings of the IRE*, Vol. 46, March 1958.

[5] Orange, China Mobile, KPN, NTT DoCoMo, Sprint, T-Mobile, Vodafone, and Telecom Italia, 'R1-070674: LTE Physical Layer Framework for Performance Verification', www.3gpp.org, 3GPP TSG RAN WG1, meeting 48, St Louis, USA, February 2007.

[6] 3GPP Technical Report 25.814, 'Physical Layer Aspects for Evolved UTRA', www.3gpp.org.

[7] NGMN, 'Next Generation Mobile Networks Radio Access Performance Evaluation Methodology', www.ngmn.org, June 2007.

[8] Ericsson, 'R1-072578: Summary of Downlink Performance Evaluation', www.3gpp.org, 3GPP TSG RAN WG1, meeting 49, Kobe, Japan, May 2007.

[9] Nokia, 'R1-072261: LTE Performance Evaluation – Uplink Summary', www.3gpp.org, 3GPP TSG RAN WG1, meeting 49, Kobe, Japan, May 2007.

[10] 3GPP Technical Report 25.913, 'Requirements for Evolved UTRA (E-UTRA) and Evolved UTRAN (E-UTRAN)', www.3gpp.org.

[11] 3GPP Technical Specification 36.307, 'Evolved Universal Terrestrial Radio Access (E-UTRA); Requirements on User Equipments (UEs) supporting a release-independent frequency band', www.3gpp.org.

[13] All web sites confirmed 1st March 2011.

[12] 3GPP Technical Specification 36.306, 'Evolved Universal Terrestrial Radio Access (E-UTRA); User Equipment (UE) radio access capabilities', www.3gpp.org.

[13] 3GPP Technical Specification 36.331, 'Evolved Universal Terrestrial Radio Access (E-UTRA); Radio Resource Control (RRC); Protocol specification', www.3gpp.org.

[14] ITU-R Report M.2134, 'Requirements related to technical performance for IMT-Advanced radio interface(s)', www.itu.int/itu-r.

Part I

Network Architecture and Protocols

2

Network Architecture

Sudeep Palat and Philippe Godin

2.1 Introduction

As mentioned in the preceding chapter, LTE has been designed to support only Packet-Switched (PS) services, in contrast to the Circuit-Switched (CS) model of previous cellular systems. It aims to provide seamless Internet Protocol (IP) connectivity between User Equipment (UE) and the Packet Data Network (PDN), without any disruption to the end users' applications during mobility. While the term 'LTE' encompasses the evolution of the radio access through the Evolved-UTRAN[1] (E-UTRAN), it is accompanied by an evolution of the non-radio aspects under the term 'System Architecture Evolution' (SAE) which includes the Evolved Packet Core (EPC) network. Together LTE and SAE comprise the Evolved Packet System (EPS).

EPS uses the concept of *EPS bearers* to route IP traffic from a gateway in the PDN to the UE. A bearer is an IP packet flow with a defined Quality of Service (QoS). The E-UTRAN and EPC together set up and release bearers as required by applications. EPS natively supports voice services over the IP Multimedia Subsystem (IMS) using Voice over IP (VoIP), but LTE also supports interworking with legacy systems for traditional CS voice support.

This chapter presents the overall EPS network architecture, giving an overview of the functions provided by the Core Network (CN) and E-UTRAN. The protocol stack across the different interfaces is then explained, along with an overview of the functions provided by the different protocol layers. Section 2.4 outlines the end-to-end bearer path including QoS aspects, provides details of a typical procedure for establishing a bearer and discusses the inter-working with legacy systems for CS voice services. The remainder of the chapter presents the network interfaces in detail, with particular focus on the E-UTRAN interfaces

[1]Universal Terrestrial Radio Access Network.

LTE – The UMTS Long Term Evolution: From Theory to Practice, Second Edition.
Stefania Sesia, Issam Toufik and Matthew Baker.

and associated procedures, including those for the support of user mobility. The network elements and interfaces used solely to support broadcast services are covered in Chapter 13, and the aspects related to UE positioning in Chapter 19.

2.2 Overall Architectural Overview

EPS provides the user with IP connectivity to a PDN for accessing the Internet, as well as for running services such as VoIP. An EPS bearer is typically associated with a QoS. Multiple bearers can be established for a user in order to provide different QoS streams or connectivity to different PDNs. For example, a user might be engaged in a voice (VoIP) call while at the same time performing web browsing or File Transfer Protocol (FTP) download. A VoIP bearer would provide the necessary QoS for the voice call, while a best-effort bearer would be suitable for the web browsing or FTP session. The network must also provide sufficient security and privacy for the user and protection for the network against fraudulent use.

Release 9 of LTE introduced several additional features. To meet regulatory requirements for commercial voice, services such as support of IMS, emergency calls and UE positioning (see Chapter 19) were introduced. Enhancements to Home cells (HeNBs) were also introduced in Release 9 (see Chapter 24).

All these features are supported by means of several EPS network elements with different roles. Figure 2.1 shows the overall network architecture including the network elements and the standardized interfaces. At a high level, the network is comprised of the CN (i.e. EPC) and the access network (i.e. E-UTRAN). While the CN consists of many logical nodes, the access network is made up of essentially just one node, the evolved NodeB (eNodeB), which connects to the UEs. Each of these network elements is inter-connected by means of interfaces which are standardized in order to allow multivendor interoperability.

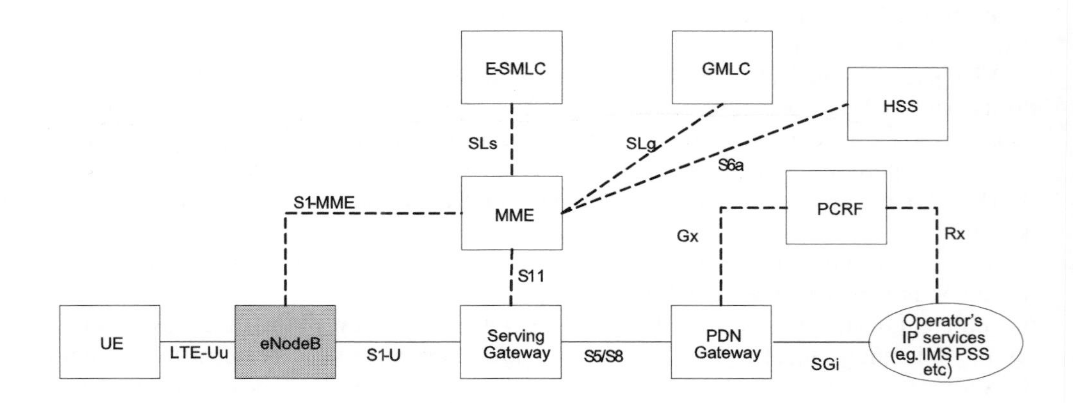

Figure 2.1: The EPS network elements.

The functional split between the EPC and E-UTRAN is shown in Figure 2.2. The EPC and E-UTRAN network elements are described in more detail below.

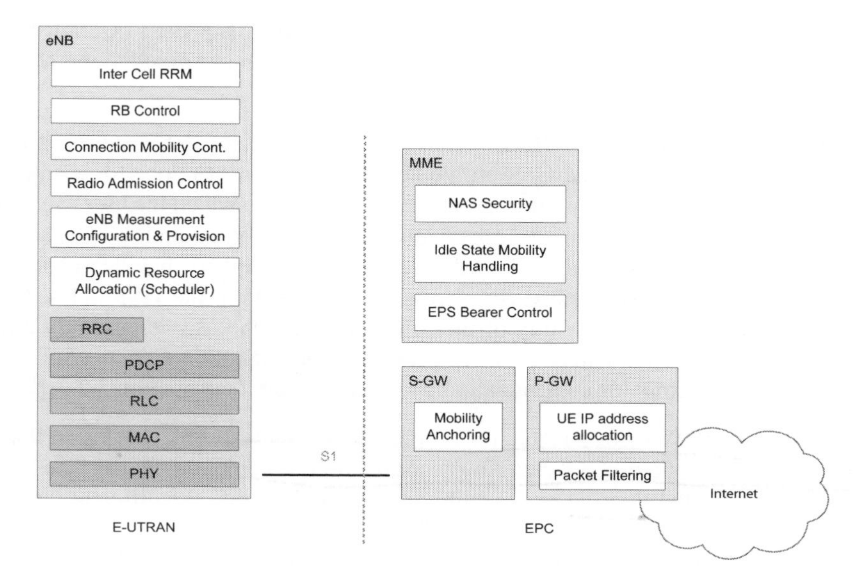

Figure 2.2: Functional split between E-UTRAN and EPC.
Reproduced by permission of © 3GPP.

2.2.1 The Core Network

The CN (called the EPC in SAE) is responsible for the overall control of the UE and the establishment of the bearers. The main logical nodes of the EPC are:

- PDN Gateway (P-GW);
- Serving GateWay (S-GW);
- Mobility Management Entity (MME) ;
- Evolved Serving Mobile Location Centre (E-SMLC).

In addition to these nodes, the EPC also includes other logical nodes and functions such as the Gateway Mobile Location Centre (GMLC), the Home Subscriber Server (HSS) and the Policy Control and Charging Rules Function (PCRF). Since the EPS only provides a bearer path of a certain QoS, control of multimedia applications such as VoIP is provided by the IMS which is considered to be outside the EPS itself. When a user is roaming outside his home country network, the user's P-GW, GMLC and IMS domain may be located in either the home network or the visited network. The logical CN nodes (specified in [1]) are shown in Figure 2.1 and discussed in more detail below.

- **PCRF.** The PCRF is responsible for policy control decision-making, as well as for controlling the flow-based charging functionalities in the Policy Control Enforcement Function (PCEF) which resides in the P-GW. The PCRF provides the QoS authorization (QoS class identifier and bit rates) that decides how a certain data flow will be treated in the PCEF and ensures that this is in accordance with the user's subscription profile.

- **GMLC.** The GMLC contains functionalities required to support LoCation Services (LCS). After performing authorization, it sends positioning requests to the MME and receives the final location estimates.

- **Home Subscriber Server (HSS).** The HSS contains users' SAE subscription data such as the EPS-subscribed QoS profile and any access restrictions for roaming (see Section 2.2.3). It also holds information about the PDNs to which the user can connect. This could be in the form of an Access Point Name (APN) (which is a label according to DNS[2] naming conventions describing the access point to the PDN), or a PDN Address (indicating subscribed IP address(es)). In addition, the HSS holds dynamic information such as the identity of the MME to which the user is currently attached or registered. The HSS may also integrate the Authentication Centre (AuC) which generates the vectors for authentication and security keys (see Section 3.2.3.1).

- **P-GW.** The P-GW is responsible for IP address allocation for the UE, as well as QoS enforcement and flow-based charging according to rules from the PCRF. The P-GW is responsible for the filtering of downlink user IP packets into the different QoS-based bearers. This is performed based on Traffic Flow Templates (TFTs) (see Section 2.4). The P-GW performs QoS enforcement for Guaranteed Bit Rate (GBR) bearers. It also serves as the mobility anchor for inter-working with non-3GPP technologies such as CDMA2000 and WiMAX networks (see Section 2.4.2 and Chapter 22 for more information about mobility).

- **S-GW.** All user IP packets are transferred through the S-GW, which serves as the local mobility anchor for the data bearers when the UE moves between eNodeBs. It also retains the information about the bearers when the UE is in idle state (known as EPS Connection Management IDLE (ECM-IDLE), see Section 2.2.1.1) and temporarily buffers downlink data while the MME initiates paging of the UE to re-establish the bearers. In addition, the S-GW performs some administrative functions in the visited network, such as collecting information for charging (e.g. the volume of data sent to or received from the user) and legal interception. It also serves as the mobility anchor for inter-working with other 3GPP technologies such as GPRS[3] and UMTS[4] (see Section 2.4.2 and Chapter 22 for more information about mobility).

- **MME.** The MME is the control node which processes the signalling between the UE and the CN. The protocols running between the UE and the CN are known as the *Non-Access Stratum* (NAS) protocols.

  The main functions supported by the MME are classified as:

  Functions related to bearer management. This includes the establishment, maintenance and release of the bearers, and is handled by the session management layer in the NAS protocol.

  Functions related to connection management. This includes the establishment of the connection and security between the network and UE, and is handled by the connection or mobility management layer in the NAS protocol layer.

---

[2]Domain Name System.

[3]General Packet Radio Service.

[4]Universal Mobile Telecommunications System.

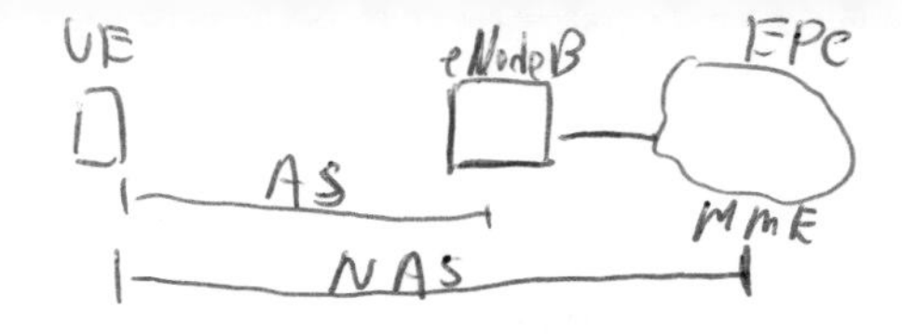

NAS control procedures are specified in [1] and are discussed in more detail in the following section.

Functions related to inter-working with other networks. This includes handing over of voice calls to legacy networks and is explained in more detail in Section 2.4.2.

- **E-SMLC.** The E-SMLC manages the overall coordination and scheduling of resources required to find the location of a UE that is attached to E-UTRAN. It also calculates the final location based on the estimates it receives, and it estimates the UE speed and the achieved accuracy. The positioning functions and protocols are explained in detail in Chapter 19.

2.2.1.1 Non-Access Stratum (NAS) Procedures

The NAS procedures, especially the connection management procedures, are fundamentally similar to UMTS. The main change from UMTS is that EPS allows concatenation of some procedures so as to enable faster establishment of the connection and the bearers.

The MME creates a *UE context*when a UE is turned on and attaches to the network. It assigns to the UE a unique short temporary identity termed the SAE-Temporary Mobile Subscriber Identity (S-TMSI) which identifies the UE context in the MME. This UE context holds user subscription information downloaded from the HSS. The local storage of subscription data in the MME allows faster execution of procedures such as bearer establishment since it removes the need to consult the HSS every time. In addition, the UE context also holds dynamic information such as the list of bearers that are established and the terminal capabilities.

To reduce the overhead in the E-UTRAN and the processing in the UE, all UE-related information in the access network can be released during long periods of data inactivity. The UE is then in the ECM-IDLE state. The MME retains the UE context and the information about the established bearers during these idle periods.

To allow the network to contact an ECM-IDLE UE, the UE updates the network as to its new location whenever it moves out of its current Tracking Area (TA); this procedure is called a 'Tracking Area Update'. The MME is responsible for keeping track of the user location while the UE is in ECM-IDLE.

When there is a need to deliver downlink data to an ECM-IDLE UE, the MME sends a paging message to all the eNodeBs in its current TA, and the eNodeBs page the UE over the radio interface. On receipt of a paging message, the UE performs a service request procedure which results in moving the UE to the ECM-CONNECTED state. UE-related information is thereby created in the E-UTRAN, and the bearers are re-established. The MME is responsible for the re-establishment of the radio bearers and updating the UE context in the eNodeB. This transition between the UE states is called an 'idle-to-active transition'. To speed up the idle-to-active transition and bearer establishment, EPS supports concatenation of the NAS and AS[5] procedures for bearer activation (see also Section 2.4.1). Some inter-relationship between the NAS and AS protocols is intentionally used to allow procedures to run simultaneously, rather than sequentially as in UMTS. For example, the bearer establishment procedure can be executed by the network without waiting for the completion of the security procedure.

[5] Access Stratum – the protocols which run between the eNodeBs and the UE.

Security functions are the responsibility of the MME for both signalling and user data. When a UE attaches with the network, a mutual authentication of the UE and the network is performed between the UE and the MME/HSS. This authentication function also establishes the security keys which are used for encryption of the bearers, as explained in Section 3.2.3.1. The security architecture for SAE is specified in [2].

The NAS also handles IMS Emergency calls, whereby UEs without regular access to the network (i.e. terminals without a Universal Subscriber Identity Module (USIM) or UEs in limited service mode) are allowed access to the network using an 'Emergency Attach' procedure; this bypasses the security requirements but only allows access to an emergency P-GW.

2.2.2 The Access Network

The access network of LTE, E-UTRAN, simply consists of a network of eNodeBs, as illustrated in Figure 2.3. For normal user traffic (as opposed to broadcast), there is no centralized controller in E-UTRAN; hence the E-UTRAN architecture is said to be flat.

Important Concept

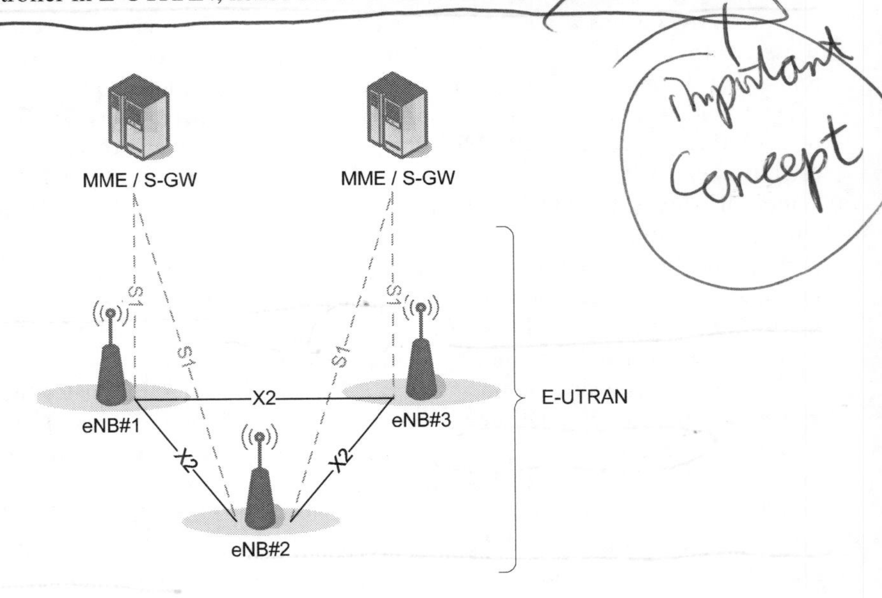

Figure 2.3: Overall E-UTRAN architecture. Reproduced by permission of © 3GPP.

The eNodeBs are normally inter-connected with each other by means of an interface known as *X2*, and to the EPC by means of the *S1 interface* – more specifically, to the MME by means of the S1-MME interface and to the S-GW by means of the S1-U interface.

The protocols which run between the eNodeBs and the UE are known as the *Access Stratum* (AS) protocols.

The E-UTRAN is responsible for all radio-related functions, which can be summarized briefly as:

- **Radio Resource Management.** This covers all functions related to the radio bearers, such as radio bearer control, radio admission control, radio mobility control, scheduling and dynamic allocation of resources to UEs in both uplink and downlink.
- **Header Compression.** This helps to ensure efficient use of the radio interface by compressing the IP packet headers which could otherwise represent a significant overhead, especially for small packets such as VoIP (see Section 4.2.2).
- **Security.** All data sent over the radio interface is encrypted (see Sections 3.2.3.1 and 4.2.3).
- **Positioning.** The E-UTRAN provides the necessary measurements and other data to the E-SMLC and assists the E-SMLC in finding the UE position (see Chapter 19).
- **Connectivity to the EPC.** This consists of the signalling towards the MME and the bearer path towards the S-GW.

On the network side, all of these functions reside in the eNodeBs, each of which can be responsible for managing multiple cells. Unlike some of the previous second- and third-generation technologies, LTE integrates the radio controller function into the eNodeB. This allows tight interaction between the different protocol layers of the radio access network, thus reducing latency and improving efficiency. Such distributed control eliminates the need for a high-availability, processing-intensive controller, which in turn has the potential to reduce costs and avoid 'single points of failure'. Furthermore, as LTE does not support soft handover there is no need for a centralized data-combining function in the network.

One consequence of the lack of a centralized controller node is that, as the UE moves, the network must transfer all information related to a UE, i.e. the UE context, together with any buffered data, from one eNodeB to another. As discussed in Section 2.3.1.1, mechanisms are therefore needed to avoid data loss during handover. The operation of the X2 interface for this purpose is explained in more detail in Section 2.6.

An important feature of the S1 interface linking the access network to the CN is known as *S1-flex*. This is a concept whereby multiple CN nodes (MME/S-GWs) can serve a common geographical area, being connected by a mesh network to the set of eNodeBs in that area (see Section 2.5). An eNodeB may thus be served by multiple MME/S-GWs, as is the case for eNodeB#2 in Figure 2.3. The set of MME/S-GW nodes serving a common area is called an *MME/S-GW pool* , and the area covered by such a pool of MME/S-GWs is called a *pool area*. This concept allows UEs in the cell(s) controlled by one eNodeB to be shared between multiple CN nodes, thereby providing a possibility for load sharing and also eliminating single points of failure for the CN nodes. The UE context normally remains with the same MME as long as the UE is located within the pool area.

2.2.3 Roaming Architecture

A network run by one operator in one country is known as a Public Land Mobile Network (PLMN). Roaming, where users are allowed to connect to PLMNs other than those to which they are directly subscribed, is a powerful feature for mobile networks, and LTE/SAE is no exception. A roaming user is connected to the E-UTRAN, MME and S-GW of the visited LTE network. However, LTE/SAE allows the P-GW of either the visited or the home network to be used, as shown in Figure 2.4. Using the home network's P-GW allows the user to access the home operator's services even while in a visited network. A P-GW in the visited network allows a 'local breakout' to the Internet in the visited network.

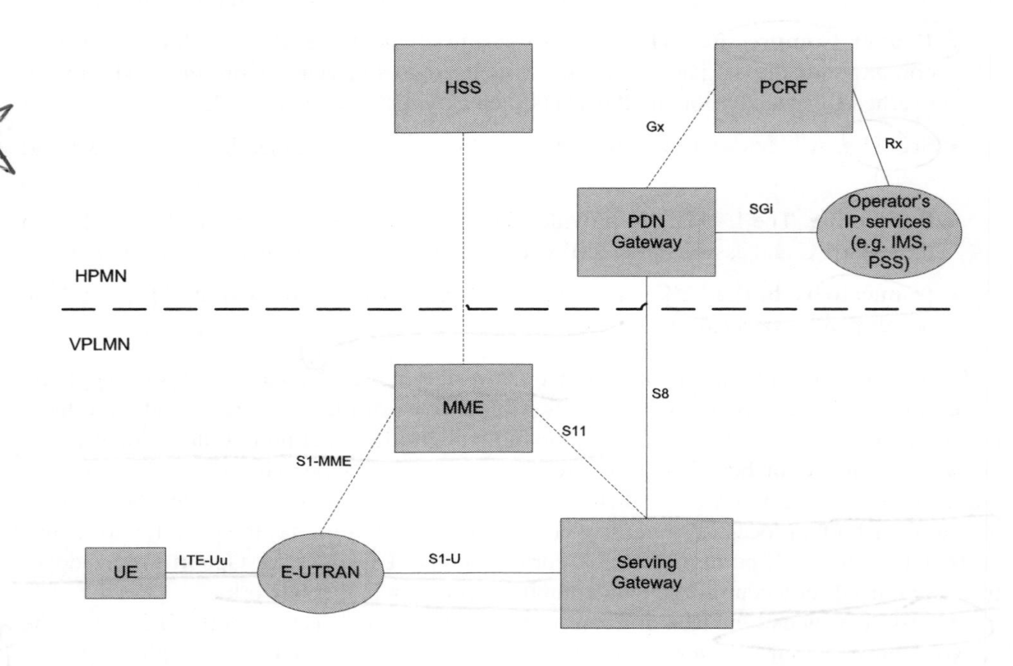

Figure 2.4: Roaming architecture for 3GPP accesses with P-GW in home network.

2.3 Protocol Architecture

We outline here the radio protocol architecture of E-UTRAN.

2.3.1 User Plane

An IP packet for a UE is encapsulated in an EPC-specific protocol and tunnelled between the P-GW and the eNodeB for transmission to the UE. Different tunnelling protocols are used across different interfaces. A 3GPP-specific tunnelling protocol called the GPRS Tunnelling Protocol (GTP) [4] is used over the core network interfaces, S1 and S5/S8.[6]

The E-UTRAN user plane protocol stack, shown greyed in Figure 2.5, consists of the Packet Data Convergence Protocol (PDCP), Radio Link Control (RLC) and Medium Access Control (MAC) sublayers which are terminated in the eNodeB on the network side. The respective roles of each of these layers are explained in detail in Chapter 4.

[6]SAE also provides an option to use Proxy Mobile IP (PMIP) on S5/S8. More details on the MIP-based S5/S8 interface can be found in [3].

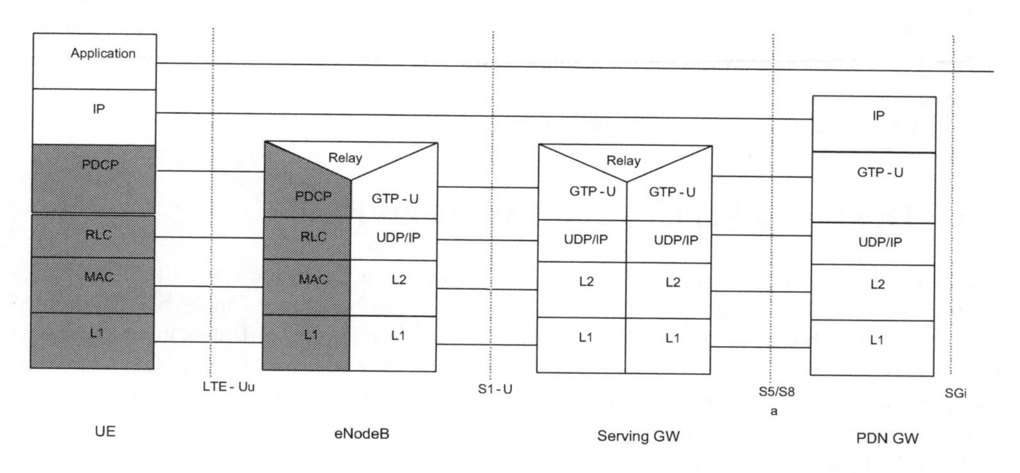

Figure 2.5: The E-UTRAN user plane protocol stack. Reproduced by permission of © 3GPP.

2.3.1.1 Data Handling During Handover

In the absence of any centralized controller node, data buffering during handover due to user mobility in the E-UTRAN must be performed in the eNodeB itself. Data protection during handover is a responsibility of the PDCP layer and is explained in detail in Section 4.2.4. The RLC and MAC layers both start afresh in a new cell after handover is completed.

2.3.2 Control Plane

The protocol stack for the control plane between the UE and MME is shown in Figure 2.6.

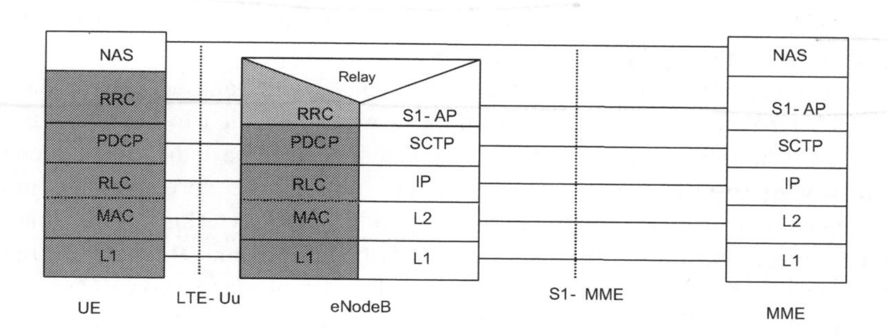

Figure 2.6: Control plane protocol stack. Reproduced by permission of © 3GPP.

The greyed region of the stack indicates the AS protocols. The lower layers perform the same functions as for the user plane with the exception that there is no header compression function for control plane.

The Radio Resource Control (RRC) protocol is known as 'Layer 3' in the AS protocol stack. It is the main controlling function in the AS, being responsible for establishing the radio bearers and configuring all the lower layers using RRC signalling between the eNodeB and the UE. These functions are detailed in Section 3.2.

2.4 Quality of Service and EPS Bearers

In a typical case, multiple applications may be running in a UE at the same time, each one having different QoS requirements. For example, a UE can be engaged in a VoIP call while at the same time browsing a web page or downloading an FTP file. VoIP has more stringent requirements for QoS in terms of delay and delay jitter than web browsing and FTP, while the latter requires a much lower packet loss rate. In order to support multiple QoS requirements, different bearers are set up within EPS, each being associated with a QoS.

Broadly, bearers can be classified into two categories based on the nature of the QoS they provide:

- **Minimum Guaranteed Bit Rate (GBR) bearers** which can be used for applications such as VoIP. These have an associated GBR value for which dedicated transmission resources are permanently allocated (e.g. by an admission control function in the eNodeB) at bearer establishment/modification. Bit rates higher than the GBR may be allowed for a GBR bearer if resources are available. In such cases, a Maximum Bit Rate (MBR) parameter, which can also be associated with a GBR bearer, sets an upper limit on the bit rate which can be expected from a GBR bearer.

- **Non-GBR bearers** which do not guarantee any particular bit rate. These can be used for applications such as web browsing or FTP transfer. For these bearers, no bandwidth resources are allocated permanently to the bearer.

In the access network, it is the eNodeB's responsibility to ensure that the necessary QoS for a bearer over the radio interface is met. Each bearer has an associated Class Identifier (QCI), and an Allocation and Retention Priority (ARP).

Each QCI is characterized by priority, packet delay budget and acceptable packet loss rate. The QCI label for a bearer determines the way it is handled in the eNodeB. Only a dozen such QCIs have been standardized so that vendors can all have the same understanding of the underlying service characteristics and thus provide the corresponding treatment, including queue management, conditioning and policing strategy. This ensures that an LTE operator can expect uniform traffic handling behaviour throughout the network regardless of the manufacturers of the eNodeB equipment. The set of standardized QCIs and their characteristics (from which the PCRF in an EPS can select) is provided in Table 2.1 (from Section 6.1.7 in [5]).

The priority and packet delay budget (and, to some extent, the acceptable packet loss rate) from the QCI label determine the RLC mode configuration (see Section 4.3.1), and how the scheduler in the MAC (see Section 4.4.2.1) handles packets sent over the bearer (e.g. in terms of scheduling policy, queue management policy and rate shaping policy). For example, a packet with a higher priority can be expected to be scheduled before a packet with lower priority. For bearers with a low acceptable loss rate, an Acknowledged Mode (AM) can be

Table 2.1: Standardized QoS Class Identifiers (QCIs) for LTE.

QCI	Resource type	Priority	Packet delay budget (ms)	Packet error loss rate	Example services
1	GBR	2	100	10^{-2}	Conversational voice
2	GBR	4	150	10^{-3}	Conversational video (live streaming)
3	GBR	5	300	10^{-6}	Non-conversational video (buffered streaming)
4	GBR	3	50	10^{-3}	Real time gaming
5	Non-GBR	1	100	10^{-6}	IMS signalling
6	Non-GBR	7	100	10^{-3}	Voice, video (live streaming), interactive gaming
7	Non-GBR	6	300	10^{-6}	Video (buffered streaming)
8	Non-GBR	8	300	10^{-6}	TCP-based (e.g. WWW, e-mail) chat, FTP, p2p file sharing, progressive video, etc.
9	Non-GBR	9	300	10^{-6}	

used within the RLC protocol layer to ensure that packets are delivered successfully across the radio interface (see Section 4.3.1.3).

The ARP of a bearer is used for call admission control – i.e. to decide whether or not the requested bearer should be established in case of radio congestion. It also governs the prioritization of the bearer for pre-emption with respect to a new bearer establishment request. Once successfully established, a bearer's ARP does not have any impact on the bearer-level packet forwarding treatment (e.g. for scheduling and rate control). Such packet forwarding treatment should be solely determined by the other bearer-level QoS parameters such as QCI, GBR and MBR.

An EPS bearer has to cross multiple interfaces as shown in Figure 2.7 – the S5/S8 interface from the P-GW to the S-GW, the S1 interface from the S-GW to the eNodeB, and the radio interface (also known as the LTE-Uu interface) from the eNodeB to the UE. Across each interface, the EPS bearer is mapped onto a lower layer bearer, each with its own bearer identity. Each node must keep track of the binding between the bearer IDs across its different interfaces.

An S5/S8 bearer transports the packets of an EPS bearer between a P-GW and an S-GW. The S-GW stores a one-to-one mapping between an S1 bearer and an S5/S8 bearer. The bearer is identified by the GTP tunnel ID across both interfaces.

An S1 bearer transports the packets of an EPS bearer between an S-GW and an eNodeB. A radio bearer [6] transports the packets of an EPS bearer between a UE and an eNodeB. An E-UTRAN Radio Access Bearer (E-RAB) refers to the concatenation of an S1 bearer and the corresponding radio bearer. An eNodeB stores a one-to-one mapping between a radio bearer ID and an S1 bearer to create the mapping between the two. The overall EPS bearer service architecture is shown in Figure 2.8.

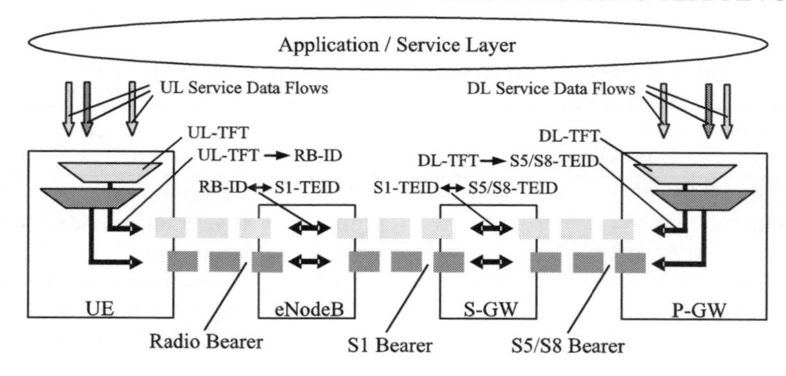

Figure 2.7: LTE/SAE bearers across the different interfaces. Reproduced by permission of © 3GPP.

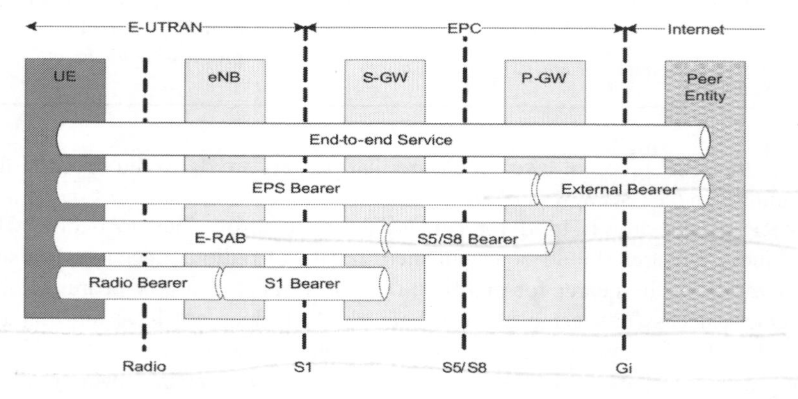

Figure 2.8: The overall EPS bearer service architecture. Reproduced by permission of © 3GPP.

IP packets mapped to the same EPS bearer receive the same bearer-level packet forwarding treatment (e.g. scheduling policy, queue management policy, rate shaping policy, RLC configuration). Providing different bearer-level QoS thus requires that a separate EPS bearer is established for each QoS flow, and user IP packets must be filtered into the different EPS bearers.

Packet filtering into different bearers is based on Traffic Flow Templates (TFTs). The TFTs use IP header information such as source and destination IP addresses and Transmission Control Protocol (TCP) port numbers to filter packets such as VoIP from web browsing traffic so that each can be sent down the respective bearers with appropriate QoS. An UpLink TFT (UL TFT) associated with each bearer in the UE filters IP packets to EPS bearers in the uplink direction. A DownLink TFT (DL TFT) in the P-GW is a similar set of downlink packet filters.

As part of the procedure by which a UE attaches to the network, the UE is assigned an IP address by the P-GW and at least one bearer is established, called the default bearer, and it remains established throughout the lifetime of the PDN connection in order to provide the UE with always-on IP connectivity to that PDN. The initial bearer-level QoS parameter values of the default bearer are assigned by the MME, based on subscription data retrieved from the HSS. The PCEF may change these values in interaction with the PCRF or according to local configuration. Additional bearers called dedicated bearers can also be established at any time during or after completion of the attach procedure. A dedicated bearer can be either GBR or non-GBR (the default bearer always has to be a non-GBR bearer since it is permanently established). The distinction between default and dedicated bearers should be transparent to the access network (e.g. E-UTRAN). Each bearer has an associated QoS, and if more than one bearer is established for a given UE, then each bearer must also be associated with appropriate TFTs. These dedicated bearers could be established by the network, based for example on a trigger from the IMS domain, or they could be requested by the UE. The dedicated bearers for a UE may be provided by one or more P-GWs.

The bearer-level QoS parameter values for dedicated bearers are received by the P-GW from the PCRF and forwarded to the S-GW. The MME only transparently forwards those values received from the S-GW over the S11 reference point to the E-UTRAN.

2.4.1 Bearer Establishment Procedure

This section describes an example of the end-to-end bearer establishment procedure across the network nodes using the functionality described in the previous sections.

A typical bearer establishment flow is shown in Figure 2.9. Each of the messages is described below.

When a bearer is established, the bearers across each of the interfaces discussed above are established.

The PCRF sends a 'PCC[7] Decision Provision' message indicating the required QoS for the bearer to the P-GW. The P-GW uses this QoS policy to assign the bearer-level QoS parameters. The P-GW then sends to the S-GW a 'Create Dedicated Bearer Request' message including the QoS and UL TFT to be used in the UE.

The S-GW forwards the Create Dedicated Bearer Request message (including bearer QoS, UL TFT and S1-bearer ID) to the MME (message 3 in Figure 2.9).

The MME then builds a set of session management configuration information including the UL TFT and the EPS bearer identity, and includes it in the 'Bearer Setup Request' message which it sends to the eNodeB (message 4 in Figure 2.9). The session management configuration is NAS information and is therefore sent transparently by the eNodeB to the UE.

The Bearer Setup Request also provides the QoS of the bearer to the eNodeB; this information is used by the eNodeB for call admission control and also to ensure the necessary QoS by appropriate scheduling of the user's IP packets. The eNodeB maps the EPS bearer QoS to the radio bearer QoS. It then signals a 'RRC Connection Reconfiguration' message (including the radio bearer QoS, session management configuration and EPS radio bearer identity) to the UE to set up the radio bearer (message 5 in Figure 2.9). The RRC Connection Reconfiguration message contains all the configuration parameters for the radio interface.

[7]Policy Control and Charging.

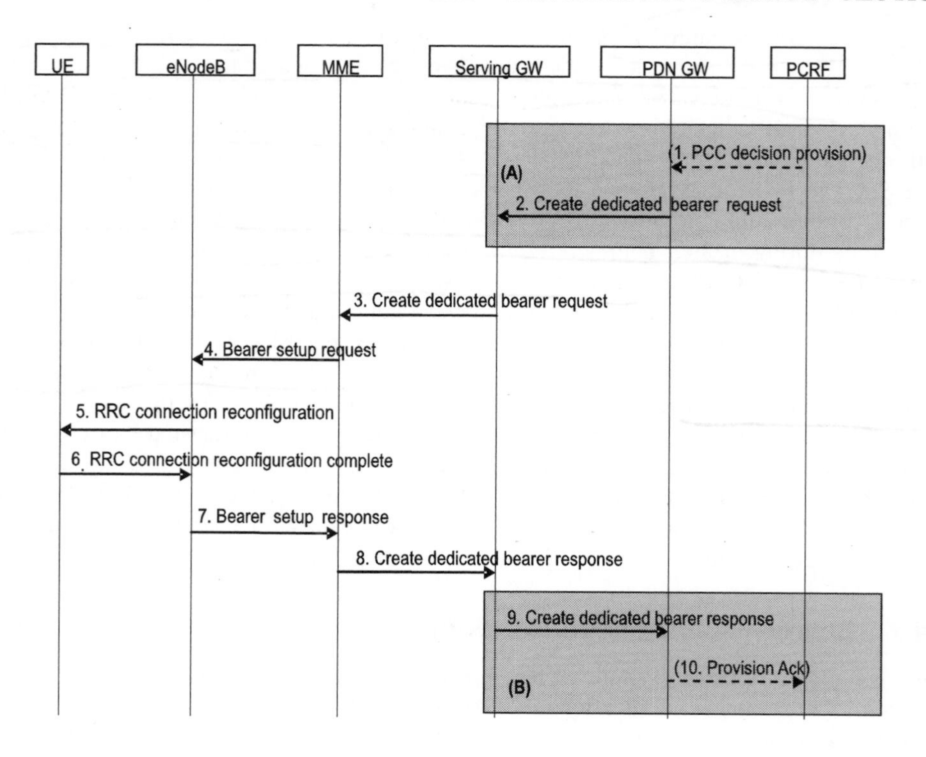

Figure 2.9: An example message flow for an LTE/SAE bearer establishment. Reproduced by permission of © 3GPP.

This is mainly for the configuration of the Layer 2 (PDCP, RLC and MAC) parameters, but also the Layer 1 parameters required for the UE to initialize the protocol stack.

Messages 6 to 10 are the corresponding response messages to confirm that the bearers have been set up correctly.

2.4.2 Inter-Working with other RATs

EPS also supports inter-working and mobility (handover) with networks using other Radio Access Technologies (RATs), notably GSM[8], UMTS, CDMA2000 and WiMAX. The architecture for inter-working with 2G and 3G GPRS/UMTS networks is shown in Figure 2.10. The S-GW acts as the mobility anchor for inter-working with other 3GPP technologies such as GSM and UMTS, while the P-GW serves as an anchor allowing seamless mobility to non-3GPP networks such as CDMA2000 or WiMAX. The P-GW may also support a Proxy Mobile Internet Protocol (PMIP) based interface. While VoIP is the primary mechanism for voice services, LTE also supports inter-working with legacy systems for CS voice services.

[8]Global System for Mobile Communications.

This is controlled by the MME and is based on two procedures outlined in Sections 2.4.2.1 and 2.4.2.2.

More details of the radio interface procedures for inter-working with other RATs are specified in [3] and covered in Sections 2.5.6.2 and 3.2.4.

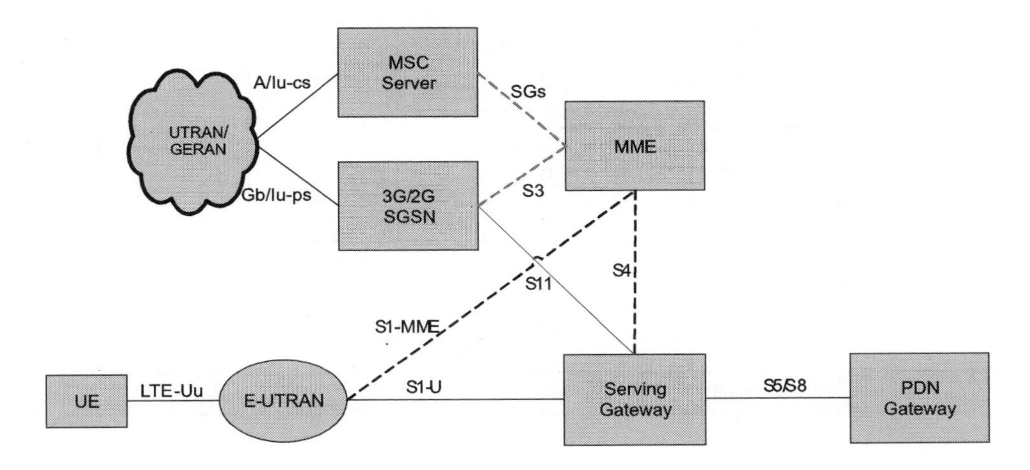

Figure 2.10: Architecture for 3G UMTS interworking.

2.4.2.1 Circuit-Switched Fall Back (CSFB)

LTE natively supports VoIP only using IMS services. However, in case IMS services are not deployed from the start, LTE also supports a Circuit-Switched FallBack (CSFB) mechanism which allows CS voice calls to be handled via legacy RATs for UEs that are camped on LTE.

CSFB allows a UE in LTE to be handed over to a legacy RAT to originate a CS voice call. This is supported by means of an interface, referred to as SGs[9], between the MME and the Mobile Switching Centre (MSC) of the legacy RAT shown in Figure 2.10. This interface allows the UE to attach with the MSC and register for CS services while still in LTE. Moreover it carries paging messages from the MSC for incoming voice calls so that UEs can be paged over LTE. The network may choose a handover, cell change order, or redirection procedure to move the UE to the legacy RAT.

Figure 2.11 shows the message flow for a CSFB call from LTE to UMTS, including paging from the MSC via the SGs interface and MME in the case of UE-terminated calls, and the sending of an Extended Service Request NAS message from the UE to the MME to trigger either a handover or redirection to the target RAT in the case of a UE-originated call. In the latter case, the UE then originates the CS call over the legacy RAT using the procedure defined in the legacy RAT specification. Further details of CSFB can be found in [7].

[9]SGs is an extension of the Gs interface between the Serving GPRS Support Node (SGSN) and the Mobile Switching Centre (MSC)

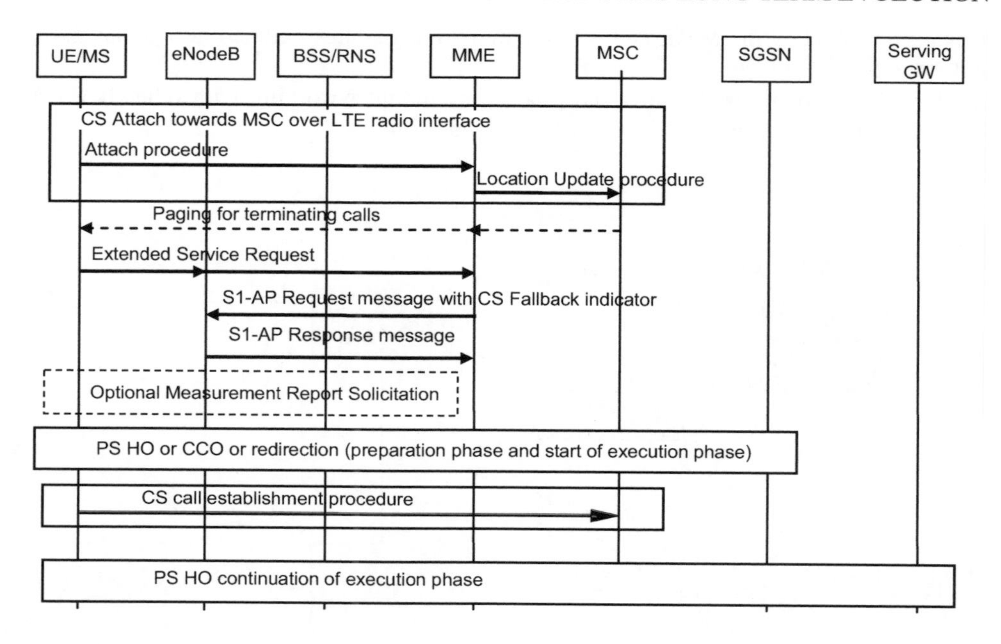

Figure 2.11: Message sequence diagram for CSFB from LTE to UMTS/GERAN.

2.4.2.2 Single Radio Voice Call Continuity (SRVCC)

If ubiquitous coverage of LTE is not available, it is possible that a UE involved in a VoIP call over LTE might then move out of LTE coverage to enter a legacy RAT cell which only offers CS voice services. The Single Radio Voice Call Continuity (SRVCC) procedure is designed for handover of a Packet Switched (PS) VoIP call over LTE to a CS voice call in the legacy RAT, involving the transfer of a PS bearer into a CS bearer.

Figure 2.12 shows an overview of the functions involved in SRVCC. The eNodeB may detect that the UE is moving out of LTE coverage and trigger a handover procedure towards the MME by means of an SRVCC indication. The MME is responsible for the SRVCC procedure and also for the transfer of the PS E-RAB carrying VoIP into a CS bearer. The MSC Server then initiates the session transfer procedure to IMS and coordinates it with the CS handover procedure to the target cell. The handover command provided to the UE to request handover to the legacy RAT also provides the information to set up the CS and PS radio bearers. The UE can continue with the call over the CS domain on completion of the handover. Further details of SRVCC can be found in [8].

2.5 The E-UTRAN Network Interfaces: S1 Interface

The S1 interface connects the eNodeB to the EPC. It is split into two interfaces, one for the control plane and the other for the user plane. The protocol structure for the S1 and the functionality provided over S1 are discussed in more detail below.

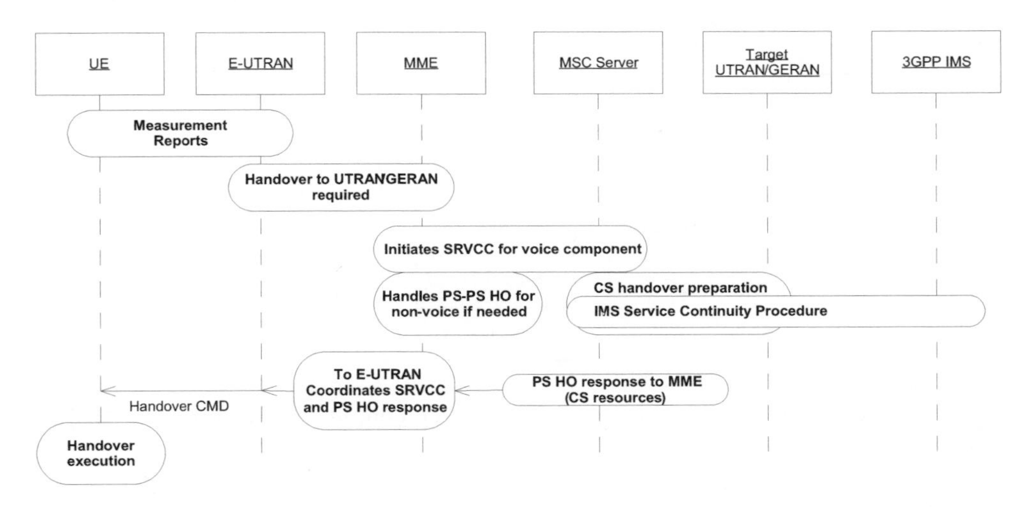

Figure 2.12: The main procedures involved in an SRVCC handover of a PS VoIP call from LTE to CS voice call in UMTS/GERAN.

2.5.1 Protocol Structure over S1

The protocol structure over S1 is based on a full IP transport stack with no dependency on legacy SS7[10] network configuration as used in GSM or UMTS networks. This simplification provides one area of potential savings on operational expenditure with LTE networks.

2.5.1.1 Control Plane

Figure 2.13 shows the protocol structure of the S1 control plane which is based on the Stream Control Transmission Protocol / IP (SCTP/IP) stack.

The SCTP protocol is well known for its advanced features inherited from TCP which ensure the required reliable delivery of the signalling messages. In addition, it makes it possible to benefit from improved features such as the handling of multistreams to implement transport network redundancy easily and avoid head-of-line blocking or multihoming (see 'IETF RFC4960' [9]).

A further simplification in LTE (compared to the UMTS Iu interface, for example) is the direct mapping of the S1-AP (S1 Application Protocol) on top of SCTP which results in a simplified protocol stack with no intermediate connection management protocol. The individual connections are directly handled at the application layer. Multiplexing takes place between S1-AP and SCTP whereby each stream of an SCTP association is multiplexed with the signalling traffic of multiple individual connections.

One further area of flexibility that comes with LTE lies in the lower layer protocols for which full optionality has been left regarding the choice of the IP version and the choice

[10] Signalling System #7 (SS7) is a communications protocol defined by the International Telecommunication Union (ITU) Telecommunication Standardization Sector (ITU-T) with a main purpose of setting up and tearing down telephone calls. Other uses include Short Message Service (SMS), number translation, prepaid billing mechanisms, and many other services.

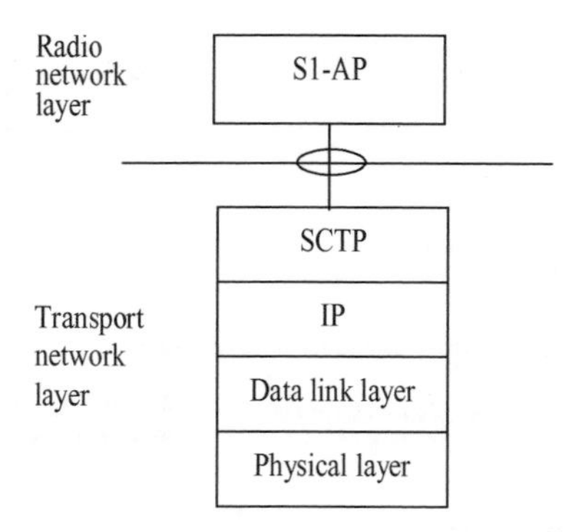

Figure 2.13: S1-MME control plane protocol stack. Reproduced by permission of © 3GPP.

of the data link layer. For example, this enables the operator to start deployment using IP version 4 with the data link tailored to the network deployment scenario.

2.5.1.2 User Plane

Figure 2.14 shows the protocol structure of the S1 user plane, which is based on the GTP/User Datagram Protocol (UDP) IP stack which is already well known from UMTS networks.

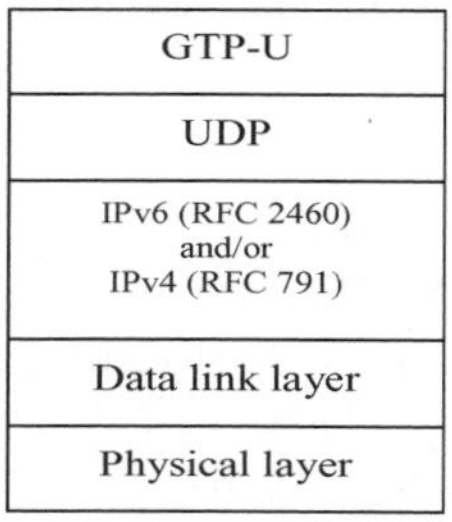

GTP-U
UDP
IPv6 (RFC 2460) and/or IPv4 (RFC 791)
Data link layer
Physical layer

Figure 2.14: S1-U user plane protocol stack. Reproduced by permission of © 3GPP.

One of the advantages of using GTP-User plane (GTP-U) is its inherent facility to identify tunnels and also to facilitate intra-3GPP mobility.

The IP version number and the data link layer have been left fully optional, as for the control plane stack.

A transport bearer is identified by the GTP tunnel endpoints and the IP address (source Tunnelling End ID (TEID), destination TEID, source IP address, destination IP address).

The S-GW sends downlink packets of a given bearer to the eNodeB IP address (received in S1-AP) associated to that particular bearer. Similarly, the eNodeB sends upstream packets of a given bearer to the EPC IP address (received in S1-AP) associated to that particular bearer.

Vendor-specific traffic categories (e.g. real-time traffic) can be mapped onto Differentiated Services (Diffserv) code points (e.g. expedited forwarding) by network O&M (Operation and Maintenance) configuration to manage QoS differentiation between the bearers.

2.5.2 Initiation over S1

The initialization of the S1-MME control plane interface starts with the identification of the MMEs to which the eNodeB must connect, followed by the setting up of the Transport Network Layer (TNL).

With the support of the S1-flex function in LTE, an eNodeB must initiate an S1 interface towards each MME node of the pool area to which it belongs. This list of MME nodes of the pool together with an initial corresponding remote IP address can be directly configured in the eNodeB at deployment (although other means may also be used). The eNodeB then initiates the TNL establishment with that IP address. Only one SCTP association is established between one eNodeB and one MME.

During the establishment of the SCTP association, the two nodes negotiate the maximum number of streams which will be used over that association. However, multiple pairs of streams[11] are typically used in order to avoid the head-of-line blocking issue mentioned above. Among these pairs of streams, one particular pair must be reserved by the two nodes for the signalling of the common procedures (i.e. those which are not specific to one UE). The other streams are used for the sole purpose of the dedicated procedures (i.e. those which are specific to one UE).

Once the TNL has been established, some basic application-level configuration data for the system operation is automatically exchanged between the eNodeB and the MME through an 'S1 SETUP' procedure initiated by the eNodeB. This procedure is one case of a Self-Optimizing Network process and is explained in detail in Section 25.3.1.

Once the S1 SETUP procedure has been completed, the S1 interface is operational.

2.5.3 Context Management over S1

Within each pool area, a UE is associated to one particular MME for all its communications during its stay in this pool area. This creates a context in this MME for the UE. This particular MME is selected by the NAS Node Selection Function (NNSF) in the first eNodeB from which the UE entered the pool.

Whenever the UE becomes active (i.e. makes a transition from idle to active mode) under the coverage of a particular eNodeB in the pool area, the MME provides the UE context information to this eNodeB using the 'INITIAL CONTEXT SETUP REQUEST' message (see Figure 2.15). This enables the eNodeB in turn to create a context and manage the UE while it is in active mode.

Even though the setup of bearers is otherwise relevant to a dedicated 'Bearer Management' procedure described below, the creation of the eNodeB context by the INITIAL CONTEXT SETUP procedure also includes the creation of one or several bearers including the default bearers.

At the next transition back to idle mode following a 'UE CONTEXT RELEASE' message sent from the MME, the eNodeB context is erased and only the MME context remains.

[11] Note that a stream is unidirectional and therefore pairs must be used.

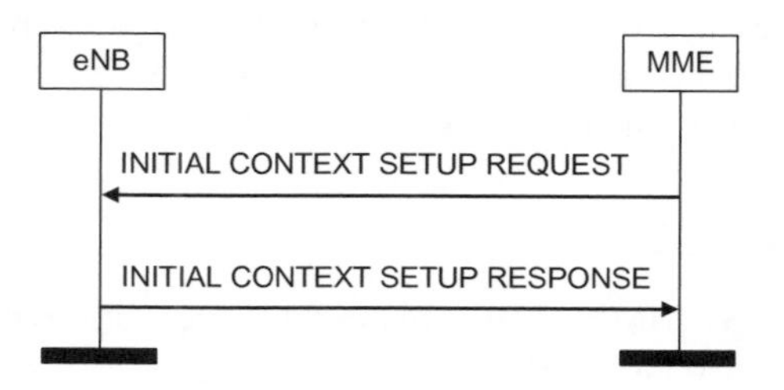

Figure 2.15: Initial context setup procedure. Reproduced by permission of © 3GPP.

2.5.4 Bearer Management over S1

LTE uses independent dedicated procedures respectively covering the setup, modification and release of bearers. For each bearer requested to be set up, the transport layer address and the tunnel endpoint are provided to the eNodeB in the 'BEARER SETUP REQUEST' message to indicate the termination of the bearer in the S-GW where uplink user plane data must be sent. Conversely, the eNodeB indicates in the 'BEARER SETUP RESPONSE' message the termination of the bearer in the eNodeB where the downlink user plane data must be sent.

For each bearer, the QoS parameters (see Section 2.4) requested for the bearer are also indicated. Independently of the standardized QCI values, it is also still possible to use extra proprietary labels for the fast introduction of new services if vendors and operators agree upon them.

2.5.5 Paging over S1

As mentioned in Section 2.5.3, in order to re-establish a connection towards a UE in idle mode, the MME distributes a 'PAGING REQUEST' message to the relevant eNodeBs based on the TAs where the UE is expected to be located. When receiving the paging request, the eNodeB sends a page over the radio interface in the cells which are contained within one of the TAs provided in that message.

The UE is normally paged using its S-TMSI. The 'PAGING REQUEST' message also contains a UE identity index value in order for the eNodeB to calculate the paging occasions at which the UE will switch on its receiver to listen for paging messages (see Section 3.4).

In Release 10, paging differentiation is introduced over the S1 interface to handle Multimedia Priority Service (MPS)[12] users. In case of MME or RAN overload, it is necessary to page a UE with higher priority during the establishment of a mobile-terminated MPS call. In case of MME overload, the MME can itself discriminate between the paging messages and discard the lower priority ones. In case of RAN overload in some cells, the eNodeB can perform this discrimination based on a new Paging Priority Indicator sent by the MME. The MME can signal up to eight such priority values to the eNodeB. In case of an IMS MPS call, the terminating UE will further set up an RRC connection with the same eNodeB that will also get automatically prioritized. In case of a CS fallback call, the eNodeB will instead

[12]MPS allows the delivery of calls or complete sessions of a high priority nature, in case for example of public safety or national security purposes, from mobile to mobile, mobile to fixed, and fixed to mobile networks during network congestion conditions.

signal to the UE that it must set the cause value 'high priority terminating call' when trying to establish the UMTS RRC Connection.

2.5.6 Mobility over S1

LTE/SAE supports mobility within LTE/SAE, and also to other systems using both 3GPP and non-3GPP technologies. The mobility procedures over the radio interface are defined in Section 3.2. These mobility procedures also involve the network interfaces. The sections below discuss the procedures over S1 to support mobility. The mobility performance requirements from the UE point of view are outlined in Chapter 22.

2.5.6.1 Intra-LTE Mobility

There are two types of handover procedure in LTE for UEs in active mode: the S1-handover procedure and the X2-handover procedure.

For intra-LTE mobility, the X2-handover procedure is normally used for the inter-eNodeB handover (described in Section 2.6.3). However, when there is no X2 interface between the two eNodeBs, or if the source eNodeB has been configured to initiate handover towards a particular target eNodeB via the S1 interface, then an S1-handover will be triggered.

The S1-handover procedure has been designed in a very similar way to the UMTS Serving Radio Network Subsystem (SRNS) relocation procedure and is shown in Figure 2.16: it consists of a preparation phase involving the core network, where the resources are first prepared at the target side (steps 2 to 8), followed by an execution phase (steps 8 to 12) and a completion phase (after step 13).

Compared to UMTS, the main difference is the introduction of the 'STATUS TRANSFER' message sent by the source eNodeB (steps 10 and 11). This message has been added in order to carry some PDCP status information that is needed at the target eNodeB in cases when PDCP status preservation applies for the S1-handover (see Section 4.2.4); this is in alignment with the information which is sent within the X2 'STATUS TRANSFER' message used for the X2-handover (see below). As a result of this alignment, the handling of the handover by the target eNodeB as seen from the UE is exactly the same, regardless of the type of handover (S1 or X2) the network had decided to use; indeed, the UE is unaware of which type of handover is used by the network.

The 'Status Transfer' procedure is assumed to be triggered in parallel with the start of data forwarding after the source eNodeB has received the 'HANDOVER COMMAND' message from the source MME. This data forwarding can be either direct or indirect, depending on the availability of a direct path for the user plane data between the source eNodeB and the target eNodeB.

The 'HANDOVER NOTIFY' message (step 13), which is sent later by the target eNodeB when the arrival of the UE at the target side is confirmed, is forwarded by the MME to trigger the update of the path switch in the S-GW towards the target eNodeB. In contrast to the X2-handover, the message is not acknowledged and the resources at the source side are released later upon reception of a 'RELEASE RESOURCE' message directly triggered from the source MME (step 17 in Figure 2.16).

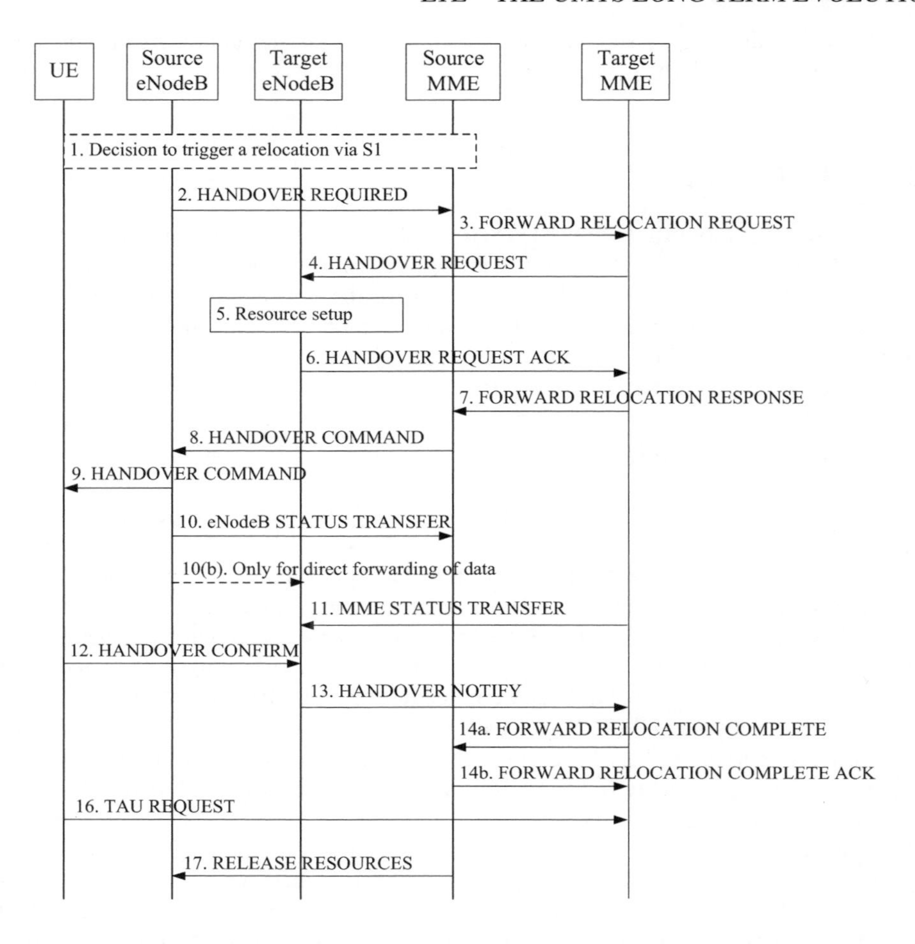

Figure 2.16: S1-based handover procedure. Reproduced by permission of © 3GPP.

2.5.6.2 Inter-RAT Mobility

One key element of the design of LTE is the need to co-exist with other Radio Access Technologies (RATs).

For mobility from LTE towards UMTS, the handover process can reuse the S1-handover procedures described above, with the exception of the 'STATUS TRANSFERŠ message which is not needed at steps 10 and 11 since no PDCP context is continued.

For mobility towards CDMA2000, dedicated uplink and downlink procedures have been introduced in LTE. They essentially aim at tunnelling the CDMA2000 signalling between the UE and the CDMA2000 system over the S1 interface, without being interpreted by the eNodeB on the way. An 'UPLINK S1 CDMA2000 TUNNELLING' message is sent from the eNodeB to the MME; this also includes the RAT type in order to identify which CDMA2000

RAT the tunnelled CDMA2000 message is associated with in order for the message to be routed to the correct node within the CDMA2000 system.

2.5.6.3 Mobility towards Home eNodeBs

Mobility towards HeNBs involves additional functions from the source LTE RAN node and the MME. In addition to the E-UTRAN Cell Global Identifier (ECGI), the source RAN node should include the Closed Subscriber Group Identity (CSG ID) and the access mode of the target HeNB in the 'HANDOVER REQUIRED' message to the MME so that the MME can perform the access control to that HeNB. If the target HeNB operates in closed access mode (see Chapter 24) and the MME fails the access control, the MME will reject the handover by sending back a 'HANDOVER PREPARATION FAILURE' message. Otherwise the MME will accept and continue the handover while indicating to the target HeNB whether the UE is a 'CSG member' if the HeNB is operating in hybrid mode. A detailed description of mobility towards the HeNB and the associated call flow is provided in Chapter 24.

2.5.7 Load Management over S1

Three types of load management procedures apply over S1: a normal 'load balancing' procedure to distribute the traffic, an 'overload' procedure to overcome a sudden peak in the loading and a 'load rebalancing' procedure to partially/fully offload an MME.

The MME load balancing procedure aims to distribute the traffic to the MMEs in the pool evenly according to their respective capacities. To achieve that goal, the procedure relies on the normal NNSF present in each eNodeB as part of the S1-flex function. Provided that suitable weight factors corresponding to the capacity of each MME node are available in the eNodeBs beforehand, a weighted NNSF done by every eNodeB in the network normally achieves a statistically balanced distribution of load among the MME nodes without further action. However, specific actions are still required for some particular scenarios:

- If a new MME node is introduced (or removed), it may be necessary temporarily to increase (or decrease) the weight factor normally corresponding to the capacity of this node in order to make it catch more (or less) traffic at the beginning until it reaches an adequate level of load.

- In case of an unexpected peak in the loading, an 'OVERLOAD' message can be sent over the S1 interface by the overloaded MME. When received by an eNodeB, this message calls for a temporary restriction of a certain type of traffic. An MME can adjust the reduction of traffic it desires by defining the number of eNodeBs to which it sends the 'OVERLOAD' message and by defining the types of traffic subject to restriction.Two new rejection types are introduced in Release 10 to combat CN Overload:

  - 'reject low priority access', which can be used by the MME to reduce access of some low-priority devices or applications such as Machine-Type Communication (MTC) devices (see Section 31.4);
  - 'permit high priority sessions', to allow access only to high-priority users and mobile-terminated services.

- Finally, if the MME wants to force rapidly the offload of part or all of its UEs, it will use the rebalancing function. This function forces the UEs to reattach to another MME by using a specific 'cause value' in the 'UE Release Command S1' message. In a first step it applies to idle mode UEs and in a second step it may also apply to UEs in connected mode (if the full MME offload is desired, e.g. for maintenance reasons).

2.5.8 Trace Function

In order to trace the activity of a UE in connected mode, two types of trace session can be started in the eNodeB:

- Signalling-Based Trace. This is triggered by the MME and is uniquely identified by a trace identity. Only one trace session can be activated at a time for one UE. The MME indicates to the eNodeB the interfaces to trace (e.g. S1, X2, Uu) and the associated trace depth. The trace depth represents the granularity of the signalling to be traced from the high-level messages down to the detailed ASN.1[13] and is comprised of three levels: minimum, medium and maximum. The MME also indicates the IP address of a Trace Collection Entity where the eNodeB must send the resulting trace record file. If an X2 handover preparation has started at the time when the eNodeB receives the order to trace, the eNodeB will signal back a TRACE FAILURE INDICATION message to the MME, and it is then up to the MME to take appropriate action based on the indicated failure reason. Signalling-based traces are propagated at X2 and S1 handover.

- Management-Based Trace. This is triggered in the eNodeB when the conditions required for tracing set by O&M are met. The eNodeB then allocates a trace identity that it sends to the MME in a CELL TRAFFIC TRACE message over S1, together with the Trace Collection Entity identity that shall be used by the MME for the trace record file (in order to assemble the trace correctly in the Trace Collection Entity). Management-based traces are propagated at X2 and S1 handover.

In Release 10, the trace function supports the Minimization of Drive Tests (MDT) feature, which is explained in Section 31.3.

2.5.9 Delivery of Warning Messages

Two types of warning message may need to be delivered with the utmost urgency over a cellular system, namely Earthquake and Tsunami Warning System (ETWS)) messages and Commercial Mobile Alert System (CMAS) messages (see Section 13.7). The delivery of ETWS messages is already supported since Release 8 via the S1 Write-Replace Warning procedure which makes it possible to carry either primary or secondary notifications over S1 for the eNodeB to broadcast over the radio. The Write-Replace Warning procedure also includes a Warning Area List where the warning message needs to be broadcast. It can be a list of cells, tracking areas or emergency area identities. The procedure also contains information on how the broadcast is to be performed (for example, the number of broadcasts requested).

[13] Abstract Syntax Notation One

In contrast to ETWS, the delivery of CMAS messages is only supported from Release 9 onwards. One difference between the two public warning systems is that in ETWS the eNodeB can only broadcast one message at a time, whereas CMAS allows the broadcast of multiple concurrent warning messages over the radio. Therefore an ongoing ETWS broadcast needs to be overwritten if a new ETWS warning has to be delivered immediately in the same cell. With CMAS, a new Kill procedure has also been added to allow easy cancellation of an ongoing broadcast when needed. This Kill procedure includes the identity of the message to be stopped and the Warning Area where it is to be stopped.

2.6 The E-UTRAN Network Interfaces: X2 Interface

The X2 interface is used to inter-connect eNodeBs. The protocol structure for the X2 interface and the functionality provided over X2 are discussed below.

2.6.1 Protocol Structure over X2

The control plane and user plane protocol stacks over the X2 interface are the same as over the S1 interface, as shown in Figures 2.17 and 2.18 respectively (with the exception that in Figure 2.17 the X2-AP (X2 Application Protocol) is substituted for the S1-AP). This also means again that the choice of the IP version and the data link layer are fully optional. The use of the same protocol structure over both interfaces provides advantages such as simplifying the data forwarding operation.

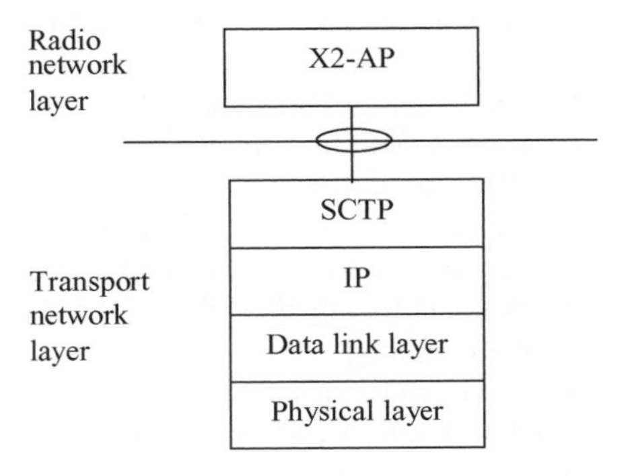

Figure 2.17: X2 signalling bearer protocol stack. Reproduced by permission of © 3GPP.

2.6.2 Initiation over X2

The X2 interface may be established between one eNodeB and some of its neighbour eNodeBs in order to exchange signalling information when needed. However, a full mesh is not mandated in an E-UTRAN network. Two types of information may typically need to be exchanged over X2 to drive the establishment of an X2 interface between two eNodeBs:

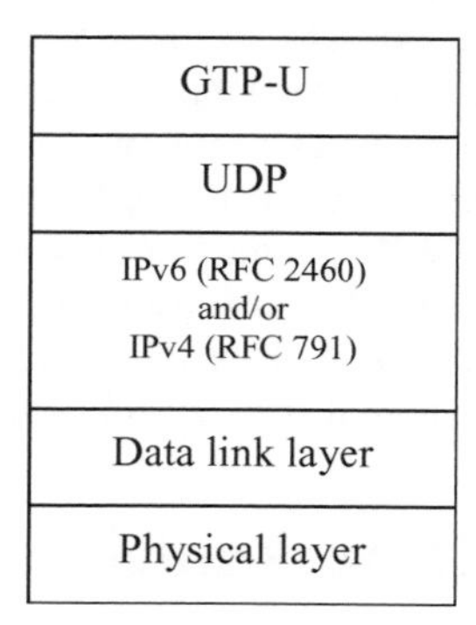

Figure 2.18: Transport network layer for data streams over X2. Reproduced by permission of © 3GPP.

load or interference related information (see Section 2.6.4) and handover related information (see mobility in Section 2.6.3).

Because these two types of information are fully independent of one another, it is possible that an X2 interface may be present between two eNodeBs for the purpose of exchanging load or interference information, even though the X2-handover procedure is not used to hand over UEs between those eNodeBs.[14]

The initialization of the X2 interface starts with the identification of a suitable neighbour followed by the setting up of the TNL.

The identification of a suitable neighbour may be done by configuration, or alternatively by a self-optimizing process known as the Automatic Neighbour Relation Function (ANRF).[15] This is described in more detail in Section 25.2.

Once a suitable neighbour has been identified, the initiating eNodeB can further set up the TNL using the transport layer address of this neighbour – either as retrieved from the network or locally configured. The automatic retrieval of the X2 IP address(es) via the network and the eNodeB Configuration Transfer procedure are described in details in Section 25.3.2.

Once the TNL has been set up, the initiating eNodeB must trigger the X2 setup procedure. This procedure enables an automatic exchange of application level configuration data relevant to the X2 interface, similar to the S1 setup procedure already described in Section 2.5.2. For example, each eNodeB reports within the 'X2 SETUP REQUESTŠ message to a neighbour eNodeB information about each cell it manages, such as the cell's physical identity, the frequency band, the tracking area identity and/or the associated PLMNs.

This automatic exchange of application-level configuration data within the X2 setup procedure is also the core of two additional SON features: automatic self-configuration of the Physical Cell Identities (PCIs) and RACH self-optimization. These features both aim to avoid conflicts between cells controlled by neighbouring eNodeBs; they are explained in detail in Sections 25.4 and 25.7 respectively.

Once the X2 setup procedure has been completed, the X2 interface is operational.

[14]In such a case, the S1-handover procedure is used instead.

[15]Under this function the UEs are requested to detect neighbour eNodeBs by reading the Cell Global Identity (CGI) contained in the broadcast information.

2.6.3 Mobility over X2

Handover via the X2 interface is triggered by default unless there is no X2 interface established or the source eNodeB is configured to use the S1-handover instead.

The X2-handover procedure is illustrated in Figure 2.19. Like the S1-handover, it is also composed of a preparation phase (steps 4 to 6), an execution phase (steps 7 to 9) and a completion phase (after step 9).

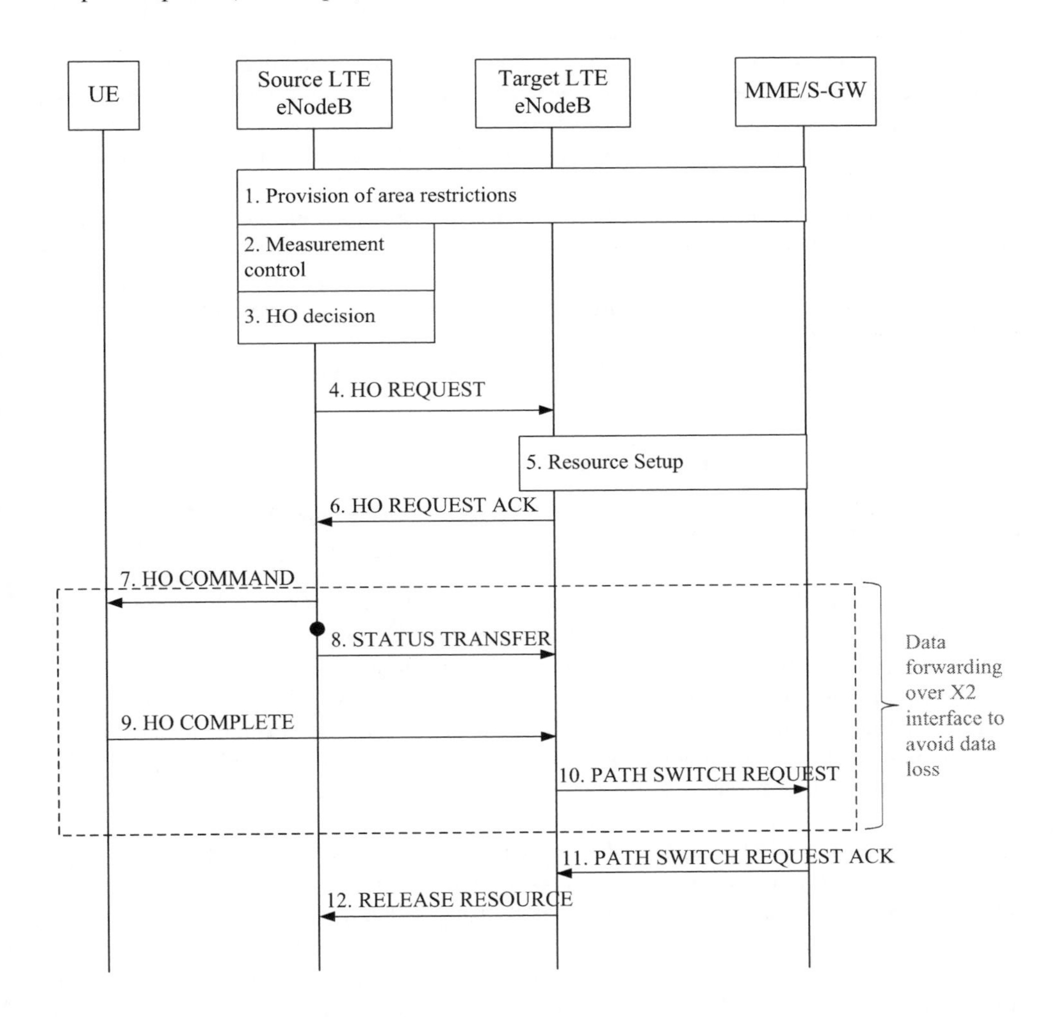

Figure 2.19: X2-based handover procedure.

The key features of the X2-handover for intra-LTE handover are:

- The handover is directly performed between two eNodeBs. This makes the preparation phase quick.

- Data forwarding may be operated per bearer in order to minimize data loss.
- The MME is only informed at the end of the handover procedure once the handover is successful, in order to trigger the path switch.
- The release of resources at the source side is directly triggered from the target eNodeB.

For those bearers for which in-sequence delivery of packets is required, the 'STATUS TRANSFER' message (step 8) provides the Sequence Number (SN) and the Hyper Frame Number (HFN) which the target eNodeB should assign to the first packet with no sequence number yet assigned that it must deliver. This first packet can either be one received over the target S1 path or one received over X2 if data forwarding over X2 is used (see below). When it sends the 'STATUS TRANSFER' message, the source eNodeB freezes its transmitter/receiver status – i.e. it stops assigning PDCP SNs to downlink packets and stops delivering uplink packets to the EPC.

Mobility over X2 can be categorized according to its resilience to packet loss: the handover can be said 'seamless' if it minimizes the interruption time during the move of the UE, or 'lossless' if it tolerates no loss of packets at all. These two modes use data forwarding of user plane downlink packets. The source eNodeB may decide to operate one of these two modes on a per-EPS-bearer basis, based on the QoS received over S1 for this bearer (see Section 2.5.4) and the service in question. These two modes are described in more detail below.

2.6.3.1 Seamless Handover

If, for a given bearer, the source eNodeB selects the seamless handover mode, it proposes to the target eNodeB in the 'HANDOVER REQUEST' message to establish a GTP tunnel to operate the downlink data forwarding. If the target eNodeB accepts, it indicates in the 'HANDOVER REQUEST ACK' message the tunnel endpoint where the forwarded data is expected to be received. This tunnel endpoint may be different from the one set up as the termination point of the new bearer established over the target S1.

Upon reception of the 'HANDOVER REQUEST ACK' message, the source eNodeB can start forwarding the data freshly arriving over the source S1 path towards the indicated tunnel endpoint in parallel with sending the handover trigger to the UE over the radio interface. This forwarded data is thus available at the target eNodeB to be delivered to the UE as early as possible.

When forwarding is in operation and in-sequence delivery of packets is required, the target eNodeB is assumed to deliver first the packets forwarded over X2 before delivering the first ones received over the target S1 path once the S1 path switch has been performed. The end of the forwarding is signalled over X2 to the target eNodeB by the reception of some 'special GTP packets' which the S-GW has inserted over the source S1 path just before switching this S1 path; these are then forwarded by the source eNodeB over X2 like any other regular packets.

2.6.3.2 Lossless Handover

If the source eNodeB selects the lossless mode for a given bearer, it will additionally forward over X2 those user plane downlink packets which it has PDCP processed but are still buffered locally because they have not yet been delivered and acknowledged by the UE. These packets are forwarded together with their assigned PDCP SN included in a GTP extension header field. They are sent over X2 prior to the freshly arriving packets from the source S1 path. The same mechanisms described above for the seamless handover are used for the GTP tunnel establishment. The end of forwarding is also handled in the same way, since in-sequence packet delivery applies to lossless handovers. In addition, the target eNodeB must ensure that all the packets – including the ones received with sequence number over X2 – are delivered in sequence at the target side. Further details of seamless and lossless handover are described in Section 4.2.

Selective retransmission. A new feature in LTE compared to previous systems is the optimization of the radio interface usage by selective retransmission. When lossless handover is operated, the target eNodeB may, however, not deliver over the radio interface some of the forwarded downlink packets received over X2 if it is informed by the UE that those packets have already been received at the source side (see Section 4.2.6). This is called downlink selective retransmission.

Similarly in the uplink, the target eNodeB may desire that the UE does not retransmit packets already received earlier at the source side by the source eNodeB, for example to avoid wasting radio resources. To operate this uplink selective retransmission scheme for one bearer, it is necessary that the source eNodeB forwards to the target eNodeB, over another new GTP tunnel, those user plane uplink packets which it has received out of sequence. The target eNodeB must first request the source eNodeB to establish this new forwarding tunnel by including in the 'HANDOVER REQUEST ACK' message a GTP tunnel endpoint where it expects the forwarded uplink packets to be received. The source eNodeB must, if possible, then indicate in the 'STATUS TRANSFER' message for this bearer the list of SNs corresponding to the forwarded packets which are to be expected. This list helps the target eNodeB to inform the UE earlier of the packets not to be retransmitted, making the overall uplink selective retransmission scheme faster (see also Section 4.2.6).

2.6.3.3 Multiple Preparation

'Multiple preparation' is another new feature of the LTE handover procedure. This feature enables the source eNodeB to trigger the handover preparation procedure towards multiple candidate target eNodeBs. Even though only one of the candidates is indicated as target to the UE, this makes recovery faster in case the UE fails on this target and connects to one of the other prepared candidate eNodeBs. The source eNodeB receives only one 'RELEASE RESOURCE' message from the final selected eNodeB.

Regardless of whether multiple or single preparation is used, the handover can be cancelled during or after the preparation phase. If the multiple preparation feature is operated, it is recommended that upon reception of the 'RELEASE RESOURCE' message the source eNodeB triggers a 'cancel' procedure towards each of the non-selected prepared eNodeBs.

2.6.3.4 Mobility Robustness Handling

In order to detect and report the cases where the mobility is unsuccessful and results in connection failures, specific messages are available over the X2 interface from Release 9 onwards to report handovers that are triggered too late or too early or to an inappropriate cell. These scenarios are explained in detail in Section 25.6.

2.6.3.5 Mobility towards Home eNodeBs via X2

In Release 10, in order to save backhaul bandwidth reduce delays, mobility between two HeNBs does not necessarily need to use S1 handover and transit via the MME but can directly use the X2 handover. This optimization is described in detail in Section 24.2.3.

2.6.4 Load and Interference Management Over X2

The exchange of load information between eNodeBs is of key importance in the flat architecture used in LTE, as there is no central Radio Resource Management (RRM) node as was the case, for example, in UMTS with the Radio Network Controller (RNC).

The exchange of load information falls into two categories depending on the purpose it serves:

- **Load balancing.** If the exchange of load information is for the purpose of load balancing, the frequency of exchange is rather low (in the order of seconds). The objective of load balancing is to counteract local traffic load imbalance between neighbouring cells with the aim of improving the overall system capacity. The mechanisms for this are explained in detail in Section 25.5.

  In Release 10, partial reporting is allowed per cell and per measurement. Therefore, if a serving eNodeB does not support some measurements, it will still report the other measurements that it does support. For each unsupported measurement, the serving eNodeB can indicate if the lack of support is permanent or temporary.

- **Interference coordination.** If the exchange of load information is to optimize RRM processes such as interference coordination, the frequency of exchange is rather high (in the order of tens of milliseconds). A special X2 'LOAD INDICATION' message is provided over the X2 interface for the exchange of load information related to interference management. For uplink interference management, two indicators can be provided within the 'LOAD INDICATION' message: a 'High Interference Indicator' and an 'Overload Indicator'. The usage of these indicators is explained in detail in Section 12.5.

The Load Indication procedure allows an eNodeB to signal to its neighbour eNodeBs new interference coordination intentions when applicable. This can either be frequency-domain interference management, as explained in Sections 12.5.1 and 12.5.2, or time-domain interference management, as explained in Section 31.2.3.

2.6.5 UE Historical Information Over X2

The provision of UE historical information is part of the X2-handover procedure and is designed to support self-optimization of the network.

Generally, the UE historical information consists of some RRM information which is passed from the source eNodeB to the target eNodeB within the 'HANDOVER REQUEST' message to assist the RRM management of a UE. The information can be partitioned into two types:

- UE RRM-related information, passed over X2 within the RRC transparent container;

- Cell RRM-related information, passed over X2 directly as an information element of the 'X2 AP HANDOVER REQUEST' message itself.

An example of such UE historical information is the list of the last few cells visited by the UE, together with the time spent in each one. This information is propagated from one eNodeB to another and can be used to determine the occurrence of ping-pong between two or three cells for instance. The length of the history information can be configured for more flexibility.

2.7 Summary

The EPS provides UEs with IP connectivity to the packet data network. In this chapter we have seen an overview of the EPS network architecture, including the functionalities provided by the E-UTRAN access network and the evolved packet core network..

It can be seen that the concept of EPS bearers, together with their associated quality of service attributes, provide a powerful tool for the provision of a variety of simultaneous services to the end user. Depending on the nature of the application, the EPS can supply the UE with multiple data flows with different QoSs. A UE can thus be engaged in a VoIP call which requires guaranteed delay and bit rate at the same time as browsing the web with a best effort QoS.

From the perspective of the network operator, the LTE system breaks new ground in terms of its degree of support for self-optimization and self-configuration of the network via the X2, S1 and Uu interfaces; these aspects are described in more detail in Chapter 25.

References[16]

[1] 3GPP Technical Specification 24.301, 'Non-Access-Stratum (NAS) protocol for Evolved Packet System (EPS); Stage 3', www.3gpp.org.

[2] 3GPP Technical Specification 33.401, 'System Architecture Evolution (SAE): Security Architecture', www.3gpp.org.

[3] 3GPP Technical Specification 23.402, 'Architecture enhancements for non-3GPP accesses', www.3gpp.org.

[4] 3GPP Technical Specification 29.060, 'General Packet Radio Service (GPRS); GPRS Tunnelling Protocol (GTP) across the Gn and Gp interface', www.3gpp.org.

[5] 3GPP Technical Specification 23.203, 'Policy and charging control architecture', www.3gpp.org.

[6] 3GPP Technical Specification 36.300, 'Evolved Universal Terrestrial Radio Access (E-UTRA) and Evolved Universal Terrestrial Radio Access Network (E-UTRAN); Overall description; Stage 2', www.3gpp.org.

[7] 3GPP Technical Specification 23.272, ' Circuit Switched (CS) fallback in Evolved Packet System (EPS); Stage 2', www.3gpp.org.

[8] 3GPP Technical Specification 23.272, 'Single Radio Voice Call Continuity (SRVCC); Stage 2', www.3gpp.org.

[9] Request for Comments 4960 The Internet Engineering Task Force (IETF), Network Working Group, 'Stream Control Transmission Protocol', http://www.ietf.org.

[16]All web sites confirmed 1st March 2011.

2.7 Summary

References

3

Control Plane Protocols

Himke van der Velde

3.1 Introduction

As introduced in Section 2.2.2, the Control Plane of the Access Stratum (AS) handles radio-specific functionalities. The AS interacts with the Non-Access Stratum (NAS), also referred to as the ‘upper layers’. Among other functions, the NAS control protocols handle Public Land Mobile Network[1] (PLMN) selection, tracking area update, paging, authentication and Evolved Packet System (EPS) bearer establishment, modification and release.

The applicable AS-related procedures largely depend on the Radio Resource Control (RRC) state of the User Equipment (UE), which can be either RRC_IDLE or RRC_CONNECTED.

A UE in RRC_IDLE performs cell selection and reselection – in other words, it decides on which cell to camp. The cell (re)selection process takes into account the priority of each applicable frequency of each applicable Radio Access Technology (RAT), the radio link quality and the cell status (i.e. whether a cell is barred or reserved). An RRC_IDLE UE monitors a paging channel to detect incoming calls, and also acquires system information. The System Information (SI) mainly consists of parameters by which the network (E-UTRAN) can control the cell (re)selection process.

In RRC_CONNECTED, the E-UTRAN allocates radio resources to the UE to facilitate the transfer of (unicast) data via shared data channels.[2] To support this operation, the UE monitors an associated control channel[3] used to indicate the dynamic allocation of the shared transmission resources in time and frequency. The UE provides the network with reports of its

[1]The network of one operator in one country.

[2]The Physical Downlink Shared CHannel (PDSCH) and Physical Uplink Shared CHannel (PUSCH)– see Sections 9.2.2 and 16.2 respectively.

[3]The Physical Downlink Control CHannel (PDCCH) – see Section 9.3.5.

LTE – The UMTS Long Term Evolution: From Theory to Practice, Second Edition.
Stefania Sesia, Issam Toufik and Matthew Baker.

buffer status and of the downlink channel quality, as well as neighbouring cell measurement information to enable E-UTRAN to select the most appropriate cell for the UE. These measurement reports include cells using other frequencies or RATs. The UE also receives SI, consisting mainly of information required to use the transmission channels. To extend its battery lifetime, a UE in RRC_CONNECTED may be configured with a Discontinuous Reception (DRX) cycle.

RRC, as specified in [1], is the protocol by which the E-UTRAN controls the UE behaviour in RRC_CONNECTED. RRC also includes the control signalling applicable for a UE in RRC_IDLE, namely paging and SI. The UE behaviour in RRC_IDLE is specified in [2].

Chapter 22 gives some further details of the UE measurements which support the mobility procedures.

Functionality related to Multimedia Broadcast/Multicast Services (MBMSs) is covered separately in Chapter 13.

3.2 Radio Resource Control (RRC)

3.2.1 Introduction

The RRC protocol supports the transfer of *common* NAS information (i.e. NAS information which is applicable to all UEs) as well as *dedicated* NAS information (which is applicable only to a specific UE). In addition, for UEs in RRC_IDLE, RRC supports notification of incoming calls (via paging).

The RRC protocol covers a number of functional areas:

- **System information** handles the broadcasting of SI, which includes NAS common information. Some of the system information is applicable only for UEs in RRC_IDLE while other SI is also applicable for UEs in RRC_CONNECTED.

- **RRC connection control** covers all procedures related to the establishment, modification and release of an RRC connection, including paging, initial security activation, establishment of Signalling Radio Bearers (SRBs) and of radio bearers carrying user data (Data Radio Bearers, DRBs), handover within LTE (including transfer of UE RRC context information[4]), configuration of the lower protocol layers,[5] access class barring and radio link failure.
- **Network controlled inter-RAT mobility** includes handover, cell change orders and redirection upon connection release, security activation and transfer of UE RRC context information.
- **Measurement configuration and reporting** for intra-frequency, inter-frequency and inter-RAT mobility, includes configuration and activation of measurement gaps.
- **Miscellaneous functions** including, for example, transfer of dedicated NAS information and transfer of UE radio access capability information.

[4]This UE context information includes the radio resource configuration including local settings not configured across the radio interface, UE capabilities and radio resource management information.

[5]Packet Data Convergence Protocol (PDCP), Radio Link Control (RLC), Medium Access Control (MAC), all of which are explained in detail in Chapter 4, and the physical layer which is explained in Chapters 5–11 and 14–18.

Dedicated RRC messages are transferred across SRBs, which are mapped via the PDCP and RLC layers onto logical channels – either the Common Control CHannel (CCCH) during connection establishment or a Dedicated Control CHannel (DCCH) in RRC_CONNECTED. System Information and Paging messages are mapped directly to logical channels – the Broadcast Control CHannel (BCCH) and Paging Control CHannel (PCCH) respectively. The various logical channels are described in more detail in Section 4.4.1.2.

SRB0 is used for RRC messages which use the CCCH, SRB1 is for RRC messages using DCCH, and SRB2 is for the (lower-priority) RRC messages using DCCH which only include NAS dedicated information.[6] All RRC messages using DCCH are integrity-protected and ciphered by the PDCP layer (after security activation) and use Automatic Repeat reQuest (ARQ) protocols for reliable delivery through the RLC layer. The RRC messages using CCCH are not integrity-protected and do not use ARQ in the RLC layer.

It should also be noted that the NAS independently applies integrity protection and ciphering.

Figure 3.1 illustrates the overall radio protocol architecture as well as the use of radio bearers, logical channels, transport channels and physical channels.

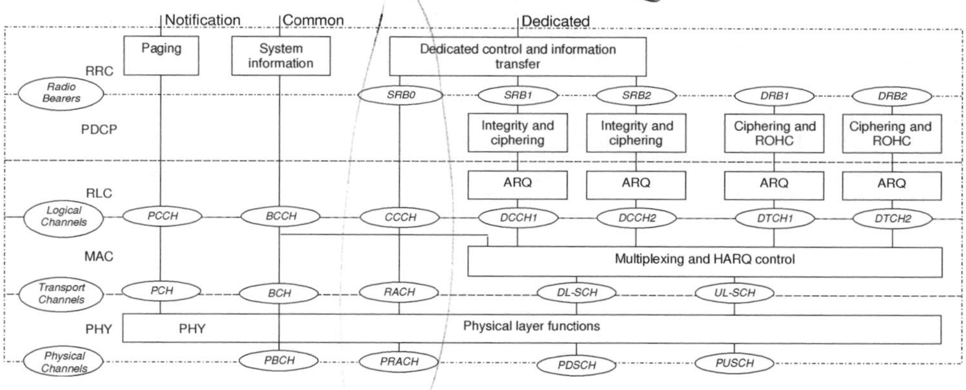

Figure 3.1: Radio architecture.

For control information for which low transfer delay is more important than reliable transfer (i.e. for which the use of ARQ is inappropriate due to the additional delay it incurs), MAC signalling is used provided that there are no security concerns (integrity protection and ciphering are not applicable for MAC signalling).

3.2.2 System Information

System information is structured by means of System Information Blocks (SIBs), each of which contains a set of functionally-related parameters. The SIB types that have been defined include:

[6]Prior to SRB2 establishment, SRB1 is also used for RRC messages which only include NAS dedicated information. In addition, SRB1 is used for higher priority RRC messages which only include NAS dedicated information.

- **The Master Information Block** (MIB), which includes a limited number of the most frequently transmitted parameters which are essential for a UE's initial access to the network.
- **System Information Block Type 1** (SIB1), which contains parameters needed to determine if a cell is suitable for cell selection, as well as information about the time-domain scheduling of the other SIBs.
- **System Information Block Type 2** (SIB2), which includes common and shared channel information.
- **SIB3–SIB8**, which include parameters used to control intra-frequency, inter-frequency and inter-RAT cell reselection.
- **SIB9**, which is used to signal the name of a Home eNodeB (HeNBs).
- **SIB10–SIB12**, which include the Earthquake and Tsunami Warning Service (ETWS) notifications and Commercial Mobile Alert System (CMAS) warning messages (See Section 13.7).
- **SIB13**, which includes MBMS related control information (See Section 13.6.3.2.

Three types of RRC message are used to transfer system information: the MIB message, the SIB1 message and SI messages. An SI message, of which there may be several, includes one or more SIBs which have the same scheduling requirements (i.e. the same transmission periodicity). Table 3.1 provides an example of a possible system information scheduling configuration, also showing which SIBs the UE has to acquire in the idle and connected states. The physical channels used for carrying the SI are explained in Section 9.2.1.

Table 3.1: Example of SI scheduling configuration.

Message	Content	Period (ms)	Applicability
MIB	Most essential parameters	40	Idle and connected
SIB1	Cell access related parameters, scheduling information	80	Idle and connected
1st SI	SIB2: Common and shared channel configuration	160	Idle and connected
2nd SI	SIB3: Common cell reselection information and intra-frequency cell reselection parameters other than the neighbouring cell information SIB4: Intra-frequency neighbouring cell information	320	Idle only
3rd SI	SIB5: Inter-frequency cell reselection information	640	Idle only
4th SI	SIB6: UTRA cell reselection information SIB7: GERAN cell reselection information	640	Idle only, depending on UE support of UMTS or GERAN

3.2.2.1 Time-Domain Scheduling of System Information

The time-domain scheduling of the MIB and SIB1 messages is fixed with a periodicities of 40 ms and 80 ms respectively, as explained in Sections 9.2.1 and 9.2.2.2.

The time-domain scheduling of the SI messages is dynamically flexible: each SI message is transmitted in a defined periodically-occurring time-domain window, while physical layer control signalling[7] indicates in which subframes[8] within this window the SI is actually scheduled. The scheduling windows of the different SI messages (referred to as SI-windows) are consecutive (i.e. there are neither overlaps nor gaps between them) and have a common length that is configurable. SI-windows can include subframes in which it is not possible to transmit SI messages, such as subframes used for SIB1, and subframes used for the uplink in TDD.

Figure 3.2 illustrates an example of the time-domain scheduling of SI, showing the subframes used to transfer the MIB, SIB1 and four SI messages. The example uses an SI-window of length 10 subframes, and shows a higher number of 'blind' Hybrid ARQ (HARQ) transmissions[9] being used for the larger SI messages.

SI messages may have different periodicities. Consequently, in some clusters of SI-windows all the SI messages are scheduled, while in other clusters only the SIs with shorter repetition periods are transmitted. For the example of Table 3.1, the cluster of SI-windows beginning at System Frame Number (SFN) 0 contains all the SI messages, the cluster starting at SFN160 contains only the first SI message, that beginning at SFN320 contains the first and second SI messages, and the one starting at SFN480 contains only the first SI message.

Note that Figure 3.2 shows a cluster of SI-windows where all the SI messages are transmitted. At occasions where a given SI is not transmitted (due to a longer repetition period), its corresponding SI-window is not used.

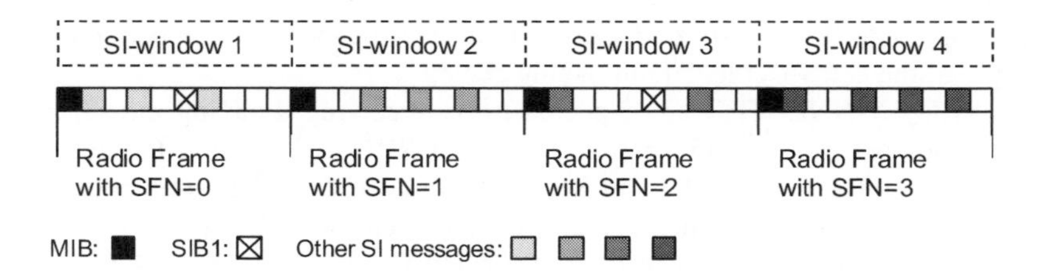

Figure 3.2: SI scheduling example.

3.2.2.2 Validity and Change Notification

SI normally changes only at specific radio frames whose System Frame Number is given by SFN mod $N = 0$, where N is configurable and defines the period between two radio frames at which a change may occur, known as the *modification period*. Prior to performing a change

[7]The Physical Downlink Control Channel – PDCCH; see Section 9.3.5.

[8]A subframe in LTE has a duration of 1 ms; see Section 6.2

[9]With blind HARQ retransmissions, there is no feedback to indicate whether the reception has been successful.

of the system information, the E-UTRAN notifies the UEs by means of a *Paging* message including a *SystemInfoModification* flag. Figure 3.3 illustrates the change of SI, with different shading indicating different content.

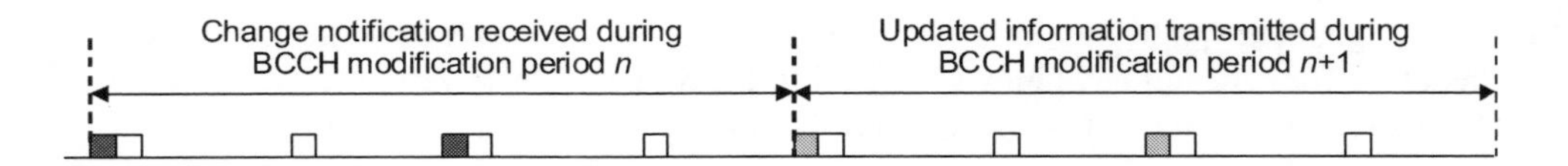

Figure 3.3: SI modification periods. Reproduced by permission of © 3GPP.

LTE provides two mechanisms for indicating that SI has changed:

1. A paging message including a flag indicating whether or not SI has changed.
2. A value tag in SIB1 which is incremented every time one or more SI message changes.

UEs in RRC_IDLE use the first mechanism, while UEs in RRC_CONNECTED can use either mechanism; the second being useful, for example, in cases when a UE was unable to receive the paging messages.

UEs in RRC_IDLE are only required to receive the paging message at their normal paging occasions – i.e. no additional wake-ups are expected to detect changes of SI. In order to ensure reliability of reception, the change notification paging message is normally repeated a number of times during the BCCH modification period preceding that in which the new system information is first transmitted. Correspondingly, the modification period is expressed as a multiple of the cell-specific default paging cycle.

UEs in RRC_CONNECTED are expected to try receiving a paging message the same number of times per modification period as UEs in RRC_IDLE using the default paging cycle. The exact times at which UEs in RRC_CONNECTED which are using this method have to try to receive a paging message are not specified; the UE may perform these tries at convenient times, such as upon wake-up from DRX, using any of the subframes which are configured for paging during the modification period. Since the eNodeB anyway has to notify all the UEs in RRC_IDLE, it has to send a paging message in all subframes which are configured for paging (up to a maximum of four subframes per radio frame) during an entire modification period. Connected mode UEs can utilize any of these subframes. The overhead of transmitting paging messages to notify UEs of a change of SI is considered marginal, since such changes are expected to be infrequent – at most once every few hours.

If the UE receives a notification of a change of SI, it starts acquiring SI from the start of the next modification period. Until the UE has successfully acquired the updated SI, it continues to use the existing parameters. If a critical parameter changes, the communication may be seriously affected, but any service interruption that may result is considered acceptable since it is short and infrequent.

If the UE returns to a cell, it is allowed to assume that the SI previously acquired from the cell remains valid if it was received less than 3 hours previously and the value tag matches.

3.2.3 Connection Control within LTE

Connection control involves:

- Security activation;
- Connection establishment, modification and release;
- DRB establishment, modification and release;
- Mobility within LTE.

3.2.3.1 Security Key Management

Security is a very important feature of all 3GPP RATs. LTE provides security in a similar way to its predecessors UMTS and GSM.

Two functions are provided for the maintenance of security: *ciphering* of both control plane (RRC) data (i.e. SRBs 1 and 2) and user plane data (i.e. all DRBs), and *integrity protection* which is used for control plane (RRC) data only. Ciphering is used in order to protect the data streams from being received by a third party, while integrity protection allows the receiver to detect packet insertion or replacement. RRC always activates both functions together, either following connection establishment or as part of the handover to LTE.

The hierarchy of keys by which the AS security keys are generated is illustrated in Figure 3.4. The process is based on a common secret key K_{ASME} (Access Security Management Entity) which is available only in the Authentication Centre in the Home Subscriber Server (HSS) (see Section 2.2.1) and in a secure part of the Universal Subscriber Identity Module (USIM) in the UE. A set of keys and checksums are generated at the Authentication Centre using this secret key and a random number. The generated keys, checksums and random number are transferred to the Mobility Management Entity (MME) (see Section 2.2.1), which passes one of the generated checksums and the random number to the UE. The USIM in the UE then computes the same set of keys using the random number and the secret key. Mutual authentication is performed by verifying the computed checksums in the UE and network using NAS protocols.

Upon connection establishment, the AS derives an *AS base-key* K_{eNB} (eNodeB-specific) and Next Hop (NH), from K_{ASME}.

K_{eNB} is used to generate three further security keys known as the *AS derived-keys*: one, called $K_{RRC\ int}$, is used for integrity protection of the RRC signalling (SRBs), one for ciphering of the RRC signalling known as $K_{RRC\ enc}$ and $K_{UP\ enc}$ used for ciphering of user data (i.e. DRBs).

NH is an intermediate key used to implement 'forward security'[10] [3]. It is derived by the UE and MME using K_{ASME} and K_{eNB} when the security context is established or using K_{ASME} and the previous NH otherwise. NH is associated with a counter called Next hop Chaining Counter (NCC) which is initially set to 0 at connection establishment.

In case of handover within E-UTRAN, a new AS base-key and new AS Derived-keys are computed from the AS base-key used in the source cell. An intermediate key, K_{eNB^*} is derived by the UE and the source eNodeB based on the Physical Cell Identity (PCI) of the target cell,

[10]Forward security refers to the property that, for an eNodeB sharing a K_{eNB} with a UE, it shall be computationally infeasible to predict any future K_{eNB}, that will be used between the same UE and another eNodeB

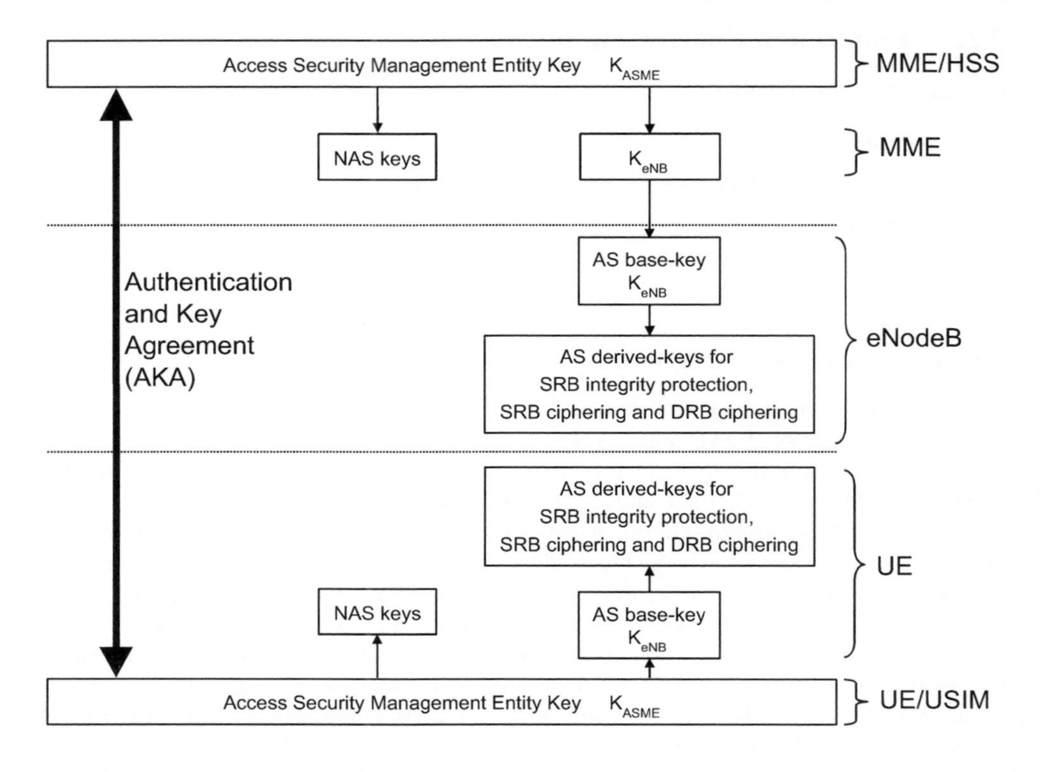

Figure 3.4: Security key derivation.

the target frequency and NH or K_{eNB}. If a fresh NH is available[11], the derivation of K_{eNB^*} is based on NH (referred to as vertical derivation). If no fresh NH is available then the K_{eNB^*} derivation is referred to as horizontal derivation and is based on K_{eNB}. K_{eNB^*} is then used at the target cell as the new K_{eNB} for RRC and data traffic.

For handover to E-UTRAN from UTRAN or GERAN, the AS base-key is derived from integrity and ciphering keys used in the UTRAN or GERAN. Handover within LTE may be used to take a new K_{ASME} into account, i.e. following a re-authentication by NAS.

The use of the security keys for the integrity protection and ciphering functions is handled by the PDCP layer, as described in Section 4.2.3.

The security functions are never deactivated, although it is possible to apply a 'NULL' ciphering algorithm. The 'NULL' algorithm may also be used in certain special cases, such as for making an emergency call without a USIM.

[11] In this case the NCC is incremented and is then larger than that of the currently active K_{eNB^*}

3.2.3.2 Connection Establishment and Release

Two levels of NAS states reflect the state of a UE in respect of connection establishment: the EPS Mobility Management (EMM) state (EMM-DEREGISTERED or EMM-REGISTERED) reflects whether the UE is registered in the MME, and the EPS Connection Management (ECM) state (ECM-IDLE or ECM-CONNECTED) reflects the connectivity of the UE with the Evolved Packet Core (EPC – see Chapter 2).

The NAS states, and their relationship to the AS RRC states, are illustrated in Figure 3.5.

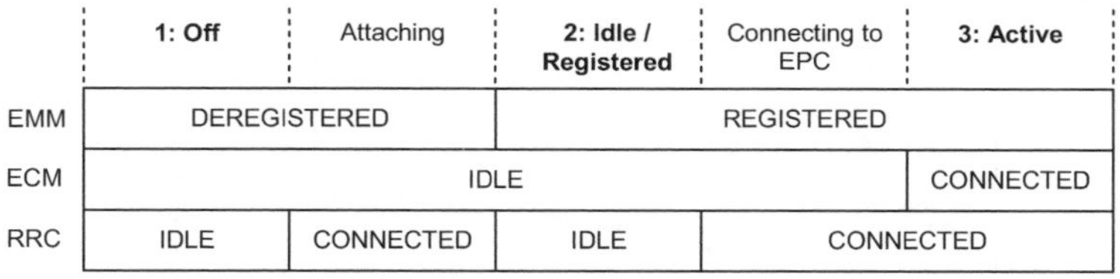

	1: Off	Attaching	2: Idle / Registered	Connecting to EPC	3: Active
EMM	DEREGISTERED		REGISTERED		
ECM	IDLE				CONNECTED
RRC	IDLE	CONNECTED	IDLE	CONNECTED	

Figure 3.5: Possible combinations of NAS and AS states.

The transition from ECM-IDLE to ECM-CONNECTED not only involves establishment of the RRC connection but also includes establishment of the S1-connection (see Section 2.5). RRC connection establishment is initiated by the NAS and is completed prior to S1-connection establishment, which means that connectivity in RRC_CONNECTED is initially limited to the exchange of control information between UE and E-UTRAN.

UEs are typically moved to ECM-CONNECTED when becoming active. It should be noted, however, that in LTE the transition from ECM-IDLE to ECM-CONNECTED is performed within 100 ms. Hence, UEs engaged in intermittent data transfer need not be kept in ECM-CONNECTED if the ongoing services can tolerate such transfer delays. In any case, an aim in the design of LTE was to support similar battery power consumption levels for UEs in RRC_CONNECTED as for UEs in RRC_IDLE.

RRC connection release is initiated by the eNodeB following release of the S1-connection between the eNodeB and the Core Network (CN).

Connection establishment message sequence. RRC connection establishment involves the establishment of SRB1 and the transfer of the initial uplink NAS message. This NAS message triggers the establishment of the S1-connection, which normally initiates a subsequent step during which E-UTRAN activates AS-security and establishes SRB2 and one or more DRBs (corresponding to the default and optionally dedicated EPS bearers).

Figure 3.6 illustrates the RRC connection establishment procedure, including the subsequent step of initial security activation and radio bearer establishment.

Step 1: Connection establishment

- Upper layers in the UE trigger connection establishment, which may be in response to paging. The UE checks if access is barred (see Section 3.3.4.6). If this is not the case, the lower layers in the UE perform a contention-based random access procedure

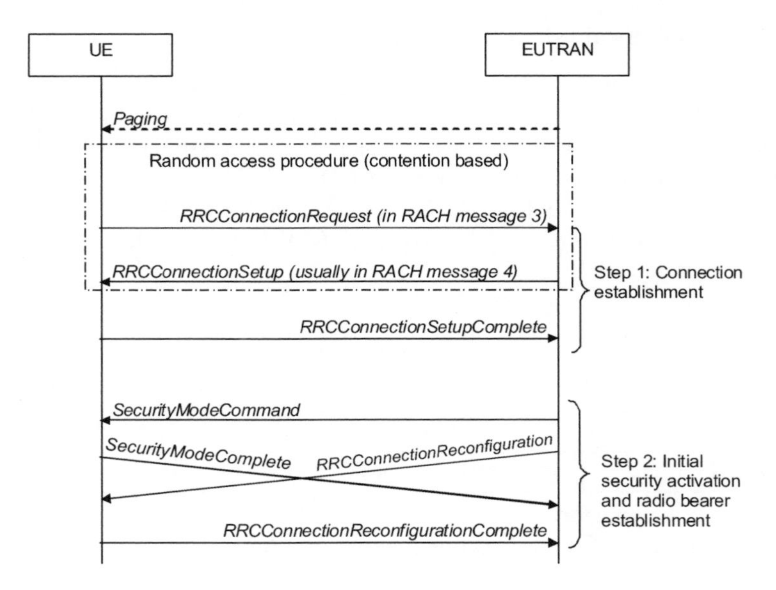

Figure 3.6: Connection establishment (see Section 17.3.1 for details of the contention-based RACH procedure).

as described in Section 17.3, and the UE starts a timer (known as *T300*) and sends the RRCConnectionRequest message. This message includes an initial identity (S-TMSI[12] or a random number) and an establishment cause.

- If E-UTRAN accepts the connection, it returns the RRCConnectionSetup message that includes the initial radio resource configuration including SRB1. Instead of signalling each individual parameter, E-UTRAN may order the UE to apply a default configuration – i.e. a configuration for which the parameter values are specified in the RRC specification [1].
- The UE returns the RRCConnectionSetupComplete message and includes the NAS message, an identifier of the selected PLMN (used to support network sharing) and, if provided by upper layers, an identifier of the registered MME. Based on the last two parameters, the eNodeB decides on the CN node to which it should establish the S1-connection.

Step 2: Initial security activation and radio bearer establishment

- E-UTRAN sends the SecurityModeCommand message to activate integrity protection and ciphering. This message, which is integrity-protected but not ciphered, indicates which algorithms shall be used.
- The UE verifies the integrity protection of the SecurityModeControl message, and, if this succeeds, it configures lower layers to apply integrity protection and ciphering to all subsequent messages (with the exception that ciphering is not applied to the

[12] S-Temporary Mobile Subscriber Identity.

response message, i.e. the SecurityModeComplete (or SecurityModeFailure) message).

- E-UTRAN sends the RRCConnectionReconfiguration message including a radio resource configuration used to establish SRB2 and one or more DRBs. This message may also include other information such as a piggybacked NAS message or a measurement configuration. E-UTRAN may send the RRCConnectionReconfiguration message prior to receiving the SecurityModeComplete message. In this case, E-UTRAN should release the connection when one (or both) procedures fail (because the two procedures result from a single S1-procedure, which does not support partial success).
- The UE finally returns the RRCConnectionReconfigurationComplete message.

A connection establishment may fail for a number of reasons, such as the following:

- Access may be barred (see Section 3.3.4.6).
- In case cell reselection occurs during connection establishment, the UE aborts the procedure and informs upper layers of the failure to establish the connection.
- E-UTRAN may temporarily reject the connection establishment by including a wait timer, in which case the UE rejects any connection establishment request until the wait time has elapsed.
- The NAS may abort an ongoing RRC connection establishment, for example upon NAS timer expiry.

3.2.3.3 DRB Establishment

To establish, modify or release DRBs, E-UTRAN applies the RRC connection reconfiguration procedure as described in Section 3.2.3.2.

When establishing a DRB, E-UTRAN decides how to transfer the packets of an EPS bearer across the radio interface. An EPS bearer is mapped (1-to-1) to a DRB, a DRB is mapped (1-to-1) to a DTCH (Dedicated Traffic CHannel – see Section 4.4.1.2) logical channel, all logical channels are mapped (n-to-1) to the Downlink or Uplink Shared Transport CHannel (DL-SCH or UL-SCH), which are mapped (1-to-1) to the corresponding Physical Downlink or Uplink Shared CHannel (PDSCH or PUSCH). This radio bearer mapping is illustrated in Figure 3.1.

The radio resource configuration covers the configuration of the PDCP, RLC, MAC and physical layers. The main configuration parameters / options include the following:

- For services using small packet sizes (e.g. VoIP), PDCP may be configured to apply indexheader!compressionheader compression to significantly reduce the signalling overhead.
- The RLC Mode is selected from those listed in Section 4.3.1. RLC Acknowledged Mode (AM) is applicable, except for services which require a very low transfer delay and for which reliable transfer is less important.
- E-UTRAN assigns priorities and Prioritized Bit-Rates (PBRs) to control how the UE divides the granted uplink resources between the different radio bearers (see Section 4.4.2.6).

- Unless the transfer delay requirements for any of the ongoing services are very strict, the UE may be configured with a DRX cycle (see Section 4.4.2.5).
- For services involving a semi-static packet rate (e.g. VoIP), semi-persistent scheduling may be configured to reduce the control signalling overhead (see Section 4.4.2.1). Specific resources may also be configured for reporting buffer status and radio link quality.
- Services tolerating higher transfer delays may be configured with a HARQ profile involving a higher average number of HARQ transmissions.

3.2.3.4 Mobility Control in RRC_IDLE and RRC_CONNECTED

Mobility control in RRC_IDLE is UE-controlled (cell-reselection), while in RRC_CONNECTED it is controlled by the E-UTRAN (handover). However, the mechanisms used in the two states need to be consistent so as to avoid ping-pong (i.e. rapid handing back and forth) between cells upon state transitions. The mobility mechanisms are designed to support a wide variety of scenarios including network sharing, country borders, home deployment and varying cell ranges and subscriber densities; an operator may, for example, deploy its own radio access network in populated areas and make use of another operator's network in rural areas.

If a UE were to access a cell which does not have the best radio link quality of the available cells on a given frequency, it may create significant interference to the other cells. Hence, as for most technologies, radio link quality is the primary criterion for selecting a cell on an LTE frequency. When choosing between cells on different frequencies or RATs the interference concern does not apply. Hence, for inter-frequency and inter-RAT cell reselection other criteria may be considered such as UE capability, subscriber type and call type. As an example, UEs with no (or limited) capability for data transmission may be preferably handled on GSM, while home customers or 'premium subscribers' might be given preferential access to the frequency or RAT supporting the highest data rates. Furthermore, in some LTE deployment scenarios, voice services may initially be provided by a legacy RAT only (as a Circuit Switching (CS) application), in which case the UE needs to be moved to the legacy RAT upon establishing a voice call (also referred to as *CS FallBack* (CSFB)).

E-UTRAN provides a list of neighbouring frequencies and cells which the UE should consider for cell reselection and for reporting of measurements. In general, such a list is referred to as a *white-list* if the UE is to consider only the listed frequencies or cells – i.e. other frequencies or cells are not available; conversely, in the case of a *black-list* being provided, a UE may consider any *un*listed frequencies or cells. In LTE, white-listing is used to indicate all the neighbouring frequencies of each RAT that the UE is to consider. On the other hand, E-UTRAN is not required to indicate all the neighbouring cells that the UE shall consider. Which cells the UE is required to detect by itself depends on the UE state as well as on the RAT, as explained below.

Note that for GERAN, typically no information is provided about individual cells. Only in specific cases, such as at country borders, is signalling[13] provided to indicate the group of cells that the UE is to consider – i.e. a white cell list.

[13]The 'NCC-permitted' parameter – see GERAN specifications.

Mobility in idle mode. In RRC_IDLE, cell reselection between frequencies is based on absolute priorities, where each frequency has an associated priority. Cell-specific default values of the priorities are provided via SI. In addition, E-UTRAN may assign UE-specific values upon connection release, taking into account factors such as UE capability or subscriber type. In case equal priorities are assigned to multiple cells, the cells are ranked based on radio link quality. Equal priorities are not applicable between frequencies of different RATs. The UE does not consider frequencies for which it does not have an associated priority; this is useful in situations such as when a neighbouring frequency is applicable only for UEs of one of the sharing networks.

Table 3.2 provides an overview of the SI parameters which E-UTRAN may use to control cell reselection . Other than the cell reselection priority of a frequency, no idle mode mobility-related parameters may be assigned via dedicated signalling. Further details of the parameters listed are provided in Section 3.3.

Mobility in connected mode. In RRC_CONNECTED, the E-UTRAN decides to which cell a UE should hand over in order to maintain the radio link. As with RRC_IDLE, E-UTRAN may take into account not only the radio link quality but also factors such as UE capability, subscriber type and access restrictions. Although E-UTRAN may trigger handover without measurement information (blind handover), normally it configures the UE to report measurements of the candidate target cells – see Section 22.3. Table 3.3 provides an overview of the frequency- and cell-specific parameters which E-UTRAN can configure for mobility-related measurement reporting.

In LTE the UE always connects to a single cell only – in other words, the switching of a UE's connection from a source cell to a target cell is a *hard* handover. The hard handover process is normally a 'backward' one, whereby the eNodeB which controls the source cell requests the target eNodeB to prepare for the handover. The target eNodeB subsequently generates the RRC message to order the UE to perform the handover, and the message is transparently forwarded by the source eNodeB to the UE. LTE also supports a kind of 'forward' handover, in which the UE by itself decides to connect to the target cell, where it then requests that the connection be continued. The UE applies this connection re-establishment procedure only after loss of the connection to the source cell; the procedure only succeeds if the target cell has been prepared in advance for the handover.

Besides the handover procedure, LTE also provides for a UE to be redirected to another frequency or RAT upon connection release. Redirection during connection establishment is not supported, since at that time the E-UTRAN may not yet be in possession of all the relevant information such as the capabilities of the UE and the type of subscriber (as may be reflected, for example, by the SPID, the Subscriber Profile ID for RAT/Frequency Priority). However, the redirection may be performed while AS-security has not (yet) been activated. When redirecting the UE to UTRAN or GERAN, E-UTRAN may provide SI for one or more cells on the relevant frequency. If the UE selects one of the cells for which SI is provided, it does not need to acquire it.

Message sequence for handover within LTE. In RRC_CONNECTED, the E-UTRAN controls mobility by ordering the UE to perform handover to another cell, which may be

Table 3.2: List of SI parameters which may be used to control cell reselection.

Parameter	Intra-Freq.	Inter-Freq.	UTRA	GERAN	CDMA2000
Common	(SIB3)	(SIB5)	(SIB6)	(SIB7)	(SIB8)
Reselection info	Q-Hyst MobilityStatePars Q-HystSF S-Search[(a)]		T-Reselect T-ReselectSF	T-Reselect T-ReselectSF	T-Reselect T-ReselectSF
Frequency list	(SIB3)	(SIB5)	(SIB6)	(SIB7)	(SIB8)
White frequency list	n/a	+	+	+	+
Frequency specific reselection info[(b)]	Priority $Thresh_{Serving\text{-}Low}$ $Thresh_{Serving\text{-}LowQ}$ T-Reselect T-ReselectSF $Thresh_{X\text{-}LowQ}$ T-Reselect T-ReselectSF	Priority Qoffset, $Thresh_{X\text{-}High}$, $Thresh_{X\text{-}Low}$ $Thresh_{X\text{-}HighQ}$,	Priority $Thresh_{X\text{-}High}$, $Thresh_{X\text{-}Low}$ $Thresh_{X\text{-}HighQ}$, $Thresh_{X\text{-}LowQ}$	Priority $Thresh_{X\text{-}High}$, $Thresh_{X\text{-}Low}$	Priority $Thresh_{X\text{-}High}$, $Thresh_{X\text{-}Low}$
Frequency specific suitability info[(c)]	Q-RxLevMin MaxTxPower Q-QualMin	Q-RxLevMin MaxTxPower Q-QualMin	Q-RxLevMin, MaxTxPower, Q-QualMin	Q-RxLevMin MaxTxPower	
Cell list	(SIB4)	(SIB5)	(SIB6)	(SIB7)	(SIB8)
White cell list	–	–	–	NCC permitted[(d)]	–
Black cell list	+	+	–	–	–
List of cells with specific info[(e)]	Qoffset	Qoffset	–	–	–

[(a)] Separate parameters for intra/ inter-frequency, both for RSRP and RSRQ.
[(b)] See Section 3.3.4.2.
[(c)] See Section 3.3.3.
[(d)] See GERAN specifications.
[(e)] See Section 3.3.4.3.

on the same frequency ('intra-frequency') or a different frequency ('inter-frequency'). Inter-frequency measurements may require the configuration of measurement gaps, depending on the capabilities of the UE (e.g. whether it has a dual receiver) – see Section 22.3.

The E-UTRAN may also use the handover procedures for completely different purposes, such as to change the security keys to a new set (see Section 3.2.3.1), or to perform a 'synchronized reconfiguration' in which the E-UTRAN and the UE apply the new configuration simultaneously.

The message sequence for the procedure for handover within LTE is shown in Figure 3.7. The sequence is as follows:

1. The UE may send a MeasurementReport message (see Section 3.2.5).

Table 3.3: Frequency- and cell-specific information which can be configured in connected mode.

Parameter	Intra-Freq.	Inter-Freq.	UTRA	GERAN	CDMA2000
Frequency list					
White frequency list	n/a	+	+	+	+
Frequency specific info	Qoffset	Qoffset	Qoffset	Qoffset	Qoffset
Cell list					
White cell list	–	–	+	NCC permitted	+
Black cell list	+	+	–	–	–
List of cells with specific info.	Qoffset	Qoffset	–	–	–

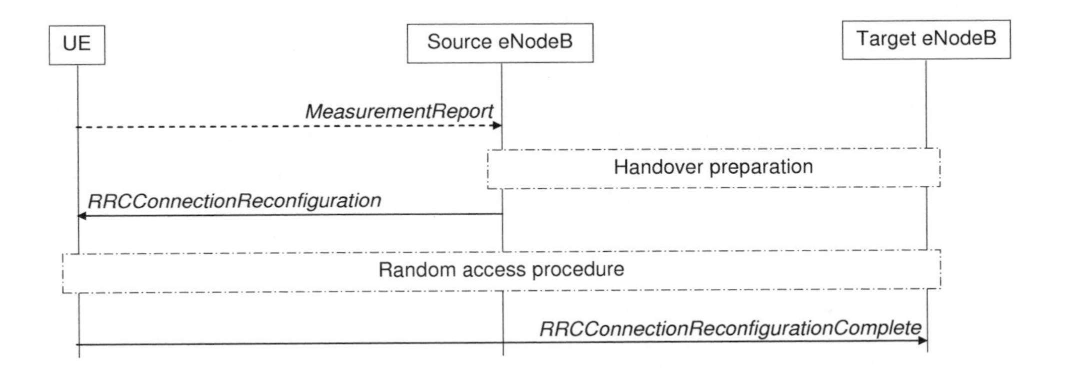

Figure 3.7: Handover within LTE.

2. Before sending the handover command to the UE, the source eNodeB requests one or more target cells to prepare for the handover. As part of this 'handover preparation request', the source eNodeB provides UE RRC context information[14] about the UE capabilities, the current AS-configuration and UE-specific Radio Resource Management (RRM) information. In response, the eNodeB controlling the target cell generates the 'handover command'. The source eNodeB will forward this command to the UE in the RRCConnectionReconfiguration message. This is done transparently (apart from performing integrity protection and ciphering) – i.e. the source eNodeB does not add or modify the protocol information contained in the message.

3. The source eNodeB sends the RRCConnectionReconfiguration message which to the UE orders it to perform handover. It includes mobility control information (namely the identity, and optionally the frequency, of the target cell) and the radio resource

[14]This UE context information includes the radio resource configuration including local settings not configured across the radio interface, UE capabilities and radio resource management information.

configuration information which is common to all UEs in the target cell (e.g. information required to perform random access. The message also includes the dedicated radio resource configuration, the security configuration and the C-RNTI[15] to be used in the target cell. Although the message may optionally include the measurement configuration, the E-UTRAN is likely to use another reconfiguration procedure for re-activating measurements, in order to avoid the RRCConnectionReconfiguration message becoming excessively large. If no measurement configuration information is included in the message used to perform inter-frequency handover, the UE stops any inter-frequency and inter-RAT measurements and deactivates the measurement gap configuration.

4. If the UE is able to comply with the configuration included in the received RRC-ConnectionReconfiguration message, the UE starts a timer, known as *T304*, and initiates a random access procedure (see Section 17.3), using the received Random Access CHannel (RACH) configuration, to the target cell at the first available occasion.[16] It is important to note that the UE does not need to acquire system information from the target cell prior to initiating random access and resuming data communication. However, the UE may be unable to use some parts of the physical layer configuration from the very start (e.g. semi-persistent scheduling (see Section 4.4.2.1), the PUCCH (see Section 16.3) and the Sounding Reference Signal (SRS) (see Section 15.6)). The UE derives new security keys and applies the received configuration in the target cell.

5. Upon successful completion of the random access procedure, the UE stops the timer T304. The AS informs the upper layers in the UE about any uplink NAS messages for which transmission may not have completed successfully, so that the NAS can take appropriate action.

For handover to cells broadcasting a Closed Subscriber Group (CSG) identity, normal measurement and mobility procedures are used to support handover. In addition, E-UTRAN may configure the UE to report that it is entering or leaving the proximity of cell(s) included in its CSG whitelist. Furthermore, E-UTRAN may request the UE to provide additional information broadcast by the handover candidate cell, for example the cell global identity, CSG identity or CSG membership status. E-UTRAN may use a indexproximity report‘proximity report’ to configure measurements and to decide whether or not to request the UE to provide additional information broadcast by the handover candidate cell. The additional information is used to verify whether or not the UE is authorized to access the target cell and may also be needed to identify handover candidate cells.[17] Further details of the mobility procedures for HeNBs can be found in Section 24.2.3.

[15]The Cell Radio Network Temporary Identifier is the RNTI to be used by a given UE while it is in a particular cell.

[16]The target cell does not specify when the UE is to initiate random access in that cell. Hence, the handover process is sometimes described as *asynchronous*.

[17]This may be the case if PCI confusion occurs, i.e. when the PCI that is included in the measurement report does not uniquely identify the cell.

3.2.3.5 Connection Re-Establishment Procedure

In a number of failure cases (e.g. radio link failure, handover failure, RLC unrecoverable error, reconfiguration compliance failure), the UE initiates the RRC connection re-establishment procedure, provided that security is active. If security is not active when one of the indicated failures occurs, the UE moves to RRC_IDLE instead.

To attempt RRC connection re-establishment, the UE starts a timer known as *T311* and performs cell selection. The UE should prioritize searching on LTE frequencies. However, no requirements are specified regarding for how long the UE shall refrain from searching for other RATs. Upon finding a suitable cell on an LTE frequency, the UE stops the timer T311, starts the timer T301 and initiates a contention based random access procedure to enable the RRCConnectionReestablishmentRequest message to be sent. In the RRCConnectionReestablishmentRequest message, the UE includes the identity used in the cell in which the failure occurred, the identity of that cell, a short Message Authentication Code and a cause.

The E-UTRAN uses the re-establishment procedure to continue SRB1 and to re-activate security without changing algorithms. A subsequent RRC connection reconfiguration procedure is used to resume operation on radio bearers other than SRB1 and to re-activate measurements. If the cell in which the UE initiates the re-establishment is not prepared (i.e. does not have a context for that UE), the E-UTRAN will reject the procedure, causing the UE to move to RRC_IDLE.

3.2.4 Connected Mode Inter-RAT Mobility

The overall procedure for the control of mobility is explained in this section; some further details can be found in Chapter 22.

3.2.4.1 Handover to LTE

The procedure for handover to LTE is largely the same as the procedure for handover within LTE, so it is not necessary to repeat the details here. The main difference is that upon handover to LTE the entire AS-configuration needs to be signalled, whereas within LTE it is possible to use 'delta signalling', whereby only the changes to the configuration are signalled.

If ciphering had not yet been activated in the previous RAT, the E-UTRAN activates ciphering, possibly using the NULL algorithm, as part of the handover procedure. The E-UTRAN also establishes SRB1, SRB2 and one or more DRBs (i.e. at least the DRB associated with the default EPS bearer).

3.2.4.2 Mobility from LTE

Generally, the procedure for mobility from LTE to another RAT supports both handover and Cell Change Order (CCO), possibly with Network Assistance (NACC – Network Assisted Cell Change). The CCO/NACC procedure is applicable only for mobility to GERAN. Mobility from LTE is performed only after security has been activated. When used for enhanced CSFB[18] to CDMA2000, the procedure includes support for parallel handover (i.e.

[18]See Section 2.4.2.1.

to both 1XRTT and HRPD), for handover to 1XRTT in combination with redirection to HRPD, and for redirection to HRPD only.

The procedure is illustrated in Figure 3.8.

1. The UE may send a MeasurementReport message (see Section 3.2.5 for further details).

2. In case of handover (as opposed to CCO), the source eNodeB requests the target Radio Access Network (RAN) node to prepare for the handover. As part of the 'handover preparation request' the source eNodeB provides information about the applicable inter-RAT UE capabilities as well as information about the currently-established bearers. In response, the target RAN generates the 'handover command' and returns this to the source eNodeB.

3. The source eNodeB sends a MobilityFromEUTRACommand message to the UE, which includes either the inter-RAT message received from the target (in case of handover), or the target cell/frequency and a few inter-RAT parameters (in case of CCO).

4. Upon receiving the MobilityFromEUTRACommand message, the UE starts the timer T304 and connects to the target node, either by using the received radio configuration (handover) or by initiating connection establishment (CCO) in accordance with the applicable specifications of the target RAT.

Upper layers in the UE are informed, by the AS of the target RAT, which bearers are established. From this, the UE can derive if some of the established bearers were not admitted by the target RAN node.

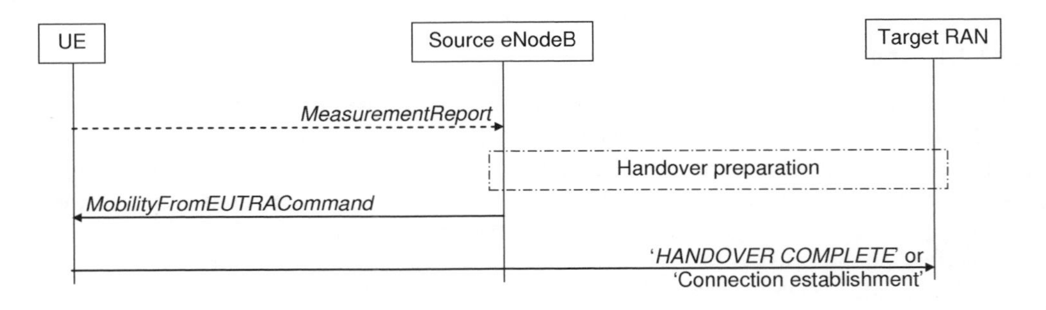

Figure 3.8: Mobility from LTE.

3.2.4.3 CDMA2000

For CDMA2000, additional procedures have been defined to support the transfer of dedicated information from the CDMA2000 upper layers, which are used to register the UE's presence in the target core network prior to performing the handover (referred to as preregistration). These procedures use SRB1.

3.2.5 Measurements

3.2.5.1 Measurement Configuration

The E-UTRAN can configure the UE to report measurement information to support the control of UE mobility. The following measurement configuration elements can be signalled via the RRCConnectionReconfiguration message.

1. **Measurement objects.** A measurement object defines on what the UE should perform the measurements – such as a carrier frequency. The measurement object may include a list of cells to be considered (white-list or black-list) as well as associated parameters, e.g. frequency- or cell-specific offsets.

2. **Reporting configurations.** A reporting configuration consists of the (periodic or event-triggered) criteria which cause the UE to send a measurement report, as well as the details of what information the UE is expected to report (e.g. the quantities, such as Received Signal Code Power (RSCP) (see Section 22.3.2.1) for UMTS or Reference Signal Received Power (RSRP) (see Section 22.3.1.1) for LTE, and the number of cells).

3. **Measurement identities.** These identify a measurement and define the applicable measurement object and reporting configuration.

4. **Quantity configurations**. The quantity configuration defines the filtering to be used on each measurement.

5. **Measurement gaps.** Measurement gaps define time periods when no uplink or downlink transmissions will be scheduled, so that the UE may perform the measurements. The measurement gaps are common for all gap-assisted measurements. Further details of the measurement gaps are discussed in Section 22.2.1.2.

The details of the above parameters depend on whether the measurement relates to an LTE, UMTS, GERAN or CDMA2000 frequency. Further details of the measurements performed by the UE are explained in Section 22.3. The E-UTRAN configures only a single measurement object for a given frequency, but more than one measurement identity may use the same measurement object. The identifiers used for the measurement object and reporting configuration are unique across all measurement types. An example of a set of measurement objects and their corresponding reporting configurations is shown in Figure 3.9.

In LTE it is possible to configure the quantity which triggers the report (RSCP or RSRP) for each reporting configuration. The UE may be configured to report either the trigger quantity or both quantities.

The RRC measurement reporting procedures include some extensions specifically to support Self-Optimizing Network (SON) functions such as the determination of Automatic Neighbour Relations (ANR) – see Section 25.2. The RRC measurement procedures also support UE positioning[19] by means of the enhanced cell identity method – see Section 19.4.

[19]See Chapter 19.

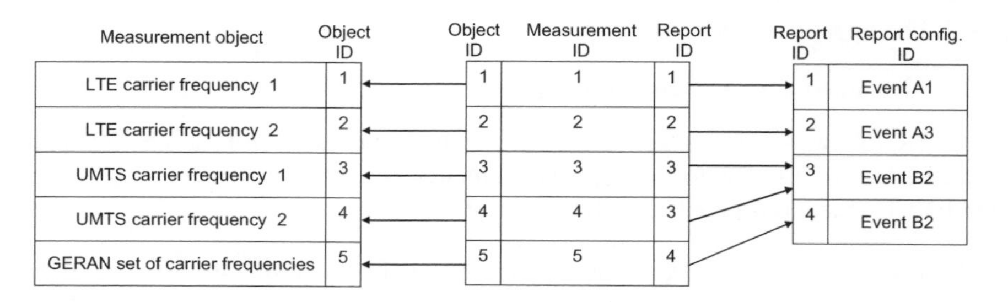

Figure 3.9: Example measurement configuration.

3.2.5.2 Measurement Report Triggering

Depending on the measurement type, the UE may measure and report any of the following:

- The serving cell;
- Listed cells (i.e. cells indicated as part of the measurement object);
- Detected cells on a listed frequency (i.e. cells which are not listed cells but are detected by the UE).

For some RATs, the UE measures and reports listed cells only (i.e. the list is a white-list), while for other RATs the UE also reports detected cells. For further details, see Table 3.3. Additionally, E-UTRAN can configure UTRAN PCI ranges for which the UE is allowed to send a measurement reports (mainly for the support of handover to UTRAN cells broadcasting a CSG identity).

For LTE, the following event-triggered reporting criteria are specified:

- **Event A1.** Serving cell becomes better than absolute threshold.
- **Event A2.** Serving cell becomes worse than absolute threshold.
- **Event A3.** Neighbour cell becomes better than an offset relative to the serving cell.
- **Event A4.** Neighbour cell becomes better than absolute threshold.
- **Event A5.** Serving cell becomes worse than one absolute threshold and neighbour cell becomes better than another absolute threshold.

For inter-RAT mobility, the following event-triggered reporting criteria are specified:

- **Event B1.** Neighbour cell becomes better than absolute threshold.
- **Event B2.** Serving cell becomes worse than one absolute threshold and neighbour cell becomes better than another absolute threshold.

The UE triggers an event when one or more cells meets a specified 'entry condition'. The E-UTRAN can influence the entry condition by setting the value of some configurable parameters used in these conditions – for example, one or more thresholds, an offset, and/or a hysteresis. The entry condition must be met for at least a duration corresponding to a 'timeToTrigger' parameter configured by the E-UTRAN in order for the event to be triggered.

The UE scales the timeToTrigger parameter depending on its speed (see Section 3.3 for further detail).

Figure 3.10 illustrates the triggering of event A3 when a timeToTrigger and an offset are configured.

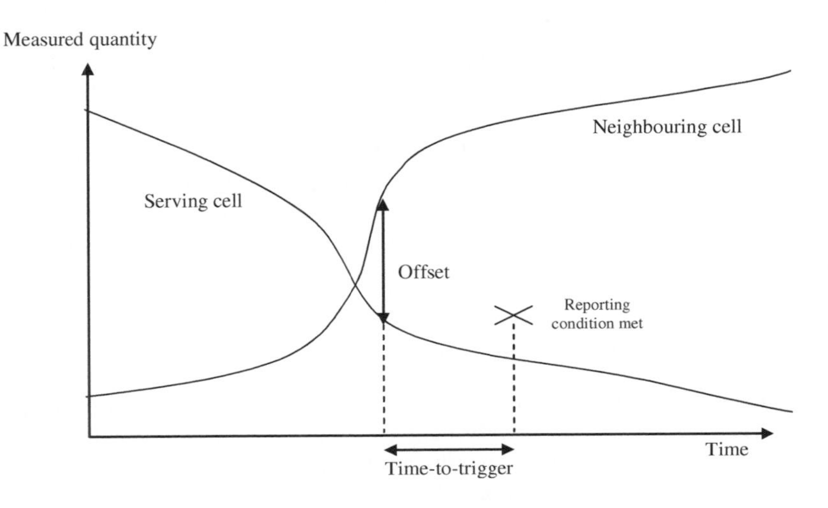

Figure 3.10: Event triggered report condition (Event A3).

The UE may be configured to provide a number of periodic reports after having triggered an event. This 'event-triggered periodic reporting' is configured by means of parameters 'reportAmount' and 'reportInterval', which specify respectively the number of periodic reports and the time period between them. If event-triggered periodic reporting is configured, the UE's count of the number of reports sent is reset to zero whenever a new cell meets the entry condition. The same cell cannot then trigger a new set of periodic reports unless it first meets a specified 'leaving condition'.

In addition to event-triggered reporting, the UE may be configured to perform periodic measurement reporting. In this case, the same parameters may be configured as for event-triggered reporting, except that the UE starts reporting immediately rather than only after the occurrence of an event.

3.2.5.3 Measurement Reporting

In a MeasurementReport message, the UE only includes measurement results related to a single measurement – in other words, measurements are not combined for reporting purposes. If multiple cells triggered the report, the UE includes the cells in order of decreasing value of the reporting quantity – i.e. the best cell is reported first. The number of cells the UE includes in a MeasurementReport may be limited by a parameter 'indexmaxReportCellsmaxReportCells'.

3.2.6 Other RRC Signalling Aspects

3.2.6.1 UE Capability Transfer

In order to avoid signalling of the UE radio access capabilities across the radio interface upon each transition from RRC_IDLE to RRC_CONNECTED, the core network stores the AS capabilities (both the E-UTRA and GERAN capabilities) while the UE is in RRC_IDLE/EMM-REGISTERED. Upon S1 connection establishment, the core network provides the capabilities to the E-UTRAN. If the E-UTRAN does not receive the (required) capabilities from the core network (e.g. due to the UE being in EMM-DEREGISTERED), it requests the UE to provide its capabilities using the UE capability transfer procedure. The E-UTRAN can indicate for each RAT (LTE, UMTS, GERAN) whether it wants to receive the associated capabilities. The UE provides the requested capabilities using a separate container for each RAT. Dynamic change of UE capabilities is not supported, except for change of the GERAN capabilities in RRC_IDLE which is supported by the tracking area update procedure.

3.2.6.2 Uplink/Downlink Information Transfer

The uplink/downlink information transfer procedures are used to transfer only upper layer information (i.e. no RRC control information is included). The procedure supports the transfer of 3GPP NAS dedicated information as well as CDMA2000 dedicated information.

In order to reduce latency, NAS information may also be included in the RRCConnection-SetupComplete and RRCConnectionReconfiguration messages. For the latter message, NAS information is only included if the AS and NAS procedures are dependent (i.e. they jointly succeed or fail). This applies for EPS bearer establishment, modification and release.

As noted earlier, some additional NAS information transfer procedures have also been defined for CDMA2000 for preregistration.

3.2.6.3 UE Information Transfer

The UE information transfer procedure was introduced in Release 9 to support SON (see Chapter 25). The procedure supports network optimization for mobility robustness by the reporting, at a later point in time, of measurement information available when a radio link failure occurs (see Section 25.6). E-UTRAN may also use the UE information transfer procedure to retrieve information regarding the last successful random access, which it may use for RACH optimization – see Section 25.7.

3.3 PLMN and Cell Selection

3.3.1 Introduction

After a UE has selected a PLMN, it performs *cell selection* – in other words, it searches for a suitable cell on which to camp (see Chapter 7). While camping on the chosen cell, the UE acquires the SI that is broadcast (see Section 9.2.1). Subsequently, the UE registers its presence in the tracking area, after which it can receive paging information which is used to notify UEs of incoming calls. The UE may establish an RRC connection, for example to establish a call or to perform a tracking area update.

When camped on a cell, the UE regularly verifies if there is a better cell; this is known as performing *cell reselection*.

LTE cells are classified according to the service level the UE obtains on them: a *suitable cell* is a cell on which the UE obtains normal service. If the UE is unable to find a suitable cell, but manages to camp on a cell belonging to another PLMN, the cell is said to be an *acceptable cell*, and the UE enters a 'limited service' state in which it can only perform emergency calls (and receive public warning messages) – as is also the case when no USIM is present in the UE. Finally, some cells may indicate via their SI that they are barred or reserved; a UE can obtain no service on such a cell.

A category called 'operator service' is also supported in LTE, which provides normal service but is applicable only for UEs with special access rights.

Figure 3.11 provides a high-level overview of the states and the cell (re)selection procedures.

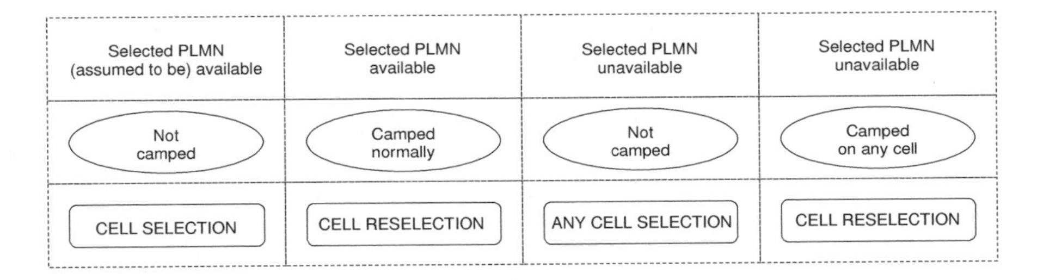

Figure 3.11: Idle mode states and procedures.

3.3.2 PLMN Selection

The NAS handles PLMN selection based on a list of available PLMNs provided by the AS. The NAS indicates the selected PLMN together with a list of equivalent PLMNs, if available. After successful registration, the selected PLMN becomes the *Registered* PLMN (R-PLMN).

The AS may autonomously indicate available PLMNs. In addition, NAS may request the AS to perform a full search for available PLMNs. In the latter case, the UE searches for the strongest cell on each carrier frequency. For these cells, the UE retrieves the PLMN identities from SI. If the quality of a cell satisfies a defined radio criterion, the corresponding PLMNs are marked as *high quality*; otherwise, the PLMNs are reported together with their quality.

3.3.3 Cell Selection

Cell selection consists of the UE searching for the strongest cell on all supported carrier frequencies of each supported RAT until it finds a suitable cell. The main requirement for cell selection is that it should not take too long, which becomes more challenging with the ever increasing number of frequencies and RATs to be searched. The NAS can speed up the search process by indicating the RATs associated with the selected PLMN. In addition, the UE may use information stored from a previous access.

The cell selection criterion is known as the *S-criterion* and is fulfilled when the cell-selection receive level and the quality level are above a given value: Srxlev > 0 and Squal > 0, where

$$\text{Srxlev} = Q_{\text{rxlevmeas}} - (Q_{\text{rxlevmin}} - Q_{\text{rxlevminoffset}})$$

$$\text{Squal} = Q_{\text{qualmeas}}(Q_{\text{qualmin}} + Q_{\text{qualminoffset}})$$

in which $Q_{\text{rxlevmeas}}$ is the measured cell receive level value, also known as the RSRP (see Section 22.3.1.1), and Q_{rxlevmin} is the minimum required receive level in the cell. Q_{qualmeas} and Q_{qualmin} are the corresponding parameters for the quality level, also known as the RSRQ. $Q_{\text{rxlevminoffset}}$ and $Q_{\text{qualminoffset}}$ are offsets which may be configured to prevent ping-pong between PLMNs, which may otherwise occur due to fluctuating radio conditions. The offsets are taken into account only when performing a periodic search for a higher priority PLMN while camped on a suitable cell in a visited PLMN.

The cell selection related parameters are broadcast within the SIB1 message.

For some specific cases, additional requirements are defined:

- Upon leaving connected mode, the UE should normally attempt to select the cell to which it was connected. However, the connection release message may include information directing the UE to search for a cell on a particular frequency.
- When performing ‘any cell selection’, the UE tries to find an acceptable cell of any PLMN by searching all supported frequencies on all supported RATs. The UE may stop searching upon finding a cell that meets the ‘high quality’ criterion applicable for that RAT.

Note that the UE only verifies the suitability of the strongest cell on a given frequency. In order to avoid the UE needing to acquire SI from a candidate cell that does not meet the S-criterion, suitability information is provided for inter-RAT neighbouring cells.

3.3.4 Cell Reselection

Once the UE camps on a suitable cell, it starts cell reselection. This process aims to move the UE to the ‘best’ cell of the selected PLMN and of its equivalent PLMNs, if any. As described in Section 3.2.3.4, cell reselection between frequencies and RATs is primarily based on absolute priorities. Hence, the UE first evaluates the frequencies of all RATs based on their priorities. Secondly, the UE compares the cells on the relevant frequencies based on radio link quality, using a ranking criterion. Finally, upon reselecting to the target cell the UE verifies the cell’s accessibility. Further rules have also been defined to allow the UE to limit the frequencies to be measured, to speed up the process and save battery power, as discussed in Section 3.3.4.1. Figure 3.12 provides a high-level overview of the cell reselection procedure.

It should be noted that the UE performs cell reselection only after having camped for at least one second on the current serving cell.

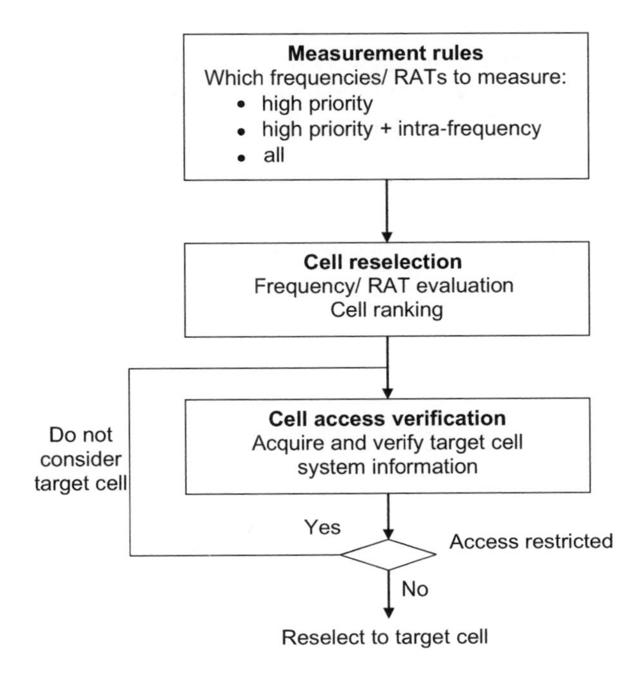

Figure 3.12: Cell reselection.

3.3.4.1 Measurement Rules

To enable the UE to save battery power, rules have been defined which limit the measurements the UE is required to perform. Firstly, the UE is required to perform intra-frequency measurements only when the quality of the serving cell is below or equal to a threshold ('SintraSearch'). Furthermore, the UE is required to measure other frequencies/RATs of lower or equal priority only when the quality of the serving cell is below or equal to another threshold ('SnonintraSearch'). The UE is always required to measure frequencies and RATs of higher priority. For both cases (i.e. intra-frequency and inter-frequency) the UE may refrain from measuring when a receive level and a quality criterion is fulfilled. The required performance (i.e. how often the UE is expected to make the measurements, and to what extent this depends on, for example, the serving cell quality) is specified in [4].

3.3.4.2 Frequency/RAT Evaluation

E-UTRAN configures an absolute priority for all applicable frequencies of each RAT. In addition to the cell-specific priorities which are optionally provided via SI, E-UTRAN can assign UE-specific priorities via dedicated signalling. Of the frequencies that are indicated in the system information, the UE is expected to consider for cell reselection only those for which it has priorities. Equal priorities are not applicable for inter-RAT cell reselection.

The UE reselects to a cell on a higher priority frequency if the S-criterion (see Section 3.3.3) of the concerned target cell exceeds a high threshold ($\text{Thresh}_{\text{X-High}}$) for longer than a certain duration $T_{\text{reselection}}$. The UE reselects to a cell on a lower-priority frequency if the S-criterion of the serving cell is below a low threshold ($\text{Thresh}_{\text{Serving-Low}}$) while the S-criterion of the target cell on a lower-priority frequency (possibly on another RAT) exceeds a low threshold ($\text{Thresh}_{\text{X-Low}}$) during the time interval $T_{\text{reselection}}$, and in the same time no cell on a higher-priority frequency is available. The UE evaluates the thresholds either based on receive level or on quality level, depending on which parameters E-UTRAN configures. Figure 3.13 illustrates the condition(s) to be met for reselecting to a cell on a higher-priority frequency (light grey bar) and to a cell on a lower priority frequency (dark grey bars).

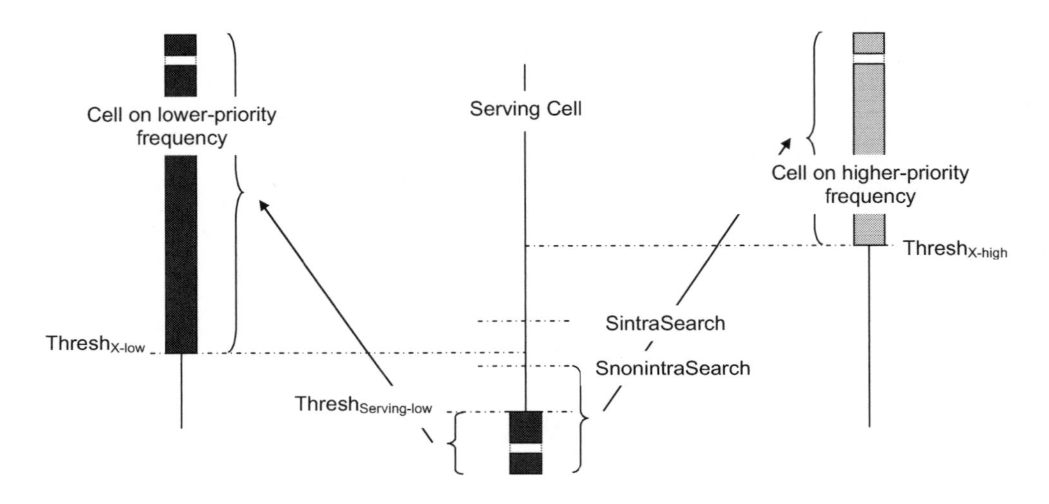

Figure 3.13: Frequency/RAT evaluation.

When reselecting to a frequency, possibly on another RAT, which has a different priority, the UE reselects to the highest-ranked cell on the concerned frequency (see Section 3.3.4.3).

Note that, as indicated in Section 3.2.3.4, thresholds and priorities are configured per frequency, while $T_{\text{reselection}}$ is configured per RAT.

From Release 8 onwards, UMTS and GERAN support the same priority-based cell reselection as provided in LTE, with a priority per frequency. Release 8 RANs will continue to handle legacy UEs by means of offset-based ranking. Likewise, Release 8 UEs should apply the ranking based on radio link quality (with offsets) unless UMTS or GERAN indicate support for priority-based reselection.

3.3.4.3 Cell Ranking

The UE ranks the intra-frequency cells and the cells on other frequencies having equal priority which fulfil the S-criterion using a criterion known as the *R-criterion*. The R-criterion generates rankings R_s and R_n for the serving cell and neighbour cells respectively:

$$\text{For the serving cell:} \quad R_s = Q_{\text{meas,s}} + Q_{\text{hyst,s}}$$

$$\text{For neighbour cells:} \quad R_n = Q_{\text{meas,n}} + Q_{\text{off s,n}}$$

where Q_{meas} is the measured cell received quality (RSRP) (see Section 22.3.1.1), $Q_{hyst,s}$ is a parameter controlling the degree of hysteresis for the ranking, and $Q_{off\,s,n}$ is an offset applicable between serving and neighbouring cells on frequencies of equal priority (the sum of the cell-specific and frequency-specific offsets).

The UE reselects to the highest-ranked candidate cell provided that it is better ranked than the serving cell for at least the duration of $T_{reselection}$. The UE scales the parameters $T_{reselection}$ and Q_{hyst}, depending on the UE speed (see Section 3.3.4.5 below).

3.3.4.4 Accessibility Verification

If the best cell on an LTE frequency is barred or reserved, the UE is required to exclude this cell from the list of cell reselection candidates. In this case, the UE may consider other cells on the same frequency unless the barred cell indicates (by means of field 'intraFreqReselection' within SIB1) that intra-frequency reselection is not allowed for a certain duration, unless the barred cell is an inaccessible Closed Subscriber Group (CSG) cell. If, however, the best cell is unsuitable for some other specific reason (e.g. because it belongs to a forbidden tracking area or to another non-equivalent PLMN), the UE is not permitted to consider any cell on the concerned frequency as a cell reselection candidate for a maximum of 300 s.

3.3.4.5 Speed Dependent Scaling

The UE scales the cell reselection parameters depending on its speed. This applies both in idle mode ($T_{reselection}$ and Q_{hyst}) and in connected mode (timeToTrigger). The UE speed is categorized by a mobility state (high, normal or low), which the UE determines based on the number of cell reselections/handovers which occur within a defined period, excluding consecutive reselections/handovers between the same two cells. The state is determined by comparing the count with thresholds for medium and high state, while applying some hysteresis. For idle and connected modes, separate sets of control parameters are used, signalled in SIB3 and within the measurement configuration respectively.

3.3.4.6 Cell Access Restrictions

The UE performs an access barring check during connection establishment (see Section 3.2.3.2). This function provides a means to control the load introduced by UE-originated traffic. There are separate means for controlling Mobile Originated (MO) calls and MO signalling. On top of the regular access class barring, Service Specific Access Control (SSAC) may be applied. SSAC facilitates separate control for MultiMedia TELephony (MMTEL) voice and video calls. Most of the SSAC functionality is handled by upper layers. In addition, separate access control exists to protect against E-UTRAN overload due to UEs accessing E-UTRAN merely to perform CSFB to CDMA2000.

Each UE belongs to an Access Class (AC) in the range 0–9. In addition, some UEs may belong to one or more high-priority ACs in the range 11–15, which are reserved for specific uses (e.g. security services, public utilities, emergency services, PLMN staff). AC10 is used for emergency access. Further details, for example regarding in which PLMN the high priority ACs apply, are provided in [5]. The UE considers access to be barred if access is barred for all its applicable ACs.

SIB2 may include a set of AC barring parameters for MO calls and/or MO signalling. This set of parameters comprises a probability factor and a barring timer for AC0–9 and a list of barring bits for AC11–15. For AC0–9, if the UE initiates a MO call and the relevant AC barring parameters are included, the UE draws a random number. If this number exceeds the probability factor, access is not barred. Otherwise access is barred for a duration which is randomly selected centred on the broadcast barring timer value. For AC11–15, if the UE initiates a MO call and the relevant AC barring parameters are included, access is barred whenever the bit corresponding to all of the UE's ACs is set. The behaviour is similar in the case of UE-initiated MO signalling.

For cell (re)selection, the UE is expected to consider cells which are neither barred nor reserved for operator or future use. In addition, a UE with an access class in the range 11–15 shall consider a cell that is (only) reserved for operator use and part of its home PLMN (or an equivalent) as a candidate for cell reselection. The UE is never allowed to (re)select a cell that is not a reselection candidate even for emergency access.

3.3.4.7 Any Cell Selection

When the UE is unable to find a suitable cell of the selected PLMN, it performs 'any cell selection'. In this case, the UE performs normal idle mode operation: monitoring paging, acquiring SI, performing cell reselection. In addition, the UE regularly attempts to find a suitable cell on other frequencies or RATs (i.e. not listed in SI). If a UE supporting voice services is unable to find a suitable cell, it should attempt to find an acceptable cell on any supported RAT regardless of the cell reselection priorities that are broadcast. The UE is not allowed to receive MBMS in this state.

3.3.4.8 Closed Subscriber Group

LTE supports the existence of cells which are accessible only for a limited set of UEs – a Closed Subscriber Group (CSG). In order to prevent UEs from attempting to register on a CSG cell on which they do not have access, the UE maintains a CSG white list, i.e. a list of CSG identities for which access has been granted to the UE. The CSG white list can be transferred to the UE by upper layers, or updated upon successful access to a CSG cell. To facilitate the latter, UEs support 'manual selection' of CSG cells which are not in the CSG white list. The manual selection may be requested by the upper layers, based on a text string broadcast by the cell. LTE also supports hybrid cells. Like CSG cells, hybrid cells broadcast a CSG identity; they are accessible as CSG cells by UEs whose CSG white lists include the CSG identity, and as normal cells by all other UEs (see Section 24.2.2).

3.4 Paging

To receive paging messages from E-UTRAN, UEs in idle mode monitor the PDCCH channel for an RNTI value used to indicate paging: the P-RNTI (see Section 9.2.2.2). The UE only needs to monitor the PDCCH channel at certain UE-specific occasions (i.e. at specific subframes within specific radio frames – see Section 6.2 for an introduction to the LTE radio frame structure.). At other times, the UE may apply DRX, meaning that it can switch off its receiver to preserve battery power.

The E-UTRAN configures which of the radio frames and subframes are used for paging. Each cell broadcasts a default paging cycle. In addition, upper layers may use dedicated signalling to configure a UE-specific paging cycle. If both are configured, the UE applies the lowest value. The UE calculates the radio frame (the Paging Frame (PF)) and the subframe within that PF (the Paging Occasion (PO)), which E-UTRAN applies to page the UE as follows:

$$\begin{aligned} \text{SFN} \bmod T &= (T/N) \times (\text{UE\_ID} \bmod N) \\ \text{i\_s} &= \lfloor \text{UE\_ID}/N \rfloor \bmod \text{Ns} \\ T &= \text{UE DRX cycle (i.e. paging cycle)} = \min(T_c, T_{ue}) \\ N &= \min(T, \text{nB}) \\ \text{Ns} &= \max(1, \text{nB}/T) \end{aligned} \tag{3.1}$$

where:

T_c is the cell-specific default paging cycle {32, 64, 128, 256} radio frames,

T_{ue} is the UE-specific paging cycle {32, 64, 128, 256} radio frames,

N is the number of paging frames within the paging cycle of the UE,

UE_ID is the IMSI[20] mod 1024, with IMSI being the decimal rather than the binary number,

i_s is an index pointing to a pre-defined table defining the corresponding subframe,

nB is the number of 'paging subframes' per paging cycle (across all UEs in the cell),

Ns is the number of 'paging subframes' in a radio frame that is used for paging.

Table 3.4 includes a number of examples to illustrate the calculation of the paging radio frames (PF) and subframes (PO).

Table 3.4: Examples for calculation of paging frames and subframes.

Case	UE_ID	T_c	T_{ue}	T	nB	N	Ns	PF	i_s	PO
A	147	256	256	256	64	64	1	76	0	9
B	147	256	128	128	32	32	1	76	0	9
C	147	256	128	128	256	128	2	19	1	4

In cases A and B in Table 3.4, one out of every four radio frames is used for paging, using one subframe in each of those radio frames. For case B, there are 32 paging frames within the UE's paging cycle, across which the UEs are distributed based on the UE-identity. In case C, two subframes in each radio frame are used for paging, i.e. Ns = 2. In this case, there are 128 paging frames within the UE's paging cycle and the UEs are also distributed across the two subframes within the paging frame. The LTE specifications include a table that indicates the subframe applicable for each combination of Ns and i_s, which is the index that follows from Equation (3.1). Figure 3.14 illustrates cases B and C. All the shaded subframes can be used for paging; the darker ones are applicable for the UE with the indicated identity.

[20]International Mobile Subscriber Identity.

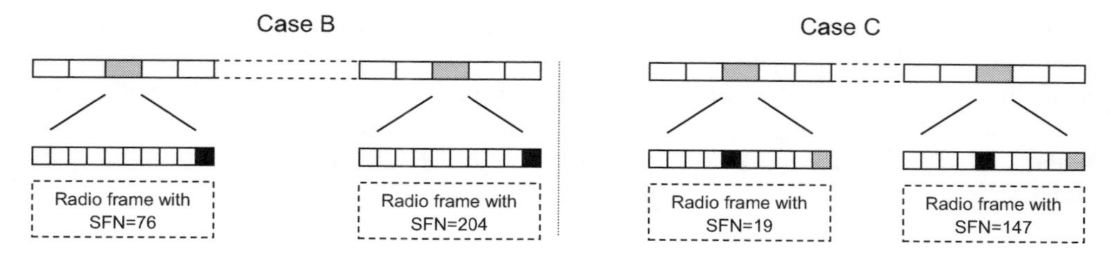

Figure 3.14: Paging frame and paging occasion examples.

3.5 Summary

The main aspects of the Control Plane protocols in LTE can be broken down into the Cell Selection and Reselection Procedures when the UE is in Idle Mode, and the RRC protocol when the UE is in Connected Mode.

The roles of these protocols include supporting security, mobility both between different LTE cells and between LTE and other radio systems, and establishment and reconfiguration of the radio bearers which carry control information and user data.

References[21]

[1] 3GPP Technical Specification 36.331, ‘Evolved Universal Terrestrial Radio Access (E-UTRA); Radio Resource Control (RRC); Protocol specification’, www.3gpp.org.

[2] 3GPP Technical Specification 36.304, ‘Evolved Universal Terrestrial Radio Access (E-UTRA); User Equipment (UE) procedures in idle mode (Release 9)’, www.3gpp.org.

[3] 3GPP Technical Specification 33.401, ‘3GPP System Architecture Evolution; Security Architecture’, www.3gpp.org.

[4] 3GPP Technical Specification 36.133, ‘Evolved Universal Terrestrial Radio Access (E-UTRA); Requirements for support of radio resource management’, www.3gpp.org.

[5] 3GPP Technical Specification 22.011, ‘Service accessibility’, www.3gpp.org.

[21] All web sites confirmed 1st March 2011.

4

User Plane Protocols

Patrick Fischer, SeungJune Yi, SungDuck Chun and YoungDae Lee

4.1 Introduction to the User Plane Protocol Stack

The LTE Layer 2 user-plane protocol stack is composed of three sublayers, as shown in Figure 4.1:

- **The Packet Data Convergence Protocol (PDCP) layer [1]:** This layer processes Radio Resource Control (RRC) messages in the control plane and Internet Protocol (IP) packets in the user plane. Depending on the radio bearer, the main functions of the PDCP layer are header compression, security (integrity protection and ciphering), and support for reordering and retransmission during handover. For radio bearers which are configured to use the PDCP layer, there is one PDCP entity per radio bearer.
- **The Radio Link Control (RLC) layer [2]:** The main functions of the RLC layer are segmentation and reassembly of upper layer packets in order to adapt them to the size which can actually be transmitted over the radio interface. For radio bearers which need error-free transmission, the RLC layer also performs retransmission to recover from packet losses. Additionally, the RLC layer performs reordering to compensate for out-of-order reception due to Hybrid Automatic Repeat reQuest (HARQ) operation in the layer below. There is one RLC entity per radio bearer.
- **The Medium Access Control (MAC) layer [3]:** This layer performs multiplexing of data from different radio bearers. Therefore there is only one MAC entity per UE. By deciding the amount of data that can be transmitted from each radio bearer and instructing the RLC layer as to the size of packets to provide, the MAC layer aims to achieve the negotiated Quality of Service (QoS) for each radio bearer. For the

LTE – The UMTS Long Term Evolution: From Theory to Practice, Second Edition.
Stefania Sesia, Issam Toufik and Matthew Baker.

uplink, this process includes reporting to the eNodeB the amount of buffered data for transmission.

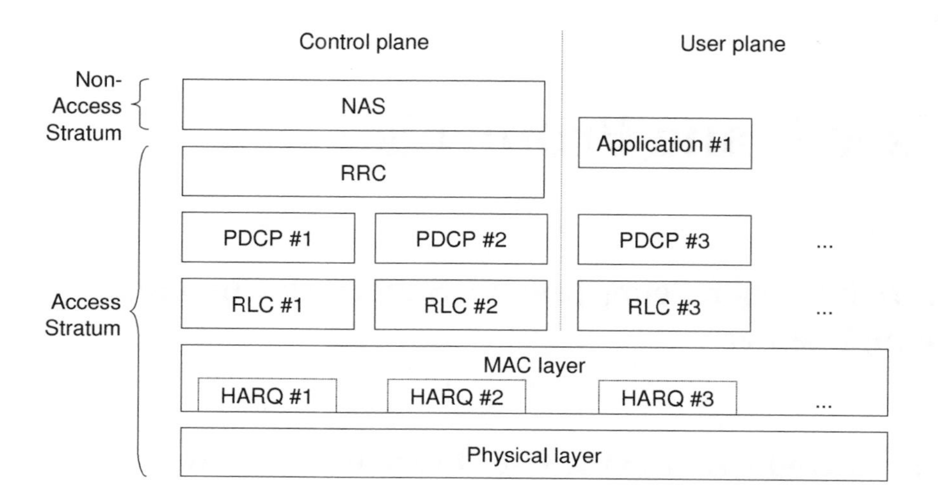

Figure 4.1: Overview of user-plane architecture.

At the transmitting side, each layer receives a Service Data Unit (SDU) from a higher layer, for which the layer provides a service, and outputs a Protocol Data Unit (PDU) to the layer below. The RLC layer receives packets from the PDCP layer. These packets are called PDCP PDUs from a PDCP point of view and represent RLC SDUs from an RLC point of view. The RLC layer creates packets which are provided to the layer below, i.e. the MAC layer. The packets provided by RLC to the MAC layer are RLC PDUs from an RLC point of view, and MAC SDUs from a MAC point of view. At the receiving side, the process is reversed, with each layer passing SDUs up to the layer above, where they are received as PDUs.

An important design feature of the LTE protocol stack is that all the PDUs and SDUs are *byte aligned*.[1] This is to facilitate handling by microprocessors, which are normally defined to handle packets in units of bytes. In order to further reduce the processing requirements of the user plane protocol stack in LTE, the headers created by each of the PDCP, RLC and MAC layers are also byte-aligned. This implies that sometimes unused padding bits are needed in the headers, and thus the cost of designing for efficient processing is that a small amount of potentially-available capacity is wasted.

[1]Byte alignment means that the lengths of the PDUs and SDUs are multiples of 8 bits.

4.2 Packet Data Convergence Protocol (PDCP)

4.2.1 Functions and Architecture

The PDCP layer performs the following functions:

- Header compression and decompression for user plane data;
- Security functions:
  - ciphering and deciphering for user plane and control plane data;
  - integrity protection and verification for control plane data;
- Handover support functions:
  - in-sequence delivery and reordering of PDUs for the layer above at handover;
  - lossless handover for user plane data mapped on RLC Acknowledged Mode (AM, see Section 4.3.1).
- Discard for user plane data due to timeout.

The PDCP layer manages data streams in the user plane, as well as in the control plane (i.e. the RRC protocol – see Section 3.2), only for the radio bearers using either a Dedicated Control CHannel (DCCH) or a Dedicated Transport CHannel (DTCH) — see Section 4.4.1.2. The architecture of the PDCP layer differs for user plane data and control plane data, as shown in Figures 4.2 and 4.3 respectively.

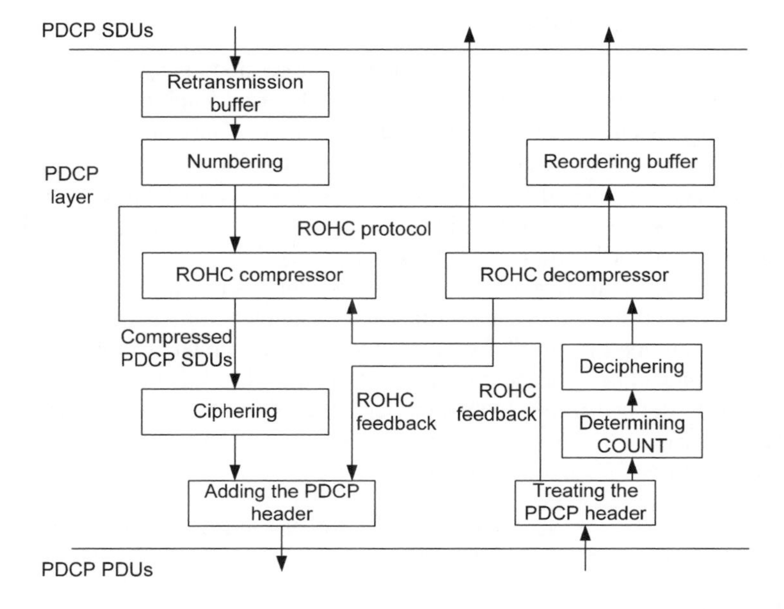

Figure 4.2: Overview of user-plane PDCP. Reproduced by permission of © 3GPP.

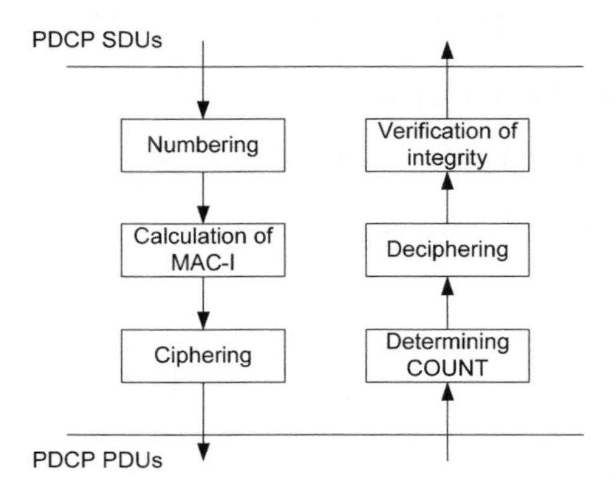

Figure 4.3: Overview of control-plane PDCP. Reproduced by permission of © 3GPP.

Each of the main functions is explained in the following subsections. Two different types of PDCP PDU are defined in LTE: PDCP Data PDUs and PDCP Control PDUs. PDCP Data PDUs are used for both control and user plane data. PDCP Control PDUs are only used to transport the feedback information for header compression, and for PDCP status reports which are used in case of handover (see Section 4.2.6) and hence are only used within the user plane.

4.2.2 Header Compression

One of the main functions of PDCP is header compression using the RObust Header Compression (ROHC) protocol defined by the IETF (Internet Engineering Task Force). In LTE, header compression is very important because there is no support for the transport of voice services via the Circuit-Switched (CS) domain.[2] Thus, in order to provide voice services on the Packet-Switched (PS) domain in a way that comes close to the efficiency normally associated with CS services it is necessary to compress the IP/UDP/RTP[3] header which is typically used for Voice over IP (VoIP) services.

The IETF specifies in 'RFC 4995'[4] a framework which supports a number of different header compression 'profiles' (i.e. sets of rules and parameters for performing the compression). The header compression profiles supported for LTE are shown in Table 4.1. This means that a UE may implement one or more of these ROHC profiles. It is important to notice that the profiles already defined in the IETF's earlier 'RFC 3095' have been redefined in RFC 4995 in order to increase robustness in some cases. The efficiency of RFC 3095 and RFC 4995 is similar, and UMTS[5] supports only RFC 3095.

[2]LTE does, however, support a CS FallBack (CSFB) procedure to allow an LTE UE to be handed over to a legacy RAT to originate a CS voice call, as well as a Single Radio Voice Call Continuity (SRVCC) procedure to hand over a Packet-Switched (PS) VoIP call to a CS voice call – see Sections 2.4.2.1 and 2.4.2.2.

[3]Internet Protocol / User Datagram Protocol / Real-Time Transport Protocol.

[4]Requests for Comments (RFCs) capture much of the output of the IETF.

[5]Universal Mobile Telecommunications System.

Table 4.1: Supported header compression protocols.

Reference	Usage
RFC 4995	No compression
RFC 3095, RFC 4815	RTP/UDP/IP
RFC 3095, RFC 4815	UDP/IP
RFC 3095, RFC 4815	ESP/IP
RFC 3843, RFC 4815	IP
RFC 4996	TCP/IP
RFC 5225	RTP/UDP/IP
RFC 5225	UDP/IP
RFC 5225	ESP/IP
RFC 5225	IP

The support of ROHC is not mandatory for the UE, except for those UEs which support VoIP. UEs which support VoIP have to support at least one profile for compression of RTP, UDP and IP.[6] The eNodeB controls by RRC signalling which of the ROHC profiles supported by the UE are allowed to be used. The ROHC compressors in the UE and the eNodeB then dynamically detect IP flows that use a certain IP header configuration and choose a suitable compression profile from the allowed and supported profiles.

ROHC header compression operates by allowing both the sender and the receiver to store the static parts of the header (e.g. the IP addresses of the sender/receiver), and to update these only when they change. Furthermore, dynamic parts (for example, the timestamp in the RTP header) are compressed by transmitting only the difference from a reference clock maintained in both the transmitter and the receiver.

As the non-changing parts of the headers are thus transmitted only once, successful decompression depends on their correct reception. Feedback is therefore used in order to confirm the correct reception of initialization information for the header decompression. Furthermore, the correct decompression of the received PDCP PDUs is confirmed periodically, depending on the experienced packet losses.

As noted above, the most important use case for ROHC is VoIP. Typically, for the transport of a VoIP packet which contains a payload of 32 bytes, the header added will be 60 bytes for the case of IPv6 and 40 bytes for the case of IPv4[7] – i.e. an overhead of 188% and 125% respectively. By means of ROHC, after the initialization of the header compression entities, this overhead can be compressed to four to six bytes, and thus to a relative overhead of 12.5–18.8%. This calculation is valid during the active periods, but during silence periods the payload size is smaller so the relative overhead is higher.

[6]ROHC is required for VoIP supported via the IP Multimedia Subsystem (IMS); in theory it could be possible to support raw IP VoIP without implementing ROHC.

[7]IPv6 is the successor to the original IPv4, for many years the dominant version of IP used on the Internet, and introduces a significantly expanded address space.

4.2.3 Security

The security architecture of LTE was introduced in Section 3.2.3.1. The implementation of security, by ciphering (of both control plane (RRC) data and user plane data) and integrity protection (for control plane (RRC) data only), is the responsibility of the PDCP layer.

A PDCP Data PDU counter (known as 'COUNT' in the LTE specifications) is used as an input to the security algorithms. The COUNT value is incremented for each PDCP Data PDU during an RRC connection; it has a length of 32 bits in order to allow an acceptable duration for an RRC connection.

During an RRC connection, the COUNT value is maintained by both the UE and the eNodeB by counting each transmitted/received PDCP Data PDU. In order to provide robustness against lost packets, each protected PDCP Data PDU includes a PDCP Sequence Number (SN) which corresponds to the least significant bits of the COUNT value.[8] Thus if one or more packets are lost, the correct COUNT value of a newly received packet can be determined using the PDCP SN. This means that the associated COUNT value is the next highest COUNT value for which the least significant bits correspond to the PDCP SN. A loss of synchronization of the COUNT value between the UE and eNodeB can then only occur if a number of packets corresponding to the maximum SN are lost consecutively. In principle, the probability of this kind of loss of synchronization occurring could be minimized by increasing the length of the SN, even to the extent of transmitting the whole COUNT value in every PDCP Data PDU. However, this would cause a high overhead, and therefore only the least significant bits are used as the SN; the actual SN length depends on the configuration and type of PDU, as explained in the description of the PDCP PDU formats in Section 4.2.6.

This use of a counter is designed to protect against a type of attack known as a *replay attack*, where the attacker tries to resend a packet that has been intercepted previously; the use of the COUNT value also provides protection against attacks which aim at deriving the used key or ciphering pattern by comparing successive patterns. Due to the use of the COUNT value, even if the same packet is transmitted twice, the ciphering pattern will be completely uncorrelated between the two transmissions, thus preventing possible security breaches.

Integrity protection is realized by adding a field known as 'Message Authentication Code for Integrity' (MAC-I)[9] to each RRC message. This code is calculated based on the Access Stratum (AS) derived keys (see Section 3.2.3.1), the message itself, the radio bearer ID, the direction (i.e. uplink or downlink) and the COUNT value.

If the integrity check fails, the message is discarded and the integrity check failure is indicated to the RRC layer so that the RRC connection re-establishment procedure can be executed (see Section 3.2.3.5).

Ciphering is realized by performing an XOR operation with the message and a ciphering stream that is generated by the ciphering algorithm based on the AS derived keys (see Section 3.2.3.1), the radio bearer ID, the direction (i.e. uplink or downlink), and the COUNT value.

Ciphering can only be applied to PDCP Data PDUs. PDCP Control PDUs (such as ROHC feedback or PDCP status reports) are neither ciphered nor integrity protected.

---

[8] In order to avoid excessive overhead, the most significant bits of the COUNT value, also referred to as the Hyper Frame Number (HFN), are not signalled but derived from counting overflows of the PDCP SN.

[9] Note that the MAC-I has no relation to the MAC layer.

Except for identical retransmissions, the same COUNT value is not allowed to be used more than once for a given security key. The eNodeB is responsible for avoiding reuse of the COUNT with the same combination of radio bearer ID, AS base-key and algorithm. In order to avoid such reuse, the eNodeB may for example use different radio bearer IDs for successive radio bearer establishments, trigger an intracell handover or trigger a UE state transition from connected to idle and back to connected again (see Section 3.2).

4.2.4 Handover

Handover is performed when the UE moves from the coverage of one cell to the coverage of another cell in RRC_CONNECTED state. Depending on the required QoS, either a seamless or a lossless handover is performed as appropriate for each user plane radio bearer, as explained in the following subsections.

4.2.4.1 Seamless Handover

Seamless handover is applied for user plane radio bearers mapped on RLC Unacknowledged Mode (UM, see Section 4.3.1). These types of data are typically reasonably tolerant of losses but less tolerant of delay (e.g. voice services). Seamless handover is therefore designed to minimize complexity and delay, but may result in loss of some SDUs.

At handover, for radio bearers to which seamless handover applies, the PDCP entities including the header compression contexts are reset, and the COUNT values are set to zero. As a new key is anyway generated at handover, there is no security reason to maintain the COUNT values. PDCP SDUs in the UE for which the transmission has not yet started will be transmitted after handover to the target cell. In the eNodeB, PDCP SDUs that have not yet been transmitted can be forwarded via the X2 interface[10] to the target eNodeB. PDCP SDUs for which the transmission has already started but that have not been successfully received will be lost. This minimizes the complexity because no context (i.e. configuration information) has to be transferred between the source and the target eNodeB at handover.

4.2.4.2 Lossless Handover

Based on the SN that is added to PDCP Data PDUs it is possible to ensure in-sequence delivery during handover, and even provide a fully lossless handover functionality, performing retransmission of PDCP SDUs for which reception has not yet been acknowledged prior to the handover. This lossless handover function is used mainly for delay-tolerant services such as file downloads where the loss of one PDCP SDU can result in a drastic reduction in the data rate due to the reaction of the Transmission Control Protocol (TCP).

Lossless handover is applied for user plane radio bearers that are mapped on RLC Acknowledged Mode (AM, see Section 4.3.1).

For lossless handover, the header compression protocol is reset in the UE because the header compression context is not forwarded from the source eNodeB to the target eNodeB. However, the PDCP SNs and the COUNT values associated with PDCP SDUs are maintained. For simplicity reasons, inter-eNodeB handover and intra-eNodeB handover are handled in the same way in LTE.

[10] For details of the X2 interface, see Section 2.6.

In normal transmission, while the UE is not handing over from one cell to another, the RLC layer in the UE and the eNodeB ensures in-sequence delivery. PDCP PDUs that are retransmitted by the RLC protocol, or that arrive out of sequence due to the variable delay in the HARQ transmission, are reordered based on the RLC SN. At handover, the RLC layer in the UE and in the eNodeB will deliver all PDCP PDUs that have already been received to the PDCP layer in order to have them decompressed before the header compression protocol is reset. Because some PDCP SDUs may not be available at this point, the PDCP SDUs that are not available in-sequence are not delivered immediately to higher layers in the UE or to the gateway in the network. In the PDCP layer, the PDCP SDUs received out of order are stored in the reordering buffer (see Figure 4.2). PDCP SDUs that have been transmitted but not yet been acknowledged by the RLC layer are stored in a retransmission buffer in the PDCP layer.

In order to ensure lossless handover in the *uplink*, the UE retransmits the PDCP SDUs stored in the PDCP retransmission buffer. This is illustrated in Figure 4.4. In this example the PDCP entity has initiated transmission for the PDCP SDUs with the sequence numbers 1 to 5; the packets with the sequence numbers 3 and 5 have not been received by the source eNodeB, for example due to the handover interrupting the HARQ retransmissions. After the handover, the UE restarts the transmission of the PDCP SDUs for which successful transmission has not yet been acknowledged to the target eNodeB. In the example in Figure 4.4 only the PDCP SDUs 1 and 2 have been acknowledged prior to the handover. Therefore, after the handover the UE will retransmit the packets 3, 4 and 5, although the network had already received packet 4.

In order to ensure in-sequence delivery in the uplink, the source eNodeB, after decompression, delivers the PDCP SDUs that are received in-sequence to the gateway, and forwards the PDCP SDUs that are received out-of-sequence to the target eNodeB. Thus, the target eNodeB can reorder the decompressed PDCP SDUs received from the source eNodeB and the retransmitted PDCP SDUs received from the UE based on the PDCP SNs which are maintained during the handover, and deliver them to the gateway in the correct sequence.

In order to ensure lossless handover in the *downlink*, the source eNodeB forwards the uncompressed PDCP SDUs for which reception has not yet been acknowledged by the UE to the target eNodeB for retransmission in the downlink. The source eNodeB receives an indication from the gateway that indicates the last packet sent to the source eNodeB. The source eNodeB also forwards this indication to the target eNodeB so that the target eNodeB knows when it can start transmission of packets received from the gateway. In the example in Figure 4.5, the source eNodeB has started the transmission of the PDCP SDUs 1 to 4; due to, for example, a handover occurring prior to the HARQ retransmissions of packet 3, packet 3 will not be received by the UE from the source eNodeB. Furthermore the UE has only sent an acknowledgment for packets 1 and 2, although packet 4 has been received by the UE. The target eNodeB then ensures that the PDCP SDUs that have not yet been acknowledged in the source eNodeB are sent to the UE. Thus, the UE can reorder the received PDCP SDUs and the PDCP SDUs that are stored in the reordering buffer, and deliver them to higher layers in sequential order.

The UE will expect the packets from the target eNodeB in ascending order of SNs. In the case of a packet not being forwarded from the source eNodeB to the target eNodeB, i.e. when one of the packets that the UE expects is missing during the handover operation, the UE can immediately conclude that the packet is lost and can forward the packets which have already been received in sequence to higher layers. This avoids the UE having to retain

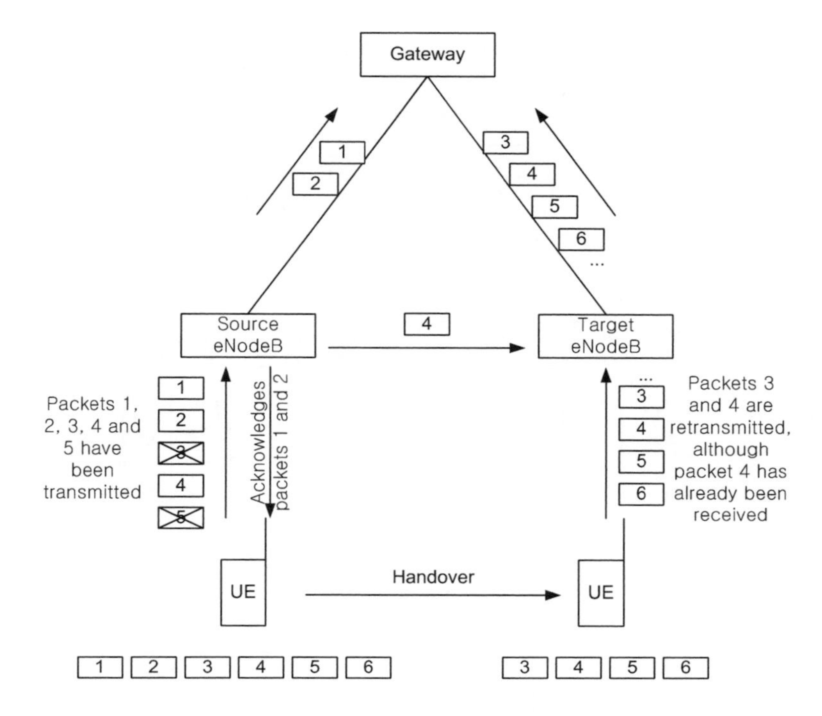

Figure 4.4: Lossless handover in the uplink.

already-received packets in order to wait for a potential retransmission. Thus the forwarding of the packets in the network can be decided without informing the UE.

In some cases it may happen that a PDCP SDU has been successfully received, but a corresponding RLC acknowledgement has not. In this case, after the handover, there may be unnecessary retransmissions initiated by the UE or the target eNodeB based on the incorrect status received by the RLC layer. In order to avoid these unnecessary retransmissions a PDCP status report can be sent from the eNodeB to the UE and from the UE to the eNodeB as described in Section 4.2.6. Additionally, a PDCP 'Status Report' can request retransmission of PDCP SDUs which were correctly received but failed in header decompression. Whether to send a PDCP status report after handover is configured independently for each radio bearer.

4.2.5 Discard of Data Packets

Typically, the data rate that is available on the radio interface is smaller than the data rate available on the network interfaces (e.g. S1[11]). Thus, when the data rate of a given service is higher than the data rate provided by the LTE radio interface, this leads to buffering in the UE and in the eNodeB. This buffering allows the scheduler in the MAC layer some freedom to vary the instantaneous data rate at the physical layer in order to adapt to the current radio

[11]For details of the S1 interface, see Section 2.5.

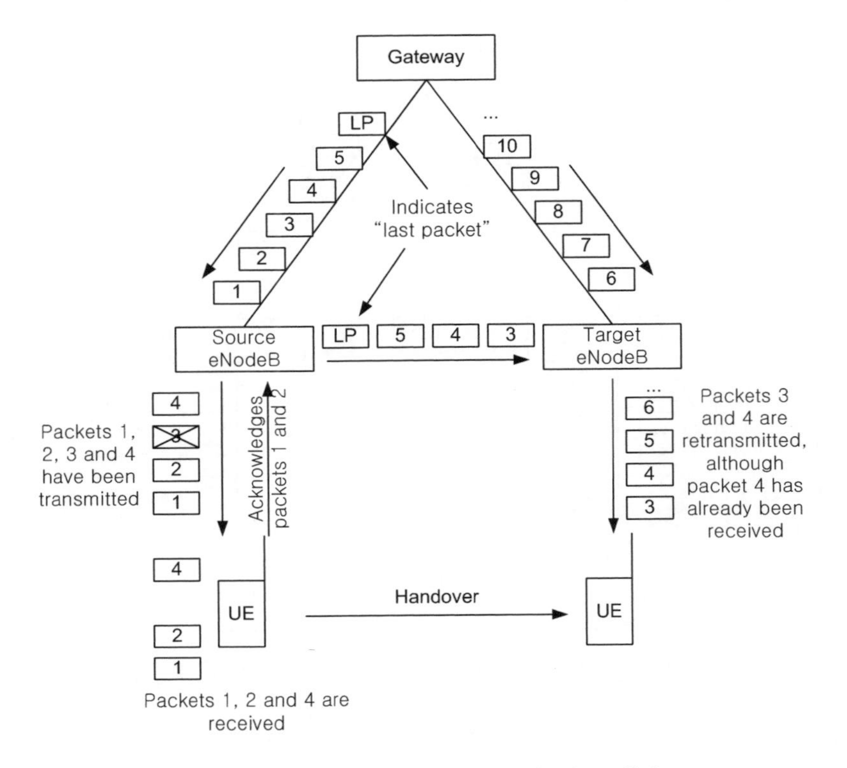

Figure 4.5: Lossless handover in the downlink.

channel conditions. Thanks to the buffering, the variations in the instantaneous data rate are then seen by the application only as some jitter in the transfer delay.

However, when the data rate provided by the application exceeds the data rate provided by the radio interface for a long period, large amounts of buffered data can result. This may lead to a large loss of data at handover if lossless handover is not applied to the bearer, or to an excessive delay for real time applications.

In the fixed internet, one of the roles typically performed by the routers is to drop packets when the data rate of an application exceeds the available data rate in a part of the internet. An application may then detect this loss of packets and adapt its data rate to the available rate. A typical example is the TCP transmit window handling, where the transmit window of TCP is reduced when a lost packet is detected, thus adapting to the available rate. Other applications such as video or voice calls via IP can also detect lost packets, for example via RTCP (Real-time Transport Control Protocol) feedback, and can adapt the data rate accordingly.

In order to allow these mechanisms to work, and to prevent excessive delay, a discard function is included in the PDCP layer for LTE. This discard function is based on a timer, where for each PDCP SDU received from the higher layers in the transmitter a timer is started, and when the transmission of the PDCP SDU has not yet been initiated in the UE at the expiry of this timer the PDCP SDU is discarded. If the timer is set to an appropriate value for the required QoS of the radio bearer, this discard mechanism can prevent excessive delay and queuing in the transmitter.

4.2.6 PDCP PDU Formats

PDCP PDUs for user plane data comprise a 'D/C' field in order to distinguish Data and Control PDUs, the formats of which are shown in Figures 4.6 and 4.7 respectively. PDCP Data PDUs comprise a 7- or 12-bit SN as shown in Table 4.2. PDCP Data PDUs for user plane data contain either an uncompressed (if header compression is not used) or a compressed IP packet.

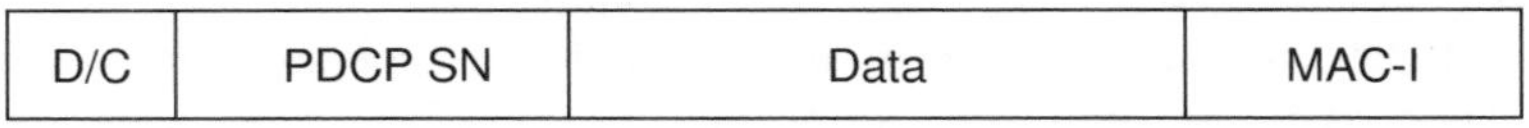

D/C	PDCP SN	Data	MAC-I

Figure 4.6: Key features of PDCP Data PDU format. See Table 4.2 for presence of D/C and MAC-I fields.

Table 4.2: PDCP Data PDU formats.

PDU type	D/C field	Sequence number length	MAC-I	Applicable RLC Modes (see Section 4.3.1)
User plane long SN	Present	12 bits	Absent	AM / UM
User plane short SN	Present	7 bits	Absent	UM
Control plane	Absent	5 bits	32 bits	AM

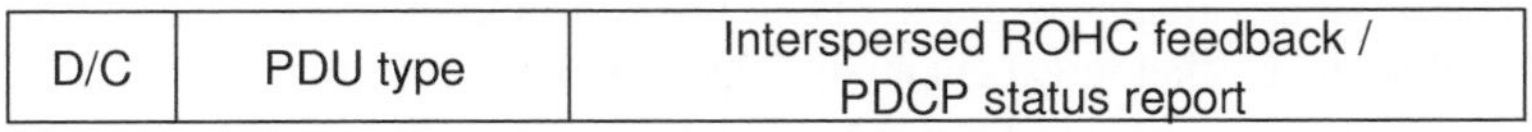

D/C	PDU type	Interspersed ROHC feedback / PDCP status report

Figure 4.7: Key features of PDCP Control PDU format.

PDCP Data PDUs for control plane data (i.e. RRC signalling) comprise a MAC-I field of 32-bit length for integrity protection. PDCP Data PDUs for control plane data contain one complete RRC message.

As can be seen in Table 4.2 there are three types of PDCP Data PDU, distinguished mainly by the length of the PDCP SN and the presence of the MAC-I field. As mentioned in Section 4.2.3, the length of the PDCP SN in relation to the data rate, the packet size and the packet inter-arrival rate determines the maximum possible interruption time without desynchronizing the COUNT value which is used for ciphering and integrity protection.

The PDCP Data PDU for user plane data using the long SN allows longer interruption times, and makes it possible, when it is mapped on RLC Acknowledged Mode (AM) (see Section 4.3.1.3), to perform lossless handover as described in Section 4.2.4, but implies a higher overhead. Therefore it is mainly used for data applications with a large IP packet size where the overhead compared to the packet size is not too significant, for example for file transfer, web browsing, or e-mail traffic.

The PDCP Data PDU for user plane data using the short SN is mapped on RLC Unacknowledged Mode (UM) (see Section 4.3.1.2) and is typically used for VoIP services, where only seamless handover is used and retransmission is not necessary.

PDCP Control PDUs are used by PDCP entities handling user plane data (see Figure 4.1). There are two types of PDCP Control PDU, distinguished by the PDU Type field in the PDCP header. PDCP Control PDUs carry either PDCP 'Status Reports' for the case of lossless handover, or ROHC feedback created by the ROHC header compression protocol. PDCP Control PDUs carrying ROHC feedback are used for user plane radio bearers mapped on either RLC UM or RLC AM, while PDCP Control PDUs carrying PDCP Status Reports are used only for user plane radio bearers mapped on RLC AM.

In order to reduce complexity, a PDCP Control PDU carrying ROHC feedback carries exactly one ROHC feedback packet – there is no possibility to transmit several ROHC feedback packets in one PDCP PDU.

A PDCP Control PDU carrying a PDCP Status Report for the case of lossless handover is used to prevent the retransmission of already-correctly-received PDCP SDUs, and also to request retransmission of PDCP SDUs which were correctly received but for which header decompression failed. This PDCP Control PDU contains a bitmap indicating which PDCP SDUs need to be retransmitted and a reference SN, the First Missing SDU (FMS). In the case that all PDCP SDUs have been received in sequence this field indicates the next expected SN, and no bitmap is included.

4.3 Radio Link Control (RLC)

The RLC layer is located between the PDCP layer (the 'upper' layer) and the MAC layer (the 'lower' layer). It communicates with the PDCP layer through a Service Access Point (SAP), and with the MAC layer via logical channels. The RLC layer reformats PDCP PDUs in order to fit them into the size indicated by the MAC layer; that is, the RLC transmitter segments and/or concatenates the PDCP PDUs, and the RLC receiver reassembles the RLC PDUs to reconstruct the PDCP PDUs.

In addition, the RLC reorders the RLC PDUs if they are received out of sequence due to the HARQ operation performed in the MAC layer. This is the key difference from UMTS, where the HARQ reordering is performed in the MAC layer. The advantage of HARQ reordering in RLC is that no additional SN and reception buffer are needed for HARQ reordering. In LTE, the RLC SN and RLC reception buffer are used for both HARQ reordering and RLC-level ARQ related operations.

The functions of the RLC layer are performed by 'RLC entities'. An RLC entity is configured in one of three data transmission modes: Transparent Mode (TM), Unacknowledged Mode (UM), and Acknowledged Mode (AM). In AM, special functions are defined to support retransmission. When UM or AM is used, the choice between the two modes is made by the eNodeB during the RRC radio bearer setup procedure (see Section 3.2.3.3), based on the QoS requirements of the EPS bearer.[12] The three RLC modes are described in detail in the following sections.

4.3.1 RLC Entities

4.3.1.1 Transparent Mode (TM) RLC Entity

As the name indicates, the TM RLC entity is transparent to the PDUs that pass through it – no functions are performed and no RLC overhead is added. Since no overhead is added, an RLC SDU is directly mapped to an RLC PDU and vice versa. Therefore, the use of TM RLC is very restricted. Only RRC messages which do not need RLC configuration can utilize the TM RLC, such as broadcast System Information (SI) messages, paging messages, and RRC messages which are sent when no Signalling Radio Bearers (SRBs) other than SRB0 (see Section 3.2.1) are available. TM RLC is not used for user plane data transmission in LTE.

TM RLC provides a unidirectional data transfer service – in other words, a single TM RLC entity is configured either as a transmitting TM RLC entity or as a receiving TM RLC entity.

4.3.1.2 Unacknowledged Mode (UM) RLC Entity

UM RLC provides a unidirectional data transfer service like TM RLC. UM RLC is mainly utilized by delay-sensitive and error-tolerant real-time applications, especially VoIP, and other delay-sensitive streaming services. Point-to-multipoint services such as MBMS (Multimedia Broadcast/Multicast Service) also use UM RLC – since no feedback path is available in the case of point-to-multipoint services, AM RLC cannot be utilized by these services.

A block diagram of the UM RLC entity is shown in Figure 4.8.

The main functions of UM RLC can be summarized as follows:

- Segmentation and concatenation of RLC SDUs;
- Reordering of RLC PDUs;
- Duplicate detection of RLC PDUs;
- Reassembly of RLC SDUs.

Segmentation and concatenation. The transmitting UM RLC entity performs segmentation and/or concatenation on RLC SDUs received from upper layers, to form RLC PDUs. The size of the RLC PDU at each transmission opportunity is decided and notified by the MAC layer depending on the radio channel conditions and the available transmission resources; therefore, the size of each transmitted RLC PDU can be different.

The transmitting UM RLC entity includes RLC SDUs into an RLC PDU in the order in which they arrive at the UM RLC entity. Therefore, a single RLC PDU can contain RLC SDUs or segments of RLC SDUs according to the following pattern:

(zero or one) SDU segment + (zero or more) SDUs + (zero or one) SDU segment.

[12]Evolved Packet System – see Section 2.

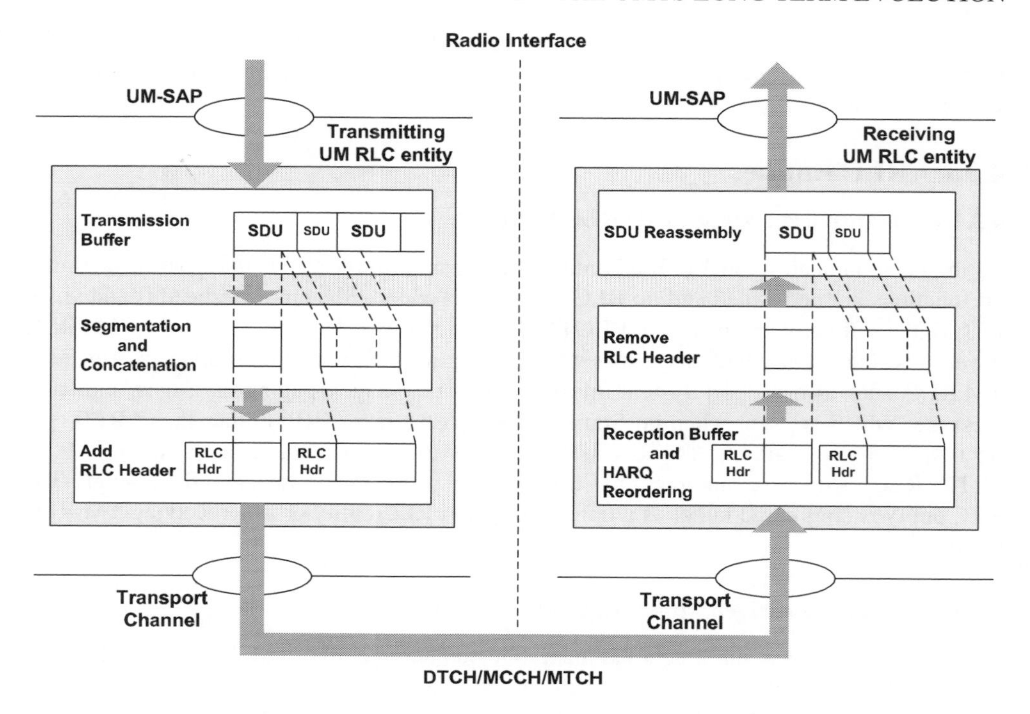

Figure 4.8: Model of UM RLC entities. Reproduced by permission of © 3GPP.

The constructed RLC PDU is always byte-aligned and has no padding.

After segmentation and/or concatenation of RLC SDUs, the transmitting UM RLC entity includes relevant UM RLC headers in the RLC PDU to indicate the sequence number[13] of the RLC PDU, and additionally the size and boundary of each included RLC SDU or RLC SDU segment.

Reordering, duplicate detection, and reassembly. When the receiving UM RLC entity receives RLC PDUs, it first reorders them if they are received out of sequence. Out-of-sequence reception is unavoidable due to the fact that the HARQ operation in the MAC layer uses multiple HARQ processes (see Section 4.4). Any RLC PDUs received out of sequence are stored in the reception buffer until all the previous RLC PDUs are received and delivered to the upper layer.

During the reordering process, any duplicate RLC PDUs received are detected by checking the SNs and discarded. This ensures that the upper layer receives upper layer PDUs only once. The most common cause of receiving duplicates is HARQ ACKs for MAC PDUs being misinterpreted as NACKs, resulting in unnecessary retransmissions of the MAC PDUs, which causes duplication in the RLC layer.

To detect reception failures and avoid excessive reordering delays, a reordering timer is used in the receiving UM RLC entity to set the maximum time to wait for the reception of RLC PDUs that have not been received in sequence. The receiving UM RLC entity starts

[13]Note that the RLC sequence number is independent from the sequence number added by PDCP.

the reordering timer when a missing RLC PDU is detected, and it waits for the missing RLC PDUs until the timer expires. When the timer expires, the receiving UM RLC entity declares the missing RLC PDUs as lost and starts to reassemble the next available RLC SDUs from the RLC PDUs stored in the reception buffer.

The reassembly function is performed on an RLC SDU basis; only RLC SDUs for which all segments are available are reassembled from the stored RLC PDUs and delivered to the upper layers. RLC SDUs that have at least one missing segment are simply discarded and not reassembled. If RLC SDUs were concatenated in an RLC PDU, the reassembly function in the RLC receiver separates them into their original RLC SDUs. The RLC receiver delivers reassembled RLC SDUs to the upper layers in increasing order of SNs.

An example scenario of a lost RLC PDU with HARQ reordering is shown in Figure 4.9. A reordering timer is started when the RLC receiver receives PDU#8. If PDU#7 has not been received before the timer expires, the RLC receiver decides that the PDU#7 is lost, and starts to reassemble RLC SDUs from the next received RLC PDU. In this example, SDU#22 and SDU#23 are discarded because they are not completely received, and SDU#24 is kept in the reception buffer until all segments are received. Only SDU#21 is completely received, so it is delivered up to the PDCP layer.

The reordering and duplicate detection functions are not applicable to RLC entities using the Multicast Control CHannel (MCCH) or Multicast Traffic CHannel (MTCH) (see Section 4.4.1.2). This is because the HARQ operation in the MAC layer is not used for these channels.

4.3.1.3 Acknowledged Mode (AM) RLC Entity

In contrast to the other RLC transmission modes, AM RLC provides a bidirectional data transfer service. Therefore, a single AM RLC entity is configured with the ability both to transmit and to receive – we refer to the corresponding parts of the AM RLC entity as the *transmitting side* and the *receiving side* respectively.

The most important feature of AM RLC is 'retransmission'. An ARQ operation is performed to support error-free transmission. Since transmission errors are corrected by retransmissions, AM RLC is mainly utilized by error-sensitive and delay-tolerant non-real-time applications. Examples of such applications include most of the interactive/background type services, such as web browsing and file downloading. Streaming-type services also frequently use AM RLC if the delay requirement is not too stringent. In the control plane, RRC messages typically utilize the AM RLC in order to take advantage of RLC acknowledgements and retransmissions to ensure reliability.

A block diagram of the AM RLC entity is shown in Figure 4.10.

Although the AM RLC block diagram looks complicated at first glance, the transmitting and receiving sides are similar to the UM RLC transmitting and receiving entities respectively, except for the retransmission-related blocks. Therefore, most of the UM RLC behaviour described in the previous section applies to AM RLC in the same manner. The transmitting side of the AM RLC entity performs segmentation and/or concatenation of RLC SDUs received from upper layers to form RLC PDUs together with relevant AM RLC headers, and the receiving side of the AM RLC entity reassembles RLC SDUs from the received RLC PDUs after HARQ reordering.

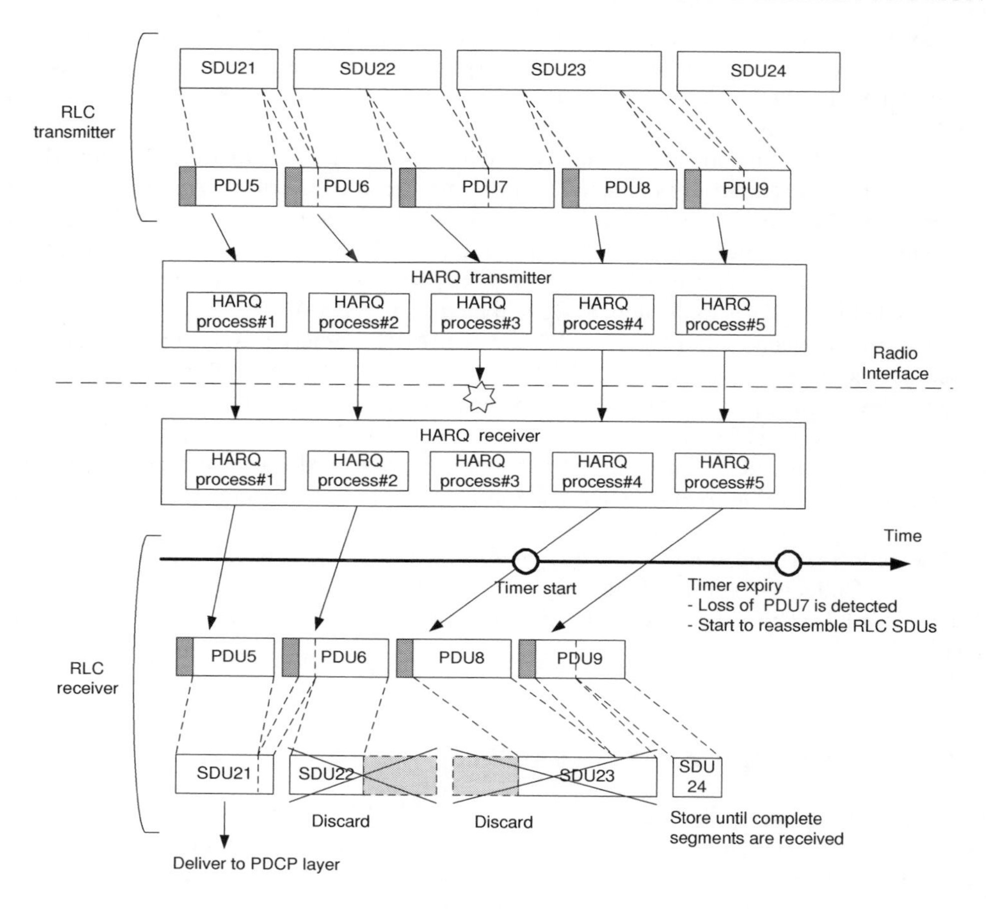

Figure 4.9: Example of PDU loss detection with HARQ reordering.

In addition to performing the functions of UM RLC, the main functions of AM RLC can be summarized as follows:

- Retransmission of RLC Data PDUs;
- Re-segmentation of retransmitted RLC Data PDUs;
- Polling;
- Status reporting;
- Status prohibit.

Retransmission and resegmentation. As mentioned before, the most important function of AM RLC is *retransmission*. In order that the transmitting side retransmits only the missing RLC PDUs, the receiving side provides a 'Status Report' to the transmitting side indicating ACK and/or NACK information for the RLC PDUs. Status reports are sent by

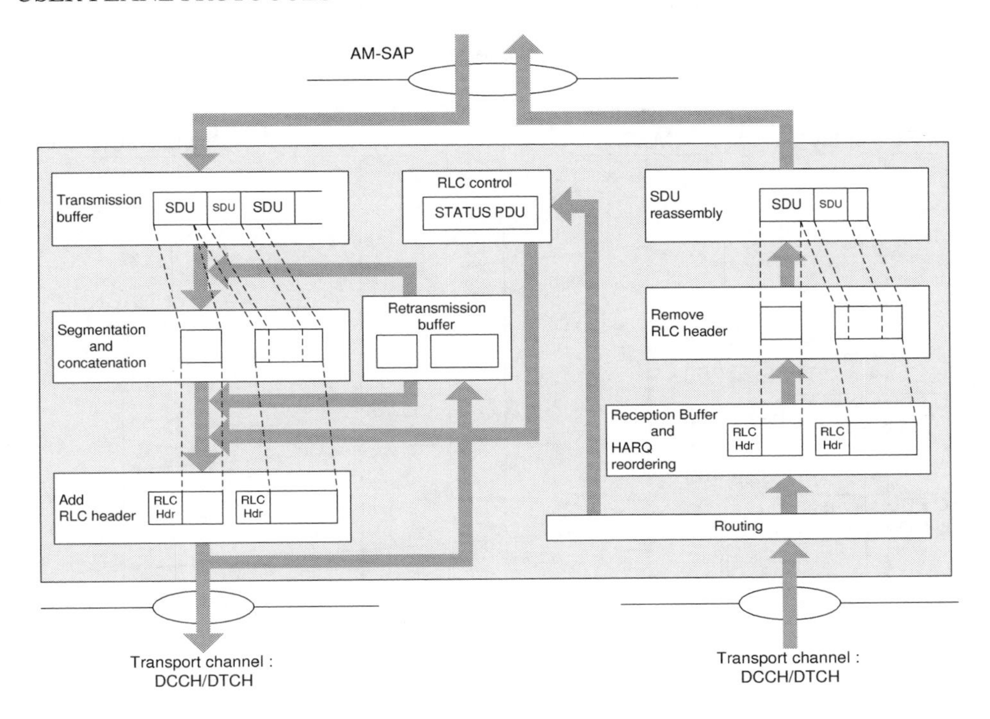

Figure 4.10: Model of AM RLC entity. Reproduced by permission of © 3GPP.

the transmitting side of the AM RLC entity whose receiving side received the corresponding RLC PDUs. Hence, the AM RLC transmitting side is able to transmit two types of RLC PDU, namely RLC Data PDUs containing data received from upper layers and RLC Control PDUs generated in the AM RLC entity itself. To differentiate between Data and Control PDUs, a 1-bit flag is included in the AM RLC header (see Section 4.3.2.3).

When the transmitting side transmits RLC Data PDUs, it stores the PDUs in the retransmission buffer for possible retransmission if requested by the receiver through a status report. In case of retransmission, the transmitter can resegment the original RLC Data PDUs into smaller PDU segments if the MAC layer indicates a size that is smaller than the original RLC Data PDU size.

An example of RLC resegmentation is shown in Figure 4.11. In this example, an original PDU of 600 bytes is resegmented into two PDU segments of 200 and 400 bytes at retransmission.

The original RLC PDU is distinguished from the retransmitted segments by another 1-bit flag in the AM RLC header: in the case of a retransmitted segment, some more fields are included in the AM RLC header to indicate resegmentation related information. The receiver can use status reports to indicate the status of individual retransmitted segments, not just full PDUs.

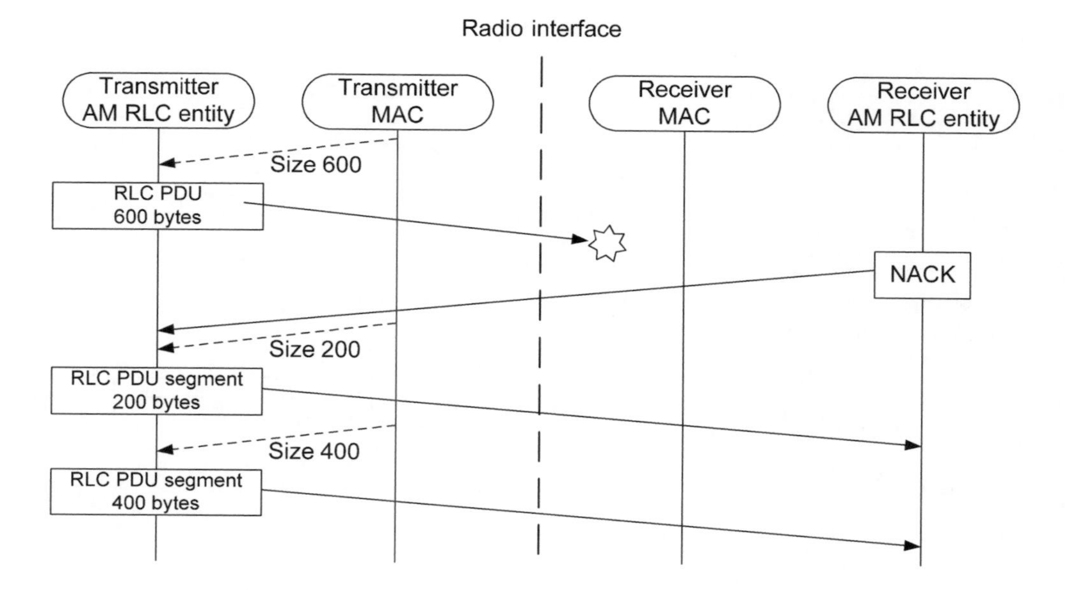

Figure 4.11: Example of RLC resegmentation.

Polling, status report and status prohibit. The transmitting side of the AM RLC entity can proactively request a status report from the peer receiving side by *polling* using a 1-bit indicator in the AM RLC header. The transmitting side can then use the status reports to select the RLC Data PDUs to be retransmitted, and manage transmission and retransmission buffers efficiently. Typical circumstances in which the transmitting side may initiate a poll include, for example, the last PDU in the transmitting side having been transmitted, or a predefined number of PDUs or data bytes having been transmitted.

When the receiving side of the AM RLC entity receives a poll from the peer transmitting side, it checks the reception buffer status and transmits a status report at the earliest transmission opportunity.

The receiving side can also generate a status report of its own accord if it detects a reception failure of an AM RLC PDU. For the detection of a reception failure, a similar mechanism is used as in the case of UM RLC in relation to the HARQ reordering delay. In AM RLC, however, the detection of a reception failure triggers a status report instead of considering the relevant RLC PDUs as permanently lost.

Note that the transmission of status reports needs to be carefully controlled according to the trade-off between transmission delay and radio efficiency. To reduce the transmission delay, status reports need to be transmitted frequently, but on the other hand frequent transmission of status reports wastes radio resources. Moreover, if further status reports are sent whilst the retransmissions triggered by a previous status report have not yet been received, unnecessary retransmissions may result, thus consuming further radio resources; in AM RLC this is in fact a second cause of duplicate PDUs occurring which have to be discarded by the duplicate-detection functionality. Therefore, to control the frequency of status reporting in an effective way, a 'status prohibit' function is available in AM RLC, whereby the transmission of new status reports is prohibited while a timer is running.

4.3.2 RLC PDU Formats

As mentioned above, the RLC layer provides two types of PDU, namely the RLC Data PDU and the RLC Control PDU. The RLC Data PDU is used to transmit PDCP PDUs and is defined in all RLC transmission modes. The RLC Control PDU delivers control information between peer RLC entities and is defined only in AM RLC. The RLC PDUs used in each RLC transmission mode are summarized in Table 4.3.

Table 4.3: PDU types used in RLC.

RLC Mode	Data PDU	Control PDU
TM	TMD (TM Data)	N/A
UM	UMD (UM Data)	N/A
AM	AMD (AM Data)/AMD segment	STATUS

In the following subsections, each of the RLC PDU formats is explained in turn.

4.3.2.1 Transparent Mode Data PDU Format

The Transparent Mode Data (TMD) PDU consists only of a data field and does not have any RLC headers. Since no segmentation or concatenation is performed, an RLC SDU is directly mapped to a TMD PDU.

4.3.2.2 Unacknowledged Mode Data PDU Format

The Unacknowledged Mode Data (UMD) PDU (Figure 4.12) consists of a data field and UMD PDU header. PDCP PDUs (i.e. RLC SDUs) can be segmented and/or concatenated into the data field. The UMD PDU header is further categorized into a fixed part (included in each UMD PDU) and an extension part (included only when the data field contains more than one SDU or SDU segment – i.e. only when the data field contains any SDU borders).

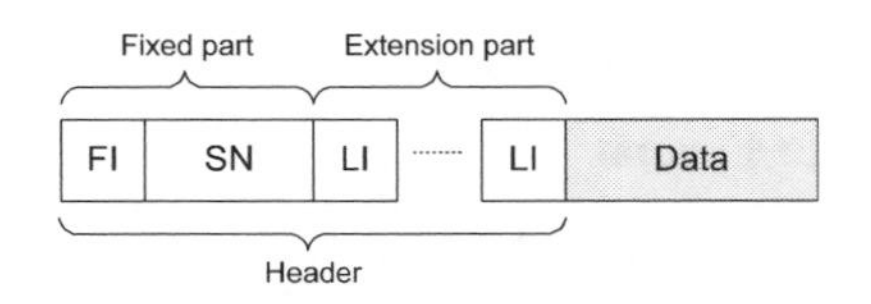

Figure 4.12: Key features of UMD PDU format.

- **Framing Info (FI).** This 2-bit field indicates whether the first and the last data field elements are complete SDUs or partial SDUs (i.e. whether the receiving RLC entity needs to receive multiple RLC PDUs in order to reassemble the corresponding SDU).
- **Sequence Number (SN).** For UMD PDUs, either a short (5 bits) or a long (10 bits) SN field can be used. This field allows the receiving RLC entity unambiguously to identify a UMD PDU, which allows reordering and duplicate-detection to take place.
- **Length Indicator (LI).** This 11-bit field indicates the length of the corresponding data field element present in the UMD PDU. There is a one-to-one correspondence between each LI and a data field element, except for the last data field element, for which the LI field is omitted because its length can be deduced from the UMD PDU size.

4.3.2.3 Acknowledged Mode Data PDU Format

In addition to the UMD PDU header fields, the Acknowledged Mode Data (AMD) PDU header (Figure 4.13) contains fields to support the RLC ARQ mechanism. The only difference in the PDU fields is that only the long SN field (10 bits) is used for AMD PDUs. The additional fields are as follows:

- **Data/Control (D/C).** This 1-bit field indicates whether the RLC PDU is an RLC Data PDU or an RLC Control PDU. It is present in all types of PDU used in AM RLC.
- **Resegmentation Flag (RF).** This 1-bit field indicates whether the RLC PDU is an AMD PDU or an AMD PDU segment.
- **Polling (P).** This 1-bit field is used to request a status report from the peer receiving side.

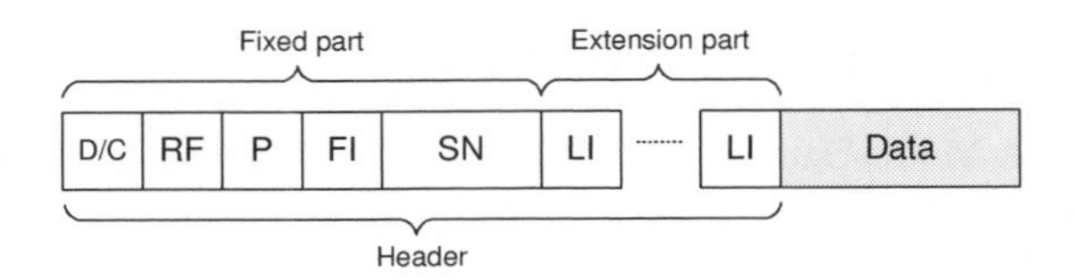

Figure 4.13: Key features of AMD PDU format.

4.3.2.4 AMD PDU Segment Format

The AMD PDU segment format (Figure 4.14) is used in case of resegmented retransmissions (when the available resource for retransmission is smaller than the original PDU size), as described in Section 4.3.1.3.

If the RF field indicates that the RLC PDU is an AMD PDU segment, the following additional resegmentation related fields are included in the fixed part of the AMD PDU header to enable correct reassembly:

- **Last Segment Flag (LSF).** This 1-bit field indicates whether or not this AMD PDU segment is the last segment of an AMD PDU.
- **Segmentation Offset (SO).** This 15-bit field indicates the starting position of the AMD PDU segment within the original AMD PDU.

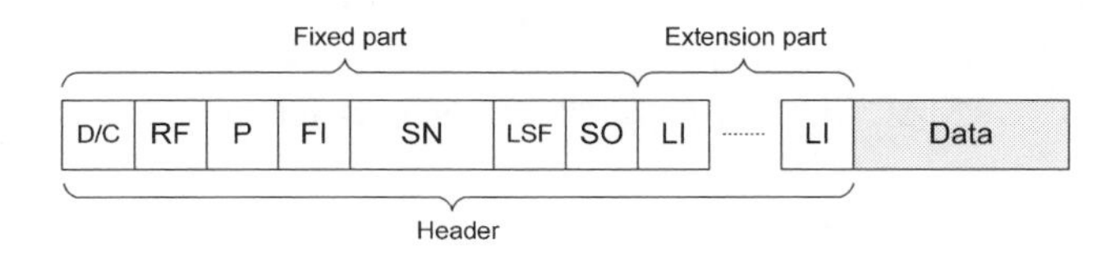

Figure 4.14: Key features of AMD PDU segment format.

4.3.2.5 STATUS PDU Format

The STATUS PDU (Figure 4.15) is designed to be very simple, as the RLC PDU error rate should normally be low in LTE due to the use of HARQ in the MAC layers. Therefore, the STATUS PDU simply lists all the missing portions of AMD PDUs by means of the following fields:

- **Control PDU Type (CPT).** This 3-bit field indicates the type of the RLC Control PDU, allowing more RLC Control PDU types to be defined in a later release of the LTE specifications. (The STATUS PDU is the only type of RLC Control PDU defined in the first version of LTE.)
- **ACK_SN.** This 10-bit field indicates the SN of the first AMD PDU which is neither received nor listed in this STATUS PDU. All AMD PDUs up to but not including this AMD PDU are correctly received by the receiver except the AMD PDUs or portions of AMD PDUs listed in the NACK_SN List.
- **NACK_SN List.** This field contains a list of SNs of the AMD PDUs that have not been completely received, optionally including indicators of which bytes of the AMD PDU are missing in the case of resegmentation.

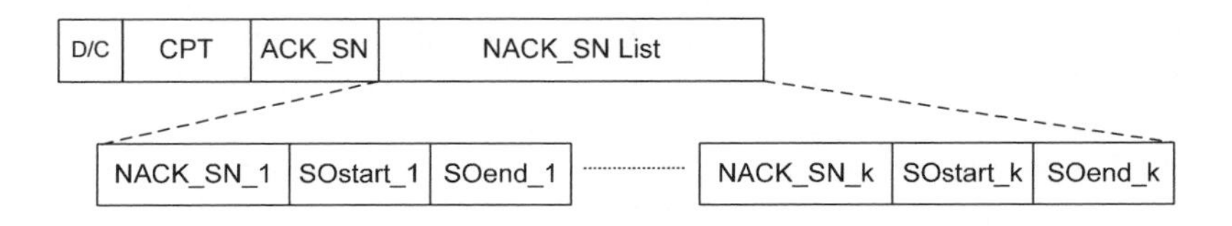

Figure 4.15: STATUS PDU format.

4.4 Medium Access Control (MAC)

The MAC layer is the lowest sublayer in the Layer 2 architecture of the LTE radio protocol stack. The connection to the physical layer below is through transport channels, and the connection to the RLC layer above is through logical channels. The MAC layer therefore performs multiplexing and demultiplexing between logical channels and transport channels: the MAC layer in the transmitting side constructs MAC PDUs, known as Transport Blocks (TBs), from MAC SDUs received through logical channels, and the MAC layer in the receiving side recovers MAC SDUs from MAC PDUs received through transport channels.

4.4.1 MAC Architecture

4.4.1.1 Overall Architecture

Figure 4.16 shows a conceptual overview of the architecture of the MAC layer.

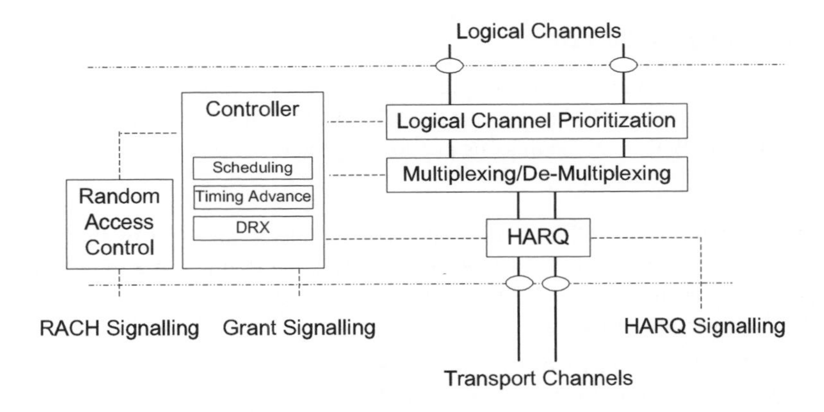

Figure 4.16: Conceptual overview of the UE-side MAC architecture.

The MAC layer consists of a HARQ entity, a multiplexing/demultiplexing entity, a logical channel prioritization entity, a random access control entity, and a controller which performs various control functions.

The HARQ entity is responsible for the transmit and receive HARQ operations. The transmit HARQ operation includes transmission and retransmission of TBs, and reception and processing of ACK/NACK signalling. The receive HARQ operation includes reception of TBs, combining of the received data and generation of ACK/NACK signalling. In order to enable continuous transmission while previous TBs are being decoded, up to eight HARQ processes in parallel are used to support multiprocess 'Stop-And-Wait' (SAW) HARQ operation.

SAW operation means that upon transmission of a TB, a transmitter stops further transmissions and awaits feedback from the receiver. When a NACK is received, or when a certain time elapses without receiving any feedback, the transmitter retransmits the TB. Such a simple SAW HARQ operation cannot on its own utilize the transmission resources during the period between the first transmission and the retransmission. Therefore multiprocess HARQ interlaces several independent SAW processes in time so that all the transmission resources can be used. Each HARQ process is responsible for a separate SAW operation and manages a separate buffer.

In general, HARQ schemes can be categorized as either *synchronous* or *asynchronous*, with the retransmissions in each case being either *adaptive* or *non-adaptive*.

In a synchronous HARQ scheme, the retransmission(s) for each process occur at predefined times relative to the initial transmission. In this way, there is no need to signal information such as HARQ process number, as this can be inferred from the transmission timing. By contrast, in an asynchronous HARQ scheme, the retransmissions can occur at any time relative to the initial transmission, so additional explicit signalling is required to indicate the HARQ process number to the receiver, so that the receiver can correctly associate each retransmission with the corresponding initial transmission. In summary, synchronous HARQ schemes reduce the signalling overhead while asynchronous HARQ schemes allow more flexibility in scheduling.

In an adaptive HARQ scheme, transmission attributes such as the Modulation and Coding Scheme (MCS) and transmission resource allocation in the frequency domain can be changed at each retransmission in response to variations in the radio channel conditions. In a non-adaptive HARQ scheme, the retransmissions are performed without explicit signalling of new transmission attributes – either by using the same transmission attributes as those of the previous transmission, or by changing the attributes according to a predefined rule. Accordingly, adaptive schemes bring more scheduling gain at the expense of increased signalling overhead.

In LTE, asynchronous adaptive HARQ is used for the downlink, and synchronous HARQ for the uplink. In the uplink, the retransmissions may be either adaptive or non-adaptive, depending on whether new signalling of the transmission attributes is provided.

The details of the HARQ incremental redundancy schemes and timing for retransmissions are explained in Section 10.3.2.5.

In the multiplexing and demultiplexing entity, data from several logical channels can be (de)multiplexed into/from one transport channel. The multiplexing entity generates MAC PDUs from MAC SDUs when radio resources are available for a new transmission, based on the decisions of the logical channel prioritization entity. The demultiplexing entity reassembles the MAC SDUs from MAC PDUs and distributes them to the appropriate RLC entities. In addition, for peer-to-peer communication between the MAC layers, control messages called 'MAC Control Elements' can be included in the MAC PDU as explained in Section 4.4.2.7 below.

The logical channel prioritization entity prioritizes the data from the logical channels to decide how much data and from which logical channel(s) should be included in each MAC PDU, as explained in Section 4.4.2.6. The decisions are delivered to the multiplexing and demultiplexing entity.

The random access control entity is responsible for controlling the Random Access CHannel (RACH) procedure (see Section 4.4.2.3). The controller entity is responsible for a number of functions including Discontinuous Reception (DRX) , the Data Scheduling procedure, and for maintaining the uplink timing alignment. These functions are explained in the following sections.

4.4.1.2 Logical Channels

The MAC layer provides a data transfer service for the RLC layer through logical channels, which are either Control Logical Channels (for the transport of control data such as RRC signalling), or Traffic Logical Channels (for user plane data).

Control logical channels.

- **Broadcast Control CHannel (BCCH).** This is a downlink channel which is used to broadcast System Information (SI) and any Public Warning System (PWS) messages (see Section 13.7). In the RLC layer, it is associated with a TM RLC entity (see Section 4.3.1).
- **Paging Control CHannel (PCCH).** This is a downlink channel which is used to notify UEs of an incoming call or a change of SI. In the RLC layer, it is associated with a TM RLC entity (see Section 4.3.1).
- **Common Control CHannel (CCCH).** This channel is used to deliver control information in both uplink and downlink directions when there is no confirmed association between a UE and the eNodeB – i.e. during connection establishment. In the RLC layer, it is associated with a TM RLC entity (see Section 4.3.1).
- **Multicast Control CHannel (MCCH).** This is a downlink channel which is used to transmit control information related to the reception of MBMS services (see Chapter 13). In the RLC layer, it is associated with a UM RLC entity (see Section 4.3.1).
- **Dedicated Control CHannel (DCCH).** This channel is used to transmit dedicated control information relating to a specific UE, in both uplink and downlink directions. It is used when a UE has an RRC connection with eNodeB. In the RLC layer, it is associated with an AM RLC entity (see Section 4.3.1).

Traffic logical channels.

- **Dedicated Traffic CHannel (DTCH).** This channel is used to transmit dedicated user data in both uplink and downlink directions. In the RLC layer, it can be associated with either a UM RLC entity or an AM RLC entity (see Section 4.3.1).
- **Multicast Traffic CHannel (MTCH).** This channel is used to transmit user data for MBMS services in the downlink (see Chapter 13). In the RLC layer, it is associated with a UM RLC entity (see Section 4.3.1).

4.4.1.3 Transport Channels

Data from the MAC layer is exchanged with the physical layer through transport channels. Data is multiplexed into transport channels depending on how it is transmitted over the air. Transport channels are classified as downlink or uplink as follows:

Downlink transport channels.

- **Broadcast CHannel (BCH).** This channel is used to transport the parts of the SI which are essential for access the Downlink Shared CHannel (DL-SCH). The transport format is fixed and the capacity is limited.
- **Downlink Shared CHannel (DL-SCH).** This channel is used to transport downlink user data or control messages. In addition, the remaining parts of the SI that are not transported via the BCH are transported on the DL-SCH.
- **Paging CHannel (PCH).** This channel is used to transport paging information to UEs, and to inform UEs about updates of the SI (see Section 3.2.2) and PWS messages (see Section 13.7).
- **Multicast CHannel (MCH).** This channel is used to transport MBMS user data or control messages that require MBSFN combining (see Chapter 13).

The mapping of the downlink transport channels onto physical channels is explained in Section 6.4.

Uplink transport channels.

- **Uplink Shared CHannel (UL-SCH).** This channel is used to transport uplink user data or control messages.
- **Random Access CHannel (RACH).** This channel is used for access to the network when the UE does not have accurate uplink timing synchronization, or when the UE does not have any allocated uplink transmission resource (see Chapter 17).

The mapping of the uplink transport channels onto physical channels is explained in Chapter 16.

4.4.1.4 Multiplexing and Mapping between Logical Channels and Transport Channels

Figures 4.17 and 4.18 show the possible multiplexing between logical channels and transport channels in the downlink and uplink respectively.

Note that in the downlink, the DL-SCH carries information from all the logical channels except the PCCH, MCCH and MTCH.

In the uplink, the UL-SCH carries information from all the logical channels.

4.4.2 MAC Functions

4.4.2.1 Scheduling

The scheduler in the eNodeB distributes the available radio resources in one cell among the UEs, and among the radio bearers of each UE. The details of the scheduling algorithm are left to the eNodeB implementation, but the signalling to support the scheduling is standardized. Some possible scheduling algorithms are discussed in Chapter 12.

In principle, the eNodeB allocates downlink or uplink radio resources to each UE based respectively on the downlink data buffered in the eNodeB and on Buffer Status Reports

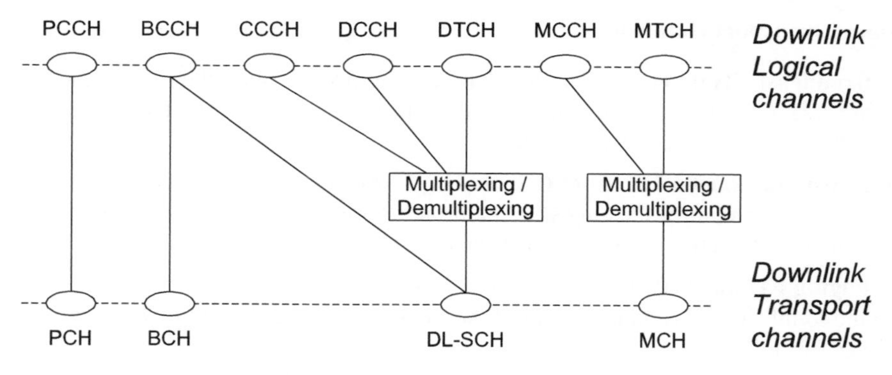

Figure 4.17: Downlink logical channel multiplexing. Reproduced by permission of © 3GPP.

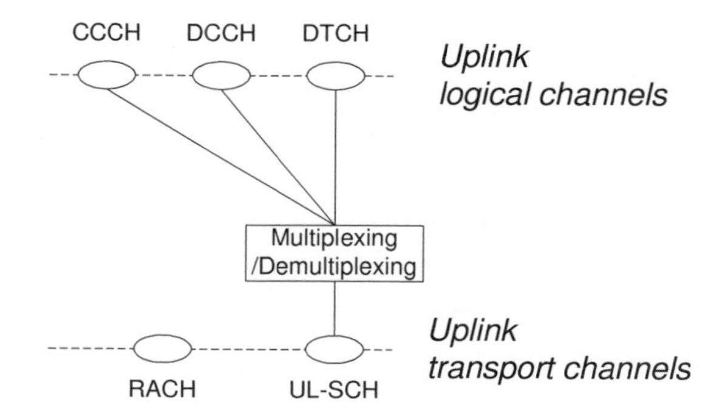

Figure 4.18: Uplink logical channel multiplexing. Reproduced by permission of © 3GPP.

(BSRs) received from the UE. In this process, the eNodeB considers the QoS requirements of each configured radio bearer, and selects the size of the MAC PDU.

The usual mode of scheduling is *dynamic scheduling*, by means of downlink assignment messages for the allocation of downlink transmission resources and uplink grant messages for the allocation of uplink transmission resources; these are valid for specific single subframes.[14] These messages also indicate whether the scheduled data is to be the first transmission of a new TB or a retransmission, by means of a 1-bit New Data Indicator (NDI); if the value of the NDI is changed relative to its previous value for the same HARQ process, the transmission is the start of a new TB. These messages are transmitted on the Physical Downlink Control CHannel (PDCCH) using a Cell Radio Network Temporary Identifier (C-RNTI) to identify the intended UE, as described in Section 9.3. This kind of scheduling is efficient for service types such as TCP or the SRBs, in which the traffic is bursty and dynamic in rate.

[14]The dynamic uplink transmission resource grants are valid for specific single subframes for initial transmissions, although they may also imply a resource allocation in later subframes for HARQ retransmissions.

In addition to the dynamic scheduling, *Semi-Persistent Scheduling* (SPS) may be used. SPS enables radio resources to be semi-statically configured and allocated to a UE for a longer time period than one subframe, avoiding the need for specific downlink assignment messages or uplink grant messages over the PDCCH for each subframe. It is useful for services such as VoIP for which the data packets are small, periodic and semi-static in size. For this kind of service the timing and amount of radio resources needed are predictable. Thus the overhead of the PDCCH is significantly reduced compared to the case of dynamic scheduling.

For the configuration of SPS, RRC signalling indicates the interval at which the radio resources are periodically assigned. Specific transmission resource allocations in the frequency domain, and transmission attributes such as the MCS, are signalled using the PDCCH. The actual transmission timing of the PDCCH messages is used as the reference timing to which the SPS interval applies. It is necessary to distinguish the PDCCH messages which apply to SPS from those used for dynamic scheduling; hence for SPS a special identity is used, known as the Semi-Persistent Scheduling C-RNTI (SPS-C-RNTI), which for each UE is different from the C-RNTI used for dynamic scheduling messages. The SPS-C-RNTI is used both for the configuration of SPS resources and for the indication of HARQ retransmissions of semi-persistently scheduled TBs. To differentiate these two cases, the NDI is used: for the configuration of the SPS resources, the SPS-C-RNTI with NDI set to 0 is used; for HARQ retransmissions, the SPS-C-RNTI with NDI set to 1 is used.

Reconfiguration of resources used for SPS can be performed for cases such as transitions between silent periods and talk spurts, or when the codec rate changes. For example, when the codec rate for a VoIP service is increased, a new downlink assignment message or uplink grant message can be transmitted to configure a larger semi-persistently scheduled radio resource for the support of bigger VoIP packets.

Allocated resources for SPS can be cancelled by an explicit scheduling message on the PDCCH using the SPS-C-RNTI indicating *SPS release*.[15] However, because there is a risk that scheduling messages can be lost in transmission, or the eNodeB's decision to release the resources may be late, an implicit mechanism to release the allocated radio resources is also specified. In the implicit mechanism, when a certain number of MAC PDUs not containing any MAC SDUs has been transmitted, the UE releases the radio resources.

Resources allocated for SPS can be temporarily overridden in a specific subframe by a scheduling message using the C-RNTI. For example, if the semi-persistently scheduled resources collide with resources configured for the Physical Random Access CHannel (PRACH) in a certain subframe, the eNodeB may choose to allocate other resources for the SPS in that subframe in order to avoid a collision with the PRACH.

Other factors potentially affecting the scheduling are uplink *Transmission Time Interval (TTI) bundling*, designed to improve uplink coverage (see Section 14.3.2), and the configuration of measurement gaps during which the UE tunes its receiver to other frequencies (see Section 22.2.1.2). In the latter case, whenever a subframe for a given HARQ process collides with a configured measurement gap, the UE can neither receive from nor transmit to a serving cell in that subframe. In such a case, if the UE cannot receive HARQ ACK/NACK feedback for an uplink TB, the UE considers that HARQ ACK is received TB and does not autonomously start a HARQ retransmission at the next transmission opportunity; to resume

[15]Explicit SPS resource release messages are positively acknowledged by the UE if they relate to downlink SPS, but not for uplink SPS.

HARQ operation, the UE has to receive a new scheduling message. If an uplink TB cannot be transmitted due to a measurement gap, the UE considers that HARQ NACK is received for that TB and transmits the TB at the next opportunity.

4.4.2.2 Scheduling Information Transfer

Buffer Status Reports (BSRs) from the UE to the eNodeB are used to assist the eNodeB's allocation of uplink radio resources. The basic assumption underlying scheduling in LTE is that radio resources are only allocated for transmission to or from a UE if data is available to be sent or received. In the downlink direction, the scheduler in the eNodeB is obviously aware of the amount of data to be delivered to each UE; however, in the uplink direction, because the scheduling decisions are performed in the eNodeB and the buffer for the data is located in the UE, BSRs have to be sent from the UE to the eNodeB to indicate the amount of data in the UE that needs to be transmitted over the UL-SCH.[16]

Two types of BSR are defined in LTE: a long BSR and a short BSR; which one is transmitted depends on the amount of available uplink transmission resources for sending the BSR, on how many groups of logical channels have non-empty buffers, and on whether a specific event is triggered at the UE. The long BSR reports the amount of data for four logical channel groups, whereas the short BSR reports the amount of data for only one logical channel group. Although the UE might actually have more than four logical channels configured, the overhead would be large if the amount of data in the UE were to be reported for every logical channel individually. Thus, grouping the logical channels into four groups for reporting purposes represents a compromise between efficiency and accuracy.

A BSR can be triggered in the following situations:

- whenever data arrives for a logical channel which has a higher priority than the logical channels whose buffers previously contained data (this is known as a Regular BSR);
- whenever data becomes available for any logical channel when there was previously no data available for transmission (a Regular BSR);
- whenever a 'retxBSR' timer expires and there is data available for transmission (a Regular BSR);
- whenever a 'periodicBSR' timer[17] expires (a Periodic BSR);
- whenever spare space in a MAC PDU can accommodate a BSR (a Padding BSR).

The 'retxBSR' timer provides a mechanism to recover from situations where a BSR is transmitted but not received. For example, if the eNodeB fails to decode a MAC PDU containing a BSR and returns a HARQ NACK, but the UE erroneously decodes the NACK as ACK, the UE will think that transmission of the BSR was successful even though it was not received by the eNodeB. In such a case, a long delay would be incurred while the UE

[16]Note that, unlike High Speed Uplink Packet Access (HSUPA), there is no possibility in LTE for a UE to transmit autonomously in the uplink by means of a transmission grant for non-scheduled transmissions. This is because the uplink transmissions from different UEs in LTE are orthogonal in time and frequency, and therefore if an uplink resource is allocated but unused, it cannot be accessed by another UE; by contrast, in HSUPA, if a UE does not use its transmission grant for non-scheduled transmissions, the resulting reduction in uplink interference can benefit other UEs. Furthermore, the short subframe length in LTE enables uplink transmission resources to be dynamically allocated more quickly than in HSUPA.

[17]Periodic BSR timer is used by RRC to control BSR reporting.

waited for an uplink resource grant that would not be forthcoming. To avoid this, the retxBSR timer is restarted whenever a uplink grant message is received; if no uplink grant is received before the timer expires, the UE transmits another BSR.

If a UE does not have enough allocated UL-SCH resources to send a BSR when a trigger for a Regular BSR occurs, the UE sends a Scheduling Request (SR) on the Physical Uplink Control CHannel (PUCCH – see Section 16.3.7) if possible; otherwise, the random access procedure (see Section 4.4.2.3) is used to request an allocation of uplink resources for sending a BSR. However, if a periodic or padding BSR is triggered when the UE does not have UL-SCH resources for a new transmission, the SR is not triggered.

Thus LTE provides suitable signalling to ensure that the eNodeB has sufficient information about the data waiting in each UE's uplink transmission buffer to allocate corresponding uplink transmission resources in a timely manner.

4.4.2.3 Random Access Procedure

The random access procedure is used when a UE is not allocated with uplink radio resources but has data to transmit, or when the UE is not time-synchronized in the uplink direction. Control of the random access procedure is an important part of the MAC layer functionality in LTE. The details are explained in Chapter 17.

4.4.2.4 Uplink Timing Alignment

Uplink timing alignment maintenance is controlled by the MAC layer and is important for ensuring that a UE's uplink transmissions arrive in the eNodeB without overlapping with the transmissions from other UEs. The details of the uplink timing advance mechanism used to maintain timing alignment are explained in Section 18.2.

The timing advance mechanism utilizes MAC Control Elements (see Section 4.4.2.7) to update the uplink transmission timing. However, maintaining the uplink synchronization in this way during periods when no data is transferred wastes radio resources and adversely impacts the UE battery life. Therefore, when a UE is inactive for a certain period of time the UE is allowed to lose uplink synchronization even in RRC_CONNECTED state. The random access procedure is then used to regain uplink synchronization when the data transfer resumes in either uplink or downlink.

4.4.2.5 Discontinuous Reception (DRX)

DRX functionality can be configured for an 'RRC_CONNECTED' UE[18] so that it does not always need to monitor the downlink channels. A DRX cycle consists of an 'On Duration' during which the UE should monitor the PDCCH and a 'DRX period' during which a UE can skip reception of downlink channels for battery saving purposes.

The parameterization of the DRX cycle involves a trade-off between battery saving and latency. On the one hand, a long DRX period is beneficial for lengthening the UE's battery life. For example, in the case of a web browsing service, it is usually a waste of resources for a UE continuously to receive downlink channels while the user is reading a downloaded

[18]Different DRX functionality applies to UEs which are in 'RRC_IDLE'. These RRC states are discussed in Chapter 3.

web page. On the other hand, a shorter DRX period is better for faster response when data transfer is resumed – for example when a user requests another web page.

To meet these conflicting requirements, two DRX cycles – a short cycle and a long cycle – can be configured for each UE, with the aim of providing a similar degree of power saving for the UE in RRC_CONNECTED as in RRC_IDLE. The transition between the short DRX cycle, the long DRX cycle and continuous reception is controlled either by a timer or by explicit commands from the eNodeB. In some sense, the short DRX cycle can be considered as a confirmation period in case a late packet arrives, before the UE enters the long DRX cycle – if data arrives at the eNodeB while the UE is in the short DRX cycle, the data is scheduled for transmission at the next wake-up time and the UE then resumes continuous reception. On the other hand, if no data arrives at the eNodeB during the short DRX cycle, the UE enters the long DRX cycle, assuming that the packet activity is finished for the time being.

Figure 4.19 shows an example of DRX operation. The UE checks for scheduling messages (indicated by its C-RNTI on the PDCCH) during the 'On Duration' period of either the long DRX cycle or the short DRX cycle depending on the currently active cycle. When a scheduling message is received during an 'On Duration', the UE starts a 'DRX Inactivity Timer' and monitors the PDCCH in every subframe while the DRX Inactivity Timer is running. During this period, the UE can be regarded as being in a continuous reception mode. Whenever a scheduling message is received while the DRX Inactivity Timer is running, the UE restarts the DRX Inactivity Timer, and when it expires the UE moves into a short DRX cycle and starts a 'DRX Short Cycle Timer'. The short DRX cycle may also be initiated by means of a MAC Control Element (see Section 4.4.2.7). When the 'DRX Short Cycle Timer' expires, the UE moves into a long DRX cycle.

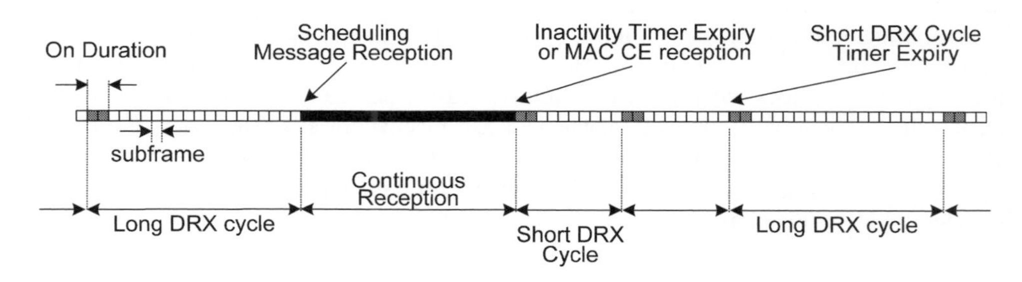

Figure 4.19: The two-level DRX procedure.

In addition to this DRX behaviour, a 'HARQ RTT (Round Trip Time) timer' is defined per downlink HARQ process with the aim of allowing the UE to sleep during the HARQ RTT. When decoding of a downlink TB for one HARQ process fails, the UE can assume that the next retransmission of the TB will occur after at least 'HARQ RTT' subframes. While the HARQ RTT timer is running, the UE does not need to monitor the PDCCH (provided that there is no other reason to be monitoring it). At the expiry of the HARQ RTT timer, if received data for a downlink HARQ process is not correctly decoded, the UE starts a 'DRX Retransmission Timer' for that HARQ process. While the timer is running, the UE monitors the PDCCH for HARQ retransmissions. The length of the DRX Retransmission Timer is related to the degree of flexibility desired for the eNodeB's scheduler. For optimal UE battery consumption, it is desirable that eNodeB schedules a HARQ retransmission as soon as the HARQ RTT timer expires. However, this requires that eNodeB always reserve some radio resources for this, and therefore the DRX Retransmission timer can be used to relax this scheduling limitation while limiting the amount of time for which the UE has to monitor the PDCCH. The HARQ RTT is illustrated in Section 10.3.2.5.

4.4.2.6 Multiplexing and Logical Channel Prioritization

Unlike the downlink, where the multiplexing and logical channel prioritization is left to the eNodeB implementation, for the uplink the process by which a UE creates a MAC PDU to transmit using the allocated radio resources is fully standardized; this is designed to ensure that the UE satisfies the QoS of each configured radio bearer in a way which is optimal and consistent between different UE implementations. Based on the uplink transmission resource grant message signalled on the PDCCH, the UE has to decide on the amount of data for each logical channel to be included in the new MAC PDU, and, if necessary, also to allocate space for a MAC Control Element.

One simple way to meet this purpose is to serve radio bearers in order of their priority. Following this principle, the data from the logical channel of the highest priority is the first to be included into the MAC PDU, followed by data from the logical channel of the next highest priority, continuing until the MAC PDU size allocated by the eNodeB is completely filled or there is no more data to transmit.

Although this kind of priority-based multiplexing is simple and favours the highest priorities, it sometimes leads to starvation of low-priority bearers. Starvation occurs when the logical channels of the lower priority cannot transmit any data because the data from higher priority logical channels always takes up all the allocated radio resources.

To avoid starvation, while still serving the logical channels according to their priorities, in LTE a Prioritized Bit Rate (PBR) is configured by the eNodeB for each logical channel. The PBR is the data rate provided to one logical channel before allocating any resource to a lower-priority logical channel.

In order to take into account both the PBR and the priority, each logical channel is served in decreasing order of priority, but the amount of data from each logical channel included into the MAC PDU is initially limited to the amount corresponding to the configured PBR. Only when all logical channels have been served up to their PBR, then if there is still room left in the MAC PDU, each logical channel is served again in decreasing order of priority. In this second round, each logical channel is served only if all logical channels of higher priority have no more data for transmission.

In most cases, a MAC Control Element has higher priority than any other logical channel because it controls the operation of a MAC entity. Thus, when a MAC PDU is composed and there is a MAC Control Element to send, the MAC Control Element is generally included first and the remaining space is used to include data from logical channels. However, since the padding BSR is used to fill up remaining space in a MAC PDU, it is included into a MAC PDU after other logical channels. Among the various types of MAC Control Element (see Section 4.4.2.7) and logical channel, the 'C-RNTI' MAC Control Element and CCCH (CCCH) have the highest priority because they are used for either contention resolution or

RRC connection management. For example, the 'RRCConnectionReestablishmentRequest' message (see Section 3.2.3.5) on the uplink CCCH is used to recover a lost RRC connection, and it is more important to complete the connection reestablishment procedure as soon as possible than to inform the eNodeB of the UE's buffer status; otherwise, the data transfer interruption time would be longer and the probability of call failure would increase due to the delayed signalling. Likewise, a BSR of an unknown user is useless until the eNodeB knows which UE transmitted the BSR. Thus, the C-RNTI MAC Control Element has higher priority than the BSR MAC Control Element.

Figure 4.20 illustrates the LTE MAC multiplexing by way of example. First, channel 1 is served up to its PBR, channel 2 up to its PBR and then channel 3 with as much data as is available (since in this example the amount of data available is less than would be permitted by the PBR configured for that channel). After that, the remaining space in the MAC PDU is filled with data from the channel 1 which is of the highest priority until there is no further room in the MAC PDU or there is no further data from channel 1. If there is still a room after serving the channel 1, channel 2 is served in a similar way.

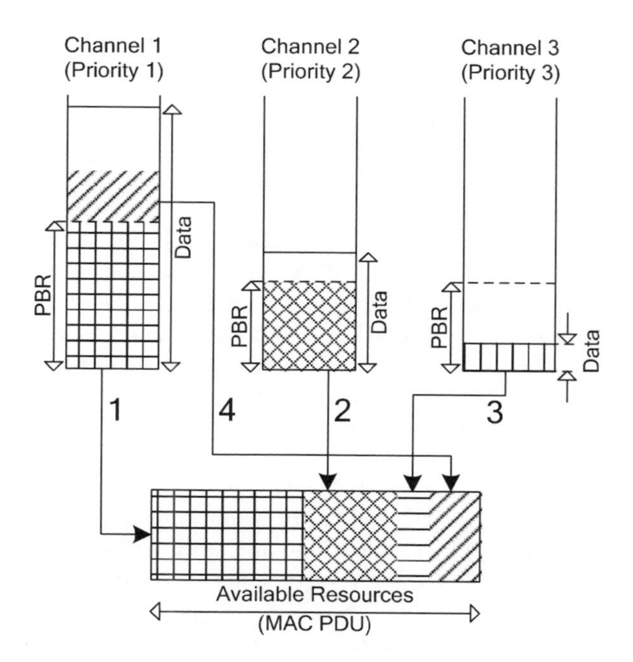

Figure 4.20: Example of MAC multiplexing.

4.4.2.7 MAC PDU Formats

When the multiplexing is done, the MAC PDU itself can be composed. The general MAC PDU format is shown in Figure 4.21. A MAC PDU primarily consists of the MAC header and the MAC payload. The MAC header is further composed of MAC subheaders, while the MAC payload is composed of MAC control elements, MAC SDUs and padding.

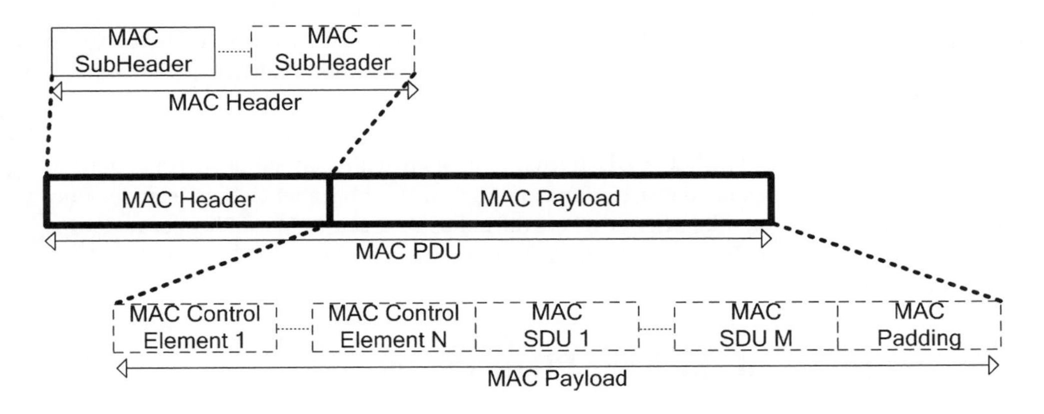

Figure 4.21: General MAC PDU format.

Each MAC subheader consists of a Logical Channel ID (LCID) and a Length (L) field. The LCID indicates whether the corresponding part of the MAC payload is a MAC Control Element, and if not, to which logical channel the related MAC SDU belongs. The L field indicates the size of the related MAC SDU or MAC Control Element.

MAC Control Elements are used for MAC-level peer-to-peer Signalling. The available types of MAC Control Element include the following:

- **Buffer Status Report MAC Control Element** for delivery of BSR information (see Section 4.4.2.2);
- **Power Headroom MAC Control Element** for the UE to report available power headroom (see Section 18.3.3);
- **DRX Command MAC Control Element** to transmit the downlink DRX commands to the UEs (see Section 4.4.2.5);
- **Timing Advance Command MAC Control Element** to transmit timing advance commands to the UEs for uplink timing alignment (see Sections 4.4.2.4 and 18.2.2);
- **C-RNTI MAC Control Element** for the UE to transmit its own C-RNTI during the random access procedure for the purpose of contention resolution (see Section 17.3.1);
- **UE Contention Resolution Identity MAC Control Element** for the eNodeB to transmit the uplink CCCH SDU that the UE has sent during the random access procedure for the purpose of contention resolution when the UE has no C-RNTI (see Section 17.3.1);
- **MBMS Dynamic Scheduling Information MAC Control Element** transmitted for each MCH to inform MBMS-capable UEs about scheduling of data transmissions on MTCH (see Section 13.6).

For each type of MAC Control Element, one special LCID is allocated.

When a MAC PDU is used to transport data from the PCCH or BCCH , the MAC PDU includes data from only one logical channel. In this case, because multiplexing is not applied, there is no need to include the LCID field in the header. In addition, if there is a one-to-one correspondence between a MAC SDU and a MAC PDU, the size of the MAC SDU can be known implicitly from the TB size. Thus, for these cases a headerless MAC PDU format is used as a transparent MAC PDU.

When a MAC PDU is used to transport the Random Access Response (RAR – see Section 17.3.1.2), a special MAC PDU format is applied with a MAC header and zero or more RARs. The MAC header consists of one or more MAC subheaders which include either a random access preamble identifier or a backoff indicator. Each MAC subheader including the Random Access Preamble IDentifier (RAPID) corresponds to one RAR in the MAC PDU (see Section 17.3.1).

4.5 Summary of the User Plane Protocols

The LTE Layer 2 protocol stack, consisting of the PDCP, RLC and MAC sublayers, acts as the interface between the radio access technology-agnostic sources of packet data traffic and the LTE physical layer. By providing functionality such as IP packet header compression, security, handover support, segmentation/concatenation, retransmission and reordering of packets, and transmission scheduling, the protocol stack enables the physical layer to be used efficiently for packet data traffic.

References[19]

[1] 3GPP Technical Specification 36.323, 'Evolved Universal Terrestrial Radio Access (E-UTRA); Packet Data Convergence Protocol (PDCP) Specification', www.3gpp.org.

[2] 3GPP Technical Specification 36.322, 'Evolved Universal Terrestrial Radio Access (E-UTRA); Radio Link Control (RLC) Protocol Specification', www.3gpp.org.

[3] 3GPP Technical Specification 36.321, 'Evolved Universal Terrestrial Radio Access (E-UTRA); Medium Access Control (MAC) Protocol Specification', www.3gpp.org.

[19] All web sites confirmed 1st March 2011.

Part II

Physical Layer for Downlink

5

Orthogonal Frequency Division Multiple Access (OFDMA)

Andrea Ancora, Issam Toufik, Andreas Bury and Dirk Slock

5.1 Introduction

The choice of an appropriate modulation and multiple-access technique for mobile wireless data communications is critical to achieving good system performance. In particular, typical mobile radio channels tend to be dispersive and time-variant, and this has generated interest in multicarrier modulation.

In general, multicarrier schemes subdivide the used channel bandwidth into a number of parallel subchannels as shown in Figure 5.1(a). Ideally the bandwidth of each subchannel is such that they are, ideally, each non-frequency-selective (i.e. having a spectrally flat gain); this has the advantage that the receiver can easily compensate for the subchannel gains individually in the frequency domain.

Orthogonal Frequency Division Multiplexing (OFDM) is a special case of multicarrier transmission where the non-frequency-selective narrowband subchannels, into which the frequency-selective wideband channel is divided, are overlapping but orthogonal, as shown in Figure 5.1(b). This avoids the need to separate the carriers by means of guard-bands, and therefore makes OFDM highly spectrally efficient. The spacing between the subchannels in OFDM is such that they can be perfectly separated at the receiver. This allows for a low-complexity receiver implementation, which makes OFDM attractive for high-rate mobile data transmission such as the LTE downlink.

It is worth noting that the advantage of separating the transmission into multiple narrowband subchannels cannot itself translate into robustness against time-variant channels if no channel coding is employed. The LTE downlink combines OFDM with channel coding

LTE – The UMTS Long Term Evolution: From Theory to Practice, Second Edition.
Stefania Sesia, Issam Toufik and Matthew Baker.

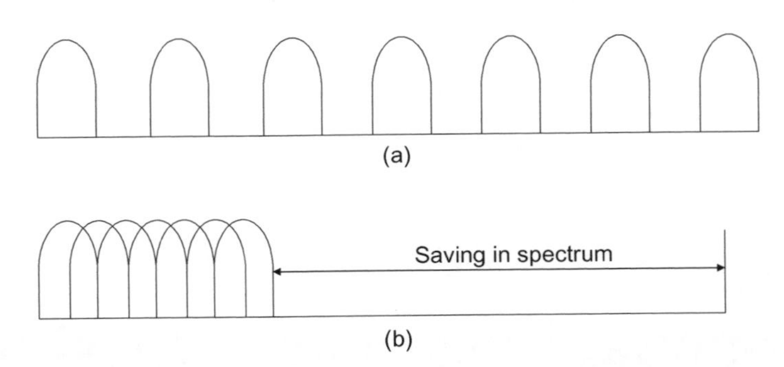

Figure 5.1: Spectral efficiency of OFDM compared to classical multicarrier modulation: (a) classical multicarrier system spectrum; (b) OFDM system spectrum.

and Hybrid Automatic Repeat reQuest (HARQ) to overcome the deep fading which may be encountered on the individual subchannels. These aspects are considered in Chapter 10 and lead to the LTE downlink falling under the category of system often referred to as 'Coded OFDM' (COFDM).

5.1.1 History of OFDM Development

Multicarrier communication systems were first introduced in the 1960s [1, 2], with the first OFDM patent being filed at Bell Labs in 1966. Initially only analogue design was proposed, using banks of sinusoidal signal generators and demodulators to process the signal for the multiple subchannels. In 1971, the use of the Discrete Fourier Transform (DFT) was proposed [3], which made OFDM implementation cost-effective. Further complexity reductions were realized in 1980 by the application of the Winograd Fourier Transform (WFT) or the Fast Fourier Transform (FFT) [4].

OFDM then became the modulation of choice for many applications for both wired systems (such as Asymmetric Digital Subscriber Line (ADSL)) and wireless systems. Wireless applications of OFDM tended to focus on broadcast systems, such as Digital Video Broadcasting (DVB) and Digital Audio Broadcasting (DAB), and relatively low-power systems such as Wireless Local Area Networks (WLANs). Such applications benefit from the low complexity of the OFDM receiver, while not requiring a high-power transmitter in the consumer terminals. This avoids one of the main disadvantages of OFDM, namely that the transmitters tend to be more expensive because of the high Peak to Average Power Ratio (PAPR); this aspect is discussed in Section 5.2.2.

The first cellular mobile radio system based on OFDM was proposed in [5]. Since then, the processing power of modern Digital Signal Processors (DSPs) has increased remarkably, paving the way for OFDM to be used in the LTE downlink. Here, the key benefits of OFDM which come to the fore are not only the low-complexity receiver but also the ability of OFDM to be adapted in a straightforward manner to operate in different channel bandwidths according to spectrum availability.

5.2 OFDM

5.2.1 Orthogonal Multiplexing Principle

A high-rate data stream typically faces the problem of having a symbol period T_s much smaller than the channel delay spread T_d if it is transmitted serially. This generates Inter-Symbol Interference (ISI) which can only be undone by means of a complex equalization procedure. In general, the equalization complexity grows with the square of the channel impulse response length.

In OFDM, the high-rate stream of data symbols is first Serial-to-Parallel (S/P) converted for modulation onto M parallel subcarriers as shown in Figure 5.2. This increases the symbol duration on each subcarrier by a factor of approximately M, such that it becomes significantly longer than the channel delay spread.

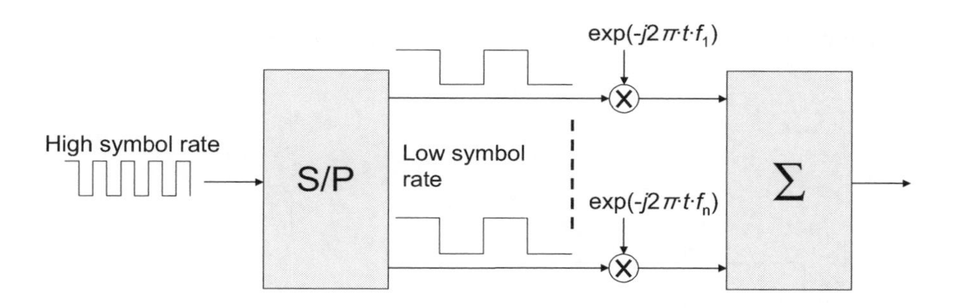

Figure 5.2: Serial-to-Parallel (S/P) conversion operation for OFDM.

This operation has the important advantage of requiring a much less complex equalization procedure in the receiver, under the assumption that the time-varying channel impulse response remains substantially constant during the transmission of each modulated OFDM symbol. Figure 5.3 shows how the resulting long symbol duration is virtually unaffected by ISI compared to the short symbol duration, which is highly corrupted.

Figure 5.4 shows the typical block diagram of an OFDM system. The signal to be transmitted is defined in the frequency domain. An S/P converter collects serial data symbols into a data block $\mathbf{S}[k] = [S_0[k], S_1[k], \ldots, S_{M-1}[k]]^T$ of dimension M, where k is the index of an OFDM symbol (spanning the M subcarriers). The M parallel data streams are first independently modulated resulting in the complex vector $\mathbf{X}[k] = [X_0[k], X_1[k], \ldots, X_{M-1}[k]]^T$. Note that in principle it is possible to use different modulations (e.g. QPSK or 16QAM) on each subcarrier; due to channel frequency selectivity, the channel gain may differ between subcarriers, and thus some subcarriers can carry higher data-rates than others. The vector $\mathbf{X}[k]$ is then used as input to an N-point Inverse FFT (IFFT) resulting in a set of N complex time-domain samples $\mathbf{x}[k] = [x_0[k], \ldots, x_{N-1}[k]]^T$. In a practical OFDM system, the number of processed subcarriers is greater than the number of modulated subcarriers (i.e. $N \geq M$), with the un-modulated subcarriers being padded with zeros.

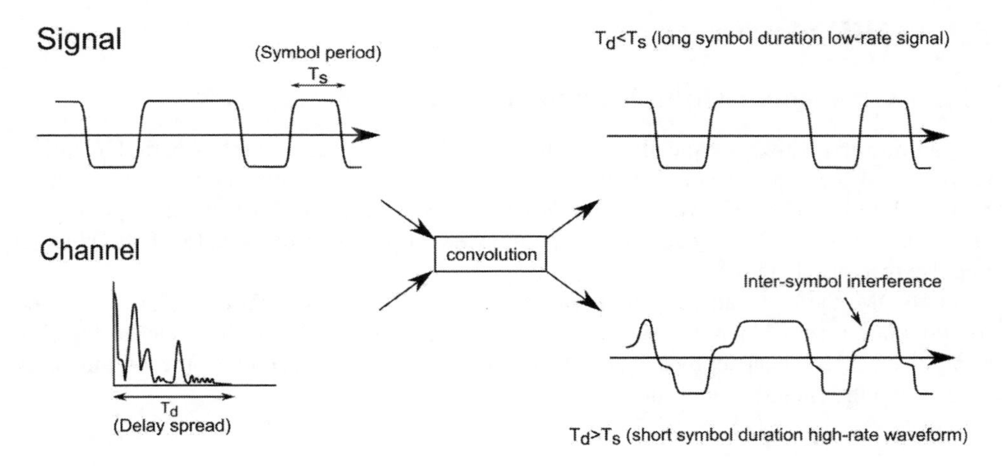

Figure 5.3: Effect of channel on signals with short and long symbol duration.

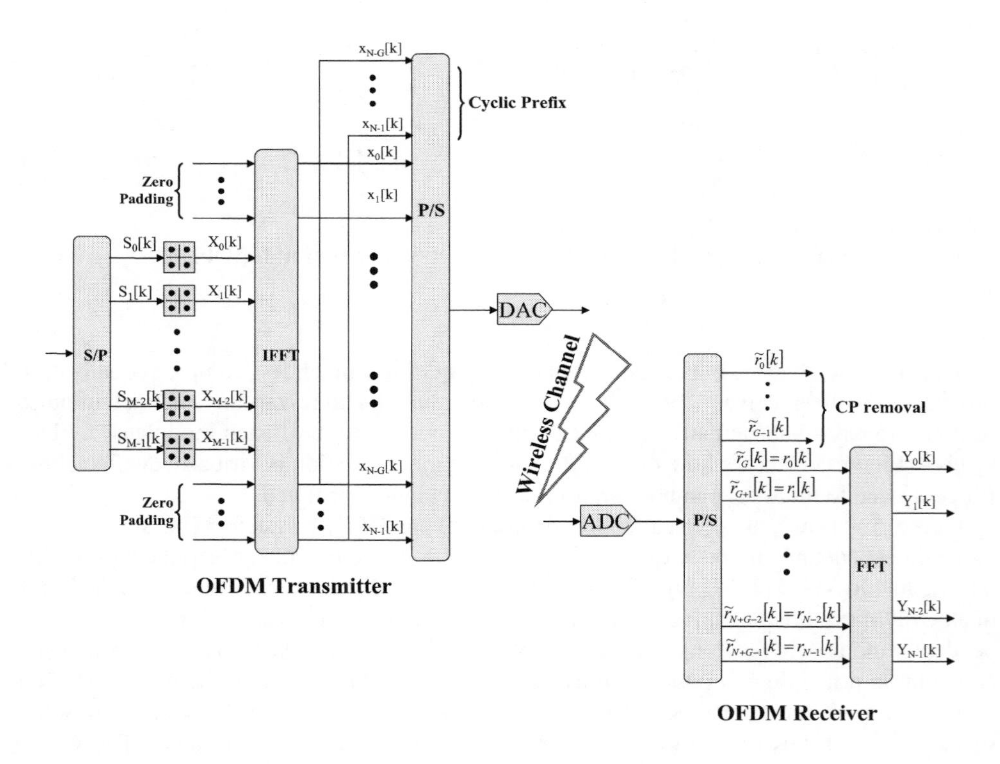

Figure 5.4: OFDM system model: (a) transmitter; (b) receiver.

The next key operation in the generation of an OFDM signal is the creation of a guard period at the beginning of each OFDM symbol $\mathbf{x}[k]$ by adding a Cyclic Prefix (CP), to eliminate the remaining impact of ISI caused by multipath propagation. The CP is generated by duplicating the last G samples of the IFFT output and appending them at the beginning of $\mathbf{x}[k]$. This yields the time domain OFDM symbol $[x_{N-G}[k], \ldots, x_{N-1}[k], x_0[k], \ldots, x_{N-1}[k]]^T$, as shown in Figure 5.5.

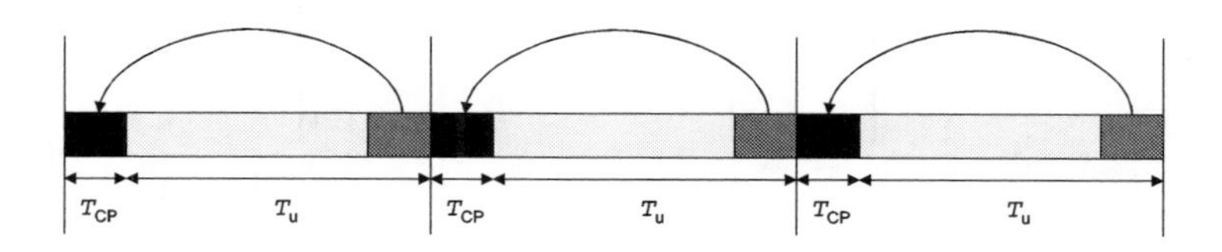

Figure 5.5: OFDM Cyclic Prefix (CP) insertion.

To avoid ISI completely, the CP length G must be chosen to be longer than the longest channel impulse response to be supported. The CP converts the linear (i.e. aperiodic) convolution of the channel into a circular (i.e. periodic) one which is suitable for DFT processing. The insertion of the CP into the OFDM symbol and its implications are explained more formally later in this section.

The output of the IFFT is then Parallel-to-Serial (P/S) converted for transmission through the frequency-selective channel.

At the receiver, the reverse operations are performed to demodulate the OFDM signal. Assuming that time- and frequency-synchronization is achieved (see Chapter 7), a number of samples corresponding to the length of the CP are removed, such that only an ISI-free block of samples is passed to the DFT. If the number of subcarriers N is designed to be a power of 2, a highly efficient FFT implementation may be used to transform the signal back to the frequency domain. Among the N parallel streams output from the FFT, the modulated subset of M subcarriers are selected and further processed by the receiver.

Let $x(t)$ be the symbol transmitted at time instant t. The received signal in a multipath environment is then given by

$$r(t) = x(t) * h(t) + z(t) \tag{5.1}$$

where $h(t)$ is the continuous-time impulse response of the channel, $*$ represents the convolution operation and $z(t)$ is the additive noise. Assuming that $x(t)$ is band-limited to $[-\frac{1}{2T_s}, \frac{1}{2T_s}]$, the continuous-time signal $x(t)$ can be sampled at sampling rate T_s such that the Nyquist criterion is satisfied. As a result of the multipath propagation, several replicas of the transmitted signals arrive at the receiver at different delays.

The received discrete-time OFDM symbol k including CP, under the assumption that the channel impulse response has a length smaller than or equal to G, can be expressed as

$$\tilde{\mathbf{r}} = \begin{bmatrix} \tilde{r}_0[k] \\ \tilde{r}_1[k] \\ \vdots \\ \tilde{r}_{G-2}[k] \\ \tilde{r}_{G-1}[k] \\ \tilde{r}_G[k] \\ \vdots \\ \tilde{r}_{N+G-1}[k] \end{bmatrix} = \mathbf{A} \cdot \begin{bmatrix} h_0[k] \\ h_1[k] \\ \vdots \\ h_{G-1}[k] \end{bmatrix} + \begin{bmatrix} z_0[k] \\ z_1[k] \\ \vdots \\ z_{G-2}[k] \\ z_{G-1}[k] \\ z_G[k] \\ \vdots \\ z_{N+G-1}[k] \end{bmatrix} \tag{5.2}$$

where

$$\mathbf{A} = \begin{bmatrix} x_{N-G}[k] & x_{N-1}[k-1] & x_{N-2}[k-1] & \cdots & x_{N-G-1}[k-1] \\ x_{N-G+1}[k] & x_{N-G}[k] & x_{N-1}[k-1] & \cdots & x_{N-G+2}[k-1] \\ \vdots & \vdots & \ddots & \ddots & \vdots \\ x_{N-2}[k] & x_{N-3}[k] & \ddots & x_{N-G}[k] & x_{N-1}[k-1] \\ x_{N-1}[k] & x_{N-2}[k] & \ddots & x_{N-G+1}[k] & x_{N-G}[k] \\ x_0[k] & x_{N-1}[k] & \ddots & x_{N-G+2}[k] & x_{N-G+1}[k] \\ \vdots & \cdots & \cdots & \cdots & \vdots \\ x_{N-1}[k] & x_{N-2}[k] & \cdots & \cdots & x_{N-G}[k] \end{bmatrix}$$

In general broadband transmission systems, one of the most complex operations the receiver has to handle is the equalization process to recover $x_n[k]$ (from Equation (5.2)).

Equation (5.2) can be written as the sum of intra-OFDM symbol interference (generated by the frequency-selective behaviour of the channel within an OFDM symbol) and the inter-OFDM symbol interference (between two consecutive OFDM block transmissions at time k and time $(k-1)$). This can be expressed as

$$\tilde{\mathbf{r}} = \mathbf{A}_{\text{Intra}} \cdot \begin{bmatrix} h_0[k] \\ h_1[k] \\ \vdots \\ h_{G-1}[k] \end{bmatrix} + \mathbf{A}_{\text{Inter}} \cdot \begin{bmatrix} h_0[k] \\ h_1[k] \\ \vdots \\ h_{G-1}[k] \end{bmatrix} + \begin{bmatrix} z_0[k] \\ z_1[k] \\ \vdots \\ z_{N+G-1}[k] \\ z_{N-G}[k] \\ \vdots \\ z_{N+G-1}[k] \end{bmatrix} \tag{5.3}$$

where

$$\mathbf{A}_{\text{Intra}} = \begin{bmatrix} x_{\text{N-G}}[k] & 0 & \cdots & 0 \\ x_{\text{N-G+1}}[1] & x_{\text{N-G}}[k] & \cdots & 0 \\ \vdots & \ddots & \ddots & \vdots \\ x_{\text{N-1}}[k] & \cdots & \ddots & x_{\text{N-G}}[k] \\ x_0[k] & x_{\text{N-1}}[k] & \cdots & x_{\text{N-G+1}}[k] \\ \vdots & \ddots & \ddots & \vdots \\ x_{\text{N-1}}[k] & x_{\text{N-2}}[k] & \cdots & x_{\text{N-G}}[k] \end{bmatrix}$$

and

$$\mathbf{A}_{\text{Inter}} = \begin{bmatrix} 0 & x_{\text{N-1}}[k-1] & x_{\text{N-2}}[k-1] & \cdots & x_{\text{N-G+1}}[k-1] \\ 0 & 0 & x_{\text{N-1}}[k-1] & \cdots & x_{\text{N-G+2}}[N-1] \\ \vdots & \ddots & \ddots & \ddots & \vdots \\ 0 & 0 & \cdots & 0 & x_{\text{N-1}}[k-1] \\ 0 & 0 & \cdots & \cdots & 0 \\ \vdots & \ddots & \ddots & \ddots & \vdots \\ 0 & 0 & \cdots & \cdots & 0 \end{bmatrix}$$

In order to suppress the inter-OFDM-symbol interference, the first G samples of the received signal are discarded, eliminating the contribution of the matrix $\mathbf{A}_{\text{Inter}}$ and the G first rows of $\mathbf{A}_{\text{Intra}}$. This can be expressed as

$$\mathbf{r}[k] = \begin{bmatrix} r_0[k] \\ r_1[k] \\ \vdots \\ r_{\text{N-1}}[k] \end{bmatrix} = \begin{bmatrix} \tilde{r}_{\text{G}}[k] \\ \tilde{r}_{\text{G+1}}[k] \\ \vdots \\ \tilde{r}_{\text{N+G-1}}[k] \end{bmatrix}$$

$$= \begin{bmatrix} x_0[k] & x_{\text{N-1}}[k] & \cdots & x_{\text{N-G+1}}[k] \\ x_1[k] & x_0[k] & \cdots & x_{\text{N-G+2}}[k] \\ \vdots & \ddots & \ddots & \vdots \\ x_{\text{N-1}}[k] & x_{\text{N-2}}[k] & \cdots & x_{\text{N-G}}[k] \end{bmatrix} \cdot \begin{bmatrix} h_0[k] \\ h_1[k] \\ \vdots \\ h_{\text{G-1}}[k] \end{bmatrix} + \begin{bmatrix} z_{\text{G}}[k] \\ z_{\text{G+1}}[k] \\ \vdots \\ z_{\text{N+G-1}}[k] \end{bmatrix}$$

Adding zeros to the channel vector can extend the signal matrix without changing the output vector. This can be expressed as

$$\begin{bmatrix} r_0[k] \\ r_1[k] \\ \vdots \\ r_{\text{N-1}}[k] \end{bmatrix} = \mathbf{B} \cdot \begin{bmatrix} h_0[k] \\ h_1[k] \\ \vdots \\ h_{\text{G-1}}[k] \\ 0 \\ \vdots \\ 0 \end{bmatrix} + \begin{bmatrix} z_{\text{G}}[k] \\ z_{\text{G+1}}[k] \\ \vdots \\ z_{\text{N+G-1}}[k] \end{bmatrix}$$

where matrix **B** is given by

$$\mathbf{B} = \begin{bmatrix} x_0[k] & x_{N-1}[k] & \cdots & x_{N-G+1}[k] & x_{N-G}[k] & \cdots & x_1[k] \\ x_1[k] & x_0[k] & \cdots & x_{N-G+2}[k] & x_{N-G+1}[k] & \cdots & x_2[k] \\ \vdots & \ddots & \ddots & \ddots & \ddots & \ddots & \vdots \\ x_{N-1}[k] & x_{N-2}[k] & \cdots & x_{N-G}[k] & x_{N-G-1}[k] & \cdots & x_0[k] \end{bmatrix}$$

The matrix **B** is circulant and thus its Fourier transform is diagonal with eigenvalues given by the FFT of its first row [6, 7]. It can then be written as $\mathbf{B} = \mathbf{F}^H\mathbf{X}\mathbf{F}$, with **X** diagonal, and the equivalent received signal can be expressed as follows:

$$\begin{bmatrix} r_0[k] \\ r_{N-G+1}[k] \\ \vdots \\ r_{N-1}[k] \end{bmatrix} = \mathbf{F}^H \cdot \begin{bmatrix} X_0[k] & 0 & \cdots & 0 \\ 0 & X_1[k] & \cdots & 0 \\ \vdots & \cdots & \ddots & \vdots \\ 0 & 0 & \cdots & X_{N-1}[k] \end{bmatrix} \cdot \mathbf{F} \cdot \begin{bmatrix} h_0[k] \\ h_1[k] \\ \vdots \\ h_{G-1}[k] \\ 0 \\ \vdots \\ 0 \end{bmatrix}$$

$$+ \begin{bmatrix} z_{N-G}[k] \\ z_{N-G+1}[k] \\ \vdots \\ z_{N+G-1}[k] \end{bmatrix} \tag{5.4}$$

(5.5)

where **F** is the Fourier transform matrix whose elements are

$$(\mathbf{F})_{n,m} = \frac{1}{\sqrt{N}} \exp\left(-\frac{j2\pi}{N}(nm)\right)$$

for $0 \leq n \leq N-1$ and $0 \leq m \leq N-1$ and N is the length of the OFDM symbol. The elements on the diagonal are the eigenvalues of the matrix **B**, obtained as

$$X_m[k] = \frac{1}{\sqrt{N}} \sum_{n=1}^{N} x_n[k] \exp\left(-2j\pi m\frac{n}{N}\right) \tag{5.6}$$

In reverse, the time-domain signal $x_n[k]$ can be obtained by

$$x_n[k] = \frac{1}{\sqrt{N}} \sum_{m=1}^{N} X_m[k] \exp\left(2j\pi m\frac{n}{N}\right) \tag{5.7}$$

By applying the Fourier transform to Equation (5.4), the equivalent received signal in the frequency domain can be obtained,

$$\begin{bmatrix} R_0[k] \\ \vdots \\ R_{N-1}[k] \end{bmatrix} = \begin{bmatrix} X_0[0] & 0 & \cdots & 0 \\ 0 & X_1[k] & \cdots & 0 \\ \vdots & \ddots & \ddots & \vdots \\ 0 & 0 & \cdots & X_{N-1}[k] \end{bmatrix} \begin{bmatrix} H_0[k] \\ H_1[k] \\ \vdots \\ H_{N-1}[k] \end{bmatrix} + \begin{bmatrix} Z_0[k] \\ \vdots \\ Z_{N-1}[k] \end{bmatrix}$$

In summary, the CP of OFDM changes the linear convolution into a circular one. The circular convolution is very efficiently transformed by means of an FFT into a multiplicative operation in the frequency domain. Hence, the transmitted signal over a frequency-selective (i.e. multipath) channel is converted into a transmission over N parallel flat-fading channels in the frequency domain:

$$R_m[k] = X_m[k] \cdot H_m[k] + Z_m[k] \tag{5.8}$$

As a result, the equalization is much simpler than for single-carrier systems and consists of just one complex multiplication per subcarrier.

5.2.2 Peak-to-Average Power Ratio and Sensitivity to Non-Linearity

While the previous section shows the advantages of OFDM, this section highlights its major drawback: the Peak-to-Average Power Ratio (PAPR).

In the general case, the OFDM transmitter can be seen as a linear transform performed over a large block of independent identically distributed (i.i.d) QAM[1]-modulated complex symbols (in the frequency domain). From the central limit theorem [8, 9], the time-domain OFDM symbol may be approximated as a Gaussian waveform. The amplitude variations of the OFDM modulated signal can therefore be very high. However, practical Power Amplifiers (PAs) of RF transmitters are linear only within a limited dynamic range. Thus, the OFDM signal is likely to suffer from non-linear distortion caused by clipping. This gives rise to out-of-band spurious emissions and in-band corruption of the signal. To avoid such distortion, the PAs have to operate with large power back-offs, leading to inefficient amplification or expensive transmitters.

The PAPR is one measure of the high dynamic range of the input amplitude, and hence a measure of the expected degradation. To analyse the PAPR mathematically, let x_n be the signal after IFFT as given by Equation (5.7) where the time index k can be dropped without loss of generality. The PAPR of an OFDM symbol is defined as the square of the peak amplitude divided by the mean power, i.e.

$$\text{PAPR} = \frac{\max_n\{|x_n|^2\}}{E\{|x_n|^2\}} \tag{5.9}$$

Under the hypothesis that the Gaussian approximation is valid, the amplitude of x_n has a Rayleigh distribution, while its power has a central chi-square distribution with two degrees of freedom. The Cumulative Distribution Function (CDF) $F_x(\alpha)$ of the normalized power is given by

$$F_x(\alpha) = \Pr\left(\frac{|x_n|^2}{E\{|x_n|^2\}} < \alpha\right) = 1 - e^{-\alpha} \tag{5.10}$$

If there is no oversampling, the time-domain samples are mutually uncorrelated and the probability that the PAPR is above a certain threshold PAPR_0 is given by[2]

$$\Pr(\text{PAPR} > \text{PAPR}_0) = 1 - F_x(\text{PAPR}_0)^N = 1 - (1 - e^{-\text{PAPR}_0})^N \tag{5.11}$$

Figure 5.6 plots the distribution of the PAPR given by Equation (5.11) for different values of the number of subcarriers N. The figure shows that a high PAPR does not occur very often. However, when it does occur, degradation due to PA non-linearities may be expected.

[1] Quadrature Amplitude Modulation.

[2] Note that the CDF of the PAPR is $F_{\text{PAPR}}(\eta) = [F_x(\eta)]^N$ for N i.i.d. samples.

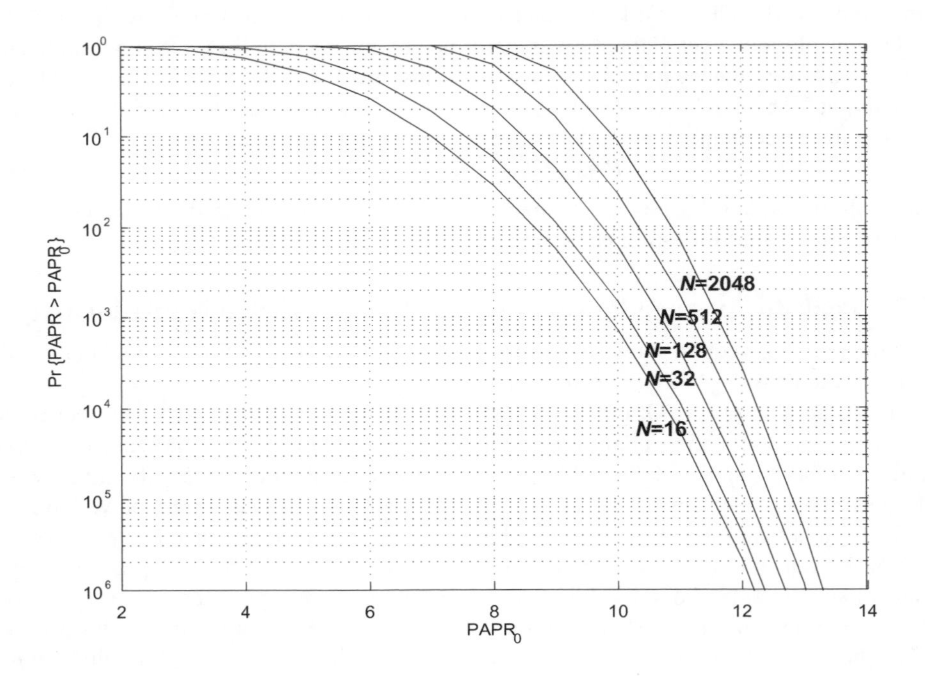

Figure 5.6: PAPR distribution for different numbers of OFDM subcarriers.

In order to evaluate the impacts of distortion on the OFDM signal reception, a useful framework for modelling non-linearities is developed in Section 21.5.2.

5.2.2.1 PAPR Reduction Techniques

Many techniques have been studied for reducing the PAPR of a transmitted OFDM signal. Although no such techniques are specified for the LTE downlink signal generation, an overview of the possibilities is provided below. In general, in LTE the cost and complexity of generating the OFDM signal with acceptable Error Vector Magnitude (EVM) (see Section 21.3.1.1) is left to the eNodeB implementation. As OFDM is not used for the LTE uplink (see Section 14.2), such considerations do not directly apply to the transmitter in the User Equipment (UE).

Techniques for PAPR reduction of OFDM signals can be broadly categorized into three main concepts: clipping and filtering [10–12], selected mapping [13] and coding techniques [14, 15]. The most potentially relevant from the point of view of LTE would be clipping and filtering, whereby the time-domain signal is clipped to a predefined level. This causes spectral leakage into adjacent channels, resulting in reduced spectral efficiency as well as in-band noise degrading the bit error rate performance. Out-of-band radiation caused by the clipping process can, however, be reduced by filtering. If discrete signals are clipped directly, the resulting clipping noise will all fall in band and thus cannot be reduced by filtering. To avoid this problem, one solution consists of oversampling the original signal by padding the input signal with zeros and processing it using a longer IFFT. The oversampled signal is clipped and then filtered to reduce the out-of-band radiation.

5.2.3 Sensitivity to Carrier Frequency Offset and Time-Varying Channels

The orthogonality of OFDM relies on the condition that transmitter and receiver operate with exactly the same frequency reference. If this is not the case, the perfect orthogonality of the subcarriers is lost, causing subcarrier leakage, also known as Inter-Carrier Interference (ICI), as can be seen in Figure 5.7.

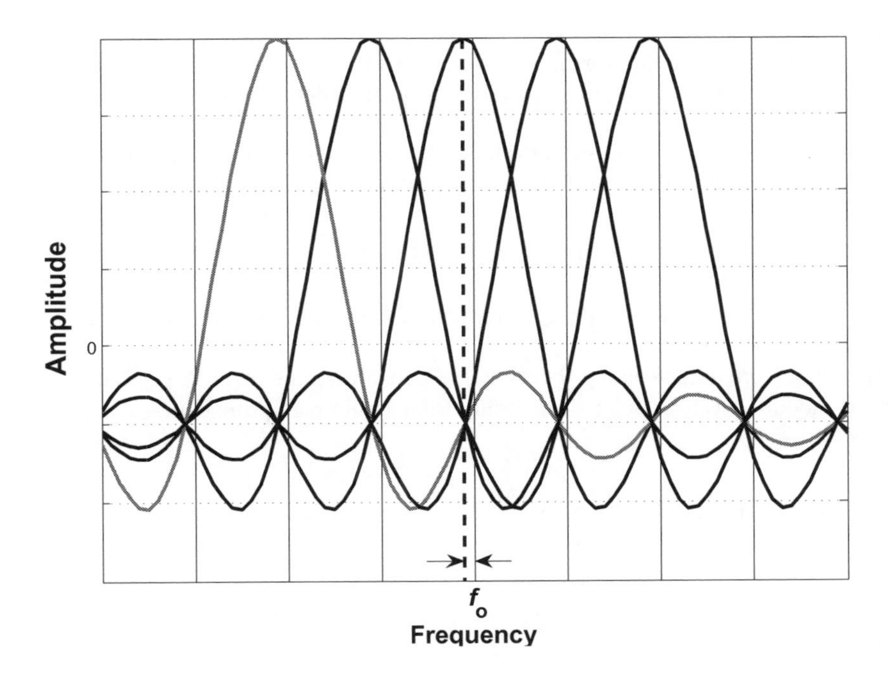

Figure 5.7: Loss of orthogonality between OFDM subcarriers due to frequency offset.

Frequency errors typically arise from a mismatch between the reference frequencies of the transmitter and the receiver local oscillators. On the receiver side in particular, due to the importance of using low-cost components in the mobile handset, local oscillator frequency drifts are usually greater than in the eNodeB and are typically a function of parameters such as temperature changes and voltage variations. This difference between the reference frequencies is widely referred to as Carrier Frequency Offset (CFO). Phase noise in the UE receiver may also result in frequency errors.

The CFO can be several times larger than the subcarrier spacing. It is usually divided into an integer part and a fractional part. Thus the frequency error can be written as

$$f_\mathrm{o} = (\Gamma + \epsilon)\Delta f \tag{5.12}$$

where Δf is the subcarrier spacing, Γ is an integer and $-0.5 < \epsilon < 0.5$. If $\Gamma \neq 0$, then the modulated data are in the wrong positions with respect to the subcarrier mapping performed at the transmitter. This simply results in a Bit Error Rate (BER) of 0.5 if the frequency offset is not compensated at the receiver, independently of the value of ϵ. In the case of $\Gamma = 0$ and $\epsilon \neq 0$, the perfect subcarrier orthogonality is lost, resulting in ICI which can degrade the BER. Typically only synchronization errors of up to a few percent of the subcarrier spacing are tolerable in OFDM systems.

Even in an ideal case where the local oscillators are perfectly aligned, the relative speed between transmitter and receiver also generates a frequency error due to Doppler.

In the case of a single-path channel, UE mobility in a constant direction with respect to the angle of arrival of the signal results in a Doppler shift f_d, while in a scattering environment this becomes a Doppler spread with spectral density $P(f)$ as discussed further in Section 8.3.1.

It can be shown [16, 17] that, for both flat and dispersive channels, the ICI power can be computed as a function of the generic Doppler spectral density $P(f)$ as follows:

$$P_\mathrm{ICI} = \int_{-f_{\mathrm{d}_\mathrm{max}}}^{f_{\mathrm{d}_\mathrm{max}}} P(f)(1 - \mathrm{sinc}^2(T_s f))\,\mathrm{d}f \tag{5.13}$$

where $f_{\mathrm{d}_\mathrm{max}}$ is the maximum Doppler frequency, and the transmitted signal power is normalized.

ICI resulting from a mismatch f_o between the transmitter and receiver oscillator frequencies can be modelled as a Doppler shift arising from single-path propagation:

$$P(f) = \delta(f - f_\mathrm{o}) \tag{5.14}$$

Hence, substituting (5.14) into (5.13), the ICI power in the case of a deterministic CFO is given by

$$P_\mathrm{ICI,CFO} = 1 - \mathrm{sinc}^2(f_\mathrm{o} T_\mathrm{s}) \tag{5.15}$$

For the classical Jakes model of Doppler spread (see Section 8.3.1), Equation (5.13) can be written as

$$P_\mathrm{ICI,Jakes} = 1 - 2\int_0^1 (1 - f)J_0(2\pi f_{\mathrm{d}_\mathrm{max}} T_\mathrm{s} f)\,\mathrm{d}f \tag{5.16}$$

where J_0 is the zero-th order Bessel function.

When no assumptions on the shape of the Doppler spectrum can be made, an upper bound on the ICI given by Equation (5.13) can be found by applying the Cauchy–Schwartz inequality, leading to [17]

$$P_\mathrm{ICI} \leq \frac{\int_0^1 [1 - \mathrm{sinc}^2(f_\mathrm{d} T_\mathrm{s} f)]^2\,\mathrm{d}f}{\int_0^1 1 - \mathrm{sinc}^2(f_\mathrm{d} T_\mathrm{s} f)\,\mathrm{d}f} \tag{5.17}$$

This upper bound on P_ICI is valid only in the case of frequency spread and does not cover the case of a deterministic CFO.

Using Equations (5.17), (5.16) and (5.15), the Signal-to-Interference Ratio (SIR) in the presence of ICI can be expressed as

$$SIR_{\mathrm{ICI}} = \frac{1 - P_{\mathrm{ICI}}}{P_{\mathrm{ICI}}} \tag{5.18}$$

Figures 5.8 and 5.9 plot these P_{ICI} and SIR_{ICI} for the cases provided. These figures show that the highest ICI is introduced by a constant frequency offset. In the case of a Doppler spread, the ICI impairment is lower. Figure 5.9 shows that, in the absence of any other impairment such as interference ($SIR = \infty$), the SIR_{ICI} rapidly decays as a function of frequency misalignments.

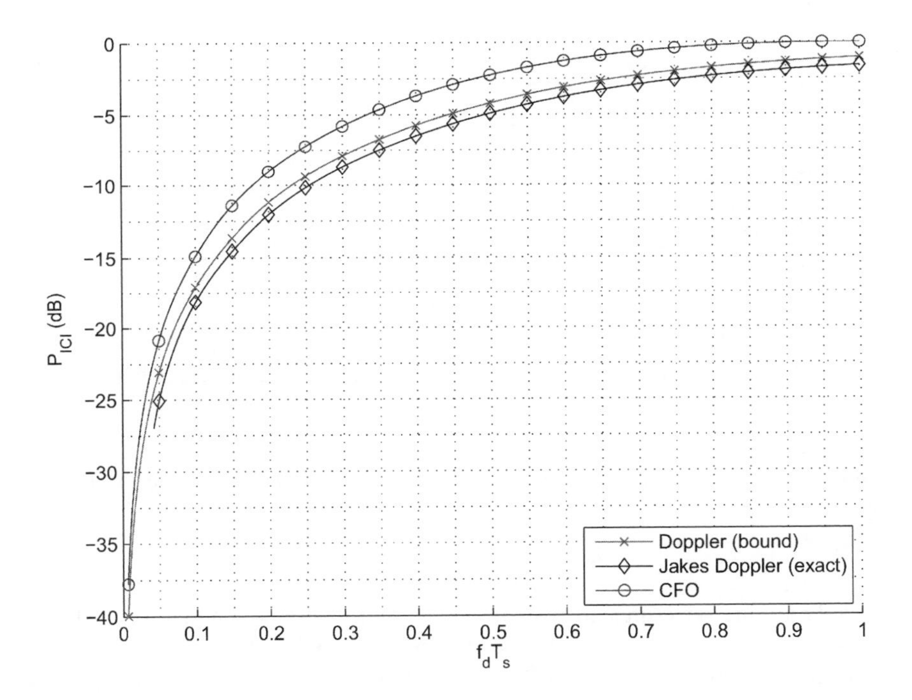

Figure 5.8: P_{ICI} for the case of a classical Doppler distribution and a deterministic CFO.

The sensitivity of the BER depends on the modulation order. It is shown in [18] that QPSK modulation can tolerate up to $\epsilon_{\mathrm{max}} = 0.05$ whereas 64-QAM requires $\epsilon \leq 0.01$.

5.2.4 Timing Offset and Cyclic Prefix Dimensioning

In the case of a memoryless channel (i.e. no delay spread), OFDM is insensitive to timing synchronization errors provided that the misalignment remains within the CP duration. In other words, if $T_{\mathrm{o}} \leq T_{\mathrm{CP}}$ (with T_{o} being the timing error), then orthogonality is maintained thanks to the cyclic nature of the CP. Any symbol timing delay only introduces a constant

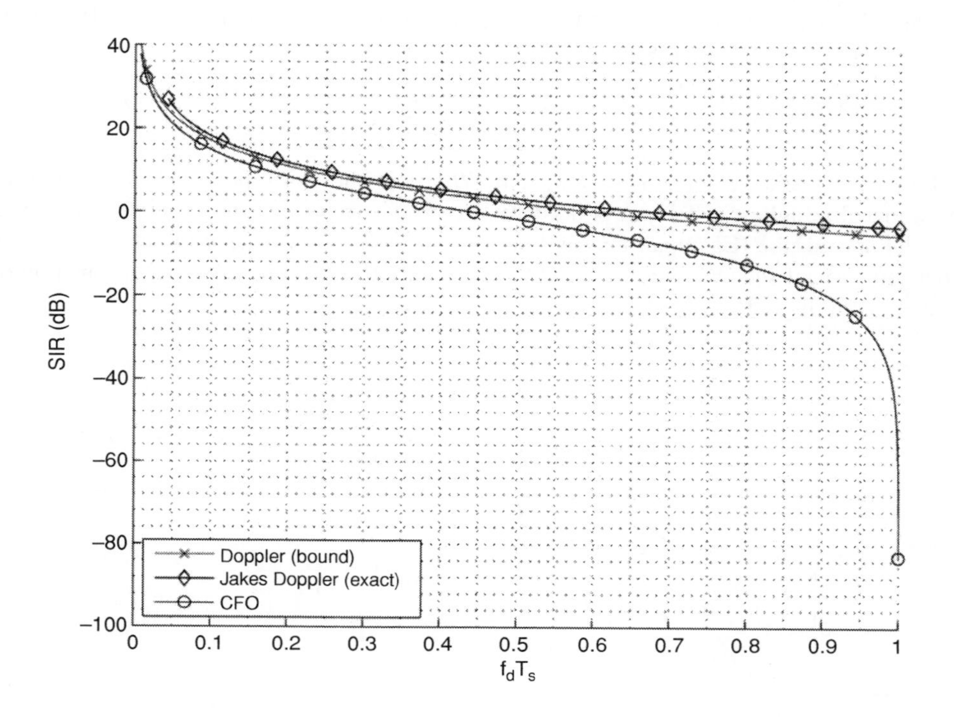

Figure 5.9: SIR_{ICI} for classical Doppler distribution and deterministic CFO.

phase shift from one subcarrier to another. The received signal at the m-th subcarrier is given by

$$R_m[k] = X_m[k] \exp\left(j2\pi \frac{dm}{N}\right) \tag{5.19}$$

where d is the timing offset in samples corresponding to a duration equal to T_o. This phase shift can be recovered as part of the channel estimation operation.

It is worth highlighting that the insensitivity to timing offsets would not hold for any kind of guard period other than a cyclic prefix; for example, zero-padding would not exhibit the same property, resulting in a portion of the useful signal power being lost.

In the general case of a channel with delay spread, for a given CP length, the maximum tolerated timing offset without degrading the OFDM reception is reduced by an amount equal to the length of the channel impulse response: $T_o \leq T_{\mathrm{CP}} - T_d$. For greater timing errors, ISI and ICI occur. The effect caused by an insufficient CP is discussed in the following section. Timing synchronization hence becomes more critical in long-delay-spread channels.

Initial timing acquisition in LTE is normally achieved by the cell-search and synchronization procedures (see Chapter 7). Thereafter, for continuous tracking of the timing-offset, two classes of approach exist, based on either CP correlation or Reference Signals (RSs). A combination of the two is also possible. The reader is referred to [19] for a comprehensive survey of OFDM synchronization techniques.

5.2.4.1 Effect of Insufficient Cyclic Prefix Length

As already explained, if an OFDM system is designed with a CP of length G samples such that $L < G$ where L is the length of the channel impulse response (in number of samples), the system benefits from turning the linear convolution into a circular one to keep the subcarriers orthogonal. The condition of a sufficient CP is therefore strictly related to the orthogonality property of OFDM.

As shown in [20], for an OFDM symbol consisting of $N + G$ samples where N is the FFT size, the power of the ICI and ISI can be computed as

$$P_{\mathrm{ICI}} = 2 \sum_{k=G}^{N+G-1} |h[k]|^2 \frac{N(k-G) - (k-G)^2}{N^2} \tag{5.20}$$

$$P_{\mathrm{ISI}} = \sum_{k=G}^{N+G-1} |h[k]|^2 \frac{(k-G)^2}{N^2} \tag{5.21}$$

The signal power P_{S} can be written as

$$P_{\mathrm{S}} = \sum_{k=0}^{G-1} |h(k)|^2 + \sum_{k=G}^{N+G-1} |h(k)|^2 \frac{(N-k+G)^2}{N^2} \tag{5.22}$$

The resulting SIR due to the CP being too short can then be written as

$$SIR_{\mathrm{og}} = \frac{P_{\mathrm{S}}}{P_{\mathrm{ISI}} + P_{\mathrm{ICI}}} \tag{5.23}$$

Figures 5.10 and 5.11 plot Equations (5.21) to (5.23) for the case of the normal CP length in LTE (see Section 5.4) assuming a channel with a uniform and normalized Power-Delay Profile (PDP) of length $L < N + G$, where the dashed line marks the boundary of the CP ($L = G$).

5.3 OFDMA

Orthogonal Frequency Division Multiple Access (OFDMA) is an extension of OFDM to the implementation of a multiuser communication system. In the discussion above, it has been assumed that a single user receives data on all the subcarriers at any given time. OFDMA distributes subcarriers to different users at the same time, so that multiple users can be scheduled to receive data simultaneously. Usually, subcarriers are allocated in contiguous groups for simplicity and to reduce the overhead of indicating which subcarriers have been allocated to each user.

OFDMA for mobile communications was first proposed in [21] based on multicarrier FDMA (Frequency Division Multiple Access), where each user is assigned to a set of randomly selected subchannels.

OFDMA enables the OFDM transmission to benefit from multi-user diversity, as discussed in Chapter 12. Based on feedback information about the frequency-selective channel conditions from each user, adaptive user-to-subcarrier assignment can be performed, enhancing considerably the total system spectral efficiency compared to single-user OFDM systems.

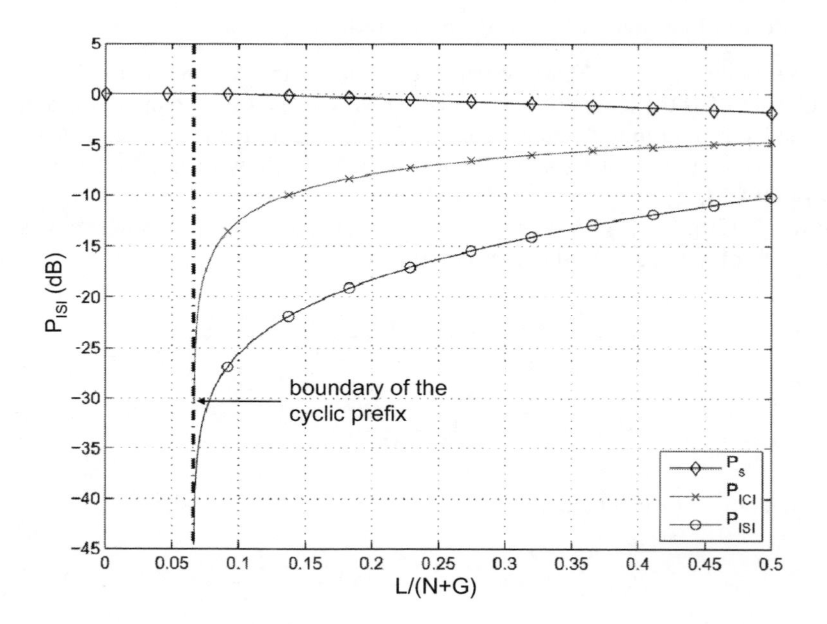

Figure 5.10: Power of signal, ICI and ISI in case of a too-short CP.

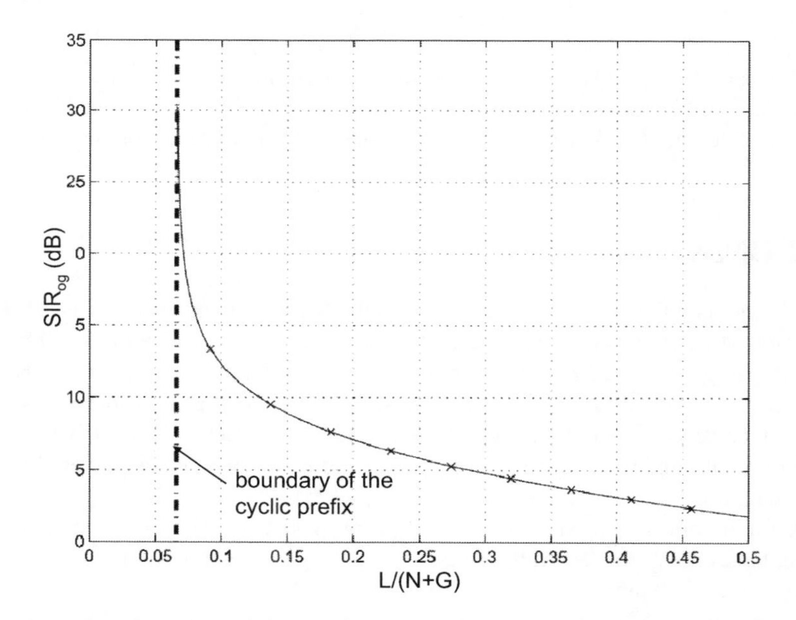

Figure 5.11: Effective SIR as a function of channel impulse response length for a given CP length.

OFDMA can also be used in combination with Time Division Multiple Access (TDMA), such that the resources are partitioned in the time-frequency plane – i.e. groups of subcarriers for a specific time duration. In LTE, such time-frequency blocks are known as Resource Blocks (RBs), as explained in Section 6.2. Figure 5.12 depicts such an OFDMA/TDMA mixed strategy as used in LTE.

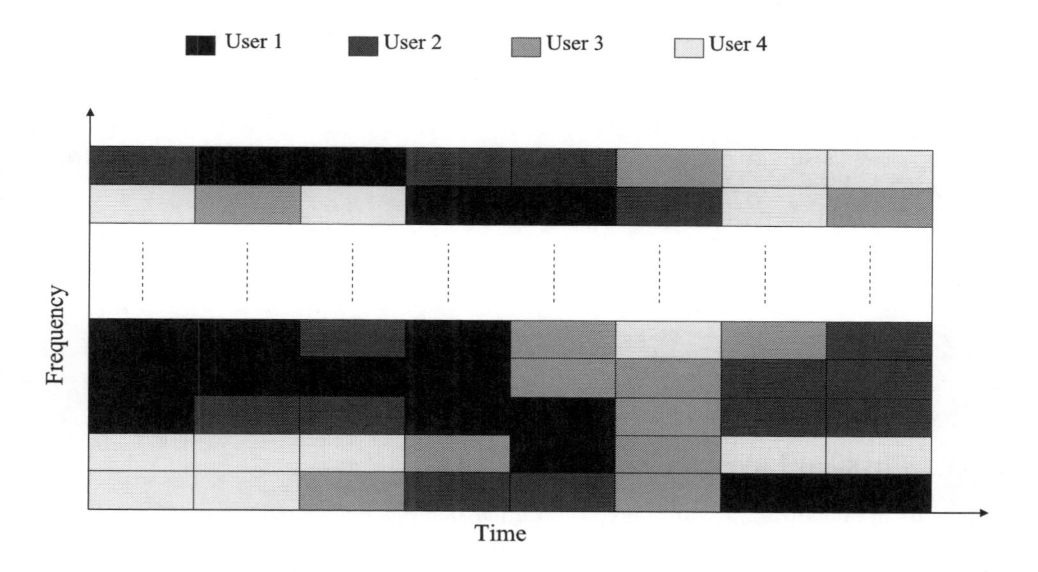

Figure 5.12: Example of resource allocation in a combined OFDMA/TDMA system.

5.4 Parameter Dimensioning

As highlighted in the previous sections, certain key parameters determine the performance of OFDM and OFDMA systems. Inevitably, some compromises have to be made in defining these parameters appropriately to maximize system spectral efficiency while maintaining robustness against propagation impairments.

For a given system, the main propagation characteristics which should be taken into account when designing an OFDM system are the expected delay spread T_{d}, the maximum Doppler frequency $f_{\mathrm{d}_{\max}}$, and, in the case of cellular systems, the targeted cell size.

The propagation characteristics impose constraints on the choice of the CP length and of the subcarrier spacing.

As already mentioned, the CP should be longer than the Channel Impulse Response (CIR) in order to ensure robustness against ISI. For cellular systems, and especially for large cells, longer delay spreads may typically be experienced than those encountered, for example, in WLAN systems, implying the need for a longer CP. On the other hand, a longer CP for a given OFDM symbol duration corresponds to a larger overhead in terms of energy per transmitted bit. Out of the $N + G$ transmitted symbols, only N convey information, leading to a rate loss. This reduction in bandwidth efficiency can be expressed as a function of the CP duration $T_{CP} = GT_s$ and the OFDM symbol period $T_u = NT_s$ (where T_s is the sampling period), as follows:

$$\beta_{overhead} = \frac{T_{CP}}{T_u + T_{CP}} \tag{5.24}$$

It is clear that to maximize spectral efficiency, T_u should be chosen to be large relative to the CP duration, but small enough to ensure that the channel does not vary within one OFDM symbol.

Further, the OFDM symbol duration T_u is related to the subcarrier spacing by $\Delta f = 1/T_u$. Choosing a large T_u leads to a smaller subcarrier separation Δf, which has a direct impact on the system sensitivity to Doppler and other sources of frequency offset, as explained in Section 5.2.3.

Thus, in summary, the following three design criteria can be identified:

$$\begin{aligned} &T_{CP} \geq T_d && \text{to prevent ISI,} \\ &\frac{f_{d_{max}}}{\Delta f} \ll 1 && \text{to keep ICI due to Doppler sufficiently low,} \\ &T_{CP}\Delta f \ll 1 && \text{for spectral efficiency.} \end{aligned} \tag{5.25}$$

5.4.1 Physical Layer Parameters for LTE

LTE aims at supporting a wide range of cellular deployment scenarios, including indoor, urban, suburban and rural situations covering both low and high UE mobility conditions (up to 350 or even 500 km/h). The cell sizes may range from home networks only a few metres across to large cells with radii of 100 kilometres or more.

Typical deployed carrier frequencies are in the range 400 MHz to 4 GHz, with bandwidths ranging from 1.4 to 20 MHz. All these cases imply different delay spreads and Doppler frequencies.

The 'normal' parameterization of the LTE downlink uses a $\Delta f = 15$ kHz subcarrier spacing with a CP length of approximately 5 μs. This subcarrier spacing is a compromise between the percentage overhead of the CP and the sensitivity to frequency offsets. A 15 kHz subcarrier spacing is sufficiently large to allow for high mobility and to avoid the need for closed-loop frequency adjustments.

In addition to the normal parameterization, it is possible to configure LTE with an extended CP of length approximately 17 μs.[3] This is designed to ensure that even in large suburban and rural cells, the delay spread should be contained within the CP. This, however, comes at the expense of a higher overhead from the CP as a proportion of the total system transmission resources. This is particularly useful for the multi-cell broadcast transmission mode supported by LTE known as Multimedia Broadcast Single Frequency Network (MBSFN) (see Section 13.4), in which the UE receives and combines synchronized signals from multiple cells. In this case the relative timing offsets from the multiple cells must all be received at the UE's receiver within the CP duration, if ISI is to be avoided, thus also requiring a long CP.

[3]The length of the extended CP is $\frac{1}{4}$ of the OFDM symbol.

In case MBSFN transmission were, in a future release of LTE, to be configured on a dedicated carrier rather than sharing the carrier with unicast data (this is not possible in current releases of LTE), a further set of parameters is defined with the subcarrier spacing halved to 7.5 kHz; this would allow the OFDM symbol length to be doubled to provide an extended CP of length approximately 33 μs while remaining $\frac{1}{4}$ of the OFDM symbol length. This would come at the expense of increasing the sensitivity to mobility and frequency errors.

These modes and their corresponding parameters are summarized in Figure 5.13. It is worth noting that when LTE is configured with the normal CP length, the CP length for the first OFDM symbol in each 0.5 ms interval is slightly longer than that of the next six OFDM symbols (i.e. 5.2 μs compared to 4.7 μs). This characteristic is due to the need to accommodate an integer number of OFDM symbols, namely 7, into each 0.5 ms interval, with assumed FFT block-lengths of 2048.

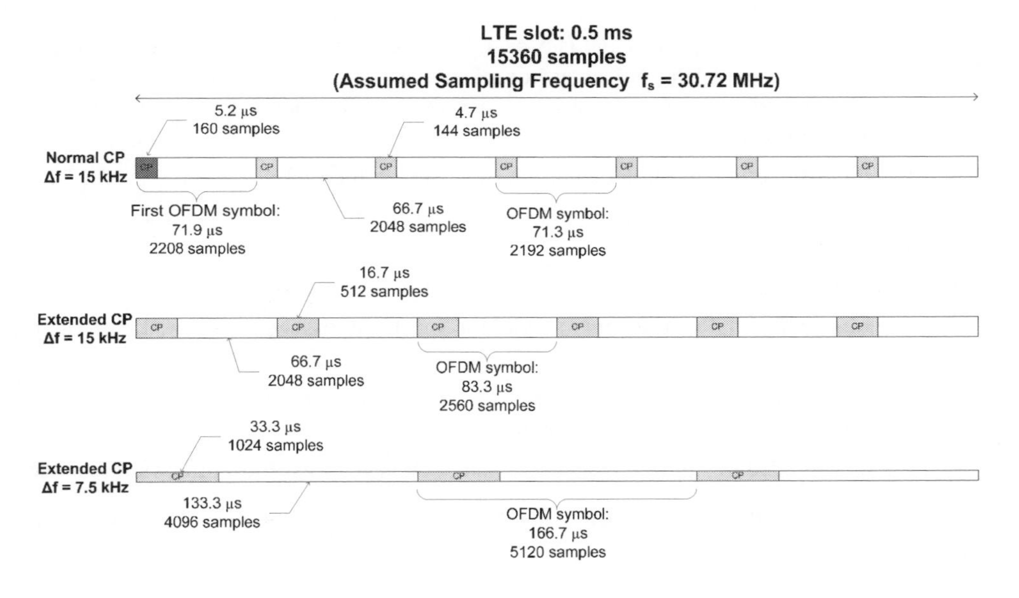

Figure 5.13: LTE OFDM symbol and CP lengths.

The actual FFT size and sampling frequency for the LTE downlink are not specified. However, the above parameterizations are designed to be compatible with a sampling frequency of 30.72 MHz. Thus, the basic unit of time in the LTE specifications, of which all other time periods are a multiple, is defined as $T_s = 1/30.72$ μs. This is itself chosen for backward compatibility with UMTS, for which the chip rate is 3.84 MHz – exactly one eighth of the assumed LTE sampling frequency.

In the case of a 20 MHz system bandwidth, an FFT order of 2048 may be assumed for efficient implementation. However, in practice the implementer is free to use other DFT sizes.

Lower sampling frequencies (and proportionally lower FFT orders) are always possible to reduce RF and baseband processing complexity for narrower bandwidth deployments: for example, for a 5 MHz system bandwidth the FFT order and sampling frequency could be scaled down to 512 and $f_s = 7.68$ MHz respectively, while only 300 subcarriers are actually modulated with data.

For the sake of simplifying terminal implementation, the direct current (d.c.) subcarrier is left unused, in order to avoid d.c. offset errors in direct conversion receivers.

The OFDMA parameters used in the downlink are defined in the 3GPP Technical Specification 36.211 [22, Section 6].

5.5 Summary

In this chapter, we have reviewed the key features, benefits and sensitivities of OFDM and OFDMA systems. In summary, it can be noted that:

- OFDM is a mature technology.
- It is already widely deployed and is especially suited for broadcast or downlink applications because of the low receiver complexity while requiring a higher transmitter complexity (expensive PA). The low receiver complexity also makes it well-suited to MIMO schemes.
- It benefits from efficient implementation by means of the FFT.
- It achieves the high transmission rates of broadband transmission, with low receiver complexity.
- It makes use of a CP to avoid ISI, enabling block-wise processing.
- It exploits orthogonal subcarriers to avoid the spectrum wastage associated with inter-subcarrier guard-bands.
- The parameterization allows the system designer to balance tolerance of Doppler and delay spread depending on the deployment scenario.
- It can be extended to a multiple-access scheme, OFDMA, in a straightforward manner.

These factors together have made OFDMA the technology of choice for the LTE downlink.

References[4]

[1] R. W. Chang, 'Synthesis of Band-limited Orthogonal Signals for Multichannel Data Transmission'. *Bell Systems Technical Journal*, Vol. 46, pp. 1775–1796, December 1966.

[2] B. R. Saltzberg, 'Performance of an Efficient Parallel Data Transmission System'. *IEEE Trans. on Communications*, Vol. 15, pp. 805–811, December 1967.

[3] S. B. Weinstein and P. M. Ebert, 'Data Transmission by Frequency-Division Multiplexing using the Discrete Fourier Transform'. *IEEE Trans. on Communications*, Vol. 19, pp. 628–634, October 1971.

[4] A. Peled and A. Ruiz, 'Frequency Domain Data Transmission using Reduced Computational Complexity Algorithms' in *Proc. IEEE International Conference on Acoustics, Speech and Signal Processing*, Vol. 5, pp. 964–967, April 1980.

[4] All web sites confirmed 1st March 2011.

[5] L. J. Cimini, 'Analysis and Simulation of Digital Mobile Channel using Orthogonal Frequency Division Multiplexing'. *IEEE Trans. on Communications*, Vol. 33, pp. 665–675, July 1985.

[6] G. H. Golub and C. F. Van Loan, *Matrix Computations*. John Hopkins University Press, 1996.

[7] R. A. Horn and C. R. Johnson, *Matrix Analysis*. Cambridge University Press, 1990.

[8] S. N. Bernstein, 'On the Work of P. L. Chebyshev in Probability Theory'. *The Scientific Legacy of P. L. Chebyshev. First Part: Mathematics*, edited by S. N. Bernstein. Academiya Nauk SSSR, Moscow-Leningrad, p. 174, 1945.

[9] T. Henk, *Understanding Probability: Chance Rules in Everyday Life*. Cambridge University Press, 2004.

[10] X. Li and L. J. Cimini, 'Effects of Clipping and Filtering on the Performance of OFDM'. *IEEE Comm. Lett.*, Vol. 2, pp. 131–133, May 1998.

[11] L. Wane and C. Tellambura, 'A Simplified Clipping and Filtering Technique for PAPR Reduction in OFDM Systems'. *IEEE Sig. Proc. Lett.*, Vol. 12, pp. 453–456, June 2005.

[12] J. Armstrong, 'Peak to Average Power Reduction for OFDM by Repeated Clipping and Frequency Domain Filtering'. *Electronics Letters*, Vol. 38, pp. 246–247, February 2002.

[13] A. Jayalath, 'OFDM for Wireless Broadband Communications (Peak Power Reduction, Spectrum and Coding)'. PhD thesis, School of Computer Science and Software Engineering, Monash University, 2002.

[14] A. Jones, T. Wilkinson, and S. Barton, 'Block Coding Scheme for Reduction of Peak to Mean Envelope Power Ratio of Multicarrier Transmission Schemes'. *Electronics Letters*, Vol. 30, pp. 2098–2099, December 1994.

[15] C. Tellambura, 'Use of M-sequence for OFDM Peak-to-Average Power Ratio Reduction'. *Electronics Letters*, Vol. 33, pp. 1300–1301, July 1997.

[16] Y. Li and L. J. Cimini, 'Bounds on the Interchannel Interference of OFDM in Time-Varying Impairments'. *IEEE Trans. on Communications*, Vol. 49, pp. 401–404, March 2001.

[17] X. Cai and G. B. Giannakis, 'Bounding Performance and Suppressing Intercarrier Interference in Wireless Mobile OFDM'. *IEEE Trans. on Communications*, Vol. 51, pp. 2047–2056, March 2003.

[18] J. Heiskala and J. Terry, *OFDM Wireless LANs: A Theoretical and Practical Guide*. SAMS Publishing, 2001.

[19] L. Hanzo and T. Keller, *OFDM And MC-CDMA: A Primer*. Wiley-IEEE Press, 2006.

[20] A. G. Burr, 'Irreducible BER of COFDM on IIR channel'. *Electronics Letters*, Vol. 32, pp. 175–176, February 1996.

[21] R. Nogueroles, M. Bossert, A. Donder, and V. Zyablov, 'Improved Performance of a Random OFDMA Mobile Communication System' in *Proc. IEEE Vehicular Technology Conference*, Vol. 3, pp. 2502–2506, May 1998.

[22] 3GPP Technical Specification 36.311, 'Evolved Universal Terrestrial Radio Access (E-UTRA); Physical Channels and Modulation', www.3gpp.org.

6

Introduction to Downlink Physical Layer Design

Matthew Baker

6.1 Introduction

The LTE downlink transmissions from the eNodeB consist of user-plane and control-plane data from the higher layers in the protocol stack (as described in Chapters 3 and 4) multiplexed with physical layer signalling to support the data transmission. The multiplexing of these various parts of the downlink signal is facilitated by the Orthogonal Frequency Division Multiple Access (OFDMA) structure described in Chapter 5, which enables the downlink signal to be subdivided into small units of time and frequency.

This subdivided structure is introduced below, together with an outline of the general steps in forming the transmitted downlink signal in the physical layer.

6.2 Transmission Resource Structure

The downlink transmission resources in LTE possess dimensions of time, frequency and space. The spatial dimension, measured in 'layers', is accessed by means of multiple 'antenna ports' at the eNodeB; for each antenna port a Reference Signal (RS) is provided to enable the User Equipment (UE) to estimate the radio channel (see Section 8.2); the techniques for using multiple antenna ports to exploit multiple spatial layers are explained in Section 11.2.

The time-frequency resources for each transmit antenna port are subdivided according to the following structure: the largest unit of time is the 10 ms radio frame, which is subdivided into ten 1 ms subframes, each of which is split into two 0.5 ms slots. Each slot comprises seven OFDM symbols in the case of the normal Cyclic Prefix (CP) length, or six if the

LTE – The UMTS Long Term Evolution: From Theory to Practice, Second Edition.
Stefania Sesia, Issam Toufik and Matthew Baker.

extended CP is configured in the cell as explained in Section 5.4.1. In the frequency domain, resources are grouped in units of 12 subcarriers (thus occupying a total of 180 kHz with a subcarrier spacing of 15 kHz), such that one unit of 12 subcarriers for a duration of one slot is termed a Resource Block (RB).[1]

The smallest unit of resource is the Resource Element (RE), which consists of one subcarrier for a duration of one OFDM symbol. An RB thus comprises 84 REs in the case of the normal cyclic prefix length, and 72 REs in the case of the extended cyclic prefix.

The detailed resource structure is shown in Figure 6.1 for the normal cyclic prefix length.

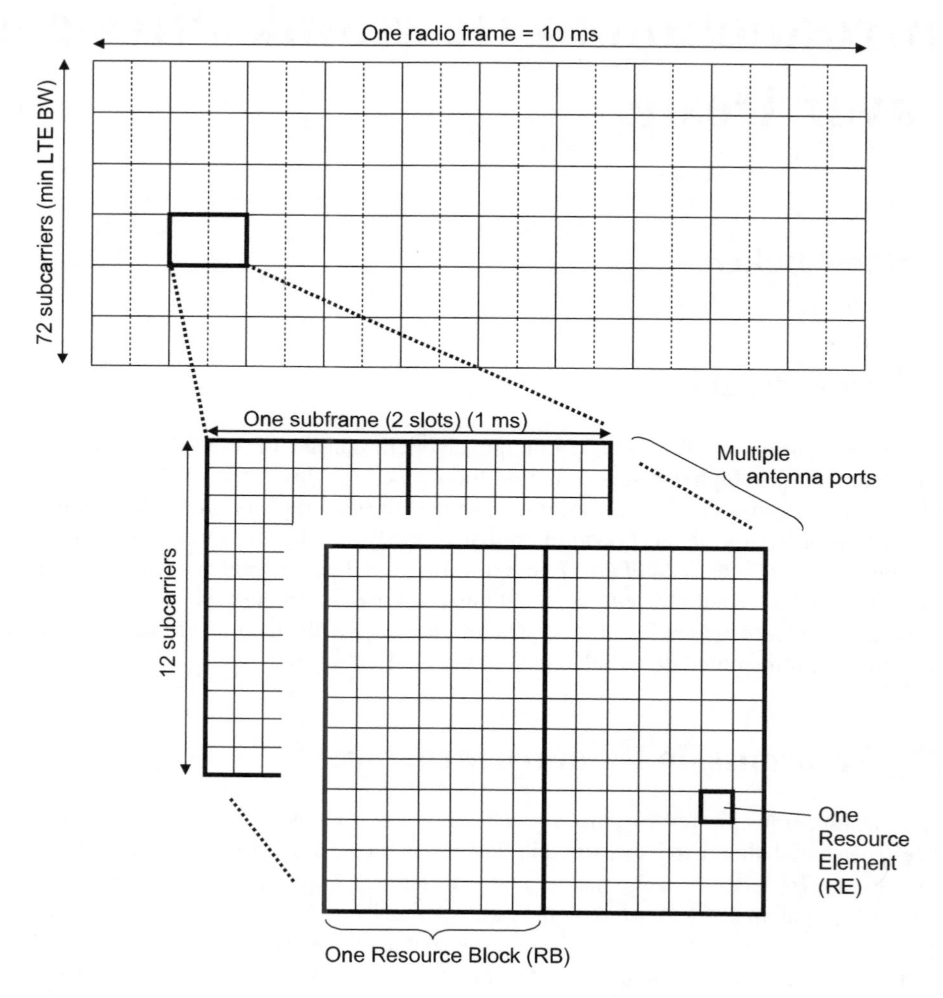

Figure 6.1: Basic time-frequency resource structure of LTE (normal cyclic prefix case).

[1]In the case of the 7.5 kHz subcarrier spacing which may be available for Multimedia Broadcast Multicast Service (MBMS) transmission in later releases of LTE (see Section 5.4.1 and Chapter 13), one RB consists of 24 subcarriers for a duration of one slot.

Within certain RBs, some REs are reserved for special purposes: synchronization signals (Chapter 7), Reference Signals (RSs – see Chapter 8), control signalling and critical broadcast system information (Chapter 9). The remaining REs are used for data transmission, and are usually allocated in pairs of RBs (the pairing being in the time domain).

The structure shown in Figure 6.1 assumes that all subframes are available for downlink transmission. This is known as 'Frame Structure Type 1' and is applicable for Frequency Division Duplexing (FDD) in paired radio spectrum, or for a standalone downlink carrier. For Time Division Duplexing (TDD) in unpaired radio spectrum, the basic structure of RBs and REs remains the same, but only a subset of the subframes are available for downlink transmission; the remaining subframes are used for uplink transmission, or for special subframes which contain a guard period to allow for switching between downlink and uplink transmission. The guard period allows the uplink transmission timing to be advanced as described in Section 18.2. This TDD structure is known as 'Frame Structure Type 2', of which seven different configurations are defined, as shown in Figure 6.2; these allow a variety of downlink-uplink ratios and switching periodicities. Further details of TDD operation using this frame structure are described in Chapter 23.

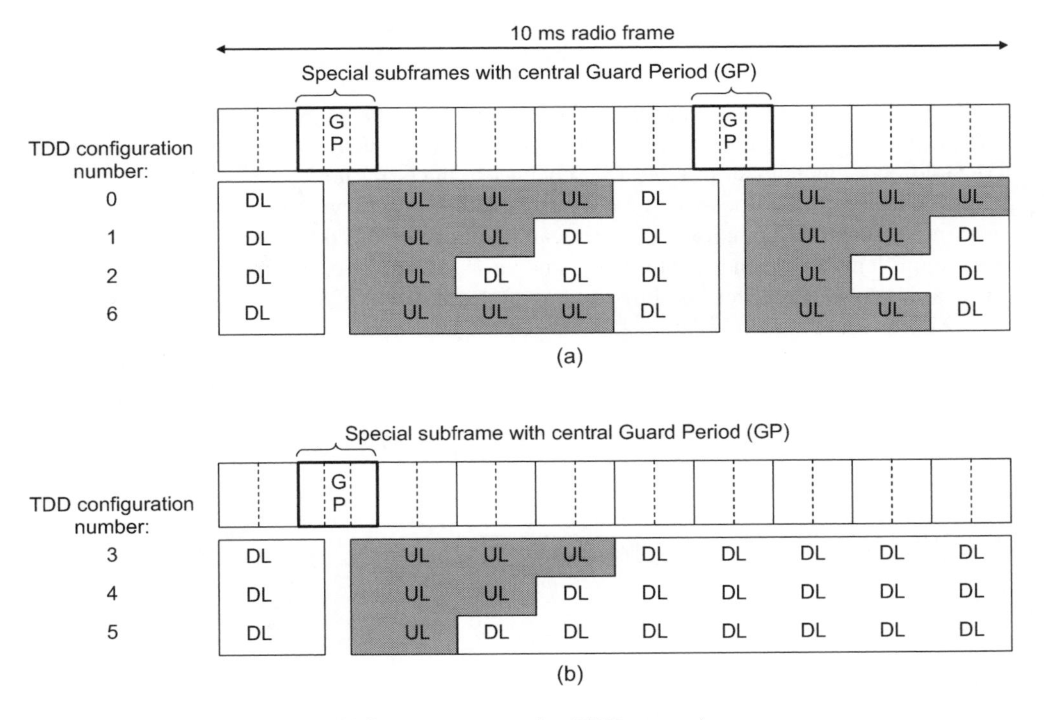

Figure 6.2: LTE subframe structure for TDD operation:
(a) configurations with 5 ms periodicity of switching from downlink (DL) to uplink (UL);
(b) configurations with 10 ms periodicity of switching from downlink (DL) to uplink (UL).

6.3 Signal Structure

The role of the physical layer is primarily to translate data into a reliable signal for transmission across the radio interface between the eNodeB and the User Equipment (UE). Each block of data is first protected against transmission errors, usually first with a Cyclic Redundancy Check (CRC), and then with channel coding, to form a *codeword* (see Chapter 10). After channel coding, the steps in the formation of the downlink LTE signal on a given carrier are illustrated in Figure 6.3.

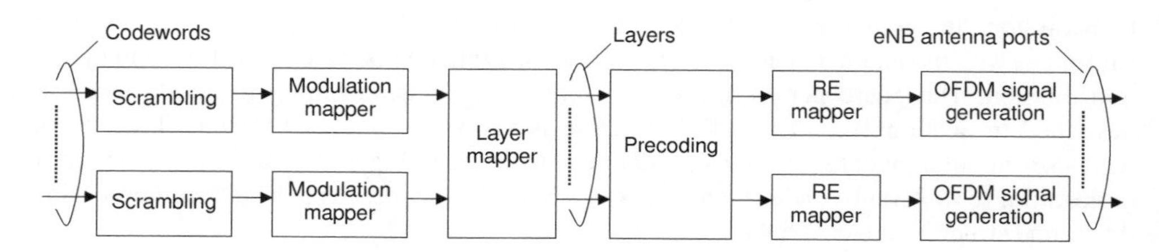

Figure 6.3: General signal structure for LTE downlink. Reproduced by permission of © 3GPP.

The initial scrambling stage is applied to all downlink physical channels, and serves the purpose of interference rejection. The scrambling sequence in all cases uses an order-31 Gold code, which can provide 2^{31} sequences which are not cyclic shifts of each other. Gold codes [1, 2] also possess the attractive feature that they can be generated with very low implementation complexity, as they can be derived from the modulo-2 addition of two maximum-length sequences (otherwise known as *M-sequences*), which can be generated from a simple shift-register.[2] A shift-register implementation of the LTE scrambling sequence generator is illustrated in Figure 6.4.

Figure 6.4: Shift-register implementation of scrambling sequence generator.

The scrambling sequence generator is re-initialized every subframe (except for the Physical Broadcast CHannel (PBCH), which is discussed in Section 9.2.1), based on the

[2]Gold Codes were also used in WCDMA (Wideband Code Division Multiple Access), for the uplink long scrambling codes.

identity of the cell, the subframe number (within a radio frame) and the UE identity. This randomizes interference between cells and between UEs. In addition, in cases where multiple data streams (codewords) are transmitted via multiple layers (see Table 11.2), the identity of the codeword is also used in the initialization.

As a useful feature for avoiding unnecessary complexity, the scrambling sequence generator described here is the same as for the pseudo-random sequence used for the RSs as described in Chapter 8, the only difference being in the method of initialization; in all cases, however, a fast-forward of 1600 places is applied at initialization, in order to ensure low cross-correlation between sequences used in adjacent cells.

Following the scrambling stage, the data bits from each channel are mapped to complex-valued modulation symbols depending on the relevant modulation scheme, then mapped to layers and precoded as explained in Chapter 11, mapped to REs, and finally translated into a complex-valued OFDM signal by means of an Inverse Fast Fourier Transform (IFFT).

6.4 Introduction to Downlink Operation

In order to communicate with an eNodeB supporting one or more cells, the UE must first identify the downlink transmission from one of these cells and synchronize with it. This is achieved by means of special synchronization signals which are embedded into the OFDM structure described above. The procedure for cell search and synchronization is described in Chapter 7.

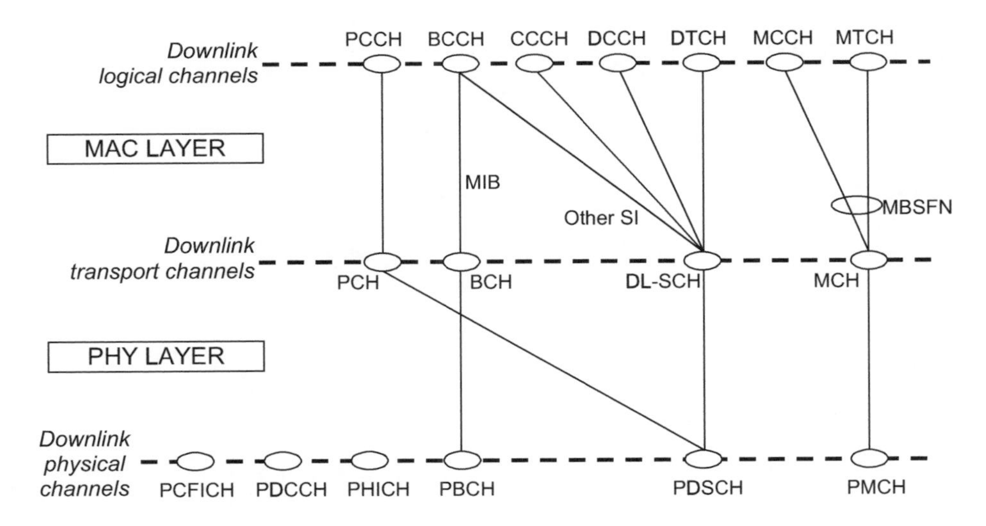

Figure 6.5: Summary of downlink physical channels and mapping to higher layers.

The next step for the UE is to estimate the downlink radio channel in order to be able to perform coherent demodulation of the information-bearing parts of the downlink signal. Some suitable techniques for channel estimation are described in Chapter 8, based on the reference signals which are inserted into the downlink signal.

Chapter 9 describes the parts of the downlink signal which carry data originating from higher protocol layers, including the PBCH, the Physical Downlink Shared Channel (PDSCH) and, in the case of MBMS transmission from Release 9 onwards, the Physical Multicast CHannel (PMCH). In addition, the design of the downlink control signalling is explained, including its implications for the ways in which downlink transmission resources may be allocated to different users for data transmission.

The downlink physical channels described in the following chapters are summarized in Figure 6.5, together with their relationship to the higher-layer channels.

The subsequent chapters explain the key techniques which enable these channels to make efficient use of the radio spectrum: channel coding and link adaptation are explained in Chapter 10, the LTE schemes for exploiting multiple antennas are covered in Chapter 11, and techniques for effective scheduling of transmission resources to multiple users are described in Chapter 12.

References[3]

[1] W. Huffman and V. Pless, *Handbook of Coding Theory, vol II*. Amsterdam: North-Holland, 1998.

[2] R. McEliece, *Finite Fields for Computer Scientists and Engineers*. Boston, MA: Kluwer Academic Publishers, 2003.

[3] All web sites confirmed 1st March 2011.

7

Synchronization and Cell Search

Fabrizio Tomatis and Stefania Sesia

7.1 Introduction

A User Equipment (UE) wishing to access an LTE cell must first undertake a *cell search procedure*. This chapter focuses on the aspects of the physical layer that are designed to facilitate cell search. The way this relates to the overall mobility functionality and protocol aspects for cell reselection and handover is explained in Chapters 3 and 22. The performance requirements related to cell search and synchronization are explained in Section 22.2.

At the physical layer, the cell search procedure consists of a series of synchronization stages by which the UE determines time and frequency parameters that are necessary to demodulate the downlink and to transmit uplink signals with the correct timing. The UE also acquires some critical system parameters.

Three major synchronization requirements can be identified in the LTE system:

1. Symbol and frame timing acquisition, by which the correct symbol start position is determined, for example to set the Discrete Fourier Transform (DFT) window position;
2. Carrier frequency synchronization, which is required to reduce or eliminate the effect of frequency errors[1] arising from a mismatch of the local oscillators between the transmitter and the receiver, as well as the Doppler shift caused by any UE motion;
3. Sampling clock synchronization.

7.2 Synchronization Sequences and Cell Search in LTE

The cell search procedure in LTE begins with a synchronization procedure which makes use of two specially designed physical signals that are broadcast in each cell: the Primary Synchronization Signal (PSS) and the Secondary Synchronization Signal (SSS). The detection of

[1]Frequency offsets may arise from factors such as temperature drift, ageing and imperfect calibration.

LTE – The UMTS Long Term Evolution: From Theory to Practice, Second Edition.
Stefania Sesia, Issam Toufik and Matthew Baker.

these two signals not only enables time and frequency synchronization, but also provides the UE with the physical layer identity of the cell and the cyclic prefix length, and informs the UE whether the cell uses Frequency Division Duplex (FDD) or Time Division Duplex (TDD).

In the case of initial synchronization (when the UE is not already camping on or connected to an LTE cell) after detecting the synchronization signals, the UE decodes the Physical Broadcast CHannel (PBCH), from which critical system information is obtained (see Section 9.2.1). In the case of neighbour cell identification, the UE does not need to decode the PBCH; it simply makes quality-level measurements based on the reference signals (see Chapter 8) transmitted from the newly detected cell and uses them for cell reselection (in RRC_IDLE state) or handover (in RRC_CONNECTED state); in the latter case, the UE reports these measurements to its serving cell.

The cell search and synchronization procedure is summarized in Figure 7.1, showing the information ascertained by the UE at each stage. The PSS and SSS structure is specifically designed to facilitate this acquisition of information.

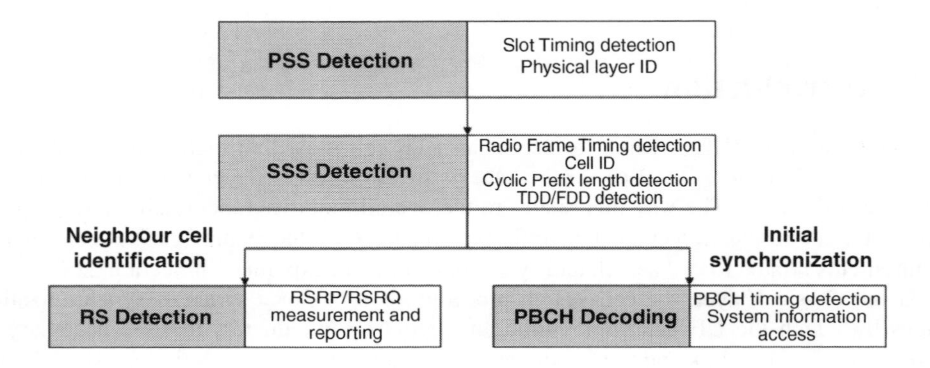

Figure 7.1: Information acquired at each step of the cell search procedure.

The PSS and SSS structure in time is shown in Figure 7.2 for the FDD case and in Figure 7.3 for TDD: the synchronization signals are transmitted periodically, twice per 10 ms radio frame. In an FDD cell, the PSS is always located in the last OFDM (Orthogonal Frequency Division Multiplexing) symbol of the first and 11th slots of each radio frame (see Chapter 6), thus enabling the UE to acquire the slot boundary timing independently of the Cyclic Prefix (CP) length. The SSS is located in the symbol immediately preceding the PSS, a design choice enabling coherent detection[2] of the SSS relative to the PSS, based on the assumption that the channel coherence duration is significantly longer than one OFDM symbol. In a TDD cell, the PSS is located in the third symbol of the 3rd and 13th slots, while the SSS is located three symbols earlier; coherent detection can be used under the assumption that the channel coherence time is significantly longer than four OFDM symbols.

The precise position of the SSS changes depending on the length of the CP configured for the cell. At this stage of the cell detection process, the CP length is unknown a priori

[2]See Section 7.3.

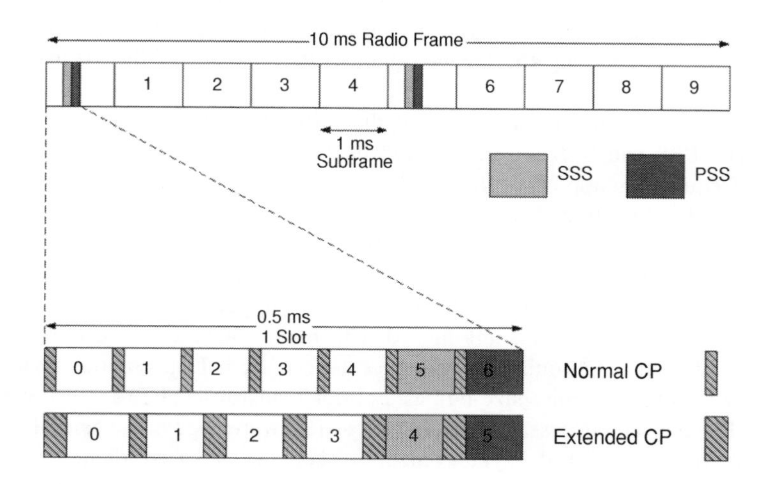

Figure 7.2: PSS and SSS frame and slot structure in time domain in the FDD case.

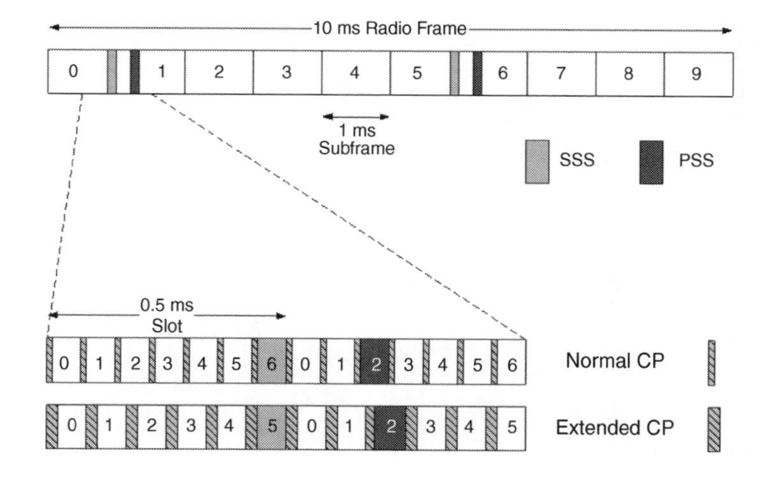

Figure 7.3: PSS and SSS frame and slot structure in time domain in the TDD case.

to the UE, and it is, therefore, blindly detected by checking for the SSS at the two possible positions.[3]

[3]Hence a total of four possible SSS positions must be checked if the UE is searching for both FDD and TDD cells.

While the PSS in a given cell is the same in every subframe in which it is transmitted, the two SSS transmissions in each radio frame change in a specific manner as described in Section 7.2.3, thus enabling the UE to establish the position of the 10 ms radio frame boundary.

In the frequency domain, the mapping of the PSS and SSS to subcarriers is shown in Figure 7.4. The PSS and SSS are transmitted in the central six Resource Blocks[4] (RBs), enabling the frequency mapping of the synchronization signals to be invariant with respect to the system bandwidth (which can in principle vary from 6 to 110 RBs to suit channel bandwidths between around 1.4 MHz and 20 MHz); this allows the UE to synchronize to the network without any a priori knowledge of the allocated bandwidth. The PSS and SSS are each comprised of a sequence of length 62 symbols, mapped to the central 62 subcarriers around the d.c. subcarrier, which is left unused. This means that the five resource elements at each extremity of each synchronization sequence are not used. This structure enables the UE to detect the PSS and SSS using a size-64 Fast Fourier Transform (FFT) and a lower sampling rate than would have been necessary if all 72 subcarriers were used in the central six resource blocks. The shorter length for the synchronization sequences also avoids the possibility in a TDD system of a high correlation with the uplink demodulation reference signals which use the same kind of sequence as the PSS (see Chapter 15).

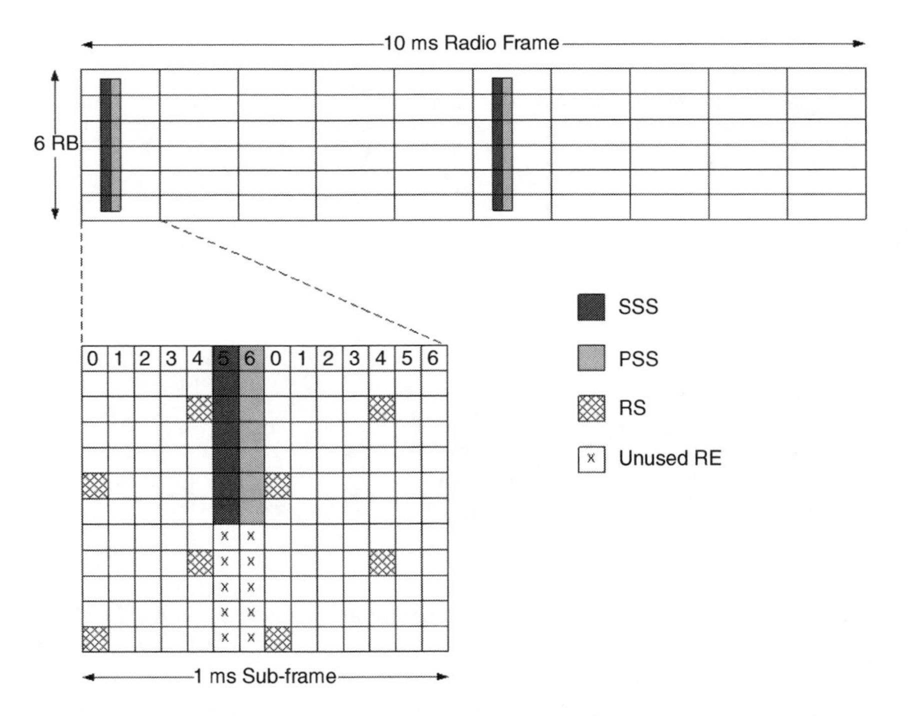

Figure 7.4: PSS and SSS frame structure in frequency and time domain for an FDD cell.

[4]Six Resource Blocks correspond to 72 subcarriers.

In the case of multiple transmit antennas being used at the eNodeB, the PSS and SSS are always transmitted from the same antenna port[5] in any given subframe, while between different subframes they may be transmitted from different antenna ports in order to benefit from time-switched antenna diversity.

The particular sequences which are transmitted for the PSS and SSS in a given cell are used to indicate the physical layer cell identity to the UE. There are 504 unique Physical Cell Identities (PCIs) in LTE, grouped into 168 groups of three identities. The three identities in a group would usually be assigned to cells under the control of the same eNodeB. Three PSS sequences are used to indicate the cell identity within the group, and 168 SSS sequences are used to indicate the identity of the group.[6]

The PSS uses sequences known as *Zadoff–Chu*. This kind of sequence is widely used in LTE, including for the uplink reference signals and random access preambles (see Chapters 15 and 17) in addition to the PSS. Therefore, Section 7.2.1 is devoted to an explanation of the fundamental principles behind Zadoff–Chu sequences, before discussing the specific constructions of the PSS and SSS sequences in the subsequent sections.

7.2.1 Zadoff–Chu Sequences

Zadoff–Chu (ZC) sequences (also known as *Generalized Chirp-Like* (GCL) sequences) are named after the authors of [1] and [2]. ZC sequences are non-binary unit-amplitude sequences [3], which satisfy a Constant Amplitude Zero Autocorrelation (CAZAC) property. CAZAC sequences are complex signals of the form $e^{j\alpha_k}$. The ZC sequence of odd-length N_{ZC} is given by

$$a_q(n) = \exp\left[-j2\pi q \frac{n(n+1)/2 + ln}{N_{\mathrm{ZC}}}\right] \tag{7.1}$$

where $q \in \{1, \ldots, N_{\mathrm{ZC}} - 1\}$ is the ZC sequence root index, $n = 0, 1, \ldots, N_{\mathrm{ZC}} - 1$, $l \in \mathbb{N}$ is any integer. In LTE, $l = 0$ is used for simplicity.

ZC sequences have the following three important properties:

Property 1. A ZC sequence has constant amplitude, and its N_{ZC}-point DFT also has constant amplitude. The constant amplitude property limits the Peak-to-Average Power Ratio (PAPR) and generates bounded and time-flat interference to other users. It also simplifies the implementation as only phases need to be computed and stored, not amplitudes.

Property 2. ZC sequences of any length have 'ideal' cyclic autocorrelation (i.e. the correlation with its circularly shifted version is a delta function). The zero autocorrelation property may be formulated as:

$$r_{kk}(\sigma) = \sum_{n=0}^{N_{\mathrm{ZC}}-1} a_k(n) a_k^*[(n+\sigma)] = \delta(\sigma) \tag{7.2}$$

where $r_{kk}(\cdot)$ is the discrete periodic autocorrelation function of a_k at lag σ. This property is of major interest when the received signal is correlated with a reference sequence and the received reference sequences are misaligned. As an example, Figure 7.5 shows the difference

[5]The concept of an antenna port in LTE is explained in Section 8.2.

[6]A subset of the available PCIs can be reserved for Closed Subscriber Group (CSG) cells (e.g. Home eNodeBs). If this is the case, the reserved PCIs are indicated in System Information Block 4 (SIB4), enabling the UE to eliminate those PCIs from its search if CSG cells are not applicable for it.

between the periodic autocorrelation of a truncated Pseudo-Noise (PN) sequence (as used in WCDMA [4]) and a ZC sequence. Both are 839 symbols long in this example. The ZC periodic autocorrelation is exactly zero for $\sigma \neq 0$ and it is non-zero for $\sigma = 0$, whereas the PN periodic autocorrelation shows significant peaks, some above 0.1, at non-zero lags.

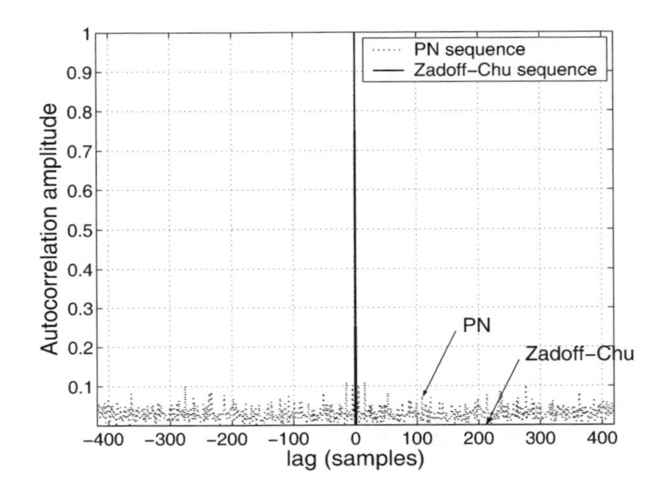

Figure 7.5: Zadoff–Chu versus PN sequence: periodic autocorrelation.

The main benefit of the CAZAC property is that it allows multiple orthogonal sequences to be generated from the same ZC sequence. Indeed, if the periodic autocorrelation of a ZC sequence provides a single peak at the zero lag, the periodic correlation of the same sequence against its cyclic shifted replica provides a peak at lag N_{CS}, where N_{CS} is the number of samples of the cyclic shift. This creates a Zero-Correlation Zone (ZCZ) between the two sequences. As a result, as long as the ZCZ is dimensioned to cope with the largest possible expected time misalignment between them, the two sequences are orthogonal for all transmissions within this time misalignment.

Property 3. The absolute value of the cyclic cross-correlation function between any two ZC sequences is constant and equal to $1/\sqrt{N_{\mathrm{ZC}}}$,[7] if $|q_1 - q_2|$ (where q_1 and q_2 are the sequence indices) is relatively prime with respect to N_{ZC} (a condition that can be easily guaranteed if N_{ZC} is a prime number). The cross-correlation of $\sqrt{N_{\mathrm{ZC}}}$ at all lags achieves the theoretical minimum cross-correlation value for any two sequences that have ideal autocorrelation.

Selecting N_{ZC} as a prime number results in $N_{\mathrm{ZC}} - 1$ ZC sequences which have the optimal cyclic cross-correlation between any pair. However, it is not always convenient to use sequences of prime length. In general, a sequence of non-prime length may be generated by either cyclic extension or truncation of a prime-length ZC sequence. A further useful property of ZC sequences is that the DFT of a ZC sequence $x_u(n)$ (in Equation (7.1)) is a weighted cyclicly shifted ZC sequence $X_{\mathrm{w}}(k)$ such that $w = -1/u \bmod N_{\mathrm{ZC}}$. This means that a ZC sequence can be generated directly in the frequency domain without the need for a DFT operation.

[7] Note that this value corresponds to the normalized cross-correlation function.

7.2.2 Primary Synchronization Signal (PSS) Sequences

The PSS is constructed from a frequency-domain ZC sequence of length 63, with the middle element punctured to avoid transmitting on the d.c. subcarrier.

The mapping of the PSS sequence to the subcarriers is shown in Figure 7.6.

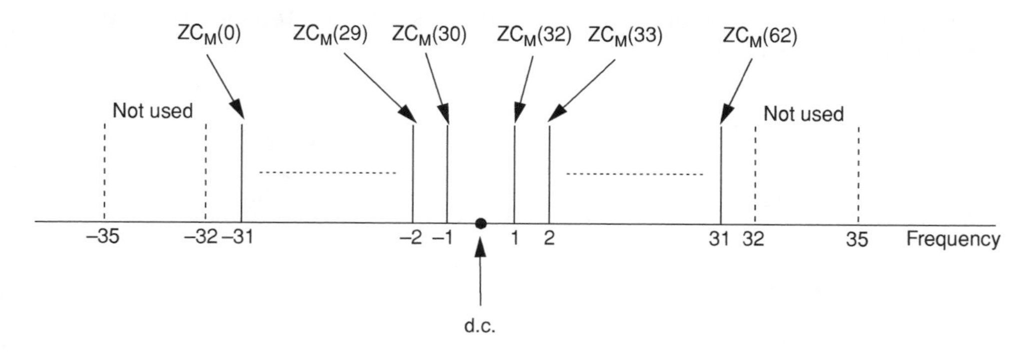

Figure 7.6: PSS sequence mapping in the frequency domain.

Three PSS sequences are used in LTE, corresponding to the three physical layer identities within each group of cells. The selected roots for the three ZC PSS sequences are $M = 29, 34, 25$, such that the frequency-domain length-63 sequence for root M is given by

$$\mathrm{ZC}_M^{63}(n) = \exp\left[-j\frac{\pi M n(n+1)}{63}\right], \quad n = 0, 1, \ldots, 62 \tag{7.3}$$

This set of roots for the ZC sequences was chosen for its good periodic autocorrelation and cross-correlation properties. In particular, these sequences have a low frequency-offset sensitivity, defined as the ratio of the maximum undesired autocorrelation peak in the time domain to the desired correlation peak computed at a certain frequency offset. This allows a certain robustness of the PSS detection during the initial synchronization, as shown in Figure 7.7.

Figures 7.8 and 7.9 show respectively the cross-correlation (for roots 29 and 25) as a function of timing and frequency offset and the autocorrelation as a function of timing offset (for root 29). It can be seen that the average and peak values of the cross-correlation are low relative to the autocorrelation. Furthermore, the ZC sequences are robust against frequency drifts as shown in Figure 7.10. Thanks to the flat frequency-domain autocorrelation property and to the low frequency offset sensitivity, the PSS can be easily detected during the initial synchronization with a frequency offset up to ±7.5 kHz.

From the UE's point of view, the selected root combination satisfies time-domain root-symmetry, in that sequences 29 and 34 are complex conjugates of each other and can be detected with a single correlator, thus allowing for some complexity reduction.

The UE must detect the PSS without any a priori knowledge of the channel, so non-coherent correlation[8] is required for PSS timing detection. A maximum likelihood detector,

[8]See Section 7.3.

as explained in Section 7.3, finds the timing offset m^*_M that corresponds to the maximum correlation, i.e.

$$m^*_M = \text{argmax}_m \left| \sum_{i=0}^{N-1} Y[i+m] S^*_M[i] \right|^2 \tag{7.4}$$

where i is time index, m is the timing offset, N is the PSS time-domain signal length, $Y[i]$ is the received signal at time instant i and $S_M[i]$ is the PSS with root M replica signal at time i as defined in Equation (7.3).

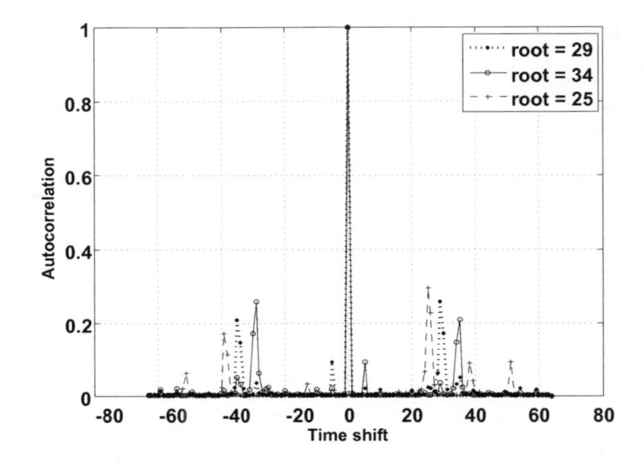

Figure 7.7: Autocorrelation profile at 7.5 kHz frequency offset for roots = 25, 29, 34.

7.2.3 Secondary Synchronization Signal (SSS) Sequences

The SSS sequences are based on maximum length sequences, known as M-sequences, which can be created by cycling through every possible state of a shift register of length n. This results in a sequence of length $2^n - 1$.

Each SSS sequence is constructed by interleaving, in the frequency domain, two length-31 BPSK[9]-modulated secondary synchronization codes, denoted here SSC1 and SSC2, as shown in Figure 7.11.

These two codes are two different cyclic shifts of a single length-31 M-sequence. The cyclic shift indices of the M-sequences are derived from a function of the PCI group (as given in Table 6.11.2.1-1 in [5]). The two codes are alternated between the first and second SSS transmissions in each radio frame. This enables the UE to determine the 10 ms radio frame timing from a single observation of an SSS, which is important for UEs handing over to LTE from another Radio Access Technology (RAT). For each transmission, SSC2 is scrambled by a sequence that depends on the index of SSC1. The sequence is then scrambled by a code that depends on the PSS. The scrambling code is one-to-one mapped to the physical layer

[9]Binary Phase Shift Keying.

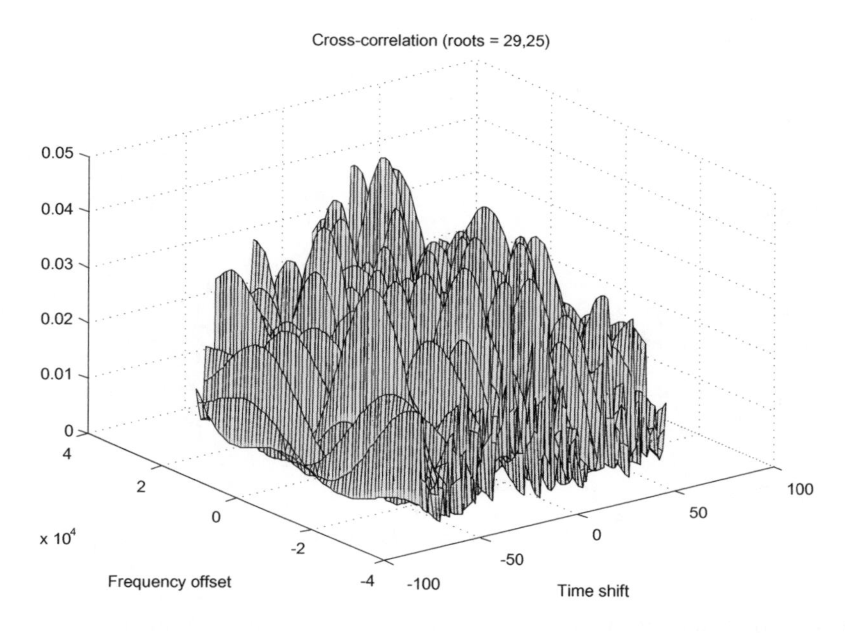

Figure 7.8: Cross-correlation of the PSS sequence pair 25 and 29.

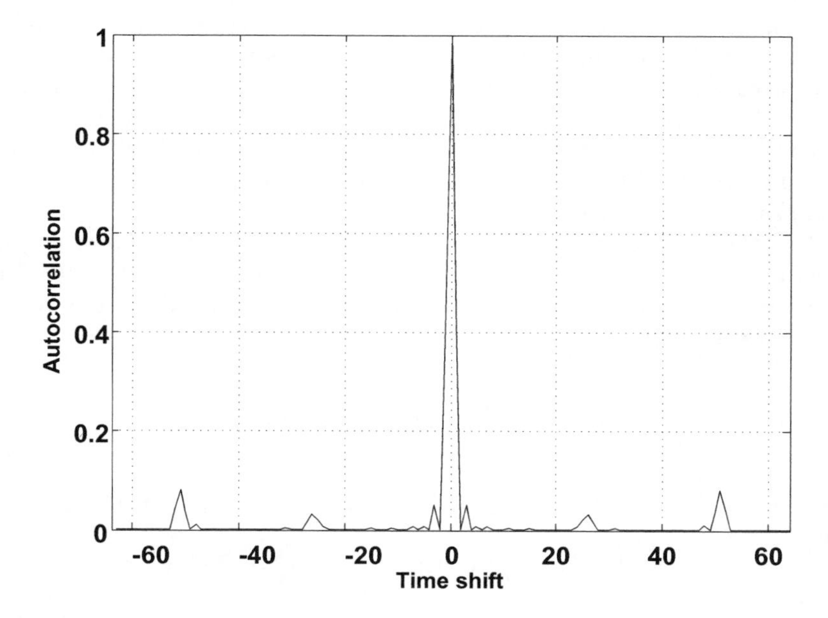

Figure 7.9: Autocorrelation of the PSS sequence 29 as a function of time offset.

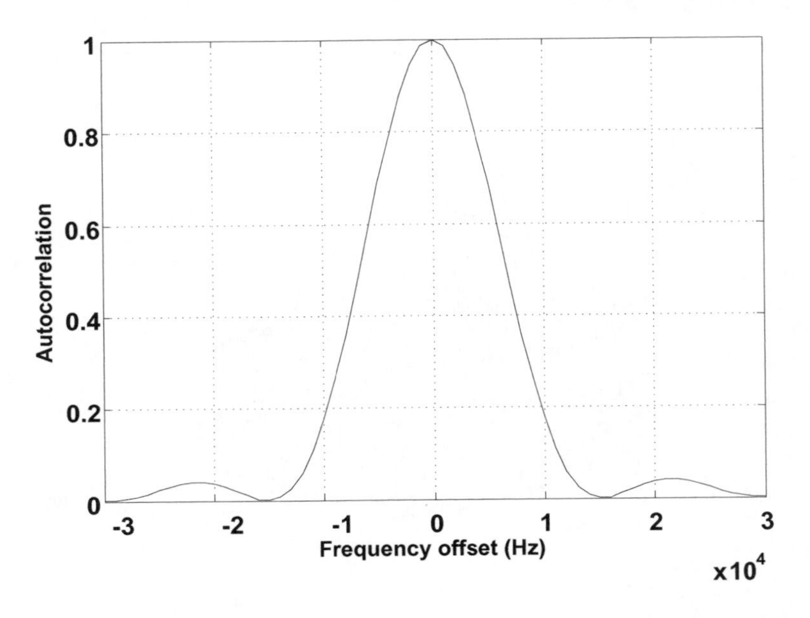

Figure 7.10: Autocorrelation of the PSS sequence as a function of frequency offset.

identity within the group corresponding to the target eNodeB. The sequence construction is illustrated in Figure 7.12; details of the scrambling operations are given in [5].

The SSS sequences have good frequency-domain properties, as shown in Figure 7.13. As in the case of the PSS, the SSS can be detected with a frequency offset up to ±7.5 kHz. In the time domain, the cross-correlation between any cyclic shifts of an SSS sequence is not as good as for classical M-sequences (for which the cross-correlation is known to be -1), owing to the effects of the scrambling operations.

From the UE's point of view, the SSS detection is done after the PSS detection, and the channel can therefore be assumed to be known (i.e. estimated based on the detected PSS sequence). It follows that a coherent detection method, as described in Section 7.3 (Equation (7.8)), can be applied:

$$\hat{\mathbf{S}}_m = \underset{\mathbf{S}}{\text{argmin}}\left(\sum_{n=1}^{N} |\mathbf{y}[n] - \mathbf{S}[n, n]\hat{h}_n|^2\right) \tag{7.5}$$

where the symbols $\mathbf{S}[n, n]$ represent the SSS sequences and $\hat{h}_n$ are the estimated channel coefficients.

However, in the case of synchronized neighbouring eNodeBs, the performance of a coherent detector can be degraded. This is because if an interfering eNodeB employs the same PSS as the one used by the target cell, the phase difference between the two eNodeBs can have an impact on the quality of the estimation of the channel coefficients. On the other hand, the performance of a non-coherent detector degrades if the coherence bandwidth of the channel is less than the six resource blocks occupied by the SSS.

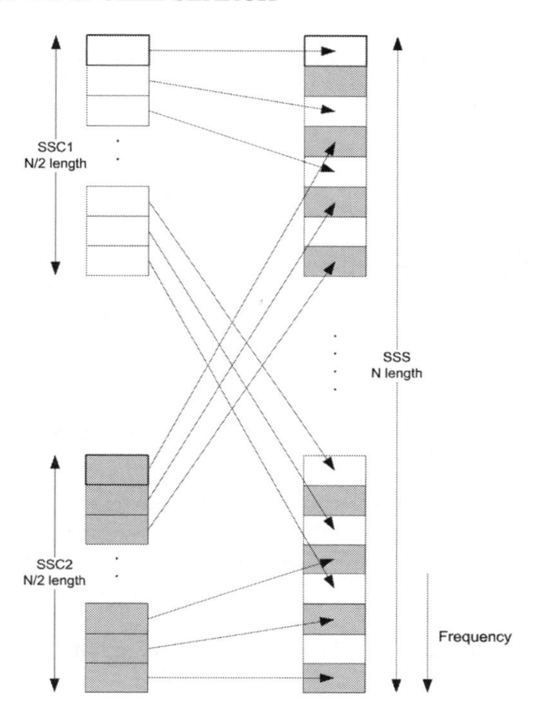

Figure 7.11: SSS sequence mapping.

In order to reduce the complexity of the SSS detector, the equivalence between M-sequence matrices and Walsh–Hadamard matrices can be exploited [6]. Using this property, a fast M-sequence transform is equivalent to a fast Walsh–Hadamard transform with index remapping. Thanks to this property the complexity of the SSS detector is reduced to $N \log_2 N$ where $N = 32$.

7.3 Coherent Versus Non-Coherent Detection

This section gives some theoretical background to the difference between coherent and non-coherent detection.

Both coherent and non-coherent detection may play a part in the synchronization procedures: in the case of the PSS, non-coherent detection is used, while for SSS sequence detection, coherent or non-coherent techniques can be used. From a conceptual point of view, a coherent detector takes advantage of knowledge of the channel, while a non-coherent detector uses an optimization metric corresponding to the average channel statistics.

We consider a generic system model where the received sequence $\mathbf{y}_m$ at time instant m is given by

$$\mathbf{y}_m = \mathbf{S}_m\mathbf{h} + \boldsymbol{\nu}_m \tag{7.6}$$

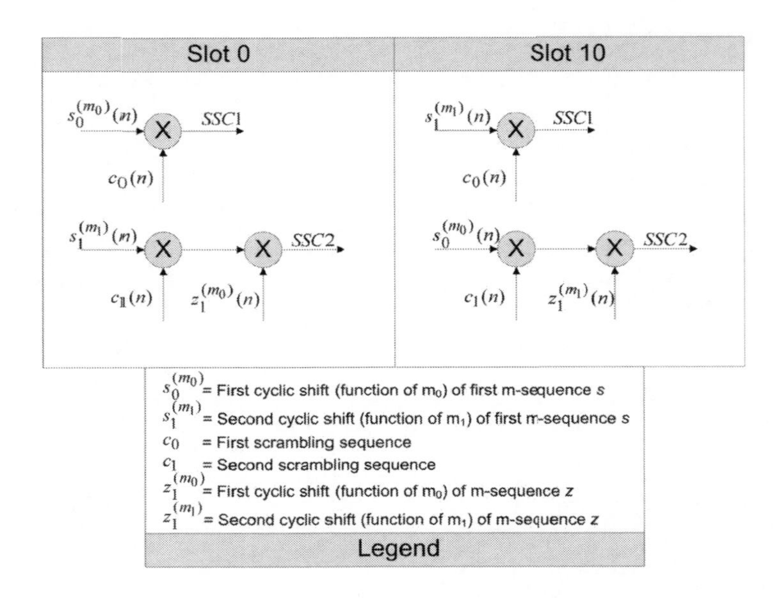

Figure 7.12: SSS sequence generation.

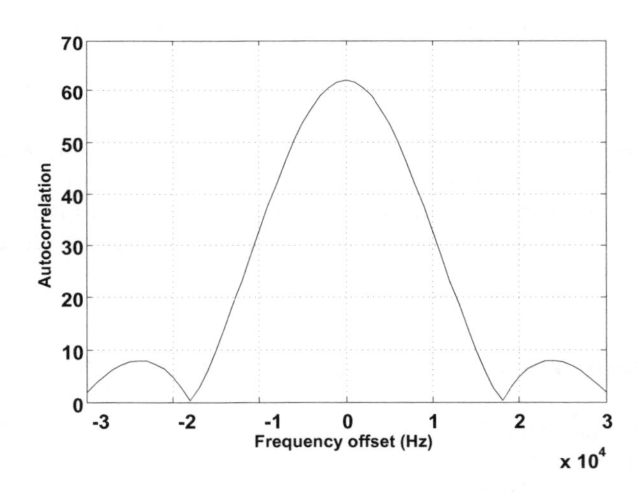

Figure 7.13: Autocorrelation of an SSS sequence as a function of frequency offset.

where the matrix $\mathbf{S}_m = \text{diag}\,[s_{m,1}, \ldots, s_{m,N}]$ represents the transmitted symbol at time instant m, $\mathbf{h} = [h_1, \ldots, h_N]^{\mathrm{T}}$ is the channel vector and $\boldsymbol{\nu}_m$ is zero-mean complex Gaussian noise with variance σ^2.

Maximum Likelihood (ML) coherent detection of a transmitted sequence consists of finding the sequence such that the probability that this sequence was transmitted, conditioned on the knowledge of the channel, is maximized. Coherent detectors therefore require channel estimation to be performed first.

Since the noise is assumed to be independently identically distributed (i.i.d.) and detection is symbol-by-symbol, we focus on one particular symbol interval, neglecting the time instant without loss of generality. The problem becomes

$$\hat{\mathbf{S}} = \underset{\mathbf{S}}{\operatorname{argmax}}\ \Pr(\mathbf{y} \mid \mathbf{S}, \mathbf{h}) = \underset{\mathbf{S}}{\operatorname{argmax}}\ \frac{1}{(\pi N_0)^N} \exp\left[-\frac{\|\mathbf{y} - \mathbf{S}\mathbf{h}\|^2}{N_0}\right] \tag{7.7}$$

$$= \underset{\mathbf{S}}{\operatorname{argmin}} \left(\sum_{n=1}^{N} |\mathbf{y}_n - \mathbf{S}_{n,n} h_n|^2\right) \tag{7.8}$$

Equation (7.8) is a minimum squared Euclidean distance rule where the symbols $\mathbf{S}_{n,n}$ are weighted by the channel coefficient h_n.

When channel knowledge is not available or cannot be exploited, non-coherent detection can be used. The ML detection problem then maximizes the following conditional probability:

$$\hat{\mathbf{S}} = \underset{\mathbf{S}}{\operatorname{argmax}}\ \Pr(\mathbf{y} \mid \mathbf{S})$$

$$= \underset{\mathbf{S}}{\operatorname{argmax}}\ \underset{\mathbf{h}}{\mathbb{E}}[\Pr(\mathbf{y} \mid \mathbf{S}, \mathbf{h})] = \int_{\mathbf{h}} \Pr(\mathbf{y} \mid \mathbf{S}, \mathbf{h}) \Pr(\mathbf{h})\, \mathrm{d}\mathbf{h}$$

$$= \int_{\mathbf{h}} \frac{1}{(\pi\, N_0)^N} \exp\left[-\frac{\|\mathbf{y}_m - \mathbf{S}_m \mathbf{h}\|^2}{N_0}\right] \Pr(\mathbf{h})\, \mathrm{d}\mathbf{h} \tag{7.9}$$

By considering the Probability Density Function (PDF) of an AWGN channel, it can be shown [9] that the ML non-coherent detector yields

$$\hat{\mathbf{S}} = \underset{\mathbf{S}}{\operatorname{argmax}}\ \{\mathbf{y}^{\mathrm{H}} \mathbf{S} (I + N_0 R_{\mathbf{h}}^{-1})^{-1} \mathbf{S}^{\mathrm{H}} \mathbf{y}\} \tag{7.10}$$

where the maximization is done over the input symbols $\mathbf{S}$, so all terms which do not depend on $\mathbf{S}$ can be discarded. Depending on the form of $R_{\mathbf{h}}$, the ML non-coherent detector can be implemented in different ways. For example, in the case of a frequency non-selective channel, the channel correlation matrix can be written as $R_{\mathbf{h}} = \sigma_{\mathbf{h}}^2 \mathbf{V}$ where $\mathbf{V}$ is the all-ones matrix.

Under this assumption, the non-coherent ML detector is thus obtained by the maximization of

$$\hat{\mathbf{S}} = \underset{\mathbf{S}}{\operatorname{argmax}} \left\{ \left| \sum_{i=1}^{N} \mathbf{S}[i, i] \mathbf{y}[i] \right|^2 \right\} \tag{7.11}$$

References[10]

[1] J. D. C. Chu, 'Polyphase Codes with Good Periodic Correlation Properties'. *IEEE Trans. on Information Theory*, Vol. 18, pp. 531–532, July 1972.

[10] All web sites confirmed 1st March 2011.

[2] R. Frank, S. Zadoff and R. Heimiller, 'Phase Shift Pulse Codes With Good Periodic Correlation Properties'. *IEEE Trans. on Information Theory*, Vol. 8, pp. 381–382, October 1962.

[3] B. M. Popovic, 'Generalized Chirp-Like Polyphase Sequences with Optimum Correlation Properties'. *IEEE Trans. on Information Theory*, Vol. 38, pp. 1406–1409, July 1992.

[4] 3GPP Technical Specification 25.213, 'Spreading and modulation (FDD) ', www.3gpp.org.

[5] 3GPP Technical Specification 36.211, 'Physical Channels and Modulation ', www.3gpp.org.

[6] M. Cohn and A. Lempel, 'On Fast M-Sequence Transforms'. *IEEE Trans. on Information Theory*, Vol. 23, pp. 135–137, January 1977.

[7] Texas Instruments, NXP, Motorola, Ericsson, and Nokia, 'R4-072215: Simulation Assumptions for Intra-frequency Cell Identification', www.3gpp.org 3GPP TSG RAN WG4, meeting 45, Jeju, Korea, November 2007.

[8] NXP, 'R4-080691: LTE Cell Identification Performance in Multi-cell Environment', www.3gpp.org 3GPP TSG RAN WG4, meeting 46bis, Shenzen, China, February 2008.

[9] D. Reader, 'Blind Maximum Likelihood Sequence Detection over Fast Fading Communication Channels', Dissertation, University of South Australia, Australia, August 1996.

8

Reference Signals and Channel Estimation

Andrea Ancora, Stefania Sesia and Alex Gorokhov

8.1 Introduction

A simple communication system can be generally modelled as in Figure 8.1, where the transmitted signal **x** passes through a radio channel **H** and suffers additive noise before being received. Mobile radio channels usually exhibit multipath fading, which causes Inter-Symbol Interference (ISI) in the received signal. In order to remove ISI, various kinds of equalization and detection algorithms can be utilized, which may or may not exploit knowledge of the Channel Impulse Response (CIR). Orthogonal Frequency Division Multiple Access (OFDMA) is particularly robust against ISI, thanks to its structure and the use of the Cyclic Prefix (CP) which allows the receiver to perform a low-complexity single-tap scalar equalization in the frequency domain, as described in Section 5.2.1.

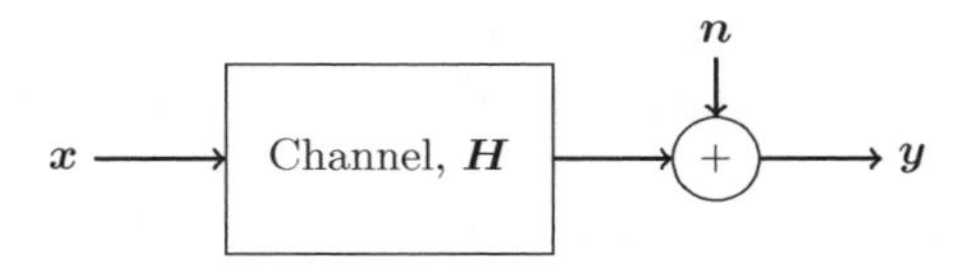

Figure 8.1: A simple transmission model.

As explained in Section 7.3, when the detection method exploits channel knowledge, it is generally said to be 'coherent'; otherwise it is called 'non-coherent'. Coherent detection can make use of both amplitude and phase information carried by the complex signals, and not of only amplitude information as with non-coherent detection. Optimal reception by coherent detection therefore typically requires accurate estimation of the propagation channel.

LTE – The UMTS Long Term Evolution: From Theory to Practice, Second Edition.
Stefania Sesia, Issam Toufik and Matthew Baker.

The main advantage of coherent detection is the simplicity of implementation compared to the more complex algorithms required by non-coherent detection for equivalent performance. However, this simplicity comes at a price, namely the overhead needed in order to be able to estimate the channel. A common and simple way to estimate the channel is to exploit known signals which do not carry any data, but which therefore cause a loss in spectral efficiency. In general, it is not an easy task to find the optimal trade-off between minimizing the spectral efficiency loss due to the overhead and providing adequate ability to track variations in the channel.

Other possible techniques for channel estimation include exploiting the correlation properties of the channel or using blind estimation.

Once synchronization between an eNodeB and a UE has been achieved, LTE (in common with earlier systems such as GSM and UMTS) is a coherent communication system, for which purpose known Reference Signals (RSs) are inserted into the transmitted signal structure.

In general, a variety of methods can be used to embed RSs into a transmitted signal. The RSs can be multiplexed with the data symbols (which are unknown at the receiver) in either the frequency, time or code domains (the latter being used in the case of the common pilot channel in the UMTS downlink). A special case of time multiplexing, known as preamble-based training, involves transmitting the RSs at the beginning of each data burst. Multiplexing-based techniques have the advantage of low receiver complexity, as the data symbol detection is decoupled from the channel estimation problem. Alternatively, RSs may be superimposed on top of the unknown data, without the two necessarily being orthogonal. Note that multiplexing RSs in the code domain is a particular type of superposition with a constraint on orthogonality between known RSs and the unknown data. A comprehensive analysis of the optimization of RS design can be found in [1, 2].

Orthogonal RS multiplexing is by far the most common technique. For example, to facilitate channel estimation in the UMTS downlink, two types of orthogonal RS are provided. The first is code-multiplexed, available to all users in a cell, and uses a specific spreading code which is orthogonal to the codes used to spread the users' data. The second type is time-multiplexed dedicated RSs, which may in some situations be inserted into the users' data streams [3].

In the LTE downlink, the OFDM transmission can be described by a two-dimensional lattice in time and frequency, as shown in Figure 6.1 and described in Chapter 6. This structure facilitates the multiplexing of the RSs, which are mapped to specific Resource Elements (REs) of the two-dimensional lattice in Figure 6.1 according to patterns explained in Section 8.2.

In order to estimate the channel as accurately as possible, all correlations between channel coefficients in time, frequency and space should be taken into account. Since RSs are sent only on particular OFDM REs (i.e. on particular OFDM symbols on particular subcarriers), channel estimates for the REs which do not bear RSs have to be computed via interpolation. The optimal interpolating channel estimator in terms of mean-squared error is based on a two-dimensional Wiener filter interpolation [4]. Due to the high complexity of such a filter, a trade-off between complexity and accuracy is achieved by using one-dimensional filters. In Sections 8.4, 8.5 and 8.6, the problem of channel estimation is approached from a theoretical point of view, and some possible solutions are described.

The work done in the field of channel estimation, and the corresponding literature available, is vast. Nevertheless many challenges still remain, and we refer the interested reader to [2, 5] for general surveys of open issues in this area.

8.2 Design of Reference Signals in the LTE Downlink

In the LTE downlink, five different types of RS are provided [6, Section 6.10]:

- Cell-specific RSs (often referred to as 'common' RSs, as they are available to all UEs in a cell and no UE-specific processing is applied to them);
- UE-specific RSs (introduced in Release 8, and extended in Releases 9 and 10), which may be embedded in the data for specific UEs (also known as DeModulation Reference Signals (DM-RSs);
- MBSFN-specific RSs, which are used only for Multimedia Broadcast Single Frequency Network (MBSFN) operation and are discussed further in Section 13.4.1;
- Positioning RSs, which from Release 9 onwards may be embedded in certain 'positioning subframes' for the purpose of UE location measurements; these are discussed in Section 19.3.1;
- Channel State Information (CSI) RSs, which are introduced in Release 10 specifically for the purpose of estimating the downlink channel state and not for data demodulation (see Section 29.1.2).

Each RS pattern is transmitted from an *antenna port* at the eNodeB. An antenna port may in practice be implemented either as a single physical transmit antenna, or as a combination of multiple physical antenna elements. In either case, the signal transmitted from each antenna port is not designed to be further deconstructed by the UE receiver: the transmitted RS corresponding to a given antenna port defines the antenna port from the point of view of the UE, and enables the UE to derive a channel estimate for all data transmitted on that antenna port – regardless of whether it represents a single radio channel from one physical antenna or a composite channel from a multiplicity of physical antenna elements together comprising the antenna port. The designations of the antenna ports available in LTE are summarized below:

- Antenna Ports 0–3: cell-specific RSs – see Section 8.2.1;
- Antenna Port 4: MBSFN – see Section 13.4.1;
- Antenna Port 5: UE-specific RSs for single-layer beamforming – see Section 8.2.2;
- Antenna Port 6: positioning RSs (introduced in Release 9) – see Section 19.3.1;
- Antenna Ports 7–8: UE-specific RSs for dual-layer beamforming (introduced in Release 9) – see Section 8.2.3;
- Antenna Ports 9–14: UE-specific RSs for multi-layer beamforming (introduced in Release 10) – see Section 29.1.1;
- Antenna Ports 15–22: CSI RSs (introduced in Release 10) – see Section 29.1.2.

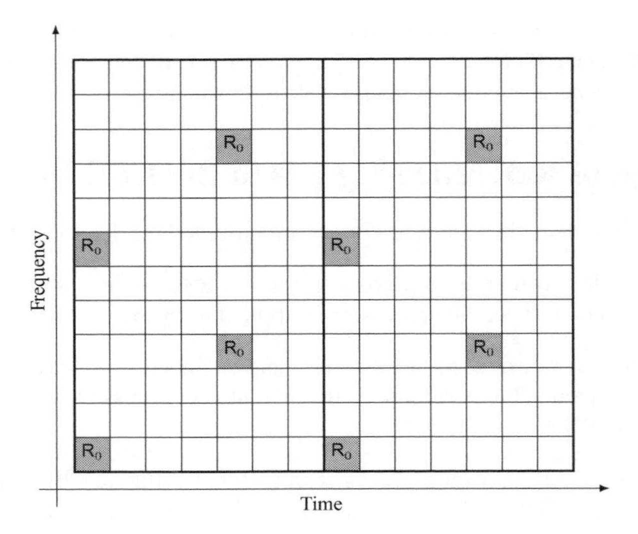

Figure 8.2: Cell-specific reference symbol arrangement in the case of normal CP length for one antenna port. Reproduced by permission of © 3GPP.

Details of the techniques provided for multi-antenna transmission in LTE can be found in Chapter 11 for Releases 8 and 9, and in Chapter 29 for the new techniques introduced in Release 10 for LTE-Advanced.

8.2.1 Cell-Specific Reference Signals

The cell-specific RSs enable the UE to determine the phase reference for demodulating the downlink control channels (see Section 9.3) and the downlink data in most transmission modes[1] of the Physical Downlink Shared Channel (PDSCH – see Section 9.2.2). If UE-specific precoding is applied to the PDSCH data symbols before transmission,[2] downlink control signalling is provided to inform the UE of the corresponding phase adjustment it should apply relative to the phase reference provided by the cell-specific RSs. The cell-specific RSs are also used by the UEs to generate Channel State Information (CSI) feedback (see Sections 10.2.1 and 11.2.2.4).

References [7, 8] show that in an OFDM-based system an equidistant arrangement of reference symbols in the lattice structure achieves the Minimum Mean-Squared Error (MMSE) estimate of the channel. Moreover, in the case of a uniform reference symbol grid, a 'diamond shape' in the time-frequency plane can be shown to be optimal.

In LTE, the arrangement of the REs on which the cell-specific RSs are transmitted follows these principles. Figure 8.2 illustrates the RS arrangement for the normal CP length.[3]

[1]Transmission modes 1 to 6.

[2]In PDSCH transmission modes 3 to 6.

[3]In the case of the extended CP, the arrangement of the reference symbols changes slightly, but the explanations in the rest of this chapter are no less valid. The detailed arrangement of reference symbols for the extended CP can be found in [6].

The LTE system is designed to work under high-mobility assumptions, in contrast to WLAN systems which are generally optimized for pedestrian-level mobility. WLAN systems typically use a preamble-based training sequence, and the degree of mobility such systems can support depends on how often the preamble is transmitted.

The required spacing in time between the reference symbols can be determined by considering the maximum Doppler spread (highest speed) to be supported, which for LTE corresponds to 500 km/h [9]. The Doppler shift is $f_\mathrm{d} = (f_\mathrm{c}\, v/c)$ where f_c is the carrier frequency, v is the UE speed in metres per second, and c is the speed of light ($3 \cdot 10^8$ m/s). Considering $f_\mathrm{c} = 2$ GHz and $v = 500$ km/h, then the Doppler shift is $f_\mathrm{d} \simeq$ 950 Hz. According to Nyquist's sampling theorem, the minimum sampling frequency needed in order to reconstruct the channel is therefore given by $T_\mathrm{c} = 1/(2f_\mathrm{d}) \simeq 0.5$ ms under the above assumptions. This implies that two reference symbols per slot are needed in the time domain in order to estimate the channel correctly.

In the frequency direction, there is one reference symbol every six subcarriers on each OFDM symbol that includes reference symbols, but the reference symbols are staggered so that within each Resource Block (RB) there is one reference symbol every threesubcarriers, as shown in Figure 8.2. This spacing is related to the expected coherence bandwidth of the channel, which is in turn related to the channel delay spread. In particular the 90% and 50% coherence bandwidths[4] are given by $B_{\mathrm{c},90\%} = 1/50\sigma_\tau$ and $B_{\mathrm{c},50\%} = 1/5\sigma_\tau$ respectively, where σ_τ is the r.m.s delay spread. In [10], the maximum r.m.s channel delay spread considered is 991 ns, corresponding to $B_{\mathrm{c},90\%} = 20$ kHz and $B_{\mathrm{c},50\%} = 200$ kHz. In LTE, the spacing between two reference symbols in frequency, in one RB, is 45 kHz, thus allowing the expected frequency-domain variations of the channel to be resolved.

Up to four cell-specific antenna ports, numbered 0 to 3, may be used by an LTE eNodeB, thus requiring the UE to derive up to four separate channel estimates.[5] For each antenna port, a different RS pattern has been designed, with particular attention having been given to the minimization of the intra-cell interference between the multiple transmit antenna ports. In Figure 8.3, R_p indicates that the RE is used for the transmission of an RS on antenna port p. When an RE is used to transmit an RS on one antenna port, the corresponding RE on the other antenna ports is set to zero to limit the interference.

From Figure 8.3, it can be noticed that the density of reference symbols for the third and fourth antenna ports is half that of the first two; this is to reduce the overhead in the system. Frequent reference symbols are useful for high-speed conditions as explained above. In cells with a high prevalence of high-speed users, the use of four antenna ports is unlikely, hence for these conditions reference symbols with lower density can provide sufficient channel estimation accuracy.

All the RSs (cell-specific, UE-specific or MBSFN-specific) are QPSK modulated – a constant modulus modulation. This property ensures that the Peak-to-Average Power Ratio (PAPR) of the transmitted waveform is kept low. The signal can be written as

$$r_{l,n_\mathrm{s}}(m) = \frac{1}{\sqrt{2}}[1 - 2c(2m)] + j\frac{1}{\sqrt{2}}[1 - 2c(2m+1)] \tag{8.1}$$

[4] $B_{\mathrm{c},x\%}$ is the bandwidth where the autocorrelation of the channel in the frequency domain is equal to $x\%$ of the peak.

[5] Any MBSFN and UE-specific RSs, if transmitted, constitute additional independent antenna ports in the LTE specifications.

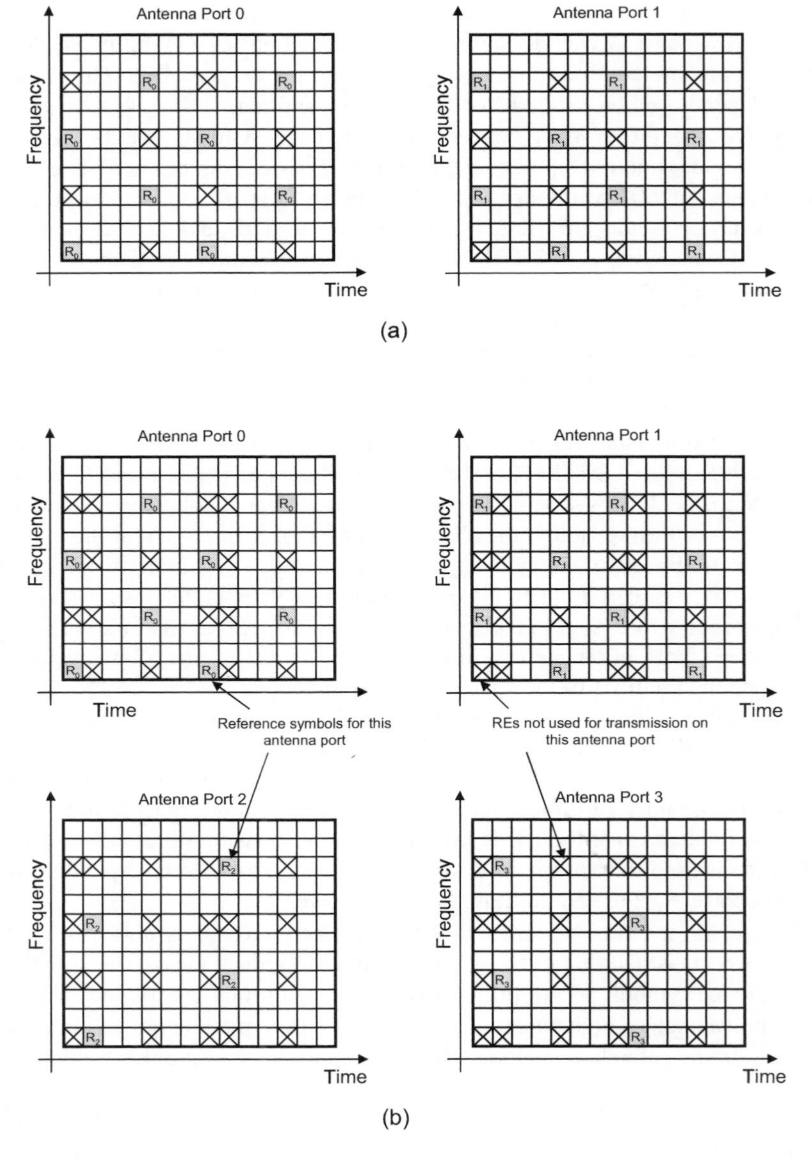

Figure 8.3: Cell-specific RS arrangement in the case of normal CP length for (a) two antenna ports, (b) four antenna ports. Reproduced by permission of © 3GPP.

where m is the index of the RS, n_s is the slot number within the radio frame and 'l' is the symbol number within the time slot. The pseudo-random sequence $c(i)$ is comprised of a length-31 Gold sequence, already introduced in Chapter 6, with different initialization values depending on the type of RSs. For the cell-specific RSs, the sequence is reinitialized at the start of each OFDM symbol, with a value that depends on the cell identity, $N_{\mathrm{ID}}^{\mathrm{cell}}$. The cell-specific RS sequence therefore carries unambiguously one of the 504 different cell identities.

A cell-specific frequency shift is applied to the patterns of reference symbols shown in Figures 8.2 and 8.3, given by $N_{\mathrm{ID}}^{\mathrm{cell}}\mathrm{mod}6$.[6] This shift helps to avoid time-frequency collisions between cell-specific RSs from up to six adjacent cells. Avoidance of collisions is particularly relevant in cases when the transmission power of the RS is boosted, as is possible in LTE up to a maximum of 6 dB relative to the surrounding data symbols. RS power-boosting is designed to improve channel estimation in the cell, but if adjacent cells transmit high-power RSs on the same REs, the resulting inter-cell interference will prevent the benefit from being realized.

8.2.2 UE-Specific Reference Signals in Release 8

In Release 8 of LTE, UE-specific RSs may be transmitted in addition to the cell-specific RSs described above if the UE is configured (by higher-layer RRC signalling) to receive its downlink PDSCH data in transmission mode 7 (see Section 9.2.2.1). The UE-specific RSs are embedded only in the RBs to which the PDSCH is mapped for those UEs. If UE-specific RSs are transmitted, the UE is expected to use them to derive the channel estimate for demodulating the data in the corresponding PDSCH RBs. The same precoding is applied to the UE-specific RSs as to the PDSCH data symbols, and therefore there is no need for signalling to inform the UE of the precoding applied. Thus the UE-specific RSs are treated as being transmitted using a distinct antenna port (number 5), with its own channel response from the eNodeB to the UE.

A typical usage of the UE-specific RSs is to enable beamforming of the data transmissions to specific UEs. For example, rather than using the physical antennas used for transmission of the other (cell-specific) antenna ports, the eNodeB may use a correlated array of physical antenna elements to generate a narrow beam in the direction of a particular UE. Such a beam will experience a different channel response between the eNodeB and UE, thus requiring the use of UE-specific RSs to enable the UE to demodulate the beamformed data coherently. The use of UE-specific beamforming is discussed in more detail in Section 11.2.2.3.

As identified in [11], the structure shown in Figure 8.4 (for the normal CP) has been chosen because there is no collision with the cell specific RSs, and hence the presence of UE-specific RSs does not affect features related to the cell-specific RSs. The UE-specific RSs have a similar pattern to that of the cell-specific RSs, which allows a UE to re-use similar channel estimation algorithms. The density is half that of the cell-specific RS, hence minimizing the overhead. Unlike the cell-specific RSs, the sequence for the UE-specific RSs is only reinitialized at the start of each subframe, as the number of REs to which the sequence is mapped in one OFDM symbol may be very small (in the event of a small number of RBs being transmitted to a UE). The initialization depends on the UE's identity.

The corresponding pattern for use in case of the extended CP being configured in a cell can be found in [6, Section 6.10.3.2].

8.2.3 UE-Specific Reference Signals in Release 9

A new design for UE-specific RSs is defined in Release 9 of the LTE specifications in order to extend UE-specific RS support to dual layer transmission. This includes transmission of

[6]The mod6 operation is used because reference symbols are spaced apart by six subcarriers in the lattice grid.

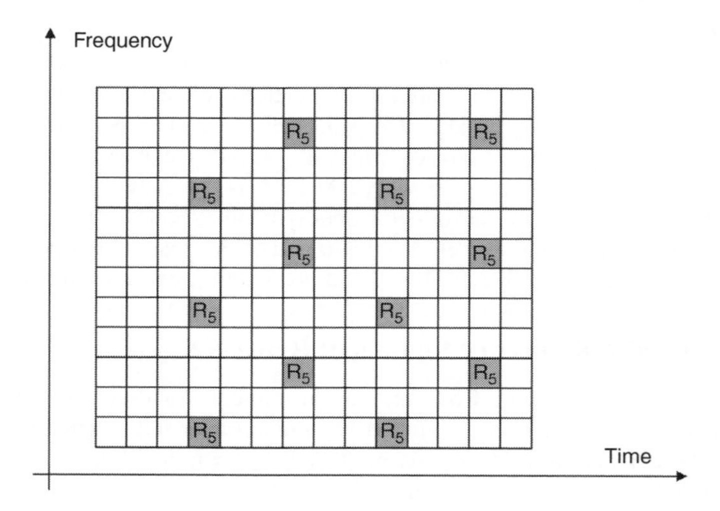

Figure 8.4: Release 8 UE-specific RS arrangement for PDSCH transmission mode 7 with normal CP. Reproduced by permission of © 3GPP.

two spatial layers to one UE, or single-layer transmission to each of two UEs as a Multi-User Multiple Input Multiple Output (MU-MIMO) transmission (see Chapter 11).

Dual layer beamforming in Release 9 is a precursor to higher-rank single- and multi-user transmissions in Release 10, and therefore scalability towards LTE-Advanced (see Section 29.1) was a key factor in the choice of design for the new RSs. This includes enabling efficient inter-cell coordination, which can be facilitated by choosing the set of REs to be non-cell-specific (i.e. without a frequency shift that depends on the cell identity).

A further consideration was backward compatibility with Release 8; this requires coexistence with the 'legacy'physical channels, and therefore the REs carrying the new UE-specific RSs have to avoid the cell-specific RSs and downlink control channels.

A final design requirement was to keep equal power spectral density of the downlink transmission in one RB. This favours designs where the UE-specific RSs for both layers are present in the same OFDM symbol.

Since UE-specific RSs are designed for time-frequency channel estimation within a given RB, the eigen-structure of the time and frequency channel covariance matrix provides insights into the optimal pattern of REs for the RSs for MMSE channel estimation. As an example, Figure 8.5 shows the eigenvalues and three dominant eigenvectors of the time-domain channel covariance matrix for a pair of RBs in a subframe for various mobile speeds assuming spatially uniform (Jakes) scattering. The same behaviour is observed in the frequency domain.

It can be seen that the three principal eigenvectors can be accurately approximated by constant, linear and quadratic functions respectively. Such a channel eigen-structure stems from the time-frequency variations across a pair of RBs being limited. Most of the energy is captured by the constant and linear components, and hence these should be the focus for optimization. While the reference symbol locations do not affect estimation accuracy for the

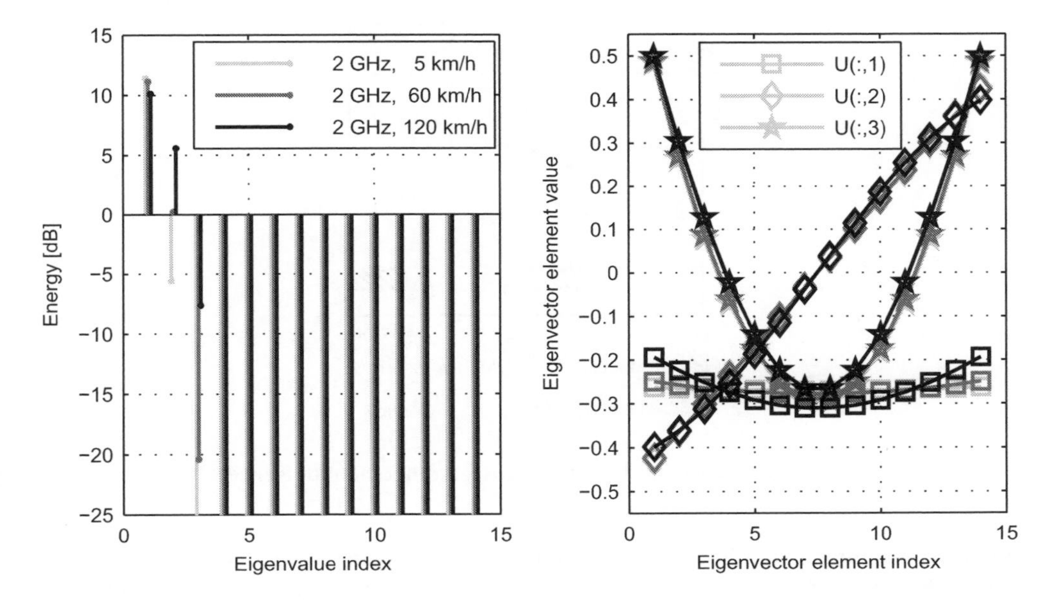

Figure 8.5: Time-domain channel covariance over a pair of RBs: eigenvalues (left) and eigenvectors (right), assuming spatially uniform (Jakes) scattering. The eigenvalue energies are normalized relative to the per-RE energy.

constant component, the best strategy for estimation of the linear component is to allocate RS energy at the edges of the measured region. Finally, the quadratic component is small but non-negligible, warranting three disjoint RS positions across the bandwidth of an RB.

Based on these observations, Figure 8.6 shows the pattern adopted for the new UE-specific RS pattern in Release 9.[7] The reference symbols are positioned in the earliest and latest available pairs of OFDM symbols that avoid collisions with the cell-specific RSs. Pairs of REs are used so that the UE-specific RSs for the two layers can be code-multiplexed. The UE-specific RSs for the two layers using this pattern are termed antenna ports 7 and 8. A UE configured to use the Release 9 dual-layer UE-specific RSs is configured in PDSCH transmission mode 8 (see Section 9.2.2.1).

Length-2 orthogonal Walsh codes are used for the code-multiplexing of the two RS ports.[8] Compared to the use of frequency-multiplexed RS ports, code multiplexing can improve the accuracy of interference estimation and potentially simplifies the implementation by maintaining the same set of RS REs regardless of the number of layers transmitted (thus facilitating dynamic switching between one and two layers).

Finally, the Release 9 UE-specific RS sequence is initialized using only the cell identity (without the UE identity used for the Release 8 UE-specific RSs), in order to enable the two orthogonally code-multiplexed UE-specific RS ports to be used for MU-MIMO (with the two RS ports assigned to different UEs). In addition, two different RS sequence initializations are

[7]Note that distributed RB mapping (see Section 9.2.2.1) is not supported in conjunction with UE-specific RSs.

[8]The mapping of the second Walsh code to each pair of REs reverses from one subcarrier to the next; this is to minimize the Peak-to-Average Power Ratio (PAPR) of the RS transmissions [12].

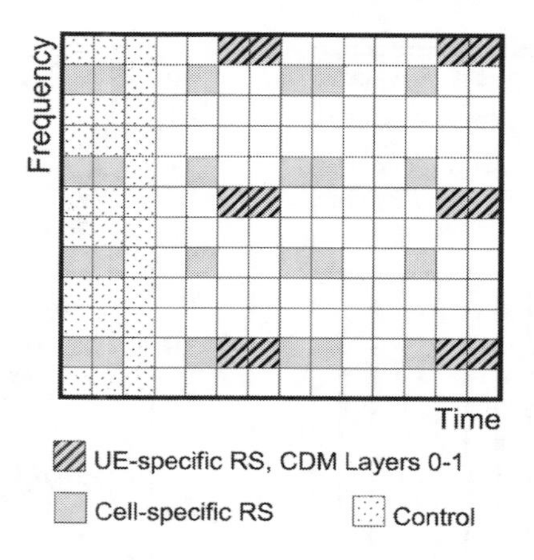

Figure 8.6: Dual-layer UE-specific RS arrangement in Release 9.

provided, enabling non-orthogonal UE-specific RS multiplexing to be used, for example for MU-MIMO with up to dual layer transmission to each UE. An example of a possible usage is shown in Figure 8.7.

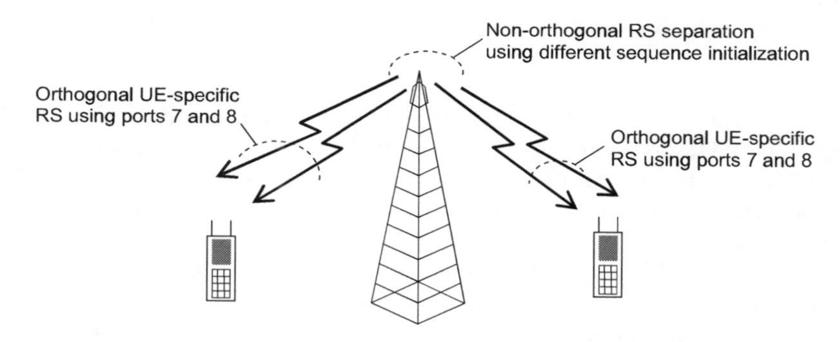

Figure 8.7: Example of usage of orthogonal and non-orthogonal UE-specific RSs in Release 9, for dual-layer transmission to each of two UEs with MU-MIMO.

8.3 RS-Aided Channel Modelling and Estimation

The channel estimation problem is related to the channel model, itself determined by the physical propagation characteristics, including the number of transmit and receive antennas, transmission bandwidth, carrier frequency, cell configuration and relative speed between eNodeB and UE receivers. In general,

- The carrier frequencies and system bandwidth mainly determine the scattering nature of the channel.

- The cell deployment governs its multipath, delay spread and spatial correlation characteristics.
- The relative speed sets the time-varying properties of the channel.

The propagation conditions characterize the channel correlation function in a three-dimensional space comprising frequency, time and spatial domains. In the general case, each MIMO multipath channel component can experience different but related spatial scattering conditions leading to a full three-dimensional correlation function across the three domains. Nevertheless, for the sake of simplicity, assuming that the multipath components of each spatial channel experience the same scattering conditions, the spatial correlation can be assumed to be independent from the other two domains and can be handled separately.

This framework might be suboptimal in general, but is nevertheless useful in mitigating the complexity of channel estimation as it reduces the general three-dimensional joint estimation problem into independent estimation problems.

For a comprehensive survey of MIMO channel estimation, the interested reader is referred to [13]. The following two subsections define the channel model and the corresponding correlation properties which are then used as the basis for an overview of channel estimation techniques.

8.3.1 Time-Frequency-Domain Correlation: The WSSUS Channel Model

The Wide-Sense Stationary Uncorrelated Scattering (WSSUS) channel model is commonly employed for the multipath channels experienced in mobile communications.

Neglecting the spatial dimension for the sake of simplicity, let $h(\tau; t)$ denote the time-varying complex baseband impulse response of a multipath channel realization at time instant t and delay τ.

Considering the channel as a random process in the time direction t, the channel is said to be delay Uncorrelated-Scattered (US) if

$$\mathbb{E}[h(\tau_\mathrm{a}; t_1)^* h(\tau_\mathrm{b}; t_2)] = \phi_h(\tau_\mathrm{a}; t_1, t_2)\delta(\tau_\mathrm{b} - \tau_\mathrm{a}) \tag{8.2}$$

where $\mathbb{E}[\cdot]$ is the expectation operator. According to the US assumption, two CIR components a and b at relative delays τ_a and τ_b are uncorrelated if $\tau_\mathrm{a} \neq \tau_\mathrm{b}$.

The channel is Wide-Sense Stationary (WSS) uncorrelated if

$$\phi_h(\tau; t_1, t_2) = \phi_h(\tau; t_2 - t_1) \tag{8.3}$$

which means that the correlation of each delay component of the CIR is only a function of the *difference* in time between each realization.

Hence, the second-order statistics of this model are completely described by its delay cross-power density $\phi_h(\tau; \Delta t)$ or by its Fourier transform, the scattering function defined as

$$S_h(\tau; f) = \int \phi_h(\tau; \Delta t) e^{-j2\pi f \Delta t}\, \mathrm{d}\Delta t \tag{8.4}$$

with f being the Doppler frequency. Other related functions of interest include the *Power Delay Profile* (PDP)

$$\psi_h(\tau) = \phi_h(\tau; 0) = \int S_h(\tau; f)\, \mathrm{d}f$$

the *time-correlation function*

$$\overline{\phi}_h(\Delta t) = \int \phi_h(\tau; \Delta t)\, \mathrm{d}\tau$$

and the *Doppler power spectrum*

$$\overline{S}_h(f) = \int S_h(\tau; f)\, \mathrm{d}\tau.$$

A more general exposition of WSSUS models is given in [14]. Classical results were derived by Clarke [15] and Jakes [16] for the case of a mobile terminal communicating with a stationary base station in a two-dimensional propagation geometry.

These well-known results state that

$$\overline{S}_h(f) = \frac{1}{\sqrt{f_\mathrm{d}^2 - f^2}} \tag{8.5}$$

for $|f| \leq f_\mathrm{d}$ with f_d being the maximum Doppler shift and

$$\overline{\phi}_h(\Delta t) = J_0(2\pi f_\mathrm{d} \Delta t) \tag{8.6}$$

where $J_0(\cdot)$ is the zeroth-order Bessel function. Figure 8.8 shows the PSD of the classical Doppler spectrum described by Clarke and Jakes [15, 16]. The Clarke and Jakes derivations are based on the assumption that the physical scattering environment is chaotic and therefore the angle of arrival of the electromagnetic wave at the receiver is a uniformly distributed random variable in the angular domain. As a consequence, the time-correlation function is strictly real-valued, the Doppler spectrum is symmetric and interestingly there is a *delay-temporal separability* property in the general bi-dimensional scattering function $S_h(\tau, \Delta t)$. In other words,

$$S_h(\tau; f) = \psi_h(\tau)\overline{S}_h(f) \tag{8.7}$$

or equivalently

$$\phi_h(\tau; \Delta t) = \psi_h(\tau)\overline{\phi}_h(\Delta t) \tag{8.8}$$

If the time and frequency aspects can be assumed to be separable, the complexity of channel estimation can be significantly reduced, as the problem is reduced to two one-dimensional operations.

The continuous-time correlation properties of the channel $h(\tau, t)$ discussed above apply equivalently to the corresponding low-pass sampled discrete-time CIR, i.e. $h[l, k]$ [17], for every l^{th} delay sampled at the kT^{th} instant where T is the sampling period. Moreover, these properties are maintained if we assume $h[l, k]$ to be well-approximated by a Finite-Impulse Response (FIR) vector $\mathbf{h}[k] = [h[0, k], \cdots, h[L-1, k]]^{\mathrm{T}}$ with a maximum delay spread of L samples. For the sake of notational simplicity and without loss of generality, the index k will be dropped in the following sections.

8.3.2 Spatial-Domain Correlation: The Kronecker Model

While the frequency and time correlations addressed in the previous section are induced by the channel delays and Doppler spread, the spatial correlation arises from the spatial scattering conditions.

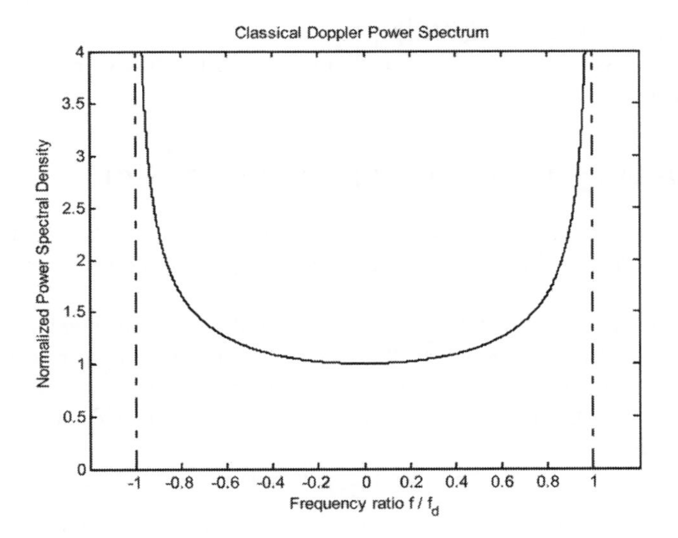

Figure 8.8: Normalized PSD for Clarke's model.

Among the possible spatial correlation models, for performance evaluation of LTE the Kronecker model is generally used [10]. Despite its simplicity, this correlation-based analytical model is widely used for the theoretical analysis of MIMO systems and yields experimentally verifiable results when limited to two or three antennas at each end of the radio link.

Let us assume the narrowband MIMO channel $N_{\mathrm{Rx}} \times N_{\mathrm{Tx}}$ matrix for a system with N_{Tx} transmitting antennas and N_{Rx} receiving antennas to be

$$\mathbf{H} = \begin{bmatrix} h_{0,0} & \cdots & h_{0,N_{\mathrm{Tx}}-1} \\ \vdots & & \vdots \\ h_{N_{\mathrm{Rx}}-1,N_{\mathrm{Tx}}-1} & \cdots & h_{N_{\mathrm{Rx}}-1,N_{\mathrm{Tx}}-1} \end{bmatrix} \tag{8.9}$$

The narrowband assumption particularly suits OFDM systems (see Section 5.2.1), where each MIMO channel component $h_{n,m}$ can be seen as the complex channel coefficient of each spatial link at a given subcarrier index.

The matrix $\mathbf{H}$ is rearranged into a vector by means of the operator $\mathrm{vec}(\mathbf{H}) = [\mathbf{h}_0^{\mathrm{T}}, \mathbf{h}_1^{\mathrm{T}}, \ldots, \mathbf{h}_{N_{\mathrm{Tx}}-1}^{\mathrm{T}}]^{\mathrm{T}}$ where $\mathbf{h}_i$ is the ith column of $\mathbf{H}$ and $\{\cdot\}^{\mathrm{T}}$ is the transpose operation. Hence, the correlation matrix can be defined as

$$\mathbf{C}_{\mathrm{S}} = \mathbb{E}[\mathrm{vec}(\mathbf{H})\mathrm{vec}(\mathbf{H})^{\mathrm{H}}] \tag{8.10}$$

where $\{\cdot\}^{\mathrm{H}}$ is the Hermitian operation. The matrix in (8.10) is the full correlation matrix of the MIMO channel.

The Kronecker model assumes that the full correlation matrix results from separate spatial correlations at the transmitter and receiver, which is equivalent to writing the full correlation matrix as a Kronecker product ($\otimes$) of the transmitter and receiver correlation matrices:

$$\mathbf{C}_{\mathrm{S}} = \mathbf{C}_{\mathrm{Tx}} \otimes \mathbf{C}_{\mathrm{Rx}} \tag{8.11}$$

with $\mathbf{C}_{\mathrm{Tx}} = \mathbb{E}[\mathbf{H}^{\mathrm{H}}\mathbf{H}]$ and $\mathbf{C}_{\mathrm{Rx}} = \mathbb{E}[\mathbf{H}\mathbf{H}^{\mathrm{H}}]$.

Typical values assumed for these correlations according to the Kronecker model are discussed in Section 20.3.5.

8.4 Frequency-Domain Channel Estimation

In this section, we address the channel estimation problem over one OFDMA symbol (specifically a symbol containing reference symbols) to exploit the frequency-domain characteristics.

In the LTE context, as for any OFDM system with uniformly distributed reference symbols [18], the Channel Transfer Function (CTF) can be estimated using a maximum likelihood approach in the frequency domain at the REs containing the RSs by de-correlating the constant modulus RS. Using a matrix notation, the CTF estimate $\widehat{\mathbf{z}}_p$ on reference symbol p can be written as

$$\widehat{\mathbf{z}}_p = \mathbf{z}_p + \widetilde{\mathbf{z}}_p = \mathbf{F}_p\mathbf{h} + \widetilde{\mathbf{z}}_p \tag{8.12}$$

for $p \in (0, \ldots, P)$ where P is the number of available reference symbols and $\mathbf{h}$ is the $L \times 1$ CIR vector. $\mathbf{F}_p$ is the $P \times L$ matrix obtained by selecting the rows corresponding to the reference symbol positions and the first L columns of the $N \times N$ Discrete Fourier Transform (DFT) matrix where N is the FFT order. $\widetilde{\mathbf{z}}_p$ is a $P \times 1$ zero-mean complex circular white noise vector whose $P \times P$ covariance matrix is given by $\mathbf{C}_{\widetilde{\mathbf{z}}_p}$. The effective channel length $L \leq L_{\mathrm{CP}}$ is assumed to be known.

8.4.1 Channel Estimate Interpolation

8.4.1.1 Linear Interpolation Estimator

The natural approach to estimate the whole CTF is to interpolate its estimate between the reference symbol positions. In the general case, let $\mathbf{A}$ be a generic interpolation filter; then the interpolated CTF estimate at subcarrier index i can be written as

$$\widehat{\mathbf{z}}_i = \mathbf{A}\widehat{\mathbf{z}}_p \tag{8.13}$$

Substituting Equation (8.12) into (8.13), the error of the interpolated CTF estimate is

$$\widetilde{\mathbf{z}}_i = \mathbf{z} - \widehat{\mathbf{z}}_i = (\mathbf{F}_L - \mathbf{A}\mathbf{F}_p)\mathbf{h} - \mathbf{A}\widetilde{\mathbf{z}}_p \tag{8.14}$$

where $\mathbf{F}_L$ is the $N \times L$ matrix obtained by taking the first L columns of the DFT matrix and $\mathbf{z} = \mathbf{F}_L\mathbf{h}$. In Equation (8.14), it can be seen that the channel estimation error is constituted of a bias term (itself dependent on the channel) and an error term.

The error covariance matrix is

$$\mathbf{C}_{\widetilde{\mathbf{z}}_i} = (\mathbf{F}_L - \mathbf{A}\mathbf{F}_p)\mathbf{C}_{\mathbf{h}}(\mathbf{F}_L - \mathbf{A}\mathbf{F}_p)^{\mathrm{H}} + \sigma^2_{\widetilde{\mathbf{z}}_p}\mathbf{A}\mathbf{A}^{\mathrm{H}} \tag{8.15}$$

where $\mathbf{C}_h = \mathbb{E}[\mathbf{h}\mathbf{h}^{\mathrm{H}}]$ is the channel covariance matrix.

Recalling Equation (8.13), linear interpolation would be the intuitive choice. Such an estimator is deterministically biased, but unbiased from the Bayesian viewpoint regardless of the structure of $\mathbf{A}$.

8.4.1.2 IFFT Estimator

As a second straightforward approach, the CTF estimate over all subcarriers can be obtained by IFFT interpolation. In this case, the matrix $\mathbf{A}$ from Equation (8.13) becomes:

$$\mathbf{A}_{\text{IFFT}} = \frac{1}{P}\mathbf{F}_L\mathbf{F}_p^{\text{H}} \tag{8.16}$$

The error of the IFFT-interpolated CTF estimate and its covariance matrix can be obtained by substituting Equation (8.16) into Equations (8.14) and (8.15).

With the approximation of $\mathbf{I}_L \approx (1/P)\mathbf{F}_p^{\text{H}}\mathbf{F}_p$, where $\mathbf{I}_L$ is the $L \times L$ identity matrix, it can immediately be seen that the bias term in Equation (8.14) would disappear, providing for better performance.

Given the LTE system parameters and the patterns of REs used for RSs, in practice $(1/P)\mathbf{F}_p^{\text{H}}\mathbf{F}_p$ is far from being a multiple of an identity matrix. The approximation becomes an equality when $K = N$, $N/W > L$ and N/W is an integer,[9] i.e. the system would have to be dimensioned without guard-bands and the RS would have to be positioned with a spacing which is an exact factor of the FFT order N, namely a power of two.

In view of the other factors affecting the design of the RS RE patterns outlined above, such constraints are impractical.

8.4.2 General Approach to Linear Channel Estimation

Compared to the simplistic approaches presented in the previous section, more elaborate linear estimators derived from both deterministic and statistical viewpoints are proposed in [19–21]. Such approaches include Least Squares (LS), Regularized LS, Minimum Mean-Squared Error (MMSE) and Mismatched MMSE. These can all be expressed under the following general formulation:

$$\mathbf{A}_{\text{gen}} = \mathbf{B}(\mathbf{G}^{\text{H}}\mathbf{G} + \mathbf{R})^{-1}\mathbf{G}^{\text{H}} \tag{8.17}$$

where $\mathbf{B}$, $\mathbf{G}$ and $\mathbf{R}$ are matrices that vary according to each estimator as expressed in Table 8.1, where $\mathbf{0}_L$ is the all-zeros $L \times L$ matrix.

The LS estimator discussed in [19] is theoretically unbiased. However, as shown in [21], it is not possible to apply the LS estimator to LTE directly, because the expression $(\mathbf{F}_p\mathbf{F}_p^{\text{H}})^{-1}$ is ill-conditioned due to the unused portion of the spectrum corresponding to the unmodulated subcarriers.

To counter this problem, the classical robust regularized LS estimator can be used instead, so as to yield a better conditioning of the matrix to be inverted. A regularization matrix $\alpha\mathbf{I}_L$ is introduced [22] where α is a constant (computed off-line) chosen to optimize the performance of the estimator in a given Signal-to-Noise Ratio (SNR) working range.

The MMSE estimator belongs to the class of statistical estimators. Unlike deterministic LS and its derivations, statistical estimators need knowledge of the second-order statistics (PDP and noise variance) of the channel in order to perform the estimation process, normally with much better performance compared to deterministic estimators. However, second-order statistics vary as the propagation conditions change and therefore need appropriate re-estimation regularly. For this reason statistical estimators are, in general, more complex due

[9] W is the spacing (in terms of number of subcarriers) between reference symbols.

Table 8.1: Linear estimators.

Interpolation method	Components of interpolation filter (see Equation (8.17)) B	G	R
Simple interpolator	$\mathbf{A}$	$\mathbf{I}_P$	$\mathbf{0}_L$
FFT	$\frac{1}{P}\mathbf{F}_L\mathbf{F}_p^{\mathrm{H}}$	$\mathbf{I}_P$	$\mathbf{0}_L$
LS	$\mathbf{F}_L$	$\mathbf{F}_p$	$\mathbf{0}_L$
Regularized LS	$\mathbf{F}_L$	$\mathbf{F}_p$	$\alpha\mathbf{I}_L$
MMSE	$\mathbf{F}_L$	$\mathbf{F}_p$	$\sigma^2_{\widetilde{\mathbf{z}}_p}\mathbf{C}_{\mathbf{h}}{}^{-1}$
Mismatched MMSE	$\mathbf{F}_L$	$\mathbf{F}_p$	$\sigma^2_{\widetilde{\mathbf{z}}_p}/\sigma^2_{\widetilde{\mathbf{h}}}\cdot\mathbf{I}_L$

to the additional burden of estimating the second-order statistics and computing the filter coefficients.

A mismatched MMSE estimator avoids the estimation of the second-order channel statistics and the consequent on-line inversion of an $L \times L$ matrix (as required in the straightforward application of MMSE) by assuming that the channel PDP is uniform[10] [20]. This estimator is, in practice, equivalent to the regularized LS estimator where the only difference lies in the fact that $\alpha = \sigma^2_{\widetilde{\mathbf{z}}_p}/\sigma^2_{\widetilde{\mathbf{h}}}$ is estimated and therefore adapted. The mismatched MMSE formulation offers the advantage that the filter coefficients can be computed to be real numbers because the uniform PDP is symmetric. Moreover, since the length of the CIR is small compared to the FFT size, the matrix A_{gen} can be considered to be 'low-density', so storing only the significant coefficients can reduce the complexity.

However, LTE does not use an exactly uniform pattern of reference symbols; for the cell-specific RSs this is the case around the d.c. subcarrier which is not transmitted. This implies that a larger number of coefficients needs to be stored.

8.4.3 Performance Comparison

For the sake of comparison between the performance of the different classes of estimator, we introduce the Truncated Normalized Mean Squared Error (TNMSE).

For each estimator, the TNMSE is computed from its covariance matrix $\mathbf{C}_{\widetilde{\mathbf{z}}}$ and the true CTF $\mathbf{z}$ as follows:

$$\mathrm{TNMSE}_{\widetilde{\mathbf{z}}} = \frac{\mathrm{Ttr}(\mathbf{C}_{\widetilde{\mathbf{z}}})}{\mathrm{Ttr}(\mathbf{F}_L\mathbf{C}_{\mathrm{h}}\mathbf{F}_L^{\mathrm{H}})} \tag{8.18}$$

where $\mathrm{Ttr}\{\cdot\}$ denotes the truncated trace operator, where the truncation is such that only the K used subcarriers are included.

Figure 8.9 shows the performance of an LTE FDD downlink with 10 MHz transmission bandwidth ($N = 1024$) and Spatial Channel Model-A (SCM-A – see Section 20.3.4).

It can be seen that the IFFT and linear interpolation methods yield the lowest performance. The regularized LS and the mismatched MMSE perform equally and the curve of the latter is therefore omitted. As expected, the optimal MMSE estimator outperforms any other estimator.

[10]This results in the second-order statistics of the channel having the structure of an identity matrix.

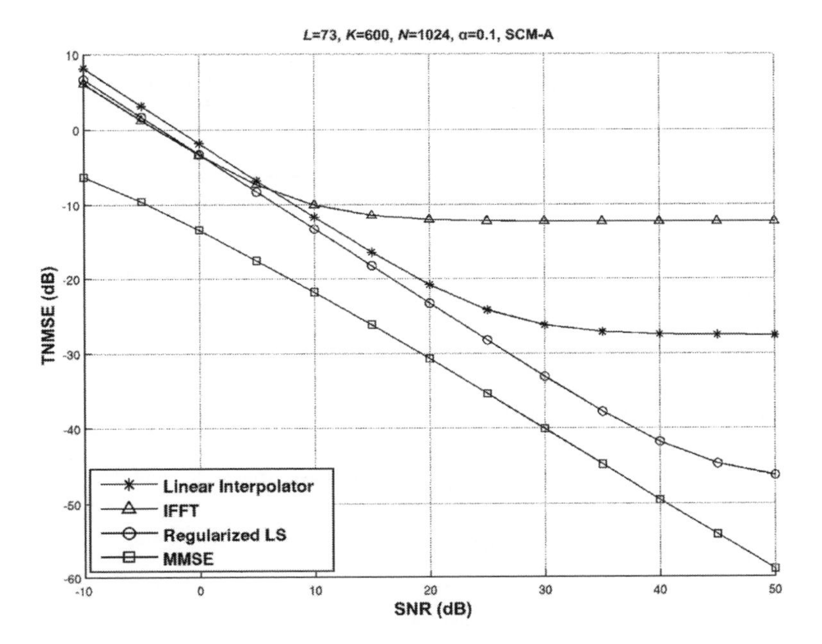

Figure 8.9: Frequency-domain channel estimation performance.

The TNMSE computed over all subcarriers actually hides the behaviour of each estimator in relation to a well-known problem of frequency-domain channel estimation techniques: the band-edge effect. This can be represented by the Gibbs [23] phenomenon in a finite-length Fourier series approximation; following this approach, Figure 8.10 shows that MMSE-based channel estimation suffers the least band-edge degradation, while all the other methods presented are highly adversely affected.

8.5 Time-Domain Channel Estimation

The main benefit of time-domain channel estimation is the possibility to enhance the channel estimate of one OFDM symbol containing reference symbols by exploiting its time correlation with the channel at previous OFDM symbols containing reference symbols. This requires sufficient memory for buffering soft values of data over several OFDM symbols while the channel estimation is carried out.

However, the correlation between consecutive symbols decreases as the UE speed increases, as expressed by Equation (8.6), and this sets a limit on the possibilities for time-domain filtering in high-mobility conditions.

Time-domain filtering is applied to the CIR estimate, rather than to the CTF estimate in the frequency domain. The use of a number of parallel scalar filters equal to the channel length L does not imply a loss of optimality, because of the WSSUS assumption.

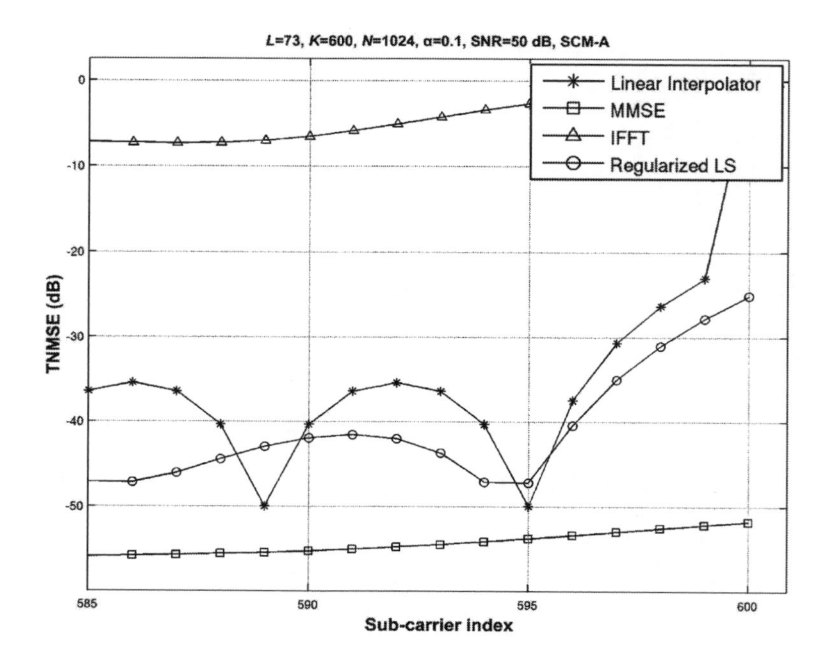

Figure 8.10: Band-edge behaviour of frequency-domain channel estimation.

8.5.1 Finite and Infinite Length MMSE

The statistical time-domain filter which is optimal in terms of Mean Squared Error (MSE) can be approximated in the form of a finite impulse response filter [24]. The channel at the lth tap position and at time instant n is estimated as

$$\hat{\hat{h}}_{l,n} = \mathbf{w}_l^H \hat{\mathbf{h}}_{l,n}^M \tag{8.19}$$

where $\hat{\hat{h}}_l$ is the smoothed CIR l^{th} tap estimate which exploits the vector $\hat{\mathbf{h}}_{l,n}^M = [\hat{h}_{l,n}, \ldots, \hat{h}_{l,n-M+1}]^{\mathrm{T}}$ of length M of the channel tap h_l across estimates at M time instants.[11] This is obtained by inverse Fourier transformation of, for example, any frequency-domain technique illustrated in Section 8.4 or even a raw estimate obtained by RS decorrelation.

The $M \times 1$ vector of Finite Impulse Response (FIR) filter coefficients $\mathbf{w}_l$ is given by

$$\mathbf{w}_l = (\mathbf{R}_h + \sigma_n^2 \mathbf{I})^{-1} \mathbf{r}_h \tag{8.20}$$

where $\mathbf{R}_h = \mathbb{E}[\mathbf{h}_l^M (\mathbf{h}_l^M)^{\mathrm{H}}]$ is the lth channel tap $M \times M$ correlation matrix, σ_n^2 the additive noise variance and $\mathbf{r}_h = \mathbb{E}[\mathbf{h}_l^M h_{l,n}^*]$ (the $M \times 1$ correlation vector between the lth tap of the current channel realization and M previous realizations including the current one).

[11] $h_{l,k}$ is the lth component of the channel vector $\mathbf{h}_k$ at time instant k. $\hat{h}_{l,k}$ is its estimate. Note that for the frequency-domain treatment, the time index was dropped.

In practical cases, the FIR filter length M is dimensioned according to a performance-complexity trade-off as a function of UE speed.

By setting M infinite, the upper bound on performance is obtained.

The MSE performance of the finite-length estimator of a channel of length L can be analytically computed as

$$\epsilon^{(M)} = 1 - \frac{1}{\sigma_h^2} \sum_{l=0}^{L-1} \mathbf{w}_l^{\mathrm{H}} \mathbf{r}_h \tag{8.21}$$

The upper bound given by an infinite-length estimator is therefore given by

$$\epsilon^{(\infty)} = \lim_{M \to \infty} \epsilon^{(M)} \tag{8.22}$$

From Equation (8.20) it can be observed that, unlike frequency-domain MMSE filtering, the size of the matrix to be inverted for a finite-length time-domain MMSE estimator is independent of the channel length L but dependent on the chosen FIR order M. Similarly to the frequency-domain counterpart, the time-domain MMSE estimator requires knowledge of the PDP, the UE speed and the noise variance.

Figure 8.11 shows the performance of time-domain MMSE channel estimation as a function of filter length M (Equation (8.21)) for a single-tap channel with a classical Doppler spectrum for low UE speed. The performance bounds derived for an infinite-length filter in each case are also indicated.

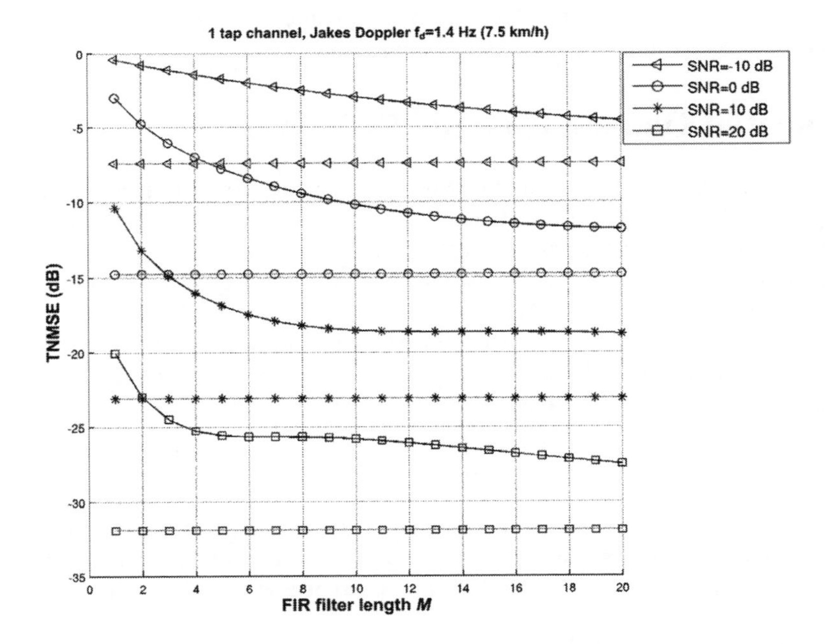

Figure 8.11: Time-domain channel estimation performance.

8.5.2 Normalized Least-Mean-Square

As an alternative to time-domain MMSE channel estimation, an adaptive estimation approach can be considered which does not require knowledge of second-order statistics of both channel and noise. A feasible solution is the Normalized Least-Mean-Square (NLMS) estimator.

It can be expressed exactly as in Equation (8.19) but with the $M \times 1$ vector of filter coefficients $\mathbf{w}$ updated according to

$$\mathbf{w}_{l,n} = \mathbf{w}_{l,n-1} + \mathbf{k}_{l,n-1} e_{l,n} \tag{8.23}$$

where M here denotes the NLMS filter order. The $M \times 1$ updated gain vector is computed according to the well-known NLMS adaptation:

$$\mathbf{k}_{l,n} = \frac{\mu}{||\hat{\mathbf{h}}_{l,n}||^2} \hat{\mathbf{h}}_{l,n}^M \tag{8.24}$$

where μ is an appropriately chosen step adaptation, $\hat{\mathbf{h}}_{l,n}^M$ is defined as for Equation (8.19) and

$$e_{l,n} = \hat{h}_{l,n} - \hat{h}_{l,n-1} \tag{8.25}$$

It can be observed that the time-domain NLMS estimator requires much lower complexity compared to time-domain MMSE because no matrix inversion or a priori statistical knowledge is required.

Other adaptive approaches could also be considered, such as Recursive Least Squares (RLS) and Kalman-based filtering. Although more complex than NLMS, the Kalman filter is a valuable candidate and is reviewed in detail in [25].

8.6 Spatial-Domain Channel Estimation

It is assumed that an LTE UE has multiple receiving antennas. Consequently, whenever the channel is correlated in the spatial domain, the correlation can be exploited to provide a further means for enhancing the channel estimate.

If it is desired to exploit spatial correlation, a natural approach is again offered by spatial domain MMSE filtering [26].

We consider here the case of a MIMO OFDM communication system with N subcarriers, N_{Tx} transmitting antennas and N_{Rx} receiving antennas. The $N_{\mathrm{Rx}} \times 1$ received signal vector at subcarrier k containing the RS sequence can be written as

$$\mathbf{r}(k) = \frac{1}{\sqrt{N_{\mathrm{Tx}}}} \mathbf{W}(k)\mathbf{s}(k) + \mathbf{n}(k) \tag{8.26}$$

where $\mathbf{W}(k)$ is the $N_{\mathrm{Rx}} \times N_{\mathrm{Tx}}$ channel frequency response matrix at RS subcarrier k, $\mathbf{s}(k)$ is the $N_{\mathrm{Tx}} \times 1$ known zero-mean and unit-variance transmitted RS sequence at subcarrier k, and $\mathbf{n}(k)$ is the $N_{\mathrm{Rx}} \times 1$ complex additive white Gaussian noise vector with zero mean and variance σ_n^2.

The CTF at subcarrier k, $\mathbf{W}(k)$, is obtained by DFT from the CIR matrix at the lth tap $\mathbf{H}_l$ as

$$\mathbf{W}(k) = \sum_{l=0}^{L-1} \mathbf{H}_l \exp\left[-j2\pi \frac{lk}{N}\right] \tag{8.27}$$

The vector $\mathbf{h}$ is obtained by rearranging the elements of all $\mathbf{H}_l$ channel tap matrices as follows:

$$\mathbf{h} = [\text{vec}(\mathbf{H}_0)^{\text{T}}, \text{vec}(\mathbf{H}_1)^{\text{T}}, \ldots, \text{vec}(\mathbf{H}_{L-1})^{\text{T}}]^{\text{T}} \tag{8.28}$$

The correlation matrix of $\mathbf{h}$ is given by

$$\mathbf{C}_h = \mathbb{E}[\mathbf{h}\mathbf{h}^{\text{H}}] \tag{8.29}$$

Using Equations (8.27) and (8.28), Equation (8.26) can now be rewritten as

$$\mathbf{r}(k) = \frac{1}{\sqrt{N_{\text{Tx}}}}\mathbf{G}(k)\mathbf{h} + \mathbf{n}(k) \tag{8.30}$$

where $\mathbf{G}(k) = [\mathbf{D}_0(k), \mathbf{D}_1(k), \ldots, \mathbf{D}_{L-1}(k)]$ and $\mathbf{D}_l(k) = \exp[-j2\pi(lk/N)]\mathbf{s}^{\text{T}}(k) \otimes \mathbf{I}_{N_{\text{Rx}}}$.

Rearranging the received signal matrix $\mathbf{G}$ and the noise matrix at all N subcarriers into a vector as follows,

$$\mathbf{r} = [\mathbf{r}^{\text{T}}(0), \mathbf{r}^{\text{T}}(1), \ldots, \mathbf{r}^{\text{T}}(N-1)]^{\text{T}} \tag{8.31}$$

Equation (8.30) can be rewritten more compactly as

$$\mathbf{r} = \frac{1}{\sqrt{N_{\text{Tx}}}}\mathbf{G}\mathbf{h} + \mathbf{n} \tag{8.32}$$

Hence, the spatial domain MMSE estimation of the rearranged channel impulse vector $\mathbf{h}$ can be simply obtained by

$$\hat{\mathbf{h}} = \mathbf{Q}\mathbf{r} \tag{8.33}$$

with

$$\mathbf{Q} = \frac{1}{\sqrt{N_{\text{Tx}}}}\mathbf{C}_h\mathbf{G}^{\text{H}}\left(\frac{1}{N_{\text{Tx}}}\mathbf{G}\mathbf{C}_h\mathbf{G}^{\text{H}} + \sigma_n^2\mathbf{I}_{N \cdot N_{\text{Rx}}}\right)^{-1} \tag{8.34}$$

Therefore the NMSE can be computed as

$$\text{NMSE}_{\text{SD-MMSE}} = \frac{\text{tr}(\mathbf{C}_{\text{SD-MMSE}})}{\text{tr}(\mathbf{C}_h)} \tag{8.35}$$

where $\mathbf{C}_{\text{SD-MMSE}}$ is the error covariance matrix.

Figure 8.12 shows the spatial domain MMSE performance given by Equation (8.35) compared to the performance obtained by the ML channel estimation on the subcarriers which carry RS.

8.7 Advanced Techniques

The LTE specifications do not mandate any specific channel estimation technique, and there is therefore complete freedom in implementation provided that the performance requirements are met and the complexity is affordable.

Particular aspects, such as channel estimation based on the UE-specific RSs, band-edge effect reduction or bursty reception, might require further improvements which go beyond the techniques described here.

Blind and semi-blind techniques are promising for some such aspects, as they try to exploit not only the a priori knowledge of the RSs but also the unknown data structure. A comprehensive analysis is available in [27] and references therein.

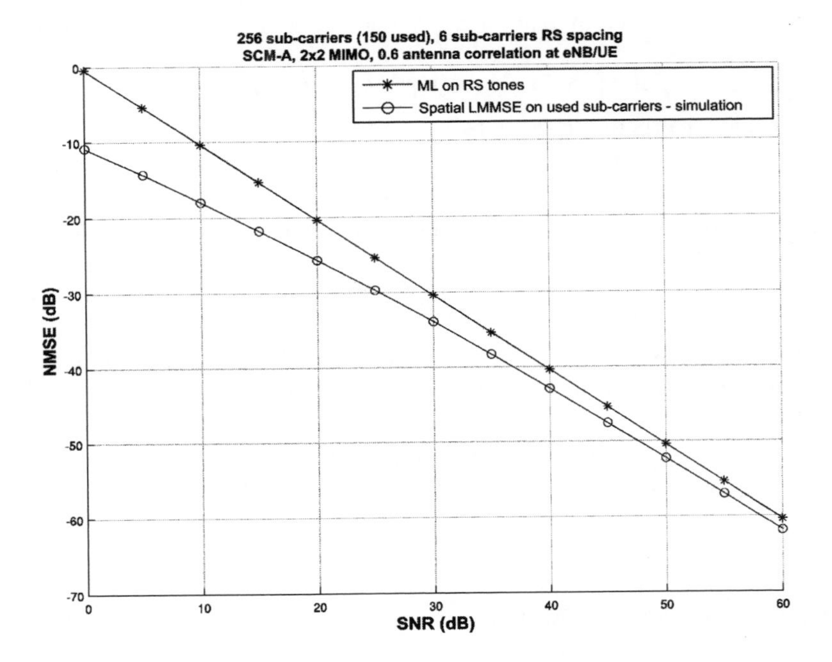

Figure 8.12: Spatial-domain channel estimation performance: CIR NMSE versus SNR.

References[12]

[1] A. Vosoughi and A. Scaglione, 'The Best Training Depends on the Receiver Architecture' in *Proc. IEEE International Conference on Acoustics, Speech, and Signal Processing (ICASSP 2004)*, (Montreal, Canada), May 2004.

[2] A. Vosoughi and A. Scaglione, 'Everything You Always Wanted to Know about Training: Guidelines Derived using the Affine Precoding Framework and the CRB'. *IEEE Trans. on Signal Processing*, Vol. 54, pp. 940–954, March 2006.

[3] 3GPP Technical Specification 25.211, 'Physical Channels and Mapping of Transport Channels onto Physical Channels (FDD)', www.3gpp.org.

[4] N. Wiener, *Extrapolation, Interpolation, and Smoothing of Stationary Time Series*. New York: John Wiley & Sons, Ltd, 1949.

[5] D. T. M. Slock, 'Signal Processing Challenges for Wireless Communications' in *Proc. International Symposium on Control, Communications and Signal Processing (ISCCSP' 2004)*, (Hammamet, Tunisia), March 2004.

[6] 3GPP Technical Specification 36.211, 'Evolved Universal Terrestrial Radio Access (E-UTRA); Physical Channels and Modulation', www.3gpp.org.

[7] R. Negi and J. Cioffi, 'Pilot Tone Selection for Channel Estimation in a Mobile OFDM System'. *IEEE Trans. on Consumer Electronics*, Vol. 44, pp. 1122–1128, August 1998.

[12] All web sites confirmed 1st March 2011.

[8] I. Barhumi, G. Leus and M. Moonen, 'Optimal Training Design for MIMO OFDM Systems in Mobile Wireless Channels'. *IEEE Trans. on Signal Processing*, Vol. 51, pp. 1615–1624, June 2003.

[9] 3GPP Technical Report 25.913, 'Requirements for Evolved UTRA (E-UTRA) and Evolved UTRAN (E-UTRAN) (Release 7)', www.3gpp.org.

[10] 3GPP Technical Specification 36.101, 'User Equipment (UE) Radio Transmission and Reception', www.3gpp.org.

[11] Motorola, Nortel, Broadcomm, Nokia, NSN, NTT DoCoMo, NEC, Mitsubishi, Alcatel-Lucent, CATT, Huawei, Sharp, Texas Instrument, ZTE, Panasonic, Philips and Toshiba, 'R1-081108: Way Forward on Dedicated Reference Signal Design for LTE Downlink with Normal CP', www.3gpp.org, 3GPP TSG RAN WG1, meeting 52, Sorrento, Italy, February 2008.

[12] NTT DoCoMo and NEC, 'R1-094337: Remaining Issues for Rel. 9 Downlink DM-RS Design', www.3gpp.org, 3GPP TSG RAN WG1, meeting 58bis, Miyazaki, Japan, Octboer 2009.

[13] P. Almers, E. Bonek, A. Burr, N. Czink, M. Debbah, V. Degli-Esposti, H. Hofstetter, P. Kyösty, D. Laurenson, G. Matz, F. Molisch, C. Oestges, and H. Ožcelik, 'Survey of Channel and Radio Propagation Models for Wireless MIMO Systems'. *EURASIP Journal on Wireless Communications and Networking*, pp. 957–1000, July 2007.

[14] J. G. Proakis, *Digital Communications*. New York: McGraw Hill, 1995.

[15] R. H. Clarke, 'A Statistical Theory of Mobile-radio Reception'. *Bell Syst. Tech. J.*, pp. 957–1000, July 1968.

[16] W. C. Jakes, *Microwave Mobile Communications*. New York: John Wiley & Sons, Ltd/Inc., 1974.

[17] P. Hoeher,'A statistical discrete-time model for the WSSUS multipath channel' in *IEEE Transactions on Vehicular Technology*, Vol 41, No. 4, pp. 461–468, 1992.

[18] J.-J. van de Beek, O. Edfors, M. Sandell, S. K. Wilson, and P. O. Börjesson, 'On Channel Estimation in OFDM Systems' in *Proc. VTC'1995, Vehicular Technology Conference (VTC 1995),* Chicago, USA, July 1995.

[19] M. Morelli and U. Mengali, 'A Comparison of Pilot-aided Channel Estimation Methods for OFDM Systems'. *IEEE Trans. on Signal Processing*, Vol. 49, pp. 3065–3073, December 2001.

[20] P. Hoher, S. Kaiser and P. Robertson, 'Pilot-Symbol-aided Channel Estimation in Time and Frequency' in *Proc. Communication Theory Mini-Conf. (CTMC) within IEEE Global Telecommun. Conf. (Globecom 1997)* (Phoenix, USA), July 1997.

[21] A. Ancora, C. Bona and D. T. M. Slock, 'Down-Sampled Impulse Response Least-Squares Channel Estimation for LTE OFDMA' in *Proc. IEEE International Conference on Acoustics, Speech and Signal Processing (ICASSP 2007),* Honolulu, Hawaii, April 2007.

[22] P. C. Hansen, 'Rank-deficient and discrete ill-posed problems: numerical aspects of linear inversion', Society for Industrial Mathematics, 1998.

[23] J. W. Gibbs, 'Fourier Series'. *Nature*, Vol. 59, p. 200, 1898.

[24] D. Schafhuber, G. Matz and F. Hlawatsch, 'Adaptive Wiener Filters for Time-varying Channel Estimation in Wireless OFDM Systems' in *Proc. IEEE International Conference on Acoustics, Speech, and Signal Processing (ICASSP 2003),* Hong Kong, April 2003.

[25] J. Cai, X. Shen and J. W. Mark, 'Robust Channel Estimation for OFDM Wireless Communication Systems. An H∞ Approach'. *IEEE Trans. Wireless Communications*, Vol. 3, November 2004.

[26] H. Zhang, Y. Li, A. Reid and J. Terry, 'Channel Estimation for MIMO OFDM in Correlated Fading Channels' in *Proc. 2005 IEEE International Conference on Communications (ICC 2005),* South Korea, May 2005.

[27] H. Bölcskei, D. Gesbert and C. Papadias, *Space-time Wireless Systems: From Array Processing to MIMO Communications*. Cambridge: Cambridge University Press, 2006.

9

Downlink Physical Data and Control Channels

Matthew Baker and Tim Moulsley

9.1 Introduction

Chapters 7 and 8 have described the signals which enable User Equipment (UEs) to synchronize with the network and estimate the downlink radio channel in order to be able to demodulate data. This chapter first reviews the downlink physical channels which transport the data and then explains the control-signalling channels; the latter support the data channels by indicating the particular time-frequency transmission resources to which the data is mapped and the format in which the data itself is transmitted.

9.2 Downlink Data-Transporting Channels

9.2.1 Physical Broadcast Channel (PBCH)

In cellular systems, the basic System Information (SI) which allows the other channels in the cell to be configured and operated is usually carried by a Broadcast CHannel (BCH). Therefore the achievable coverage for reception of the BCH is crucial to the successful operation of such cellular communication systems; LTE is no exception. As already noted in Chapter 3, the broadcast (SI) is divided into two categories:

LTE – The UMTS Long Term Evolution: From Theory to Practice, Second Edition.
Stefania Sesia, Issam Toufik and Matthew Baker.

- The 'Master Information Block' (MIB), which consists of a limited number of the most frequently transmitted parameters essential for initial access to the cell,[1] and is carried on the Physical Broadcast CHannel (PBCH).
- The other System Information Blocks (SIBs) which, at the physical layer, are multiplexed with unicast data transmitted on the Physical Downlink Shared CHannel (PDSCH) as discussed in Section 9.2.2.2.

This section focuses in particular on the PBCH, the design of which reflects some specific requirements:

- Detectability without prior knowledge of the system bandwidth;
- Low system overhead;
- Reliable reception right to the edge of the LTE cells;
- Decodability with low latency and low impact on UE battery life.

The resulting overall PBCH structure is shown in Figure 9.1.

Detectability without the UE having prior knowledge of the system bandwidth is achieved by mapping the PBCH only to the central 72 subcarriers of the OFDM[2] signal (which corresponds to the minimum possible LTE system bandwidth of 6 Resource Blocks (RBs)), regardless of the actual system bandwidth. The UE will have first identified the system centre-frequency from the synchronization signals as described in Chapter 7.

Low system overhead for the PBCH is achieved by deliberately keeping the amount of information carried on the PBCH to a minimum, since achieving stringent coverage requirements for a large quantity of data would result in a high system overhead. The size of the MIB is therefore just 14 bits, and, since it is repeated every 40 ms, this corresponds to a data rate on the PBCH of just 350 bps.

The main mechanisms employed to facilitate reliable reception of the PBCH in LTE are time diversity, Forward Error Correction (FEC) coding and antenna diversity.

Time diversity is exploited by spreading out the transmission of each MIB on the PBCH over a period of 40 ms. This significantly reduces the likelihood of a whole MIB being lost in a fade in the radio propagation channel, even when the mobile terminal is moving at pedestrian speeds.

The FEC coding for the PBCH uses a convolutional coder, as the number of information bits to be coded is small; the details of the convolutional coder are explained in Section 10.3.3. The basic code rate is 1/3, after which a high degree of repetition of the systematic (i.e. information) bits and parity bits is used, such that each MIB is coded at a very low code rate (1/48 over a 40 ms period) to give strong error protection.

Antenna diversity may be utilized at both the eNodeB and the UE. The UE performance requirements specified for LTE assume that all UEs can achieve a level of decoding performance commensurate with dual-antenna receive diversity (although it is recognized that in low-frequency deployments, such as below 1 GHz, the advantage obtained from receive antenna diversity is reduced due to the correspondingly higher correlation between

[1]The MIB information consists of the downlink system bandwidth, the PHICH size (Physical Hybrid ARQ Indicator CHannel, see Section 9.3.4), and the most-significant eight bits of the System Frame Number (SFN) – the remaining two bits being gleaned from the 40 ms periodicity of the PBCH.

[2]Orthogonal Frequency Division Multiplexing.

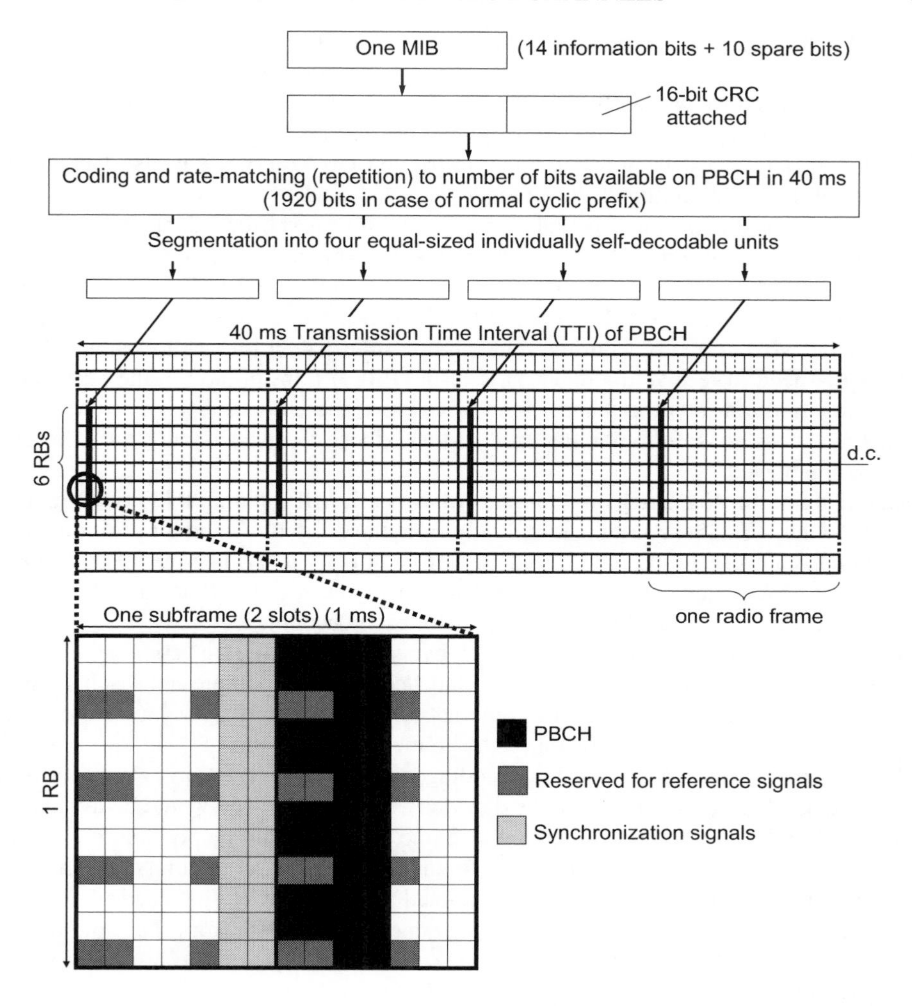

Figure 9.1: PBCH structure.

the antennas); this enables LTE system planners to rely on this level of performance being common to all UEs, thereby enabling wider cell coverage to be achieved with fewer cell sites than would otherwise be possible. Transmit antenna diversity may be also employed at the eNodeB to further improve coverage, depending on the capability of the eNodeB; eNodeBs with two or four transmit antenna ports transmit the PBCH using a Space-Frequency Block Code (SFBC), details of which are explained in Section 11.2.2.1.

The precise set of Resource Elements (REs) used by the PBCH is independent of the number of transmit antenna ports used by the eNodeB; any REs which may be used for Reference Signal (RS) transmission are avoided by the PBCH, irrespective of the actual number of transmit antenna ports deployed at the eNodeB. The number of transmit antenna ports used by the eNodeB must be ascertained blindly by the UE, by performing the decoding for each SFBC scheme corresponding to the different possible numbers of transmit antenna ports (namely one, two or four). This discovery of the number of transmit antenna ports is further facilitated by the fact that the Cyclic Redundancy Check (CRC) on each MIB is masked with a codeword representing the number of transmit antenna ports.

Finally, achieving low latency and a low impact on UE battery life is also facilitated by the design of the coding outlined above: the low code rate with repetition enables the full set of coded bits to be divided into four subsets, each of which is self-decodable in its own right. Each of these subsets of the coded bits is then transmitted in a different one of the four radio frames during the 40 ms transmission period, as shown in Figure 9.1. This means that if the Signal to Interference Ratio (SIR) of the radio channel is sufficiently good to allow the UE to decode the MIB correctly from the transmission in less than four radio frames, then the UE does not need to receive the other parts of the PBCH transmission in the remainder of the 40 ms period; on the other hand, if the SIR is low, the UE can receive further parts of the MIB transmission, soft-combining each part with those received already, until successful decoding is achieved.

The timing of the 40 ms transmission interval for each MIB on the PBCH is not indicated explicitly to the UE; it is ascertained implicitly from the scrambling and bit positions, which are re-initialized every 40 ms. The UE can therefore initially determine the 40 ms timing by performing four separate decodings of the PBCH using each of the four possible phases of the PBCH scrambling code, checking the CRC for each decoding.

When a UE initially attempts to access a cell by reading the PBCH, a variety of approaches may be taken to carry out the necessary blind decodings. A simple approach is always to perform the decoding using a soft combination of the PBCH over four radio frames, advancing a 40 ms sliding window one radio frame at a time until the window aligns with the 40 ms period of the PBCH and the decoding succeeds. However, this would result in a 40–70 ms delay before the PBCH can be decoded. A faster approach would be to attempt to decode the PBCH from the first single radio frame, which should be possible provided the SIR is sufficiently high; if the decoding fails for all four possible scrambling code phases, the PBCH from the first frame could be soft-combined with the PBCH bits received in the next frame – there is a 3-in-4 chance that the two frames contain data from the same transport block. If decoding still fails, a third radio frame could be combined, and failing that a fourth. It is evident that the latter approach may be much faster (potentially taking only 10 ms), but on the other hand requires slightly more complex logic.

9.2.2 Physical Downlink Shared CHannel (PDSCH)

The Physical Downlink Shared CHannel (PDSCH) is the main data-bearing downlink channel in LTE. It is used for all user data, as well as for broadcast system information which is not carried on the PBCH, and for paging messages – there is no specific physical layer paging channel in LTE. The use of the PDSCH for user data is explained in Section 9.2.2.1; the use of the PDSCH for system information and paging is covered in Section 9.2.2.2.

Data is transmitted on the PDSCH in units known as *Transport Blocks* (TBs), each of which corresponds to a Medium Access Control (MAC) layer Protocol Data Unit (PDU) as described in Section 4.4. Transport blocks may be passed down from the MAC layer to the physical layer once per Transmission Time Interval (TTI), where a TTI is 1 ms, corresponding to the subframe duration.

9.2.2.1 General Use of the PDSCH

When employed for user data, one or, at most, two TBs can be transmitted per UE per subframe, depending on the transmission mode selected for the PDSCH for each UE. The transmission mode configures the multi-antenna transmission scheme usually applied:[3]

Transmission Mode 1: Transmission from a single eNodeB antenna port;

Transmission Mode 2: Transmit diversity (see Section 11.2.2.1);

Transmission Mode 3: Open-loop spatial multiplexing (see Section 11.2.2.2);

Transmission Mode 4: Closed-loop spatial multiplexing (see Section 11.2.2.2);

Transmission Mode 5: Multi-User Multiple-Input Multiple-Output (MU-MIMO) (see Section 11.2.3);

Transmission Mode 6: Closed-loop rank-1 precoding (see Section 11.2.2.2);

Transmission Mode 7: Transmission using UE-specific RSs with a single spatial layer (see Sections 8.2 and 11.2.2.3);

Transmission Mode 8: Introduced in Release 9, transmission using UE-specific RSs with up to two spatial layers (see Sections 8.2.3 and 11.2.2.3);

Transmission Mode 9: Introduced in Release 10, transmission using UE-specific RSs with up to eight spatial layers (see Sections 29.1 and 29.3).

With the exception of transmission modes 7, 8 and 9, the phase reference for demodulating the PDSCH is given by the cell-specific Reference Signals (RSs) described in Section 8.2.1, and the number of eNodeB antenna ports used for transmission of the PDSCH is the same as the number of antenna ports used in the cell for the PBCH. In transmission modes 7, 8 and 9, UE-specific RSs (see Sections 8.2.2, 8.2.3 and 29.1.1 respectively) provide the phase reference for the PDSCH. The configured transmission mode also controls the formats of the associated downlink control signalling messages, as described in Section 9.3.5.1, and the modes of channel quality feedback from the UE (see Section 10.2.1).

After channel coding (see Section 10.3.2) and mapping to spatial layers according to the selected transmission mode, the coded PDSCH data bits are mapped to modulation symbols depending on the modulation scheme selected for the current radio channel conditions and required data rate.

The modulation order may be selected between two bits per symbol (using QPSK (Quadrature Phase Shift Keying)), four bits per symbol (using 16QAM (Quadrature Amplitude Modulation)) and six bits per symbol (using 64QAM); constellation diagrams for these modulation schemes are illustrated in Figure 9.2. Support for reception of 64QAM modulation is mandatory for all classes of LTE UE.

The REs used for the PDSCH can be any which are not reserved for other purposes (i.e. RSs, synchronization signals, PBCH and control signalling). Thus when the control

[3]In addition to the transmission schemes listed here for each mode, transmission modes 3 to 9 also support the use of transmit diversity as a 'fallback' technique; this is useful, for example, when radio conditions are temporarily inappropriate for the usual scheme, or to ensure that a common scheme is available during reconfiguration of the transmission mode.

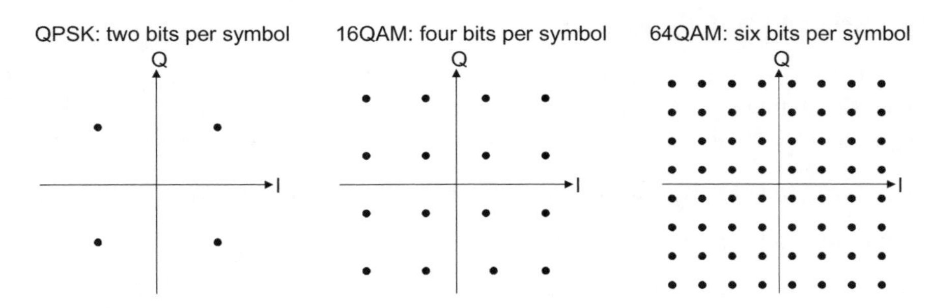

Figure 9.2: Constellations of modulation schemes applicable to PDSCH transmission.

signalling informs a UE that a particular pair of RBs[4] in a subframe are allocated to that UE, it is only the *available* REs within those RBs which actually carry PDSCH data.

Normally the allocation of pairs of RBs to PDSCH transmission for a particular UE is signalled to the UE by means of dynamic control signalling transmitted at the start of the relevant subframe using the Physical Downlink Control Channel (PDCCH), as described in Section 9.3.

The mapping of data to physical RBs can be carried out in one of two ways: *localized mapping* and *distributed mapping*.[5]

Localized resource mapping entails allocating all the available REs in a pair of RBs to the same UE. This is suitable for most scenarios, including the use of dynamic channel-dependent scheduling according to frequency-specific channel quality information reported by the UE (see Sections 10.2.1 and 12.4).

Distributed resource mapping entails separating in frequency the two physical RBs comprising each pair, with a frequency-hop occurring at the slot boundary in the middle of the subframe, as shown in Figure 9.3. This is a useful means of obtaining frequency diversity for small amounts of data which would otherwise be constrained to a narrow part of the downlink bandwidth and would therefore be more susceptible to narrow-band fading. An example of a typical use for this transmission mode could be a Voice-over-IP (VoIP) service, where, in order to minimize overhead, certain frequency resources may be 'semi-persistently scheduled' (see Section 4.4.2.1) – in other words, certain RBs in the frequency domain are allocated on a periodic basis to a specific UE by Radio Resource Control (RRC) signalling rather than by dynamic PDCCH signalling. This means that the transmissions are not able to benefit from dynamic channel-dependent scheduling, and therefore the frequency diversity which is achieved through distributed mapping is a useful tool to improve performance. Moreover, as the amount of data to be transmitted per UE for a VoIP service is small (typically sufficient to occupy only one or two pairs of RBs in a given subframe), the degree of frequency diversity obtainable via localized scheduling is very limited.

[4]The term 'pair of RBs' here means a pair of resource blocks which occupy the same set of 12 subcarriers and are contiguous in time, thus having a duration of one subframe.

[5]Distributed mapping is not supported in conjunction with UE-specific RSs in transmission modes 8 and 9.

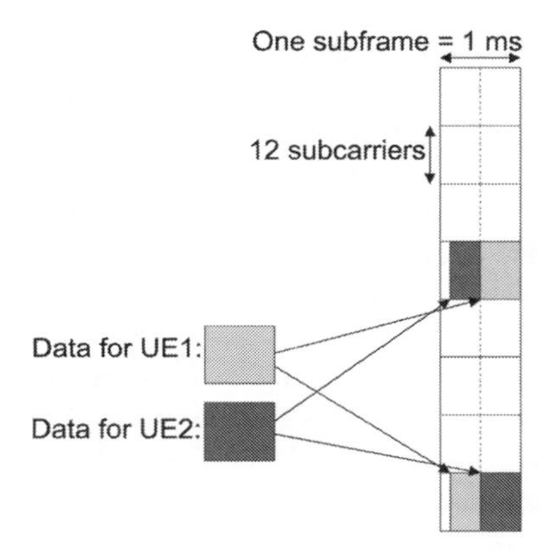

Figure 9.3: Frequency-distributed data mapping in LTE downlink.

The potential increase in the number of VoIP users which can be accommodated in a cell as a result of using distributed resource mapping as opposed to localized resource mapping is illustrated by way of example in Figure 9.4.

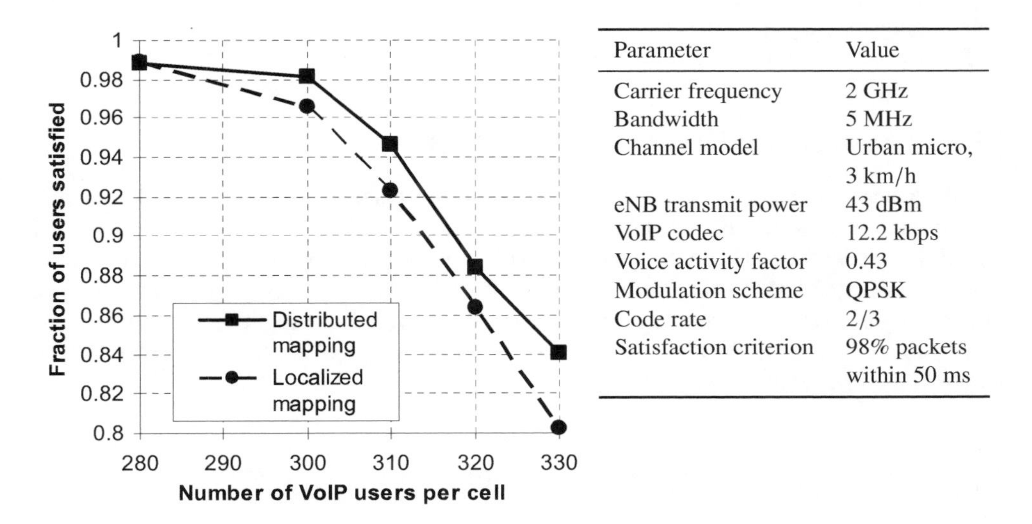

Parameter	Value
Carrier frequency	2 GHz
Bandwidth	5 MHz
Channel model	Urban micro, 3 km/h
eNB transmit power	43 dBm
VoIP codec	12.2 kbps
Voice activity factor	0.43
Modulation scheme	QPSK
Code rate	2/3
Satisfaction criterion	98% packets within 50 ms

Figure 9.4: Example of increase in VoIP capacity arising from frequency-distributed resource mapping.

9.2.2.2 Special Uses of the PDSCH

As noted above, the PDSCH is used for some special purposes in addition to normal user data transmission.

One such use is for the broadcast system information (i.e. SIBs) that is not carried on the PBCH. The RBs used for broadcast data of this sort are indicated by signalling messages on the PDCCH in the same way as for other PDSCH data, except that the identity indicated on the PDCCH is not the identity of a specific UE but is, rather, a designated broadcast identity known as the System Information Radio Network Temporary Identifier (SI-RNTI), which is fixed in the specifications (see Section 7.1 of [1]) and therefore known a priori to all UEs. Some constraints exist as to which subframes may be used for SI messages on the PDSCH; these are explained in Section 3.2.2.

Another special use of the PDSCH is paging, as no separate physical channel is provided in LTE for this purpose. In previous systems such as WCDMA,[6] a special 'Paging Indicator Channel' was provided, which was specially designed to enable the UE to wake up its receiver periodically for a very short period of time, in order to minimize the impact on battery life; on detecting a paging indicator (typically for a group of UEs), the UE would then keep its receiver switched on to receive a longer message indicating the exact identity of the UE being paged. By contrast, in LTE the PDCCH signalling is already very short in duration, and therefore the impact on UE battery life of monitoring the PDCCH from time to time is low. Therefore the normal PDCCH signalling can be used to carry the equivalent of a paging indicator, with the detailed paging information being carried on the PDSCH in RBs indicated by the PDCCH. In a similar way to broadcast data, paging indicators on the PDCCH use a single fixed identifier called the Paging RNTI (P-RNTI). Rather than providing different paging identifiers for different groups of UEs, different UEs monitor different subframes for their paging messages, as described in Section 3.4. Paging messages may be received in subframes 0, 4, 5 or 9 in each radio frame.

9.2.3 Physical Multicast Channel (PMCH)

In LTE Release 9, the use of a third data-transporting channel became available, namely the Physical Multicast CHannel (PMCH), designed to carry data for Multimedia Broadcast and Multicast Services (MBMS). The PMCH can only be transmitted in certain specific subframes known as MBSFN (Multimedia Broadcast Single Frequency Network) subframes, indicated in the system information carried on the PDSCH. A Release 8 UE must be aware of the possible existence of MBSFN subframes, but is not required to decode the PMCH. The details of the PMCH are explained in Section 13.4.1.

9.3 Downlink Control Channels

9.3.1 Requirements for Control Channel Design

The control channels in LTE are provided to support efficient data transmission. In common with other wireless systems, the control channels convey physical layer messages which cannot be carried sufficiently efficiently, quickly or conveniently by higher layer protocols. The design of the control channels in the LTE downlink aims to balance a number of somewhat conflicting requirements, the most important of which are discussed below.

[6]Wideband Code Division Multiple Access.

9.3.1.1 Physical Layer Signalling to Support the MAC Layer

The general requirement to support MAC operation is very similar to that in WCDMA, but there are a number of differences of detail, mainly arising from the frequency domain resource allocation supported in the LTE multiple access schemes.

The use of the uplink transmission resources on the Physical Uplink Shared Channel (PUSCH) is determined dynamically by an uplink scheduling process in the eNodeB, and therefore physical layer signalling must be provided to indicate to UEs which time/frequency resources they have been granted permission to use.

The eNodeB also schedules downlink transmissions on the PDSCH, and therefore similar physical layer messages from the eNodeB are needed to indicate which resources in the frequency domain contain the downlink data transmissions intended for particular UEs, together with parameters such as the modulation scheme and code rate used for the data. Explicit signalling of this kind avoids the considerable additional complexity which would arise if UEs had to search for their data among all the possible combinations of data packet size, format and resource allocation.

In order to facilitate efficient operation of HARQ,[7] further physical layer signals are needed to convey acknowledgements of uplink data packets received by the eNodeB, and power control commands are needed to ensure that uplink transmissions are made at appropriate power levels (as explained in Section 18.3).

9.3.1.2 Flexibility, Overhead and Complexity

The LTE physical layer specification is intended to allow operation in any system bandwidth from six resource blocks (1.08 MHz) to 110 resource blocks (19.8 MHz). It is also designed to support a range of scenarios including, for example, just a few users in a cell each demanding high data rates, or very many users with low data rates. Considering the possibility that both uplink resource grants and downlink resource allocations could be required for every UE in each subframe, the number of control channel messages carrying resource information could be as many as a couple of hundred if every resource allocation were as small as one resource block. Since every additional control channel message implies additional overhead which consumes downlink resources, it is desirable that the control channel is designed to minimize unnecessary overhead for any given signalling load, whatever the system bandwidth.

Similar considerations apply to the signalling of HARQ acknowledgements for each uplink packet transmission.

Furthermore, as in any mobile communication system, the complexity and power consumption of the terminals are important considerations for LTE. Therefore, the control signalling must be designed so that the necessary scalability and flexibility is achieved without undue decoding complexity.

9.3.1.3 Coverage and Robustness

In order to achieve good coverage it must be possible to configure the system so that the control channels can be received with sufficient reliability over a substantial part of every cell. As an example, if a message indicating resource allocation is not received correctly, then the corresponding data transmission will also fail, with a direct and proportionate impact

[7]Hybrid Automatic Repeat reQuest – see Sections 4.4 and 10.3.2.5.

on throughput efficiency. Techniques such as channel coding and frequency diversity can be used to make the control channels more robust. However, in order to make good use of system resources, it is desirable to be able to adapt the transmission parameters of the control signalling for different UEs or groups of UEs, so that lower code rates and higher power levels are only applied for those UEs for which it is necessary (e.g. near the cell border, where signal levels are likely to be low and interference from other cells high).

Also, it is desirable to avoid unintended reception of control channels from other cells, by applying cell-specific randomization.

9.3.1.4 System-Related Design Aspects

Since the different parts of LTE are intended to provide a complete system, some aspects of control channel design cannot be considered in isolation.

A basic design decision in LTE is that a control channel message is intended to be transmitted to a particular UE (or, in some cases, a group of UEs). Therefore, in order to reach multiple UEs in a cell within a subframe, it must be possible to transmit multiple control channels within the duration of a single subframe. However, in cases where the control channel messages are intended for reception by more than one UE (for example, when relating to the transmission of a SIB on the PDSCH), it is more efficient to arrange for all the UEs to receive a single transmission rather than to transmit the same information to each UE individually. This requires that both common and dedicated control channel messages be supported.

Finally, some scenarios may be characterized by the data arriving at regular intervals, as is typical for VoIP traffic. It is then possible to predict in advance when resources will need to be assigned in the downlink or granted in the uplink, and the number of control channel messages which need to be sent can be reduced by means of 'Semi-Persistent Scheduling' (SPS) as discussed in Section 4.4.2.1.

9.3.2 Control Channel Structure

Three downlink physical control channels are provided in LTE: the Physical Control Format Indicator CHannel (PCFICH), the Physical HARQ Indicator CHannel (PHICH) and the Physical Downlink Control CHannel (PDCCH). In general, the downlink control channels can be configured to occupy the first 1, 2 or 3 OFDM symbols in a subframe, extending over the entire system bandwidth as shown in Figure 9.5.

This flexibility allows the control channel overhead to be adjusted according to the particular system configuration, traffic scenario and channel conditions. There are two special cases: in subframes containing MBSFN transmissions (see Sections 9.2.3 and 13.4.1) there may be 0, 1 or 2 OFDM symbols for control signalling, while for narrow system bandwidths (less than 10 RBs) the number of control symbols is increased and may be 2, 3 or 4 to ensure sufficient coverage at the cell border.

9.3.3 Physical Control Format Indicator CHannel (PCFICH)

The PCFICH carries a Control Format Indicator (CFI) which indicates the number of OFDM symbols (i.e. normally 1, 2 or 3) used for transmission of control channel information in each subframe. In principle, the UE could deduce the value of the CFI without a channel such as

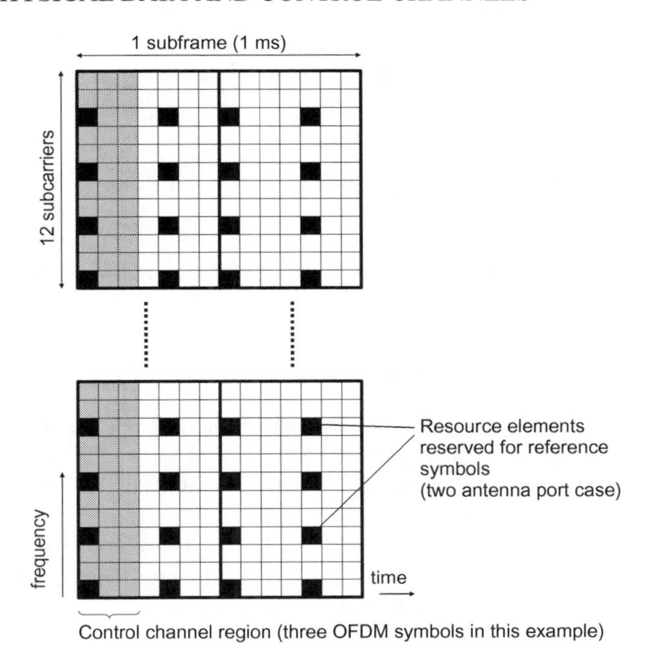

Figure 9.5: The time-frequency region used for downlink control signalling.

the PCFICH, for example by multiple attempts to decode the control channels assuming each possible number of symbols, but this would result in significant additional processing load. Three different CFI values are used in LTE. In order to make the CFI sufficiently robust, each codeword is 32 bits in length, mapped to 16 REs using QPSK modulation. These 16 REs are arranged in groups of 4, known as Resource Element Groups (REGs). The REs occupied by RSs are not included within the REGs, which means that the total number of REGs in a given OFDM symbol depends on whether or not cell-specific RSs are present. The concept of REGs (i.e. mapping in groups of four REs) is also used for the other downlink control channels (the PHICH and PDCCH).

The PCFICH is transmitted on the same set of antenna ports as the PBCH, with transmit diversity being applied if more than one antenna port is used.

In order to achieve frequency diversity, the 4 REGs carrying the PCFICH are distributed across the frequency domain. This is done according to a predefined pattern in the first OFDM symbol in each downlink subframe (see Figure 9.6), so that the UEs can always locate the PCFICH information, which is a prerequisite to being able to decode the rest of the control signalling.

To minimize the possibility of confusion with PCFICH information from a neighbouring cell, a cell-specific frequency offset is applied to the positions of the PCFICH REs; this offset depends on the Physical Cell ID (PCI), which is deduced from the Primary and Secondary Synchronization Signals (PSS and SSS) as explained in Section 7.2. In addition, a cell-specific scrambling sequence (again a function of the PCI) is applied to the CFI codewords, so that the UE can preferentially receive the PCFICH from the desired cell.

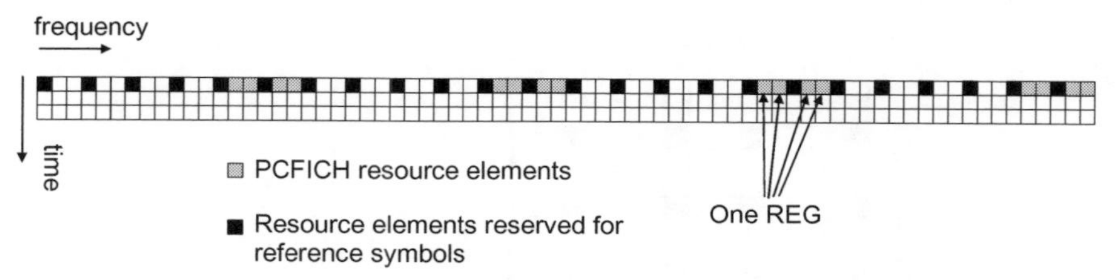

Figure 9.6: PCFICH mapping to Resource Element Groups (REGs).

9.3.4 Physical Hybrid ARQ Indicator Channel (PHICH)

The PHICH carries the HARQ ACK/NACK, which indicates whether the eNodeB has correctly received a transmission on the PUSCH. The HARQ indicator is set to 0 for a positive ACKnowledgement (ACK) and 1 for a Negative ACKnowledgement (NACK). This information is repeated in each of three BPSK[8] symbols.

Multiple PHICHs are mapped to the same set of REs. These constitute a PHICH group, where different PHICHs within the same PHICH group are separated through different complex orthogonal Walsh sequences. Each PHICH is uniquely identified by a PHICH index, which indicates both the group and the sequence. The sequence length is four for the normal cyclic prefix (or two in the case of the extended cyclic prefix). As the sequences are complex, the number of PHICHs in a group (i.e. the number of UEs receiving their acknowledgements on the same set of downlink REs) can be up to twice the sequence length. A cell-specific scrambling sequence is applied.

Factor-3 repetition coding is applied for robustness, resulting in three instances of the orthogonal Walsh code being transmitted for each ACK or NACK. The error rate on the PHICH is intended to be of the order of 10^{-2} for ACKs and as low as 10^{-4} for NACKs. The resulting PHICH construction, including repetition and orthogonal spreading, is shown in Figure 9.7.

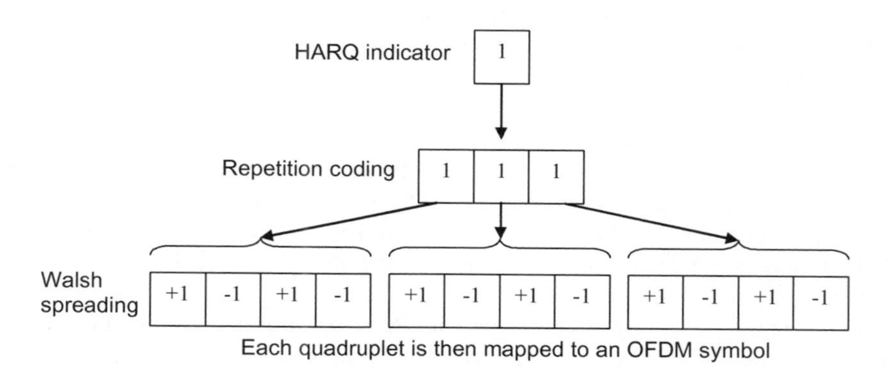

Figure 9.7: An example of PHICH signal construction.

[8]Binary Phase Shift Keying.

The PHICH duration, in terms of the number of OFDM symbols used in the time domain, is configurable (by an indication transmitted on the PBCH), normally to either one or three OFDM symbols.[9] As the PHICH cannot extend into the PDSCH transmission region, the duration configured for the PHICH puts a lower limit on the size of the control channel region at the start of each subframe (as signalled by the PCFICH).

Finally, each of the three instances of the orthogonal code of a PHICH transmission is mapped to an REG on one of the first three OFDM symbols of each subframe,[10] in such a way that each PHICH is partly transmitted on each of the available OFDM symbols. This mapping is illustrated in Figure 9.8 for each possible PHICH duration.

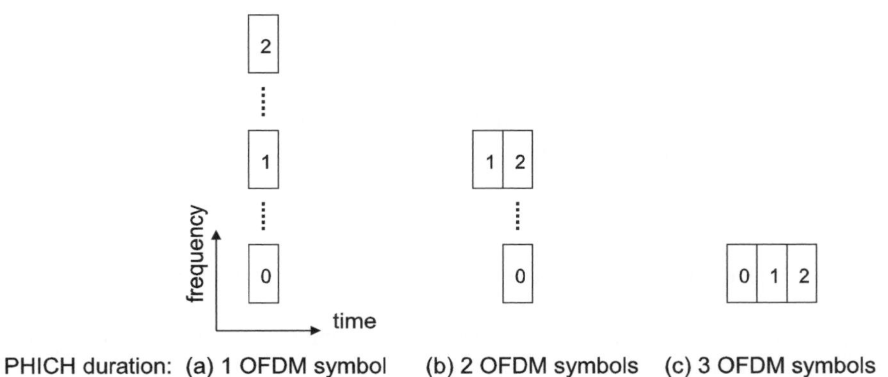

Figure 9.8: Examples of the mapping of the three instances of a PHICH orthogonal code to OFDM symbols, depending on the configured PHICH duration.

The PBCH also signals the number of PHICH groups configured in the cell, which enables the UEs to deduce to which remaining REs in the control region the PDCCHs are mapped.[11]

In order to obviate the need for additional signalling to indicate which PHICH carries the ACK/NACK response for each PUSCH transmission, the PHICH index is implicitly associated with the index of the lowest uplink RB used for the corresponding PUSCH transmission. This relationship is such that adjacent PUSCH RBs are associated with PHICHs in different PHICH groups, to enable some degree of load balancing. However, this mechanism alone is not sufficient to enable multiple UEs to be allocated the same RBs for a PUSCH transmission, as occurs in the case of uplink multi-user MIMO (see Section 16.6); in this case, different cyclic shifts of the uplink demodulation RSs are configured for the different UEs which are allocated the same time-frequency PUSCH resources, and the same

[9]In some special cases, the three-OFDM-symbol duration is reduced to two OFDM symbols; these cases are (i) MBSFN subframes on mixed carriers supporting MBSFN and unicast data, and (ii) the second and seventh subframes in case of frame structure type 2 for Time Division Duplex (TDD) operation.

[10]The mapping avoids REs used for reference symbols or PCFICH.

[11]For Frequency Division Duplex (FDD) operation with Frame Structure Type 1 (see Section 6.2), the configured number of PHICH groups is the same in all subframes; for TDD operation with Frame Structure Type 2, the number of PHICH groups is 0, 1 or 2 times the number signalled by the PBCH, according to the correspondence with uplink subframes.

cyclic shift index is then used to shift the PHICH allocations in the downlink so that each UE receives its ACK or NACK on a different PHICH. This mapping of the PHICH allocations is illustrated in Figure 9.9.

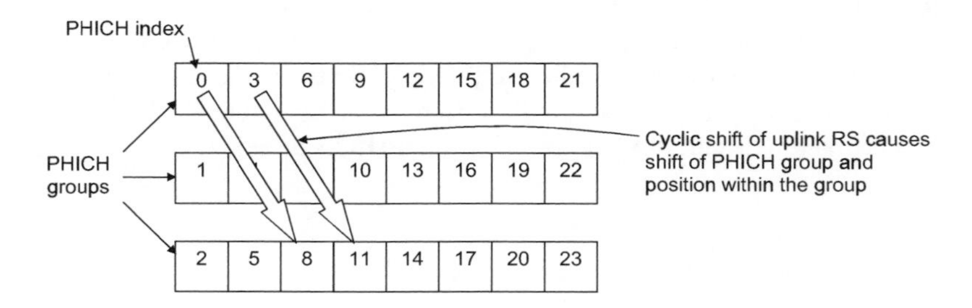

Figure 9.9: Indexing of PHICHs within PHICH groups, and shifting in the case of cyclic shifting of the uplink demodulation reference signals.

The PHICH indexing for the case of uplink MIMO in Release 10 is explained in Section 29.4.1. The use of the PHICH in the case of aggregation of multiple carriers in Release 10 is explained in Section 28.3.1.3.

The PHICHs are transmitted on the same set of antenna ports as the PBCH, and transmit diversity is applied if more than one antenna port is used.

9.3.5 Physical Downlink Control CHannel (PDCCH)

Each PDCCH carries a message known as Downlink Control Information (DCI), which includes resource assignments and other control information for a UE or group of UEs. In general, several PDCCHs can be transmitted in a subframe.

Each PDCCH is transmitted using one or more *Control Channel Elements* (CCEs), where each CCE corresponds to nine REGs. Four QPSK symbols are mapped to each REG.

Four PDCCH formats are supported, as listed in Table 9.1.

Table 9.1: PDCCH formats.

PDCCH format	Number of CCEs (n)	Number of REGs	Number of PDCCH bits
0	1	9	72
1	2	18	144
2	4	36	288
3	8	72	576

CCEs are numbered and used consecutively, and, to simplify the decoding process, a PDCCH with a format consisting of n CCEs may only start with a CCE with a number equal to a multiple of n.

The number of CCEs aggregated for transmission of a particular PDCCH is known as the 'aggregation level' and is determined by the eNodeB according to the channel conditions. For example, if the PDCCH is intended for a UE with a good downlink channel (e.g. close to the eNodeB), then one CCE is likely to be sufficient. However, for a UE with a poor channel (e.g. near the cell border) then eight CCEs may be required in order to achieve sufficient robustness. In addition, the power level of a PDCCH may be adjusted to match the channel conditions.

9.3.5.1 Formats for Downlink Control Information (DCI)

The required content of the control channel messages depends on the system deployment and UE configuration. For example, if the infrastructure does not support MIMO, or if a UE is configured in a transmission mode which does not involve MIMO, there is no need to signal the parameters that are only required for MIMO transmissions. In order to minimize the signalling overhead, it is therefore desirable that several different message formats are available, each containing the minimum payload required for a particular scenario. On the other hand, to avoid too much complexity in implementation and testing, it is desirable not to specify too many formats. The set of DCI message formats in Table 9.2 is specified in LTE; Format 2B was added in Release 9, and Formats 2C and 4 were added in Release 10. Additional formats may be defined in future.

Table 9.2: Supported DCI formats.

DCI format	Purpose	Applicable PDSCH transmission mode(s)
0	PUSCH grants	All
1	PDSCH assignments with a single codeword	1,2,7
1A	PDSCH assignments using a compact format	All
1B	PDSCH assignments for rank-1 transmission	6
1C	PDSCH assignments using a very compact format	n/a
1D	PDSCH assignments for multi-user MIMO	5
2	PDSCH assignments for closed-loop MIMO operation	4
2A	PDSCH assignments for open-loop MIMO operation	3
2B	PDSCH assignments for dual-layer beamforming	8
2C	PDSCH assignments for up to 8-layer spatial multiplexing	9
3	Transmit Power Control (TPC) commands for multiple users for PUCCH and PUSCH with 2-bit power adjustments	n/a
3A	Transmit Power Control (TPC) commands for multiple users for PUCCH and PUSCH with 1-bit power adjustments	n/a
4	PUSCH grants for up to 4-layer spatial multiplexing	All (if configured for PUSCH transmission mode 2)

The information content of the different DCI message formats is listed below for Frequency Division Duplex (FDD) operation. Some small differences exist for Time Division Duplex (TDD), and these are outlined afterwards.

Format 0. DCI Format 0 is used for the transmission of resource grants for the PUSCH. The following information is transmitted:

- Flag to differentiate between Format 0 and Format 1A;
- Resource assignment and frequency hopping flag;
- Modulation and Coding Scheme (MCS);
- New Data Indicator (NDI);
- HARQ information and Redundancy Version (RV);
- Power control command for scheduled PUSCH;
- Cyclic shift for uplink Demodulation RS;
- Request for transmission of an aperiodic CQI report (see Sections 10.2.1 and 28.3.2.3).

Format 1. DCI Format 1 is used for the transmission of resource assignments for single codeword PDSCH transmissions (transmission modes 1, 2 and 7 (see Section 9.2.2.1)). The following information is transmitted:

- Resource allocation type (see Section 9.3.5.4);
- RB assignment;
- MCS;
- HARQ information and RV;
- Power control command for Physical Uplink Control CHannel (PUCCH).

Format 1A. DCI Format 1A is used for compact signalling of resource assignments for single codeword PDSCH transmissions for any PDSCH transmission mode. It is also used to allocate a dedicated preamble signature to a UE to trigger contention-free random access (see Section 17.3.2); in this case the PDCCH message is known as a *PDCCH order*. The following information is transmitted:

- Flag to differentiate between Format 0 and Format 1A;
- Flag to indicate that the distributed mapping mode (see Section 9.2.2.1) is used for the PDSCH transmission (otherwise the allocation is a contiguous set of physical RBs);
- RB assignment;
- MCS;
- HARQ information and RV;
- Power control command for PUCCH.

Format 1B. DCI Format 1B is used for compact signalling of resource assignments for PDSCH transmissions using closed-loop precoding with rank-1 transmission (transmission mode 6). The information transmitted is the same as in Format 1A, but with the addition of an indicator of the precoding vector applied for the PDSCH transmission.

Format 1C. DCI Format 1C is used for very compact transmission of PDSCH assignments. When format 1C is used, the PDSCH transmission is constrained to using QPSK modulation. This is used, for example, for signalling paging messages and some broadcast system information messages (see Section 9.2.2.2), and for notifying UEs of a change of MBMS control information on the Multicast Control Channel (MCCH – see Section 13.6.3.2). The following information is transmitted:

- RB assignment;
- Coding scheme.

The RV is not signalled explicitly, but is deduced from the SFN (see [1, Section 5.3.1]).

Format 1D. DCI Format 1D is used for compact signalling of resource assignments for PDSCH transmissions using multi-user MIMO (transmission mode 5). The information transmitted is the same as in Format 1B, but, instead of one of the bits of the precoding vector indicators, there is a single bit to indicate whether a power offset is applied to the data symbols. This is needed to show whether the transmission power is shared between two UEs.

Format 2. DCI Format 2 is used for the transmission of resource assignments for PDSCH for closed-loop MIMO operation (transmission mode 4). The following information is transmitted:

- Resource allocation type (see Section 9.3.5.4);
- RB assignment;
- Power control command for PUCCH;
- HARQ information and RV for each codeword;
- MCS for each codeword;
- A flag to indicate if the mapping from transport blocks to codewords is reversed;
- Number of spatial layers;
- Precoding information and indication of whether one or two codewords are transmitted on the PDSCH.

Format 2A. DCI Format 2A is used for the transmission of resource assignments for PDSCH for open-loop MIMO operation (transmission mode 3). The information transmitted is the same as for Format 2, except that if the eNodeB has two transmit antenna ports, there is no precoding information, and, for four antenna ports, two bits are used to indicate the transmission rank.

Format 2B. DCI Format 2B is introduced in Release 9 and is used for the transmission of resource assignments for PDSCH for dual-layer beamforming (transmission mode 8). The information transmitted is similar to Format 2A, except that no precoding information is included and the bit in Format 2A for indicating reversal of the transport block to codeword mapping is replaced in Format 2B by a bit indicating the scrambling code applied to the UE-specific RSs for the corresponding PDSCH transmission (see Section 8.2.3).

Format 2C. DCI Format 2C is introduced in Release 10 and is used for the transmission of resource assignments for PDSCH for closed-loop single-user or multi-user MIMO operation with up to 8 layers (transmission mode 9). The information transmitted is similar to Format 2B; full details are given in Section 29.3.2.

Formats 3 and 3A. DCI Formats 3 and 3A are used for the transmission of power control commands for PUCCH and PUSCH, with 2-bit or 1-bit power adjustments respectively. These DCI formats contain individual power control commands for a group of UEs.

Format 4. DCI Format 4 is introduced in Release 10 and is used for the transmission of resource grants for the PUSCH when the UE is configured in PUSCH transmission mode 2 for uplink single-user MIMO. The information transmitted is similar to Format 0, with the addition of MCS and NDI information for a second transport block, and precoding information; full details are given in Section 29.4.

DCI Formats for TDD. In TDD operation, the DCI formats contain the same information as for FDD, but with some additions (see Section 23.4.3 for an explanation of the usage of these additions):

- Uplink index (in DCI Formats 0 and 4, uplink-downlink configuration 0 only);
- Downlink Assignment Index (DAI) (in DCI Formats 0, 1, 1A, 1B, 1D, 2, 2A 2B, 2C and 4, uplink-downlink configurations 1–6 only); see Section 23.4.3 for details of DAI usage.

DCI Format modifications in Release 10. In the case of aggregation of multiple carriers in Release 10, DCI Formats 0, 1, 1A, 1B, 1D, 2, 2A, 2B, 2C and 4 can be configured to include a carrier indicator for cross-carrier scheduling; this is explained in detail in Section 28.3.1.1. In DCI Formats 0 and 4, additional fields are included to request transmission of an aperiodic Sounding Reference Signal (SRS) (see Section 29.2.2) and to indicate whether the uplink PRB allocation is contiguous or multi-clustered (see Section 28.3.6.2 for details). In TDD operation, DCI Formats 2B and 2C may also be configured to include an additional field to request transmission of an aperiodic SRS.

9.3.5.2 PDCCH CRC Attachment

In order that the UE can identify whether it has received a PDCCH transmission correctly, error detection is provided by means of a 16-bit CRC appended to each PDCCH. Furthermore, it is necessary that the UE can identify which PDCCH(s) are intended for it. This could in theory be achieved by adding an identifier to the PDCCH payload; however, it turns out to be more efficient to scramble the CRC with the 'UE identity', which saves the additional payload but at the cost of a small increase in the probability of falsely detecting a PDCCH intended for another UE.

In addition, for UEs which support antenna selection for uplink transmissions (see Section 16.6), the requested antenna may be indicated using Format 0 by applying an antenna-specific mask to the CRC. This has the advantage that the same size of DCI message can be used, irrespective of whether antenna selection is used.

9.3.5.3 PDCCH Construction

In general, the number of bits required for resource assignment depends on the system bandwidth, and therefore the message sizes also vary with the system bandwidth. The numbers of payload bits for each DCI format (including information bits and CRC) are summarized in Table 9.3, for each of the supported values of system bandwidth. In addition, padding bits are added if necessary in the following cases:

- To ensure that Formats 0 and 1A are the same size, even in the case of different uplink and downlink bandwidths, in order to avoid additional complexity at the UE receiver;
- To ensure that Formats 3 and 3A are the same size as Formats 0 and 1A, likewise to avoid additional complexity at the UE receiver;
- To avoid potential ambiguity in identifying the correct PDCCH location as described in Section 9.3.5.5;
- To ensure that Format 1 has a different size from Formats 0/1A, so that these formats can be easily distinguished at the UE receiver;
- To ensure that Format 4 has a different size from Formats 1/2/2A/2B/2C, again so that this format can be easily distinguished.

In Release 10, some optional additional information bits may be configured in some of the DCI formats; these are not included in Table 9.3, but are explained below the table. Because their presence may affect the number of padding bits, any such additional bits do not necessarily increase the transmitted size of the DCI format by the same amount.

In order to provide robustness against transmission errors, the PDCCH information bits are coded as described in Section 10.3.3. The set of coded and rate-matched bits for each PDCCH are then scrambled with a cell-specific scrambling sequence; this reduces the possibility of confusion with PDCCH transmissions from neighbouring cells. The scrambled bits are mapped to blocks of four QPSK symbols (REGs). Interleaving is applied to these symbol blocks, to provide frequency diversity, followed by mapping to the available physical REs on the set of OFDM symbols indicated by the PCFICH. This mapping process excludes the REs reserved for RSs and the other control channels (PCFICH and PHICH).

The PDCCHs are transmitted on the same set of antenna ports as the PBCH, and transmit diversity is applied if more than one antenna port is used.

9.3.5.4 Resource Allocation

Conveying indications of physical layer resource allocation is one of the major functions of the PDCCHs. While the exact use of the PDCCHs depends on the algorithms implemented in the eNodeB, it is nevertheless possible to outline some general principles of typical operation.

In each subframe, PDCCHs indicate the frequency-domain resource allocations. As discussed in Section 9.2.2.1, resource allocations are normally localized, meaning that a Physical RB (PRB) in the first half of a subframe is paired with the PRB at the same frequency in the second half of the subframe. For simplicity, the explanation here is in terms of the first half subframe only.

The main design challenge for the signalling of frequency-domain resource allocations (in terms of a set of RBs) is to find a good compromise between flexibility and signalling overhead. The most flexible, and arguably the simplest, approach is to send each UE a

Table 9.3: DCI format payload sizes (in bits), without padding, for different FDD system bandwidths.

	Bandwidth (PRBs)					
	6	15	25	50	75	100
Format 0	35	37	39	41	42	43
Format 1	35	38	43	47	49	55
Format 1A	36	38	40	42	43	44
Format 1B/1D (2 transmit antenna ports)	38	40	42	44	45	46
Format 1C	24	26	28	29	30	31
Format 2 (2 transmit antenna ports)	47	50	55	59	61	67
Format 2A (2 transmit antenna ports)	44	47	52	56	58	64
Format 2B (2 or 4 transmit antenna ports)	44	47	52	56	58	64
Format 2C	46	49	54	58	60	66
Format 4 (2 UE transmit antennas)	46	47	50	52	53	54
Format 1B/1D (4 transmit antenna ports)	40	42	44	46	47	48
Format 2 (4 transmit antenna ports)	50	53	58	62	64	70
Format 2A (4 transmit antenna ports)	46	49	54	58	60	66
Format 4 (4 UE transmit antennas)	49	50	53	55	56	57

Note that for Release 10 UEs:

- DCI Format 0 is extended by 1 bit for a multi-cluster resource allocation flag and may be further extended by 1 or 2 bits for aperiodic CQI request (see Section 28.3.2.3) and aperiodic SRS request (see Section 29.2.2), depending on configuration.
- DCI Format 1A may be extended by 1 bit for aperiodic SRS request, depending on configuration.
- In TDD operation, DCI Formats 2B and 2C may be extended by 1 bit for aperiodic SRS request, depending on configuration.
- DCI Format 4 always includes 2 bits for requesting aperiodic SRS (see Section 29.2.2), and 1 bit for aperiodic CQI request, but may be extended by an additional 1 bit for aperiodic CQI request in case of carrier aggregation (see Section 28.3.2.3).
- DCI Formats 0, 1, 1A, 1B, 1D, 2, 2A, 2B, 2C and 4 can be configured to be extended by 3 bits for a carrier indicator field (see Section 28.3.1.1).

bitmap in which each bit indicates a particular PRB. This would work well for small system bandwidths, but for large system bandwidths (i.e. up to 110 PRBs) the bitmap would need 110 bits, which would be a prohibitive overhead – particularly for small packets, where the PDCCH message could be larger than the data packet! One possible solution would be to send a combined resource allocation message to all UEs, but this was rejected on the grounds of the high power needed to reach all UEs reliably, including those at the cell edges. The approaches adopted in LTE Releases 8 and 9 are listed in Table 9.4, and further details are given below.

Resource allocation Type 0. In resource allocations of Type 0, a bitmap indicates the Resource Block Groups (RBGs) which are allocated to the scheduled UE, where an RBG is a set of consecutive PRBs. The RBG size (P) is a function of the system bandwidth as shown in Table 9.5. The total number of RBGs (N_{RBG}) for a downlink system bandwidth of

Table 9.4: Methods for indicating Resource Block (RB) allocation.

Method	UL/DL	Description	Number of bits required (see text for definitions)
Direct bitmap	DL	The bitmap comprises 1 bit per RB. This method is the only one applicable when the bandwidth is less than 10 resource blocks.	$N_{\mathrm{RB}}^{\mathrm{DL}}$
Bitmap: 'Type 0'	DL	The bitmap addresses Resource Block Groups (RBGs), where the group size (2, 3 or 4) depends on the system bandwidth.	$\lceil N_{\mathrm{RB}}^{\mathrm{DL}}/P \rceil$
Bitmap: 'Type 1'	DL	The bitmap addresses individual RBs in a subset of RBGs. The number of subsets (2, 3, or 4) depends on the system bandwidth. The number of bits is arranged to be the same as for Type 0, so the same DCI format can carry either type of allocation.	$\lceil N_{\mathrm{RB}}^{\mathrm{DL}}/P \rceil$
Contiguous allocations: 'Type 2'	DL or UL	Any possible arrangement of contiguous RB allocations can be signalled in terms of a starting position and number of RBs.	$\lceil \log_2(N_{\mathrm{RB}}^{\mathrm{DL}}(N_{\mathrm{RB}}^{\mathrm{DL}} + 1)) \rceil$ or $\lceil \log_2(N_{\mathrm{RB}}^{\mathrm{UL}}(N_{\mathrm{RB}}^{\mathrm{UL}} + 1)) \rceil$
Distributed allocations	DL	In the downlink, a limited set of resource allocations can be signalled where the RBs are scattered across the frequency domain and shared between two UEs. The number of bits is arranged to be the same as for contiguous allocations Type 2, so the same DCI format can carry either type of allocation.	$\lceil \log_2(N_{\mathrm{RB}}^{\mathrm{DL}}(N_{\mathrm{RB}}^{\mathrm{DL}} + 1)) \rceil$

$N_{\mathrm{RB}}^{\mathrm{DL}}$ PRBs is given by $N_{\mathrm{RBG}} = \lceil N_{\mathrm{RB}}^{\mathrm{DL}}/P \rceil$. An example for the case of $N_{\mathrm{RB}}^{\mathrm{DL}} = 25$, $N_{\mathrm{RBG}} = 13$ and $P = 2$ is shown in Figure 9.10, where each bit in the bitmap indicates a pair of PRBs (i.e. two PRBs which are adjacent in frequency).

Table 9.5: RBG size for Type 0 resource allocation.

Downlink bandwidth $N_{\mathrm{RB}}^{\mathrm{DL}}$	RBG size (P)
$0 \leq 10$	1
11–26	2
27–63	3
64–110	4

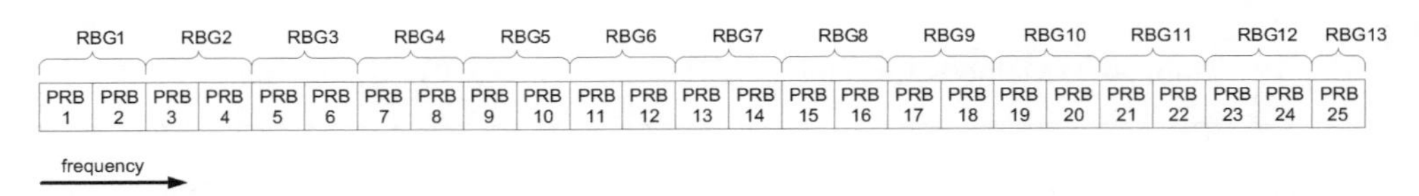

Figure 9.10: PRB addressed by a bitmap Type 0, each bit addressing a complete RBG.

Resource allocation Type 1. In resource allocations of Type 1, individual PRBs can be addressed (but only within a subset of the PRBs available). The bitmap used is slightly smaller than for Type 0, since some bits are used to indicate which subset of the RBG is addressed, and a shift in the position of the bitmap. The total number of bits (including these additional flags) is the same as for Type 0. An example for the case of $N_{\mathrm{RB}}^{\mathrm{DL}} = 25$, $N_{\mathrm{RBG}} = 11$ and $P = 2$ is shown in Figure 9.11.

The motivation for providing this method of resource allocation is flexibility in spreading the resources across the frequency domain to exploit frequency diversity.

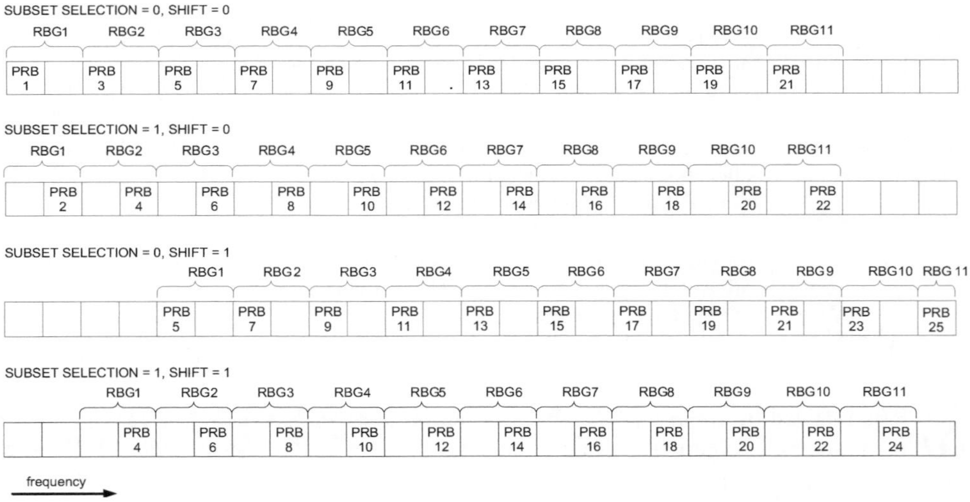

Figure 9.11: PRBs addressed by a bitmap Type 1, each bit addressing a subset of an RBG, depending on a subset selection and shift value.

Resource allocation Type 2. In resource allocations of Type 2, the resource allocation information indicates a contiguous set of PRBs, using either localized or distributed mapping (see Section 9.2.2.1) as indicated by a 1-bit flag in the resource allocation message. PRB allocations may vary from a single PRB up to a maximum number of PRBs spanning the system bandwidth. A Type 2 resource allocation field consists of a Resource Indication

Value (RIV) corresponding to a starting RB (RB_{START}) and a length in terms of contiguously-allocated RBs (L_{CRBs}). The resource indication value is defined by

$$\text{if} \quad (L_{CRBs} - 1) \leq \lfloor N_{RB}^{DL}/2 \rfloor \quad \text{then} \quad RIV = N_{RB}^{DL}(L_{CRBs} - 1) + RB_{START}$$

$$\text{else} \quad RIV = N_{RB}^{DL}(N_{RB}^{DL} - L_{CRBs} + 1) + (N_{RB}^{DL} - 1 - RB_{START})$$

An example of a method for reversing the mapping to derive the resource allocation from the RIV can be found in [2].

Resource allocation in Release 10. In addition to the above methods, Release 10 supports a non-contiguous resource allocation method for the uplink, allowing two separate contiguous sets of PRBs to be assigned, as explained in detail in Section 28.3.6.2.

9.3.5.5 PDCCH Transmission and Blind Decoding

The previous discussion has covered the structure and possible contents of an individual PDCCH message, and transmission by an eNodeB of multiple PDCCHs in a subframe. This section addresses the question of how these transmissions are organized so that a UE can locate the PDCCHs intended for it, while at the same time making efficient use of the resources allocated for PDCCH transmission.

A simple approach, at least for the eNodeB, would be to allow the eNodeB to place any PDCCH anywhere in the PDCCH resources (or CCEs) indicated by the PCFICH. In this case, the UE would need to check all possible PDCCH locations, PDCCH formats and DCI formats, and act on the messages with correct CRCs (taking into account that the CRC is scrambled with a UE identity). Carrying out such a 'blind decoding' of all the possible combinations would require the UE to make many PDCCH decoding attempts in every subframe. For small system bandwidths, the computational load would be reasonable, but for large system bandwidths, with a large number of possible PDCCH locations, it would become a significant burden, leading to excessive power consumption in the UE receiver. For example, blind decoding of 100 possible CCE locations for PDCCH Format 0 would be equivalent to continuously receiving a data rate of around 4 Mbps.

The alternative approach adopted for LTE is to define for each UE a limited set of CCE locations where a PDCCH may be placed. Such a constraint may lead to some limitations as to which UEs can be sent PDCCHs within the same subframe, which thus restricts the UEs to which the eNodeB can grant resources. Therefore it is important for good system performance that the set of possible PDCCH locations available for each UE is not too small.

The set of CCE locations in which the UE may find its PDCCHs can be considered as a 'search space'. In LTE, the search space is a different size for each PDCCH format. Moreover, separate *UE-specific* and *common* search spaces are defined; a UE-specific search space is configured for each UE individually, whereas all UEs are aware of the extent of the common search space. Note that the UE-specific and common search spaces may overlap for a given UE. The sizes of the common and UE-specific search spaces in Releases 8 and 9 are listed in Table 9.6.

With such small search spaces, it is quite possible in a given subframe that the eNodeB cannot find CCE resources to send PDCCHs to all the UEs that it would like to, because having assigned some CCE locations, the remaining ones are not in the search space of a

Table 9.6: Search spaces for PDCCH formats in Releases 8 and 9.

PDCCH format	Number of CCEs (n)	Number of candidates in common search space	Number of candidates in UE-specific search space
0	1	—	6
1	2	—	6
2	4	4	2
3	8	2	2

Note: The search space sizes in Release 10 are discussed in Section 28.3.1.1.

particular UE. To minimize the possibility of such blocking persisting into the next subframe, a UE-specific hopping sequence (derived from the UE identity) is applied to the starting positions of the UE-specific search spaces from subframe to subframe.

In order to keep under control the computational load arising from the total number of blind decoding attempts, the UE is not required to search for all the defined DCI formats simultaneously. Typically, in the UE-specific search space, the UE will always search for Formats 0 and 1A, which are the same size and are distinguished by a flag in the message. In addition, a UE may be required to receive a further format (i.e. 1, 1B, 1D, 2, 2A or 2B, depending on the PDSCH transmission mode configured by the eNodeB).

In the common search space, the UE will typically search for Formats 1A and 1C. In addition, the UE may be configured to search for Formats 3 or 3A, which have the same size as Formats 0 and 1A, and may be distinguished by having the CRC scrambled by a different (common) identity, rather than a UE-specific one.

Considering the above, a Release 8/9 UE is required to carry out a maximum of 44 blind decodings in any subframe (12 in the common search space and 32 in the UE-specific search space). This does not include checking the same message with different CRC values, which requires only a small additional computational complexity. The number of blind decodings required for a Release 10 UE is discussed in Section 28.3.1.1.

Finally, the PDCCH structure is also adapted to avoid cases where a PDCCH CRC 'pass' might occur for multiple positions in the configured search-spaces due to repetition in the channel coding (for example, if a PDCCH was mapped to a high number of CCEs with a low code rate, then the CRC could pass for an overlapping smaller set of CCEs as well, if the channel coding repetition was thus aligned). Such cases are avoided by adding a padding bit to any PDCCH messages having a size which could result in this problem occurring.

9.3.6 PDCCH Scheduling Process

To summarize the arrangement of the PDCCH transmissions in a given subframe, a typical sequence of steps carried out by the eNodeB is depicted in Figure 9.12.

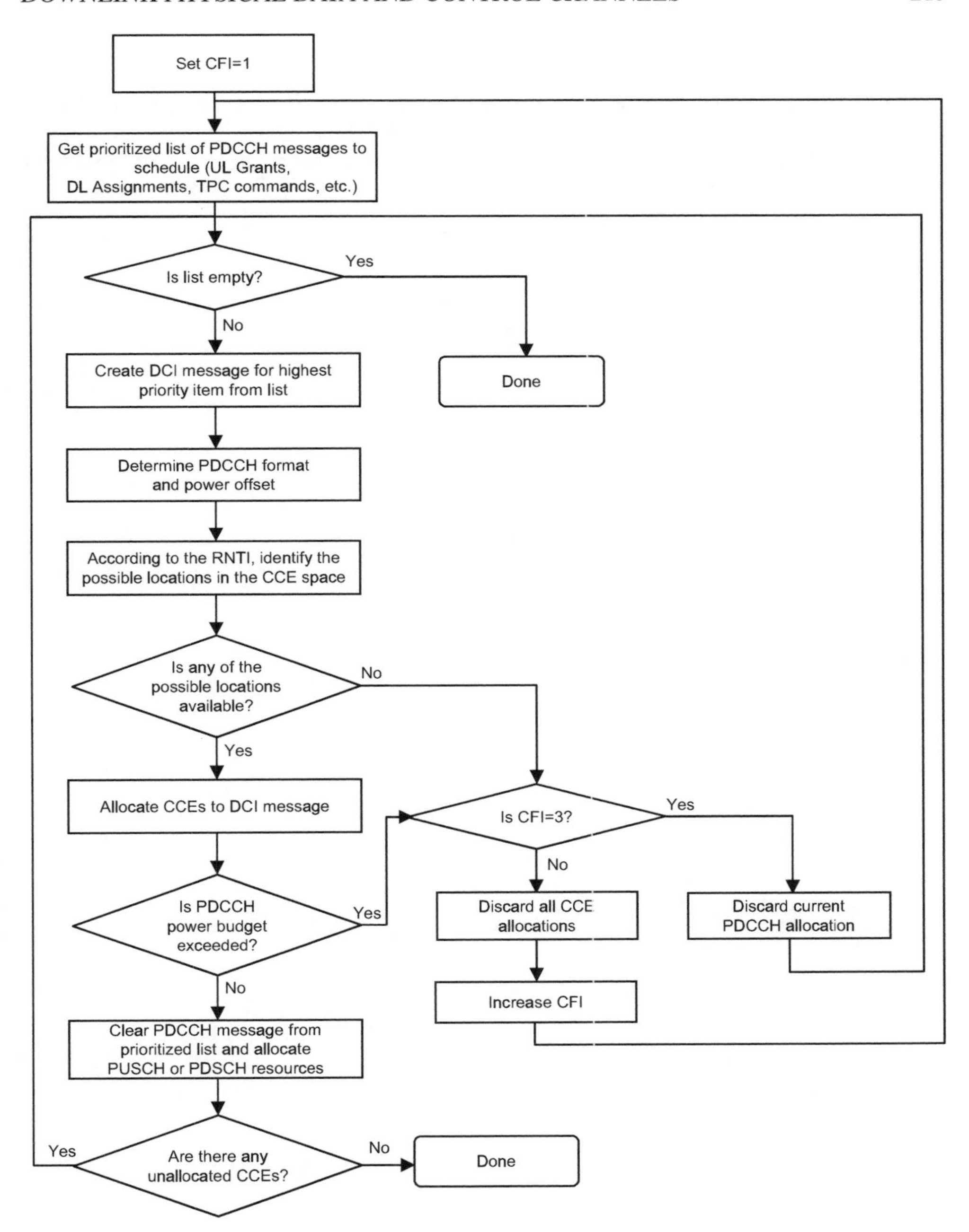

Figure 9.12: A typical sequence of PDCCH scheduling operations in a subframe.

References[12]

[1] 3GPP Technical Specification 36.321, 'Evolved Universal Terrestrial Radio Access (E-UTRA); Medium Access Control (MAC) protocol specification', www.3gpp.org.

[2] NEC, 'R1-072119: DL Unicast Resource Allocation Signalling', www.3gpp.org, 3GPP TSG RAN WG1, meeting 49, Kobe, Japan, May 2007.

[12]All web sites confirmed 1st March 2011.

10

Link Adaptation and Channel Coding

Brian Classon, Ajit Nimbalker, Stefania Sesia and Issam Toufik

10.1 Introduction

The principle of link adaptation is fundamental to the design of a radio interface which is efficient for packet-switched data traffic. Unlike the early versions of UMTS (Universal Mobile Telecommunication System), which used fast closed-loop power control to support circuit-switched services with a roughly constant data rate, link adaptation in HSPA (High Speed Packet Access) and LTE adjusts the transmitted information data rate (modulation scheme and channel coding rate) dynamically to match the prevailing radio channel capacity for each user. Link adaptation is therefore very closely related to the design of the channel coding scheme used for forward error correction.

For the downlink data transmissions in LTE, the eNodeB typically selects the modulation scheme and code rate depending on a prediction of the downlink channel conditions. An important input to this selection process is the Channel Quality Indicator (CQI) feedback transmitted by the User Equipment (UE) in the uplink. CQI feedback is an indication of the data rate which can be supported by the channel, taking into account the Signal-to-Interference-plus-Noise Ratio (SINR) and the characteristics of the UE's receiver. Section 10.2 explains the principles of link adaptation as applied in LTE; it also shows how the eNodeB can select different CQI feedback modes to trade off the improved downlink link adaptation enabled by CQI against the uplink overhead caused by the CQI itself.

The LTE specifications are designed to provide the signalling necessary for interoperability between the eNodeB and the UEs so that the eNodeB can optimize the link adaptation,

LTE – The UMTS Long Term Evolution: From Theory to Practice, Second Edition.
Stefania Sesia, Issam Toufik and Matthew Baker.

but the exact methods used by the eNodeB to exploit the information that is available are left to the manufacturer's choice of implementation.

In general, in response to the CQI feedback the eNodeB can select between QPSK, 16QAM and 64QAM[1] schemes and a wide range of code rates. As discussed further in Section 10.2.1, the optimal switching points between the different combinations of modulation order and code rate depend on a number of factors, including the required quality of service and cell throughput.

The channel coding scheme for forward error correction, on which the code rate adaptation is based, was the subject of extensive study during the standardization of LTE. The chapter therefore continues with a review of the key theoretical aspects of the types of channel coding studied for LTE: convolutional codes, turbo codes with iterative decoding, and a brief introduction of Low-Density Parity Check (LDPC) codes. The theory of channel coding has seen intense activity in recent decades, especially since the discovery of turbo codes offering near-Shannon limit performance, and the development of iterative processing techniques in general. 3GPP was an early adopter of these advanced channel coding techniques, with the turbo code being standardized in the first version of the UMTS as early as 1999. Later releases of UMTS (in HSPA) added more advanced channel coding features with the introduction of link adaptation, including Hybrid Automatic Repeat reQuest (HARQ), a combination of ARQ and channel coding which provides more robustness against fading; these schemes include incremental redundancy, whereby the code rate is progressively reduced by transmitting additional parity information with each retransmission. However, the underlying turbo code from the first version of UMTS remained untouched. Meanwhile, the academic and research communities were generating new insights into code design, iterative decoding and the implementation of decoders. Section 10.3.2 explains how these developments impacted the design of the channel coding for LTE, and in particular the decision to enhance the turbo code from UMTS by means of a new contention-free interleaver, rather than to adopt a new LDPC code.

For the LTE uplink transmissions, the link adaptation process is similar to that for the downlink, with the selection of modulation and coding schemes also being under the control of the eNodeB. An identical channel coding structure is used for the uplink, while the modulation scheme may be selected between QPSK, 16QAM and, for the highest category of UE, also 64QAM. The main difference from the downlink is that instead of basing the link adaptation on CQI feedback, the eNodeB can directly make its own estimate of the supportable uplink data rate by channel sounding, for example using the Sounding Reference Signals (SRSs) which are described separately in Section 15.6.

A final important aspect of link adaptation is its use in conjunction with multi-user scheduling in time and frequency, which enables the radio transmission resources to be shared efficiently between users as the channel capacity to individual users varies. The CQI can therefore be used not only to adapt the modulation and coding rate to the channel conditions, but also for the optimization of time/frequency selective scheduling and for inter-cell interference management as discussed in Chapter 12.

[1]Quadrature Phase Shift Keying (QPSK) and Quadrature Amplitude Modulation (QAM).

10.2 Link Adaptation and CQI Feedback

In cellular communication systems, the quality of the signal received by a UE depends on the channel quality from the serving cell, the level of interference from other cells, and the noise level. To optimize system capacity and coverage for a given transmission power, the transmitter should try to match the information data rate for each user to the variations in received signal quality (see, for example, [1, 2] and references therein). This is commonly referred to as link adaptation and is typically based on Adaptive Modulation and Coding (AMC).

The degrees of freedom for the AMC consist of the modulation and coding schemes:

- **Modulation Scheme.** Low-order modulation (i.e. few data bits per modulated symbol, e.g. QPSK) is more robust and can tolerate higher levels of interference but provides a lower transmission bit rate. High-order modulation (i.e. more bits per modulated symbol, e.g. 64QAM) offers a higher bit rate but is more prone to errors due to its higher sensitivity to interference, noise and channel estimation errors; it is therefore useful only when the SINR is sufficiently high.

- **Code rate.** For a given modulation, the code rate can be chosen depending on the radio link conditions: a lower code rate can be used in poor channel conditions and a higher code rate in the case of high SINR. The adaptation of the code rate is achieved by applying puncturing (to increase the code rate) or repetition (to reduce the code rate) to the output of a mother code, as explained in Section 10.3.2.4.

A key issue in the design of the AMC scheme for LTE was whether all Resource Blocks (RBs) allocated to one user in a subframe should use the same Modulation and Coding Scheme (MCS) (see, for example, [3–6]) or whether the MCS should be frequency-dependent within each subframe. It was shown that, in general, only a small throughput improvement arises from a frequency-dependent MCS compared to an RB-common MCS in the absence of transmission power control, and therefore the additional control signalling overhead associated with frequency-dependent MCS is not justified. Therefore in LTE the modulation and channel coding rates are constant over the allocated frequency resources for a given user, and time-domain channel-dependent scheduling and AMC are supported instead. In addition, when multiple transport blocks are transmitted to one user in a given subframe using multistream Multiple-Input Multiple-Output (MIMO) (as discussed in Chapter 11), each transport block can use an independent MCS.

In LTE, the UE can be configured to report CQIs to assist the eNodeB in selecting an appropriate MCS to use for the downlink transmissions. The CQI reports are derived from the downlink received signal quality, typically based on measurements of the downlink reference signals (see Section 8.2). It is important to note that, like HSDPA, the reported CQI is not a direct indication of SINR. Instead, the UE reports the highest MCS that it can decode with a BLER (BLock Error Rate, computed on the transport blocks) probability not exceeding 10%. Thus the information received by the eNodeB takes into account the characteristics of the UE's receiver, and not just the prevailing radio channel quality. Hence a UE that is designed with advanced signal processing algorithms (for example, using interference cancellation techniques) can report a higher channel quality and, depending on the characteristics of the eNodeB's scheduler, can receive a higher data rate.

A simple method by which a UE can choose an appropriate CQI value could be based on a set of BLER thresholds, as illustrated in Figure 10.1. The UE would report the CQI value corresponding to the MCS that ensures BLER $\leq 10^{-1}$ based on the measured received signal quality.

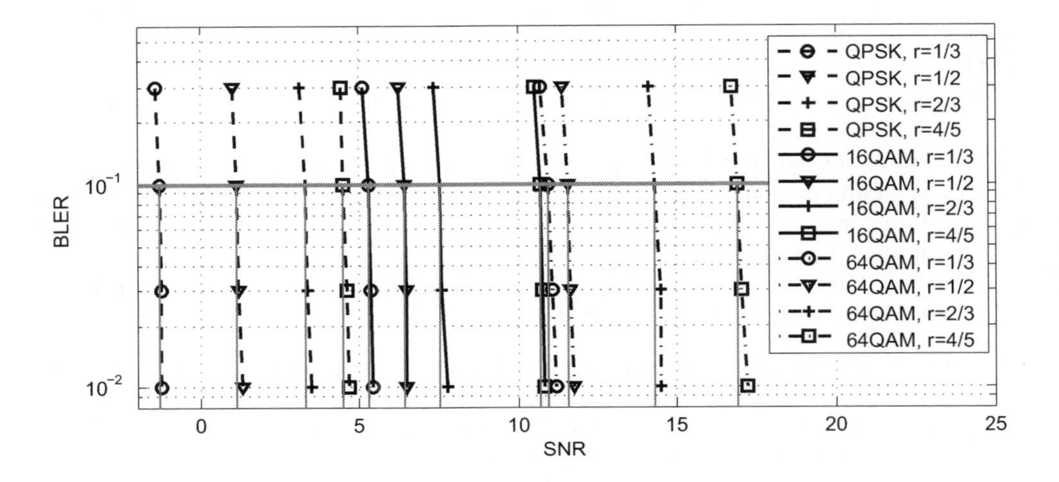

Figure 10.1: Typical BLER versus Signal-to-Noise Ratio (SNR) for different MCSs. From left to right, the curves in this example correspond to QPSK, 16QAM and 64QAM, rates 1/3, 1/2, 2/3 and 4/5.

The list of modulation schemes and code rates which can be signalled by means of a CQI value is shown in Table 10.1.

AMC can exploit the UE feedback by assuming that the channel fading is sufficiently slow. This requires the channel coherence time to be at least as long as the time between the UE's measurement of the downlink reference signals and the subframe containing the correspondingly adapted downlink transmission on the Physical Downlink Shared CHannel (PDSCH). This time is typically 7–8 ms (equivalent to a UE speed of ∼16 km/h at 1.9 GHz).

However, a trade-off exists between the amount of CQI information reported by the UEs and the accuracy with which the AMC can match the prevailing conditions. Frequent reporting of the CQI in the time domain allows better matching to the channel and interference variations, while fine resolution in the frequency domain allows better exploitation of frequency-domain scheduling. However, both lead to increased feedback overhead in the uplink. Therefore, the eNodeB can configure both the time-domain update rate and the frequency-domain resolution of the CQI, as discussed in the following section.

10.2.1 CQI Feedback in LTE

The periodicity and frequency resolution to be used by a UE to report CQI are both controlled by the eNodeB. In the time domain, both periodic and aperiodic CQI reporting are supported.

Table 10.1: CQI values. Reproduced by permission of © 3GPP.

CQI index	Modulation	Approximate code rate	Efficiency (information bits per symbol)
0	'Out-of-range'	—	—
1	QPSK	0.076	0.1523
2	QPSK	0.12	0.2344
3	QPSK	0.19	0.3770
4	QPSK	0.3	0.6016
5	QPSK	0.44	0.8770
6	QPSK	0.59	1.1758
7	16QAM	0.37	1.4766
8	16QAM	0.48	1.9141
9	16QAM	0.6	2.4063
10	64QAM	0.45	2.7305
11	64QAM	0.55	3.3223
12	64QAM	0.65	3.9023
13	64QAM	0.75	4.5234
14	64QAM	0.85	5.1152
15	64QAM	0.93	5.5547

The Physical Uplink Control CHannel (PUCCH, see Section 16.3.1) is used for periodic CQI reporting only while the Physical Uplink Shared CHannel (PUSCH, see Section 16.2) is used for aperiodic reporting of the CQI, whereby the eNodeB specifically instructs the UE to send an individual CQI report embedded into a resource which is scheduled for uplink data transmission.

The frequency granularity of the CQI reporting is determined by defining a number of sub-bands (N), each comprised of k contiguous Physical Resource Blocks (PRBs).[2] The value of k depends on the type of CQI report considered and is a function of the system bandwidth. In each case, the number of sub-bands spans the whole system bandwidth and is given by $N = \lceil N_{\mathrm{RB}}^{\mathrm{DL}}/k \rceil$, where $N_{\mathrm{RB}}^{\mathrm{DL}}$ is the number of RBs across the system bandwidth. The CQI reporting modes can be Wideband CQI, eNodeB-configured sub-band feedback, or UE-selected sub-band feedback. These are explained in detail in the following sections. In addition, in the case of multiple transmit antennas at the eNodeB, CQI value(s) may be reported for a second codeword.

For some downlink transmission modes, additional feedback signalling consisting of Precoding Matrix Indicators (PMI) and Rank Indications (RI) is also transmitted by the UE, as explained in Section 11.2.2.4.

10.2.1.1 Aperiodic CQI Reporting

Aperiodic CQI reporting on the PUSCH is scheduled by the eNodeB by setting a CQI request bit in an uplink resource grant sent on the Physical Downlink Control CHannel (PDCCH).

[2]Note that the last sub-band may have less than k contiguous PRBs.

The type of CQI report is configured by the eNodeB by RRC signalling. Table 10.2 summarizes the relationship between the configured downlink transmission mode (see Section 9.2.2.1) and the possible CQI reporting type.

Table 10.2: Aperiodic CQI feedback types on PUSCH for each PDSCH transmission mode.

PDSCH transmission mode (See Section 9.2.2.1)	PDSCH transmission scheme assumed by UE for deriving CQI	CQI type		
		Wideband only	eNodeB-configured sub-bands	UE-selected sub-bands
Mode 1	Single antenna port		X	X
Mode 2	Transmit diversity		X	X
Mode 3	Transmit diversity if RI=1, otherwise large-delay CDD[a]		X	X
Mode 4	Closed-loop spatial multiplexing	X	X	X
Mode 5	Multi-user MIMO		X	
Mode 6	Closed-loop spatial multiplexing using a single transmission layer	X	X	X
Mode 7	Single antenna port if one PBCH antenna port, otherwise transmit diversity		X	X
Mode 8[b] with PMI/RI configured	Closed-loop spatial multiplexing	X	X	X
Mode 8[b] without PMI/RI configured	Single antenna port if one PBCH antenna port, otherwise transmit diversity		X	X
Mode 9[c] with PMI/RI configured and >1 CSI-RS port[d]	Closed-loop spatial multiplexing	X	X	X
Mode 9[c] otherwise	Single antenna port if one PBCH antenna port, otherwise transmit diversity		X	X

[a] Cyclic Delay Diversity. [b] Introduced in Release 9. [c] Introduced in Release 10. [d] See Section 29.1.2.

The CQI reporting types are as follows:

- **Wideband feedback.** The UE reports one wideband CQI value for the whole system bandwidth.

- **eNodeB-configured sub-band feedback.** The UE reports a wideband CQI value for the whole system bandwidth. In addition, the UE reports a CQI value for each sub-band, calculated assuming transmission only in the relevant sub-band. Sub-band CQI reports are encoded differentially with respect to the wideband CQI using 2-bits as follows:

  Sub-band differential CQI offset = Sub-band CQI index − Wideband CQI index

  Possible sub-band differential CQI offsets are $\{\leq -1, 0, +1, \geq +2\}$. The sub-band size k is a function of system bandwidth as summarized in Table 10.3.

Table 10.3: Sub-band size (k) versus system bandwidth for eNodeB-configured aperiodic CQI reports. Reproduced by permission of © 3GPP.

System bandwidth (RBs)	Sub-band size (k RBs)
6–7	(Wideband CQI only)
8–10	4
11–26	4
27–63	6
64–110	8

- **UE-selected sub-band feedback.** The UE selects a set of M preferred sub-bands of size k (where k and M are given in Table 10.4 for each system bandwidth range) within the whole system bandwidth. The UE reports one wideband CQI value and one CQI value reflecting the average quality of the M selected sub-bands. The UE also reports the positions of the M selected sub-bands using a combinatorial index r defined as

  $$r = \sum_{i=0}^{M-1} \left\langle \begin{matrix} N - s_i \\ M - i \end{matrix} \right\rangle$$

  where the set $\{s_i\}_{i=0}^{M-1}$, $1 \leq s_i \leq N$, $s_i < s_{i+1}$ contains the M sorted sub-band indices and

  $$\left\langle \begin{matrix} x \\ y \end{matrix} \right\rangle = \begin{cases} \binom{x}{y} & \text{if } x \geq y \\ 0 & \text{if } x < y \end{cases}$$

  is the extended binomial coefficient, resulting in a unique label $r \in \{0, \dots, \binom{N}{M} - 1\}$. Some possible algorithms for deriving the combinatorial index r in the UE and extracting the information from it in the eNodeB can be found in [7, 8].

  The CQI value for the M selected sub-bands for each codeword is encoded differentially using two bits relative to its respective wideband CQI as defined by

  Differential CQI

  = Index for average of M preferred sub-bands − Wideband CQI index

  Possible differential CQI values are $\{\leq +1, +2, +3, \geq +4\}$.

Table 10.4: Sub-band size k and number of preferred sub-bands (M) versus downlink system bandwidth for aperiodic CQI reports for UE-selected sub-band feedback. Reproduced by permission of © 3GPP.

System bandwidth (RBs)	Sub-band size (k RBs)	Number of preferred sub-bands (M)
6–7	(Wideband CQI only)	(Wideband CQI only)
8–10	2	1
11–26	2	3
27–63	3	5
64–110	4	6

10.2.1.2 Periodic CQI Reporting

If the eNodeB wishes to receive periodic reporting of the CQI, the UE will transmit the reports using the PUCCH.[3]

Only wideband and UE-selected sub-band feedback is possible for periodic CQI reporting, for all downlink (PDSCH) transmission modes. As with aperiodic CQI reporting, the type of periodic reporting is configured by the eNodeB by RRC signalling. For wideband periodic CQI reporting, the period can be configured to[4]):

$$\{2, 5, 10, 16, 20, 32, 40, 64, 80, 128, 160\}\,\text{ms, or Off.}$$

While the wideband feedback mode is similar to that sent via the PUSCH, the 'UE-selected sub-band' CQI using PUCCH is different. In this case, the total number of sub-bands N is divided into J fractions called *bandwidth parts*. The value of J depends on the system bandwidth as summarized in Table 10.5. In this case, one CQI value is computed and reported for a single selected sub-band from each bandwidth part, along with the corresponding sub-band index.

Table 10.5: Periodic CQI reporting with UE-selected sub-bands: sub-band size (k) and bandwidth parts (J) versus downlink system bandwidth. Reproduced by permission of © 3GPP.

System bandwidth (RBs)	Sub-band size (k RBs)	Number of bandwidth parts (J)
6–7	(Wideband CQI only)	1
8–10	4	1
11–26	4	2
27–63	6	3
64–110	8	4

[3]If PUSCH transmission resources are scheduled for the UE in one of the periodic subframes, the periodic CQI report is sent on the PUSCH instead, as explained in Section 16.4 and Figure 16.15.

[4]These values apply to FDD operation; for TDD, see [9, Section 7.2.2]

10.2.1.3 CQI Reporting for Spatial Multiplexing

If the UE is configured in PDSCH transmission modes 3, 4, 8 or 9,[5] the eNodeB may use spatial multiplexing to transmit two codewords simultaneously to the UE with independently selected MCSs. To support this, the following behaviour is defined for the UE's CQI reports in these modes:

- If the UE is not configured to send RI feedback, or if the reported RI is equal to 1, or in any case in transmission mode 3,[6] the UE feeds back only one CQI report corresponding to a single codeword.
- If RI feedback is configured and the reported RI is greater than 1 in transmission modes 4 or 8:
  - For aperiodic CQI reporting, each CQI report (whether wideband or sub-band) comprises two independent CQI reports for the two codewords.
  - For periodic CQI reporting, one CQI report is fed back for the first codeword, and a second three-bit differential CQI report is fed back for the second codeword (for both wideband and sub-band reporting). The differential CQI report for the second codeword can take the following values relative to the CQI report for the first codeword: $\leq -4, -3, -2, -1, 0, +1, +2, \geq +3$.

10.3 Channel Coding

Channel coding, and in particular the channel decoder, has retained its reputation for being the dominant source of complexity in the implementation of wireless communications, in spite of the relatively recent prevalence of advanced antenna techniques with their associated complexity.

Section 10.3.1 introduces the theory behind the families of channel codes of relevance to LTE. This is followed in Sections 10.3.2 and 10.3.3 by an explanation of the practical design and implementation of the channel codes used in LTE for data and control signalling respectively.

10.3.1 Theoretical Aspects of Channel Coding

This section first explains convolutional codes, as not only do they remain relevant for small data blocks but also an understanding of them is a prerequisite for understanding turbo codes. The turbo-coding principle and the Soft-Input Soft-Output (SISO) decoding algorithms are then discussed. The section concludes with a brief introduction to LDPC codes.

10.3.1.1 From Convolutional Codes to Turbo Codes

A convolutional encoder $C(k, n, m)$ is composed of a shift register with m stages. At each time instant, k information bits enter the shift register and k bits in the last position of

[5] Mode 8 from Release 9 onwards; mode 9 from Release 10 onwards.

[6] If RI feedback is configured in transmission mode 3 and the RI fed back is greater than 1, although only one CQI corresponding to the first codeword is fed back, its value is adapted on the assumption that a second codeword will also be transmitted.

the shift register are dropped. The set of n output bits is a linear combination of the content of the shift register. The *rate* of the code is defined as $R_c = k/n$. Figure 10.2 shows the convolutional encoder used in LTE [10] with $m = 6$, $n = 3$, $k = 1$ and rate $R_c = 1/3$. The linear combinations are defined via n generator sequences $\mathbf{G} = [\mathbf{g}_0, \ldots, \mathbf{g}_{n-1}]$ where $\mathbf{g}_\ell = [g_{\ell,0}, g_{\ell,1}, \ldots, g_{\ell,m}]$. The generator sequences used in Figure 10.2 are

$$\mathbf{g}_0 = [1\,0\,1\,1\,0\,1\,1] = [133](\text{oct}),$$

$$\mathbf{g}_1 = [1\,1\,1\,1\,0\,0\,1] = [171](\text{oct}),$$

$$\mathbf{g}_2 = [1\,1\,1\,0\,1\,0\,1] = [165](\text{oct}).$$

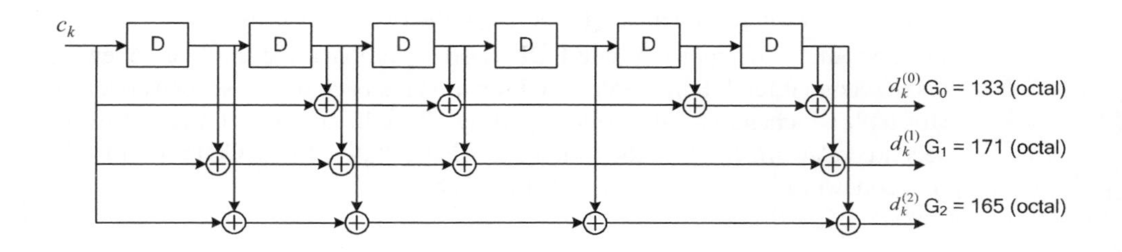

Figure 10.2: Rate 1/3 convolutional encoder used in LTE with $m = 6$, $n = 3$, $k = 1$ [10]. Reproduced by permission of © 3GPP.

A convolutional encoder can be described by a trellis diagram [11], which is a representation of a finite state machine including the time dimension.

Consider an input block with L bits encoded with a rate $1/n$ (i.e. $k = 1$) convolutional encoder, resulting in a codeword of length $(L + m) \times n$ bits, including m trellis termination bits (or *tail bits*) which are inserted at the end of the information block to drive the shift register contents back to all zeros at the end of the encoding process. Note that using tail bits is just one possible way of terminating an input sequence. Other trellis termination methods include simple truncation (i.e. no tail bits appended) and so-called *tail-biting* [12]. In the tail-biting approach, the initial and final states of the convolutional encoder are required to be identical. Usually, tail-biting for feed-forward convolutional encoders is achieved by initializing the shift register contents with the last m information bits in the input block. Tail-biting encoding facilitates uniform protection of the information bits and suffers no rate-loss owing to the tail bits. Tail-biting convolutional codes can be decoded using, for example, the *Circular Viterbi Algorithm* (CVA) [13, 14].

Let the received sequence $\mathbf{y}$ be expressed as

$$\mathbf{y} = \sqrt{E_b}\mathbf{x} + \mathbf{n} \tag{10.1}$$

where $\mathbf{n} = [n_0, n_1, \ldots, n_\ell, \ldots, n_{(L+m)\times(n-1)}]$ and $n_\ell \sim N(0, N_0)$ is the Additive White Gaussian Noise (AWGN) and E_b is the energy per bit. The transmitted codeword is $\mathbf{x} = [\mathbf{x}_0, \mathbf{x}_1, \ldots, \mathbf{x}_\ell, \ldots, \mathbf{x}_{L+m-1}]$ where $\mathbf{x}_\ell$ is the convolutional code output sequence at time instant ℓ for the input information bit i_ℓ, given by $\mathbf{x}_\ell = [x_{\ell,0}, \ldots, x_{\ell,n-1}]$. Equivalently, $\mathbf{y} = [\mathbf{y}_0, \mathbf{y}_1, \ldots, \mathbf{y}_\ell, \ldots, \mathbf{y}_{L+m-1}]$ where $\mathbf{y}_\ell = [y_{\ell,0}, \ldots, y_{\ell,n-1}]$ is the noisy received version of $\mathbf{x}$. $(L + m)$ is the total trellis length.

10.3.1.2 Soft-Input Soft-Output (SISO) Decoders

In order to define the performance of a communication system, the codeword error probability or bit error probability can be considered. The minimization of the bit error probability is in general more complicated and requires the maximization of the a posteriori bit probability (MAP symbol-by-symbol). The minimization of the codeword/sequence error probability is in general easier and is equivalent to the maximization of A Posteriori Probability (APP) for each codeword; this is expressed by the MAP sequence detection rule, whereby the estimate $\hat{\mathbf{x}}$ of the transmitted codeword is given by

$$\hat{\mathbf{x}} = \underset{\mathbf{x}}{\operatorname{argmax}}\, P(\mathbf{x} \mid \mathbf{y}) \tag{10.2}$$

When all codewords are equiprobable, the MAP criterion is equivalent to the Maximum Likelihood (ML) criterion which selects the codeword that maximizes the probability of the received sequence $\mathbf{y}$ conditioned on the estimated transmitted sequence $\mathbf{x}$, i.e.

$$\hat{\mathbf{x}} = \underset{\mathbf{x}}{\operatorname{argmax}}\, P(\mathbf{y} \mid \mathbf{x}) \tag{10.3}$$

Maximizing Equation (10.3) is equivalent to maximizing the logarithm of $P(\mathbf{y} \mid \mathbf{x})$, as $\log(\cdot)$ is a monotonically increasing function. This leads to simplified processing.[7] The log-likelihood function for a memoryless channel can be written as

$$\log P(\mathbf{y} \mid \mathbf{x}) = \sum_{i=0}^{L+m-1} \sum_{j=0}^{n-1} \log P(y_{i,j} \mid x_{i,j}) \tag{10.4}$$

For an AWGN channel, the conditional probability in Equation (10.4) is $P(y_{i,j} \mid x_{i,j}) \sim N(\sqrt{E_b}x_{i,j}, N_0)$, hence

$$\log P(\mathbf{y} \mid \mathbf{x}) \propto \|\mathbf{y}_i - \sqrt{E_b}\mathbf{x}_i\|^2 \tag{10.5}$$

The maximization of the metric in Equation (10.5) yields a codeword that is closest to the received sequence in terms of the Euclidean distance [15]. This maximization can be performed in an efficient manner by operating on the trellis.

As an example, Figure 10.3 shows a simple convolutional code with generator polynomials $\mathbf{g}_0 = [1\ 0\ 1]$ and $\mathbf{g}_1 = [1\ 1\ 1]$ and Figure 10.4 represents the corresponding trellis diagram. Each edge in the trellis corresponds to a transition from a state s to a state s', which can be obtained for a particular input information bit. In Figure 10.4, the edges are parametrized with the notation $i_\ell / x_{\ell,0}\, x_{\ell,1}$, i.e. the input/output of the convolutional encoder. The shift registers of the convolutional code are initialized to the all-zero state.

In Figure 10.4, $M(\mathbf{y}_i \mid \mathbf{x}_i) = \sum_{j=0}^{n-1} \log P(y_{i,j} \mid x_{i,j})$ denotes the branch metric at the ith trellis step (i.e. the cost of choosing a branch at trellis step i), given by Equation (10.5). The Viterbi Algorithm (VA) selects the best path through the trellis by computing at each step the best

[7] The processing is simplified because the multiplication operation can be transformed to the simpler addition operation in the logarithmic domain.

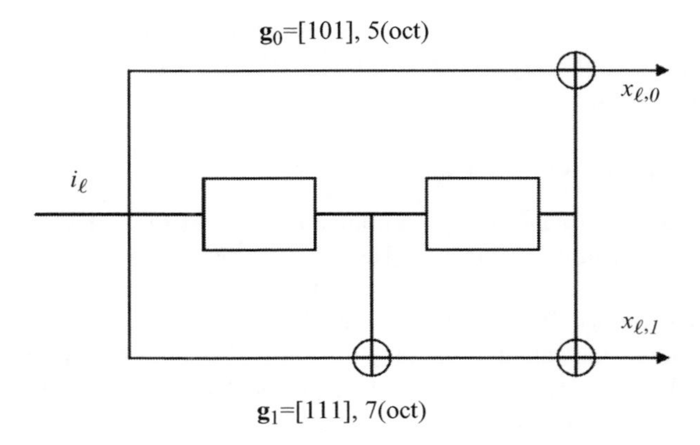

Figure 10.3: Rate $\frac{1}{2}$ convolutional encoder with $m = 2$, $m = 2$, $k = 1$, corresponding to generator polynomials $\mathbf{g}_0 = [1\ 0\ 1]$ and $\mathbf{g}_1 = [1\ 1\ 1]$.

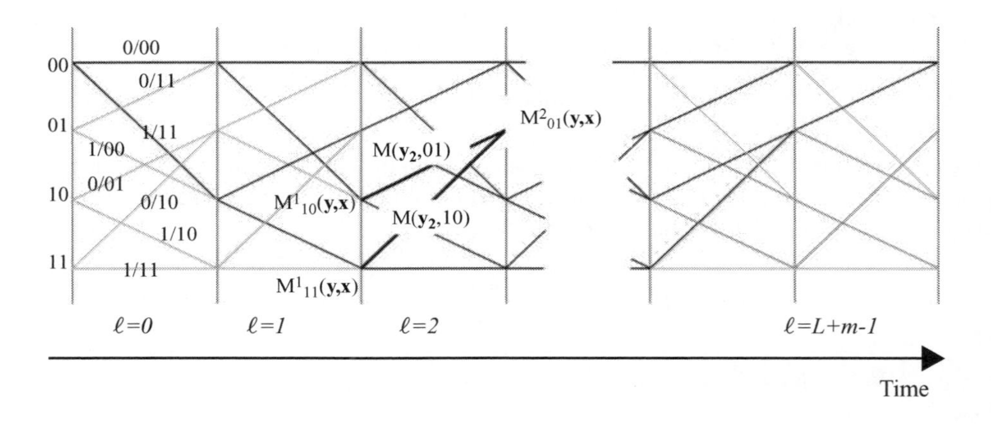

Figure 10.4: Trellis corresponding to convolutional code with generator polynomials $\mathbf{g}_0 = [1\ 0\ 1]$ and $\mathbf{g}_1 = [1\ 1\ 1]$.

partial metric (the accumulated cost of following a specific path until the ℓth transition) and selecting at the end the best total metric [16, 17]. It then traces back the selected path in the trellis to provide the estimated input sequence.

Although the original VA outputs a hard-decision estimate of the input sequence, the VA can be modified to output soft information[8] along with the hard-decision estimate of the input sequence [18]. The reliability of coded or information bits can be obtained via the computation of the a posteriori probability.

Let i_ℓ be the information bit which enters the shift register of the convolutional code at time ℓ. Assuming BPSK (Binary Phase Shift Keying) modulation ($0 \rightarrow +1$, and $1 \rightarrow -1$), the Log Likelihood Ratio (LLR) of an information symbol (or bit) i_ℓ is

[8]A soft decision gives additional information about the reliability of the decision [15].

$$\Lambda(i_\ell) = \Lambda(x_{\ell,0}) = \log \frac{\text{APP}(i_\ell = 1)}{\text{APP}(i_\ell = -1)} = \log \frac{P(i_\ell = 1 \mid \mathbf{y})}{P(i_\ell = -1 \mid \mathbf{y})} \tag{10.6}$$

The decoder can make a decision by comparing $\Lambda(i_\ell)$ to zero. Thus, the sign of the LLR gives an estimate of the information bit i_ℓ (LLR $\geq 0 \rightarrow 0$, and LLR $< 0 \rightarrow 1$), and the magnitude indicates the reliability of the estimate of the bit.

The a posteriori LLR $\Lambda(i_\ell)$ can be computed via the BCJR algorithm (named after its inventors, Bahl, Cocke, Jelinek and Raviv [19]). The BCJR algorithm is a SISO decoding algorithm that uses two Viterbi-like recursions going forwards and backwards in the trellis to compute Equation (10.6) efficiently. For this reason it is also referred to as a 'forward-backward' algorithm.

In order to explain the BCJR algorithm, it is better to write the APP from Equation (10.6) in terms of the joint probability of a transition in the trellis from the state s_ℓ at time instant ℓ, to the state $s_{\ell+1}$ at time instant $\ell + 1$,

$$\Lambda(i_\ell) = \log\left(\frac{\sum_{S^+} \text{P}(s_\ell = s', s_{\ell+1} = s, \mathbf{y})}{\sum_{S^-} \text{P}(s_\ell = s', s_{\ell+1} = s, \mathbf{y})}\right) = \log\left(\frac{\sum_{S^+} p(s', s, \mathbf{y})}{\sum_{S^-} p(s', s, \mathbf{y})}\right) \tag{10.7}$$

where $s, s' \in S$ are possible states of the convolutional encoder, S^+ is the set of ordered pairs (s', s) such that a transition from state s' to state s at time ℓ is caused by the input bit $i_\ell = 0$. Similarly, S^- is the set of transitions caused by $i_\ell = 1$. The probability $p(s', s, \mathbf{y})$ can be decomposed as follows:

$$p(s', s, \mathbf{y}) = p(s', \mathbf{y}_{t<\ell})p(s, \mathbf{y}_\ell \mid s')p(\mathbf{y}_{t>\ell} \mid s) \tag{10.8}$$

where $\mathbf{y}_{t<\ell} \triangleq [\mathbf{y}_0, \ldots, \mathbf{y}_{\ell-1}]$ and $\mathbf{y}_{t>\ell} \triangleq [\mathbf{y}_\ell, \ldots, \mathbf{y}_{L+m-1}]$.

By defining

$$\alpha_\ell(s') \triangleq p(s', \mathbf{y}_{t<\ell}) \tag{10.9}$$

$$\beta_{\ell+1}(s') \triangleq p(\mathbf{y}_{t>\ell} \mid s) \tag{10.10}$$

$$\gamma_\ell(s', s) \triangleq p(s, \mathbf{y}_\ell \mid s') \tag{10.11}$$

the APP in Equation (10.7) takes the usual form [19], i.e.

$$\Lambda(i_\ell) = \log\left(\frac{\sum_{S^+} \alpha_\ell(s')\gamma_\ell(s', s)\beta_{\ell+1}(s)}{\sum_{S^-} \alpha_\ell(s')\gamma_\ell(s', s)\beta_{\ell+1}(s)}\right) \tag{10.12}$$

The probability $\alpha_\ell(s')$ is computed iteratively in a forward recursion

$$\alpha_{\ell+1}(s) = \sum_{s' \in S} \alpha_\ell(s')\gamma_\ell(s', s) \tag{10.13}$$

with initial condition $\alpha_0(0) = 1$, $\alpha_0(s) = 0$ for $s \neq 0$, assuming that the initial state of the encoder is 0. Similarly, the probability $\beta_\ell(s')$ is computed via a backward recursion

$$\beta_\ell(s') = \sum_{s \in S} \beta_{\ell+1}(s)\gamma_\ell(s', s) \tag{10.14}$$

with initial condition $\beta_{L+m-1}(0) = 1$, $\beta_{L+m-1}(s) = 0$ for $s \neq 0$ assuming trellis termination is employed using tail bits. The transition probability can be computed as

$$\gamma_\ell(s', s) = \mathrm{P}(\mathbf{y}_\ell \mid \mathbf{x}_\ell)P(i_\ell)$$

The MAP algorithm consists of initializing $\alpha_0(s')$, $\beta_{L+m-1}(s)$, computing the branch metric $\gamma_\ell(s', s)$, and continuing the forward and backward recursion in Equations (10.13) and (10.14) to compute the updates of $\alpha_{\ell+1}(s)$ and $\beta_\ell(s')$.

The complexity of the BCJR is approximately three times that of the Viterbi decoder. In order to reduce the complexity a log-MAP decoder can be considered, where all operations are performed in the logarithmic domain. Thus, the forward and backward recursions $\alpha_{\ell+1}(s)$, $\beta_\ell(s')$ and $\gamma_\ell(s, s')$ are replaced by $\alpha^*_{\ell+1}(s) = \log[\alpha_{\ell+1}(s)]$, $\beta^*_\ell(s') = \log[\beta_\ell(s')]$ and $\gamma^*_\ell(s, s') = \log[\gamma_\ell(s, s')]$. This gives advantages for implementation and can be shown to be numerically more stable. By defining

$$\max{}^*(z_1, z_2) \triangleq \log(e^{z_1} + e^{z_2}) = \max(z_1, z_2) + \log(1 + e^{-|z_1 - z_2|}) \tag{10.15}$$

it can be shown that the recursion in Equations (10.13) and (10.14) becomes

$$\begin{aligned} \alpha^*_{\ell+1}(s) &= \log\bigg(\sum_{s' \in S} e^{\alpha^*_\ell(s') + \gamma^*_\ell(s', s)}\bigg) \\ &= \max_{s' \in S}{}^*\{\alpha^*_\ell(s') + \gamma^*_\ell(s', s)\} \end{aligned} \tag{10.16}$$

with initial condition $\alpha^*_0(0) = 0$, $\alpha^*_0(s) = -\infty$ for $s \neq 0$. Similarly,

$$\beta^*_\ell(s') = \max_{s' \in S}{}^*\{\beta^*_{\ell+1}(s) + \gamma^*_\ell(s', s)\} \tag{10.17}$$

with initial condition $\beta^*_{L+m}(0) = 0$, $\beta^*_{L+m}(s) = -\infty$ for $s \neq 0$. Figure 10.5 shows a schematic representation of the forward and backward recursion.

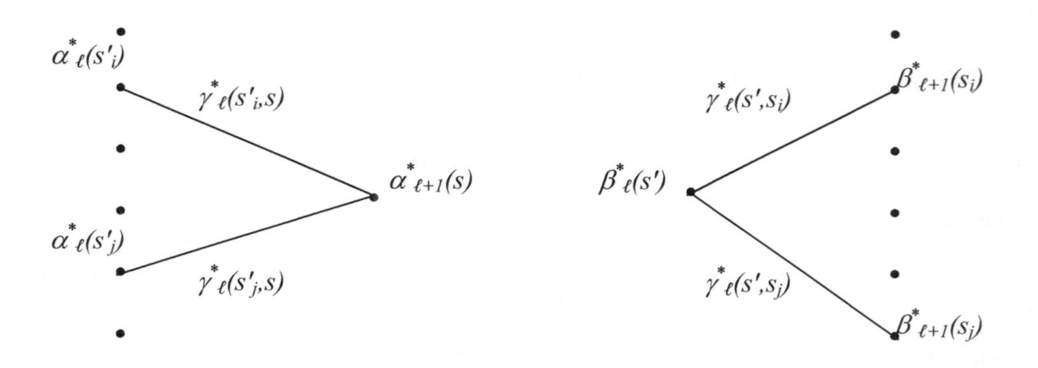

Figure 10.5: Forward and backward recursion for the BCJR decoding algorithm.

The APP in Equation (10.12) becomes

$$\Lambda(i_\ell) = \max^*_{(s',s)\in S^+}\{\alpha^*_\ell(s') + \gamma^*_\ell(s',s) + \beta^*_{\ell+1}(s)\} - \max^*_{(s',s)\in S^-}\{\alpha^*_\ell(s') + \gamma^*_\ell(s',s) + \beta^*_{\ell+1}(s)\} \tag{10.18}$$

In order to reduce the complexity further, the *max-log-MAP* approximation can be used, where the $\max^* = \log(e^x + e^y)$ is approximated by $\max(x, y)$. The advantage is that the algorithm is simpler and faster, with the complexity of the forward and backward passes being equivalent to a Viterbi decoder. The drawback is the loss of accuracy arising from the approximation.

Convolutional codes are the most widely used family of error-correcting codes owing to their reasonably good performance and the possibility of extremely fast decoders based on the VA, as well as their flexibility in supporting variable input codeword sizes. However, it is well known that there remains a significant gap between the performance of convolutional codes and the theoretical limits set by Shannon.[9] In the early 1990s, an encoding and decoding algorithm based on convolutional codes was proposed [20], which exhibited performance within a few tenths of a deciBel (dB) from the Shannon limit – the turbo code family was born. Immediately after turbo codes were discovered, Low-Density Parity Check (LDPC) codes [23–25] that also provided near-Shannon limit performance were rediscovered.

10.3.1.3 Turbo Codes

Berrou, Glavieux and Thitimajshima introduced turbo codes and the concept of iterative decoding to achieve near-Shannon limit performance [20, 26]. Some of the reasoning regarding probabilistic processing can be found in [18, 27], as recognized by Berrou in [28].

A turbo encoder consists of a concatenation of two convolutional encoders linked by an interleaver. For instance, the turbo encoder adopted in UMTS and LTE [10] is schematically represented in Figure 10.6 with two identical convolutional codes with generator polynomial given by $\mathbf{G} = [1, \mathbf{g}_0/\mathbf{g}_1]$ where $\mathbf{g}_0 = [1011]$ and $\mathbf{g}_1 = [1101]$. Thus, a turbo code encodes the input block twice (with and without interleaving) to generate two distinct set of parity bits. Each constituent encoder may be terminated to the all-zero state by using tail bits. The nominal code rate of the turbo code shown in Figure 10.6 is $1/3$.

Like convolutional codes, the optimal decoder for turbo codes would ideally be the MAP or ML decoder. However the number of states in the trellis of a turbo code is significantly larger due to the interleaver Π, thus making a true ML or MAP decoder intractable (except for trivial block sizes). Therefore, Berrou *et al.* [26] proposed the principle of iterative decoding. This is based on a suboptimal approach using a separate optimal decoder for each constituent convolutional coder, with the two constituent decoders iteratively exchanging information via a (de)interleaver. The classical decoding structure is shown in Figure 10.7.

The SISO decoder for each constituent convolutional code can be implemented via the BCJR algorithm (or the MAP algorithm), which was briefly described in the previous section. The SISO decoder outputs an a posteriori LLR value for each information bit, and this can be used to obtain a hard decision estimate as well as a reliability estimate. In addition, the

[9] 'Shannon's theorem', otherwise known as the 'noisy-channel coding theorem', states that if and only if the information rate is less than the channel capacity, then there exists a coding-decoding scheme with vanishing error probability when the code's block length tends to infinity [21, 22].

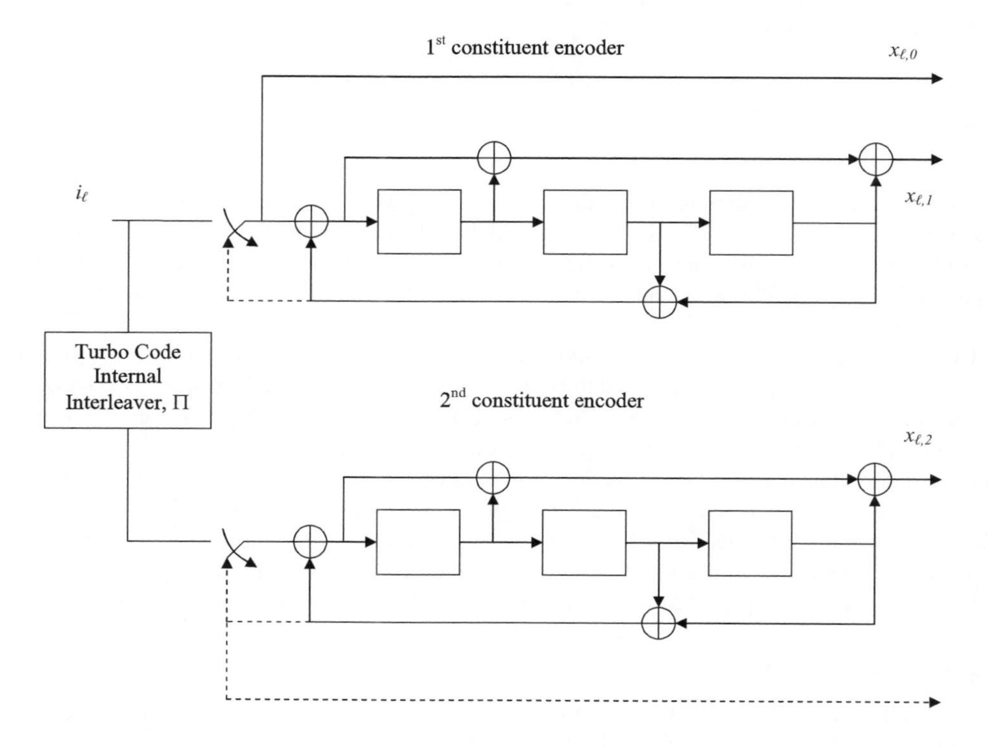

Figure 10.6: Schematic view of parallel turbo code used in LTE and UMTS [10]. Reproduced by permission of © 3GPP.

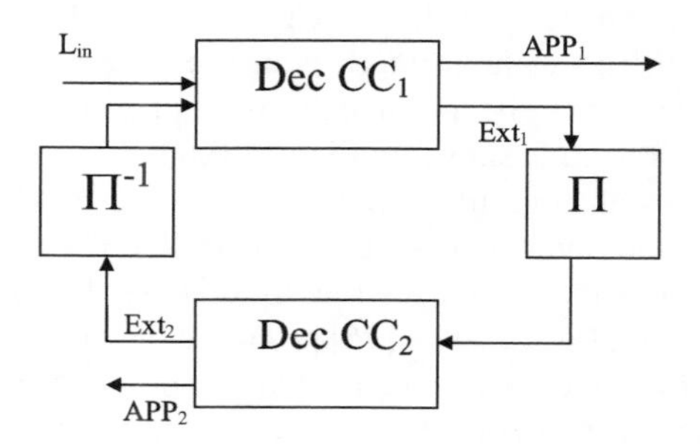

Figure 10.7: Schematic turbo decoder representation.

SISO decoder generates extrinsic LLRs (i.e. the extrinsic information of the bit x_ℓ is the LLR obtained from all the bits of the convolutional code except the bit corresponding to x_ℓ) for each information/coded bit that is utilized by the other decoder as a priori information after suitable (de)interleaving. The soft input to the first SISO decoder is the so-called *channel observation* or channel likelihood, obtained as

$$\mathrm{L_{in}}(x_{\ell,k}) = \frac{P(y_{\ell,k} \mid x_{\ell,k} = 1)}{P(y_{\ell,k} \mid x_{\ell,k} = -1)}$$

Thus the two decoders cooperate by iteratively exchanging the extrinsic information via the (de)interleaver. After a certain number of iterations (after convergence), the a posteriori LLR output can be used to obtain final hard decision estimates of the information bits.

10.3.1.4 Low-Density Parity-Check (LDPC) Codes

Low-Density Parity-Check (LDPC) codes were first studied by Gallager in his doctoral thesis [23] and have been extensively analysed by many researchers in recent years. LDPC codes are linear parity-check codes with a parity-check equation given by $H\mathbf{c}^T = 0$, where H is the $(n-k) \times n$ parity-check matrix of the code $C(k, n)$ and $\mathbf{c}$ is a length-n valid codeword belonging to the code C. Similarly to turbo codes, ML decoding for LDPC codes becomes too complex as the block size increases. In order to approximate ML decoding, Gallager in [23] introduced an iterative decoding technique which can be considered as the forerunner of message-passing algorithms [24, 29], which lead to an efficient LDPC decoding algorithm with a complexity which is linear with respect to the block size.

The term 'low-density' refers to the fact that the parity-check matrix entries are mostly zeros – in other words, the density of ones is low. The parity-check matrix of an LDPC code can be represented by a 'Tanner graph', in which two types of node, variable and check, are interconnected. The variable nodes and check nodes correspond to the codeword bits and the parity-check constraints respectively. A variable node v_j is connected to a check node c_i if the corresponding codeword bit participates in the parity-check equation, i.e. if $H(i, j) = 1$. Thus, the Tanner graph is an excellent tool by which to visualize the code constraints and to describe the decoding algorithms. Since an LDPC code has a low density of ones in H, the number of interconnections in the Tanner graph is small (and typically linear with respect to the codeword size). Figure 10.8 shows an example of a Tanner graph of a (10, 5) code together with its parity-check matrix H.[10]

As mentioned above, LDPC codes are decoded using message-passing algorithms, such as Belief Propagation (BP) or the Sum-Product Algorithm (SPA). The idea of BP is to calculate approximate marginal a posteriori LLRs by applying Bayes' rule locally and iteratively at the nodes in the Tanner graph. The variable nodes and check nodes in the Tanner graph exchange LLR messages along their interconnections in an iterative fashion, thus cooperating with each other in the decoding process. Only extrinsic messages are passed along the interconnections to ensure that a node does not receive repeats of information which it already possesses, analogous to the exchange of extrinsic information in turbo decoding.

It is important to note that the number of operations per iteration in the BP algorithm is linear with respect to the number of message nodes. However, the BP decoding of LDPC codes is also suboptimal, like the turbo codes, owing to the cycles in the Tanner graph.[11] Cycles increase the dependencies between the extrinsic messages being received at each node during the iterative process. However, the performance loss can be limited by avoiding

[10]This particular matrix is not actually 'low density', but it is used for the sake of illustration.

[11]A cycle of length P in a Tanner graph is a set of P connected edges that starts and ends at the same node. A cycle verifies the condition that no node (except the initial and final nodes) appears more than once.

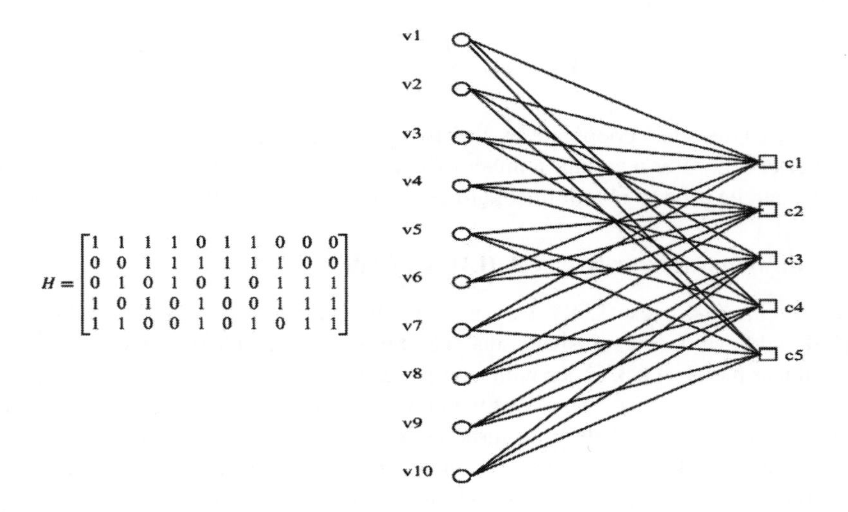

Figure 10.8: A Tanner graph representing the parity-check matrix of a (3, 6)-regular LDPC code $C(10, 5)$ and its associated parity-check matrix H.

excessive occurrences of short cycles. After a certain number of iterations, the a posteriori LLRs for the codeword bits are computed to obtain the final hard decision estimates. Typically, more iterations are required for LDPC codes to converge than are required for turbo codes. In practice, a variation of BP known as the Layered BP (LBP) algorithm is preferred, as it requires half as many iterations as the conventional BP algorithm [30].

From an encoding perspective, a straightforward algorithm based on the parity-check matrix would require a number of operations that is proportional to the square of the codeword size. However, there exist several structured LDPC code designs (including those used in the IEEE 802.16e, IEEE 802.11n and DVB-S2 standards) which have linear encoding complexity and extremely good performance. With structured or vectorized LDPC codes, the parity-check matrices are constructed using submatrices which are all-zero or circulant (shifted identity matrices), and such codes require substantially lower encoding/decoding complexity without sacrificing performance. Numerous references exist for LDPC code design and analysis, and the interested reader is referred, for example, to [23, 24, 29, 31–33] and references therein.

10.3.2 Channel Coding for Data Channels in LTE

Turbo codes found a home in UMTS relatively rapidly after their publication in 1993, with the benefits of their near-Shannon limit performance outweighing the associated costs of memory and processing requirements. Once the turbo code was defined in UMTS there was little incentive to reopen the specification as long as it was functional.

Therefore, although in UMTS Releases 5, 6 and 7 the core turbo code was not changed, it was enhanced by the ability to select different redundancy versions for HARQ retransmissions. However, for LTE, with data rates of 100 Mbps to 300 Mbps in view, the UMTS turbo code needed to be re-examined, especially in terms of decoding complexity.

Sections 10.3.2.1 and 10.3.2.2 explain the eventual decision not to use LDPC codes for LTE, while replacing the turbo interleaver with a 'contention-free' interleaver. The following sections explain the specific differences between LTE channel coding and UMTS as summarized in Table 10.6.

Table 10.6: Major features of UMTS and LTE channel coding schemes.

Channel coding	UMTS	LTE
Constituent code	Tailed, eight states, $R = 1/3$ mother code	Same
Turbo interleaver	Row/column permutation	Contention-free quadratic permutation polynomial (QPP) interleaver
Rate matching	Performed on concatenated code blocks	Virtual Circular Buffer (CB) rate matching, performed per code block
Hybrid ARQ	Redundancy Versions (RVs) defined, Chase operation allowed	RVs defined on virtual CB, Chase operation allowed
Control channel	256-state tailed convolutional code	64-state tail-biting convolutional code, CB rate matching
Per-code-block operations	Turbo coding only	CRC attachment, turbo coding, rate matching, modulation mapping

10.3.2.1 The Lure of the LDPC Code

The draw of LDPC codes is clear: performance almost up to the Shannon limit, with claims of up to eight times less complexity than turbo codes. LDPC codes had also recently been standardized as an option in IEEE 802.16-2005. The complexity angle is all-important in LTE, provided that the excellent performance of the turbo code is maintained. However, LDPC proposals put forward for LTE claimed widely varying complexity benefits, from no benefit to 2.4 times [35] up to 7.35 times [36]. In fact, it turns out that the complexity benefit is code-rate dependent, with roughly a factor-of-two reduction in the operations count at code rate 1/2 [37]. This comparison assumes that the LDPC code is decoded with the Layered Belief Propagation (LBP) decoder [30] while the turbo code is decoded with a log-MAP decoder.

On operations count alone, LDPC would be the choice for LTE. However, at least two factors curbed enthusiasm for LDPC. First, LDPC decoders have significant implementation complexity for memory and routing, which makes simple operation counts unrepresentative. Second, turbo codes were already standardized in UMTS, and a similar lengthy standardization process was undesired when perhaps a relatively simple enhancement to the known turbo code would suffice. It was therefore decided to keep the same turbo code constituent encoders as in the UMTS turbo code, including the tailing method, but to enhance it using a new contention-free interleaver that enables efficient decoding at the high data rates targeted for LTE. Table 10.7 gives a comparison of some of the features of turbo and LDPC codes which are explained in the following sections.

Table 10.7: Comparison of turbo and LDPC codes.

	Turbo codes	LDPC codes
Standards	UMTS, LTE, WiMAX	IEEE 802.16e, IEEE 802.11n DVB-S2
Encoding	Linear time	Linear time with certain designs (802.16e, 802.11n)
Decoding	log-MAP and variants	Scheduling: Layered Belief Propagation, turbo-decoding Node processing: Min-Sum, normalized Min-Sum
Main decoding concern	Computationally intensive (more operations)	Memory intensive (more memory); routing network
Throughput	No inherent parallelism; Parallelism obtained via contention-free interleaver	Inherently parallelizable; Structured LDPC codes for flexible design
Performance	Comparable for information block sizes 10^3–10^4; Slightly better than LDPC at small block sizes; Four to eight iterations on average	Can be slightly better than turbo codes at large block lengths (10^4 or larger), iterative decoding threshold analysis; Very low error floor (e.g. at BLER around 10^{-7}); 10 to 15 iterations on average
HARQ	Simple for both Chase and IR (via mother code puncturing)	Simple for Chase; IR possible using model matrix extension or puncturing from a mother code
Complexity comparison (operations count)	Slightly higher operations count than LDPC at high code rates	Slightly lower operations count than turbo codes at high code rates

10.3.2.2 Contention-Free Turbo Decoding

The key to high data-rate turbo decoding is to parallelize the turbo decoder efficiently while maintaining excellent performance. The classical turbo decoder has two MAP decoders (usually realized via a single hardware instantiation) that exchange extrinsic information in an iterative fashion between the two component codes, as described in Section 10.3.1. Thus, the first consideration is whether the parallelism is applied internally to the MAP decoder (single codeword parallelization) or externally by employing multiple turbo decoders (multiple codeword parallelizations with no exchange of extrinsic information between codewords).

If external parallelism is adopted, an input block may be split into X pieces, resulting in X codewords, and the increase in processing speed is obtained by operating up to X turbo decoders in parallel on the X codewords. In this case, in addition to a performance loss due to the smaller turbo interleaver size in each codeword, extra cost is incurred for memory and ancillary gate counts for forward-backward decoders. For this reason, one larger codeword with internal MAP parallelization is preferred. With one larger codeword, the MAP algorithm is run in parallel on each of the X pieces and the pieces can exchange extrinsic information, thus benefiting from the coding gains due to the large interleaver size. Connecting the pieces is most easily done with forward and backward state initialization based on the output of the previous iterations, although training within adjacent pieces during the current iteration is also possible [38]. As long as the size of each piece is large enough (e.g. 32 bits or greater), performance is essentially unaffected by the parallel processing.

While it is clear that the MAP algorithm can be parallelized, the MAP is not the entire turbo decoder. Efficient handling of the extrinsic message exchange (i.e. the (de)interleaving operation) is required as well. Since multiple MAP processors operate in parallel, multiple extrinsic values need to be read from or written to the memories concurrently. The memory accesses depend on the interleaver structure.

The existing UMTS interleaver has a problem with memory access contentions. Contentions result in having to read from or write to the same memory at the same time (as shown in Figure 10.9).

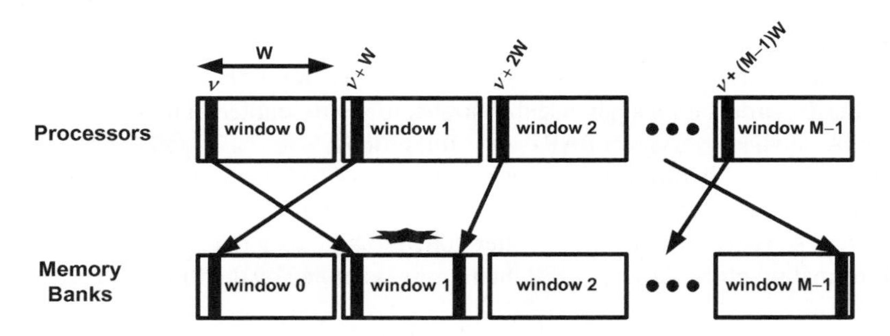

Figure 10.9: Memory access contention in window 1 due to concurrent access by Processors 0 and 2.

Contention resolution is possible (see, e.g., [39,40]), but at a cost of extra hardware, and the resolution time (cycles) may vary for every supported interleaver size. Complex memory management may also be used as contention resolution for any arbitrary interleaver [41].

A new, Contention-Free (CF) interleaver solves the problem by ensuring that no contentions occur in the first place. An interleaver $\pi(i)$, $0 \leq i < K$, is said to be contention-free for a window size W when it satisfies the following for both interleaver $\psi = \pi$ and deinterleaver

$\psi = \pi^{-1}$ [42]:

$$\left\lfloor \frac{\psi(u_1 W + v)}{W} \right\rfloor \neq \left\lfloor \frac{\psi(u_2 W + v)}{W} \right\rfloor \tag{10.19}$$

where $0 \leq vW$, $u_1 \geq 0$ and $u_2 < M$ for all $u_1 \neq u_2$. The terms on both sides of Equation (10.19) are the memory bank indices that are accessed by the M processors on the vth step. Inequality (10.19) requires that, for any given position v of each window, the memory banks accessed be unique between any two windows, thus eliminating access contentions. For significant hardware savings, instead of using M physically separate memories, it is better to use a single physical memory and fetch or store M values on each cycle from a single address. This requires the CF interleaver also to satisfy a vectorized decoding property where the intra-window permutation is the same for each window:

$$\pi(uW + v) \bmod W = \pi(v) \bmod W \tag{10.20}$$

for all $1 \leq u < M$ and $0 \leq v < W$.

Performance, implementation complexity and flexibility are concerns. However, even a simple CF interleaver composed of look-up tables (for each block size) and a bit-reversal permutation [43] can be shown to have excellent performance. In terms of flexibility, a maximally contention-free interleaver can have a parallelism order (number of windows) that is any factor of the block size. A variety of possible parallelism factors provides freedom for each individual manufacturer to select the degree of parallelism based on the target data rates for different UE categories.

After consideration of performance, available flexible classes of CF interleavers and complexity benefits, a new contention-free interleaver was selected for LTE.

10.3.2.3 The LTE Contention-Free Interleaver

The main choices for the CF interleaver included Almost Regular Permutation (ARP) [44] and Quadratic Permutation Polynomial (QPP) [45] interleavers. The ARP and QPP interleavers share many similarities and they are both suitable for LTE turbo coding, offering flexible parallelism orders for each block size, low-complexity implementation, and excellent performance. A detailed overview of ARP and QPP proposals for LTE (and their comparison with the UMTS Release-99 turbo interleaver) is given in [46]. Of these two closely competing designs, the QPP interleaver was selected for LTE as it provides more parallelism factors M and requires less storage, thus making it better-suited to high data rates.

For an information block size K, a QPP interleaver is defined by the following polynomial:

$$\pi(i) = (f_1 i + f_2 i^2) \bmod K \tag{10.21}$$

where i is the output index ($0 \leq i \leq K - 1$), $\pi(i)$ is the input index and f_1 and f_2 are the coefficients that define the permutation with the following conditions:

- f_1 is relatively prime to block size K;
- all prime factors of K also factor f_2.

The inverses of QPP interleavers can also be described via permutation polynomials but they are not necessarily quadratic, as such a requirement may result in inferior turbo code performance for certain block sizes. Therefore, it is better to select QPP interleavers with low-degree inverse polynomials (the maximum degree of the inverse polynomial is equal to four for LTE QPP interleavers) without incurring a performance penalty. In general, permutation polynomials (quadratic and non-quadratic) are implementation-friendly as they can be realized using only adders in a recursive fashion. A total of 188 interleavers are defined for LTE, of which 153 have quadratic inverses while the remaining 35 have degree-3 and degree-4 inverses. The performance of the LTE QPP interleavers is shown alongside the UMTS turbo code in Figure 10.10.

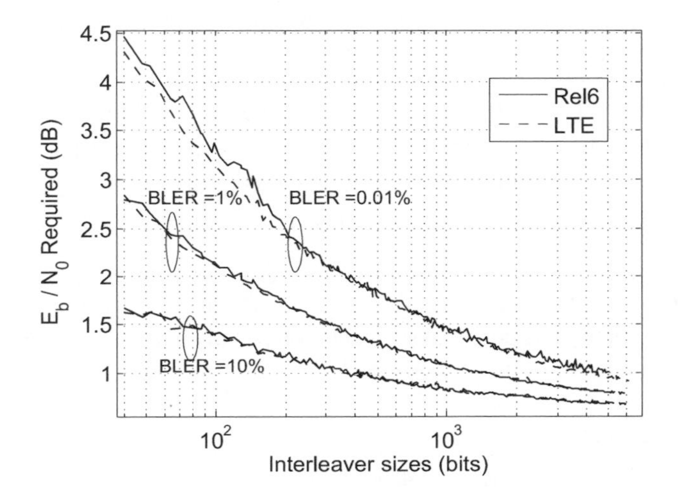

Figure 10.10: Performance of LTE QPP interleaver design versus UMTS turbo interleaver in static AWGN with QPSK modulation and eight iterations of a max-log-MAP decoder.

An attractive (and perhaps the most important) feature of QPP interleavers is that they are 'maximum contention-free', which means that they provide the maximum flexibility in supported parallelism, i.e. every factor of K is a supported parallelism factor. For example, for $K = 1024$, supported parallelism factors include $\{1, 2, 4, 8, 16, 32, 64, 128, 256, 512, 1024\}$, although factors that result in a window size less than 32 may not be required in practice.

The QPP interleavers also have the 'even-even' property whereby even and odd indices in the input are mapped to even and odd indices respectively in the output; this enables the encoder and decoder to process two information bits per clock cycle, facilitating radix-4 implementations (i.e. all log-MAP decoding operations can process two bits at a time, including forward and backward recursions, extrinsic LLR generation and memory read and write).

One key impact of the decision to use a CF interleaver is that not all input block sizes are natively supported. As the amounts of parallelism depend on the factorization of the block size, certain block sizes (e.g. prime sizes) are not natively supportable by the turbo code. Since the input can be any size, filler bits are used whenever necessary to pad the input to the nearest QPP interleaver size. The QPP sizes are selected such that:

- The number of interleavers is limited (fewer interleavers implies more filler bits).
- The fraction of filler bits is roughly the same as the block size increases (spacing increases as block size increases).
- Multiple parallelism values are available (block sizes are spaced an integer number of bytes apart).

For performance within approximately 0.1 dB of the baseline UMTS interleaver, as few as 45 interleaver sizes [47] are feasible, although the percentage of filler bits may be high (nearly 12%). For LTE, the following 188 byte-aligned interleaver sizes spaced in a semi-log manner were selected with approximately 3% filler bits [10]:

$$K = \begin{cases} 40 + 8t & \text{if } 0 \leq t \leq 59 \quad \text{(40–512 in steps of 8 bits)} \\ 512 + 16t & \text{if } 0 < t \leq 32 \quad \text{(528–1024 in steps of 16 bits)} \\ 1024 + 32t & \text{if } 0 < t \leq 32 \quad \text{(1056–2048 in steps of 32 bits)} \\ 2048 + 64t & \text{if } 0 < t \leq 64 \quad \text{(2112–6144 in steps of 64 bits)} \end{cases} \tag{10.22}$$

The maximum turbo interleaver size was also increased from 5114 in UMTS to 6144 in LTE, such that a 1500-byte TCP/IP packet would be segmented into only two segments rather than three, thereby minimizing the potential segmentation penalty and (marginally) increasing the turbo interleaver gain.

10.3.2.4 Rate-Matching

The Rate-Matching (RM) algorithm selects bits for transmission from the rate 1/3 turbo coder output via puncturing and/or repetition. Since the number of bits for transmission is determined based on the available physical resources, the RM should be capable of generating puncturing patterns for arbitrary rates. Furthermore, the RM should send as many new bits as possible in retransmissions to maximize the Incremental Redundancy (IR) HARQ gains (see Section 4.4 for further details about the HARQ protocol).

The main contenders for LTE RM were to use the same (or a similar) algorithm as HSPA, or to use Circular Buffer (CB) RM as in CDMA2000 1xEV and WiMAX. The primary advantage of the HSPA RM is that while it appears complex, it has been well studied and is well understood. However, a key drawback is that there are some severe performance degradations at higher code rates, especially near code rates 0.78 and 0.88 [48]. Therefore, circular buffer RM was selected for LTE, as it generates puncturing patterns simply and flexibly for any arbitrary code rate, with excellent performance.

In the CB approach, as shown in Figure 10.11, each of the three output streams of the turbo coder (systematic part, parity0, and parity1) is rearranged with its own interleaver (called a sub-block interleaver). In LTE, the 12 tail bits are distributed equally into the three streams as well, resulting in the sub-block size $K_s = K + 4$, where K is the QPP interleaver size. Then, an output buffer is formed by concatenating the rearranged systematic bits with the interlacing of the two rearranged parity streams. For any desired code rate, the coded bits for transmission are simply read out serially from a certain starting point in the buffer, wrapping around to the beginning of the buffer if the end of the buffer is reached.

A Redundancy Version (RV) specifies a starting point in the circular buffer to start reading out bits. Different RVs are specified by defining different starting points to enable HARQ

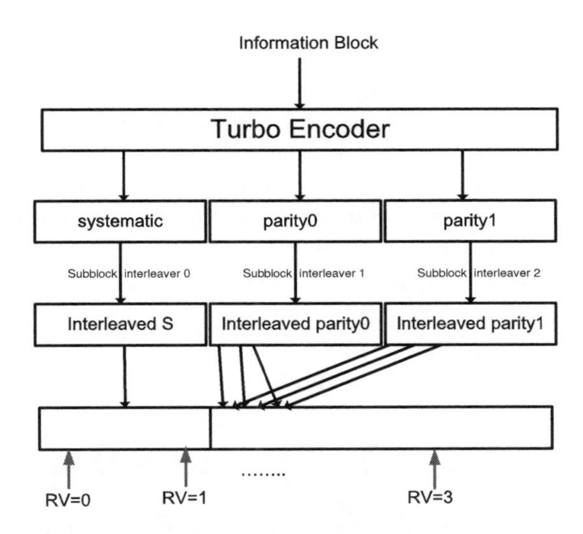

Figure 10.11: Rate-matching algorithm based on the CB. RV = 0 starts at an offset relative to the beginning of the CB to enable systematic bit puncturing on the first transmission.

operation. Usually RV = 0 is selected for the initial transmission to send as many systematic bits as possible. The scheduler can choose different RVs on transmissions of the same packet to support both IR and Chase combining HARQ.

The turbo code tail bits are uniformly distributed into the three streams, with all streams the same size. Each sub-block interleaver is based on the traditional row-column interleaver with 32 columns (for all block size) and a simple length-32 intra-column permutation.

- The bits of each stream are written row-by-row into a matrix with 32 columns (the number of rows is determined by the stream size), with dummy bits padded to the front of each stream to completely fill the matrix.
- A length-32 column permutation is applied and the bits are read out column-by-column to form the output of the sub-block interleaver. This sub-block interleaver has the property that it naturally first puts all the even indices and then all the odd indices into the rearranged sub-block. The permutation order is as follows:
  [0, 16, 8, 24, 4, 20, 12, 28, 2, 18, 10, 26, 6, 22, 14, 30, 1, 17, 9, 25, 5, 21, 13, 29, 3, 19, 11, 27, 7, 23, 15, 31]

Given the even-even property of the QPP permutations and the above property of the sub-block interleaver, the sub-block interleaver of the second parity stream is offset by an odd value $\delta = 1$ to ensure that the odd and even input indices have equal levels of protection. Thus, for index i, if $\pi_{\text{sys}}(i)$ denotes the permutation of the systematic bit sub-block interleaver, then the permutation of the two parity sub-block interleavers are $\pi_{\text{par0}}(i) = \pi_{\text{sys}}(i)$, and $\pi_{\text{par1}}(i) = (\pi_{\text{sys}}(i) + \delta) \bmod K_s$, where K_s is the sub-block size. With the offset $\delta = 1$ used in LTE, the first K_s parity bits in the interlaced parity portion of the circular buffer correspond to the K_s systematic bits, thus ensuring equal protection to all systematic bits [49]. Moreover, the offset enables the systematic bit puncturing feature whereby a small percentage of systematic bits are punctured in an initial transmission to enhance performance at high code rates. With the offset, RV = 0 results in partially systematic codes that are self-decodable at high coding rates, i.e. avoiding the 'catastrophic' puncturing patterns which have been shown to exist at some code rates in UMTS.

A two-dimensional interpretation of the circular buffer (with a total of 96 columns) is shown in Figure 10.12 (where different shadings indicate bits from different streams and black cells indicate dummy bits). The one-dimensional CB may be formed by reading bits out column-by-column from the two-dimensional CB. Bits are read column-by-column starting from a column top RV location, and the dummy bits are discarded during the output bit generation. Although the dummy bits can be discarded during sub-block interleaving, in LTE the dummy bits are kept to allow a simpler implementation.

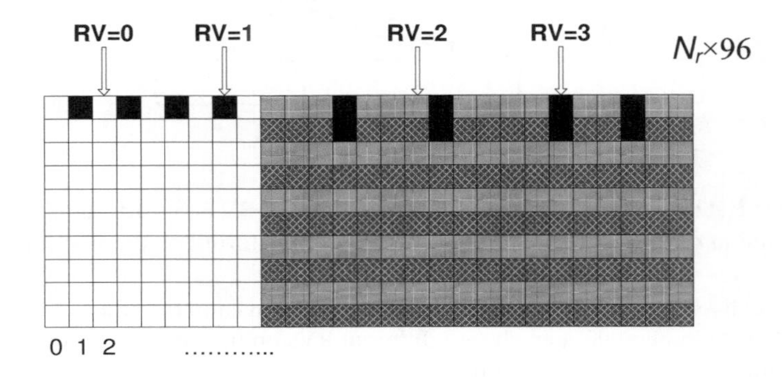

Figure 10.12: Two-dimensional visualization of the VCB. The starting points for the four RVs are the top of the selected columns.

This leads to the foremost advantage of the LTE CB approach, in that it enables efficient HARQ operation, since the CB operation can be performed without requiring an intermediate step of forming any actual physical buffer. In other words, for any combination of the 188 stream sizes and four RV values, the desired codeword bits can be equivalently obtained directly from the output of the turbo encoder using simple addressing based on sub-block permutation.

Therefore the term 'Virtual Circular Buffer' (VCB) is more appropriate in LTE. The LTE VCB operation also allows Systematic Bit Puncturing (SBP) by defining RV = 0 to skip the first two systematic columns of the CB, leading to approximately 6% punctured systematic bits (with no wrap around). Thus, with systematic bit puncturing and uniform spaced RVs, the four RVs start at the top of columns 2, 26, 50 and 74.

10.3.2.5 HARQ in LTE

The physical layer in LTE supports HARQ on the physical downlink and uplink shared channels, with separate control channels to send the associated acknowledgement feedback. In Frequency Division Duplex (FDD) operation, eight Stop-And-Wait (SAW) HARQ processes are available in both downlink and uplink with a typical Round-Trip Time (RTT) of 8 ms (see Figures 10.13 and 10.14 respectively). Each HARQ process requires a separate soft buffer allocation in the receiver for the purpose of combining the retransmissions. In

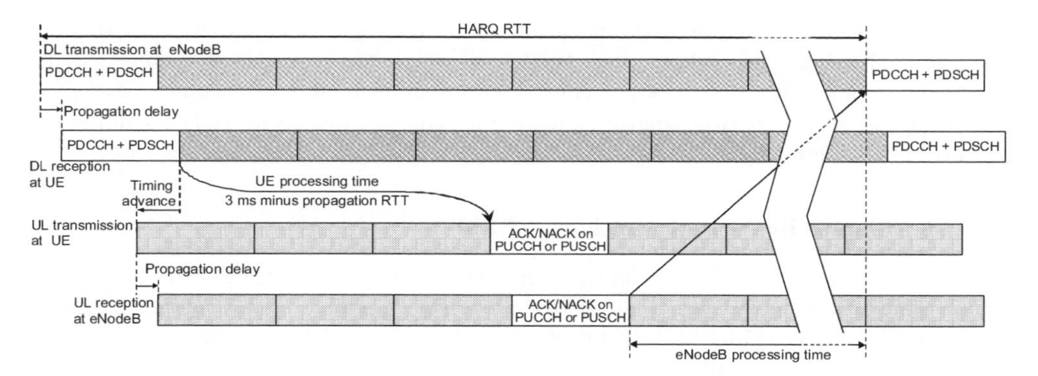

Figure 10.13: Timing diagram of the downlink HARQ (SAW) protocol.

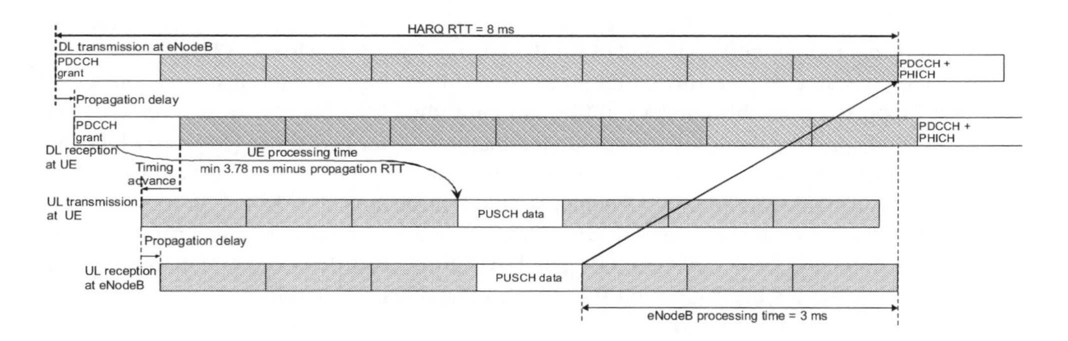

Figure 10.14: Timing diagram of the uplink HARQ (SAW) protocol.

FDD, the HARQ process to which a transport block belongs is identified by a unique three-bit HARQ process IDentifier (HARQ ID). In TDD, the number of HARQ processes depends on the uplink/downlink configuration as explained in Section 23.4.3, and four bits are used to identify the process.

There are several fields in the downlink control information (see Section 9.3.5.1) to aid the HARQ operation:

- New Data Indicator (NDI): toggled whenever a new packet transmission begins;
- Redundancy Version (RV): indicates the RV selected for the transmission or retransmission;
- MCS: Modulation and Coding Scheme.

As explained in Section 4.4.1.1, the LTE downlink HARQ is asynchronous and adaptive, and therefore every downlink transmission is accompanied by explicit signalling of control information. The uplink HARQ is synchronous, and either non-adaptive or adaptive. The uplink non-adaptive HARQ operation requires a predefined RV sequence 0, 2, 3, 1, 0, 2, 3, 1, ... for successive transmissions of a packet due to the absence of explicit control signalling. For the adaptive HARQ operation, the RV is explicitly signalled. There are also other uplink modes in which the redundancy version (or the modulation) is combined with other control information to minimize control signalling overhead. Aspects of HARQ control signalling specifically related to TDD operation are discussed in Section 23.4.3.

10.3.2.6 Limited Buffer Rate Matching (LBRM)

A major contributor to the UE implementation complexity is the UE HARQ soft buffer size, which is the total memory (over all the HARQ processes) required for LLR storage to support HARQ operation. The aim of Limited Buffer Rate Matching (LBRM) is therefore to reduce the required UE HARQ soft buffer sizes while maintaining the peak data rates and having minimal impact on the system performance. [50] shows that there is a negligible performance impact even with a 50% reduction in the soft buffer size. Therefore, for LTE, up to 50% soft buffer reduction is enabled by means of LBRM for the higher UE categories (3, 4 and 5 – see Section 1.3.4) while it is not applied to the lower UE categories (1 and 2). This reduction corresponds to a mother code rate of 2/3 for the largest Transport Block[12] (TB).

For each UE category, the soft buffer size is determined based on the instantaneous peak data rate supported per subframe (see Table 1.2) multiplied by eight (i.e. using eight HARQ processes) and applying the soft buffer reduction factor as described above [51]. For FDD, the soft buffer is split into eight equal partitions (one partition per TB) if the UE is configured to receive Physical Downlink Shared CHannel (PDSCH) transmissions based on transmission modes other than 3 or 4 (i.e. one TB per HARQ process), or sixteen equal partitions for transmission modes 3 or 4 (i.e. two TBs per HARQ process). It is important to note that all transmission modes are supported by all UE categories.[13] Therefore, a category 1 UE could be in transmission modes 3 or 4 but may be limited to rank-1 operation as it can support only one layer in spatial multiplexing (see Section 11.2.2). For simplicity and to facilitate dual-mode UE development, the soft buffer sizes for TDD were chosen to be the same as those for FDD. However, the number of downlink HARQ processes in TDD varies between 4 and 15 according to the downlink/uplink configuration (see Section 6.2); hence the soft buffer is adapted and split into $\min(M_{\text{DL\_HARQ}}, M_{\text{limit}})$ equal partitions if the UE is configured to receive PDSCH transmissions based on transmission modes other than 3 or 4, or $2 \cdot \min(M_{\text{DL\_HARQ}}, M_{\text{limit}})$ equal partitions for transmission modes 3 or 4, where $M_{\text{DL\_HARQ}}$ is the number of downlink HARQ processes, and M_{limit} is equal to 8. Thus, when the number of downlink HARQ processes is greater than 8, statistical soft buffer management can be used for efficient UE implementation [52].

A soft buffer size per code block segment is derived from the soft buffer size per TB, and LBRM then simply shortens the length of the VCB of the code block segments for certain larger sizes of TB, with the RV spacing being compressed accordingly. With LBRM, the effective mother code rate for a TB becomes a function of the TB size and the allocated UE soft buffer size. Since the eNodeB knows the soft buffer capability of the UE, it only transmits those code bits out of the VCB that can be stored in the UE's HARQ soft buffer for all (re)transmissions of a given TB.

[12] A transport block is equivalent to a MAC PDU – see Section 4.4.

[13] The only exceptions are that PDSCH transmission modes 7 and 8 (see Section 9.2.2.1) are optional for FDD UEs, and PUSCH transmission mode 2 (see Section 29.4.1) is optional for UE categories 1–7.

Soft buffer management in case of carrier aggregation in Release 10 is discussed in Section 28.4.2.

10.3.2.7 Overall Channel Coding Chain for Data

The overall flow diagram of the turbo coded channels in LTE is summarized in Figure 10.15.

The physical layer first attaches a 24-bit Cyclic Redundancy Check (CRC) to each TB received from the MAC layer. This is used at the receiver to verify correct reception and to generate the HARQ ACK/NACK feedback.

The TB is then segmented into 'code blocks' according to a rule which, given an arbitrary TB size, is designed to minimize the number of filler bits needed to match the available QPP sizes. This is accomplished by allowing two adjacent QPP sizes to be used when segmenting a TB, rather than being restricted to a single QPP size. The filler bits would then be placed in the first segment. However, in LTE the set of possible TB sizes is restricted such that the segmentation rule described above always results in a single QPP size for each segment with no filler bits.

Following segmentation, a further 24-bit CRC is attached to each code block if the TB was split into two or more code blocks. This code-block-level CRC can be utilized to devise

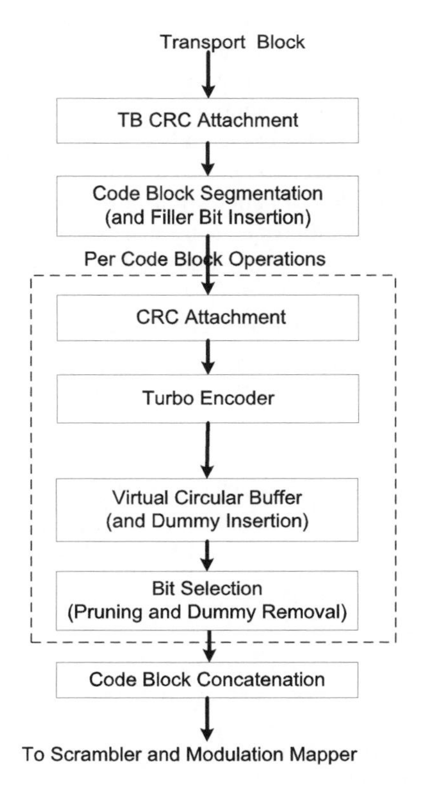

Figure 10.15: Flow diagram for turbo coded channels in LTE.

rules for early termination of the turbo decoding iterations to reduce decoding complexity. It is worth noting that the polynomial used for the code-block-level CRC is different from the polynomial used for the transport block CRC. This is a deliberate design feature in order to avoid increasing the probability of failing to detect errors as a result of the use of individual CRCs per code block; if all the code block CRCs pass, the decoder should still check the transport block CRC, which, being based on a different polynomial, is likely to detect an error which was not detected by a code-block CRC.

A major difference between LTE and HSDPA is that in LTE most of the channel coding operations on the PDSCH are performed at a code-block level, whereas in HSDPA only turbo coding is performed at the code block level. Although in the LTE specifications [10] the code block concatenation for transmission is done prior to the scrambler and modulation mapper, each code block is associated with a distinct set of modulation symbols, implying that the scrambling and modulation mapping operations may in fact be done individually for each code block, which facilitates an efficient pipelined implementation.

10.3.3 Channel Coding for Control Channels in LTE

Unlike the data, control information (such as is sent on the Physical Downlink Control CHannel (PDCCH) and Physical Broadcast CHannel (PBCH)!channel coding) is coded with a convolutional code, as the code blocks are significantly smaller and the additional complexity of the turbo coding operation is therefore not worthwhile.

The PDCCH is especially critical from a decoding complexity point of view, since a UE must decode a large number of potential control channel locations as discussed in Section 9.3.5.5. Another relevant factor in the code design for the PDCCH and the PBCH is that they both carry a relatively small number of bits, making the tail bits a more significant overhead. Therefore, it was decided to adopt a tail-biting convolutional code for LTE but, in order to limit the decoding complexity, using a new convolutional code with only 64 states instead of the 256-state convolutional code used in UMTS. These key differences are summarized in Table 10.8.

The LTE convolutional code, shown in Figure 10.2, offers slightly better performance for the target information block sizes, as shown in Figure 10.16. The initial value of the shift register of the encoder is set to the values corresponding to the last six information

Table 10.8: Differences between LTE and HSPA convolutional coding.

Property	LTE convolutional code	HSPA convolutional code
Number of states	64	256
Tailing method	Tail-biting	Tailed
Generators	[133, 171, 165](oct) $R = 1/3$	[561, 753](oct) $R = 1/2$ [557, 663, 711](oct) $R = 1/3$
Normalized decoding complexity	1/2 (assuming two iterations of decoding)	1
Rate matching	Circular buffer	Algorithmic calculation of rate-matching pattern

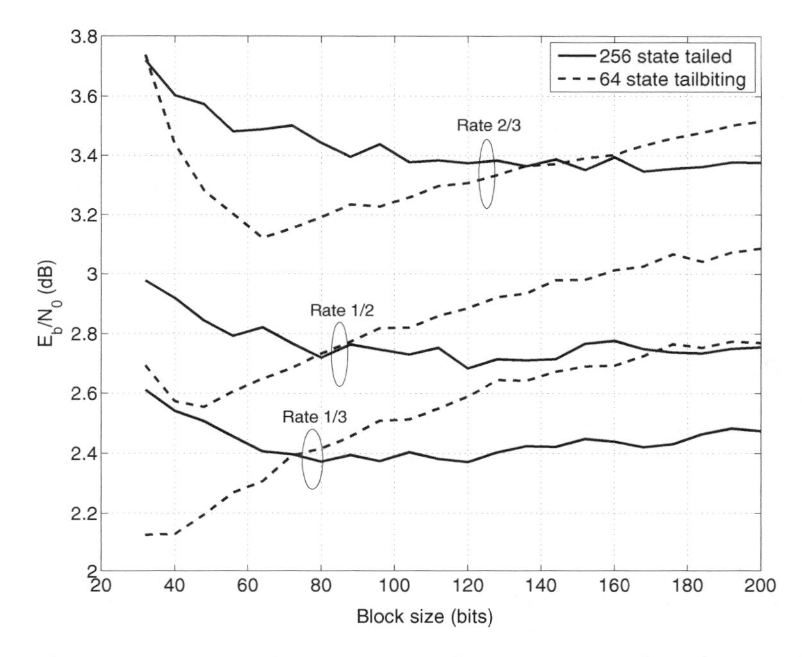

Figure 10.16: E_b/N_0 (dB) versus block size (bits) for BLER target of 1% for convolutional codes with rate 1/3, 1/2, and 2/3. The HSPA convolutional code has 256-state code with tail bits and the 64-state LTE code is tail-biting.

bits in the input stream so that the initial and final states of the shift register are the same. The decoder can utilize a Circular Viterbi Algorithm [14] or MAP algorithm [53], with decoding complexity approximately twice that of the Viterbi decoder with two iterations (passes through the trellis). With only a quarter the number of states, the overall complexity of the LTE convolutional code can therefore be argued to be half that of the HSPA code, provided that only two iterations are used.

The rate-matching for the convolutional code in LTE uses a similar CB method as for the turbo code. A 32-column interleaver is used, with no interlacing in the CB (the three parity streams are concatenated in the CB). This structure gives good performance at higher code rates as well as lower code rates, and therefore LTE has no need for an additional (different) $R = 1/2$ generator polynomial as used in UMTS.

Other even smaller items of control information use block codes (for example, a Reed–Muller code for CQI, or a simple list of codewords for the Physical Control Format Indicator CHannel (PCFICH) – see Section 9.3.3). With small information words, block codes lend themselves well to a Maximum Likelihood (ML) decoding approach.

10.4 Conclusions

The LTE channel coding is a versatile design that has benefited from the decades of research and development in the area of iterative processing. Although the turbo codes used in LTE and UMTS are of the same form as Berrou's original scheme, the LTE turbo code with its contention-free interleaver provides hardware designers with sufficient design flexibility to support the high data rates offered by LTE and LTE-Advanced. However, with increased support for parallelism comes the cost of routing the extrinsic values to and from the memory. The routing complexity in the turbo decoder with a large number of processors (e.g. $M = 64$) may in fact become comparable to that of an LDPC code with similar processing capability. Therefore, it is possible that the cost versus performance trade-offs between turbo and LDPC codes may be reinvestigated in the future. Nevertheless, it is clear that the turbo code will continue to shine for a long time to come.

References[14]

[1] A. J. Goldsmith and S. G. Chua, 'Adaptive Coded Modulation for Fading Channels'. *IEEE Trans. on Communications*, Vol. 46, pp. 595–602, May 1998.

[2] J. Hayes, 'Adaptive feedback communications'. *IEEE Trans. on Communication Technologies*, Vol. 16, pp. 15–22, February 1968.

[3] NTT DoCoMo, Fujitsu, Mitsubishi Electric Corporation, NEC, QUALCOMM Europe, Sharp, and Toshiba Corporation, 'R1-060039: Adaptive Modulation and Channel Coding Rate Control for Single-antenna Transmission in Frequency Domain Scheduling in E-UTRA Downlink', www.3gpp.org, 3GPP TSG RAN WG1, LTE ad-hoc meeting, Helsinki, Finland, January 2006.

[4] LG Electronics, 'R1-060051: Link Adaptation in E-UTRA Downlink', www.3gpp.org, 3GPP TSG RAN WG1, LTE ad-hoc meeting, Helsinki, Finland, January 2006.

[5] NTT DoCoMo, Fujitsu, Mitsubishi Electric Corporation, NEC, Sharp, and Toshiba Corporation, 'R1-060040: Adaptive Modulation and Channel Coding Rate Control for MIMO Transmission with Frequency Domain Scheduling in E-UTRA Downlink', www.3gpp.org, 3GPP TSG RAN WG1, LTE ad-hoc meeting, Helsinki, Finland, January 2006.

[6] Samsung, 'R1-060076: Adaptive Modulation and Channel Coding Rate', www.3gpp.org, 3GPP TSG RAN WG1, LTE ad-hoc meeting, Helsinki, Finland, January 2006.

[7] Huawei, 'R1-080555: Labelling Complexity of UE Selected Subbands on PUSCH', www.3gpp.org, 3GPP TSG RAN WG1, meeting 51bis, Sevilla, Spain, January 2008.

[8] Huawei, 'R1-080182: Labelling of UE-selected Subbands on PUSCH', www.3gpp.org, 3GPP TSG RAN WG1, meeting 51bis, Sevilla, Spain, January 2008.

[9] 3GPP Technical Specification 36.213, 'Evolved Universal Terrestrial Radio Access (E-UTRA); Physical Layer Procedures', www.3gpp.org.

[10] 3GPP Technical Specification 36.212, 'Evolved Universal Terrestrial Radio Access (E-UTRA); Multiplexing and Channel Coding (FDD)', www.3gpp.org.

[11] J. G. Proakis, *Digital Communications*. New York: McGraw-Hill, 1995.

[12] J. H. Ma and J. W. Wolf, 'On Tail-Biting Convolutional Codes'. *IEEE Trans. on Communications*, Vol. 34, pp. 104–111, February 1986.

[13] R. V. Cox and C. V. Sundberg, 'An Efficient Adaptive Circular Viterbi Algorithm for Decoding Generalized Tailbiting Convolutional Codes'. *IEEE Trans. on Vehicular Technology*, Vol. 43, pp. 57–68, February 1994.

[14] H. Ma and W. J. Wolf, 'On Tail-biting Convolutional Codes'. *IEEE Trans. on Communications*, Vol. 34, pp. 104–111, February 1986.

[14] All web sites confirmed 1st March 2011.

[15] S. Lin and D. J. Costello, *Error and Control Coding: Fundamentals and Applications*. Second Edition, Prentice Hall: Englewood Cliffs, NJ, 2004.

[16] A. J. Viterbi, 'Error Bounds for Convolutional Codes and an Asymptotically Optimum Decoding Algorithm'. *IEEE Trans. on Information Theory*, Vol. 13, pp. 260–269, April 1967.

[17] G. D. Forney Jr., 'The Viterbi Algorithm', in *Proc. of the IEEE*, March 1973.

[18] J. Haguenauer and P. Hoeher, 'A Viterbi algorithm with soft-decision outputs and its applications', in *Proc. of IEEE Globecom*, Dallas, TX, 1989.

[19] L. R. Bahl, J. Cocke, F. Jelinek and J. Raviv, 'Optimal Decoding of Linear Codes for Minimizing Symbol Error Rate'. *IEEE Trans. on Information Theory*, Vol. 20, pp. 284–287, March 1974.

[20] C. Berrou and A. Glavieux, 'Near Optimum Error Correcting Coding and Decoding: Turbo-codes'. *IEEE Trans. on Communications*, Vol. 44, pp. 1261–1271, October 1996.

[21] C. E. Shannon, 'Communication in the Presence of Noise', in *Proc. IRE*, Vol. 37, pp. 10–21, January 1949.

[22] C. E. Shannon, *A Mathematical Theory of Communication*. Urbana, IL: University of Illinois Press, 1998.

[23] R. G. Gallager, 'Low-Density Parity-Check Codes'. Ph.D. Thesis, Cambridge, MA: MIT Press, 1963.

[24] S. Y. Chung, 'On the Construction of Some Capacity-Approaching Coding Scheme'. Ph.D. Thesis, Massachusetts Institute of Technology, September 2000.

[25] D. J. C. MacKay and R. M. Neal, 'Near Shannon Limit Performance of Low Density Parity Check Codes'. *IEEE Electronics Letters*, Vol. 32, pp. 1645–1646, August 1996.

[26] C. Berrou, P. Glavieux and A. Thitimajshima, 'Near Shannon Limit Error-correcting Coding and Decoding', in *Proc. IEEE International Conference on Communications*, Geneva, Switzerland, 1993.

[27] G. Battail, 'Coding for the Gaussian Channel: the Promise of Weighted-Output Decoding'. *International Journal of Satellite Communications*, Vol. 7, pp. 183–192, 1989.

[28] C. Berrou, 'The Ten-Year Old Turbo Codes Are Entering into Service'. *IEEE Communication Magazine*, Vol. 41, no. 8, 2003.

[29] T. J. Richardson and R. L Urbanke, 'The Capacity of Low-density Parity-check Codes Under Message-passing Decoding'. *IEEE Trans. on Information Theory*, Vol. 47, pp. 599–618, February 2001.

[30] M. M. Mansour and N. Shanbhag, 'High-throughput LDPC Decoders'. *IEEE Trans. on Very Large Scale Integration (VLSI) Systems*, Vol. 11, pp. 976–996, October 2003.

[31] T. J. Richardson, M. A Shokrollahi and R. L Urbanke, 'Design of Capacity-approaching Irregular Low-density Parity-check Codes'. *IEEE Trans. on Information Theory*, Vol. 47, pp. 619–637, February 2001.

[32] S-Y. Chung, G. D., Jr. Forney, T. J. Richardson and R. Urbanke, 'On the Design of Low-density Parity-check Codes within 0.0045 dB of the Shannon Limit'. *IEEE Communication Letters*, Vol. 5, pp. 58–60, February 2001.

[33] J. Chen, R. M. Tanner, J. Zhang and M. P. C Fossorier, 'Construction of Irregular LDPC Codes by Quasi-Cyclic Extension'. *IEEE Trans. on Information Theory*, Vol. 53, pp. 1479–1483, 2007.

[34] M. Bickerstaff, L. Davis, C. Thomas, D. Garrett and C. Nicol, 'A 24 Mb/s Radix-4 LogMAP Turbo Decoder for 3GPP-HSDPA Mobile Wireless', in *Proc. of 2003 IEEE International Solid State Circuits Conference*, San Francisco, CA, 2003.

[35] Motorola, 'R1-060384: LDPC Codes for EUTRA', www.3gpp.org, 3GPP TSG RAN WG1, meeting 44, Denver, USA, February 2006.

[36] Samsung, 'R1-060334: LTE Channel Coding', www.3gpp.org, 3GPP TSG RAN WG1, meeting 44, Denver, USA, February 2006.

[37] Intel, ITRI, LG, Mitsubishi, Motorola, Samsung, and ZTE, 'R1-060874: Complexity Comparison of LDPC Codes and Turbo Codes', www.3gpp.org, 3GPP TSG RAN WG1, meeting 44bis, Athens, Greece, March 2006.

[38] T. K. Blankenship, B. Classon and V. Desai, 'High-Throughput Turbo Decoding Techniques for 4G', in *Proc. of International Conference 3G and Beyond*, pp. 137–142, 2002.

[39] Panasonic, 'R1-073357: Interleaver for LTE turbo code', www.3gpp.org, 3GPP TSG RAN WG1, meeting 47, Riga, Latvia, November 2006.

[40] Nortel and Samsung, 'R1-073265: Parallel Decoding Method for the Current 3GPP Turbo Interleaver', www.3gpp.org, 3GPP TSG RAN WG1, meeting 47, Riga, Latvia, November 2006.

[41] A. Tarable, S. Benedetto and G. Montorsi, 'Mapping Interleaving Laws to Parallel Turbo and LDPC Decoder Architectures'. *IEEE Trans. on Information Theory*, Vol. 50, pp. 2002–2009, September 2004.

[42] A. Nimbalker, T. K. Blankenship, B. Classon, T. Fuja, and D. J. Costello Jr, 'Contention-free Interleavers', in *Proc. IEEE International Symposium on Information Theory*, Chicago, IL, 2004.

[43] A. Nimbalker, T. K. Blankenship, B. Classon, T. Fuja and D. J. Costello Jr, 'Inter-window Shuffle Interleavers for High Throughput Turbo Decoding' in *Proc. of 3rd International Symposium on Turbo Codes and Related Topics*, Brest, France, 2003.

[44] C. Berrou, Y. Saouter, C. Douillard, S. Kerouedan and M. Jezequel, 'Designing Good Permutations for Turbo Codes: Towards a Single Model', in *Proc. IEEE International Conference on Communications*, Paris, France, 2004.

[45] O. Y. Takeshita, 'On Maximum Contention-free Interleavers and Permutation Polynomials over Integer Rings'. *IEEE Trans. on Information Theory*, Vol. 52, pp. 1249–1253, March 2006.

[46] A. Nimbalker, Y Blankenship, B. Classon, T. Fuja and T. K. Blankenship, 'ARP and QPP Interleavers for LTE Turbo Coding', in *Proc. IEEE Wireless Communications and Networking Conference*, 2008.

[47] Motorola, 'R1-063061: A Contention-free Interleaver Design for LTE Turbo Codes', www.3gpp.org, 3GPP TSG RAN WG1, meeting 47, Riga, Latvia, November 2006.

[48] Siemens, 'R1-030421: Turbo Code Irregularities in HSDPA', www.3gpp.org, 3GPP TSG RAN WG1, meeting 32, Marne La Vallée, Paris, May 2003.

[49] Motorola, 'R1-071795: Parameters for Turbo Rate-Matching', www.3gpp.org, 3GPP TSG RAN WG1, meeting 48bis, Malta, March 2007.

[50] Motorola, 'R1-080058: Limited Buffer Rate Matching Ű System performance', www.3gpp.org, 3GPP TSG RAN WG1, meeting 51bis, Seville, Spain, January 2008.

[51] Motorola, Nokia Siemens Networks 'R1-082123: Adjustments to UE Downlink Soft Buffer Sizes based on LTE transport block sizes', www.3gpp.org, 3GPP TSG RAN WG1, meeting 53, Kansas City, USA, May 2008.

[52] Ericsson, 'R1-082018: On soft buffer usage for LTE TDD', www.3gpp.org, 3GPP TSG RAN WG1, meeting 53, Kansas City, USA, May 2008.

[53] R. Y. Shao, S. Lin, and M. Fossorier, 'An Iterative Bidirectional Decoding Algorithm for Tail Biting Codes', in *Proc. of IEEE Information Theory Workshop (ITW)*, Kruger National Park, South Africa, 1999.

11

Multiple Antenna Techniques

Thomas Sälzer, David Gesbert, Cornelius van Rensburg, Filippo Tosato, Florian Kaltenberger and Tetsushi Abe

11.1 Fundamentals of Multiple Antenna Theory

11.1.1 Overview

The value of multiple antenna systems as a means to improve communications was recognized in the very early ages of wireless transmission. However, most of the scientific progress in understanding their fundamental capabilities has occurred only in the last 20 years, driven by efforts in signal and information theory, with a key milestone being achieved with the invention of so-called Multiple-Input Multiple-Output (MIMO) systems in the mid-1990s.

Although early applications of beamforming concepts can be traced back as far as 60 years in military applications, serious attention has been paid to the utilization of multiple antenna techniques in mass-market commercial wireless networks only since around 2000. Today, the key role which MIMO technology plays in the latest wireless communication standards for Personal, Wide and Metropolitan Area Networks (PANs, WANs and MANs) testifies to its importance. Aided by rapid progress in the areas of computation and circuit integration, this trend culminated in the adoption of MIMO for the first time in a cellular mobile network standard in the Release 7 version of HSDPA (High Speed Downlink Packet Access); soon after, the development of LTE broke new ground in being the first global mobile cellular system to be designed with MIMO as a key component from the start.

This chapter first provides the reader with the theoretical background necessary for a good understanding of the role and advantages promised by multiple antenna techniques in

LTE – The UMTS Long Term Evolution: From Theory to Practice, Second Edition.
Stefania Sesia, Issam Toufik and Matthew Baker.

wireless communications in general.[1] The chapter focuses on the intuition behind the main technical results and show how key progress in information theory yields practical lessons in algorithm and system design for cellular communications. As can be expected, there is still a gap between the theoretical predictions and the performance achieved by schemes that must meet the complexity constraints imposed by commercial considerations.

We distinguish between single-user MIMO and multi-user MIMO, although a common set of concepts captures the essential MIMO benefits in both cases. Single-user MIMO techniques dominated the algorithms selected for the first version of LTE, with multi-user MIMO becoming more established in Releases 9 and 10.

The second part of the chapter describes the actual methods adopted for LTE, paying particular attention to the combinations of theoretical principles and system design constraints that led to these choices.

While traditional wireless communications (Single-Input Single-Output (SISO)) exploit time- or frequency-domain pre-processing and decoding of the transmitted and received data respectively, the use of additional antenna elements at either the base station (eNodeB) or User Equipment (UE) side (on the downlink or uplink) opens an extra spatial dimension to signal precoding and detection. Space-time processing methods exploit this dimension with the aim of improving the link's performance in terms of one or more possible metrics, such as the error rate, communication data rate, coverage area and spectral efficiency (expressed in bps/Hz/cell).

Depending on the availability of multiple antennas at the transmitter and/or the receiver, such techniques are classified as Single-Input Multiple-Output (SIMO), Multiple-Input Single-Output (MISO) or MIMO. Thus in the scenario of a multi-antenna-enabled base station communicating with a single antenna UE, the uplink and downlink are referred to as SIMO and MISO respectively. When a multi-antenna terminal is involved, a full MIMO link may be obtained, although the term MIMO is sometimes also used in its widest sense, thus including SIMO and MISO as special cases. While a point-to-point multiple-antenna link between a base station and one UE is referred to as Single-User MIMO (SU-MIMO), Multi-User MIMO (MU-MIMO) features several UEs communicating simultaneously with a common base station using the same frequency- and time-domain resources.[2] By extension, considering a multicell context, neighbouring base stations sharing their antennas in virtual MIMO fashion [4] to communicate with the same set of UEs in different cells comes under the term Coordinated MultiPoint (CoMP) transmission/reception. This latter scenario is not supported in the first versions of LTE but it is being studied for possible inclusion in later releases of LTE-Advanced as described in Section 29.5. The overall evolution of MIMO concepts, from the simplest diversity setup to the advanced joint-processing CoMP technique, is illustrated in Figure 11.1.

Despite their variety and sometimes perceived complexity, single-user and multi-user MIMO techniques tend to revolve around just a few fundamental principles, which aim at leveraging some key properties of multi-antenna radio propagation channels. As introduced

[1]For more exhaustive tutorial information on MIMO systems, the reader is referred, for example, to [1–3].

[2]Note that, in LTE, a single eNodeB may in practice control multiple cells; in such a case, we consider each cell as an independent base station for the purpose of explaining the MIMO techniques; the simultaneous transmissions in the different cells address different UEs and are typically achieved using different fixed directional physical antennas; they are therefore not classified as multi-user MIMO.

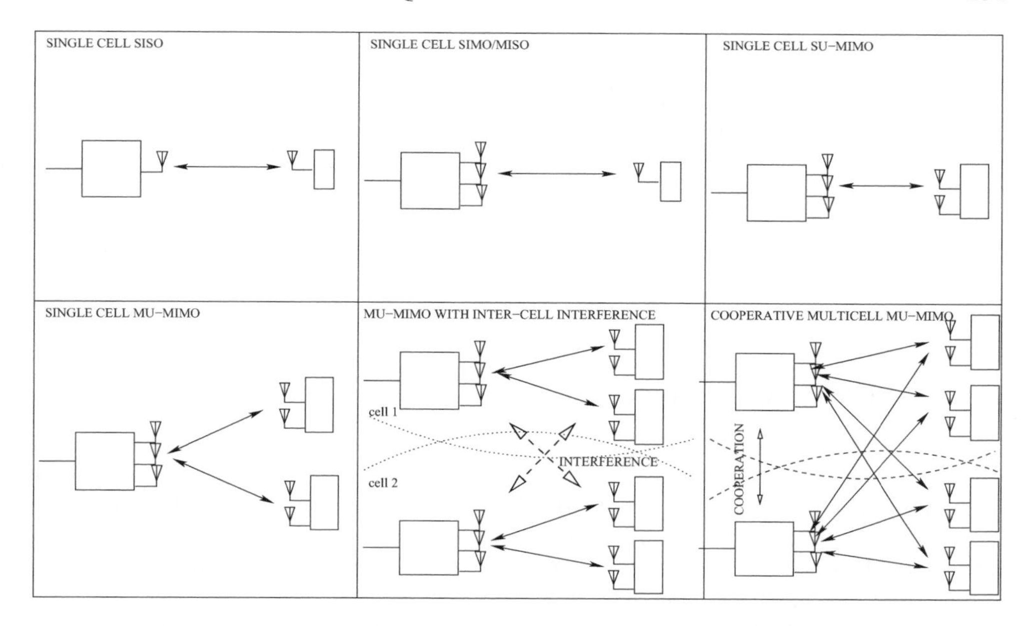

Figure 11.1: The evolution of MIMO technology, from traditional single antenna communication, to multi-user MIMO scenarios, to the possible multicell MIMO networks of the future.

in Section 1.3, there are basically three advantages associated with such channels (over their SISO counterparts):

- Diversity gain.
- Array gain.
- Spatial multiplexing gain.

Diversity gain corresponds to the mitigation of the effect of multipath fading, by means of transmitting or receiving over multiple antennas at which the fading is sufficiently decorrelated. It is typically expressed in terms of an *order*, referring either to the number of effective independent diversity branches or to the slope of the Bit Error Rate (BER) curve as a function of the Signal-to-Noise Ratio (SNR) (or possibly in terms of an SNR gain in the system's link budget).

While diversity gain is fundamentally related to improvement of the *statistics* of instantaneous SNR in a fading channel, array gain and multiplexing gain are of a different nature, rather being related to the geometry and the theory of vector spaces. Array gain corresponds to a spatial version of the well-known matched-filter gain in time-domain receivers, while multiplexing gain refers to the ability to send multiple data streams in parallel and to separate them on the basis of their spatial signature. The latter is much akin to the multiplexing of users separated by orthogonal spreading codes, timeslots or frequency assignments, with the great advantage that, unlike Code, Time or Frequency Division Multiple Access (CDMA, TDMA or FDMA), MIMO multiplexing does not come at the cost of bandwidth expansion; it does, however, suffer the expense of added antennas and signal processing complexity and its gains strongly depend on the spatial characteristics of the propagation channel (see Chapter 20).

These aspects are analysed further in the following sections, considering first theoretically optimal transmission schemes and then popular MIMO approaches. First a base station to single-user communication model is considered (to cover single-user MIMO); then the techniques are generalized to multi-user MIMO.

The schemes adopted in LTE Releases 8 and 9, specifically for the downlink, are addressed subsequently. We focus on the Frequency Division Duplex (FDD) case. Discussion of some aspects of MIMO which are specific to Time Division Duplex (TDD) operation can be found in Section 23.5.

11.1.2 MIMO Signal Model

Let $\mathbf{Y}$ be a matrix of size $N \times T$ denoting the set of (possibly precoded) signals being transmitted from N distinct antennas over T symbol durations (or, in the case of some frequency-domain systems, T subcarriers), where T is a parameter of the MIMO algorithm (defined below). Thus the nth row of $\mathbf{Y}$ corresponds to the signal emitted from the nth transmit antenna. Let $\mathbf{H}$ denote the $M \times N$ channel matrix modelling the propagation effects from each of the N transmit antennas to any one of M receive antennas, over an arbitrary subcarrier whose index is omitted here for simplicity. We assume $\mathbf{H}$ to be invariant over T symbol durations. The matrix channel is represented by way of example in Figure 11.2. Then the $M \times T$ signal $\mathbf{R}$ received over T symbol durations over this subcarrier can be conveniently written as

$$\mathbf{R} = \mathbf{H}\mathbf{Y} + \mathbf{N} \tag{11.1}$$

where $\mathbf{N}$ is the additive noise matrix of dimension $M \times T$ over all M receiving antennas. We will use $\mathbf{h}_i$ to denote the ith column of $\mathbf{H}$, which will be referred to as the *receive spatial signature* of (i.e. corresponding to) the ith transmitting antenna. Likewise, the jth row of $\mathbf{H}$ can be termed the *transmit spatial signature* of the jth receiving antenna.

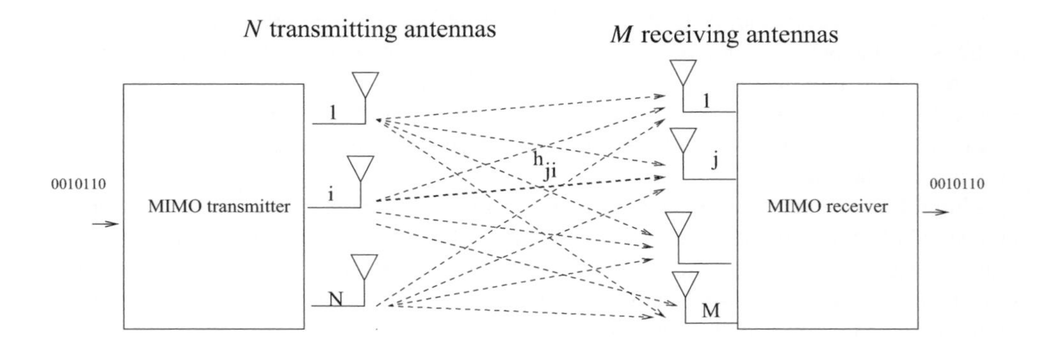

Figure 11.2: Simplified transmission model for a MIMO system with N transmit antennas and M receive antennas, giving rise to an $M \times N$ channel matrix, with MN links.

Let $\mathbf{X} = (x_1, x_2, \ldots, x_P)$ be a group of P QAM[3] symbols to be sent to the receiver over the T symbol durations; these symbols must be *mapped* to the transmitted signal $\mathbf{Y}$ before launching into the air. The choice of this mapping function $\mathbf{X} \rightarrow \mathbf{Y}(\mathbf{X})$ determines which one out of several possible MIMO transmission methods results, each yielding a different combination of the diversity, array and multiplexing gains. Meanwhile, the so-called *spatial* rate of the chosen MIMO transmission method is given by the ratio P/T.

Note that, in the most general case, the considered transmit (or receive) antennas may be attached to a single transmitting (or receiving) device (base station or UE), or distributed over different devices. The symbols in $(x_1, x_2, \ldots, x_P)$ may also correspond to the data of one or possibly multiple users, giving rise to the single-user MIMO or multi-user MIMO models.

11.1.3 Single-User MIMO Techniques

Several classes of SU-MIMO transmission methods are discussed below, both optimal and suboptimal.

11.1.3.1 Optimal Transmission over MIMO Systems

The optimal way of communicating over the MIMO channel involves a channel-dependent precoder, which fulfils the roles of both transmit beamforming and power allocation across the transmitted streams, and a matching receive beamforming structure. Full channel knowledge is therefore required at the transmit side for this method to be applicable. Consider a set of $P = NT$ symbols to be sent over the channel. The symbols are separated into N streams (or layers) of T symbols each. Stream i consists of symbols $[x_{i,1}, x_{i,2}, \ldots, x_{i,T}]$. Note that in an ideal setting, each stream may adopt a distinct code rate and modulation. This is discussed in more detail below. The transmitted signal can now be written as

$$\mathbf{Y}(\mathbf{X}) = \mathbf{V}\mathbf{P}\bar{\mathbf{X}} \tag{11.2}$$

where

$$\bar{\mathbf{X}} = \begin{pmatrix} x_{1,1} & x_{1,2} & \cdots & x_{1,T} \\ \vdots & \vdots & & \vdots \\ x_{N,1} & x_{N,2} & \cdots & x_{N,T} \end{pmatrix} \tag{11.3}$$

$\mathbf{V}$ is an $N \times N$ transmit beamforming matrix and $\mathbf{P}$ is a $N \times N$ diagonal power-allocation matrix with $\sqrt{p_i}$ as its ith diagonal element, where p_i is the power allocated to the ith stream. Of course, the power levels must be chosen so as not to exceed the available transmit power, which can often be conveniently expressed as a constraint on the total normalized transmit power P_t.[4] Under this model, the information-theoretic capacity of the MIMO channel in bps/Hz can be obtained as [3]

$$C_{\text{MIMO}} = \log_2 \det(I + \rho \mathbf{H}\mathbf{V}\mathbf{P}^2\mathbf{V}^{\text{H}}\mathbf{H}^{\text{H}}) \tag{11.4}$$

where $\{\cdot\}^{\text{H}}$ denotes the Hermitian operator for a matrix or vector and ρ is the so-called transmit SNR, given by the ratio of the transmit power over the noise power.

[3]Quadrature Amplitude Modulation.

[4]In practice, there may be a limit on the maximum transmission power from each antenna.

The optimal (capacity-maximizing) precoder (**VP**) in Equation (11.4) is obtained by the concatenation of *singular vector beamforming* and the so-called *waterfilling power allocation*.

Singular vector beamforming means that **V** should be a unitary matrix (i.e. $\mathbf{V}^{\mathrm{H}}\mathbf{V}$ is the identity matrix of size N) chosen such that $\mathbf{H} = \mathbf{U}\boldsymbol{\Sigma}\mathbf{V}^{\mathrm{H}}$ is the Singular-Value Decomposition[5] (SVD) of the channel matrix **H**. Thus the ith right singular vector of **H**, given by the ith column of **V**, is used as a transmit beamforming vector for the ith stream. At the receiver side, the optimal beamformer for the ith stream is the ith left singular vector of **H**, obtained as the ith row of $\mathbf{U}^{\mathrm{H}}$:

$$\mathbf{u}_i^{\mathrm{H}}\mathbf{R} = \lambda_i \sqrt{p_i}[x_{i,1}, x_{i,2}, \ldots, x_{i,T}] + \mathbf{u}_i^{\mathrm{H}}\mathbf{N} \tag{11.5}$$

where λ_i is the ith singular value of **H**.

Waterfilling power allocation is the optimal power allocation and is given by

$$p_i = [\mu - 1/(\rho\lambda_i^2)]_+ \tag{11.6}$$

where $[x]_+$ is equal to x if x is positive and zero otherwise. μ is the so-called 'water level', a positive real variable which is set such that the total transmit power constraint is satisfied.

Thus the optimal SU-MIMO multiplexing scheme uses SVD-based transmit and receive beamforming to decompose the MIMO channel into a number of parallel non-interfering subchannels, dubbed 'eigen-channels', each one with an SNR being a function of the corresponding singular value λ_i and chosen power level p_i.

Contrary to what would perhaps be expected, the philosophy of optimal power allocation across the eigen-channels is *not* to equalize the SNRs, but rather to render them more unequal, by 'pouring' more power into the better eigen-channels, while allocating little power (or even none at all) to the weaker ones because they are seen as not contributing enough to the total capacity. This waterfilling principle is illustrated in Figure 11.3.

The underlying information-theoretic assumption here is that the information rate on each stream can be adjusted finely to match the eigen-channel's SNR. In practice, this is done by selecting a suitable Modulation and Coding Scheme (MCS) for each stream.

11.1.3.2 Beamforming with Single Antenna Transmitter or Receiver

In the case where either the receiver or the transmitter is equipped with only a single antenna, the MIMO channel exhibits only one active eigen-channel, and hence multiplexing of more than one data stream is not possible.

In *receive* beamforming, $N = 1$ and $M > 1$ (assuming a single stream). In this case, one symbol is transmitted at a time, such that the symbol-to-transmit-signal mapping function is characterized by $P = T = 1$, and $\mathbf{Y}(\mathbf{X}) = \mathbf{X} = x$, where x is the one QAM symbol to be sent. The received signal vector is given by

$$\mathbf{R} = \mathbf{H}x + \mathbf{N} \tag{11.7}$$

The receiver combines the signals from its M antennas through the use of weights $\mathbf{w} = [w_1, \ldots, w_M]$. Thus the received signal after antenna combining can be expressed as

$$z = \mathbf{wR} = \mathbf{wH}s + \mathbf{wN} \tag{11.8}$$

[5]The reader is referred to [5] for an explanation of generic matrix operations and terminology.

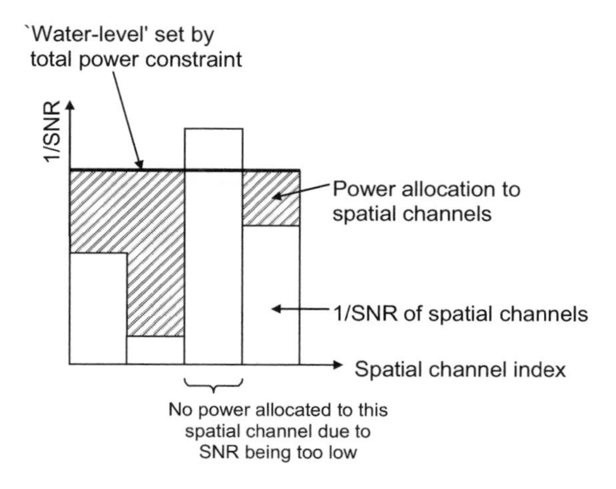

Figure 11.3: The waterfilling principle for optimal power allocation.

After the receiver has acquired a channel estimate (as discussed in Chapter 8), it can set the beamforming vector $\mathbf{w}$ to its optimal value to maximize the received SNR. This is done by aligning the beamforming vector with the UE's channel, via the so-called Maximum Ratio Combining (MRC) $\mathbf{w} = \mathbf{H}^{\mathrm{H}}$, which can be viewed as a spatial version of the well-known matched filter. Note that cancellation of an interfering signal can also be achieved, by selecting the beamforming vector to be orthogonal to the channel from the interference source. These simple concepts are illustrated in Figure 11.4.

The maximum ratio combiner provides a factor of M improvement in the received SNR compared to the $M = N = 1$ case – i.e. an array gain of $10\log_{10}(M)$ dB in the link budget.

In *transmit* beamforming, $M = 1$ and $N > 1$. The symbol-to-transmit-signal mapping function is characterized by $P = T = 1$, and $\mathbf{Y}(\mathbf{X}) = \mathbf{w}x$, where x is the one QAM symbol to be sent and $\mathbf{w}$ is the transmit beamforming vector of size $N \times 1$, computed based on channel knowledge, which is itself often obtained via a receiver-to-transmitter feedback link.[6] Assuming perfect channel knowledge at the transmitter side, the SNR-maximizing solution is given by the transmit MRC, which can be seen as a matched prefilter,

$$\mathbf{w} = \frac{\mathbf{H}^{\mathrm{H}}}{||\mathbf{H}||} \tag{11.9}$$

where the normalization by $||\mathbf{H}||$ enforces a total power constraint across the transmit antennas. The transmit MRC pre-filter provides a similar gain as its receive counterpart, namely $10\log_{10}(N)$ dB in average SNR improvement.

[6]In some situations, other techniques such as receive/transmit channel reciprocity may be used, as discussed in Section 23.5.2.

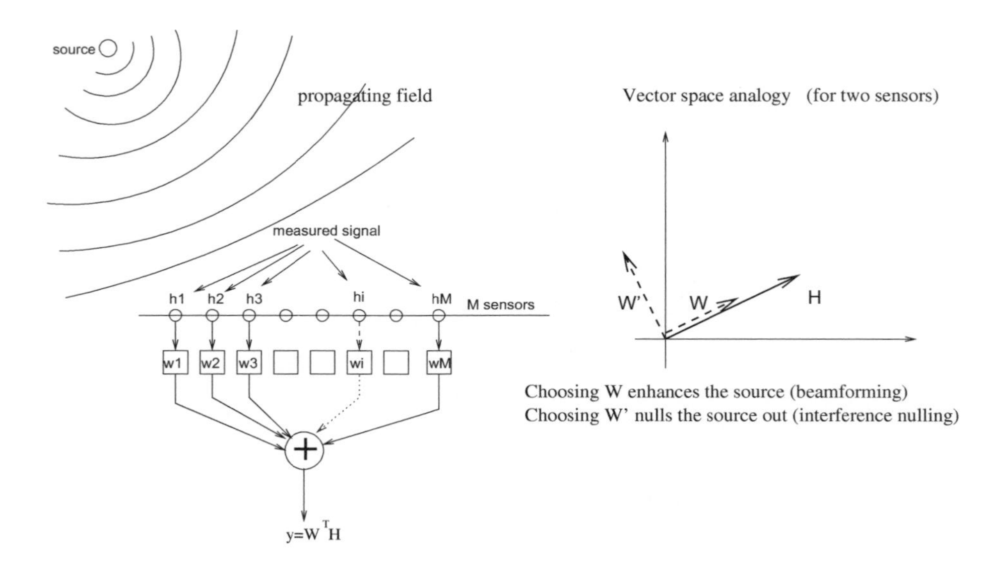

Figure 11.4: The beamforming and interference cancelling concepts.

11.1.3.3 Spatial Multiplexing without Channel Knowledge at the Transmitter

When $N > 1$ and $M > 1$, multiplexing of up to $\min(M, N)$ streams is theoretically possible even without channel knowledge at the transmitter. Assume for instance that $M \geq N$. In this case, one considers N streams, each transmitted using a different transmit antenna. As the transmitter does not have knowledge of the matrix $\mathbf{H}$, the precoder is simply the identity matrix. In this case, the symbol-to-transmit-signal mapping function is characterized by $P = NT$ and by

$$\mathbf{Y}(\mathbf{X}) = \bar{\mathbf{X}} \tag{11.10}$$

At the receiver, a variety of linear and non-linear detection techniques may be implemented to recover the symbol matrix $\bar{\mathbf{X}}$. A low-complexity solution is offered by the linear case, whereby the receiver superposes N beamformers $\mathbf{w}_1, \mathbf{w}_2, \ldots, \mathbf{w}_N$ so that the ith stream $[x_{i,1}, x_{i,2}, \ldots, x_{i,T}]$ is detected as follows,

$$\mathbf{w}_i\mathbf{R} = \mathbf{w}_i\mathbf{H}\bar{\mathbf{X}} + \mathbf{w}_i\mathbf{N} \tag{11.11}$$

The design criterion for the beamformer $\mathbf{w}_i$ can be interpreted as a compromise between single-stream beamforming and cancelling of interference (created by the other $N - 1$ streams). Inter-stream interference is fully cancelled by selecting the Zero-Forcing (ZF) receiver given by

$$\mathbf{W} = \begin{pmatrix} \mathbf{w}_1 \\ \mathbf{w}_2 \\ \vdots \\ \mathbf{w}_N \end{pmatrix} (\mathbf{H}^{\text{H}}\mathbf{H})^{-1}\mathbf{H}^{\text{H}}. \tag{11.12}$$

However, for optimal performance, $\mathbf{w}_i$ should strike a balance between alignment with respect to $\mathbf{h}_i$ and orthogonality with respect to all other signatures $\mathbf{h}_k$, $k \neq i$. Such a balance is achieved by, for example, a Minimum Mean-Squared Error (MMSE) receiver.

Beyond classical linear detection structures such as the ZF or MMSE receivers, more advanced but non-linear detectors can be exploited which provide a better error rate performance at the chosen SNR operating point, at the cost of extra complexity. Examples of such detectors include the Successive Interference Cancellation (SIC) detector and the Maximum Likelihood Detector (MLD) [3]. The principle of the SIC detector is to treat individual streams, which are channel-encoded, as layers which are 'peeled off' one by one by a processing sequence consisting of linear detection, decoding, remodulating, re-encoding and subtraction from the total received signal **R**. On the other hand, the MLD attempts to select the most likely set of all streams, simultaneously, from **R**, by an exhaustive search procedure or a lower-complexity equivalent such as the sphere-decoding technique [3].

Multiplexing gain

The multiplexing gain corresponds to the multiplicative factor by which the spectral efficiency is increased by a given scheme. Perhaps the single most important requirement for MIMO multiplexing gain to be achieved is for the various transmit and receive antennas to experience a sufficiently different channel response. This translates into the condition that the spatial signatures of the various transmitters (the $\mathbf{h}_i$'s) (or receivers) be sufficiently decorrelated and linearly independent to make the channel matrix **H** invertible (or, more generally, well-conditioned). An immediate consequence of this condition is the limitation to $\min(M, N)$ of the number of independent streams which may be multiplexed into the MIMO channel, or, more generally, to rank(**H**) streams. As an example, single-user MIMO communication between a four-antenna base station and a dual antenna UE can, at best, support multiplexing of two data streams, and thus a doubling of the UE's data rate compared with a single stream.

11.1.3.4 Diversity Techniques

Unlike the basic multiplexing scenario in Equation (11.10), where the design of the transmitted signal matrix **Y** exhibits no redundancy between its entries, a diversity-oriented design will feature some level of repetition between the entries of **Y**. For 'full diversity', each transmitted symbol $x_1, x_2, \ldots, x_P$ must be assigned to each of the transmit antennas at least once in the course of the T symbol durations. The resulting symbol-to-transmit-signal mapping function is called a Space-Time Block Code (STBC). Although many designs of STBC exist, additional properties such as the orthogonality of matrix **Y** allow improved performance and easy decoding at the receiver. Such properties are realized by the Alamouti space-time code [6], explained in Section 11.2.2.1. The total diversity order which can be realized in the N to M MIMO channel is MN when entries of the MIMO channel matrix are statistically uncorrelated. The intuition behind this is that $MN - 1$ represents the number of SISO links simultaneously in a state of severe fading which the system can sustain while still being able to convey the information to the receiver. The diversity order is equal to this number plus one. As in the previous simple multiplexing scheme, an advantage of diversity-oriented transmission is that the transmitter does not need knowledge of the channel **H**, and therefore no feedback of this parameter is necessary.

Diversity versus multiplexing trade-off

A fundamental aspect of the benefits of MIMO lies in the fact that any given multiple antenna configuration has a limited number of degrees of freedom. Thus there exists a compromise between reaching full beamforming gain in the detection of a desired stream of data and the perfect cancelling of undesired, interfering streams. Similarly, there exists a trade-off between the number of streams that may be multiplexed across the MIMO channel and the amount of diversity that each one of them will enjoy. Such a trade-off can be formulated from an information-theoretic point of view [7]. In the particular case of spatial multiplexing of N streams over an N to M antenna channel, with $M \geq N$, and using a linear detector, it can be shown that each stream can enjoy a maximum diversity order of $M - N + 1$.

To some extent, increasing the spatial load of MIMO systems (i.e. the number of spatially multiplexed streams) is akin to increasing the user load in CDMA systems. This correspondence extends to the fact that an optimal load level exists for a given target error rate in both systems.

11.1.4 Multi-User MIMO Techniques

11.1.4.1 Comparing Single-User and Multi-User MIMO

The set of MIMO techniques featuring data streams being communicated to (or from) antennas located on distinct UEs is referred to as Multi-User MIMO (MU-MIMO), which is one of the more recent developments of MIMO technology.

Although the MU-MIMO situation is just as well described by our model in Equation (11.1), the MU-MIMO scenario differs in a number of crucial ways from its single-user counterpart. We first explain these differences qualitatively, and then present a brief survey of the most important MU-MIMO transmission techniques.

In MU-MIMO, K UEs are selected for simultaneous communication over the same time-frequency resource, from a set of U active UEs in the cell. Typically, K is much smaller than U. Each UE is assumed to be equipped with J antennas, so the selected UEs together form a set of $M = KJ$ UE-side antennas. Since the number of streams that may be communicated over an N to M MIMO channel is limited to $\min(M, N)$ (if complete interference suppression is intended using linear combining of the antennas), the upper bound on the number of streams in MU-MIMO is typically dictated by the number of base station antennas N. The number of streams which may be allocated to each UE is limited by the number of antennas J at that UE. For instance, with single-antenna UEs, up to N streams can be multiplexed, thus achieving the maximum multiplexing gain, with a distinct stream being allocated to each UE. This is in contrast to SU-MIMO, where the transmission of N streams necessitates that the UE be equipped with at least N antennas. Therefore a great advantage of MU-MIMO over SU-MIMO is that the MIMO multiplexing benefits are preserved even in the case of low-cost UEs with a small number of antennas. As a result, it is generally assumed that, in MU-MIMO, it is the base station which bears the burden of spatially separating the UEs, be it on the uplink or the downlink. Thus the base station performs receive beamforming from several UEs on the uplink and transmit beamforming towards several UEs on the downlink.

Another fundamental contrast between SU-MIMO and MU-MIMO comes from the difference in the underlying channel model. While in SU-MIMO the decorrelation between the spatial signatures of the antennas requires rich multipath propagation or the use of orthogonal polarizations, in MU-MIMO the decorrelation between the signatures of the different UEs occurs naturally due to fact that the separation between such UEs is typically large relative to the wavelength.

However, all these benefits of MU-MIMO depend on the level of Channel State Information at the Transmitter (CSIT) that the eNodeB receives from each UE. In the case of SU-MIMO, it has been shown that even a small number of feedback bits per antenna can be very beneficial in steering the transmitted energy accurately towards the UE's antenna(s) [8–11]. More precisely, in SU-MIMO channels, the accuracy of CSIT only causes an SNR offset, but does not affect the slope of the cell throughput-versus-SNR curve (i.e. the multiplexing gain). Yet for the MU-MIMO downlink, the level of Channel State Information (CSI) available at the transmitter does affect the multiplexing gain, because a MU-MIMO system with finite-rate feedback is essentially interference-limited after the crucial interference rejection processing has been carried out by the transmitter. Hence, providing accurate channel feedback is considerably more important for MU-MIMO than for SU-MIMO. On the other hand, in a system with a large number of UEs, the uplink resources required for accurate CSI feedback can become large and must be carefully controlled in the system design.

11.1.4.2 Techniques for Single-Antenna UEs

In considering the case of MU-MIMO for single-antenna UEs, it is worth noting that the number of antennas available to a UE for transmission is typically less than the number available for reception. We therefore examine first the uplink scenario, followed by the downlink.

With a single antenna at each UE, the MU-MIMO uplink scenario is very similar to the one described by Equation (11.10): because the UEs in mobile communication systems such as LTE typically cannot cooperate and do not have knowledge of the uplink channel coefficients, no precoding can be applied and each UE simply transmits an independent message. Thus, if K UEs are selected for transmission in the same time-frequency resource, with each UE k transmitting symbol s_k, the received signal at the base station, over a single $T = 1$ symbol period, is written as

$$\mathbf{R} = \mathbf{H}\bar{\mathbf{X}} + \mathbf{N} \tag{11.13}$$

where $\bar{\mathbf{X}} = (x_1, \ldots, x_K)^T$. In this case, the columns of $\mathbf{H}$ correspond to the receive spatial signatures of the different UEs. The base station can recover the transmitted symbol information by applying beamforming filters, for example using MMSE or ZF solutions. Note that no more than N UEs can be served (i.e. $K \leq N$) if inter-user interference is to be suppressed fully.

MU-MIMO in the uplink is sometimes referred to as 'Virtual MIMO' as, from the point of view of a given UE, there is no knowledge of the simultaneous transmissions of the other UEs. This transmission mode and its implications for LTE are discussed in Section 16.6.2.

On the downlink, which is illustrated in Figure 11.5, the base station must resort to transmit beamforming in order to separate the data streams intended for the various UEs. Over a single $T = 1$ symbol period, the signal received by UEs 1 to K can be written compactly as

$$\mathbf{R} = (r_1, \ldots, r_K)^T = \mathbf{HVP}\bar{\mathbf{X}} + \mathbf{N} \tag{11.14}$$

This time, the *rows* of $\mathbf{H}$ correspond to the transmit spatial signatures of the various UEs. $\mathbf{V}$ is the transmit beamforming matrix and $\mathbf{P}$ is the (diagonal) power allocation matrix selected such that it fulfils the total normalized transmit power constraint P_t. To cancel out fully the inter-user interference when $K \leq N$, a transmit ZF beamforming solution may be employed (although this is not optimal due to the fact that it may require a high transmit power if the channel is ill-conditioned). Such a solution would be given by $\mathbf{V} = \mathbf{H}^{\mathrm{H}}(\mathbf{H}\mathbf{H}^{\mathrm{H}})^{-1}$.

Note that regardless of the channel realization, the power allocation must be chosen to satisfy any power constraints at the base station, for example such that trace($\mathbf{V}\mathbf{P}\mathbf{P}^{\mathrm{H}}\mathbf{V}^{\mathrm{H}}$) = P_t. It is important to note that ZF precoding requires accurate channel knowledge at the transmitter, which in most cases necessitates terminal feedback in the uplink.

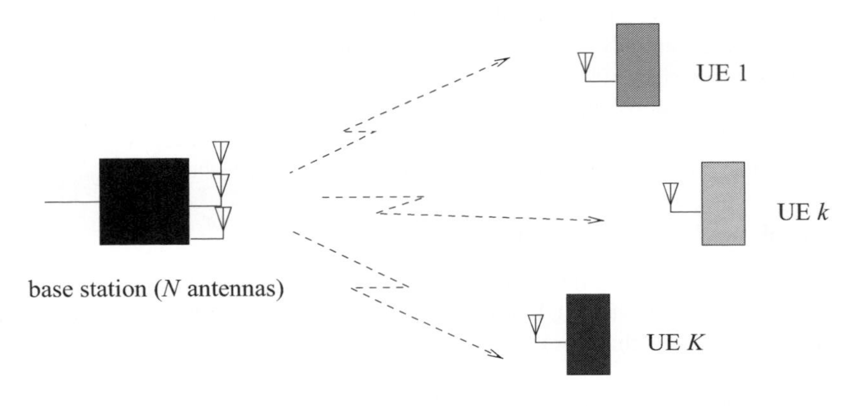

Figure 11.5: An MU-MIMO scenario in the downlink with single-antenna UEs: the base station transmits to K selected UEs simultaneously. Their signals are separated by multiple-antenna precoding at the base station side, based on channel knowledge.

11.1.4.3 Techniques for Multiple-Antenna UEs

The principles presented above for single-antenna UEs can be generalized to the case of multiple-antenna UEs. There could, in theory, be essentially two ways of exploiting the additional antennas at the UE side. In the first approach, the multiple antennas are simply treated as multiple *virtual* UEs, allowing high-capability terminals to receive or transmit more than one stream, while at the same time spatially sharing the channel with other UEs. For instance, a four-antenna base station could theoretically communicate in a MU-MIMO fashion with two UEs equipped with two antennas each, allowing two streams per UE, resulting in a total multiplexing gain of four. Another example would be that of two single-antenna UEs, receiving one stream each, and sharing access with another two-antenna UE, the latter receiving two streams. Again, the overall multiplexing factor remains limited to the number of base-station antennas.

The second approach for making use of additional UE antennas is to treat them as extra degrees of freedom for the purpose of strengthening the link between the UE and the base station. Multiple antennas at the UE may then be combined in MRC fashion in the case of the downlink, or, in the case of the uplink, space-time coding could be used. Antenna selection is another way of extracting more diversity out of the uplink channel, as discussed in Section 16.6.

11.1.4.4 Comparing Single-User and Multi-User Capacity

To illustrate the gains of multi-user multiplexing over single-user transmission, we compare the sum-rate achieved by both types of system from an information-theoretic standpoint, for single antenna UEs. We compare the Shannon capacity in single-user and multi-user scenarios both for an idealized synthetic channel and for a channel obtained from real measurement data.

The idealized channel model assumes that the entries of the channel matrix **H** in Equation (11.13) are independently and identically distributed (i.i.d.) Rayleigh fading. For the measured channel case, a channel sounder[7] was used to perform real-time wideband channel measurements synchronously for two UEs moving at vehicular speed in an outdoor semi-urban hilly environment with Line-Of-Sight (LOS) propagation predominantly present. The most important parameters of the platform are summarized in Table 11.1.

Table 11.1: Parameters of the measured channel for SU-MIMO/MU-MIMO comparison. More details can be found in [12, 13].

Parameter	Value
Centre frequency	1917.6 MHz
Bandwidth	4.8 MHz
Base station transmit power	30 dBm
Number of antennas at base station	4 (2 cross polarized)
Number of UEs	2
Number of antennas at UE	1
Number of subcarriers	160

The sum-rate capacity of a two-UE MU-MIMO system (calculated assuming a ZF precoder as described in Section 11.1.4) is compared with the capacity of an equivalent MISO system serving a single UE at a time (i.e. in TDMA), employing beamforming (see Section 11.1.3.2). The base station has four antennas and the UE has a single antenna. Full CSIT is assumed in both cases.

Figure 11.6 shows the ergodic (i.e. average) sum-rate of both schemes in both channels. The mean is taken over all frames and all subcarriers and subsequently normalized to bps/Hz. It can be seen that in both the ideal and the measured channels, MU-MIMO yields a higher sum-rate than SU-MISO in general. In fact, at high SNR, the multiplexing gain of the MU-MIMO system is two while it is limited to one for the SU-MISO case.

However, for low SNR, the SU-MISO TDMA and MU-MIMO schemes perform very similarly. This is because a sufficiently high SNR is required to excite more than one MIMO

[7]The Eurecom MIMO OpenAir Sounder (EMOS) [12].

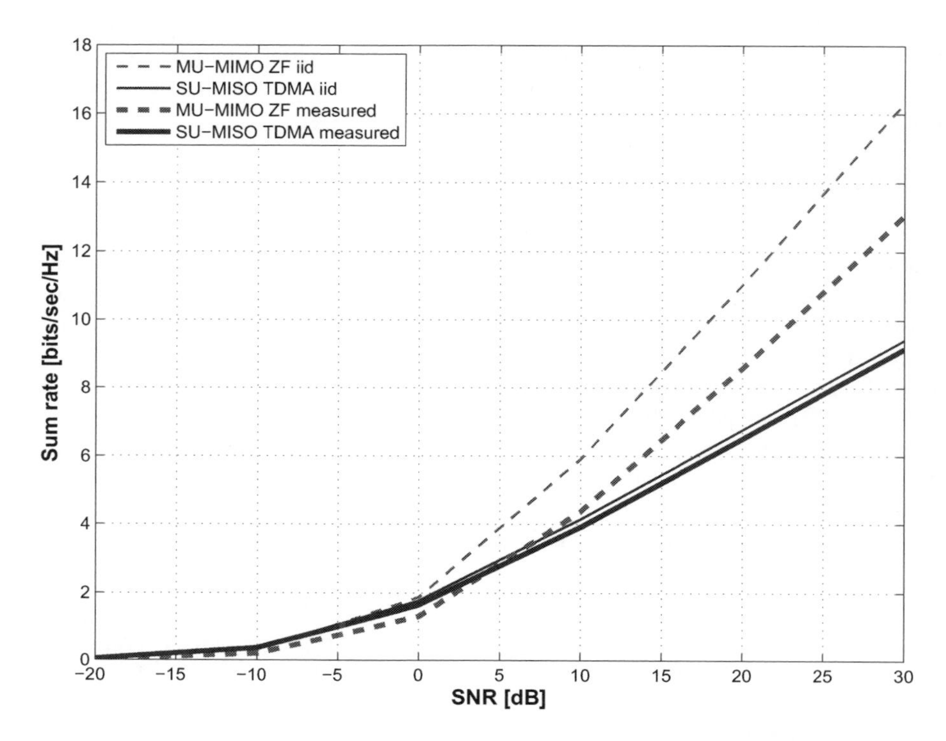

Figure 11.6: Ergodic sum-rate capacity of SU-MISO TDMA and MU-MIMO with two UEs, for an i.i.d. Rayleigh fading channel and for a measured channel.

transmission mode. Interestingly, the performance of both SU-MISO TDMA and MU-MIMO is slightly worse in the measured channels than in the idealized i.i.d. channels. This can be attributed to the correlation of the measured channel in time (due to the relatively slow movement of the users), in frequency (due to the LOS propagation), and in space (due to the transmit antenna correlation). In the MU-MIMO case, the difference between the i.i.d. and the measured channel is much higher than in the single-user TDMA case, since these correlation effects result in a rank-deficient channel matrix.

11.2 MIMO Schemes in LTE

Building on the theoretical background of the previous section, the MIMO schemes adopted for LTE Releases 8 and 9 are now reviewed and explained. These schemes relate to the downlink unless otherwise stated. The MIMO enhancements introduced for LTE-Advanced in Release 10 are explained in Chapter 29.

11.2.1 Practical Considerations

First, a few important practical constraints are briefly reviewed which affect the real-life performance of the theoretical MIMO systems considered above. Such constraints include practical deployment and antenna configurations, propagation conditions, channel knowledge, and implementation complexity. These aspects are often decisive for assessing the performance of a particular transmission strategy in a given propagation and system setting and were fundamental to the selection of MIMO schemes for LTE.

It was argued above that the full MIMO benefits (array gain, diversity gain and multiplexing gain) assume ideally decorrelated antennas and full-rank MIMO channel matrices. In this regard, the propagation environment and the antenna design (e.g. the spacing and polarization) play a significant role. If the antenna separation at the base station and the UE is small relative to the wavelength, strong correlation will be observed between the spatial signatures (especially in a LOS situation), limiting the usefulness of spatial multiplexing. However, this can to some extent be circumvented by means of dual-polarized antennas, whose design itself provides orthogonality even in LOS situations.[8] Beyond such antennas, the condition of spatial signature independence with SU-MIMO can only be satisfied with the help of rich random multipath propagation. In diversity-oriented schemes, the invertibility of the channel matrix is not required, yet the entries of the channel matrix should be statistically decorrelated. Although a greater LOS to non-LOS energy ratio will tend to correlate the fading coefficients on the various antennas, this effect will be counteracted by the reduction in fading delivered by the LOS component.

Another source of discrepancy between theoretical MIMO gains and practically achieved performance lies in the (in-)ability of the receiver and, whenever needed, the transmitter, to acquire reliable knowledge of the propagation channel. At the receiver, channel estimation is typically performed over a finite sample of Reference Signals (RSs), as discussed in Chapter 8. In the case of transmit beamforming and MIMO precoding, the transmitter then has to acquire this channel knowledge from the receiver usually through a limited feedback link, which causes further degradation to the available CSIT.

The potential advantages of MU-MIMO over SU-MIMO include robustness with respect to the propagation environment and the preservation of spatial multiplexing gain even in the case of UEs with small numbers of antennas. However, these advantages rely on the ability of the base station to compute the required transmit beamformer, which requires accurate CSIT; if no CSIT is available and the fading statistics are identical for all the UEs, the MU-MIMO gains disappear and the SU-MIMO strategy becomes optimal [14]. As a consequence, one of the most difficult challenges in making MU-MIMO practical for cellular applications, and particularly for an FDD system, is devising feedback mechanisms that allow for accurate CSI to be delivered efficiently by the UEs to the base station. This requires the use of appropriate *codebooks* for quantization (see Section 11.2.2.3). A recent account of the literature on this subject may be found in [15].

Another issue which arises for practical implementations of MIMO schemes is the interaction between the physical layer and the scheduling protocol. As noted in Section 11.1.4.1, in both uplink and downlink cases the number of UEs which can be served in a MU-MIMO fashion is typically limited to $K = N$, assuming linear combining structures. Often

[8] Horizontal and vertical polarizations, or, better, +45° and −45° polarizations, give a twofold multiplexing capability even in LOS.

one may even decide to limit K to a value strictly less than N to preserve some degrees of freedom for per-user diversity. As the number of active users U will typically exceed K, a selection algorithm must be implemented to identify which set of users will be scheduled for simultaneous transmission over a particular time-frequency Resource Block (RB). This algorithm is not specified in LTE and various approaches are possible; as discussed in Chapter 12, a combination of rate maximization and QoS constraints will typically be considered. It is important to note that the choice of UEs that will maximize the sum-rate (the sum over the K individual rates for a given subframe) is one that favours UEs exhibiting not only good instantaneous SNR but also spatial separability among their signatures.

11.2.2 Single-User Schemes

In this section, we examine the downlink multi-antenna schemes adopted in LTE for individual UEs. We consider first the diversity schemes used on the transmit side, then the single-user spatial multiplexing transmission modes, and finally beamforming using precoded UE-specific RSs (including the combination of beamforming and spatial multiplexing as introduced in Release 9).

11.2.2.1 Transmit Diversity Schemes

The theoretical aspects of transmit diversity were discussed in Section 11.1. Here we discuss the two main transmit diversity techniques defined in LTE. In LTE, transmit diversity is only defined for 2 and 4 transmit antennas and one data stream, referred to in LTE as one *codeword* since one transport block CRC[9] is used per data stream. To maximize diversity gain, the antennas typically need to be uncorrelated, so they need to be well separated relative to the wavelength or have different polarization. Transmit diversity is valuable in a number of scenarios, including low SNR, or for applications with low delay tolerance. Diversity schemes are also desirable for channels for which no uplink feedback signalling is available (e.g. Multimedia Broadcast/Multicast Services (MBMS) described in Chapter 13, Physical Broadcast CHannel (PBCH) in Chapter 9 and Synchronization Signals in Chapter 7), or when the feedback is not sufficiently accurate (e.g. at high speeds); transmit diversity is therefore also used as a fallback scheme in LTE when transmission with other schemes fails.

In LTE the multiple-antenna scheme is independently configured for the control channels and the data channels, and in the case of the data channels (Physical Downlink Shared CHannel – PDSCH) it is assigned independently per UE.

In this section we discuss in detail the transmit diversity techniques of Space-Frequency Block Codes (SFBCs) and Frequency-Switched Transmit Diversity (FSTD), as well as the combination of these schemes as used in LTE. These transmit diversity schemes may be used in LTE for the PBCH and Physical Downlink Control CHannel (PDCCH), and also for the PDSCH if it is configured in transmit diversity mode[10] for a UE.

Another transmit diversity technique which is commonly associated with OFDM is Cyclic Delay Diversity (CDD). CDD is not used in LTE as a diversity scheme in its own right but

[9]Cyclic Redundancy Check.

[10]PDSCH transmission mode 2 – see Section 9.2.2.1. As noted above, transmit diversity may also be used as a fallback in other transmission modes.

rather as a precoding scheme for spatial multiplexing on the PDSCH; we therefore introduce it in Section 11.2.2.2 in the context of spatial multiplexing.

Space-Frequency Block Codes (SFBCs)

If a physical channel in LTE is configured for transmit diversity operation using two eNodeB antennas, pure SFBC is used. SFBC is a frequency-domain version of the well-known Space-Time Block Codes (STBCs), also known as Alamouti codes [6]. This family of codes is designed so that the transmitted diversity streams are orthogonal and achieve the optimal SNR with a linear receiver. Such orthogonal codes only exist for the case of two transmit antennas.

STBC is used in UMTS, but in LTE the number of available OFDM symbols in a subframe is often odd while STBC operates on pairs of adjacent symbols in the time domain. The application of STBC is therefore not straightforward for LTE, while the multiple subcarriers of OFDM lend themselves well to the application of SFBC.

For SFBC transmission in LTE, the symbols transmitted from the two eNodeB antenna ports on each pair of adjacent subcarriers are defined as follows:

$$\begin{bmatrix} y^{(0)}(1) & y^{(0)}(2) \\ y^{(1)}(1) & y^{(1)}(2) \end{bmatrix} = \begin{bmatrix} x_1 & x_2 \\ -x_2^* & x_1^* \end{bmatrix} \tag{11.15}$$

where $y^{(p)}(k)$ denotes the symbols transmitted from antenna port p on the kth subcarrier.

Since no orthogonal codes exist for antenna configurations beyond 2×2, SFBC has to be modified in order to apply it to the case of 4 transmit antennas. In LTE, this is achieved by combining SFBC with FSTD.

Frequency Switched Transmit Diversity (FSTD) and its Combination with SFBC

General FSTD schemes transmit symbols from each antenna on a different set of subcarriers. In LTE, FSTD is only used in combination with SFBC for the case of 4 transmit antennas, in order to provide a suitable transmit diversity scheme where no orthogonal rate-1 block code exists. The LTE scheme is in fact a combination of two 2×2 SFBC schemes mapped to independent subcarriers as follows:

$$\begin{bmatrix} y^{(0)}(1) & y^{(0)}(2) & y^{(0)}(3) & y^{(0)}(4) \\ y^{(1)}(1) & y^{(1)}(2) & y^{(1)}(3) & y^{(1)}(4) \\ y^{(2)}(1) & y^{(2)}(2) & y^{(2)}(3) & y^{(2)}(4) \\ y^{(3)}(1) & y^{(3)}(2) & y^{(3)}(3) & y^{(3)}(4) \end{bmatrix} = \begin{bmatrix} x_1 & x_2 & 0 & 0 \\ 0 & 0 & x_3 & x_4 \\ -x_2^* & x_1^* & 0 & 0 \\ 0 & 0 & -x_4^* & x_3^* \end{bmatrix} \tag{11.16}$$

where, as previously, $y^{(p)}(k)$ denotes the symbols transmitted from antenna port p on the kth subcarrier. Note that the mapping of symbols to antenna ports is different in the 4 transmit-antenna case compared to the 2 transmit-antenna SFBC scheme. This is because the density of cell-specific RSs on the third and fourth antenna ports is half that of the first and second antenna ports (see Section 8.2.1), and hence the channel estimation accuracy may be lower on the third and fourth antenna ports. Thus, this design of the transmit diversity scheme avoids concentrating the channel estimation losses in just one of the SFBC codes, resulting in a slight coding gain.

11.2.2.2 Spatial Multiplexing Schemes

We begin by introducing some terminology used to describe spatial multiplexing in LTE:

- A spatial *layer* is the term used in LTE for one of the different streams generated by spatial multiplexing as described in Section 11.1. A layer can be described as a mapping of symbols onto the transmit antenna ports.[11] Each layer is identified by a precoding vector of size equal to the number of transmit antenna ports and can be associated with a radiation pattern.

- The *rank* of the transmission is the number of layers transmitted.[12]

- A *codeword* is an independently encoded data block, corresponding to a single Transport Block (TB) delivered from the Medium Access Control (MAC) layer in the transmitter to the physical layer, and protected with a CRC.

For ranks greater than 1, two codewords can be transmitted. Note that the number of codewords is always less than or equal to the number of layers, which in turn is always less than or equal to the number of antenna ports.

In principle, a SU-MIMO spatial multiplexing scheme can either use a single codeword mapped to all the available layers, or multiple codewords each mapped to one or more different layers. The main benefit of using only one codeword is a reduction in the amount of control signalling required, both for Channel Quality Indicator (CQI) reporting, where only a single value would be needed for all layers, and for HARQ ACK/NACK[13] feedback, where only one ACK/NACK would have to be signalled per subframe per UE. In such a case, the MLD receiver is optimal in terms of minimizing the BER. At the opposite extreme, a separate codeword could be mapped to each of the layers. The advantage of such a mapping is that significant gains are possible by using SIC, albeit at the expense of more signalling being required. An MMSE-SIC receiver can be shown to approach the Shannon capacity [7]. Note that an MMSE receiver is viable for both transmitter structures. For LTE, a middle-way was adopted whereby at most two codewords are used, even if four layers are transmitted. The codeword-to-layer mapping is static, since only minimal gains were shown for a dynamic mapping method. The mappings are shown in Table 11.2. Note that in LTE all RBs belonging to the same codeword use the same MCS, even if a codeword is mapped to multiple layers.

Precoding

The PDSCH transmission modes for open-loop spatial multiplexing[14] and closed-loop spatial multiplexing[15] use precoding from a defined 'codebook' to form the transmitted layers. Each codebook consists of a set of predefined precoding matrices, with the size of the set being a trade-off between the number of signalling bits required to indicate a particular matrix in the codebook and the suitability of the resulting transmitted beam direction.

[11] See Section 8.2 for an explanation of the concept of an antenna port.

[12] Note that the rank of a matrix is the number of linearly independent rows or columns, or equivalently the dimension of the subspace generated by the rows or columns of the matrix. Hence the rank of the channel matrix may be higher than the number of transmitted layers.

[13] Hybrid Automatic Repeat reQuest ACKnowledgement/Negative ACKnowledgement.

[14] PDSCH transmission mode 3 – see Section 9.2.2.1.

[15] PDSCH transmission modes 4 and 6 – see Section 9.2.2.1.

Table 11.2: Codeword-to-layer mapping in LTE.

Transmission rank	Codeword 1	Codeword 2
Rank 1	Layer 1	
Rank 2	Layer 1	Layer 2
Rank 3	Layer 1	Layer 2 and Layer 3
Rank 4	Layer 1 and Layer 2	Layer 3 and Layer 4

In the case of closed-loop spatial multiplexing, a UE feeds back to the eNodeB the index of the most desirable entry from a predefined codebook. The preferred precoder is the matrix which would maximize the capacity based on the receiver capabilities. In a single-cell, interference-free environment the UE will typically indicate the precoder that would result in a transmission with an effective SNR following most closely the largest singular values of its estimated channel matrix.

Some important properties of the LTE codebooks are as follows:

- **Constant modulus property.** LTE uses precoders which mostly comprise pure phase corrections – that is, with no amplitude changes. This is to ensure that the Power Amplifier (PA) connected to each antenna is loaded equally. The one exception (which still maintains the constant modulus property) is using an identity matrix as the precoder. However, although the identity precoder may completely switch off one antenna on one layer, since each layer is still connected to one antenna at constant power the net effect across the layers is still constant modulus to the PA.

- **Nested property.** The nested property is a method of arranging the codebooks of different ranks so that the lower rank codebook is comprised of a subset of the higher rank codebook vectors. This property simplifies the CQI calculation across different ranks. It ensures that the precoded transmission for a lower rank is a subset of the precoded transmission for a higher rank, thereby reducing the number of calculations required for the UE to generate the feedback. For example, if a specific index in the codebook corresponds to columns 1, 2 and 3 from the precoder $\mathbf{W}$ in the case of a rank 3 transmission, then the same index in the case of rank 2 transmission must consist of either columns 1 and 2 or columns 1 and 3 from $\mathbf{W}$.

- **Minimal 'complex' multiplications.** The 2-antenna codebook consists entirely of a QPSK[16] alphabet, which eliminates the need for any complex multiplications since all codebook multiplications use only ± 1 and $\pm j$. The 4-antenna codebook does contain some QPSK entries which require a $\sqrt{2}$ magnitude scaling as well; it was considered that the performance gain of including these precoders justified the added complexity.

The 2-antenna codebook in LTE comprises one 2×2 identity matrix and two DFT (Discrete Fourier Transform) matrices:

$$\begin{bmatrix} 1 & 0 \\ 0 & 1 \end{bmatrix}, \quad \begin{bmatrix} 1 & 1 \\ 1 & -1 \end{bmatrix} \quad \text{and} \quad \begin{bmatrix} 1 & 1 \\ j & -j \end{bmatrix} \tag{11.17}$$

[16]Quadrature Phase Shift Keying.

Here the columns of the matrices correspond to the layers.

The 4 transmit antenna codebook in LTE uses a Householder generating function:

$$\mathbf{W}_{\mathrm{H}} = \mathbf{I} - 2\mathbf{u}\mathbf{u}^{\mathrm{H}}/\mathbf{u}^{\mathrm{H}}\mathbf{u}, \tag{11.18}$$

which generates unitary matrices from input vectors **u**, which are defined in [16, Table 6.3.4.2.3-2]. The advantage of using precoders generated with this equation is that it simplifies the CQI calculation (which has to be carried out for each individual precoder in order to determine the preferred precoder) by reducing the number of matrix inversions. This structure also reduces the amount of control signalling required, since the optimum rank-1 version of $\mathbf{W}_{\mathrm{H}}$ is always the first column of $\mathbf{W}_{\mathrm{H}}$; therefore the UE only needs to indicate the preferred precoding matrix, and not also the individual vector within it which would be optimal for the case of rank-1 transmission.

Note that both DFT and Householder matrices are unitary.

Cyclic Delay Diversity (CDD)

In the case of *open*-loop spatial multiplexing,[17] the feedback from the UE indicates only the *rank* of the channel, and not a preferred precoding matrix. In this mode, if the rank used for PDSCH transmission is greater than 1 (i.e. more than one layer is transmitted), LTE uses Cyclic Delay Diversity (CDD) [17]. CDD involves transmitting the same set of OFDM symbols on the same set of OFDM subcarriers from multiple transmit antennas, with a different delay on each antenna. The delay is applied before the Cyclic Prefix (CP) is added, thereby guaranteeing that the delay is cyclic over the Fast Fourier Transform (FFT) size. This gives CDD its name.

Adding a time delay is identical to applying a phase shift in the frequency domain. As the same time delay is applied to all subcarriers, the phase shift will increase linearly across the subcarriers with increasing subcarrier frequency. Each subcarrier will therefore experience a different beamforming pattern as the non-delayed subcarrier from one antenna interferes constructively or destructively with the delayed version from another antenna. The diversity effect of CDD therefore arises from the fact that different subcarriers will pick out different spatial paths in the propagation channel, thus increasing the frequency-selectivity of the channel. The channel coding, which is applied to a whole transport block across the subcarriers, ensures that the whole transport block benefits from the diversity of spatial paths.

Although this approach does not optimally exploit the channel in the way that ideal precoding would (by matching the precoding to the eigenvectors of the channel), it does help to ensure that any destructive fading is constrained to individual subcarriers rather than affecting a whole transport block. This can be particularly beneficial if the channel information at the transmitter is unreliable, for example due to the feedback being limited or the UE velocity being high.

The general principle of the CDD technique is illustrated in Figure 11.7. The fact that the delay is added before the CP means that any delay value can be used without increasing the overall delay spread of the channel. By contrast, if the delay had been added after the addition of the CP, then the usable delays would have had to be kept small in order to ensure that the delay spread of the delayed symbol is no more than the maximum channel delay spread, in order to obviate any need to increase the CP length.

[17] PDSCH transmission mode 3 – see Section 9.2.2.1.

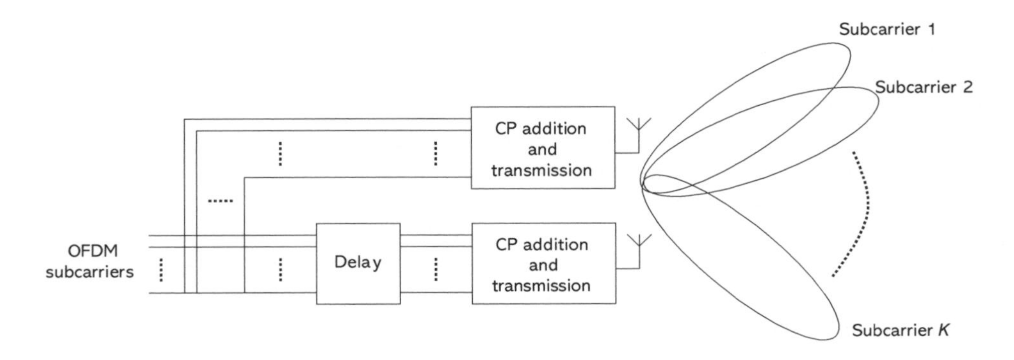

Figure 11.7: Principle of Cyclic Delay Diversity.

The time-delay/phase-shift equivalence means that the CDD operation can be implemented as a frequency-domain precoder for the affected antenna(s), where the precoder phase changes on a per-subcarrier basis according to a fixed linear function. In general, the implementation designer can choose whether to implement CDD in the time domain or the frequency domain. However, one advantage of a frequency-domain implementation is that it is not limited to delays corresponding to an integer number of samples.

As an example of CDD for a case with 2 transmit antenna ports, we can express mathematically the received symbol r_k on the kth subcarrier as

$$r_k = h_{1k}x_k + h_{2k}e^{j\phi k}x_k \tag{11.19}$$

where h_{pk} is the channel from the pth transmit antenna and $e^{j\phi k}$ is the phase shift on the kth subcarrier due to the delay operation. We can see clearly that on some subcarriers the symbols from the second transmit antenna will add constructively, while on other subcarriers they will add destructively. Here, $\phi = 2\pi d_{\text{cdd}}/N$, where N is the FFT size and d_{cdd} is the delay in samples.

The number of peaks and troughs created by CDD in the received signal spectrum across the subcarriers therefore depends on the length of the delay d_{cdd}: as d_{cdd} is increased, the number of peaks and troughs in the spectrum also increases.

In LTE, the frequency-domain phase shift values ϕ are π, $2\pi/3$ and $\pi/2$ for 2-, 3- and 4-layer transmission respectively. For a size-2048 FFT, for example, these values correspond to $d_{\text{cdd}} = 1024,\ 682.7,\ 512$ sample delays respectively.

For the application of CDD in LTE, the eNodeB transmitter combines CDD delay-based phase shifts with additional precoding using fixed unitary DFT-based precoding matrices.

The application of precoding in this way is useful when the channel coefficients of the antenna ports are correlated, since then *virtual* antennas (formed by fixed, non-channel-dependent precoding) will typically be uncorrelated. On its own, the benefit of CDD is reduced by antenna correlation. The use of uncorrelated virtual antennas created by fixed precoding can avoid this problem in correlated channels, while not degrading the performance if the individual antenna ports are uncorrelated.

For ease of explanation, the above discussion of CDD has been in terms of a rank-1 transmission – i.e. with a single layer. However, in practice, CDD is only applied in LTE when the rank used for PDSCH transmission is greater than 1. In such a case, each layer benefits independently from CDD in the same way as for a single layer. For example, for a rank-2 transmission, the transmission on the second antenna port is delayed relative to the first antenna port for each layer. This means that symbols transmitted on both layers will experience the delay and hence the increased frequency selectivity.

For multilayer CDD operation, the mapping of the layers to antenna ports is carried out using precoding matrices selected from the spatial multiplexing codebooks described earlier. As the UE does not indicate a preferred precoding matrix in the open loop spatial multiplexing transmission mode in which CDD is used, the particular spatial multiplexing matrices selected from the spatial multiplexing codebooks in this case are predetermined.

In the case of 2 transmit antenna ports, the predetermined spatial multiplexing precoding matrix $\mathbf{W}$ is always the same (the first entry in the 2 transmit antenna port codebook, which is the identity matrix). Thus, the transmitted signal can be expressed as follows:

$$\begin{bmatrix} y^{(0)}(k) \\ y^{(1)}(k) \end{bmatrix} = \mathbf{W}\mathbf{D}_2\mathbf{U}_2\mathbf{x} = \frac{1}{\sqrt{2}}\begin{bmatrix} 1 & 0 \\ 0 & 1 \end{bmatrix}\frac{1}{\sqrt{2}}\begin{bmatrix} 1 & 0 \\ 0 & e^{j\phi_1 k} \end{bmatrix}\begin{bmatrix} 1 & 1 \\ 1 & -1 \end{bmatrix}\begin{bmatrix} x^{(0)}(i) \\ x^{(1)}(i) \end{bmatrix} \tag{11.20}$$

In the case of 4 transmit antenna ports, v different precoding matrices are used from the 4 transmit antenna port codebook where v is the transmission rank. These v precoding matrices are applied in turn across groups of v subcarriers in order to provide additional decorrelation between the spatial streams.

11.2.2.3 Beamforming Schemes

The theoretical aspects of beamforming were described in Section 11.1. This section explains how it is implemented in LTE Release 8 and Release 9.

In LTE, two beamforming techniques are supported for the PDSCH:

- **Closed-loop rank-1 precoding.**[18] This mode amounts to beamforming since only a single layer is transmitted, exploiting the gain of the antenna array. However, it can also be seen as a special case of the SU-MIMO spatial multiplexing using codebook based precoding discussed in Section 11.2.2.2. In this mode, the UE feeds channel state information back to the eNodeB to indicate suitable precoding to apply for the beamforming operation. This feedback information is known as Precoding Matrix Indicators (PMIs) and is described in detail in Section 11.2.2.4.

- **Beamforming with UE-specific RSs.**[19] In this case, the MIMO precoding is not restricted to a predefined codebook, so the UE cannot use the cell-specific RS for PDSCH demodulation and precoded UE-specific RS are therefore needed.

In this section we focus on the latter case which supports a single transmitted layer in LTE Release 8 and was extended to support dual-layer beamforming in Release 9. These modes are primarily mechanisms to extend cell coverage by concentrating the eNodeB power in the directions in which the radio channel offers the strongest path(s) to reach the UE. They are

[18]PDSCH transmission mode 6 – see Section 9.2.2.1.

[19]PDSCH transmission mode 7 (from Release 8) and 8 (introduced in Release 9) – see Section 9.2.2.1.

typically implemented with an array of closely spaced antenna elements (or pairs of cross-polarized elements) for creating directional transmissions. Due to the fact that no precoding codebook is used, an arbitrary number of transmit antennas is theoretically supported; in practical deployments up to 8 antennas are typically used (e.g. an array of 4 cross-polarized pairs). The signals from the correlated antenna elements are phased appropriately so that they add up constructively at the location where the UE is situated. The UE is not really aware that it is receiving a precoded directional transmission rather than a cell-wide transmission (other than by the fact that the UE is directed to use the UE-specific RS as the phase reference for demodulation). To the UE, the phased array of antennas 'appears' as just one antenna port.

In LTE Release 8, only a single UE-specific antenna port and associated RSs is defined (see Section 8.2.2). As a consequence the beamformed PDSCH can only be transmitted with a single spatial layer. The support of two UE-specific antenna ports was introduced in Release 9 (see Section 8.2.3), thus enabling beamforming with two spatial layers with a direct one-to-one mapping between the layers and antenna ports. The UE-specific RS patterns for the two Release 9 antenna ports were designed to be orthogonal in order to facilitate separation of the spatial layers at the receiver. The two associated PDSCH layers can then be transmitted to a single user in good propagation conditions to increase its data rate, or to multiple users (MU-MIMO) to increase the system capacity. LTE Release 10 further extends the support for beamforming with UE-specific RSs to higher numbers of spatial layers, as described in Chapter 29.

When two layers are transmitted over superposed beams they share the available transmit power of the eNodeB, so the increase in data rate comes at the expense of a reduction of coverage. This makes fast rank adaptation (which is also used in the codebook-based spatial multiplexing transmission mode) an important mechanism for the dual-layer beamforming mode. To support this, it is possible to configure the UE to feed back a Rank Indicator (RI) together with the PMI (see Section 11.2.2.4).

It should also be noted that beamforming in LTE can only be applied to the PDSCH and not to the downlink control channels, so although the range of a given data rate on the PDSCH can be extended by beamforming, the overall cell coverage may still be limited by the range of the control channels unless other measures are taken. The trade-off between throughput and coverage with beamforming in LTE is illustrated in Figure 11.8.

Like any precoding technique, these beamforming modes require reliable CSIT which can, in principle, be obtained either from feedback from the UE or from estimation of the uplink channel.

In the single-layer beamforming mode of Release 8, the UE does not feed back any precoding-related information, and the eNodeB deduces this information from the uplink, for example using Direction Of Arrival (DOA) estimations. It is worth noting that in this case calibration of the eNodeB RF paths may be necessary, as discussed in Section 23.5.2.

The dual-layer beamforming mode of Release 9 provides the possibility for the UE to help the eNodeB to derive the beamforming precoding weights by sending PMI feedback in the same way as for codebook-based closed-loop spatial multiplexing. For beamforming, this is particularly suitable for FDD deployments where channel reciprocity cannot be exploited as effectively as in the case of TDD. Nevertheless, this PMI feedback can only provide partial CSIT, since it is based on measurements of the common RSs which, in beamforming deployments, are usually transmitted from multiple antenna elements with broad beam patterns; furthermore, the PMI feedback uses the same codebook as for closed-loop spatial

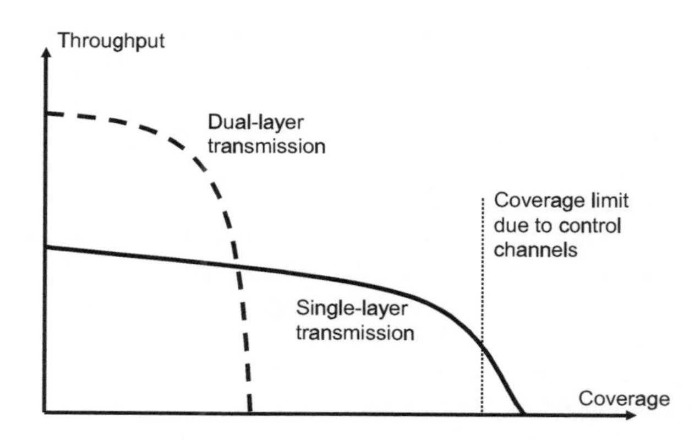

Figure 11.8: Trade-off between throughput and coverage with beamforming.

multiplexing and is therefore constrained by quantization while the beamforming precoding is not. As a consequence, additional uplink channel estimates are still required if the full benefits of beamforming are to be realized.

For the Channel Quality Indicator (CQI) feedback for link adaptation, the eNodeB must take into account the fact that in the beamforming modes the channel quality experienced by the UE on the beamformed PDSCH resources will typically be different (hopefully better) than on the cell-specific RSs from which the CQI is derived. The UE-specifc RSs cannot be used for CQI estimation because they are only present in the RBs in which beamforming transmission is used; UE-specific RSs that are precoded for a different UE are obviously not useful for channel estimation. The eNodeB has to adjust the MCS used for transmission relative to the CQI reports by estimating the expected beamforming gain depending on the number of transmitted layers; the HARQ acknowledgement rate can be useful in deriving such an adjustment.

It is worth mentioning that beamforming also affects the interference caused to neighbouring cells. If the beamforming is intermittent, it can result in a problem often known as the 'flash-light effect', where strong intermittent interference may disturb the accuracy of the UEs' CQI reporting in adjacent cells. This effect was shown to have the potential to cause throughput reductions in HSDPA [18]. In LTE the possibility for frequency-domain scheduling (as discussed in Section 12.5) provides an additional degree of freedom to avoid such issues. In addition, beamforming may even be coordinated to take advantage of the spatial interference structure created by beamforming in multiple cells, especially among the cells controlled by one eNodeB.

11.2.2.4 Feedback Computation and Signalling

In order to support MIMO operation. the UE can be configured to report Precoding Matrix Indicators (PMIs) and Rank Indicators(RIs), in addition to CQI reporting as discussed in Section 10.2.1.

In the case of codebook-based spatial multiplexing transmissions, the precoding at the eNodeB transmitter is applied relative to the transmitted phase of the cell-specific RSs for each antenna port. Thus if the UE knows the precoding matrices that could be applicable (as defined in the configured codebook) and it knows the transfer function of the channels from the different antenna ports (by making measurements on the RSs), it can determine which **W** is most suitable under the current radio conditions and signal this to the eNodeB. The preferred **W**, whose index constitutes the PMI report, is the precoder that maximizes the aggregate number of data bits which could be received across all layers. The number of RBs to which each PMI corresponds in the frequency domain depends on the feedback mode configured by the eNodeB:

- **Wideband PMI.** The UE reports a single PMI corresponding to the preferred precoder assuming transmission over the whole system bandwidth. This is applicable for PMI feedback that is configured to be sent periodically (on the PUCCH[20] unless the PUSCH[21] is transmitted), and also for PMI feedback that is sent aperiodically in conjunction with UE-selected sub-band CQI reports on the PUSCH (see Section 10.2.1).
- **Sub-band PMI.** The UE reports one PMI for each sub-band across the whole system bandwidth, where the sub-band size is between 4 and 6 RBs depending on the system bandwidth (as shown in Table 10.3). This is applicable for PMI feedback that is sent aperiodically on the PUSCH in conjunction with a wideband CQI report. It is well suited to a scenario where the eNodeB wishes to schedule a wideband data transmission to a UE while the channel direction varies across the bandwidth; the precoder used by the eNodeB can change from sub-band to sub-band within one transport block.
- **UE-selected sub-band PMI.** The UE selects a set of M preferred sub-bands, each of size k RBs (where M and k are the same as for UE-selected sub-band CQI feedback as given in Table 10.4) and reports a single PMI corresponding to the preferred precoder assuming transmission over all of the M selected sub-bands. This is applicable in conjunction with UE-selected sub-band CQI reports on the PUSCH.

The eNodeB is not bound to use the precoder requested by the UE, but clearly if the eNodeB chooses another precoder then the eNodeB will have to adjust the MCS accordingly since the reported CQI could not be assumed to be applicable.

The eNodeB may also restrict the set of precoders which the UE may evaluate and report. This is known as *codebook subset restriction.* It enables the eNodeB to prevent the UE from reporting precoders which are not useful, for example in some eNodeB antenna configurations or in correlated fading scenarios. In the case of open-loop spatial multiplexing, codebook subset restriction amounts simply to a restriction on the rank which the UE may report.

For each PDSCH transmission to a UE, the eNodeB indicates via a 'Transmitted Precoding Matrix Indicator' (TPMI) in the downlink assignment message[22] on the PDCCH whether it is applying the UE's preferred precoder, and if not, which precoder is used. This enables the UE to derive the correct phase reference relative to the cell-specific RSs in order to

[20]Physical Uplink Control CHannel.

[21]Physical Uplink Shared CHannel.

[22]The TPMI is sent in DCI Formats 2, 2A and 2B – see Section 9.3.5.1.

demodulate the PDSCH data. The ability to indicate that the eNodeB is applying the UE's preferred precoder is particularly useful in the case of sub-band PMI reporting, as it enables multiple precoders to be used across the bandwidth of a wideband transmission to a UE without incurring a high signalling overhead to notify the UE of the set of precoders used.

The UE can also be configured to report the channel rank via a RI, which is calculated to maximize the capacity over the entire bandwidth, jointly selecting the preferred precoder(s) to maximize the capacity on the assumption of the selected transmission rank. The CQI values reported by the UE correspond to the preferred transmission rank and precoders, to enable the eNodeB to perform link adaptation and multi-user scheduling as discussed in Sections 10.2 and 12.1. The number of CQI values reported normally corresponds to the number of codewords supported by the preferred transmission rank. Further, the CQI values themselves will depend on the assumed transmission rank: for example, the precoding matrix for layer 1 will usually be different depending on whether or not the UE is assuming the presence of a second layer.

Although the UE indicates the transmission rank that would maximize the downlink data rate, the eNodeB can also indicate to the UE that a different transmission rank is being used for a PDSCH transmission. This gives flexibility to the eNodeB, since the UE does not know the amount of data in the downlink buffer; if the amount of data in the buffer is small, the eNodeB may prefer to use a lower rank with higher reliability.

11.2.3 Multi-User MIMO

The main focus of MIMO in the first release of LTE was on achieving transmit antenna diversity or single-user multiplexing gain, neither of which requires such a sophisticated CSI feedback mechanism as MU-MIMO. As pointed out in Section 11.2.1, the availability of accurate CSIT is the main challenge in making MU-MIMO schemes attractive for cellular applications.

Nevertheless, basic support for MU-MIMO was provided in Release 8, and enhanced schemes were added in Releases 9 and 10.

The MU-MIMO scheme supported in LTE Release 8[23] is based on the same codebook-based scheme as is defined for SU-MIMO (see Section 11.2.2.2). In particular the same 'implicit' feedback calculation mechanism is used, whereby the UE makes hypotheses on the precoder the eNodeB would use for transmission on the PDSCH and reports a recommended precoding matrix (PMI) corresponding to the best hypothesis – i.e. an *implicit* representation of the CSI. The recommended precoder typically depends on the type of receiver implemented by the UE, e.g. MRC, linear MMSE or MMSE Decision Feedback Equalizer (DFE). However, in the same way as for SU-MIMO, the eNodeB is not bound to the UE's recommendation and can override the PMI report by choosing a different precoder from the codebook and signalling it back to the scheduled UEs using a TPMI on the PDCCH.

In the light of the similarities between implicit channel feedback by precoder recommendation and explicit channel vector quantization, it is instructive to consider the implications of the fact that the LTE 2-antenna codebook and the first eight entries in the 4-antenna codebook are derived from a DFT matrix (as noted in Section 11.2.2.2). The main reasons for this choice are as follows:

[23]PDSCH transmission mode 5 – see Section 9.2.2.1.

- Simplicity in the PMI/CQI calculation, which can be done in some cases without complex multiplications;
- A DFT matrix nicely captures the characteristics of a highly correlated MISO channel.[24]

In LTE Release 8, a UE configured in MU-MIMO mode assumes that an eNodeB transmission on the PDSCH would be performed on one layer using one of the rank-1 codebook entries defined for two or four antenna ports. Therefore, the MU-MIMO transmission is limited to a single layer and single codeword per scheduled UE, and the RI and PMI feedback are the same as for SU-MIMO rank 1.

The eNodeB can then co-schedule multiple UEs on the same RB, typically by pairing UEs that report orthogonal PMIs. The eNodeB can therefore signal to those UEs (using one bit in the DCI format 1D message on the PDCCH – see Section 9.3.5.1) that the PDSCH power is reduced by 3 dB relative to the cell-specific RSs. This assumes that the transmission power of the eNodeB is equally divided between the two codewords and implies that in practice a maximum of two UEs may be co-scheduled in Release 8 MU-MIMO mode.

Apart from this power offset signalling, MU-MIMO in Release 8 is fully transparent to the scheduled UEs, in that they are not explicitly aware of how many other UEs are co-scheduled or of which precoding vectors are used for any co-scheduled UEs.

CQI reporting for MU-MIMO in Release 8 does not take into account any interference from transmissions to co-scheduled users. The CQI values reported are therefore equivalent to those that would be reported for rank-1 SU-MIMO. It is left to the eNodeB to estimate a suitable adjustment to apply to the MCS if multiple UEs are co-scheduled.

In Release 9, the support for MU-MIMO is enhanced by the introduction of dual-layer beamforming in conjunction with precoded UE-specific RSs, as discussed in Sections 8.2.3 and 11.2.2.3. This removes the constraint on the eNodeB to use the feedback codebook for precoding, and gives flexibility to utilize other approaches to MU-MIMO user separation, for example according to a zero-forcing beamforming criterion. Note that in the limit of a large UE population, a zero-forcing beamforming solution converges to a unitary precoder because, with high probability, there will be N UEs with good channel conditions reporting orthogonal channel signatures.

The use of precoded UE-specific RSs removes the need to signal explicitly the details of the chosen precoder to the UEs. Instead, the UEs estimate the effective channel (i.e. the combination of the precoding matrix and physical channel), from which they can derive the optimal antenna combiner for their receivers.

The UE-specific RS structure described in Section 8.2.3 allows up to 4 UEs to be scheduled for MU-MIMO transmissions from Release 9.

Furthermore, the precoded UE-specific RSs mean that for Release 9 MU-MIMO it is not necessary to signal explicitly a power offset depending on the number of co-scheduled UEs. As the UE-specific RSs for two UEs are code-division multiplexed, the eNodeB transmission power is shared between the two antenna ports in the same way as the transmitted data, so the UE can use the UE-specific RSs as both the phase reference and the amplitude reference for demodulation. Thus the MU-MIMO scheduling is fully transparent in Release 9.

[24]It is not difficult to see that if the first N rows are taken from a DFT matrix of size N_q, each of the N_q column vectors comprises the spatial signature of the ith element of a uniform linear antenna array with spacing l, such that $\frac{l}{\lambda}\sin\theta = \frac{i}{N_q}$ [7].

11.2.4 MIMO Performance

The multiple antenna schemes supported by LTE contribute much to its spectral efficiency improvements compared to UMTS.

Figure 11.9 shows a typical link throughput performance comparison between the open-loop SU-MIMO spatial multiplexing mode of LTE Release 8 with rank feedback but no PMI feedback, and closed-loop SU-MIMO spatial multiplexing with PMI feedback. Four transmit and two receive antennas are used. Both modes make use of spatial multiplexing gain to enhance the peak data rate. Closed-loop MIMO provides gain from the channel-dependent precoding, while open-loop MIMO offers additional robustness by SFBC for single-layer transmissions and CDD for more than one layer.

The gain of closed-loop precoding can be observed in low mobility scenarios, while the benefit of open-loop diversity can clearly be seen in the case of high mobility.

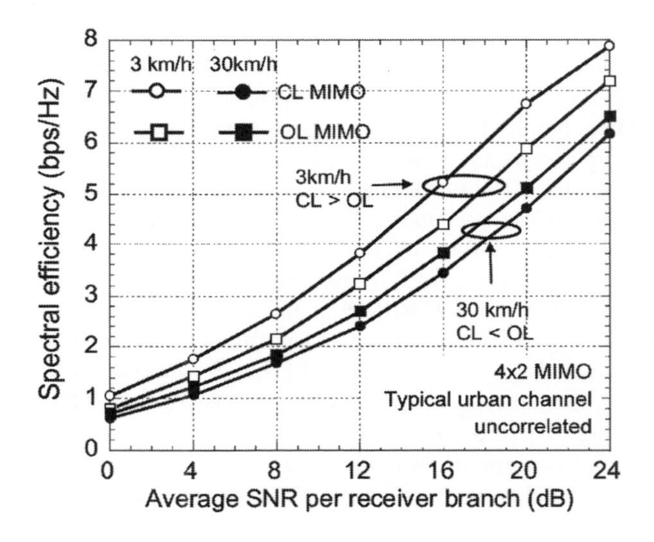

Figure 11.9: Example of closed-loop and open-loop MIMO performance.

11.3 Summary

In this chapter we have reviewed the predominant families of MIMO techniques (both single-user and multi-user) of relevance to LTE. LTE breaks new ground in drawing on such approaches to harness the power of MIMO systems not just for boosting the peak per-user data rate but also for improving overall system capacity and spectral efficiency. Single-User MIMO techniques are well-developed in Release 8, providing the possibility to benefit from precoding while avoiding a high control signalling overhead. Support for beamforming and Multi-User MIMO is significantly enhanced in Release 9. LTE-Advanced further develops and enhances these techniques, as outlined in Chapter 29.

References[25]

[1] D. Gesbert, M. Shafi, P. Smith, D. Shiu and A. Naguib, 'From Theory to Practice: An Overview of MIMO Space-time Coded Wireless Systems'. *IEEE Journal on Selected Areas in Communications*. Special Issue on MIMO systems, guest edited by the authors, April 2003.

[2] H. Bölcskei, D. Gesbert, C. Papadias and A. J. van der Veen (eds), *Space-Time Wireless Systems: From Array Processing to MIMO Communications*. Cambridge: Cambridge University Press, 2006.

[3] A. Goldsmith, *Wireless Communications*. Cambridge: Cambridge University Press, 2005.

[4] D. Gesbert, S. Hanly, H. Huang, S. Shamai, O. Simeone and W. Yu, 'Multi-Cell MIMO Cooperative Networks: A New Look at Interference', to appear in *IEEE Journal on Selected Areas in Communications*, December 2010.

[5] G. Golub and C. van Loan, *Matrix Computations*. Balitmore, MA: John Hopkins University Press, 1996.

[6] S. M. Alamouti, 'A Simple Transmitter Diversity Technique for Wireless Communications'. *IEEE Journal on Selected Areas in Communications*, Vol. 16, pp. 1451–1458, October 1998.

[7] D. Tse and P. Viswanath, *Fundamentals of Wireless Communication*. Cambridge: Cambridge University Press, 2004.

[8] A. Narula, M. J. Lopez, M. D. Trott and G. W. Wornell, 'Efficient Use of Side Information in Multiple Antenna Data Transmission over Fading Channels'. *IEEE Journal on Selected Areas in Communications*, Vol. 16, pp. 1223–1436, October 1998.

[9] D. Love, R. Heath Jr and T. Strohmer, 'Grassmannian Beamforming for Multiple-input Multiple-output Wireless Systems'. *IEEE Trans. on Information Theory*, Vol. 49, pp. 2735–2747, October 2003.

[10] D. Love, R. Heath Jr, W. Santipach, and M. Honig, 'What is the Value of Limited Feedback for MIMO Channels?'. *IEEE Communication Magazine*, Vol. 42, pp. 54–59, October 2003.

[11] N. Jindal, 'MIMO Broadcast Channels with Finite Rate Feedback'. *IEEE Trans. Information Theory*, Vol. 52, pp. 5045–5059, November 2006.

[12] F. Kaltenberger, L. S. Cardoso, M. Kountouris, R. Knopp and D. Gesbert, 'Capacity of Linear Multi-user MIMO Precoding Schemes with Measured Channel Data', in *Proc. IEEE Workshop on Sign. Proc. Adv. in Wireless Comm. (SPAWC)*, Recife, Brazil, July 2008.

[13] F. Kaltenberger, M. Kountouris, D. Gesbert and R. Knopp, 'Correlation and Capacity of Measured Multi-user MIMO Channels', in *Proc. IEEE International Symposium on Personal, Indoor and Mobile Radio Communications*, Cannes, France, September 2008.

[14] Caire, G. and Shamai, S. (Shitz), 'On the Achievable Throughput of a Multi-antenna Gaussian Broadcast Channel'. *IEEE Trans. on Information Theory*, Vol. 49, pp. 1691–1706, July 2003.

[15] R. Heath, V. Lau, D. Love, D. Gesbert, B. Rao and A. Andrews, (eds), 'Exploiting Limited Feedback in Tomorrow's Wireless Communications Networks'. Special issue of *IEEE Journal on Selected Areas in Communications*, October 2008.

[16] 3GPP Technical Specification 36.211, 'Evolved Universal Terrestrial Radio Access (E-UTRA); Physical Channels and Modulation', www.3gpp.org.

[17] Y. Li, J. Chuang and N. Sollenberger, 'Transmitter Diversity for OFDM Systems and Its Impact on High-Rate Data Wireless Networks'. *IEEE Journal on Selected Areas in Communications*, Vol. 17, pp. 1233–1243, July 1999.

[18] A. Osseiran and A. Logothetis, 'Closed Loop Transmit Diversity in WCDMA HS-DSCH', in *Proc. IEEE Vehicular Technology Conference*, May 2005.

[25] All web sites confirmed 1st March 2011.

12

Multi-User Scheduling and Interference Coordination

Issam Toufik and Raymond Knopp

12.1 Introduction

The eNodeB in an LTE system is responsible, among other functions, for managing resource scheduling for both uplink and downlink channels. The ultimate aim of this function is typically to fulfil the expectations of as many users of the system as possible, taking into account the Quality of Service (QoS) requirements of their respective applications.

A typical single-cell cellular radio system is shown in Figure 12.1, comprising K User Equipments (UEs) communicating with one eNodeB over a fixed total bandwidth B. In the uplink, each UE has several data queues corresponding to different uplink logical channel groups, each with different delay and rate constraints. In the same way, in the downlink the eNodeB may maintain several buffers per UE containing dedicated data traffic, in addition to queues for broadcast services. The different traffic queues in the eNodeB would typically have different QoS constraints. We further assume for the purposes of this analysis that only one user is allocated a particular Resource Block (RB) in any subframe, although in practice multi-user MIMO schemes may result in more than one UE per RB, as discussed in Sections 11.2.3 and 16.6.2 for the downlink and uplink respectively.

The goal of a resource scheduling algorithm in the eNodeB is to allocate the RBs and transmission powers for each subframe in order to optimize a function of a set of performance metrics, for example maximum/minimum/average throughput, maximum/minimum/average delay, total/per-user spectral efficiency or outage probability. In the downlink the resource allocation strategy is constrained by the total transmission power of the eNodeB, while in the uplink the main constraints on transmission power in different RBs arises from a multicell view of inter-cell interference and the power headroom of the UEs.

In this chapter we provide an overview of some of the key families of scheduling algorithms which are relevant for an LTE system and highlight some of the factors that an eNodeB can advantageously take into account.

LTE – The UMTS Long Term Evolution: From Theory to Practice, Second Edition.
Stefania Sesia, Issam Toufik and Matthew Baker.

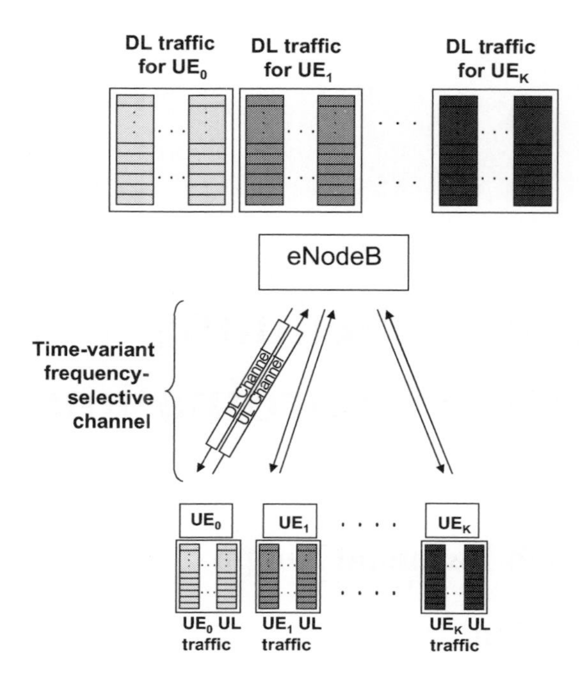

Figure 12.1: A typical single-cell cellular radio system.

12.2 General Considerations for Resource Allocation Strategies

The generic function of a resource scheduler, as shown for the downlink case in Figure 12.2, is to schedule data to a set of UEs on a shared set of physical resources.

It is worth noting that the algorithm used by the resource scheduler is also tightly coupled with the Adaptive Modulation and Coding (AMC) scheme and the retransmission protocol (Hybrid Automatic Repeat reQuest, HARQ, see Sections 4.4 and 10.3.2.5). This is due firstly to the fact that, in addition to dynamic physical resource allocation, the channel measurements are also used to adapt the Modulation and Coding Scheme (MCS) (i.e. transmission spectral-efficiency) as explained in Section 10.2. Secondly, the queue dynamics, which impact the throughput and delay characteristics of the link seen by the application, depend heavily on the HARQ protocol and transport block sizes. Moreover, the combination of channel coding and retransmissions provided by HARQ enables the spectral efficiency of an individual transmission in one subframe to be traded off against the number of subframes in which retransmissions take place. Well-designed practical scheduling algorithms will necessarily consider all these aspects.

In general, scheduling algorithms can make use of two types of measurement information to inform the scheduling decisions, namely Channel State Information (CSI) and traffic measurements (volume and priority). These are obtained either by direct measurements at the eNodeB, or via feedback signalling channels, or a combination of both. The amount of feedback used is an important consideration, since the availability of accurate CSI and traffic information helps to maximize the data rate in one direction at the expense of more overhead in the other. This fundamental trade-off, which is common to all feedback-based resource scheduling schemes, is particularly important in Frequency Division Duplex (FDD) operation where uplink-downlink reciprocity of the radio channels cannot be assumed (see Chapter 20). For Time Division Duplex (TDD) systems, coherence between the uplink and downlink channels may be used to assist the scheduling algorithm, as discussed in Section 23.5.

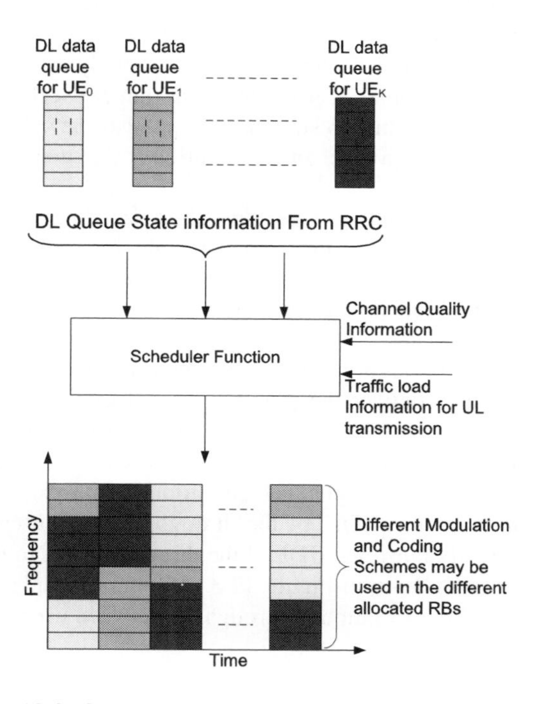

Figure 12.2: Generic view of a wideband resource scheduler.

Based on the available measurement information, the eNodeB resource scheduler must manage the differing requirements of all the UEs in the cells under its control to ensure that sufficient radio transmission resources are allocated to each UE within acceptable latencies to meet their QoS requirements in a spectrally efficient way. The details of this process are not standardized as it is largely internal to the eNodeB, allowing for vendor-specific algorithms to be developed which can be optimized for specific scenarios in conjunction with network operators. However, the key inputs available to the resource scheduling process are common.

In general, two extremes of scheduling algorithm may be identified: *opportunistic scheduling* and *fair scheduling*. The former is typically designed to maximize the sum of the transmitted data rates to all users by exploiting the fact that different users experience different channel gains and hence will experience good channel conditions at different times and frequencies. A fundamental characteristic of mobile radio channels is the fading effects arising from the mobility of the UEs in a multipath propagation environment, and from variations in the surrounding environment itself (see Chapter 20). In [1–3] it is shown that, for a multi-user system, significantly more information can be transmitted across a fading channel than a non-fading channel for the same average signal power at the receiver. This principle is known as *multi-user diversity*. With proper dynamic scheduling, allocating the channel at each given time instant only to the user with the best channel condition in a particular part of the spectrum can yield a considerable increase in the total throughput as the number of active users becomes large.

The main issue arising from opportunistic resource allocation schemes is the difficulty of ensuring fairness and the required QoS. Users' data cannot always wait until the channel conditions are sufficiently favourable for transmission, especially in slowly varying channels. Furthermore, as explained in Chapter 1, it is important that network operators can provide reliable wide area coverage, including to stationary users near the cell edge – not just to the users which happen to experience good channel conditions by virtue of their proximity to the eNodeB.

The second extreme of scheduling algorithm, fair scheduling, therefore pays more attention to latency and achieving a minimum data rate for each user than to the total data rate achieved. This is particularly important for real-time applications such as Voice-over-IP (VoIP) or video-conferencing, where a certain minimum rate must be guaranteed independently of the channel state.

In practice, most scheduling algorithms fall between the two extremes outlined above, including elements of both to deliver the required mix of QoS. A variety of metrics can be used to quantify the degree of fairness provided by a scheduling algorithm, the general objective being to avoid heavily penalizing the cell-edge users in an attempt to give high throughputs to the users with good channel conditions. One example is based on the Cumulative Distribution Function (CDF) of the throughput of all users, whereby a system may be considered sufficiently fair if the CDF of the throughput lies to the right-hand side of a particular line, such as that shown in Figure 12.3. Another example is the *Jain index* [4], which gives an indication of the variation in throughput between users. It is calculated as:

$$J = \frac{\overline{T}^2}{\overline{T}^2 + \mathrm{var}(T)}, \quad 0 \leq J \leq 1, \tag{12.1}$$

where $\overline{T}$ and $\mathrm{var}(T)$ are the mean and variance respectively of the average user throughputs. The more similar the average user throughputs ($\mathrm{var}(T) \rightarrow 0$), the higher the value of J.

Other factors also need to be taken into account, especially the fact that, in a coordinated deployment, individual cells cannot be considered in isolation – nor even the individual set of cells controlled by a single eNodeB. The eNodeBs should take into account the interference generated by co-channel cells, which can be a severe limiting factor, especially for cell-edge users. Similarly, the performance of the system as a whole can be enhanced if each eNodeB also takes into account the impact of the transmissions of its own cells on the neighbouring cells. These inter-cell aspects, and the corresponding inter-eNodeB signalling mechanisms provided in LTE, are discussed in detail in Section 12.5.

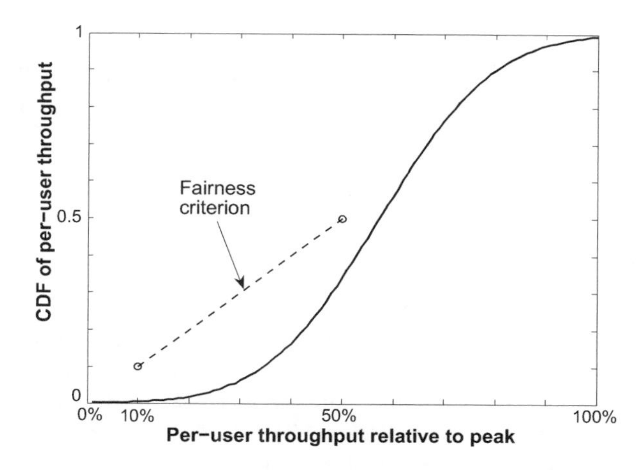

Figure 12.3: An example of a metric based on the throughput CDF over all users for scheduler fairness evaluation (10–50 metric).

12.3 Scheduling Algorithms

Multi-user scheduling finds its basis at the interface between information theory and queueing theory, in the theory of capacity-maximizing resource allocation. Before establishing an algorithm, a capacity-related metric is first formulated and then optimized across all possible resource allocation solutions satisfying a set of predetermined constraints. Such constraints may be physical (e.g. bandwidth and total power) or QoS-related.

Information theory offers a range of possible capacity metrics which are relevant in different system operation scenarios. Two prominent examples are explained here, namely the so-called *ergodic capacity* and *delay-limited capacity*, corresponding respectively to soft and hard forms of rate guarantee for the user.

12.3.1 Ergodic Capacity

The ergodic capacity (also known as the Shannon capacity) is defined as the maximum data rate which can be sent over the channel with asymptotically small error probability, averaged over the fading process. When CSI is available at the Transmitter (i.e. CSIT), the transmit power and mutual information[1]between the transmitter and the receiver can be varied depending on the fading state in order to maximize the average rates. The ergodic capacity metric considers the long-term average data rate which can be delivered to a user when the user does not have any latency constraints.

[1]The mutual information bewteen two random variables X and Y with probability density function $p(X)$ and $p(Y)$ respectively is defined as $I(X; Y) = \mathbb{E}_{X,Y}\,[\log p(X; Y)/(p(X)p(Y))]$, where $\mathbb{E}[.]$ is the expectation function [5].

12.3.1.1 Maximum Rate Scheduling

It has been shown in [3, 6] that the maximum total ergodic sum rate $\sum_{k=1}^{K} R_k$, where R_k is the total rate allocated to user k, is achieved with a transmission power given by

$$P_k(m,f) = \begin{cases} \left[\dfrac{1}{\lambda_k} - \dfrac{P_N}{|H_k(m,f)|^2}\right]^+ & \text{if } |H_k(m,f)|^2 \geq \dfrac{\lambda_k}{\lambda_{k'}}|H_{k'}(m,f)|^2 \\ 0 & \text{otherwise} \end{cases} \tag{12.2}$$

where $[x]^+ = \max(0, x)$, $H_k(m, f)$ is the channel gain of user k in RB m of subframe f and λ_k are constants which are chosen in order to satisfy an average per-user power constraint.

This approach is known in the literature as *maximum sum rate* scheduling. The result in Equation (12.2) shows that the maximum sum rate is achieved by orthogonal multiplexing where in each subchannel (i.e. each RB in LTE) only the user with the best channel gain is scheduled. This orthogonal scheduling property is in line with, and thus justifies, the philosophy of the LTE multiple-access schemes. The input power spectra given by Equation (12.2) are water-filling formulae in both frequency and time (i.e. allocating more power to a scheduled user when his channel gain is high and less power when it is low).[2]

A variant of this resource allocation strategy with no power control (relevant for cases where the power control dynamic range is limited or zero) is considered in [2]. This allocation strategy is called 'maximum-rate constant-power' scheduling, where only the user with the best channel gain is scheduled in each RB, but with no adaptation of the transmit power. It is shown in [2] that most of the performance gains offered by the maximum rate allocation in Equation (12.2) are due to multi-user diversity and not to power control, so an on-off power allocation can achieve comparable performance to maximum sum rate scheduling. This not only allows some simplification of the scheduling algorithm but also is well suited to a downlink scenario where the frequency-domain dynamic range of the transmitted power in a given subframe is limited by constraints such as the dynamic range of the UE receivers and the need to transmit wideband reference signals for channel estimation.

12.3.1.2 Proportional Fair Scheduling

The ergodic sum rate corresponds to the optimal rate for traffic which has no delay constraint. This results in an unfair sharing of the channel resources. When the QoS required by the application includes latency constraints, such a scheduling strategy is not suitable and other fairer approaches need to be considered. One such approach is the well-known Proportional Fair Scheduling (PFS) algorithm. PFS schedules a user when its instantaneous channel quality is high relative to its own average channel condition over time. It allocates user $\hat{k}_m$ in RB m in any given subframe f if [7]

$$\hat{k}_m = \underset{k=1,\dots,K}{\operatorname{argmax}} \frac{R_k(m,f)}{T_k(f)} \tag{12.3}$$

where $T_k(f)$ denotes the long-term average throughput of user k computed in subframe f and $R_k(m, f) = \log(1 + \text{SNR}_k(m, f))$ is the achievable rate by user k in RB m and subframe f.

[2]Water-filling strategies are discussed in more detail in the context of MIMO in Section 11.1.3.

The long-term average user throughputs are recursively computed by

$$T_k(f) = \left(1 - \frac{1}{t_c}\right)T_k(f-1) + \frac{1}{t_c}\sum_{m=1}^{M} R_k(m,f)\mathcal{I}\{\hat{k}_m = k\} \tag{12.4}$$

where t_c is the time window over which fairness is imposed and $\mathcal{I}\{\cdot\}$ is the indicator function equal to one if $\hat{k}_m = k$ and zero otherwise. A large time window t_c tends to maximize the total average throughput; in fact, in the limit of a very long time window, PFS and maximum-rate constant-power scheduling result in the same allocation of resources. For small t_c, the PFS tends towards a round-robin[3] scheduling of users [7].

Several studies of PFS have been conducted in the case of WCDMA/HSDPA[4], yielding insights which can be applied to LTE. In [8] the link level system performance of PFS is studied, taking into account issues such as link adaptation dynamic range, power and code resources, convergence settings, signalling overhead and code multiplexing. [9] analyses the performance of PFS in the case of VoIP, including a comparison with round-robin scheduling for different delay budgets and packet-scheduling settings. A study of PFS performance in the case of video streaming in HSDPA can be found in [10].

12.3.2 Delay-Limited Capacity

Even though PFS introduces some fairness into the system, this form of fairness may not be sufficient for applications which have a very tight latency constraint. For these cases, a different capacity metric is needed; one such example is referred to as the 'delay-limited capacity'. The delay-limited capacity (also known as zero-outage capacity) is defined as the transmission rate which can be guaranteed in all fading states under finite long-term power constraints. In contrast to the ergodic capacity, where mutual information between the transmitter and the receiver vary with the channel, the powers in delay-limited capacity are coordinated between users and RBs with the objective of maintaining constant mutual information independently of fading states. The delay-limited capacity is relevant to traffic classes where a given data rate must be guaranteed throughout the connection time, regardless of the fading dips.

The delay-limited capacity for a flat-fading multiple-access channel is characterized in [11]. The wideband case is analysed in [7, 12].

It is shown in [7] that guaranteeing a delay-limited rate incurs only a small throughput loss in high Signal to Noise Ratio (SNR) conditions, provided that the number of users is large. However, the solution to achieve the delay-limited capacity requires non-orthogonal scheduling of the users, with successive decoding in each RB. This makes this approach basically unsuitable for LTE.

It is, however, possible to combine orthogonal multiple access with hard QoS requirements. Examples of algorithms achieving such a compromise can be found in [12, 13]. It can be shown that even under hard fairness constraints it is possible to achieve performance which is very close to the optimal unfair policy; thus hard fairness constraints do not necessarily introduce a significant throughput degradation, even with orthogonal resource allocation, provided that the number of users and the system bandwidth are large. This may be a relevant

[3] A round-robin approach schedules each user in turn, irrespective of their prevailing channel conditions.

[4] Wideband Code Division Multiple Access / High Speed Downlink Packet Access.

scenario for a deployment targeting VoIP users, where densities of several hundred users per cell are foreseen (as mentioned in Section 1.2), with a latency constraint typically requiring each packet to be successfully delivered within 50 ms.

In general this discussion of scheduling strategies assumes that all users have an equal and infinite queue length – often referred to as a full-buffer traffic model. In practice this is not the case, especially for real-time services, and information on users' queue lengths is necessary to guarantee system stability. If a scheduling algorithm can be found which keeps the average queue length bounded, the system is said to be stabilized [14]. One approach to ensuring that the queue length remains bounded is to use the queue length to set the priority order in the allocation of RBs. This generally works for lightly loaded systems. In [15] and [16] it is shown that in wideband frequency-selective channels, low average packet delay can be achieved even if the fading is very slow.

12.4 Considerations for Resource Scheduling in LTE

In LTE, each logical channel has a corresponding QoS description which should influence the behaviour of the eNodeB resource scheduling algorithm. Based on the evolution of the radio and traffic conditions, this QoS description could potentially be updated for each service in a long-term fashion. The mapping between the QoS descriptions of different services and the resource scheduling algorithm in the eNodeB can be a key differentiating factor between radio network equipment manufacturers.

An important constraint for the eNodeB scheduling algorithm is the accuracy of the eNodeB's knowledge of the channel quality for the active UEs in the cell. The manner in which such information is provided to the scheduler in LTE differs between uplink and downlink transmissions. In practice, for the downlink this information is provided through the feedback of Channel Quality Indicators (CQIs)[5] by UEs as described in Chapter 10, while for the uplink the eNodeB may use Sounding Reference Signals (SRSs) or other signals transmitted by the UEs to estimate the channel quality, as discussed in Chapter 15. The frequency with which CQI reports and SRS are transmitted is configurable by the eNodeB, allowing for a trade-off between the signalling overhead and the availability of up-to-date channel information. If the most recent CQI report or SRS was received a significant time before the scheduling decision is taken, the performance of the scheduling algorithm can be significantly degraded.

In order to perform frequency-domain scheduling, the information about the radio channel needs to be frequency-specific. For this purpose, the eNodeB may configure the CQI reports to relate to specific sub-bands to assist the downlink scheduling, as explained in Section 10.2.1. Uplink frequency-domain scheduling can be facilitated by configuring the SRS to be transmitted over a large bandwidth. However, for cell-edge UEs the wider the transmitted bandwidth the lower the available power per RB; this means that accurate frequency-domain scheduling may be more difficult for UEs near the cell edge. Limiting the SRS to a subset of the system bandwidth will improve the channel quality estimation on these RBs but restrict the ability of the scheduler to find an optimal scheduling solution for all users. In general, provided that the bandwidth over which the channel can be estimated for

[5]More generally, in the context of user selection for MU-MIMO, full CSI is relevant, to support user pairing as well as rate scheduling.

scheduling purposes is greater than the intended scheduling bandwidth for data transmission by a sufficient factor, a useful element of multi-user diversity gain may still be achievable.

As noted above, in order to support QoS- and queue-aware scheduling, it is necessary for the scheduler to have not only information about the channel quality, but also information on the queue status. In the LTE downlink, knowledge of the amount of buffered data awaiting transmission to each UE is inherently available in the eNodeB; for the uplink, Section 4.4.2.2 explains the buffer status reporting mechanisms available to transfer such information to the eNodeB.

12.5 Interference Coordination and Frequency Reuse

One limiting aspect for system throughput performance in cellular networks is inter-cell interference, especially for cell-edge users. Careful management of inter-cell interference is particularly important in systems such as LTE which are designed to operate with a frequency reuse factor of one.

The scheduling strategy of the eNodeB may therefore include an element of Inter-Cell Interference Coordination (ICIC), whereby interference from and to the adjacent cells is taken into account in order to increase the data rates which can be provided for users at the cell edge. This implies for example imposing restrictions on what resources in time and/or frequency are available to the scheduler, or what transmit power may be used in certain time/frequency resources.

The impact of interference on the achievable data rate for a given user can be expressed analytically. If a user k is experiencing no interference, then its achievable rate in an RB m of subframe f can be expressed as

$$R_{k,\text{no-Int}}(m, f) = W \log\left(1 + \frac{P^s(m, f)|H_k^s(m, f)|^2}{P_N}\right) \tag{12.5}$$

where $H_k^s(m, f)$ is the channel gain from the serving cell s to user k, $P^s(m, f)$ is the transmit power from cell s, P_N is the noise power and W is the bandwidth of one RB (i.e. 180 kHz). If neighbouring cells are transmitting in the same time-frequency resources, then the achievable rate for user k reduces to

$$R_{k,\text{Int}}(m, f) = W \log\left(1 + \frac{P^s(m, f)|H_k^s(m, f)|^2}{P_N + \sum_{i \neq s} P^i(m, f)|H_k^i(m, f)|^2}\right) \tag{12.6}$$

where the indices i denote interfering cells.

The rate loss for user k can then be expressed as

$$\begin{aligned} R_{k,\text{loss}}(m, f) &= R_{k,\text{no-Int}}(m, f) - R_{k,\text{Int}}(m, f) \\ &= W \log\left(\frac{1 + \text{SNR}}{1 + \left[\frac{1}{\text{SNR}} + \frac{\sum_{i \neq s} P^i(m,f)|H_k^i(m,f)|^2}{P^s(m,f)|H_k^s(m,f)|^2}\right]^{-1}}\right) \end{aligned} \tag{12.7}$$

Figure 12.4 plots the rate loss for user k as a function of the total inter-cell interference to signal ratio $\alpha = \left(\sum_{i \neq s} P^i(m, f)|H_k^i(m, f)|^2\right) / \left(P^s(m, f)|H_k^s(m, f)|^2\right)$, with SNR = 0 dB. It can

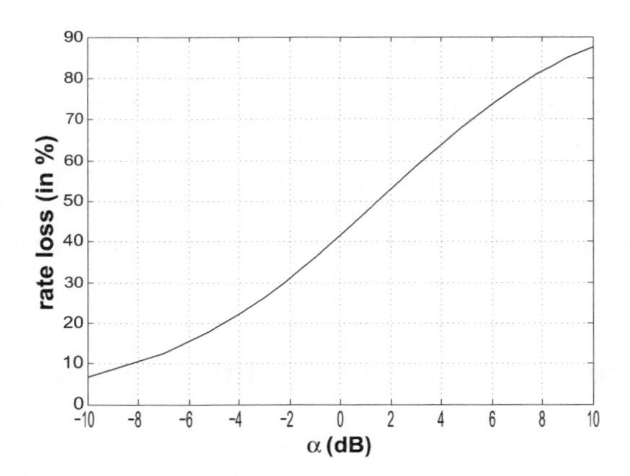

Figure 12.4: User's rate loss due to interference.

easily be seen that for a level of interference equal to the desired signal level (i.e. $\alpha \approx 0$ dB), user k experiences a rate loss of approximately 40%.

In order to demonstrate further the significance of interference and power allocation depending on the system configuration we consider two examples of a cellular system with two cells (s_1 and s_2) and one active user per cell (k_1 and k_2 respectively). Each user receives the wanted signal from its serving cell, while the inter-cell interference comes from the other cell.

In the first example, each user is located near its respective eNodeB (see Figure 12.5(a)). The channel gain from the interfering cell is small compared to the channel gain from the serving cell ($|H_{k_1}^{s_1}(m, f)| \gg |H_{k_1}^{s_2}(m, f)|$ and $|H_{k_2}^{s_2}(m, f)| \gg |H_{k_2}^{s_1}(m, f)|$). In the second example (see Figure 12.5(b)), we consider the same scenario but with the users now located close to the edge of their respective cells. In this case the channel gain from the serving cell and the interfering cell are comparable ($|H_{k_1}^{s_1}(m, f)| \approx |H_{k_1}^{s_2}(m, f)|$ and $|H_{k_2}^{s_2}(m, f)| \approx |H_{k_2}^{s_1}(m, f)|$).

The capacity of the system with two eNodeBs and two users can be written as

$$R_{\text{Tot}} = W\left(\log\left(1 + \frac{P^{s_1}|H_{k_1}^{s_1}(m, f)|^2}{P_N + P^{s_2}|H_{k_1}^{s_2}(m, f)|^2}\right) + \log\left(1 + \frac{P^{s_2}|H_{k_2}^{s_2}(m, f)|^2}{P_N + P^{s_1}|H_{k_2}^{s_1}(m, f)|^2}\right)\right) \tag{12.8}$$

From this equation, it can be noted that the optimal transmit power operating point in terms of maximum achievable throughput is different for the two considered cases. In the first scenario, the maximum throughput is achieved when both eNodeBs transmit at maximum power, while in the second the maximum capacity is reached by allowing only one eNodeB to transmit. It can in fact be shown that the optimal power allocation for maximum capacity for this situation with two base stations is binary in the general case; this means that either both base stations should be operating at maximum power in a given RB, or one of them should be turned off completely in that RB [17].

From a practical point of view, this result can be exploited in the eNodeB scheduler by treating users in different ways depending on whether they are cell-centre or cell-edge users. Each cell can then be divided into two parts – inner and outer. In the inner part, where users experience a low level of interference and also require less power to communicate with the

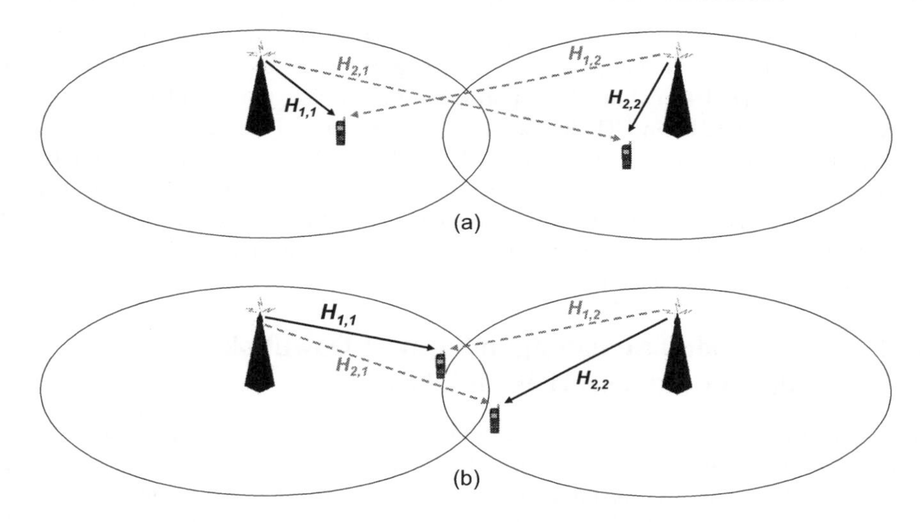

Figure 12.5: System configuration: (a) users close to eNodeBs; (b) users at the cell edge.

serving cell, a frequency reuse factor of 1 can be adopted. For the outer part, scheduling restrictions are applied: when the cell schedules a user in a given part of band, the system capacity is optimized if the neighbouring cells do not transmit at all; alternatively, they may transmit only at low power (probably to users in the inner parts of the neighbouring cells) to avoid creating strong interference to the scheduled user in the first cell. This effectively results in a higher frequency reuse factor at the cell-edge; it is often referred to as 'partial frequency reuse' or 'soft frequency reuse', and is illustrated in Figure 12.6.

In order to coordinate the scheduling in different cells in such a way, communication between neighbouring cells is required. If the neighbouring cells are managed by the

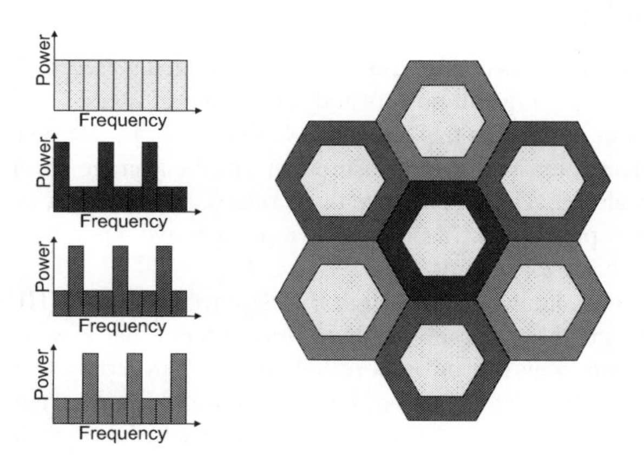

Figure 12.6: Partial frequency reuse.

same eNodeB, a coordinated scheduling strategy can be followed without the need for standardized signalling. However, where neighbouring cells are controlled by different eNodeBs, standardized signalling is important, especially in multivendor networks. The main mechanism for ICIC in LTE Releases 8 and 9 is normally assumed to be frequency-domain-based, at least for the data channels, and the Release 8/9 inter-eNodeB ICIC signalling explained in the following two sections is designed to support this. In Release 10, additional time-domain mechanisms are introduced, aiming particularly to support ICIC for the PDCCH and for heterogeneous networks comprising both macrocells and small cells; these mechanisms are explained in Section 31.2.

12.5.1 Inter-eNodeB Signalling to Support Downlink Frequency-Domain ICIC in LTE

In relation to the downlink transmissions, a bitmap termed the Relative Narrowband Transmit Power (RNTP[6]) indicator can be exchanged between eNodeBs over the X2 interface. Each bit of the RNTP indicator corresponds to one RB in the frequency domain and is used to inform the neighbouring eNodeBs if a cell is planning to keep the transmit power for the RB below a certain upper limit or not. The value of this upper limit, and the period for which the indicator is valid into the future, are configurable. This enables the neighbouring cells to take into account the expected level of interference in each RB when scheduling UEs in their own cells. The reaction of the eNodeB in case of receiving an indication of high transmit power in an RB in a neighbouring cell is not standardized (thus allowing some freedom of implementation for the scheduling algorithm); however, a typical response could be to avoid scheduling cell-edge UEs in such RBs. In the definition of the RNTP indicator, the transmit power per antenna port is normalized by the maximum output power of a base station or cell. The reason for this is that a cell with a smaller maximum output power, corresponding to smaller cell size, can create as much interference as a cell with a larger maximum output power corresponding to a larger cell size.

12.5.2 Inter-eNodeB Signalling to Support Uplink Frequency-Domain ICIC in LTE

For the uplink transmissions, two messages may be exchanged between eNodeBs to facilitate some coordination of their transmit powers and scheduling of users:

A reactive indicator, known as the 'Overload Indicator' (OI), can be exchanged over the X2 interface to indicate physical layer measurements of the average uplink interference plus thermal noise for each RB. The OI can take three values, expressing low, medium, and high levels of interference plus noise. In order to avoid excessive signalling load, it cannot be updated more often than every 20 ms.

A proactive indicator, known as the 'High Interference Indicator' (HII), can also be sent by an eNodeB to its neighbouring eNodeBs to inform them that it will, in the near future, schedule uplink transmissions by one or more cell-edge UEs in certain parts of the bandwidth, and therefore that high interference might occur in those frequency regions. Neighbouring cells may then take this information into consideration in scheduling their own users to limit the interference impact. This can be achieved either by deciding not to schedule their own

[6]RNTP is defined in [18, Section 5.2.1].

cell-edge UEs in that part of the bandwidth and only considering the allocation of those resources for cell-centre users requiring less transmission power, or by not scheduling any user at all in the relevant RBs. The HII is comprised of a bitmap with one bit per RB and, like the OI, is not sent more often than every 20 ms. The HII bitmap is addressed to specific neighbour eNodeBs.

In addition to frequency-domain scheduling in the uplink, the eNodeB also controls the degree to which each UE compensates for the path-loss when setting its uplink transmission power. This enables the eNodeB to trade off fairness for cell-edge UEs against inter-cell interference generated towards other cells, and can also be used to maximize system capacity. This is discussed in more detail in Section 18.3.2.

12.5.3 Static versus Semi-Static ICIC

In general, ICIC may be static or semi-static, with different levels of associated communication required between eNodeBs.

For *static interference coordination*, the coordination is associated with cell planning, and reconfigurations are rare. This largely avoids signalling on the X2 interface, but it may result in some performance limitation since it cannot adaptively take into account variations in cell loading and user distributions.

Semi-static interference coordination typically refers to reconfigurations carried out on a time-scale of the order of seconds or longer. The inter-eNodeB communication methods over the X2 interface can be used as discussed above. Other types of information such as traffic load information may also be used, as discussed in Section 2.6.4. Semi-static interference coordination may be more beneficial in cases of non-uniform load distributions in eNodeBs and varying cell sizes across the network.

12.6 Summary

In summary, it can be observed that a variety of resource scheduling algorithms may be applied by the eNodeB depending on the optimization criteria required. The prioritization of data will typically take into account the corresponding traffic classes, especially in regard to balancing throughput maximization for delay-tolerant applications against QoS for delay-limited applications in a fair way.

It can be seen that multi-user diversity is an important factor in all cases, and especially if the user density is high, in which case the multi-user diversity gain enables the scheduler to achieve a high capacity even with tight delay constraints. Finally, it is important to note that individual cells, and even individual eNodeBs, cannot be considered in isolation. System optimization requires some degree of coordination between cells and eNodeBs, in order to avoid inter-cell interference becoming the limiting factor. Considering the system as a whole, the best results are in many cases realized by simple 'on-off' allocation of resource blocks, whereby some eNodeBs avoid scheduling transmissions in certain resource blocks which are used by neighbouring eNodeBs for cell-edge users.

LTE Releases 8 and 9 support standardized signalling between eNodeBs to facilitate frequency-domain data-channel ICIC algorithms for both downlink and uplink. Additional time-domain ICIC mechanisms of particular relevance to the control signalling and to heterogeneous networks are introduced in Release 10 and explained in Section 31.2.

References[7]

[1] K. Knopp and P. A. Humblet, 'Information Capacity and Power Control in Single-cell Multiuser Communications', in *Proc. IEEE International Conference on Communications*, Seattle, WA, 1995.

[2] R. Knopp and P. A. Humblet, 'Multiple-Accessing over Frequency-selective Fading Channels', in *Proc. IEEE International Symposium on Personal, Indoor and Mobile Radio Communications*, September 1995.

[3] D. N. C. Tse and S. D. Hanly, 'Multiaccess Fading Channels. I: Polymatroid Structure, Optimal Resource Allocation and Throughput Capacities'. *IEEE Trans. on Information Theory*, Vol. 44, pp. 2796–2815, November 1998.

[4] R. Jain, W. Hawe and D. Chiu, 'A quantitative measure of fairness and discrimination for resource allocation in Shared Computer Systems', Digital Equipment Corporation Technical Report 301, September 1984.

[5] T. M. Cover and J. Thomas. *Elements of Information Theory*. Wiley, New York, 1991.

[6] R. Knopp and P. A. Hamblet, 'Multiuser Diversity', Technical Report, Eurecom Institute, Sophia Antipolis, France, 2002.

[7] G. Caire, R. Müller and R. Knopp, 'Hard Fairness Versus Proportional Fairness in Wireless Communications: The Single-cell Case'. *IEEE Trans. on Information Theory*, Vol. 53, pp. 1366–1385, April 2007.

[8] T. E. Kolding, 'Link and System Performance Aspects of Proportional Fair Scheduling in WCDMA/HSDPA', in *Proc. IEEE Vehicular Technology Conference*, Orlando, FL, October 2003.

[9] B. Wang, K. I. Pedersen, P. E. Kolding and T. E. Mogensen, 'Performance of VoIP on HSDPA', in *Proc. IEEE Vehicular Technology Conference*, Stockholm, May 2005.

[10] V. Vukadinovic and G. Karlsson, 'Video Streaming in 3.5G: On Throughput-Delay Performance of Proportional Fair Scheduling', in *IEEE International Symposium on Modeling, Analysis, and Simulation of Computer and Telecommunication Systems*, Monterey, CA, September 2006.

[11] S. D. Hanly and D. N. C. Tse, 'Multiaccess Fading Channels. II: Delay-limited Capacities'. *IEEE Trans. on Information Theory*, Vol. 44, pp. 2816–2831, November 1998.

[12] I. Toufik, 'Wideband Resource Allocation for Future Cellular Networks'. Ph.D. Thesis, Eurecom Institute, 2006.

[13] I. Toufik and R. Knopp, 'Channel Allocation Algorithms for Multi-carrier Systems', in *Proc. IEEE Vehicular Technology Conference*, Los Angeles, CA, September 2004.

[14] M. J. Neely, E. Modiano and C. E. Rohrs, 'Power Allocation and Routing in Multi-beam Satellites with Time Varying Channels'. *IEEE Trans. on Networking*, Vol. 11, pp. 138–152, February 2003.

[15] M. Realp, A. I. Pérez-Neira and R. Knopp, 'Delay Bounds for Resource Allocation in Wideband Wireless Systems', in *Proc. IEEE International Conference on Communications*, Istanbul, June 2006.

[16] M. Realp, R. Knopp and A. I. Pérez-Neira, 'Resource Allocation in Wideband Wireless Systems' in *Proc. IEEE International Symposium on Personal, Indoor and Mobile Radio Communications*, Berlin, September 2005.

[17] A. Gjendemsjoe, D. Gesbert, G. Oien and S. Kiani, 'Binary Power Control for Sum Rate Maximization over Multiple Interfering Links'. *IEEE Trans. Wireless Communications*, August 2008.

[18] 3GPP Technical Specification 36.213, 'Evolved Universal Terrestrial Radio Access (E-UTRA); Physical Layer Procedures', www.3gpp.org.

[7] All web sites confirmed 1st March 2011.

13

Broadcast Operation

Himke van der Velde, Olivier Hus and Matthew Baker

13.1 Introduction

The Multimedia Broadcast/Multicast Service (MBMS) aims to provide an efficient mode of delivery for both broadcast and multicast services over the core network. MBMS was introduced in the second release of the LTE specifications (Release 9), although the initial Release 8 physical layer specifications were already designed to support MBMS by including essential components to ensure forward-compatibility.

The LTE MBMS feature is largely based on that which was already available in UTRAN[1] (from Release 6) and GERAN[2] with both simplifications and enhancements.

This chapter first describes general aspects including the MBMS reference architecture, before reviewing the MBMS features supported by LTE in more detail. In particular, we describe how LTE is able to benefit from the new OFDM downlink radio interface to achieve radically improved transmission efficiency and coverage by means of multicell 'single frequency network' operation. The control signalling and user plane content synchronization mechanisms are also explained.

13.2 Broadcast Modes

In the most general sense, broadcasting is the distribution of content to an audience of multiple users; in the case of mobile multimedia services an efficient transmission system for the simultaneous delivery of content to large groups of mobile users. Typical broadcast content can include newscasts, weather forecasts or live mobile television.

Figure 13.1 illustrates the reception of a video clip showing a highlight of a sporting event. There are three possible types of transmission to reach multiple users:

[1] UMTS Terrestrial Radio Access Network.
[2] GSM Edge Radio Access Network.

LTE – The UMTS Long Term Evolution: From Theory to Practice, Second Edition.
Stefania Sesia, Issam Toufik and Matthew Baker.

Figure 13.1: Receiving football game highlights as a mobile broadcast service. Adapted from image provided courtesy of Philips 'Living Memory' project.

- **Unicast:** A bidirectional point-to-point transmission between the network and each of the multiple users; the network provides a dedicated connection to each terminal, and the same content is transmitted multiple times – i.e. separately to each individual user receiving the service.
- **Broadcast:** A downlink-only point-to-multipoint connection from the network to multiple terminals; the content is transmitted once to all terminals in a geographical area, and users are free to choose whether or not to receive it.
- **Multicast:** A downlink-only point-to-multipoint connection from the network to a managed group of terminals; the content is transmitted once to the whole group, and only users belonging to the managed user group can receive it.

Point-to-multipoint transmission typically becomes more efficient than point-to-point when there are more than around three to five users per cell, as it supports feedback which can improve the radio link efficiency; the exact switching point depends on the nature of the transfer mechanisms. At higher user densities, point-to-multipoint transmission decreases the total amount of data transmitted in the downlink, and it may also reduce the control overhead especially in the uplink.

The difference between broadcast and multicast modes manifests itself predominantly in the upper layers (i.e. the Non-Access Stratum (NAS) – see Section 3.1). Multicast includes additional procedures for subscription, authorization and charging, to ensure that the services are available only to specific users. Multicast also includes joining and leaving procedures, which aim to provide control information only in areas in which there are users who are subscribed to the MBMS multicast service. For UTRAN, both modes were developed; however, for simplicity, LTE Release 9 MBMS includes only a broadcast mode.

A point-to-multipoint transfer can involve either single or multicell transmission. In the latter case, multiple cells transmit exactly the same data in a synchronized manner so as to appear as one transmission to the UE. Whereas UTRAN MBMS supports point-to-point, single cell point-to-multipoint and multicell point-to-multipoint modes, LTE MBMS supports a single transfer mode: multicell point-to-multipoint, using a 'single frequency network' transmission as described in Section 13.4.1. The number of cells participating in the multicell transmission can, however, be limited to one.

UTRAN MBMS includes counting procedures that enable UTRAN to count the number of UEs interested in receiving a service and hence to select the optimal transfer mode (point-to-point or point-to-multipoint), as well as to avoid transmission of MBMS user data in cells in which there are no users interested in receiving the session. LTE MBMS did not include counting procedures in Release 9, so both the transfer mode and the transmission area were fixed (i.e. semi-statically configured). However, a counting procedure is introduced in Release 10 of LTE, as explained in Section 13.6.5.

13.3 Overall MBMS Architecture

13.3.1 Reference Architecture

Figure 13.2 shows the reference architecture for MBMS broadcast mode in the Evolved Packet System (EPS – see Section 2.1), with both E-UTRAN and UTRAN. E-UTRAN includes a Multicell Coordination Entity (MCE), which is a logical entity – in other words, its functionality need not be placed in a separate physical node but could, for example, be integrated with an eNodeB.

The main nodes and interfaces of the MBMS reference architecture are described in the following sections.

13.3.2 Content Provision

Sources of MBMS content vary, but are normally assumed to be external to the Core Network (CN); one example of external content providers would be television broadcasters for mobile television. The MBMS content is generally assumed to be IP based, and by design the MBMS system is integrated with the IP Multimedia Subsystem (IMS) [2] service infrastructure based on the Internet Engineering Task Force (IETF) Session Initiation Protocol (SIP) [3].

The Broadcast-Multicast Service Centre (BM-SC) is the interface between external content providers and the CN. The main functions provided by the BM-SC are:

- Reception of MBMS content from external content providers;
- Providing application and media servers for the mobile network operator;
- Announcement and scheduling of MBMS services and delivery of MBMS content into the core network (including traffic shaping and content synchronization).

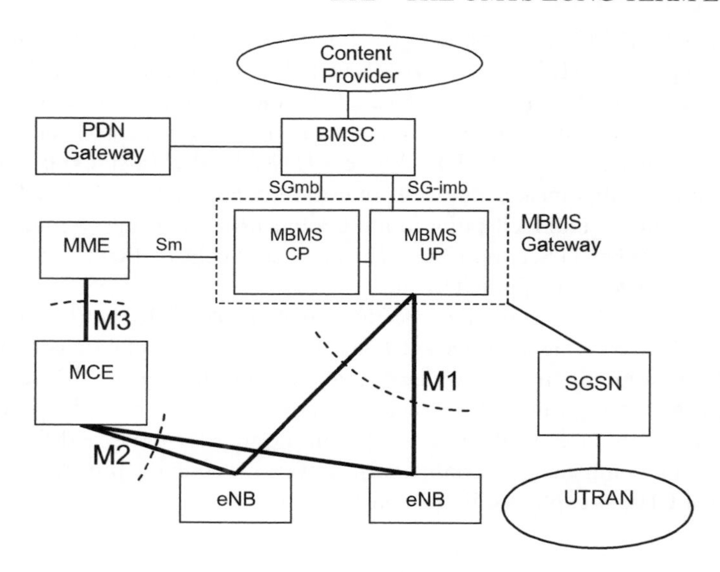

Figure 13.2: EPS reference architecture for MBMS broadcast mode [1]. Reproduced by permission of © 3GPP.

13.3.3 Core Network

In LTE/SAE,[3] the MBMS Gateway (MBMS GW) is the CN entry point for MBMS control signalling and traffic. The MBMS GW distributes session control signalling via the Mobility Management Entity (MME – see Section 2.2.1) to E-UTRAN and handles the establishment and release of user plane bearers using IP multicast traffic.

The MME is involved with MBMS session control (start/modification/stop) and reliable transmission of control messages to the E-UTRAN nodes in the 'MBMS service area'.

13.3.4 Radio Access Network – E-UTRAN/UTRAN/GERAN and UE

The Radio Access Network (RAN), whether E-UTRAN, UTRAN or GERAN, is responsible for delivering MBMS data efficiently within the designated MBMS service area.

The MCE deals with session control and manages the subframe allocation and radio resource configuration to ensure that all eNodeBs participating in an MBMS transmission within a semi-statically configured area (see Section 13.4.2) use exactly the same configuration. The MCE also handles admission control.

The eNodeB handles the transfer of MBMS control, session start notifications and transfer of MBMS data.

The final element in the MBMS content and service chain is the MBMS receiver itself: the UE.

[3] System Architecture Evolution.

13.3.5 MBMS Interfaces

For MBMS in LTE, two new control plane interfaces have been defined (M3 and M2), as well as one new user plane interface (M1), as shown in Figure 13.2.

13.3.5.1 M3 Interface (MCE – MME)

An Application Part (M3AP) is defined for this interface between the MME and MCE. The M3AP primarily specifies procedures for starting, stopping and updating MBMS sessions. Upon start or modification of an MBMS session, the MME provides the details of the MBMS bearer while the MCE verifies if the MBMS service (or modification of it) can be supported. Point-to-point signalling transport is applied, using the Stream Control Transmission Protocol (SCTP) [4].

13.3.5.2 M2 Interface (MCE – eNodeB)

Like the M3 interface, an application part (M2AP) is defined for this interface between the MCE and eNodeB, again primarily specifying procedures for starting, stopping and updating MBMS sessions. Upon start or modification of an MBMS session, the MCE provides the details of the radio resource configuration that all participating eNodeBs shall apply. In particular, the MCE provides the updated control information to be broadcast by the eNodeBs. SCTP is again used for the signalling transport.

13.3.5.3 M1 Interface (MBMS GW – eNodeB)

Similarly to the S1 and X2 interfaces (see Sections 2.5 and 2.6 respectively), the GTP-U[4] protocol over UDP[5] over IP is used to transport MBMS data streams over the M1 interface. IP multicast is used for point-to-multipoint delivery.

13.4 MBMS Single Frequency Network Transmission

One of the targets for MBMS in LTE is to support a cell edge spectral efficiency in an urban or suburban environment of 1 bps/Hz – equivalent to the support of at least 16 mobile TV channels at around 300 kbps per channel in a 5 MHz carrier. This is only achievable by exploiting the special features of the LTE OFDM[6] air interface to transmit multicast or broadcast data as a multicell transmission over a synchronized 'single frequency network': this is known as *Multimedia Broadcast Single Frequency Network (MBSFN)* operation.

13.4.1 Physical Layer Aspects

In MBSFN operation, MBMS data is transmitted simultaneously over the air from multiple tightly time-synchronized cells. A UE receiver will therefore observe multiple versions of the signal with different delays due to the multicell transmission. Provided that the transmissions from the multiple cells are sufficiently tightly synchronized for each to arrive at the UE within

[4]GPRS Tunnelling Protocol – User Plane [5].
[5]User Datagram Protocol.
[6]Orthogonal Frequency Division Multiplexing.

the Cyclic Prefix (CP) at the start of the symbol, there will be no Inter-Symbol Interference (ISI). In effect, this makes the MBSFN transmission appear to a UE as a transmission from a single large cell, and the UE receiver may treat the multicell transmissions in the same way as multipath components of a single-cell transmission without incurring any additional complexity. This is illustrated in Figure 13.3. The UE does not even need to know how many cells are transmitting the signal.

The method of achieving the required tight synchronization between MBSFN transmissions from different eNodeBs is not defined in the LTE specifications; this is left to the implementation of the eNodeBs. Typical implementations are likely to use satellite-based solutions (e.g. GPS[7]) or possibly synchronized backhaul protocols (e.g. [6]).

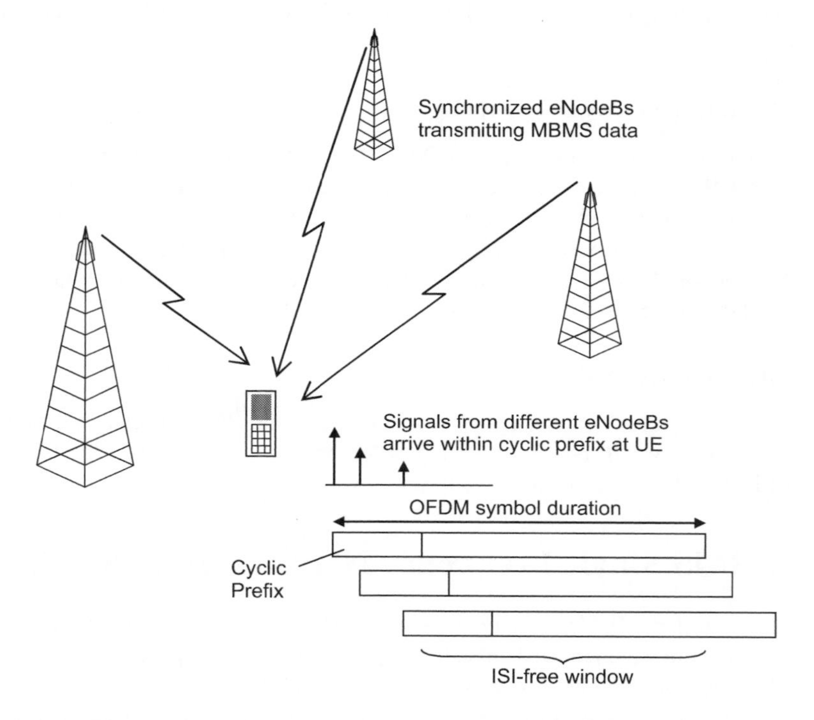

Figure 13.3: ISI-free operation with MBSFN transmission.

This single frequency network reception leads to significant improvements in spectral efficiency compared to UMTS Release 6 MBMS, as the MBSFN transmission greatly enhances the Signal-to-Interference-plus-Noise Ratio (SINR). This is especially true at the cell edge, where transmissions which would otherwise have constituted inter-cell interference are translated into useful signal energy – hence the received signal power is increased at the same time as the interference power being largely removed.

An example of the improvement in performance achievable using MBSFN transmission compared to single-cell point-to-multipoint transmission is shown in Figure 13.4. In this

[7]Global Positioning System.

example, the probability of a randomly located UE not being in outage (defined as MBMS packet loss rate < 1%) is plotted against spectral efficiency of the MBMS data transmissions (a measure of MBMS data rate in a given bandwidth). A hexagonal cell-layout is assumed, with the MBSFN area comprising one, two or three rings around a central cell for which the performance is evaluated. It can be seen that the achievable data rates increase markedly as the size of the MBSFN area is increased and hence the surrounding inter-cell interference is reduced. A 1 km cell radius is assumed, with 46 dBm eNodeB transmission power, 15 m eNodeB antenna height and 2 GHz carrier frequency.

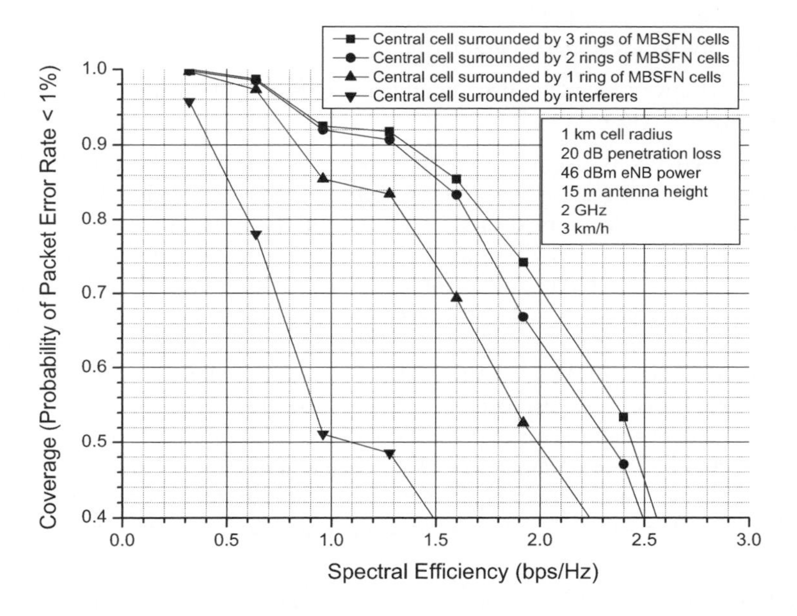

Figure 13.4: Reduction in total downlink resource usage achievable using MBSFN transmission [7]. Reproduced by permission of © 2007 Motorola.

MBSFN data transmission takes place via the Multicast CHannel (MCH) transport channel, which is mapped to the Physical Multicast CHannel (PMCH) introduced in Section 9.2.3.

The basic structure of the Physical Multicast Channel (PMCH) is very similar to the Physical Downlink Shared Channel (PDSCH). However, as the channel in MBSFN operation is in effect a composite channel from multiple cells, it is necessary for the UE to perform a separate channel estimate for MBSFN reception from that performed for reception of data from a single cell. Therefore, in order to avoid the need to mix normal Reference Signals (RSs) and RSs for MBSFN in the same subframe, frequency-division multiplexing of the PMCH and PDSCH is not permitted within a given subframe; instead, time-division multiplexing of unicast and MBSFN data is used – certain subframes are specifically designated as MBSFN subframes, and it is in these subframes that the PMCH would be transmitted.[8] Certain subframes are not allowed to be used for MBSFN transmission: in a Frequency Division Duplex (FDD) system, subframes 0, 4, 5 and 9 in each 10 ms radio frame are reserved for unicast transmission in order to avoid disrupting the synchronization signals or paging occasions, and to ensure that sufficient cell-specific RSs are available for decoding the broadcast system information; in a Time Division Duplex (TDD) system, subframes 0, 1, 2, 5 and 6 cannot be MBSFN subframes.

The key differences from PDSCH in respect of the PMCH are as follows:

- The dynamic control signalling (PDCCH and PHICH[9] – see Section 9.3) cannot occupy more than two OFDM symbols in an MBSFN subframe. The scheduling of MBSFN data on the PMCH is carried out by higher-layer signalling, so the PDCCH is used only for uplink resource grants and not for the PMCH.
- The PMCH always uses the first redundancy version (see Section 10.3.2.4) and does not use Hybrid ARQ.
- The extended CP is used (∼17 μs instead of ∼5 μs) (see Section 5.4.1). As the differences in propagation delay from multiple cells will typically be considerably greater than the delay spread in a single cell, the longer CP helps to ensure that the signals remain within the CP at the UE receivers, thereby reducing the likelihood of ISI. This avoids introducing the complexity of an equalizer in the UE receiver, at the expense of a small loss in peak data rate due to the additional overhead of the longer CP. Note, however, that if the non-MBSFN subframes use the normal CP, then the normal CP is used in the OFDM symbols used for the control signalling at the start of each MBSFN subframe. This results in there being some spare time samples whose usage is unspecified between the end of the last control signalling symbol and the first PMCH symbol, the PMCH remaining aligned with the end of the subframe; the eNodeB may transmit an undefined signal (e.g. a cyclic extension) during this time or it may switch off its transmitter – the UE cannot assume anything about the transmitted signal during these samples.
- The Reference Signal (RS) pattern embedded in the PMCH (designated 'antenna port 4') is different from non-MBSFN data transmission, as shown in Figure 13.5. The RSs are spaced more closely in the frequency domain than for non-MBSFN transmission, reducing the separation to every other subcarrier instead of every sixth subcarrier. This improves the accuracy of the channel estimate that can be achieved for the longer delay spreads. The channel estimate obtained by the UE from the MBSFN RS is a composite channel estimate, representing the composite channel from the set of cells transmitting the MBSFN data. (Note, however, that the cell-specific RS pattern embedded in the OFDM symbols carrying control signalling at the start of each MBSFN subframe remains the same as in the non-MBSFN subframes.)

The latter two features are designed so that a UE making measurements of a neighbouring cell does not need to know in advance the allocation of MBSFN and non-MBSFN subframes. The UE can take advantage of the fact that the first two OFDM symbols in all subframes use the same CP duration and RS pattern.

[8]LTE does not currently support dedicated MBMS carriers in which all subframes would be used for MBSFN transmission.

[9]Physical Downlink Control Channel and Physical HARQ Indicator Channel.

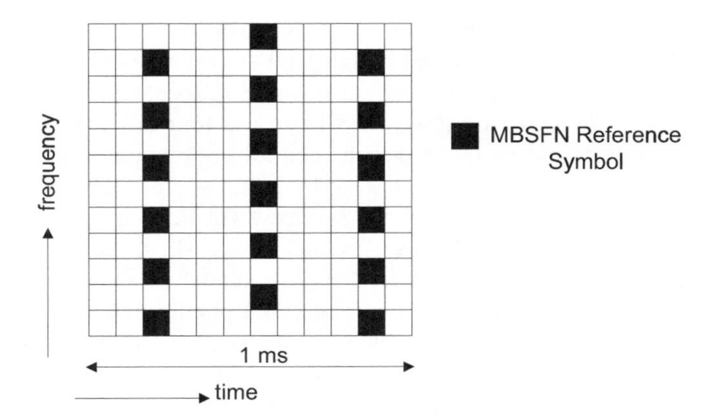

Figure 13.5: MBSFN RS pattern for 15 kHz subcarrier spacing.
Reproduced by permission of © 3GPP.

In addition to these enhancements for MBSFN transmission, a second OFDM parameterization is provided in the LTE physical layer specifications, designed for downlink-only multicast/broadcast transmissions. However, this parameterization is not supported in current releases of LTE. As discussed in Section 5.4.1, this parameterization has an even longer CP, double the length of the extended CP, resulting in approximately 33 μs. This is designed to cater in the future for deployments with very large differences in propagation delay between the signals from different cells (e.g. 10 km). This would be most likely to occur for deployments at low carrier frequencies and large inter-site distances. In order to avoid further increasing the overhead arising from the CP in this case, the number of subcarriers per unit bandwidth is also doubled, giving a subcarrier spacing of 7.5 kHz. The cost of this is an increase in Inter-Carrier Interference (ICI), especially in high-mobility scenarios with a large Doppler spread. There is therefore a trade-off between support for wide-area coverage and support for high mobile velocities. It should be noted, however, that the maximum Doppler shift is lower at the low carrier frequencies that would be likely to be used for a deployment with a 7.5 kHz subcarrier spacing. The absolute frequency spacing of the reference symbols for the 7.5 kHz parameterization is the same as for the 15 kHz subcarrier spacing MBSFN pattern, resulting in a reference symbols on every fourth subcarrier.

13.4.2 MBSFN Areas

A geographical area of the network where MBMS can be transmitted is called an *MBMS Service Area*.

A geographical area where all eNodeBs can be synchronized and can perform MBSFN transmissions is called an *MBSFN Synchronization Area*.

The area within an MBSFN Synchronization Area, covered by a group of cells participating in an MBSFN transmission, is called an *MBSFN Area*. An MBSFN Synchronization Area may support several MBSFN Areas. Moreover, a cell within an MBSFN Synchronization Area may form part of multiple MBSFN Areas, each characterized by different transmitted content and a different set of participating cells.

Figure 13.6 illustrates an example of the usage of MBSFN areas in a network providing two nationwide MBMS services (N1 and N2) as well as two regional MBMS services, R1 and R2 (i.e. with different regional content), across three different regions.

The most natural way to deploy this would be to use four different MBSFN areas, as shown in the figure. It should be noted that the control information related to a service is transmitted by the same set of cells that transmit the data – e.g. MBSFN area N provides both control and data for the two nationwide services. This kind of approach implies that within a geographical area the resources allocated to MBMS need to be semi-statically divided between the MBSFN areas, and UEs that are potentially interested in receiving both nationwide and regional services need to monitor the control information of multiple MBSFN areas. Indeed, UEs are expected to be capable of monitoring control information of multiple MBSFN areas, although simultaneous reception of multiple services is optional regardless of whether they are part of the same MBSFN area. Further details of UE capabilities can be found in Section 13.5.2.

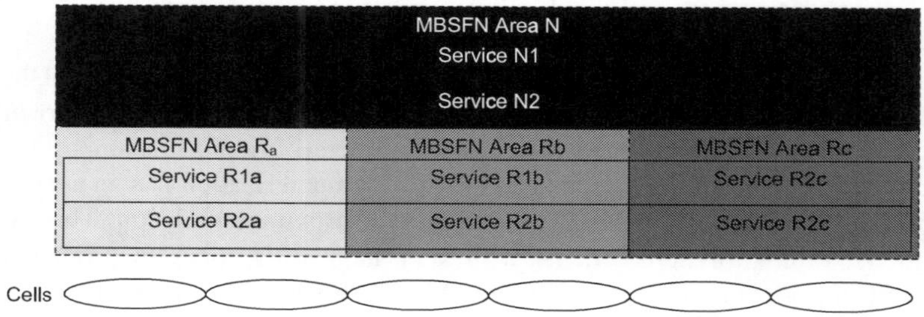

Figure 13.6: Example scenario with overlapping MBSFN areas.

Figure 13.7 illustrates how the same services can be provided without overlapping MBSFN areas. It should be noted that for the nationwide services this approach results in reduced MBSFN transmission gains and potential service interruptions upon change of MBSFN area.

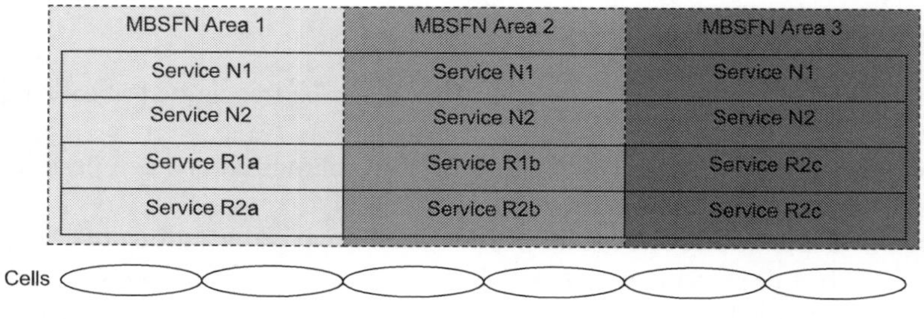

Figure 13.7: Example scenario without overlapping MBSFN areas.

Current releases of LTE do not support dynamic change of the MBSFN area; the counting procedure introduced in Release 10 (see Section 13.6.5) is used only to decide whether or not to use MBSFN transmission for an MBMS service but not to change set of cells used for an MBSFN transmission.

13.5 MBMS Characteristics

13.5.1 Mobility Support

Mobility for MBMS is purely based on the procedures defined for unicast operation. No additional mechanisms are provided to direct UEs to the carrier frequency providing MBMS, nor to reduce MBMS service interruption when moving from one MBSFN area to another.

E-UTRAN can assign the highest cell reselection priority to the carrier frequency providing MBMS. However, in connected mode, when E-UTRAN controls the mobility, the network is aware neither of whether the UE supports MBMS nor of whether it is interested in actually receiving any particular MBMS service.

13.5.2 UE Capabilities and Service Prioritization

All UEs need to support some MBMS-related behaviour, including those that do not support MBMS reception and UEs that conform to the initial Release 8 version of LTE (which does not include MBMS). These UEs must be aware of the MBSFN subframes that are configured, since the RS pattern is different. Release 8/9 UEs not supporting MBMS may also benefit from knowing that no regular downlink resource assignments will occur for them in the MBSFN subframes.

As unicast and MBMS transmissions are time-division multiplexed, parallel reception of unicast and MBMS on the same carrier should be relatively straightforward. Nevertheless, a UE may of course have other limitations preventing parallel support, for example due to memory or display restrictions. Consequently, there is no requirement for a UE that supports MBMS to support simultaneous reception of unicast, nor to support simultaneous reception of multiple MBMS services.

In the case of a UE that cannot receive all the services in which it is interested, it should be able to receive the service (unicast or MBMS) that it prioritizes most. This implies that a UE that is receiving a service should at least be able to detect the start of other services (unless the currently received service is always considered as having the highest priority).

The UE itself performs the prioritization of MBMS services for reception. The E-UTRAN is neither aware of which MBMS services a UE is interested in receiving (a UE may, for example, have successfully received an earlier transmission of a repeated session), nor about the priority ascribed to reception of each service by the UE. The details of prioritization are not specified, being left instead to UE implementation. A UE may, for example, choose to terminate a unicast service at the application layer if it prevents reception of an MBMS service that it has prioritized.

13.6 Radio Access Protocol Architecture and Signalling

13.6.1 Protocol Architecture

The LTE radio protocols have been extended in Release 9 to accommodate MBMS:

- Two new types of logical channels are introduced in Release 9: the Multicast Control CHannel (MCCH) and the Multicast Traffic CHannel (MTCH) for MBMS control and user data respectively.
- A new transport channel, the Multicast Channel (MCH), is introduced, mapped onto the PMCH to support the multicell MBSFN transmission mode.

User data for an MBMS session may employ one or more radio bearers, each of which corresponds to an MTCH logical channel. The Medium Access Control (MAC) layer may multiplex information from multiple logical channels (MCCH and one or more MTCH) together in one transport block that is carried on the MCH.

The MBMS control information is mainly carried by the MCCH, of which there is one per MBSFN area. The MCCH primarily carries the MBSFNAreaConfiguration message, which provides the radio bearer and (P)MCH configuration details, as well as the list of ongoing MBMS sessions. The remainder of the MBMS control information, consisting of the MCCH configuration and parameters related to the notification of changes in the MCCH information, clearly cannot be carried by the MCCH itself and is instead carried by the Broadcast Control CHannel (BCCH), in SIB13 (see Section 3.2.2). In Release 10, a further MCCH message was introduced for the counting procedure, as explained in Section 13.6.5.

Figure 13.8 illustrates the radio protocol extensions introduced to accommodate MBMS.

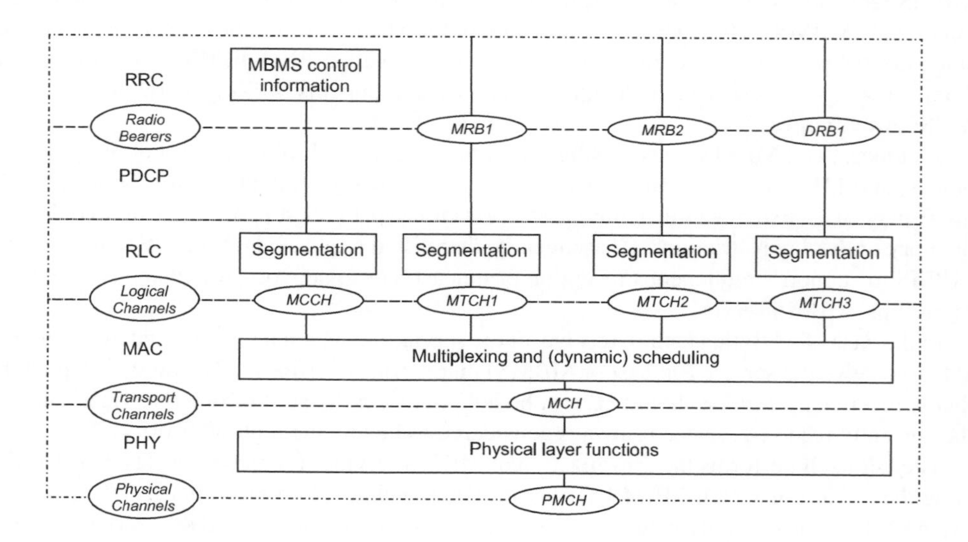

Figure 13.8: LTE protocol extensions to support MBMS.

13.6.2 Session Start Signalling

As a result of the relative simplicity of MBMS in LTE Release 9, the main signalling procedures are for 'session start' and 'session stop'. The absence of multicast support means that procedures for subscription and joining/leaving are not required; furthermore, service announcement is assumed to be an application-specific procedure, not part of the main MBMS signalling procedures.

Figure 13.9 provides an overview of the session start message sequence.

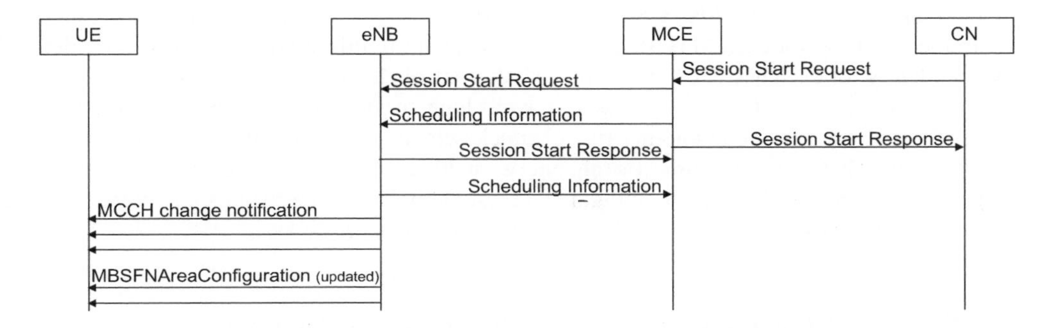

Figure 13.9: The MBMS session start message sequence.

The BM-SC may initiate multiple subsessions with different content, distinguished by flow identifiers. The 'Session Start Request' message from the BM-SC includes the following parameters:

- **QoS:** Used for the admission control and radio resource configuration in the MCE.
- **Service area:** Used to decide to which eNodeBs the session start signalling should be forwarded.
- **Session duration:** Used by admission control in the MCE.
- **Time to data transfer:** Used by the MCE to ensure all UEs receive the control information in time.
- **Access indicator:** Indicates in which Radio Access Technologies (RATs) the service will be broadcast.

Upon receiving the session start request, the MCE performs admission control, allocates IP multicast addresses, and allocates and configures the radio resources. The MCE provides the relevant session parameters (e.g. the multicast address) to the eNodeB by means of the 'Session Start Request' message. Likewise, the MCE transfers the radio resource configuration within a 'scheduling information' message to the eNodeB. Upon receiving these messages, the eNodeB notifies the UEs about a change of MCCH information and subsequently provides the updated MBMS radio resource configuration information within the MBSFNAreaConfiguration message. The eNodeB also joins the transport network IP multicast address.

The actual data transfer starts after a (configurable) duration, such as 10 s.

13.6.3 Radio Resource Control (RRC) Signalling Aspects

13.6.3.1 Scheduling

MBMS does not use the PDCCH for dynamic scheduling. In an MBSFN subframe, the MCH uses all the resources in the frequency domain, so MCH-related scheduling only relates to the subframe allocation in the time domain.

The subframes that carry MCCH are indicated semi-statically by SIB13. For MTCH, the scheduling of which subframes are used for a particular MBMS bearer is somewhat more dynamic: once per 'MCH Scheduling Period' (MSP), E-UTRAN indicates which subframes are used for each MTCH. This 'MCH Scheduling Information' (MSI) is provided independently for each MCH by means of a MAC control element (see Section 4.4.2.7).

One design goal for LTE MBMS was to avoid long service interruptions when the user switches from one service to another (i.e. channel switching). Assuming that the UE stores the complete MCCH information, the channel switching time is mainly determined by the length of the MSP; if E-UTRAN configures a reasonably low value for this parameter (e.g. 320 ms), users can smoothly switch channels without noticeable delays.

13.6.3.2 MBMS Control Information Validity and Change Notification

As mentioned above, the MBMS control information is mainly carried by the MCCH; the BCCH merely carries the control information needed to acquire the MCCH and to detect MCCH information changes.

Transmission of MBMS control information on the BCCH (i.e. in SIB13) is performed according to the usual procedures for SIBs (see Section 3.2.2.2). The transfer procedures for MBMS control information on the MCCH are similar: control information can only change at specific radio frames with System Frame Number (SFN) given by SFN mod $m = 0$, where m is the modification period. The MBSFNAreaConfiguration message is repeated a configurable number of times within the modification period.

When an MBMS session is started, E-UTRAN notifies the UEs about an MCCH information change. This change notification is provided a configurable number of times per modification period, by means of a message on the PDCCH using Downlink Control Information (DCI) Format 1C (see Section 9.3.5.1), using a special identifier, the MBMS Radio Network Temporary Identifier (M-RNTI). This message indicates which of the configured MCCHs will change. UEs that are interested in receiving an MBMS session that has not yet started should try receiving the change notification a minimum number of times during the modification period. If the UE knows which MBSFN area(s) will be used to provide the MBMS session(s) it is interested in receiving, it has to meet this requirement only for the corresponding MCCH(s).

Figure 13.10 illustrates a change of MBMS control information, affecting both the BCCH and the MCCH. In this example, E-UTRAN starts to use the updated configuration at an SFN at which the BCCH and MCCH modification period boundaries coincide. The figure illustrates the change notifications provided by E-UTRAN in such a case. Initially, the change notification indicates only that MCCH-2, which has a modification period of 10.24 s, will change. Later, the notification indicates that both MCCH-1 (which has a modification period of 5.12 s) and MCCH-2 will change.

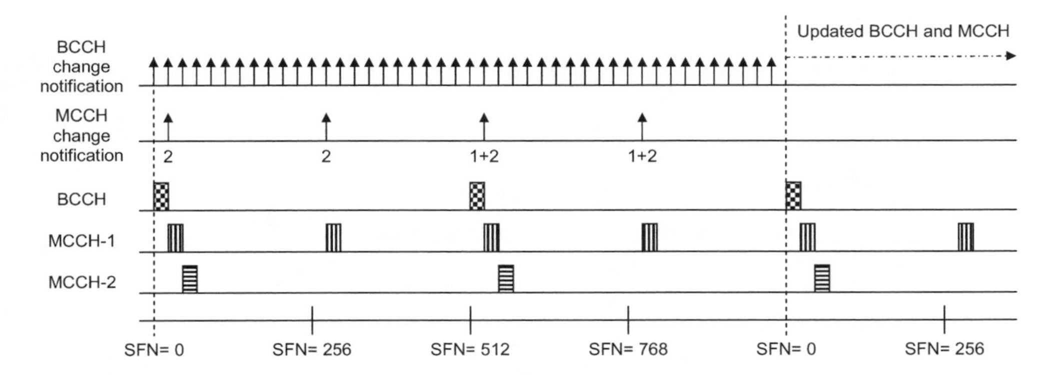

Figure 13.10: An example of a change of MBMS control information.

The change notification procedure is not used to inform UEs about modifications of ongoing sessions. UEs that are receiving an MBMS session are therefore required to acquire, every modification period, the MBMS control information of the MBSFN area used for that session.

13.6.3.3 Radio Resource Allocation

The network broadcasts in SIB13 a bitmap indicating to the UEs the set of subframes that is allocated to the MCCH. This bitmap covers a single radio frame. It is possible to configure a more robust Modulation and Coding Scheme (MCS) for the subframes that include MCCH.

Semi-static resource allocation signalling indicates which subframes are allocated to a particular MBSFN area; this is known as the Common Subframe Allocation (CSA).[10] The CSA is defined by up to eight patterns. Each pattern specifies the allocated subframes by a bitmap, that covers one or four radio frames, and repeats with a periodicity of up to 32 radio frames. The MCCH subframes are included in the CSA.

Next, the subframes assigned to the MBSFN area are allocated to each of the MCHs using the MBSFN area. Within a CSA period, the MCHs are time-multiplexed, with each MCH using one or more subframes which are consecutive within the set of subframes in the CSA; as the subframes are consecutive in this way, only the last allocated subframe is signalled for each MCH. This is known as the MCH Subframe Allocation (MSA).

The MBMS resource allocation is illustrated in Figure 13.11 for an example with two MBSFN areas:

- In even-numbered radio frames, subframes 3 and 6 are allocated, while in odd-numbered radio frames only subframe 6 is allocated;
- Within a period of 80 ms, the MBSFN areas are not interleaved – MBSFN area 1 takes the first 4.5 radio frames while area 2 takes the next 3.5;
- Both MBSFN areas use two MCHs, one with an MSP of 160 ms and one with an MSP of 320 ms;

[10]The term 'common' here indicates that the CSA subframes are shared by all the MCHs of the MBSFN area.

- Both MBSFN areas use a CSA period of 160 ms;
- Two subframes are allocated to MCCH.

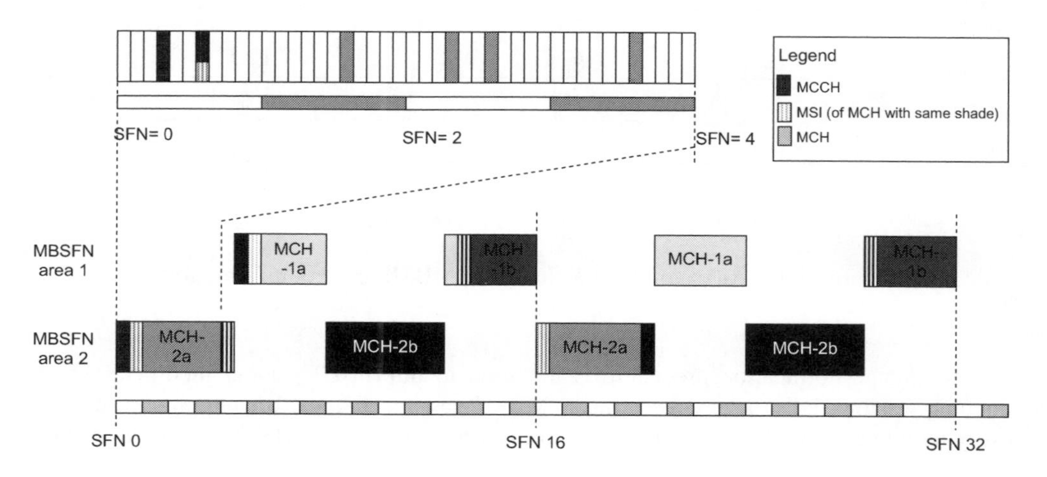

Figure 13.11: An example of MBMS resource allocation.

Every MSP, a UE that is receiving, for example, MCH-2b via MBSFN area 2 acquires the initial subframes to acquire MCCH (if present in this MSP), the MSI MAC control element (see Section 13.6.3.1) and subsequently the subframes dynamically allocated to the service in question.

13.6.4 Content Synchronization

As explained in Section 13.4, the basic property of MBSFN transmission is that all participating cells transmit exactly the same content in a synchronized manner so it appears as one transmission to the UE. Mechanisms are therefore provided to ensure synchronization of the MBMS content – i.e. to ensure that all participating eNodeBs include the same MBMS control information and data in the corresponding time-synchronized subframes.

For the MBMS control information, whenever the MCE updates the control information it indicates the modification period from which the updated control information applies by means of a parameter called 'MCCH update time'.

For the synchronization of MBMS user data, a specific protocol was designed: the SYNC protocol as specified in [8]. Each MBMS bearer operates a separate instance of the SYNC protocol, which uses a timestamp to indicate the time at which the eNodeB should start the transmission of a first packet belonging to a synchronization sequence. The timestamp is based on a common time reference and should be set taking into account transfer delays that may occur. All packets that the BM-SC allocates to a synchronization sequence are given the same timestamp. The MSP is the same as, or a multiple of, the synchronization sequence values applicable for the services carried by the MCH in question.

Before considering the details of the SYNC protocol, it is important to understand the overall architecture, which can be summarized as follows:

- The BM-SC performs traffic shaping: it discards packets as necessary to ensure that, for each synchronization sequence, the actual bit rate of each MBMS bearer does not exceed the guaranteed bit rate (see Section 2.4).The BM-SC also decides in which synchronization sequence each packet should be transmitted.[11]
- The MCE semi-statically configures which services are multiplexed together, as well as the radio resources allocated to the relevant MCH. The MCE also semi-statically configures the order in which the services are scheduled.
- The participating eNodeBs first buffer all the packets of an entire MSP. The eNodeBs then dynamically schedule the services, taking into account the order pre-defined by the MCE. If the BM-SC allocates more packets to an MSP than actually fit into the allocated radio resources, the eNodeBs discard some of the packets, starting with the last packets of the service scheduled last according to the pre-defined order (so-called tail-dropping). Finally the eNodeBs compile the MSI. The resulting MAC control element is transmitted at the start of the MSP.

Figure 13.12 illustrates the SYNC protocol by means of a typical sequence of the different SYNC Protocol Data Unit (PDU) types.

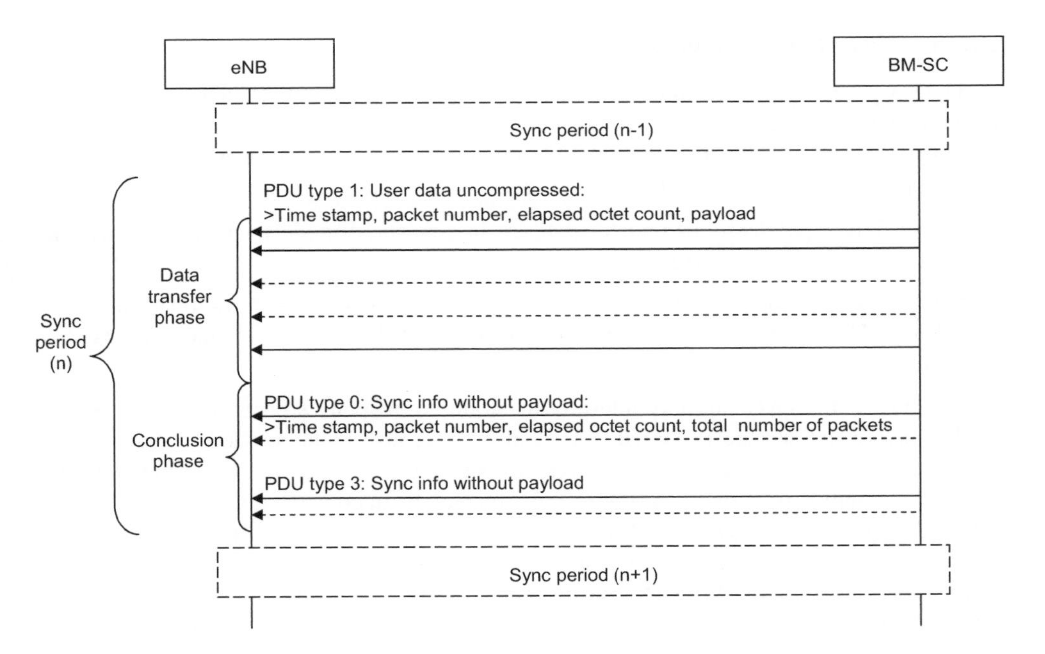

Figure 13.12: An example of a SYNC protocol PDU sequence.

[11] In a future release of LTE, it may become possible for the BM-SC to perform 'radio-aware traffic shaping' – i.e. knowing which services share a given radio resource and their respective bit rates, the BM-SC may handle temporary rate fluctuations by moving some packets to a previous or later scheduling MSP.

The BM-SC transfers the data allocated to one synchronization sequence by means of a number of PDUs of 'type 1'. These PDUs all include the same timestamp, a packet number that is reset at the start of the synchronization sequence and increases for every subsequent packet, and an elapsed octet counter that indicates the number of octets transferred since the start of the synchronization sequence.

After transferring all the user data, the BM-SC may provide additional information by means of a PDU 'type 0' or 'type 3'. These PDUs may be repeated to increase the likelihood the PDU is received correctly. PDU type 0 not only includes the regular timestamp, the packet number and the elapsed octet counter but also indicates the total number of packets and the total elapsed octets within the synchronization sequence. PDU type 3 provides the same information, but in addition provides the length of each packet in the synchronization sequence.

The eNodeB has a buffer corresponding to one MSP, in which it inserts the received packets in accordance with the pre-defined service order. If one or more non-consecutive packets are lost, the SYNC protocol provides the eNodeB with all the information required to continue filling the transmission buffer correctly. In such cases, the eNodeB only needs to mute the subframes affected by the lost packets (rather than muting all subsequent subframes until the end of the MSP). The eNodeB may be unable to do this if consecutive packets are lost (unless a PDU type 3 is received), as the size of the layer 2 headers depends on the precise construction of the transport blocks, which depends on the packet sizes.

13.6.5 Counting Procedure

A counting procedure was introduced in Release 10 that enables E-UTRAN to determine how many UEs are receiving, or are interested in receiving, an MBMS service via an MBSFN Radio Bearer (MRB) (i.e. using the MBSFN mode of operation). Compared to UMTS, the counting procedure is limited in the following respects:

- Only UEs of Release 10 or later that are in connected mode respond to a counting request from the network;
- The counting request does not include the session identity, with the result that UEs respond to a counting request even if they have already received the upcoming session (in case of session repetition).

Moreover, E-UTRAN can only interpret the counting responses properly when it knows whether or not the service is available via unicast, because if the service is available, all interested UEs should be in connected mode to receive the service, whereas if the service is not available, only a fraction of the interested UEs may be in connected mode.

Based on the counting results, E-UTRAN may decide to start or stop MBSFN transmission of a given MBMS service. Dynamic change of the MBSFN area (i.e. dynamic control of the cells that participate in the MBSFN transmission of a service) is not supported.

Figure 13.13 provides an overview of a typical message sequence for MBMS counting. The sequence shown is based on the following assumptions:

- Whenever a session of an MBMS service is scheduled to take place (e.g. according to the Electronic Programme Guide (EPG)), it is accessible via unicast transmission;
- Upon start of the session's schedule, the BM-SC allocates a Temporary Mobile Group Identifier (TMGI) and provides a session start indication to E-UTRAN;

- E-UTRAN waits a certain time (e.g. 10 s) before initiating the counting procedure; this delay allows UEs that are interested in receiving the service to enter connected mode to receive the service via unicast.

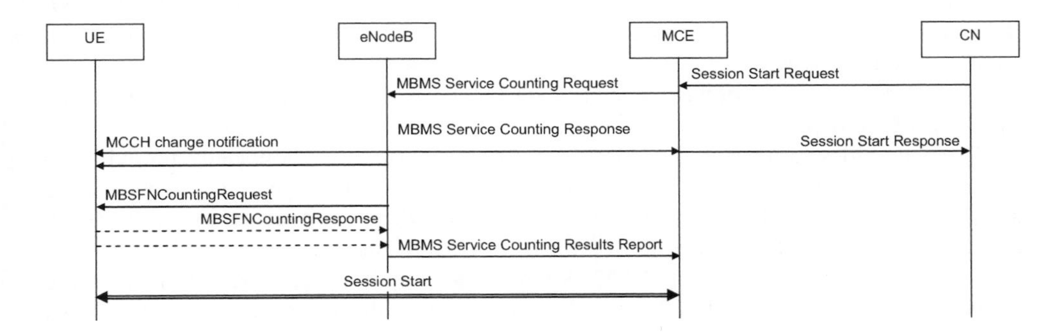

Figure 13.13: MBMS counting upon session start.

The counting request messages relate to a single MBSFN area and can cover up to 16 services. E-UTRAN can issue one 'MBMSCountingRequest' message per MBSFN area in a modification period, and UEs only respond once in a modification period. In the event of E-UTRAN repeating the same counting request (i.e. including the same services) in a subsequent modification period, the UEs respond again.

Counting may also be performed to deactivate the service in case, at a later point in time, the number of interested UEs has dropped. In such a case, the MBSFN transmission of the service may be suspended temporarily; if later the number of interested UEs increases again, the MBSFN transmission of the service may be resumed. This is illustrated in Figure 13.14.

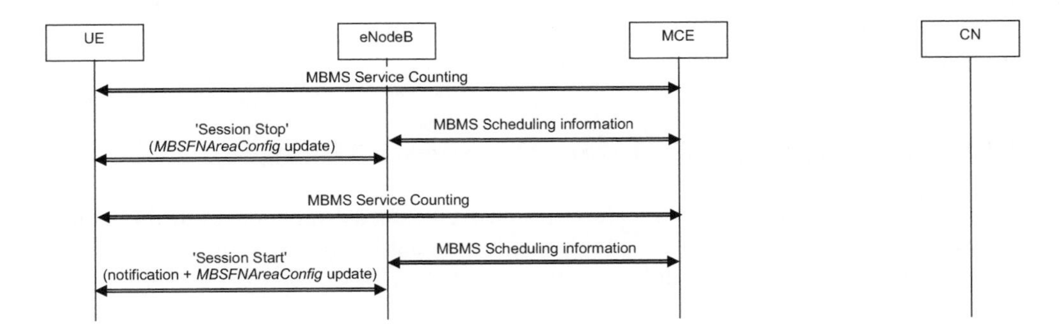

Figure 13.14: Suspension and resumption of MBSFN transmission based on counting.

It should be noted that normally an eNodeB does not initiate the MBMS change notification procedure without a session start; MBMS service resumption is an exception.

13.7 Public Warning Systems

Cellular networks are commonly used to broadcast public warnings and emergency information during events such as natural disasters (e.g. earthquakes, tsunamis or floods). There are two Public Warning System (PWS) variants: the Earthquake and Tsunami Warning System (ETWS), predominantly used in Japan, and the Commercial Mobile Alert Service (CMAS), mainly used in the Americas.

As MBMS was not part of the first release of LTE, it was agreed to support public warnings by means of the BCCH.

ETWS provides 'primary notifications', which are short notifications delivered within 4 seconds, as well as 'secondary notifications' that are less time-critical and provide more detailed information such as where to get help. Both the primary and secondary notifications, of which there may be multiple, are optional. In case of an emergency, E-UTRAN does not need to delay the scheduling of ETWS notifications until the next BCCH modification period. It notifies UEs (in both idle and connected modes) about the presence of an ETWS notification by means of a paging message including a field named 'etws-Indication'. Upon detecting this indicator, an ETWS-capable UE attempts to receive SIB10, which is used to transfer the ETWS primary notification. The UE also checks if SIB11 is scheduled, which is used to carry an ETWS secondary notification (or a segment of one). Upon receiving the primary and/or secondary ETWS notifications (depending on which is scheduled), the UE may stop acquisition of system information for ETWS.

CMAS only employs short text messages, for a single type of warning message. In an emergency situation, multiple CMAS alerts may need to be broadcast. However, E-UTRAN only schedules a subsequent CMAS alert after completely transmitting all segments of the previous CMAS warning message. In the same way as for ETWS, E-UTRAN need not delay the scheduling of CMAS warning messages until the next BCCH modification period, and paging is used for notification. Unlike ETWS, however, SIB12 is used to transfer a CMAS warning message (or a segment of one), and the UE should continue SIB12 acquisition upon receiving a complete CMAS warning message.

Both ETWS and CMAS incorporate security mechanisms, including means to verify the integrity of the warning message and to authenticate the message source. The use of these mechanisms is subject to regulatory requirements.

13.8 Comparison of Mobile Broadcast Modes

MBMS is not the only technology which is capable of efficient delivery of multimedia content to mobile consumers, and it is therefore instructive to consider briefly some of the relative strengths and weaknesses of the different general approaches.

13.8.1 Delivery by Cellular Networks

The key feature of using a cellular network for the content delivery is that the services can be deployed using a network operator's existing network infrastructure. Furthermore, it does not necessarily require the allocation or acquisition of additional spectrum beyond that to which the operator already has access. However, the advantage of reusing the cellular network and spectrum for multimedia broadcasting is also its main drawback: it reduces the bandwidth available for other mobile services.

It can be noted that mobile operators have been providing mobile television over UTRAN networks for a number of years already; this has however generally been based on point-to-point unicast connections; this is feasible in lightly loaded networks, but, as explained in Section 13.2, it does not make the most efficient use of the radio resources. MBMS transmission from a single source entity to multiple recipients on a single delivery channel allows radio resources to be shared, thus solving the capacity issues associated with the use of unicast transmissions for streaming of bandwidth-hungry services such as mobile television.

In general, the use of a cellular network for MBMS also has the advantage of bringing the capability to send 'personalized' content to groups of a few users, as well as the possibility of a built-in application-layer return channel, which is useful for the provision of interactive services. However, such functionality is not supported in current releases of LTE.

13.8.2 Delivery by Broadcast Networks

Mobile broadcast services may alternatively be provided by standalone broadcast systems such as Digital Video Broadcasting-Terrestrial (DVB-T) and Digital Audio Broadcasting (DAB). Originally developed for fixed receivers, specific mobile versions of these standards have been developed, namely Digital Video Broadcasting-Handheld (DVB-H) and Digital Mobile Broadcasting (DMB) respectively.

These systems typically assume the use of a relatively small number of higher-powered transmitters designed to cover a wide geographical area. For broadcast service provision, user capacity constraints do not require the use of small cells.

Such systems can therefore achieve relatively high data rates with excellent wide-area coverage. However, they clearly also require dedicated spectrum in which to operate, the construction and operation of a network of transmitters additional to that provided for the cellular network, and the presence of additional hardware in the mobile terminals.

Standalone broadcast systems are also, by definition, 'downlink-only': they do not provide a direct return channel for interactive services, although it may be possible to use the cellular network for this purpose.

13.8.3 Services and Applications

Both cellular delivery (MBMS) and standalone broadcast methods enable the delivery of a variety of multimedia services and applications, the most fashionable being mobile television. This gives network operators the opportunity to differentiate their service offerings.

In some ways, cellular delivery (MBMS) and standalone broadcast (e.g. DVB-H) systems can be seen as complementary. The mobile television industry and market are changing rapidly, and may be better described as part of the mobile entertainment industry, where new market models prevail. An example of this is the concept of the 'long tail of content' [9] which describes the consumption of products or services across a large population, as illustrated in Figure 13.15: given a large availability of choice, a few very popular products or services will dominate the market for the majority of consumers; a large number of other products or services will only find a niche market.

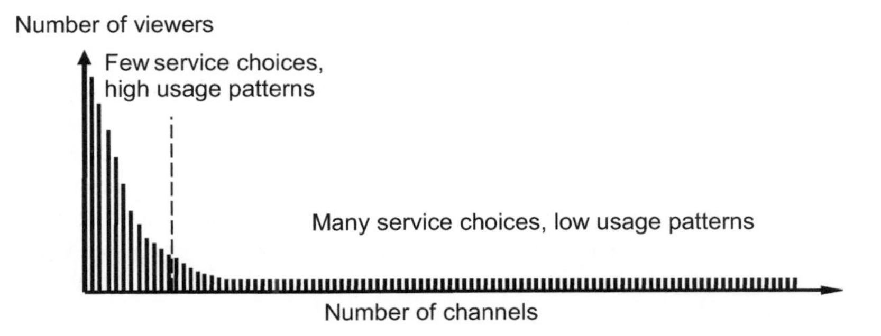

Figure 13.15: The 'Long Tail of Content'.

This is a very appropriate representation of the usage of broadcast television, whether fixed or mobile: viewing figures for a few main channels (typically national broadcasters) are high, while usage is low for a profusion of specialist channels that gather a limited yet dedicated audience.

The deployment of mobile broadcast technology worldwide (whether by cellular MBMS or standalone broadcast) is very much dependent on the availability of spectrum and on the offer of services in each country, while infrastructure costs and content agreements between network operators and broadcasters are also significant factors.

What matters ultimately to the user is the quality of reception on their terminal, and mobility. Deployment of MBMS in LTE networks can be a highly efficient solution, especially by exploiting the use of Single Frequency Network operation based on OFDM – which is also the transmission method used by DVB-H and DMB.

References[12]

[1] 3GPP Technical Specification 23.246, 'Multimedia Broadcast/Multicast Service (MBMS); Architecture and Functional Description (Release 6)', www.3gpp.org.

[2] 3GPP Technical Specification 23.228, 'IP Multimedia Subsystem (IMS); Stage 2 (Release 8)', www.3gpp.org.

[3] Internet Engineering Task Force, 'RFC3261 Session Initiation Protocol', www.ietf.org.

[4] Internet Engineering Task Force, 'RFC4960 Stream Control Transmission Protocol', www.ietf.org.

[5] 3GPP Technical Specification 29.060, 'General Packet Radio Service (GPRS); GPRS Tunnelling Protocol (GTP) across the Gn and Gp interface (Release 8)', www.3gpp.org.

[6] IEEE1588, 'IEEE 1588 Standard for a Precision Clock Synchronization Protocol for Networked Measurement and Control Systems', http://ieee1588.nist.gov.

[7] Motorola, 'R1-070051: Performance of MBMS Transmission Configurations', www.3gpp.org, 3GPP TSG RAN WG1, meeting 47bis, Sorrento, Italy, January 2007.

[8] 3GPP Technical Specification 25.446, 'MBMS synchronisation protocol (SYNC)', www.3gpp.org.

[9] D. Anderson, *The Long Tail: Why the Future of Business is Selling Less of More*. New York, USA: Hyperion, 2006.

[12] All web sites confirmed 1st March 2011.

Part III

Physical Layer for Uplink

14

Uplink Physical Layer Design

Robert Love and Vijay Nangia

14.1 Introduction

While many of the requirements for the design of the LTE uplink physical layer and multiple-access scheme are similar to those of the downlink, the uplink also poses some unique challenges. Some of the desirable attributes for the LTE uplink include:

- Orthogonal uplink transmission by different User Equipment (UEs), to minimize intra-cell interference and maximize capacity;
- Flexibility to support a wide range of data rates, and to enable the data rate to be adapted to the Signal-to-Interference-plus-Noise Ratio (SINR);
- Sufficiently low Peak-to-Average Power Ratio (PAPR) (or Cubic Metric (CM) – see Section 21.3.3) of the transmitted waveform, to avoid excessive cost, size and power consumption of the UE Power Amplifier (PA);
- Ability to exploit the frequency diversity afforded by the wideband channel (up to 20 MHz), even when transmitting at low data rates;
- Support for frequency-selective scheduling;
- Support for advanced multiple-antenna techniques, to exploit spatial diversity and enhance uplink capacity.

The multiple-access scheme selected for the LTE uplink so as to fulfil these principle characteristics is Single Carrier-Frequency Division Multiple Access (SC-FDMA).

A major advantage of SC-FDMA over the Direct-Sequence Code Division Multiple Access (DS-CDMA) scheme used in UMTS is that it achieves intra-cell orthogonality even in frequency-selective channels. SC-FDMA avoids the high level of intra-cell interference associated with DS-CDMA which significantly reduces system capacity and limits the use

LTE – The UMTS Long Term Evolution: From Theory to Practice, Second Edition.
Stefania Sesia, Issam Toufik and Matthew Baker.

of adaptive modulation. A code-multiplexed uplink also suffers the drawback of an increased CM/PAPR if multi-code transmission is used from a single UE.

The use of OFDMA (Orthogonal Frequency Division Multiple Access) for the LTE uplink would have been attractive due to the possibility for full uplink-downlink commonality. In principle, an OFDMA scheme similar to the LTE downlink could satisfy all the uplink design criteria listed above, except for low CM/PAPR. As outlined in Section 5.2.2.1, much research has been conducted in recent years on methods to reduce the CM/PAPR of OFDM; however, in general the effectiveness of these methods requires careful practical evaluation against their associated complexity and/or overhead (for example, in terms of additional signalling or additional transmission bandwidth used to achieve the CM/PAPR reduction).

SC-FDMA combines the desirable characteristics of OFDM outlined in Section 5.2 with the low CM/PAPR of single-carrier transmission schemes.

Like OFDM, SC-FDMA divides the transmission bandwidth into multiple parallel subcarriers, with the orthogonality between the subcarriers being maintained in frequency-selective channels by the use of a Cyclic Prefix (CP) or guard period. The use of a CP prevents Inter-Symbol Interference (ISI) between SC-FDMA information blocks. It transforms the linear convolution of the multipath channel into a circular convolution, enabling the receiver to equalize the channel simply by scaling each subcarrier by a complex gain factor as explained in Chapter 5.

However, unlike OFDM, where the data symbols directly modulate each subcarrier independently (such that the amplitude of each subcarrier at a given time instant is set by the constellation points of the digital modulation scheme), in SC-FDMA the signal modulated onto a given subcarrier is a linear combination of all the data symbols transmitted at the same time instant. Thus in each symbol period, all the transmitted subcarriers of an SC-FDMA signal carry a component of each modulated data symbol. This gives SC-FDMA its crucial single-carrier property, which results in the CM/PAPR being significantly lower than pure multicarrier transmission schemes such as OFDM.

14.2 SC-FDMA Principles

14.2.1 SC-FDMA Transmission Structure

An SC-FDMA signal can, in theory, be generated in either the time domain or the frequency domain [1]. Although the two techniques are duals and 'functionally' equivalent, in practice, the time-domain generation is less bandwidth-efficient due to time-domain filtering and associated requirements for filter ramp-up and ramp-down times [2, 3]. Nevertheless, we describe both approaches here to facilitate understanding of the principles of SC-FDMA in both domains.

14.2.2 Time-Domain Signal Generation

Time-domain generation of an SC-FDMA signal is shown in Figure 14.1.

The input bit stream is mapped into a single-carrier stream of QPSK or QAM[1] symbols, which are grouped into symbol-blocks of length M. This may be followed by an optional repetition stage, in which each block is repeated L times, and a user-specific frequency shift,

[1]QPSK: Quadrature Phase Shift Keying; QAM: Quadrature Amplitude Modulation.

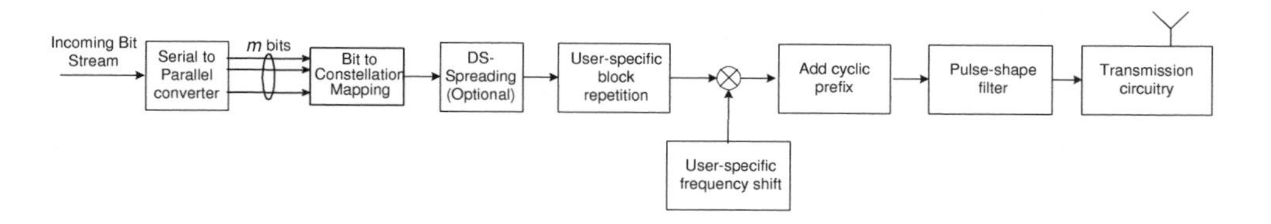

Figure 14.1: SC-FDMA time-domain transmit processing.

by which each user's transmission may be translated to a particular part of the available bandwidth. A CP is then inserted. After filtering (e.g. with a Root-Raised Cosine (RRC) pulse-shaping filter), the resulting signal is transmitted.

The repetition of the symbol blocks results in the spectrum of the transmitted signal only being non-zero at certain subcarrier frequencies, i.e. every L^{th} subcarrier as shown in the example in Figure 14.2.

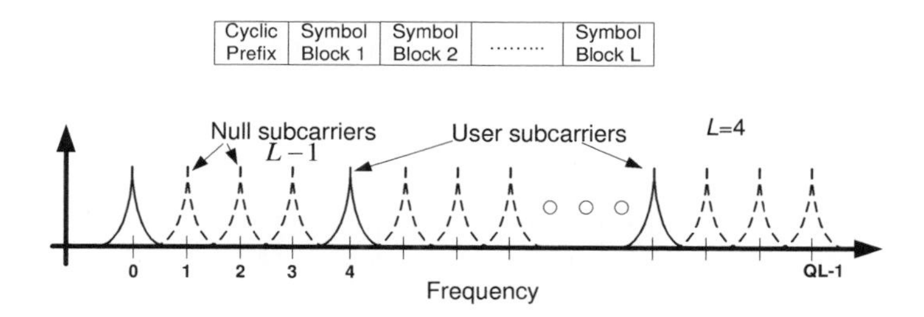

Figure 14.2: Distributed transmission with equal spacing between occupied subcarriers.

Since such a signal occupies only one in every L subcarriers, the transmission is said to be 'distributed' and is one way of providing a frequency-diversity gain.

When no symbol-block repetition is performed ($L = 1$), the signal occupies consecutive subcarriers[2] and the transmission is said to be 'localized'. Localized transmissions are beneficial for supporting frequency-selective scheduling or inter-cell interference coordination (as explained in Section 12.5). Localized transmission may also provide frequency diversity if the set of consecutive subcarriers is hopped in the frequency domain.

Different users' transmissions, using different repetition factors or bandwidths, remain orthogonal on the uplink when the following conditions are met:

- The users occupy different sets of subcarriers. This may in general be accomplished either by introducing a user-specific frequency shift (for localized transmissions) or alternatively by arranging for different users to occupy interleaved sets of subcarriers

[2]The occupied bandwidth then depends on the symbol rate.

(for distributed transmissions). The latter method is known in the literature as Interleaved Frequency Division Multiple Access (IFDMA) [4].

- The received signals are properly synchronized in time and frequency.
- The CP is longer than the sum of the delay spread of the channel and any residual timing synchronization error between the users.

14.2.3 Frequency-Domain Signal Generation (DFT-S-OFDM)

Generation of an SC-FDMA signal in the frequency domain uses a Discrete Fourier Transform-Spread-OFDM (DFT-S-OFDM) structure [5–7] as shown in Figure 14.3.

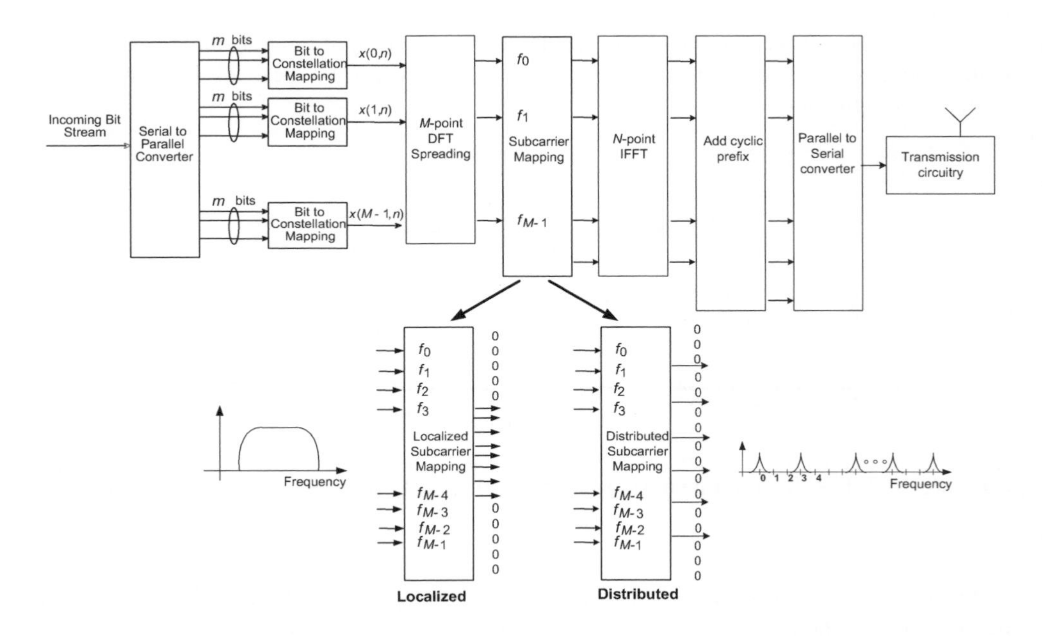

Figure 14.3: SC-FDMA frequency-domain transmit processing (DFT-S-OFDM) showing localized and distributed subcarrier mappings.

The first step of DFT-S-OFDM SC-FDMA signal generation is to perform an M-point DFT operation on each block of M QAM data symbols. Zeros are then inserted among the outputs of the DFT in order to match the DFT size to an N-subcarrier OFDM modulator (typically an Inverse Fast Fourier Transform (IFFT)). The zero-padded DFT output is mapped to the N subcarriers, with the positions of the zeros determining to which subcarriers the DFT-precoded data is mapped.

Usually N is larger than the maximum number of occupied subcarriers, thus providing for efficient oversampling and 'sinc' ($\sin(x)/x$) pulse-shaping. The equivalence of DFT-S-OFDM and a time-domain-generated SC-FDMA transmission can readily be seen by considering the case of $M = N$, where the DFT operation cancels the IFFT of the OFDM modulator resulting in the data symbols being transmitted serially in the time domain. However, this simplistic construction would not provide any oversampling or pulse-shape filtering.

As with the time-domain approach, DFT-S-OFDM is capable of generating both localized and distributed transmissions:

- **Localized transmission.** The subcarrier mapping allocates a group of M adjacent subcarriers to a user. $M < N$ results in zeros being appended to the output of the DFT spreader resulting in an upsampled/interpolated version of the original M QAM data symbols at the IFFT output of the OFDM modulator. The transmitted signal is thus similar to a narrowband single carrier with a CP (equivalent to time-domain generation with repetition factor $L = 1$) and 'sinc' pulse-shaping filtering.
- **Distributed transmission.** The subcarrier mapping allocates M equally spaced subcarriers (e.g. every L^{th} subcarrier). $(L - 1)$ zeros are inserted between the M DFT outputs, and additional zeros are appended to either side of the DFT output prior to the IFFT $(ML < N)$. As with the localized case, the zeros appended on either side of the DFT output provide upsampling or 'sinc' interpolation, while the zeros inserted between the DFT outputs produce waveform repetition in the time domain. This results in a transmitted signal similar to time-domain IFDMA with repetition factor L and 'sinc' pulse-shaping filtering.

Like the time-domain SC-FDMA signal generation (in Section 14.2.2), orthogonality between different users with different data rate requirements can be achieved by assigning each user a unique set of subcarriers. The CP structure is the same as for the time-domain signal generation, and therefore the same efficient FDE techniques can be employed at the receiver [8, 9].

It is worth noting that, in principle, any unitary matrix can be used instead of the DFT for the spreading operation with similar performance [10]. However, the use of non-DFT spreading would result in increased CM/PAPR since the transmitted signal would no longer have the single carrier characteristic.

14.3 SC-FDMA Design in LTE

Having outlined the key principles of SC-FDMA transmission, we now explain the application of SC-FDMA to the LTE uplink.

14.3.1 Transmit Processing for LTE

Although the frequency-domain generation of SC-FDMA (DFT-S-OFDM) is functionally equivalent to the time-domain SC-FDMA signal generation, each technique requires a slightly different parameterization for efficient signal generation [3]. The pulse-shaping filter used in the time domain SC-FDMA generation approach in practice has a non-zero excess bandwidth, resulting in bandwidth efficiency which is smaller than that achievable with the frequency-domain method with its inherent 'sinc' (zero excess bandwidth) pulse-shaping filter which arises from the zero padding and IFFT operation. The non-zero excess bandwidth pulse-shaping filter in the time-domain generation also requires ramp-up and ramp-down times of 3–4 samples duration, while for DFT-S-OFDM there is no explicit pulse-shaping filter, resulting in a much shorter ramp time similar to OFDM. However, the pulse-shaping filter in the time-domain generation does provide the benefit of reduced CM by approximately 0.25–0.5 dB compared to DFT-S-OFDM, as shown in Figure 14.4. Thus there is a trade-off between bandwidth efficiency and CM/PAPR reduction between the time- and frequency-domain SC-FDMA generation methods.

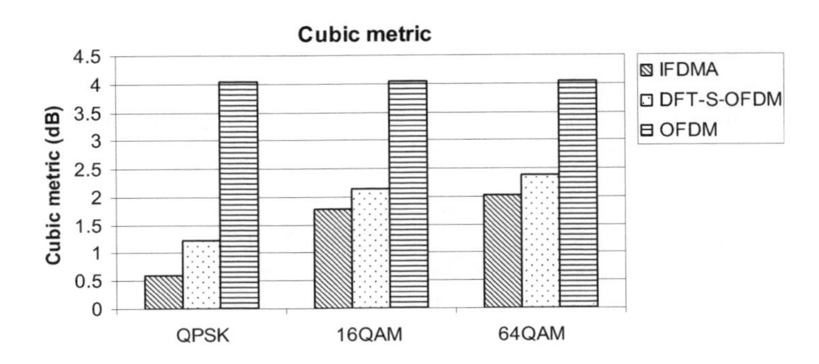

Figure 14.4: Cubic metric comparison of time-domain SC-FDMA generation (IFDMA), frequency-domain SC-FDMA generation (DFT-S-OFDM) and OFDM.

Frequency-domain signal generation for the LTE uplink has a further benefit in that it allows a very similar parameterization to be adopted as for the OFDM downlink, including the same subcarrier spacing, number of occupied subcarriers in a given bandwidth, and CP lengths. This provides maximal commonality between uplink and downlink, including for example the same clock frequency.

For these reasons, the SC-FDMA parameters chosen for the LTE uplink have been optimized under the assumption of frequency-domain DFT-S-OFDM signal generation.

An important feature of the LTE SC-FDMA parameterization is that the numbers of subcarriers which can be allocated to a UE for transmission are restricted such that the DFT size in LTE can be constructed from multiples of 2, 3 and/or 5. This enables efficient, low-complexity mixed-radix FFT implementations.

14.3.2 SC-FDMA Parameters for LTE

The same basic transmission resource structure is used for the uplink as for the downlink: a 10 ms radio frame is divided into ten 1 ms subframes each consisting of two 0.5 ms slots. As LTE SC-FDMA is based on the same fundamental processing as OFDM, it uses the same 15 kHz subcarrier spacing as the downlink. The uplink transmission resources are also defined in the frequency domain (i.e. before the IFFT), with the smallest unit of resource being a Resource Element (RE), consisting of one SC-FDMA data block length on one subcarrier. As in the downlink, a Resource Block (RB) comprises 12 REs in the frequency domain for a duration of 1 slot, as detailed in Section 6.2. The LTE uplink SC-FDMA physical layer parameters for Frequency Division Duplex (FDD) and Time Division Duplex (TDD) deployments are detailed in Table 14.1.

Table 14.1: LTE uplink SC-FDMA physical layer parameters.

Parameter	Value	Comments
Subframe duration	1 ms	
Slot duration	0.5 ms	
Subcarrier spacing	15 kHz	
SC-FDMA symbol duration	66.67 μs	
CP duration	Normal CP:	5.2 μs first symbol in each slot, 4.69 μs all other symbols
	Extended CP:	16.67 μs all symbols
Number of symbols per slot	Normal CP: 7 Extended CP: 6	
Number of subcarriers per RB	12	

Two CP durations are supported – a normal CP of duration 4.69 μs and an extended CP of 16.67 μs, as in the downlink (see Section 5.4.1). The extended CP is beneficial for deployments with large channel delay-spread characteristics, and for large cells.

The 1 ms subframe allows a 1 ms scheduling interval (or Transmission Time Interval (TTI)), as for the downlink, to enable low latency. However, one difference from the downlink is that the uplink coverage is more likely to be limited by the maximum transmission power of the UE. In some situations, this may mean that a single Voice-over-IP (VoIP) packet, for example, cannot be transmitted in a 1 ms subframe with an acceptable error rate. One solution to this is to segment the VoIP packet at higher layers to allow it to be transmitted over several subframes. However, such segmentation results in additional signalling overhead for each segment (including resource allocation signalling and Hybrid ARQ acknowledgement signalling). A more efficient technique for improving uplink VoIP coverage at the cell edge is to use the so-called *TTI bundling*, where a single transport block from the MAC layer is transmitted repeatedly in multiple consecutive subframes, with only one set of signalling messages for the whole transmission. The LTE uplink allows groups of 4 TTIs to be 'bundled' in this way, in addition to the normal 1 ms TTI.

In practice in LTE Releases 8 and 9, all the uplink data transmissions are localized, using contiguous blocks of subcarriers. This simplifies the transmission scheme. Frequency-diversity can still be exploited by means of frequency hopping, which can occur both within one subframe (at the boundary between the two slots) and between subframes. In the case of frequency hopping within a subframe, the channel coding spans the two transmission frequencies, and therefore the frequency-diversity gain is maximized through the channel decoding process. The only instance of distributed transmission in the LTE uplink (using an IFDMA-like structure) is for the Sounding Reference Signals (SRSs) which are transmitted to enable the eNodeB to perform uplink frequency-selective scheduling; these are discussed in Section 15.6.

Like the downlink, the LTE uplink supports scalable system bandwidths from approximately 1.4 MHz up to 20 MHz with the same subcarrier spacing and symbol duration for all bandwidths. The uplink scaling for the bandwidths supported in LTE is shown in

Table 14.2: LTE Uplink SC-FDMA parametrization for selected carrier bandwidths.

	Carrier bandwidth (MHz)					
	1.4	3	5	10	15	20
FFT size	128	256	512	1024	1536	2048
Sampling rate: M/N × 3.84 MHz	1/2	1/1	2/1	4/1	6/1	8/1
Number of subcarriers	72	180	300	600	900	1200
Number of RBs	6	15	25	50	75	100
Bandwidth efficiency (%)	77.1	90	90	90	90	90

Table 14.2. Note that the sampling rates resulting from the indicated FFT sizes are designed to be small rational multiples of the UMTS 3.84 MHz chip rate, for ease of implementation in a multimode UE.

Note that in the OFDM downlink parameter specification, the d.c. subcarrier is unused in order to support direct conversion (zero IF[3]) architectures. In contrast, no unused d.c. subcarrier is possible for SC-FDMA (as shown in Table 14.2) as it can affect the low CM/PAPR property of the transmit signal.

14.3.3 d.c. Subcarrier in SC-FDMA

Direct conversion transmitters and receivers can introduce distortion at the carrier frequency (zero frequency or d.c. in baseband), for example arising from local oscillator leakage [11]. In the LTE downlink, this is addressed by the inclusion of an unused d.c. subcarrier. However, for the uplink, the same solution would cause significant degradation to the low CM property of the transmitted signal. In order to minimize the impact of such distortion on the packet error rate and the CM/PAPR, the d.c. subcarrier of the SC-FDMA signal is modulated in the same way as all the other subcarriers but the subcarriers are all frequency-shifted by half a subcarrier spacing (±7.5 kHz), resulting in an offset of 7.5 kHz relative to d.c. Thus two subcarriers straddle the d.c. location, and hence the amount of distortion affecting any individual RB is halved. This is illustrated in Figure 14.5 for deployments with even and odd numbers of RBs across the system bandwidth.

14.3.4 Pulse Shaping

As explained in Sections 14.2.3 and 14.3.1, one of the benefits of frequency-domain processing for SC-FDMA is that, in principle, there is no need for explicit pulse-shaping thanks to the implicit 'sinc' pulse-shaping. Nevertheless, an additional explicit pulse-shaping filter can further reduce the CM/PAPR, but at the expense of spectral efficiency (due to the resulting non-zero excess filter bandwidth similar to that of the time domain SC-FDMA signal generation described in Section 14.2.2).

The use of pulse-shaping filters, such as RRC[4] [12] or Kaiser window [13], was considered for LTE in order to reduce CM/PAPR, especially for lower-order modulations such as QPSK

[3]Intermediate Frequency.

[4]Roor-Raised-Cosine.

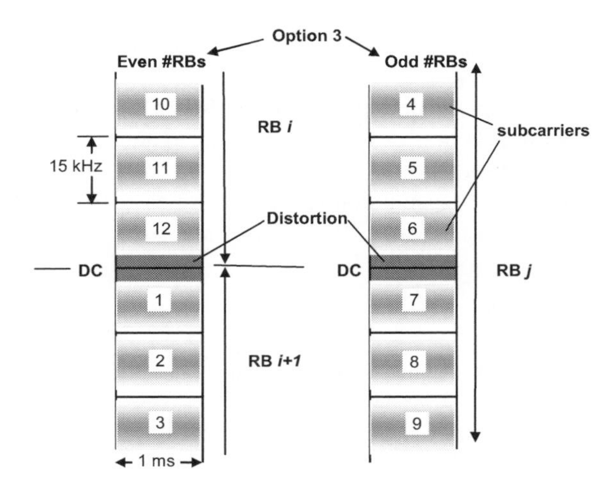

Figure 14.5: 7.5 kHz subcarrier shift for LTE system bandwidths with even or odd numbers of RBs.

to enhance uplink coverage. However, the bandwidth expansion caused by pulse-shaping filtering would require the code rate to be increased for a given data rate, and, in order to provide a net performance benefit, power boosting would be required with the size of the boost being approximately equal to the CM reduction achieved by the pulse-shaping.

Simulation results show that the CM benefits obtained from the use of pulse shaping (RRC or Kaiser window) are small. Moreover, pulse-shaping could not be applied to reference symbols, in order to prevent degradation of the channel estimation.

Finally, as discussed in Chapter 21, CM is not a problem for UEs operating at a maximum power level for resource allocations of up to 10 RBs. This means that there is no benefit possible from pulse-shaping given that UEs at the cell edge would often have less than 10 RBs allocated.

Therefore, no pulse-shaping is used in LTE.

14.4 Summary

The important properties of the SC-FDMA transmission scheme used for the LTE uplink are derived from its multicarrier OFDM-like structure with single-carrier characteristic. The multicarrier-based structure gives the LTE uplink the same robustness against ISI as the LTE downlink, with low-complexity frequency-domain equalization being facilitated by the CP. At the same time, the DFT-based precoding ensures that the LTE uplink possesses the low CM required for efficient UE design. Crucially, the LTE uplink is designed to be orthogonal in the frequency domain between different UEs, thus virtually eliminating the intra-cell interference associated with CDMA.

The parameters of the LTE uplink are designed to ensure maximum commonality with the downlink, and to facilitate frequency-domain DFT-S-OFDM signal generation.

The localized resource allocation scheme of the LTE uplink allows both frequency-selective scheduling and the exploitation of frequency diversity, the latter being achieved by means of frequency hopping.

The following chapters explain the application of SC-FDMA in LTE in more detail, showing how reference signals for channel estimation, data transmissions and control signalling are multiplexed together.

Enhancements to the uplink transmission scheme in Release 10 for LTE-Advanced are described in Section 28.3.6.

References[5]

[1] Motorola, 'R1-050397: EUTRA Uplink Numerology and Frame Structure', www.3gpp.org, 3GPP TSG RAN WG1, meeting 41, Athens, Greece, May 2005.

[2] Motorola, 'R1-050584: EUTRA Uplink Numerology and Design', www.3gpp.org, 3GPP TSG RAN WG1, meeting 41bis, Sophia Antipolis, France, June 2005.

[3] Motorola, 'R1-050971: Single Carrier Uplink Options for E-UTRA', www.3gpp.org, 3GPP TSG RAN WG1, meeting 42, London, UK, August 2005.

[4] U. Sorger, I. De Broeck and M. Schnell, 'Interleaved FDMA – A New Spread-Spectrum Multiple-Access Scheme', in *Proc. IEEE International Conference on Communications*, pp. 1013–1017, 1998.

[5] K. Bruninghaus and H. Rohling, 'Multi-carrier Spread Spectrum and its Relationship to Single-carrier Transmission', in *Proc. IEEE Vehicular Technology Conference*, May 1998.

[6] D. Galda and H. Rohling, 'A Low Complexity Transmitter Structure for OFDM-FDMA Uplink Systems', in *Proc. IEEE Vehicular Technology Conference*, May 2002.

[7] R. Dinis, D. Falconer, C. T. Lam and M. Sabbaghian, 'A Multiple Access Scheme for the Uplink of Broadband Wireless Systems', in *Proc. IEEE Global Telecommunications Conference*, November 2004.

[8] H. Sari, G. Karam and I. Jeanclaude, 'Frequency-Domain Equalization of Mobile Radio and Terrestrial Broadcast Channels', in *Proc. IEEE Global Telecommunications Conference*, November 1994.

[9] D. Falconer, S. L. Ariyavisitakul, A. Benyamin-Seeyar and B. Eidson, 'Frequency Domain Equalization for Single-carrier Broadband Wireless Systems'. *IEEE Communications Magazine*, Vol. 40, pp. 58–66, April 2002.

[10] V. Nangia and K. L. Baum, 'Experimental Broadband OFDM System – Field Results for OFDM and OFDM with Frequency Domain Spreading', in *Proc. IEEE International Conference on Vehicular Technology*, Vancouver, BC, Canada, September 2002.

[11] Motorola, 'R1-062061: Uplink DC Subcarrier Distortion Considerations in LTE', www.3gpp.org, 3GPP TSG RAN WG1, meeting 46, Tallinn, Estonia, August 2006.

[12] NTT DoCoMo, NEC, and SHARP, 'R1-050702: DFT-spread OFDM with Pulse Shaping Filter in Frequency Domain in Evolved UTRA Uplink', www.3gpp.org, 3GPP TSG RAN WG1, meeting 42, London, UK, August 2005.

[13] Huawei, 'R1-051434: Optimum family of spectrum-shaping functions for PAPR reduction in SC-FDMA', www.3gpp.org, 3GPP TSG RAN WG1, meeting 43, Seoul, Korea, November 2005.

[5] All web sites confirmed 1st March 2011.

15

Uplink Reference Signals

Robert Love and Vijay Nangia

15.1 Introduction

As in the downlink, the LTE Single-Carrier Frequency Division Multiple Access (SC-FDMA) uplink incorporates Reference Signals (RSs) for data demodulation and channel sounding. In this chapter, the design principles behind these RSs are explained, including in particular features related to interference randomization and coordination, and the flexible configuration of channel sounding.

The roles of the uplink RSs include enabling channel estimation to aid coherent demodulation, channel quality estimation for uplink scheduling, power control, timing estimation and direction-of-arrival estimation to support downlink beamforming. Two types of RS are supported on the uplink:

- **DeModulation RS** (DM-RS), associated with transmissions of uplink data on the Physical Uplink Shared CHannel (PUSCH) and/or control signalling on the Physical Uplink Control CHannel (PUCCH).[1] These RSs are primarily used for channel estimation for coherent demodulation.
- **Sounding RS** (SRS), not associated with uplink data and/or control transmissions, and primarily used for channel quality determination to enable frequency-selective scheduling on the uplink.

The uplink RSs are time-multiplexed with the data symbols. The DM-RSs of a given UE[2] occupy the same bandwidth (i.e. the same Resource Blocks (RBs)) as its PUSCH/PUCCH data transmission. Thus, the allocation of orthogonal (in frequency) sets of RBs to different

[1] See Chapter 16 for details of the uplink physical channels.
[2] User Equipment.

LTE – The UMTS Long Term Evolution: From Theory to Practice, Second Edition.
Stefania Sesia, Issam Toufik and Matthew Baker.

UEs for data transmission automatically ensures that their DM-RSs are also orthogonal to each other.

If configured by higher-layer signalling, SRSs are transmitted on the last SC-FDMA symbol in a subframe; SRS can occupy a bandwidth different from that used for data transmission, in order to allow wider bandwidth sounding. UEs transmitting SRSs in the same subframe can be multiplexed via either Frequency or Code Division Multiplexing (FDM or CDM respectively), as explained in Section 15.6.

Desirable characteristics for the uplink RSs include:

- Constant amplitude in the frequency domain for equal excitation of all the allocated subcarriers for unbiased channel estimates;
- Low Cubic Metric (CM) in the time domain (at worst no higher than that of the data transmissions; a lower CM for RSs than the data can be beneficial in enabling the transmission power of the RSs to be boosted at the cell-edge);
- Good autocorrelation properties for accurate channel estimation;
- Good cross-correlation properties between different RSs to reduce interference from RSs transmitted on the same resources in other (or, in some cases, the same) cells.

The following sections explain how these characteristics are achieved in LTE.

15.2 RS Signal Sequence Generation

The uplink RSs in LTE are mostly based on Zadoff–Chu (ZC) sequences [1, 2]. The fundamental structure and properties of these sequences are described in Section 7.2.1.

These sequences satisfy the desirable properties for RSs mentioned above, exhibiting 0 dB CM, ideal cyclic autocorrelation and optimal cross-correlation. The cross-correlation property results in the impact of an interfering signal being spread evenly in the time domain after time-domain correlation of the received signal with the desired sequence; this results in more reliable detection of the significant channel taps. However, in practice the CM of a ZC sequence is degraded at the Nyquist sampling rate because of the presence of unused guard subcarriers at each end of the sequence (due to the number of occupied RS subcarriers being less than the IFFT[3] size of the OFDM[4] modulator) and results in the ZC sequence effectively being oversampled in the time domain.

The RS sequence length, N_p, is equal to the number of assigned subcarriers $M_{\mathrm{sc}}^{\mathrm{RS}}$, which is a multiple of the number of subcarriers per RB, $N_{\mathrm{sc}}^{\mathrm{RB}} = 12$, i.e.

$$N_p = M_{\mathrm{sc}}^{\mathrm{RS}} = m \cdot N_{\mathrm{sc}}^{\mathrm{RB}}, \quad 1 \leq m \leq N_{\mathrm{RB}}^{\mathrm{UL}} \tag{15.1}$$

where $N_{\mathrm{RB}}^{\mathrm{UL}}$ is the uplink system bandwidth in terms of RBs.

The length-N_p RS sequence is directly applied (without Discrete Fourier Transform (DFT) spreading) to N_p RS subcarriers at the input of the IFFT as shown in Figure 15.1 [3]. Recall that a ZC sequence of odd-length N_{ZC} is given by

$$a_q(n) = \exp\left[-j2\pi q \frac{n(n+1)/2 + ln}{N_{\mathrm{ZC}}}\right] \tag{15.2}$$

[3] Inverse Fast Fourier Transform

[4] Orthogonal Frequency Division Multiplexing

where $q = 1, \ldots, N_{ZC} - 1$ is the ZC sequence index (also known as the root index), $n = 0, 1, \ldots, N_{ZC} - 1$, and $l = 0$ in LTE.

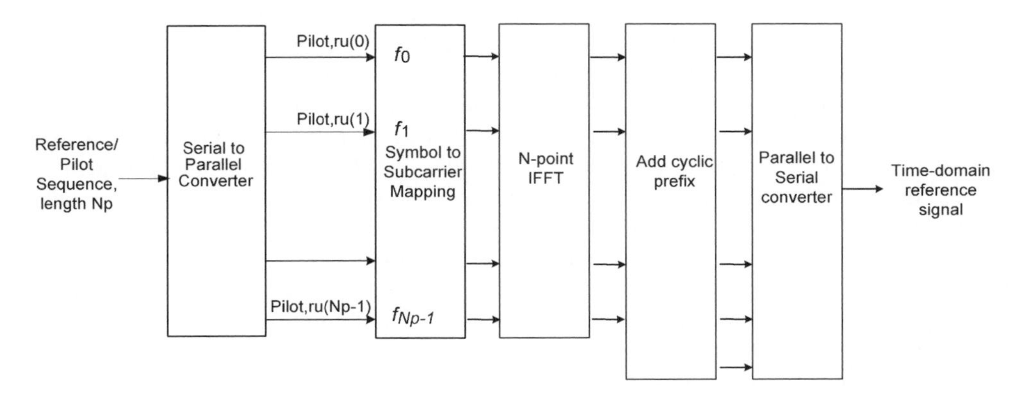

Figure 15.1: Transmitter structure for SC-FDMA reference signals. Note that no DFT spreading is applied to the RS sequence.

In LTE, N_{ZC} is selected to be the largest prime number smaller than or equal to N_p. The ZC sequence of length N_{ZC} is then cyclically extended to the target length N_p as follows:

$$\bar{r}_q(n) = a_q(n \bmod N_{ZC}), \quad n = 0, 1, \ldots, N_p - 1 \tag{15.3}$$

The cyclic extension in the frequency domain preserves the constant amplitude property. Cyclic extension of the ZC sequences is used rather than truncation, as in general it provides better CM characteristics. For sequence lengths of three or more RBs, this provides at least 30 sequences with CM smaller than or close to that of QPSK.

However, for the shortest sequence lengths, suitable for resource allocations of just one or two RBs, only a small number of low-CM extended ZC sequences is available (six and 12 sequences respectively with CM less than that of QPSK). Therefore, in order to provide at least as large a number of sequences as for the 3 RB case, 30 special RS sequences are defined in LTE for resource allocations of one or two RBs. These special sequences are QPSK rather than ZC-based sequences and were obtained from computer searches so as to have constant modulus in the frequency-domain, low CM, low memory and complexity requirements and good cross-correlation properties.

The QPSK RS sequences are given by,

$$\bar{r}(n) = e^{j\varphi(n)\pi/4}, \quad n = 0, 1, \ldots, M_{sc}^{RS} - 1, \tag{15.4}$$

where M_{sc}^{RS} is the number of subcarriers to which the sequence is mapped, and the values of $\varphi(n)$ can be found in [4, Tables 5.5.1.2-1 and 5.5.1.2-2].

15.2.1 Base RS Sequences and Sequence Grouping

In order for a cell to support uplink transmissions of different bandwidths, it is necessary to assign a cell at least one *base RS sequence* for each possible RB allocation size. Multiple RS sequences for each allocation size are then derived from each base sequence by means of different cyclic time shifts, as explained in Section 15.2.2.

The smallest number of available base sequences is for resource allocations of three RBs, where, as noted above, only 30 extended ZC sequences exist. As a result, the complete set of available base sequences across all RB allocation sizes is divided into 30 non-overlapping *sequence-groups*. A cell is then assigned one of the sequence-groups for uplink transmissions. For each resource allocation size up to and including five RBs, each of the 30 sequence-groups contains only one base sequence, since for five RBs (i.e. sequences of length 60) only 58 extended ZC-sequences are available. For sequence lengths greater than five RBs, more extended ZC-sequences are available, and therefore each of the 30 sequence-groups contains two base sequences per resource allocation size; this is exploited in LTE to support *sequence hopping* (within the sequence-group) between the two slots of a subframe.

The base sequences for resource allocations larger than three RBs are selected such that they are the sequences with high cross-correlation to the single 3 RB base sequence in the sequence-group [5]. Since the cross-correlation between the 3 RB base sequences of different sequence-groups is low due to the inherent properties of the ZC sequences, such a method for assigning the longer base sequences to sequence-groups helps to ensure that the cross-correlation between sequence-groups is kept low, thus reducing inter-cell interference.

The v base RS sequences of length 3 RBs or larger (i.e. $M_{\text{sc}}^{\text{RS}} \geq 36$) assigned to a sequence-group u are given by,

$$\bar{r}_{u,v}(n) = a_q(n \bmod N_{\text{ZC}}), \quad n = 0, 1, \ldots, M_{\text{sc}}^{\text{RS}} - 1 \tag{15.5}$$

where $u \in \{0, 1, \ldots, 29\}$ is the sequence-group number, v is the index of the base sequence of length $M_{\text{sc}}^{\text{RS}}$ within the sequence-group u, and is given by

$$v = \begin{cases} 0, 1 & \text{for } M_{\text{sc}}^{\text{RS}} \geq 72 \\ 0 & \text{otherwise} \end{cases} \tag{15.6}$$

N_{ZC} is the largest prime number smaller than $M_{\text{sc}}^{\text{RS}}$, and q is the root ZC sequence index (defined in [4, Section 5.5.1.1]).

15.2.2 Orthogonal RS via Cyclic Time-Shifts of a Base Sequence

UEs which are assigned to different RBs transmit RSs in these RBs and hence achieve separability of the RSs via FDM. However, in certain cases, UEs can be assigned to transmit on the same RBs, for example in the case of uplink multi-user MIMO[5] (sometimes also referred to as Spatial Division Multiple Access (SDMA) or 'Virtual MIMO'). In these cases, the RSs can interfere with each other, and some means of separating the RSs from the different transmitters is required. Using different base sequences for different UEs transmitting in the same RBs is not ideal due to the non-zero cross-correlation between the base sequences which can degrade the channel estimation at the eNodeB. It is preferable that

[5]Multiple-Input Multiple-Output.

the RS signals from the different UEs are fully orthogonal. In theory, this could be achieved by FDM of the RSs within the same RBs, although this would reduce the RS sequence length and the number of different RS sequences available; this would be particularly undesirable for low-bandwidth transmissions.

Therefore in LTE, orthogonality between RSs occupying the same RBs is instead provided by exploiting the fact that the correlation of a ZC sequence with any Cyclic Shift (CS) of the same sequence is zero (see Section 7.2.1). As the channel impulse response is of finite duration, different transmitters can use different cyclic time shifts of the same base RS sequence, with the RSs remaining orthogonal provided that the cyclic shifts are longer than the channel impulse response.

If the RS SC-FDMA symbol duration is T_p and the channel impulse response duration is less than T_{cs}, then up to T_p/T_{cs} different transmitters can transmit in the same symbol, with different cyclic shift values, with separable channel estimates at the receiver. For example, Figure 15.2 shows that if $T_p/T_{\mathrm{cs}} = 4$ and there are four transmitters, then each transmitter $t \in \{1, \dots, 4\}$ can use a cyclic $(t-1)T_p/4$ of the same base sequence. At the eNodeB receiver, by correlating the received composite signal from the different transmitters occupying the same RBs with the base sequence, the channel estimates from the different transmitters are separable in the time domain [6].

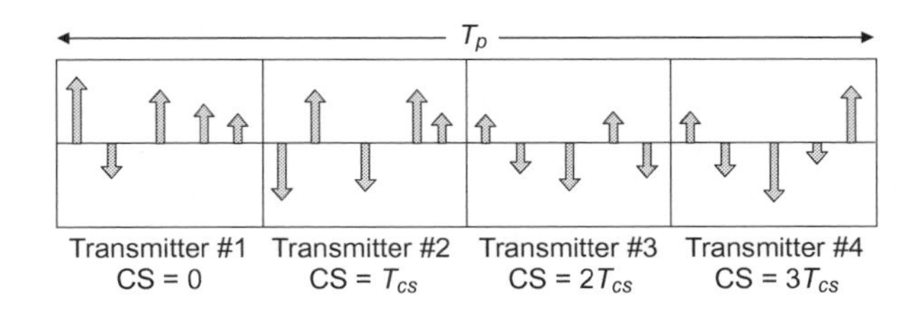

Figure 15.2: Illustration of cyclic time shift orthogonality of RS signals.

Since a cyclic time shift is equivalent to applying a phase ramp in the frequency domain, the representation of a base sequence with cyclic shift, α, in the frequency-domain is given by

$$r_{u,v}^{(\alpha)}(n) = e^{j\alpha n}\bar{r}_{u,v}(n) \tag{15.7}$$

where $\bar{r}_{u,v}(n)$ is the base (or unshifted) sequence of sequence-group u, with base sequence index v within the sequence-group, $\alpha = 2\pi n_t/P$ with n_t the cyclic time shift index for transmitter t, and P is the number of equally spaced cyclic time shifts supported.

In LTE, 12 equally spaced cyclic time shifts are defined for the DM-RS on the PUSCH and PUCCH. This allows for delay spreads up to 5.55 μs.

The degree of channel estimate separability at the receiver between different cyclic time shifts depends in practice on the (circular) distance between the shifts (as well as the received power differences between the transmitters). Cyclic time shifts spaced the furthest apart experience the least cross-talk between the channel estimates (for example, arising from practical issues such as channel estimation filtering and finite sampling granularity). Thus, when the number of UEs using different cyclic time shifts is less than the number of cyclic time shifts supported (P), it is beneficial to assign cyclic time shifts with the largest possible (circular) separation; this is supported in LTE.

In LTE the cyclic time shifts also always hop between the two slots in a subframe, for inter-cell interference randomization (see Section 15.4).

15.3 Sequence-Group Hopping and Planning

LTE supports both *RS sequence-group hopping* and *RS sequence-group planning* modes of system deployment [7, 8]; the mode is configurable by Radio Resource Control (RRC) signalling.

The sequence-group assigned to a cell is a function of its physical-layer cell identity (cell-ID) and can be different for PUSCH and PUCCH transmissions. A UE acquires knowledge of the physical layer cell-ID from its downlink synchronization signals, as described in Section 7.2.

15.3.1 Sequence-Group Hopping

The sequence-group hopping mode of deployment can be enabled in a cell by a 1-bit broadcast signalling parameter called 'groupHoppingEnabled'. This mode actually consists of a combination of hopping and shifting of the sequence-group according to one of 504 sequence-group hopping/shifting patterns corresponding to the 504 unique cell-IDs [9]. Since there are 30 base sequence-groups, $17(=\lceil 504/30 \rceil)$ unique sequence-group hopping patterns of length 20 are defined (corresponding to the duration of a radio frame with 20 slots), each of which can be offset by one of 30 sequence-group shift offsets. The sequence-group number u depends on the sequence-group hopping pattern f_{gh} and the sequence-group shift offset f_{ss} as defined in [4, Section 5.5.1.3]. The sequence-group hopping pattern changes u from slot to slot in a pseudo-random manner, while the shift offset is fixed in all slots. Both f_{gh} and f_{ss} depend on the cell-ID.

The sequence-group hopping pattern f_{gh} is obtained from a length-31 Gold sequence generator (see Section 6.3) [10,11], of which the second constituent M-sequence is initialized at the beginning of each radio frame by the sequence-group hopping pattern index of the cell. Up to 30 cell-IDs can have the same sequence-group hopping pattern (e.g. part of a planned coordinated cell cluster), with different sequence-group shift offsets being used to minimize RS collisions and inter-cell interference. The same sequence-group hopping pattern is used for PUSCH DM-RS, SRS and PUCCH DM-RS transmissions. The sequence-group shift offset can be different for PUSCH and PUCCH.

For PUSCH, it should be possible to assign cell-IDs such that the same sequence-group hopping pattern and sequence-group shift offset, and hence the same base sequences, are used in adjacent cells. This can enable the RSs from UEs in adjacent cells (for example, the cells of the same eNodeB) to be orthogonal to each other by using different cyclic time shifts of the same base sequence. Therefore in LTE the sequence-group shift offset for PUSCH is explicitly configured by a cell-specific 5-bit broadcast signalling parameter, 'groupAssignmentPUSCH'. As the sequence-group shift offset is a function of the cell-ID, the overhead for signalling one of the 504 sequence-group hopping patterns for PUSCH is reduced from nine to five bits, such that $f_{\text{ss}} = (\text{cell-ID} \bmod 30 + \Delta_{\text{ss}}) \bmod 30$, where $\Delta_{\text{ss}} \in \{0, \ldots, 29\}$ is indicated by 'groupAssignmentPUSCH'.

For PUCCH transmissions, as described in Section 16.3, the same RBs at the edge of the system bandwidth are normally used by all cells. Thus, in order to randomize interference on the PUCCH between neighbouring cells which are using the same sequence-group hopping pattern, the sequence-shift offset for PUCCH is simply given by cell-ID mod 30. Similarly, for interference randomization on the SRS transmissions which occur in the same SC-FDMA symbol (see Section 15.6), the same sequence-group shift offset as PUCCH is used.

In Section 15.2.1, it was explained that there are two base sequences per sequence-group for each RS sequence length greater than 60 (5 RBs), with the possibility of interference randomization by sequence-hopping between the two base sequences at the slot boundary in the middle of each subframe. If sequence-group hopping is used, the base sequence automatically changes between each slot, and therefore additional sequence hopping within the sequence group is not needed; hence only the first base sequence in the sequence group is used if sequence-group hopping is enabled.

15.3.2 Sequence-Group Planning

If sequence-group hopping is disabled, the same sequence-group number u is used in all slots of a radio frame and is simply obtained from the sequence group shift offset, $u = f_{ss}$.

Since 30 sequence-groups are defined (see Section 15.2.1), planned sequence-group assignment is possible for up to 30 cells in LTE. This enables neighbouring cells to be assigned sequence groups with low cross-correlation to reduce RS interference, especially for small RB allocations.[6]

An example of sequence-group planning with a conventional six-sequence-group reuse plan is shown in Figure 15.3 [12]. The same sequence-group number (and hence base sequences) are used in the three cells of each eNodeB, with different cyclic time shifts assigned to each cell (only 3 cyclic time shifts, D1, D2 and D3 are shown in this example).

With sequence-group planning, sequence hopping within the sequence group between the two slots of a subframe for interference randomization can be enabled by a 1-bit cell-specific parameter, 'sequenceHoppingEnabled'. The base sequence index for $M_{sc}^{RS} \geq 72$ used in slot n_s is then obtained from the length-31 Gold sequence generator. In order to enable the use of the same base RS sequence (and hopping pattern) in adjacent cells for PUSCH, the pseudo-random sequence generator is initialized at the beginning of each radio frame by the sequence-group hopping pattern index (based on part of the cell-ID), offset by the PUSCH sequence-group shift index of the cell (see [4, Section 5.5.1.4]).

The same hopping pattern within the sequence-group is used for PUSCH DM-RS, SRS, and PUCCH DM-RS.

Further interference randomization is provided by the cyclic time shift hopping, which is always enabled in LTE as discussed in the following section.

15.4 Cyclic Shift Hopping

Cyclic time-shift hopping is always enabled for inter-cell interference randomization for PUSCH and PUCCH transmissions. For PUSCH with $P = 12$ evenly spaced cyclic time shifts, the hopping is between the two slots in a subframe, with the cyclic shift (α in

[6]Power-limited cell-edge UEs are likely to have small RB allocations.

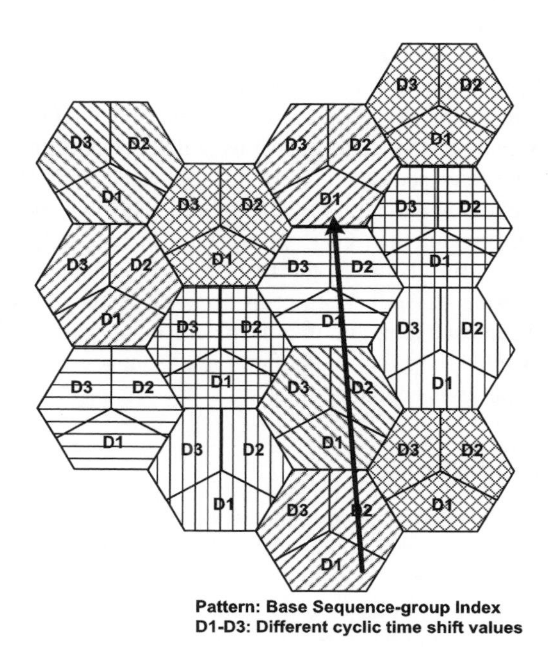

Figure 15.3: Example of RS sequence-group planning – six-sequence-group reuse plan.

Equation (15.7)) for a UE being derived in each slot from a combination of a 3-bit cell-specific broadcast cyclic time shift offset parameter, a 3-bit cyclic time shift offset indicated in each uplink scheduling grant and a pseudo-random cyclic shift offset obtained from the output of the length-31 Gold sequence generator (see [4, Section 5.5.2.1.1]).

As mentioned in Section 15.3, it should be possible to use different cyclic time shifts in adjacent cells with the same sequence-group (e.g. the cells of the same eNodeB), in order to support orthogonal RS transmissions from UEs in different cells. This requirement is similar to the case for initialization of the pseudo-random sequence generator for sequence hopping within a sequence-group, where the same hopping patterns are needed in neighbouring cells. Thus, the initialization of the PUSCH DM-RS cyclic shift pseudo-random sequence generator is the same as that for the sequence hopping pattern generator in Section 15.3.2, initialized every radio frame (see [4, Section 5.5.1.4]).

In the case of PUCCH transmission, cyclic time-shift hopping (among the $P = 12$ evenly spaced cyclic time shifts) is performed per SC-FDMA symbol, with the cyclic shift α for a given SC-FDMA symbol in a given slot being derived (as specified in [4, Section 5.4]) from a combination of the assigned PUCCH resource index (see Section 16.3) and the output of the length-31 Gold sequence generator. In order to randomize interference on the PUCCH arising from the fact that the same band-edge RBs are used for PUCCH transmissions in all cells (see Section 16.3), the pseudo-random sequence generator is initialized at the beginning of each radio frame by the cell-ID.

In addition, to achieve intra-cell interference randomization for the PUCCH DM-RS, the cyclic time shift used in the second slot is hopped such that UEs which are assigned adjacent cyclic time shifts in the first slot use non-adjacent cyclic time shifts (with large separation) in the second slot [13]. A further benefit of using a different cyclic time shift in each slot is that the non-ideal cross-correlation between different base RS sequences is averaged (as the cross-correlation is not constant for all time lags).

15.5 Demodulation Reference Signals (DM-RS)

The DM-RSs associated with uplink PUSCH data or PUCCH control transmissions from a UE are primarily provided for channel estimation for coherent demodulation and are therefore present in every transmitted uplink slot.

RSs could in theory be concentrated in one position in each slot, or divided up and positioned in multiple locations in each slot. Two alternatives were considered in the design of the DM-RSs in LTE, as shown in Figure 15.4:

- One RS symbol per slot, having the same duration as a data SC-FDMA symbol (sometimes referred to as a 'Long Block' (LB)), with the RS symbol having the same subcarrier spacing as the data SC-FDMA symbols;

- Two RS symbols per slot, each of half the duration of a data SC-FDMA symbol (sometimes referred to as a 'Short Block' (SB)), with the subcarrier spacing in the RS symbols being double that of the data SC-FDMA symbols (i.e. resulting in only six subcarriers per RB in the RS symbols).

As can be seen from Figure 15.4, it would be possible to support twice as many SB RSs in a slot in the time domain compared to the LB RS structure with the same number of data SC-FDMA symbols. However, the frequency resolution of the SB RS would be half that of the LB RS due to the subcarrier spacing being doubled.

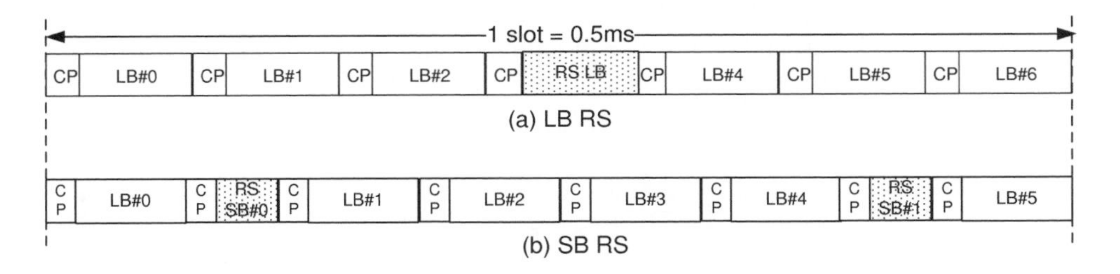

Figure 15.4: Example of slot formats considered with (a) one Long Block RS per slot and (b) two Short Block RSs per slot.

Figures 15.5 and 15.6 show a performance comparison of LB and SB RSs for a resource allocation bandwidth of 1 RB for a duration of one subframe (1 ms), for UE speeds of 30 km/h and 250 km/h respectively. The performance of LB RS is similar to that for SB

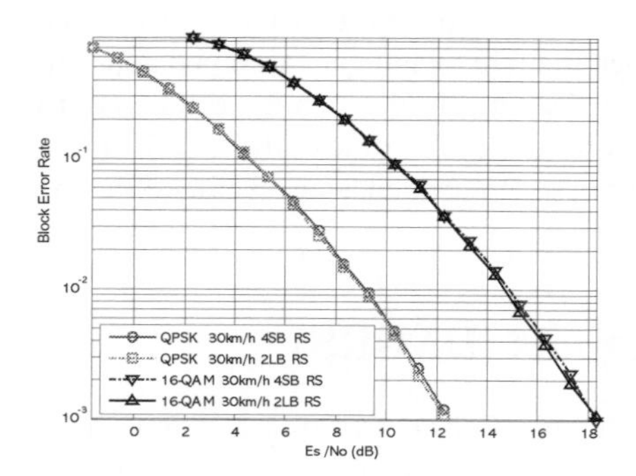

Figure 15.5: Demodulation performance comparison for Long Block RS and Short Block RS structure. Code rate $r = 1/2$, GSM Typical Urban (TU) channel model, 30 km/h, 2 GHz carrier frequency.

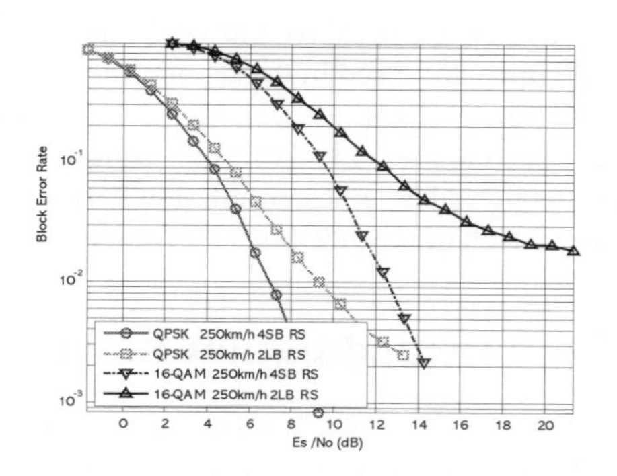

Figure 15.6: Demodulation performance comparison for Long Block RS and Short Block RS structure. Code rate $r = 1/2$, GSM Typical Urban (TU) channel model, 250 km/h, 2 GHz carrier frequency.

RS for medium speeds, with some degradation at high speeds. However, with LB RS the signal parameterization remains the same as that of the downlink OFDM. LB RSs also have the advantage of providing longer sequences for a given bandwidth allocation (due to there being twice as many RS subcarriers), and thus a larger number of RS sequences with desirable characteristics.

Therefore the LB RS structure of Figure 15.4(a) was adopted in LTE for the PUSCH DM-RS.

The exact position of the single PUSCH DM-RS symbol in each uplink slot depends on whether the normal or extended CP is used, as shown in Figure 15.7. For the case of the normal CP with seven SC-FDMA symbols per slot, the PUSCH DM-RS occupies the centre (i.e. fourth) SC-FDMA symbol. With six SC-FDMA symbols per slot in the case of the extended CP, the third SC-FDMA symbol is used. For PUCCH transmission, the position and number of DM-RS depend on the type of uplink control information being transmitted, as discussed in Section 16.3.

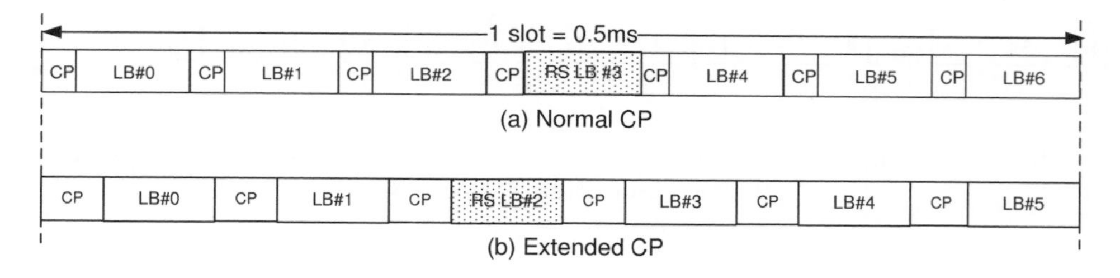

Figure 15.7: LTE uplink subframe configuration for PUSCH DM RS: (a) normal CP; (b) extended CP.

The DM-RS occupies the same RBs as the RB allocation for the uplink PUSCH data or PUCCH control transmission. Thus, the RS sequence length, M_{sc}^{RS}, is equal to the number of subcarriers allocated to the UE for PUSCH or PUCCH transmissions. Further, since the PUSCH RB allocation size is limited to multiples of two, three and/or five RBs (as explained in Section 14.3.1), the DM-RS sequence lengths are also restricted to the same multiples.

As discussed in Section 15.2.2, the DM-RS SC-FDMA symbol supports 12 cyclic time shifts with a spacing of 5.55 μs.

To support inter-cell interference randomization, cyclic time-shift hopping is always enabled for DM-RS as detailed in Section 15.4.

Release 10 enhancements to the uplink DM-RS to support non-contiguous PUSCH resource allocation and uplink Single-User MIMO are explained in Sections 28.3.6.2 and 29.2.1 respectively.

15.6 Uplink Sounding Reference Signals (SRS)

The SRSs, which are not associated with uplink data and/or control transmission, are primarily used for channel quality estimation to enable frequency-selective scheduling on the uplink. However, they can be used for other purposes, such as to enhance power control or to support various start-up functions for UEs not recently scheduled. Some examples include initial Modulation and Coding Scheme (MCS) selection, initial power control for data transmissions, timing advance, and 'frequency semi-selective scheduling' in which the frequency resource is assigned frequency-selectively for the first slot of a subframe and hops pseudo-randomly to a different frequency in the second slot [14].

15.6.1 SRS Subframe Configuration and Position

The subframes in which SRSs are transmitted by any UE within the cell are indicated by cell-specific broadcast signalling. A 4-bit cell-specific 'srsSubframeConfiguration' parameter indicates 15 possible sets of subframes in which SRS may be transmitted within each radio frame (see [4, Section 5.5.3.3]). This configurability provides flexibility in adjusting the SRS overhead depending on the deployment scenario. A 16th configuration switches the SRS off completely in the cell, which may for example be appropriate for a cell serving primarily high-speed UEs.

The SRS transmissions are always in the last SC-FDMA symbol in the configured subframes, as shown in Figure 15.8. Thus the SRS and DM-RS are located in different SC-FDMA symbols. PUSCH data transmission is not permitted on the SC-FDMA symbol designated for SRS, resulting in a worst-case sounding overhead (with an SRS symbol in every subframe) of around 7%.

15.6.2 Duration and Periodicity of SRS Transmissions

The eNodeB in LTE may either request an individual SRS transmission from a UE or configure a UE to transmit SRS periodically until terminated; a 1-bit UE-specific signalling parameter, 'duration', indicates whether the requested SRS transmission is single or periodic. If periodic SRS transmissions are configured for a UE, the periodicity may be any of 2, 5, 10, 20, 40, 80, 160 or 320 ms; the SRS periodicity and SRS subframe offset within the period

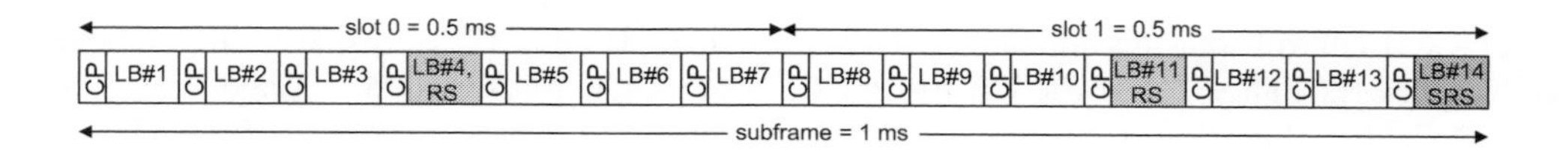

Figure 15.8: Uplink subframe configuration with SRS symbol.

in which the UE should transmit its SRS are configured by a 10-bit UE-specific dedicated signalling parameter called 'srs-ConfigIndex'.

In Release 10, a mechanism for dynamically triggering an aperiodic SRS tranmission by means of the PDCCH is introduced; this is explained in Section 29.2.2.

15.6.3 SRS Symbol Structure

In order to support frequency-selective scheduling between multiple UEs, it is necessary that SRS from different UEs with different sounding bandwidths can overlap. In order to support this, Interleaved FDMA (IFDMA, introduced in Section 14.2) is used in the SRS SC-FDMA symbol, with a RePetition Factor (RPF) of 2. The (time-domain) RPF is equivalent to a frequency-domain decimation factor, giving the spacing between occupied subcarriers of an SRS signal with a comb-like spectrum. Thus, RPF = 2 implies that the signal occupies every 2nd subcarrier within the allocated sounding bandwidth as shown by way of example in Figure 15.9. Using a larger RPF could in theory have provided more flexibility in how the bandwidth could be allocated between UEs, but it would have reduced the sounding sequence length (for a given sounding bandwidth) and the number of available SRS sequences (similar to the case for DM-RS). Therefore the RPF is limited to 2.

Due to the IFDMA structure of the SRS symbol, a UE is assigned, as part of its configurable SRS parameters, the 'transmissionComb' index (0 or 1) on which to transmit the SRS. The RS sequences used for the SRS are the same as for the DM-RS, resulting in the SRS sequence length being restricted to multiples of two, three and/or five times the RB size. In addition, the SRS bandwidth (in RBs) must be an even number, due to the RPF of 2 and the minimum SRS sequence length being 12. Therefore, the possible SRS bandwidths, $N_{\text{RB}}^{\text{SRS}}$ (in number of RBs), and the SRS sequence length, $M_{\text{sc}}^{\text{SRS}}$, are respectively given by,

$$\begin{aligned} N_{\text{RB}}^{\text{SRS}} &= 2^{(1+\alpha_2)} \cdot 3^{\alpha_3} \cdot 5^{\alpha_5} \\ M_{\text{sc}}^{\text{SRS}} &= \tfrac{1}{2} \cdot N_{\text{RB}}^{\text{SRS}} \cdot 12 \end{aligned} \tag{15.8}$$

where α_2, α_3, α_5 is a set of positive integers. Similarly to the DM-RS, simultaneous SRS can be transmitted from multiple UEs using the same RBs and the same offset of the comb, using different cyclic time shifts of the same base sequence to achieve orthogonal separation (see Section 15.2.2). For the SRS, eight (evenly spaced) cyclic time shifts per SRS comb are supported (see [4, Section 5.5.3.1]), with the cyclic shift being configured individually for each UE.

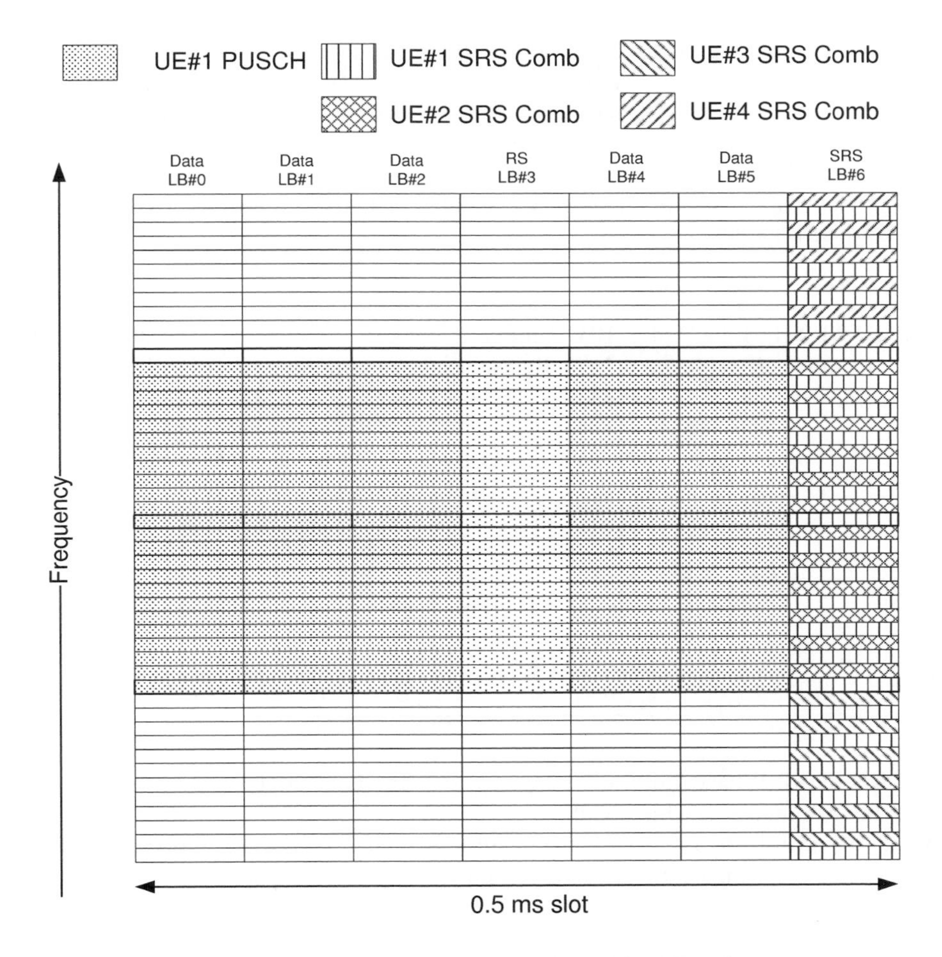

Figure 15.9: SRS symbol structure with RPF = 2.

15.6.3.1 SRS Bandwidths

Some of the factors which affect the SRS bandwidth are the maximum power of the UE, the number of supportable sounding UEs, and the sounding bandwidth needed to benefit from uplink channel-dependent scheduling. Full bandwidth sounding provides the most complete channel information when the UE is sufficiently close to the eNodeB, but degrades as the path-loss increases when the UE cannot further increase its transmit power to maintain the transmission across the full bandwidth. Full bandwidth transmission of SRS also limits the number of simultaneous UEs whose channels can be sounded, due to the limited number of cyclic time shifts (eight cyclic time shifts per SRS comb as explained above).

To improve the SNR and support a larger number of SRSs, up to four SRS bandwidths can be simultaneously supported in LTE depending on the system bandwidth. To provide flexibility with the values for the SRS bandwidths, eight sets of four SRS bandwidths are defined for each possible system bandwidth. RRC signalling indicates which of the eight sets is applicable in the cell by means of a 3-bit cell-specific parameter 'srs-BandwidthConfig'. This allows some variability in the maximum SRS bandwidths, which is important as the SRS region does not include the PUCCH region near the edges of the system bandwidth (see Section 16.3), which is itself variable in bandwidth. An example of the eight sets of four SRS bandwidths applicable to uplink system bandwidths in the range 40–60 RBs is shown in Table 15.1 (see [4, Table 5.5.3.2-2]).

Table 15.1: SRS BandWidth (BW) configurations for system bandwidths 40–60 RBs (see [4, Table 5.5.3.2-2]). Reproduced by permission of © 3GPP.

	Number of RBs			
Configuration	SRS-BW 0	SRS-BW 1	SRS-BW 2	SRS-BW 3
0	48	24	12	4
1	48	16	8	4
2	40	20	4	4
3	36	12	4	4
4	32	16	8	4
5	24	4	4	4
6	20	4	4	4
7	16	4	4	4

The specific SRS bandwidth to be used by a given UE is configured by a further 2-bit UE-specific parameter, 'srs-Bandwidth'.

As can be seen from Table 15.1, the smallest sounding bandwidth supported in LTE is 4 RBs. A small sounding bandwidth of 4 RBs provides for higher-quality channel information from a power-limited UE. The sounding bandwidths are constrained to be multiples of each other, i.e. following a tree-like structure, to support frequency hopping of the different narrowband SRS bandwidths (see [4, Section 5.5.3.2]). Frequency hopping can be enabled or disabled for an individual UE based on the value of the parameter 'freqDomainPosition'. The tree structure of the SRS bandwidths limits the possible starting positions for the different SRS bandwidths, reducing the overhead for signalling the starting position to 5 bits (signalled to each UE by the parameter 'freqDomainPosition').

Table 15.2 summarizes the various SRS configurable parameters which are signalled to a UE [15].

15.7 Summary

The uplink reference signals provided in LTE fulfil an important function in facilitating channel estimation and channel sounding. The ZC-based sequence design can be seen to be a good match to this role, with constant amplitude in the frequency domain and the ability to provide a large number of sequences with zero or low correlation. This enables both interference randomization and interference coordination techniques to be employed in LTE

Table 15.2: Uplink SRS configurable parameters.

Sounding RS parameter name	Significance	Signalling type
srs-BandwidthConfig	Maximum SRS bandwidth in the cell	Cell-specific
srs-SubframeConfig	Sets of subframes in which SRS may be transmitted in the cell	Cell-specific
srs-Bandwidth	SRS transmission bandwidth for a UE	UE-specific
freqDomainPosition	Frequency-domain position	UE-specific
srs-HoppingBandwidth	Frequency hop size	UE-specific
duration	Single SRS or periodic	UE-specific
srs-ConfigIndex	Periodicity and subframe offset	UE-specific
transmissionComb	Transmission comb offset	UE-specific
$n_{\text{SRS}}^{\text{cs}}$	Cyclic shift	UE-specific

system deployments, as appropriate to the scenario. A high degree of flexibility is provided for configuring the reference signals, especially for the sounding reference signals, where the overhead arising from their transmission can be traded off against the improvements in system efficiency which may be achievable from frequency-selective uplink scheduling.

Enhancements to the uplink RSs for LTE-Advanced are outlined in Sections 28.3.6.2 and 29.2.1 for the DM-RS and in Section 29.2.2 for the SRS.

References[7]

[1] D. C. Chu, 'Polyphase Codes With Good Periodic Correlation Properties'. *IEEE Trans. on Information Theory*, pp. 531–532, July 1972.

[2] B. M Popovic, 'Generalized Chirp-like Polyphase Sequences with Optimal Correlation Properties'. *IEEE Trans. on Information Theory*, Vol. 38, pp. 1406–1409, July 1992.

[3] Motorola, 'R1-060878: EUTRA SC-FDMA Uplink Pilot/Reference Signal Design & TP', www.3gpp.org, 3GPP TSG RAN WG1, meeting 44bis, Athens, Greece, March 2006.

[4] 3GPP Technical Specification 36.211, 'Evolved Universal Terrestrial Radio Access (E-UTRA); Physical Channels and Modulation', www.3gpp.org.

[7] All web sites confirmed 1st March 2011.

[5] Huawei, LG Electronics, NTT DoCoMo, and Panasonic, 'R1-080576: Way Forward on the Sequence Grouping for UL DM RS', www.3gpp.org, 3GPP TSG RAN WG1, meeting 51bis, Sevilla, Spain, January 2008.

[6] K. Fazel and G. P. Fettweis, *Multi-Carrier Spread-Spectrum*. Kluwer Academic Publishers, Dordrecht, Holland, 1997.

[7] Alcatel-Lucent, Ericsson, Freescale, Huawei, LGE, Motorola, Nokia, Nokia-Siemens Networks, NTT DoCoMo, Nortel, Panasonic, Qualcomm, and TI, 'R1-072584: Way Forward for PUSCH RS', www.3gpp.org, 3GPP TSG RAN WG1, meeting 49, Kobe, Japan, May 2007.

[8] Alcatel-Lucent, Ericsson, Freescale, Huawei, LGE, Motorola, Nokia, Nokia-Siemens Networks, NTT DoCoMo, Nortel, Panasonic, Qualcomm, and TI, 'R1-072585: Way forward for PUCCH RS', www.3gpp.org, 3GPP TSG RAN WG1, meeting 49, Kobe, Japan, May 2007.

[9] NTT DoCoMo, 'R1-074278: Hopping and Planning of Sequence Groups for Uplink RS', www.3gpp.org, 3GPP TSG RAN WG1, meeting 50bis, Shanghai, China, October 2007.

[10] Motorola, 'R1-080719: Hopping Patterns for UL RS', www.3gpp.org, 3GPP TSG RAN WG1, meeting 52, Sorrento, Italy, February 2008.

[11] Qualcomm Europe, Ericsson, Motorola, Samsung, Panasonic, and NTT DoCoMo, 'R1-081133: WF on UL DM-RS Hopping Pattern Generation', www.3gpp.org, 3GPP TSG RAN WG1, meeting 52, Sorrento, Italy, February 2008.

[12] Motorola, 'R1-071341: Uplink Reference Signal Planning Aspects', www.3gpp.org, 3GPP TSG RAN WG1, meeting 48bis, St. Julian's, Malta, March 2007.

[13] Panasonic, Samsung, and ETRI, 'R1-080983: Way forward on the Cyclic Shift Hopping for PUCCH', www.3gpp.org, 3GPP TSG RAN WG1, meeting 52, Sorrento, Italy, February 2008.

[14] Motorola, 'R1-073756: Benefit of Non-Persistent UL Sounding for Frequency Hopping PUSCH', www.3gpp.org, 3GPP TSG RAN WG1, meeting 50, Athens, Greece, August 2007.

[15] Ericsson, 'R1-082199: Physical-layer parameters to be configured by RRC', www.3gpp.org, 3GPP TSG RAN WG1, meeting 53, Kansas City, USA, May 2008.

16

Uplink Physical Channel Structure

Robert Love and Vijay Nangia

16.1 Introduction

The LTE Single-Carrier Frequency Division Multiple Access (SC-FDMA) uplink provides separate physical channels for the transmission of data and control signalling, the latter being predominantly to support the downlink data transmissions. The detailed structure of these channels, as explained in this chapter, is designed to make efficient use of the available frequency-domain resources and to support effective multiplexing between data and control signalling.

LTE Release 8 also incorporates uplink multiple antenna techniques, including closed-loop antenna selection and Spatial Division Multiple Access (SDMA) or Multi-User Multiple-Input Multiple-Output (MU-MIMO). More advanced multiple-antenna techniques, including closed-loop spatial multiplexing for Single-User MIMO (SU-MIMO) and transmit diversity for control signalling are introduced in Release 10 for LTE-Advanced, as explained in Section 29.4.

The physical layer transmissions of the LTE uplink comprise three physical channels and two signals:

- PRACH – Physical Random Access CHannel (see Chapter 17);
- PUSCH – Physical Uplink Shared CHannel (see Section 16.2);
- PUCCH – Physical Uplink Control CHannel (see Section 16.3);
- DM-RS – DeModulation Reference Signal (see Section 15.5);
- SRS – Sounding Reference Signal (see Section 15.6).

LTE – The UMTS Long Term Evolution: From Theory to Practice, Second Edition.
Stefania Sesia, Issam Toufik and Matthew Baker.

The uplink physical channels, and their relationship to the higher-layer channels, are summarized in Figure 16.1.

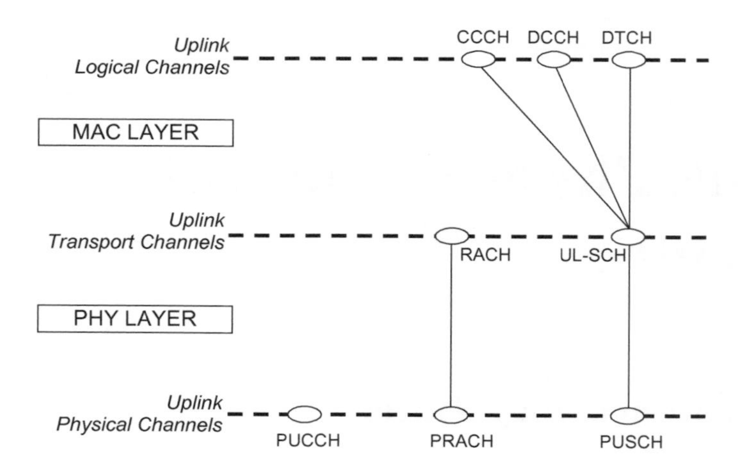

Figure 16.1: Uplink physical channels and their mapping to higher layers.

16.2 Physical Uplink Shared Data Channel Structure

The Physical Uplink Shared CHannel (PUSCH), which carries data from the Uplink Shared Channel (UL-SCH) transport channel, uses DFT-Spread OFDM (DFT-S-OFDM), as described in Chapter 14. The transmit processing chain is shown in Figure 16.2. As explained in Chapter 10, the information bits are first channel-coded with a turbo code of mother code rate $r = 1/3$, which is adapted to a suitable final code rate by a rate-matching process. This is followed by symbol-level channel interleaving which follows a simple 'time-first' mapping [1] – in other words, adjacent data symbols end up being mapped first to adjacent SC-FDMA symbols in the time domain, and then across the subcarriers (see [2, Section 5.2.2.8]). The coded and interleaved bits are then scrambled by a length-31 Gold code (as described in Section 6.3) prior to modulation mapping, DFT-spreading, subcarrier mapping[1] and OFDM modulation. The signal is frequency-shifted by half a subcarrier prior to transmission, to avoid the distortion caused by the d.c. subcarrier being concentrated in one Resource Block (RB), as described in Section 14.3.3. The modulations supported are QPSK, 16QAM and 64QAM (the latter being only for the highest categories of User Equipment (UE) – Categories 5 and 8 (see Sections 1.3.4 and 27.5)).

[1]Only localized mapping (i.e. to contiguous sets of Resource Blocks (RBs)) is supported for PUSCH and PUCCH transmissions in Releases 8 and 9. In Release 10, mapping to two clusters of RBs is also supported – see Section 28.3.6.2.

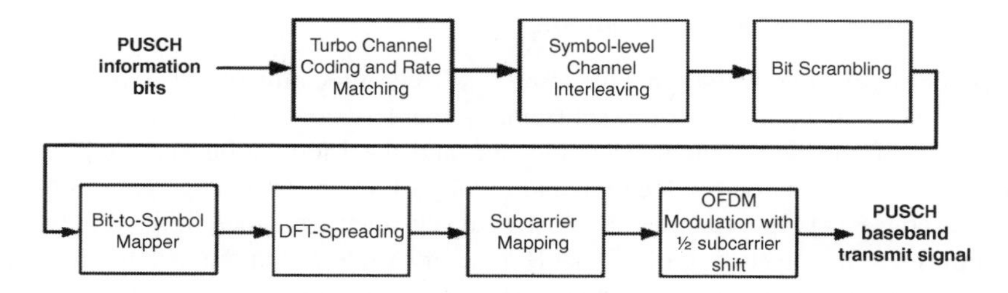

Figure 16.2: Uplink physical data channel processing.

The baseband SC-FDMA transmit signal for SC-FDMA symbol ℓ is given by the following expression (see [3, Section 5.6]),

$$s_\ell(t) = \sum_{k=-\lfloor N_{\text{RB}}^{\text{UL}} N_{\text{sc}}^{\text{RB}}/2 \rfloor}^{k=-\lceil N_{\text{RB}}^{\text{UL}} N_{\text{sc}}^{\text{RB}}/2 \rceil -1} a_{k^-,\ell} \exp[j2\pi(k+1/2)\Delta f(t - N_{\text{CP},\ell} T_s)] \tag{16.1}$$

for $0 \leq t < (N_{\text{CP},\ell} + N)T_s$, where $N_{\text{CP},\ell}$ is the number of samples of the Cyclic Prefix (CP) in SC-FDMA symbol ℓ (see Section 14.3), $N = 2048$ is the Inverse Fast Fourier Transform (IFFT) size, $\Delta f = 15$ kHz is the subcarrier spacing, $T_\text{s} = 1/(N \cdot \Delta f)$ is the sampling interval, $N_{\text{RB}}^{\text{UL}}$ is the uplink system bandwidth in RBs, $N_{\text{sc}}^{\text{RB}} = 12$ is the number of subcarriers per RB, $k^{(-)} = k + \lfloor N_{\text{RB}}^{\text{UL}} N_{\text{sc}}^{\text{RB}}/2 \rfloor$ and $a_{k,\ell}$ is the content of subcarrier k on symbol ℓ. For the PUSCH, the SC-FDMA symbol $a_{k,\ell}$ is obtained by DFT-spreading the QAM data symbols, $[d_{0,\ell}, d_{1,\ell}, \ldots, d_{M_{\text{SC}}^{\text{PUSCH}}-1,\ell}]$ to be transmitted on SC-FDMA symbol ℓ (see [3, Section 5.3.3]),

$$a_{k,\ell} = \frac{1}{\sqrt{M_{\text{sc}}^{\text{PUSCH}}}} \sum_{i=0}^{M_{\text{sc}}^{\text{PUSCH}}-1} d_{i,\ell} e^{-j2\pi ik/M_{\text{sc}}^{\text{PUSCH}}} \tag{16.2}$$

for $k = 0, 1, 2, \ldots, M_{\text{sc}}^{\text{PUSCH}} - 1$, where $M_{\text{sc}}^{\text{PUSCH}} = M_{\text{RB}}^{\text{PUSCH}} \cdot N_{\text{sc}}^{\text{RB}}$ and $M_{\text{RB}}^{\text{PUSCH}}$ is the allocated PUSCH bandwidth in RBs.

As explained in Section 4.4.1, a Hybrid Automatic Repeat reQuest (HARQ) scheme is used, which in the uplink is synchronous, using N-channel stop and wait. This means that retransmissions occur in specific periodically occurring subframes (HARQ channels). Further details of the HARQ operation are given in Section 10.3.2.5.

16.2.1 Scheduling on PUSCH

In the LTE uplink, both frequency-selective scheduling and non-frequency-selective scheduling are supported. The former is based on the eNodeB exploiting available channel knowledge to schedule a UE to transmit using specific RBs in the frequency domain where the UEs experience good channel conditions. The latter does not make use of frequency-specific channel knowledge, but rather aims to benefit from frequency diversity during the transmission of each transport block. The possible techniques supported in LTE are discussed in more detail below. Intermediate approaches are also possible.

16.2.1.1 Frequency-Selective Scheduling

With frequency-selective scheduling, the same localized[2] allocation of transmission resources is typically used in both slots of a subframe – there is no frequency hopping during a subframe. The frequency-domain RB allocation and the Modulation and Coding Scheme (MCS) are chosen based on the location and quality of an above-average gain in the uplink channel response [4]. In order to enable frequency-selective scheduling, timely channel quality information is needed at the eNodeB. One method for obtaining such information in LTE is by uplink channel sounding using the SRS described in Section 15.6. The performance of frequency-selective scheduling using the SRS depends on the sounding bandwidth and the quality of the channel estimate, the latter being a function of the transmitted power spectral density used for the SRS. With a large sounding bandwidth, link quality can be evaluated on a larger number of RBs. However, this is likely to lead to the SRS being transmitted at a lower power density, due to the limited total UE transmit power, and this reduces the accuracy of the estimate for each RB within the sounding bandwidth especially for cell-edge UEs. Conversely, sounding a smaller bandwidth can improve channel estimation on the sounded RBs but results in missing channel information for certain parts of the channel bandwidth, thus risking exclusion of the best quality RBs. As an example, it is shown in [5] that, at least for a bandwidth of 5 MHz, frequency-selective scheduling based on full-band sounding outperforms narrower bandwidth sounding.

16.2.1.2 Frequency-Diverse or Non-Selective Scheduling

There are cases when no, or limited, frequency-specific channel quality information is available, for example because of SRS overhead constraints or high Doppler conditions. In such cases, it is preferable to exploit the frequency diversity of LTE's wideband channel.

In LTE, frequency hopping of a localized transmission is used to provide frequency-diversity.[3] Two hopping modes are supported – hopping only between subframes (inter-subframe hopping) or hopping both between and within subframes (inter- and intra-subframe hopping). These modes are illustrated in Figure 16.3. Cell-specific broadcast signalling is used to configure the hopping mode via the parameter 'Hopping-mode' (see [6, Section 8.4]).

In case of intra-subframe hopping, a frequency hop occurs at the slot boundary in the middle of a subframe; this provides frequency diversity within a codeword (i.e. within a single transmission of transport block). On the other hand, inter-subframe hopping provides frequency diversity between HARQ retransmissions of a transport block, as the frequency allocation hops every allocated subframe.

Two methods are defined for the frequency hopping allocation (see [6, Section 8.4]): either a pre-determined pseudo-random frequency hopping pattern (see [3, Section 5.3.4]) or an explicit hopping offset signalled in the UL resource grant on the Physical Downlink Control CHannel (PDCCH). For uplink system bandwidths less than 50 RBs, the size of the hopping offset (modulo the system bandwidth) is approximately half the number of RBs available for PUSCH transmissions (i.e. $\lfloor N_{\text{RB}}^{\text{PUSCH}}/2\rfloor$), while for uplink system bandwidths of 50 RBs or more, the possible hopping offsets are $\lfloor N_{\text{RB}}^{\text{PUSCH}}/2\rfloor$ and $\pm\lfloor N_{\text{RB}}^{\text{PUSCH}}/4\rfloor$ (see [6, Section 8.4]).

[2]Localized means that allocated RBs are consecutive in the frequency domain.

[3]Note that there are upper limits on the size of resource allocation with which frequency hopping can be used – see [6, Section 8.4].

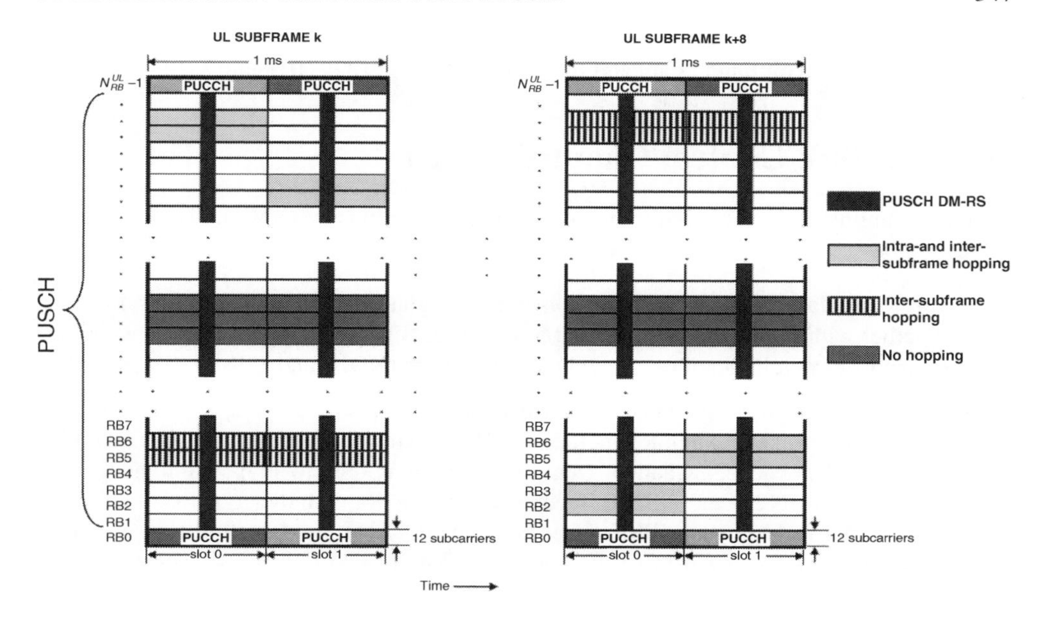

Figure 16.3: Uplink physical data channel processing.

Signalling the frequency hop via the uplink resource grant can be used for frequency semi-selective scheduling [7], in which the frequency resource is assigned selectively for the first slot of a subframe and frequency diversity is also achieved by hopping to a different frequency in the second slot. In some scenarios this may yield intermediate performance between that of fully frequency-selective and fully non-frequency-selective scheduling; this may be seen as one way to reduce the sounding overhead typically needed for fully frequency-selective scheduling.

In Release 10, another method of frequency-diverse scheduling is introduced, using dual-cluster PUSCH resource allocations; this is explained in detail in Section 28.3.6.2.

16.2.2 PUSCH Transport Block Sizes

The transport block size for a PUSCH data transmission is signalled in the corresponding resource grant on the PDCCH (DCI Format 0, or, for uplink SU-MIMO in Release 10, DCI Format 4 – see Section 9.3.5.1). Together with the indicated modulation scheme, the transport block size implies the code rate. The available transport block sizes are given in [6, Section 7.1.7.2].

In most cases, a generally linear range of code rates is available for each resource allocation size. One exception is an index which allows a transport block size of 328 bits in a single RB allocation with QPSK modulation, which corresponds to a code rate greater than unity. This is primarily designed to support cell-edge Voice-over-IP (VoIP) transmissions: by using only one RB per subframe, the UE's power spectral density is maximized for good coverage; 'TTI bundling' (see Section 14.3.2), whereby the transmission is repeated in four consecutive subframes, together with typically three retransmissions at 16 ms intervals, then enables both Chase combining gain and Incremental Redundancy (IR) gain to be achieved.

16.3 Uplink Control Channel Design

In general, uplink control signalling in mobile communications systems can be divided into two categories:

- **Data-associated control signalling** is control signalling which is always transmitted together with uplink data and is used in the processing of that data. Examples include transport format indications, 'New Data' Indicators (NDIs) and MIMO parameters.
- **Control signalling not associated with data** is transmitted independently of any uplink data packet. Examples include HARQ Acknowledgements (ACK/NACK) for downlink data packets, Channel Quality Indicators (CQIs) to support link adaptation (see Section 10.2), and MIMO feedback such as Rank Indicators (RIs) and Pre-coding Matrix Indicators (PMIs) (see Section 11.2.2.4) for downlink transmissions. Scheduling Requests (SRs) for uplink transmissions also fall into this category (see Section 4.4.2.2).

In LTE, the low signalling latency afforded by the short subframe duration of 1 ms, together with the orthogonal nature of the uplink multiple access scheme which necessitates centralized resource allocation, make it appropriate for the eNodeB to be in full control of the uplink transmission parameters. Consequently uplink data-associated control signalling is not necessary in LTE, as the relevant information is already known to the eNodeB. Therefore only data-non-associated control signalling exists in the LTE uplink.

When simultaneous uplink PUSCH data and control signalling are scheduled, the control signalling is normally multiplexed together with the data prior to the DFT spreading[4], in order to preserve the single-carrier low Cubic Metric (CM) property of the uplink transmission. The uplink control channel, PUCCH, is used by a UE to transmit any necessary control signalling only in subframes in which the UE has not been allocated any RBs for PUSCH transmission. In the design of the PUCCH, special consideration was given to maintaining a low CM [8].

16.3.1 Physical Uplink Control Channel (PUCCH) Structure

The control signalling on the PUCCH is transmitted in a frequency region that is normally configured to be on the edges of the system bandwidth.

In order to minimize the resources needed for transmission of control signalling, the PUCCH in LTE is designed to exploit frequency diversity: each PUCCH transmission in one subframe comprises a single (0.5 ms) RB at or near one edge of the system bandwidth, followed (in the second slot of the subframe) by a second RB at or near the opposite edge of the system bandwidth, as shown in Figure 16.4; together, the two RBs are referred to as a *PUCCH region*. This design can achieve a frequency diversity benefit of approximately 2 dB compared to transmission in the same RB throughout the subframe.

[4]In Release 10, the possibility of simultaneous transmission of PUSCH and PUCCH is introduced; this is explained in Section 28.3.2.

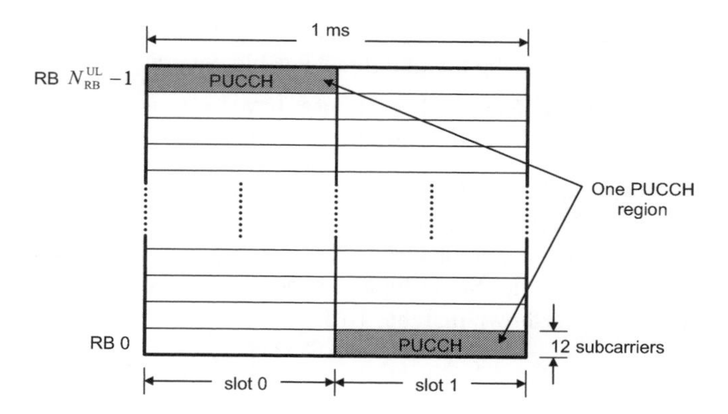

Figure 16.4: A PUCCH region.

At the same time, the narrow bandwidth of the PUCCH in each slot (only a single RB) maximizes the power per subcarrier for a given total transmission power (see Figure 16.5) and therefore helps to fulfil stringent coverage requirements.

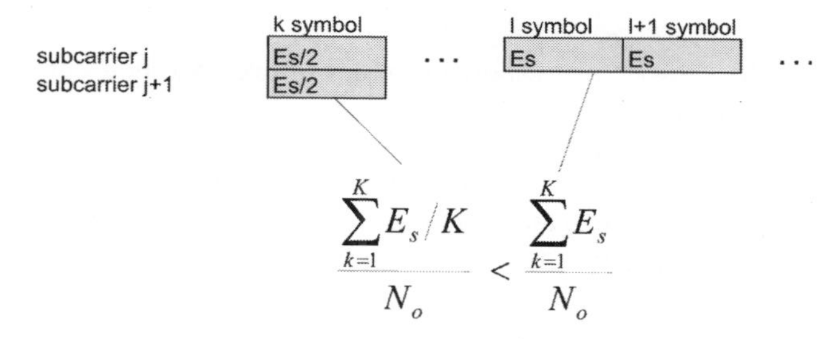

Figure 16.5: The link budget of a two-slot narrowband transmission exceeds that of a one-slot wider-band transmission, given equal coding gain.

Positioning the control regions at the edges of the system bandwidth has a number of advantages, including the following:

- The frequency diversity achieved through frequency hopping is maximized by allowing hopping from one edge of the band to the other.
- Out-Of-Band (OOB) emissions are smaller if a UE is only transmitting on a single RB per slot compared to multiple RBs. The PUCCH regions can serve as a kind of guard band between the wider-bandwidth PUSCH transmissions of adjacent carriers and can therefore improve coexistence [9].
- Using control regions on the band edges maximizes the achievable PUSCH data rate, as the entire central portion of the band can be allocated to a single UE. If the control regions were in the central portion of a carrier, a UE bandwidth allocation would be

limited to one side of the control region in order to maintain the single-carrier nature of the signal, thus limiting the maximum achievable data rate.

- Control regions on the band edges impose fewer constraints on the uplink data scheduling, both with and without inter-/intra-subframe frequency hopping.

The number of RBs (in each slot) that can be used for PUCCH transmission within the cell is $N_{\mathrm{RB}}^{\mathrm{HO}}$ (parameter 'pusch-HoppingOffset'). This is indicated to the UEs in the cell through broadcast signalling. Note that the number of PUCCH RBs per slot is the same as the number of PUCCH regions per subframe. Some typical expected numbers of PUCCH regions for different LTE bandwidths are shown in Table 16.1.

Table 16.1: Typical numbers of PUCCH regions.

Bandwidth (MHz)	Number of RBs per subframe	Number of PUCCH regions
1.4	2	1
3	4	2
5	8	4
10	16	8
20	32	16

Figures 16.6 and 16.7 respectively show examples of even and odd numbers of PUCCH regions being configured in a cell.

In the case of an even number of PUCCH regions (Figure 16.6), both RBs of each RB-pair (e.g. RB-pair 2 and RB-pair $N_{\mathrm{RB}}^{\mathrm{UL}} - 3$) are used for PUCCH transmission. However, for the case of an odd number of PUCCH regions (Figure 16.7), one RB of an RB-pair in each slot is not used for PUCCH (e.g. one RB of RB-pair 2 and RB-pair $N_{\mathrm{RB}}^{\mathrm{UL}} - 3$ is unused); the eNodeB

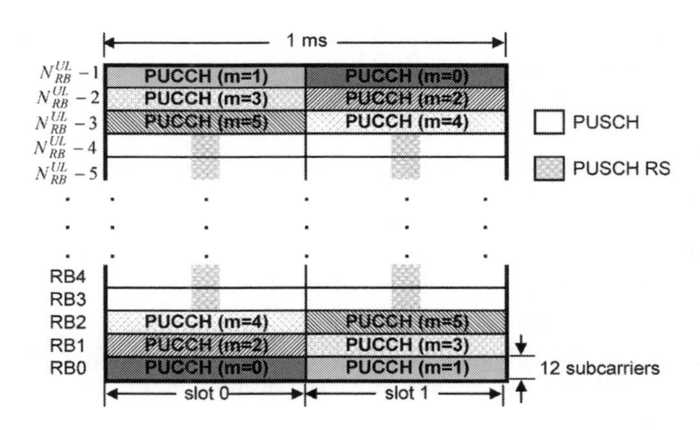

Figure 16.6: PUCCH uplink control structure with an even number of 'PUCCH Control Regions' ($N_{\mathrm{RB}}^{\mathrm{PUCCH}} = 6$).

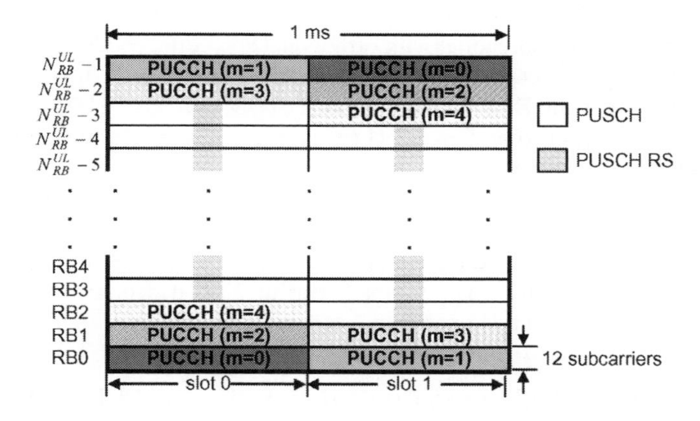

Figure 16.7: Example of an odd number of PUCCH regions ($N_{\mathrm{RB}}^{\mathrm{HO}} = 5$).

may schedule a UE with an intra-subframe frequency hopping PUSCH allocation to exploit these unused RBs.

Alternatively, a UE can be assigned a localized allocation which includes the unused RB-pair (e.g. RB-pair 2 or RB-pair $N_{\mathrm{RB}}^{\mathrm{UL}} - 3$). In this case, the UE will transmit PUSCH data on both RBs of the RB-pair, assuming that neither of the RBs are used for PUCCH by any UE in the subframe. Thus, the eNodeB scheduler can appropriately schedule PUSCH transmission on the PUCCH RBs when they are under-utilized.

The eNodeB may also choose to schedule low-power PUSCH transmission (e.g. from UEs close to the eNodeB) in the outer RBs of the configured PUCCH region, while the inner PUCCH region is used for PUCCH signalling. This can provide further reduction in OOB emissions which is necessary in some frequency bands, by moving higher-power PUCCH transmission (e.g. those from cell-edge UEs) slightly away from the edge of the band.

16.3.1.1 Multiplexing of UEs within a PUCCH Region

Control signalling from multiple UEs can be multiplexed into a single PUCCH region using orthogonal Code Division Multiplexing (CDM). In some scenarios this can have benefits over a pure Frequency Division Multiplexing (FDM) approach, as it reduces the need to limit the power differentials between the PUCCH transmissions of different UEs. One technique to provide orthogonality between UEs is by using cyclic time shifts of a sequence with suitable properties, as explained in Section 15.2.2. In a given SC-FDMA symbol, different cyclic time shifts of a waveform (e.g. a Zadoff–Chu (ZC) sequence as explained in Section 7.2.1) are modulated with a UE-specific QAM symbol carrying the necessary control signalling information, with the supported number of cyclic time shifts determining the number of UEs which can be multiplexed per SC-FDMA symbol. As the PUCCH RB spans 12 subcarriers, and assuming the channel is approximately constant over the RB (i.e. a single-tap channel), the LTE PUCCH supports up to 12 cyclic shifts per PUCCH RB.

For control information transmissions with a small number of control signalling bits, such as 1- or 2-bit positive/negative acknowledgements (ACK/NACK), orthogonality is achieved between UEs by a combination of cyclic time shifts within an SC-FDMA symbol and SC-FDMA symbol time-domain spreading with orthogonal spreading codes, i.e. modulating the SC-FDMA symbols by elements of an orthogonal spreading code [10]. CDM of multiple UEs is used rather than Time Domain Multiplexing (TDM) because CDM enables the time duration of the transmission to be longer, which increases the total transmitted energy per signalling message in the case of a power-limited UE.

Thus, the LTE PUCCH control structure uses frequency-domain code multiplexing (different cyclic time shifts of a base sequence) and/or time-domain code multiplexing (different orthogonal block spreading codes), thereby providing an efficient, orthogonal control channel which supports small payloads (up to 22 coded bits) from multiple UEs simultaneously, together with good operational capability at low SNR.

16.3.2 Types of Control Signalling Information and PUCCH Formats

The Uplink Control Information (UCI) can consist of:

- Scheduling Requests (SRs) (see Section 4.4.2.2).
- HARQ ACK/NACK in response to downlink data packets on the Physical Downlink Shared CHannel(PDSCH); one ACK/NACK bit is transmitted in the case of single-codeword downlink transmission while two ACK/NACK bits are used in the case of two-codeword downlink transmission.
- Channel State Information (CSI), which includes Channel Quality Indicators (CQIs) as well as the MIMO-related feedback consisting of RIs and PMI. 20 bits per subframe are used for the CSI.

The amount of UCI a UE can transmit in a subframe depends on the number of SC-FDMA symbols available for transmission of control signalling data (i.e. excluding SC-FDMA symbols used for RS transmission for coherent detection of the PUCCH). The PUCCH supports seven different formats (with an eighth added in Release 10), depending on the information to be signalled. The mapping between PUCCH formats and UCI is shown in Table 16.2 (see [6, Section 10.1] and [3, Table 5.4-1]).

Table 16.2: Supported uplink control information formats on PUCCH.

PUCCH Format	Uplink Control Information (UCI)
Format 1	Scheduling request (SR) (unmodulated waveform)
Format 1a	1-bit HARQ ACK/NACK with/without SR
Format 1b	2-bit HARQ ACK/NACK with/without SR
Format 2	CSI (20 coded bits)
Format 2	CSI and 1- or 2-bit HARQ ACK/NACK for extended CP only
Format 2a	CSI and 1-bit HARQ ACK/NACK (20 + 1 coded bits)
Format 2b	CSI and 2-bit HARQ ACK/NACK (20 + 2 coded bits)
Format 3	Multiple ACK/NACKs for carrier aggregation: up to 20 ACK/NACK bits plus optional SR, in 48 coded bits; see Section 28.3.2.1 for details.

The physical mapping of the PUCCH formats to the PUCCH regions is shown in Figure 16.8.

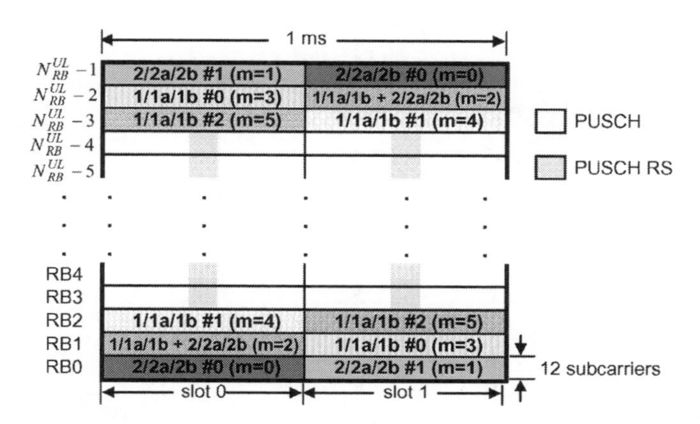

Figure 16.8: Physical mapping of PUCCH formats to PUCCH RBs or regions.

It can be seen that the PUCCH CSI formats 2/2a/2b are mapped and transmitted on the band-edge RBs (e.g. PUCCH region $m = 0, 1$) followed by a mixed PUCCH RB (if present, e.g. region $m = 2$) of CSI format 2/2a/2b and SR/HARQ ACK/NACK format 1/1a/1b, and then by PUCCH SR/HARQ ACK/NACK format 1/1a/1b (e.g. region $m = 4, 5$). The number of PUCCH RBs available for use by CSI format 2/2a/2b, N_{RB}^{2}, is indicated to the UEs in the cell by broadcast signalling.

16.3.3 Channel State Information Transmission on PUCCH (Format 2)

The PUCCH CSI channel structure (Format 2) for one slot with normal CP is shown in Figure 16.9. SC-FDMA symbols 1 and 5 are used for DM-RS transmissions in the case of the normal CP (while in the case of the extended CP only one RS is transmitted, on SC-FDMA symbol 3).

The number of RS symbols per slot results from a trade-off between channel estimation accuracy and the supportable code rate for the UCI bits. For a small number of UCI bits with a low SNR operating point (for a typical 1% target error rate), improving the channel estimation accuracy by using more RS symbols is more beneficial than being able to use a lower channel code rate. However, with larger numbers of UCI bits the required SNR operating point increases, and the higher code rate resulting from a larger overhead of RS symbols becomes more critical, thus favouring fewer RS symbols. In view of these factors, two RS symbols per slot (in case of normal CP) was considered to provide the best trade-off in terms of performance and RS overhead, given the payload sizes required.

10 CSI bits are channel coded with a rate 1/2 punctured (20, k) Reed-Muller code (see [2, Section 5.2.3.3]) to give 20 coded bits, which are then scrambled (in a similar

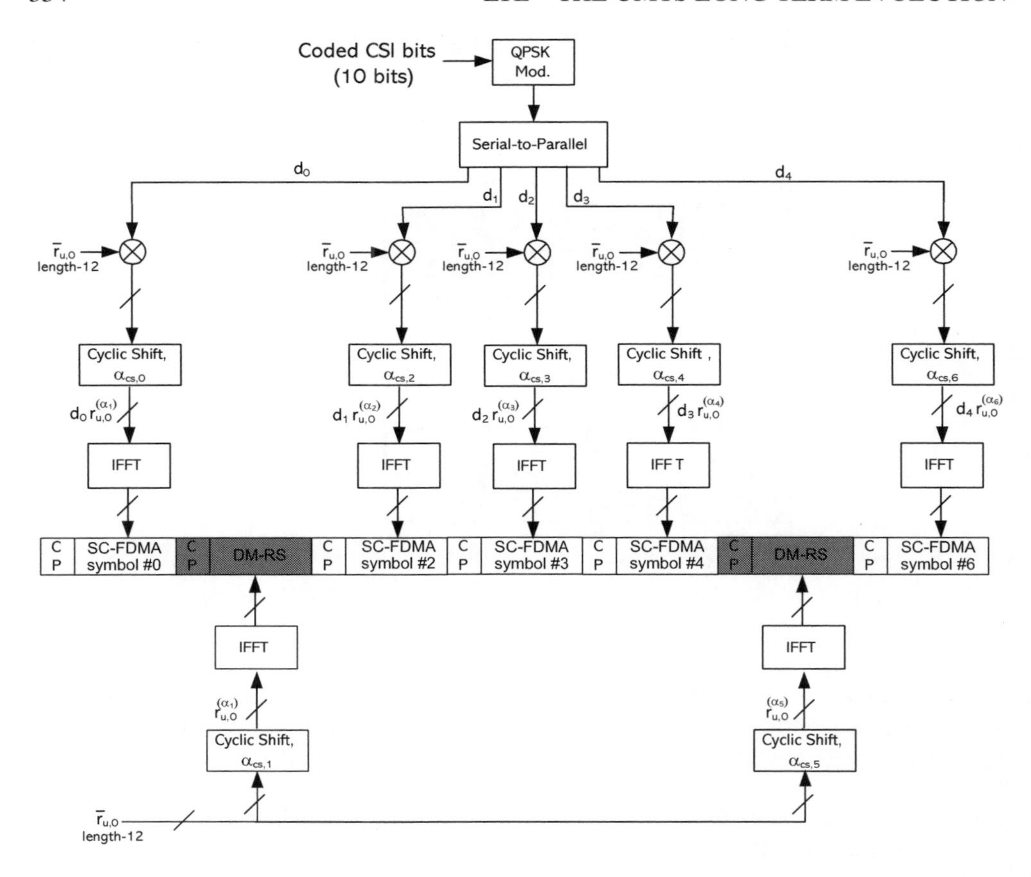

Figure 16.9: CSI channel structure for PUCCH format 2/2a/2b with normal CP for one slot.

way to PUSCH data with a length-31 Gold sequence) prior to QPSK constellation mapping. One QPSK modulated symbol is transmitted on each of the 10 SC-FDMA symbols in the subframe by modulating a cyclic time shift of the base RS sequence of length 12 prior to OFDM modulation. The 12 equally spaced cyclic time shifts allow 12 different UEs to be orthogonally multiplexed on the same PUCCH format 2 RB.

The DM-RS signal sequence (on the 2nd and 6th SC-FDMA symbols for the normal CP and the 4th symbol for the extended CP) is similar to the frequency-domain CSI signal sequence but without the CSI data modulation.

In order to provide inter-cell interference randomization, cell-specific symbol-level cyclic time-shift hopping is used, as described in Section 15.4. For example, the PUCCH cyclic time-shift index on SC-FDMA symbol l in even slots n_s is obtained by adding (modulo-12) a pseudo-random cell-specific PUCCH cyclic shift offset to the assigned cyclic time shift $n_{\text{RS}}^{\text{PUCCH}}$. Intra-cell interference randomization is achieved by cyclic time-shift remapping in the second slot as explained in Section 15.4.

A UE is semi-statically configured by higher layer signalling to report periodically different CQI, PMI, and RI types (see Section 10.2.1) on the CSI PUCCH using a PUCCH *resource index* $n^{(2)}_{\text{PUCCH}}$, which indicates both the PUCCH region and the cyclic time shift to be used. The PUCCH region m used for the PUCCH format 2/2a/2b transmission (see Figure 16.8) is given by (see [3, Section 5.4.3])

$$m = \left\lfloor \frac{n^{(2)}_{\text{PUCCH}}}{12} \right\rfloor \tag{16.3}$$

and the assigned cyclic time shift, $n^{\text{PUCCH}}_{\text{RS}}$, is given by

$$n^{\text{PUCCH}}_{\text{RS}} = n^{(2)}_{\text{PUCCH}} \bmod 12. \tag{16.4}$$

16.3.4 Multiplexing of CSI and HARQ ACK/NACK from a UE on PUCCH

The simultaneous transmission of HARQ ACK/NACK and CSI on the PUCCH can be enabled by UE-specific higher layer signalling.[5] If simultaneous transmission is not enabled and the UE needs to transmit HARQ ACK/NACK on the PUCCH in the same subframe in which a CSI report has been configured, the CSI is dropped and only HARQ ACK/NACK is transmitted using the transmission structure detailed in Section 16.3.5.

In subframes where the eNodeB scheduler allows for simultaneous transmission of CSI and HARQ ACK/NACK on the PUCCH from a UE, the CSI and the 1- or 2-bit HARQ ACK/NACK information needs to be multiplexed in the same PUCCH RB, while maintaining the low CM single-carrier property of the signal. The method used to achieve this is different for the cases of normal CP and extended CP, as described in the following sections.

16.3.4.1 Normal CP (Format 2a/2b)

The transmission structure for CSI is the same as described in Section 16.3.3. In order to transmit a 1- or 2-bit HARQ ACK/NACK together with CSI, the HARQ ACK/NACK bits (which are not scrambled) are BPSK/QPSK modulated as shown in Figure 16.10, resulting in a single HARQ ACK/NACK modulation symbol, d_{HARQ}. A positive acknowledgement (ACK) is encoded as a binary '1' and a negative acknowledgement (NACK) is encoded as a binary '0' (see [2, Section 5.2.3.4]).

The single HARQ ACK/NACK modulation symbol, d_{HARQ}, is then used to modulate the second RS symbol (SC-FDMA symbol 5) in each slot – i.e. ACK/NACK is signalled using the RS. This results in PUCCH formats 2a/2b. It can be seen from Figure 16.10 that the modulation mapping is such that a NACK (or NACK, NACK in the case of two downlink MIMO codewords) is mapped to +1, resulting in a default NACK in case neither ACK nor NACK is transmitted (so-called Discontinuous Transmission (DTX)), as happens if the UE fails to detect the downlink grant on the PDCCH. In other words, a DTX (no RS modulation) is interpreted as a NACK by the eNodeB, triggering a downlink retransmission.

As one of the RSs in a slot is modulated by the HARQ ACK/NACK symbol, a variety of ACK/NACK and CSI detection schemes are possible. In low-Doppler environments with

[5]Note that Release 10 also provides the option of configuring simultaneous PUCCH and PUSCH transmission from a UE – see Section 28.3.2.

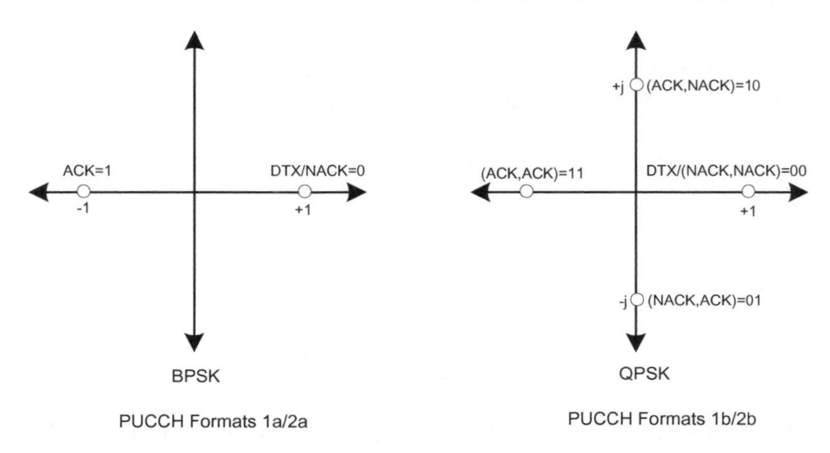

Figure 16.10: Constellation mapping for HARQ ACK/NACK.

little channel variation over the 0.5 ms slot, coherent detection of ACK/NACK and CSI can be achieved by using only the first RS symbol in the slot as the phase reference. Alternatively, to improve the channel estimation quality for CSI detection, an estimate of the HARQ ACK/NACK symbol can be used to undo the modulation on the second RS in the slot so that both RS symbols can be used for channel estimation and demodulation of CSI. In high-Doppler environments in which significant channel variations occur over a slot, relying on a single RS symbol for coherent detection degrades the performance of the ACK/NACK and CSI demodulation. In such cases, blind decoding or multiple hypothesis testing of the different ACK/NACK combinations can be used to decode the ACK/NACK and CSI, selecting the hypothesis that maximizes the correlation between the received signal and the estimated CSI [11] (i.e. a Maximum Likelihood detection).

16.3.4.2 Extended CP (Format 2)

In the case of the extended CP (with one RS symbol per slot), the 1- or 2-bit HARQ ACK/NACK is jointly encoded with the CSI resulting in a (20, $k_{\mathrm{CQI}} + k_{\mathrm{ACK/NACK}}$) Reed-Muller-based block code. A 20-bit codeword is transmitted on the PUCCH using the channel structure described in Section 16.3.3. The joint coding of the ACK/NACK and CSI is performed as shown in Figure 16.11. The largest number of information bits supported by the block code is 13, corresponding to $k_{\mathrm{CSI}} = 11$ CSI bits and $k_{\mathrm{ACK/NACK}} = 2$ bits (for two-codeword transmission in the downlink).

16.3.5 HARQ ACK/NACK Transmission on PUCCH (Format 1a/1b)

The PUCCH channel structure for HARQ ACK/NACK transmission with no CSI is shown in Figure 16.12 for one slot with normal CP. Three (two in case of extended CP) SC-FDMA symbols are used in the middle of the slot for RS transmission, with the remaining four SC-FDMA symbols being used for ACK/NACK transmission. Due to the small number of ACK/NACK bits, three RS symbols are used to improve the channel estimation accuracy for a lower SNR operating point than for the CSI structure described in Section 16.3.3.

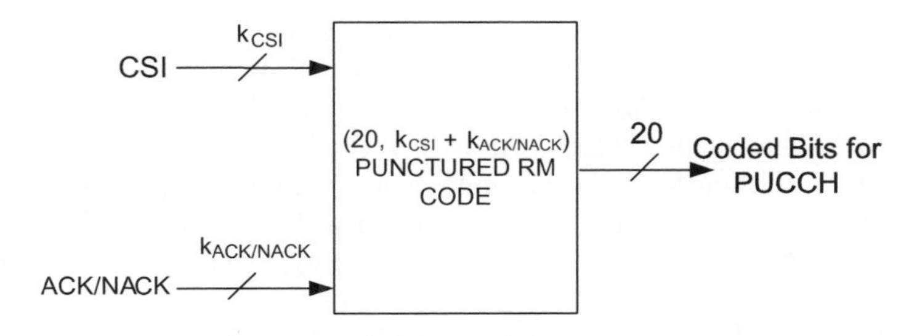

Figure 16.11: Joint coding of HARQ ACK/NACK and CSI for extended CP.

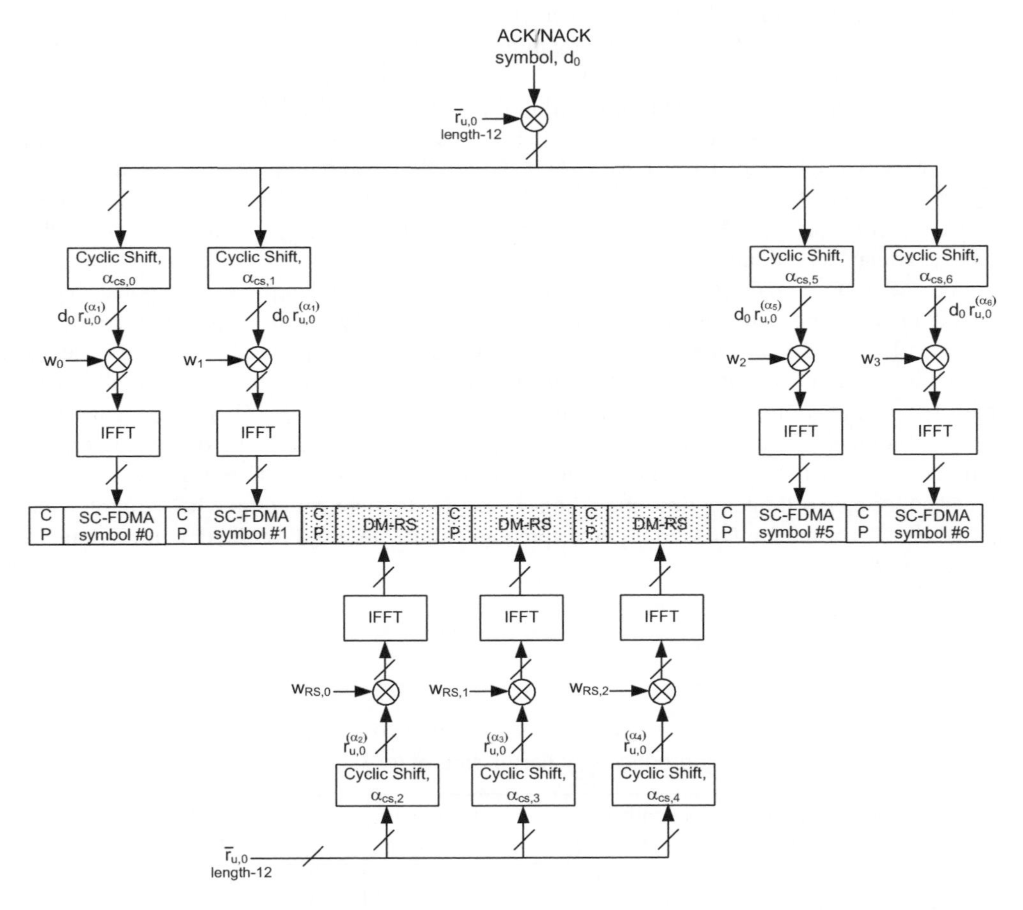

Figure 16.12: ACK/NACK structure – users are multiplexed using different cyclic shifts and time-domain spreading.

Both 1- and 2-bit acknowledgements are supported using BPSK and QPSK modulation respectively. The HARQ ACK/NACK bits are BPSK/QPSK modulated according to the modulation mapping shown in Figure 16.10 (see [3, Table 5.4.1-1]) resulting in a single HARQ ACK/NACK modulation symbol. An ACK is encoded as a binary '1' and a NACK as a binary '0' (see [2, Section 5.2.3.4]). The modulation mapping is the same as the mapping for 1- or 2-bit HARQ ACK/NACK when multiplexed with CSI for PUCCH formats 2a/2b. The modulation symbol is scrambled on a per-slot basis by either 1 or $e^{(-j\pi/2)}$ depending on the PUCCH resource index.

As in the case of CSI transmission, the one BPSK/QPSK modulated symbol (which is phase-rotated by 90 degrees in the second slot) is transmitted on each SC-FDMA data symbol by modulating a cyclic time shift of the base RS sequence of length 12 (i.e. frequency-domain CDM) prior to OFDM modulation. In addition, as mentioned in Section 16.3.1.1, time-domain spreading with orthogonal (Walsh–Hadamard or DFT) spreading codes is used to code-division-multiplex UEs. Thus, a large number of UEs (data and RSs) can be multiplexed on the same PUCCH RB using frequency-domain and time-domain code multiplexing. The RSs from the different UEs are multiplexed in the same way as the data SC-FDMA symbols.

For the cyclic time-shift multiplexing, the number of cyclic time shifts supported in an SC-FDMA symbol for PUCCH HARQ ACK/NACK RBs is configurable by a cell-specific higher-layer signalling parameter $\Delta_{\text{shift}}^{\text{PUCCH}} \in \{1, 2, 3\}$, indicating 12, 6, or 4 shifts respectively (see [3, Section 5.4.1]). The value selected by the eNodeB for $\Delta_{\text{shift}}^{\text{PUCCH}}$ can be based on the expected delay spread in the cell.

For the time-domain spreading CDM, the length-2 and length-4 orthogonal block spreading codes are based on Walsh–Hadamard codes, and the length-3 spreading codes are based on DFT codes as shown in Table 16.3. The number of supported spreading codes is limited by the number of RS symbols, as the multiplexing capacity of the RSs is smaller than that of the data symbols due to the smaller number of RS symbols. Therefore a subset of size s orthogonal spreading codes of a particular length L ($s \leq L$) is used depending on the number of RS SC-FDMA symbols. For the normal CP with four data SC-FDMA symbols and three supportable orthogonal time spreading codes (due to there being three RS symbols), the indices 0, 1, 2 of the length-4 orthogonal spreading codes are used for the data time-domain block spreading.

Table 16.3: Time-domain orthogonal spreading code sequences. Reproduced by permission of © 3GPP.

Orthogonal code sequence index	Length 2 Walsh–Hadamard	Length 3 DFT	Length 4 Walsh–Hadamard
0	$[+1 \quad +1]$	$[+1 \quad +1 \quad +1]$	$[+1 \quad +1 \quad +1 \quad +1]$
1	$[+1 \quad -1]$	$[1 \quad e^{j2\pi/3} \quad e^{j4\pi/3}]$	$[+1 \quad -1 \quad +1 \quad -1]$
2	N/A	$[1 \quad e^{j4\pi/3} \quad e^{j2\pi/3}]$	$[+1 \quad -1 \quad -1 \quad +1]$
3	N/A	N/A	$[+1 \quad +1 \quad -1 \quad -1]$

Similarly, for the extended CP case with four data SC-FDMA symbols but only two RS symbols, orthogonal spreading code indices 0 and 2 of length 4 are used for the data block spreading codes. For the length-4 orthogonal codes, the code sequences used are such that subsets of the code sequences result in the minimum inter-code interference in high-Doppler conditions where generally the orthogonality between the code sequences breaks down [12]. Table 16.4 summarizes the time-domain orthogonal spreading code lengths (i.e. spreading factors) for data and RS. The number of supportable orthogonal spreading codes is equal to the number of RS SC-FDMA symbols, $N_{\mathrm{RS}}^{\mathrm{PUCCH}}$.

Table 16.4: Spreading factors for time-domain orthogonal spreading codes for data and RS for PUCCH formats 1/1a/1b for normal and extended CP.

	Normal CP		Extended CP	
	Data, $N_{\mathrm{SF}}^{\mathrm{PUCCH}}$	RS, $N_{\mathrm{RS}}^{\mathrm{PUCCH}}$	Data, $N_{\mathrm{SF}}^{\mathrm{PUCCH}}$	RS, $N_{\mathrm{RS}}^{\mathrm{PUCCH}}$
Spreading factor	4	3	4	2

It should be noted that it is possible for the transmission of HARQ ACK/NACK and SRS to be configured in the same subframe. If this occurs, the eNodeB can also configure (by cell-specific broadcast signalling) the way in which these transmissions are to be handled by the UE. One option is for the ACK/NACK to take precedence over the SRS, such that the SRS is not transmitted and only HARQ ACK/NACK is transmitted in the relevant subframe, according to the PUCCH ACK/NACK structure in Figure 16.12. The alternative is for the eNodeB to configure the UEs to use a shortened PUCCH transmission in such subframes, whereby the last SC-FDMA symbol of the ACK/NACK (i.e. the last SC-FDMA symbol in the second slot of the subframe is not transmitted; this is shown in Figure 16.13).

This maintains the low CM single-carrier property of the transmitted signal, by ensuring that a UE never needs to transmit both HARQ ACK/NACK and SRS symbols simultaneously, even if both signals are configured in the same subframe. If the last symbol of the ACK/NACK is not transmitted in the second slot of the subframe, this is known as a *shortened* PUCCH format, as shown in Figure 16.14.[6] For the shortened PUCCH, the length of the time-domain orthogonal block spreading code is reduced by one (compared to the first slot shown in Figure 16.12). Hence, it uses the length-3 DFT basis spreading codes in Table 16.3 in place of the length-4 Walsh–Hadamard codes.

The frequency-domain HARQ ACK/NACK signal sequence on SC-FDMA symbol n is defined in [3, Section 5.4.1].

The number of HARQ ACK/NACK resource indices $N_{\mathrm{PUCCH,RB}}^{(1)}$ corresponding to cyclic-time-shift/orthogonal-code combinations that can be supported in a PUCCH RB is given by

$$N_{\mathrm{PUCCH,RB}}^{(1)} = c \cdot P, \quad c = \begin{cases} 3 & \text{normal cyclic prefix} \\ 2 & \text{extended cyclic prefix} \end{cases} \tag{16.5}$$

[6]Note that configuration of SRS in the same subframe as channel quality information or SR is not valid. Therefore the shortened PUCCH formats are only applicable for PUCCH formats 1a and 1b.

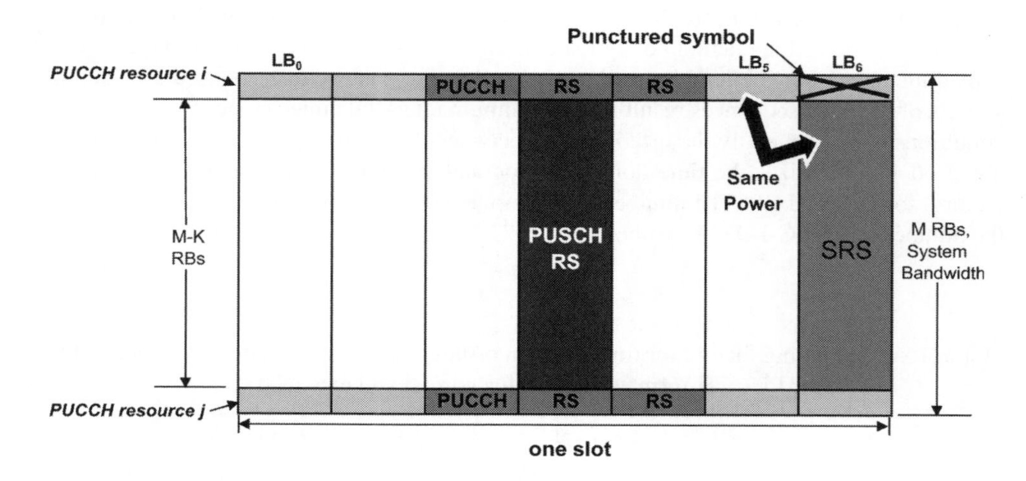

Figure 16.13: A UE may not simultaneously transmit on SRS and PUCCH or PUSCH, in order to avoid violating the single-carrier nature of the signal. Therefore, a PUCCH or PUSCH symbol may be punctured if SRS is transmitted.

where c is equal to the number of RS symbols, $P = 12/\Delta_{\text{shift}}^{\text{PUCCH}}$, and $\Delta_{\text{shift}}^{\text{PUCCH}} \in \{1, 2, 3\}$ is the number of equally spaced cyclic time shifts supported. For example, with the normal CP and $\Delta_{\text{shift}}^{\text{PUCCH}} = 2$, ACK/NACKs from 18 different UEs can be multiplexed in one RB.

As in the case of CSI (see Section 16.3.3), cyclic time shift hopping (described in Section 15.4) is used to provide inter-cell interference randomization.

In the case of semi-persistently scheduled downlink data transmissions on the PDSCH (see Section 4.4.2.1) without a corresponding downlink grant on the control channel PDCCH, the PUCCH ACK/NACK resource index $n_{\text{PUCCH}}^{(1)}$ to be used by a UE is semi-statically configured by higher layer signalling. This PUCCH ACK/NACK resource is used for ACK/NACK transmission corresponding to initial HARQ transmission. For dynamically-scheduled downlink data transmissions (including HARQ retransmissions for semi-persistent data) on the PDSCH (indicated by downlink assignment signalling on the PDCCH), the PUCCH HARQ ACK/NACK resource index $n_{\text{PUCCH}}^{(1)}$ is implicitly determined based on the index of the first Control Channel Element (CCE, see Section 9.3) of the PDCCH message.

The PUCCH region m used for the HARQ ACK/NACK with format 1/1a/1b transmission for the case with no mixed PUCCH region (shown in Figure 16.8), is given by

$$m = \left\lfloor \frac{n_{\text{PUCCH}}^{(1)}}{N_{\text{PUCCH, RB}}^{(1)}} \right\rfloor + N_{\text{RB}}^{(2)} \tag{16.6}$$

where $N_{\text{RB}}^{(2)}$ is the number of RBs that are available for PUCCH formats 2/2a/2b and is a cell-specific broadcast parameter (see [3, Section 5.4.3]).

The PUCCH resource index $n^{(1)}(n_s)$, corresponding to a combination of a cyclic time shift and orthogonal code ($n_{\text{RS}}^{\text{PUCCH}}$ and n_{oc}), within the PUCCH region m in even slots is given by

$$n^{(1)}(n_s) = n_{\text{PUCCH}}^{(1)} \bmod N_{\text{PUCCH, RB}}^{(1)} \quad \text{for } n_s \bmod 2 = 0 \tag{16.7}$$

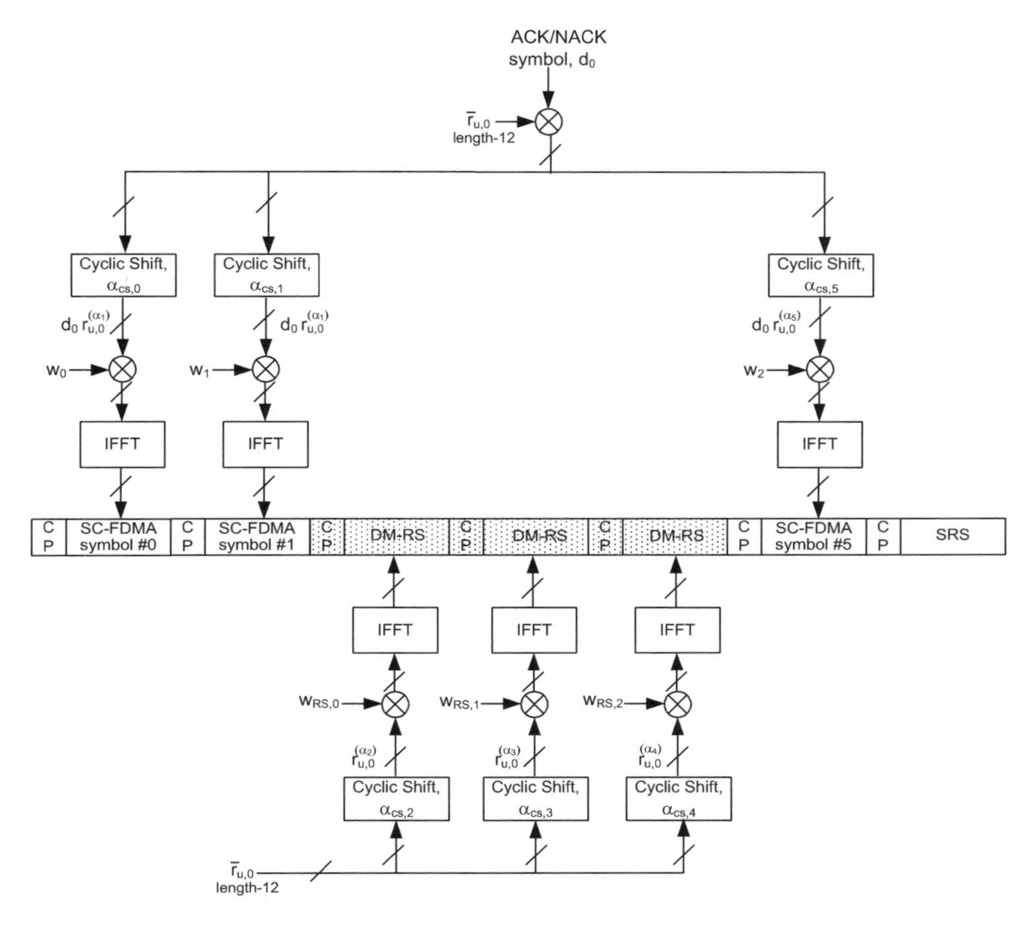

Figure 16.14: Shortened PUCCH ACK/NACK structure when simultaneous SRS and ACK/NACK is enabled in the cell.

The PUCCH resource index ($n_{\mathrm{RS}}^{\mathrm{PUCCH}}$, n_{oc}) allocation within a PUCCH RB format 1/1a/1b, is shown in Tables 16.5, 16.6 and 16.7, for $\Delta_{\mathrm{shift}}^{\mathrm{PUCCH}} \in \{1, 2, 3\}$ with 36, 18, and 12 resource indices respectively for the normal CP case [13]. For the extended CP, with two time-domain orthogonal spreading code sequences, only the first two columns of the orthogonal code sequence index, $n_{\mathrm{oc}} = 1, 2$ are used, resulting in 24, 12 and 8 resource indices for $\Delta_{\mathrm{shift}}^{\mathrm{PUCCH}} \in \{1, 2, 3\}$ respectively.

The PUCCH resources are first indexed in the cyclic time-shift domain, followed by the orthogonal time spreading code domain.

The cyclic time shifts used on *adjacent* orthogonal codes can also be staggered, providing the opportunity to separate the channel estimates prior to de-spreading. As high-Doppler breaks down the orthogonality between the spread blocks, offsetting the cyclic time shift

Table 16.5: PUCCH RB format 1/1a/1b resource index allocation, $\Delta_{\text{shift}}^{\text{PUCCH}} = 1$, 36 resource indices, normal CP.

	Orthogonal code sequence index, n_{oc}		
Cyclic shift index, $n_{\text{RS}}^{\text{PUCCH}}$	$n_{\text{oc}} = 0$	$n_{\text{oc}} = 1$	$n_{\text{oc}} = 2$
0	0	12	24
1	1	13	25
2	2	14	26
3	3	15	27
4	4	16	28
5	5	17	29
6	6	18	30
7	7	19	31
8	8	20	32
9	9	21	33
10	10	22	34
11	11	23	35

Table 16.6: PUCCH RB format 1/1a/1b resource index allocation, $\Delta_{\text{shift}}^{\text{PUCCH}} = 2$, 18 resource indices, normal CP.

	Orthogonal code sequence index, n_{oc}		
Cyclic shift index, $n_{\text{RS}}^{\text{PUCCH}}$	$n_{\text{oc}} = 0$	$n_{\text{oc}} = 1$	$n_{\text{oc}} = 2$
0	0		12
1		6	
2	1		13
3		7	
4	2		14
5		8	
6	3		15
7		9	
8	4		16
9		10	
10	5		17
11		11	

values within each SC-FDMA symbol can restore orthogonality at moderate delay spreads. This can enhance the tracking of high-Doppler channels [14].

In order to randomize intra-cell interference, PUCCH resource index remapping is used in the second slot [15]. Index remapping includes both cyclic shift remapping and orthogonal block spreading code remapping (similar to the case of CSI – see Section 16.3.3).

The PUCCH resource index remapping function in an odd slot is based on the PUCCH resource index in the even slot of the subframe, as defined in [3, Section 5.4.1].

Table 16.7: PUCCH RB format 1/1a/1b resource index allocation, $\Delta_{\text{shift}}^{\text{PUCCH}} = 3$, 12 resource indices, normal CP.

	Orthogonal code sequence index, n_{oc}		
Cyclic shift index, $n_{\text{RS}}^{\text{PUCCH}}$	$n_{\text{oc}} = 0$	$n_{\text{oc}} = 1$	$n_{\text{oc}} = 2$
0	0		
1		4	
2			7
3	1		
4		5	
5			8
6	2		
7			
8			
9	3		
10		6	
11			9

16.3.6 Multiplexing of CSI and HARQ ACK/NACK in the Same (Mixed) PUCCH RB

The multiplexing of CSI and HARQ ACK/NACK in different PUCCH RBs can in general simplify the system. However, in the case of small system bandwidths, such as 1.4 MHz, the control signalling overhead can become undesirably high with separate CSI and ACK/NACK RB allocations (two out of a total of 6 RBs for control signalling in 1.4 MHz system bandwidths). Therefore, multiplexing of CSI and ACK/NACK from different UEs in the same mixed PUCCH RB is supported in LTE to reduce the total control signalling overhead.

The ZC cyclic time-shift structure facilitates the orthogonal multiplexing of channel quality information and ACK/NACK signals with different numbers of RS symbols. This is achieved by assigning different sets of adjacent cyclic time shifts to CSI and ACK/NACK signals [16] as shown in Table 16.8. As can be seen from this table, $N_{\text{cs}}^{(1)} \in \{0, 1, \ldots, 7\}$ cyclic time shifts are used for PUCCH ACK/NACK formats 1/1a/1b in the mixed PUCCH RB case, where $N_{\text{cs}}^{(1)}$ is a cell-specific broadcast parameter (see [3, Section 5.4]) restricted to integer multiples of $\Delta_{\text{shift}}^{\text{PUCCH}}$. A guard cyclic time shift is used between the ACK/NACK and CSI cyclic shift resources to improve orthogonality and channel separation between UEs transmitting CSI and those transmitting ACK/NACK. To avoid mixing of the cyclic time shifts for ACK/NACK and CSI, the cyclic time shift (i.e. the PUCCH resource index remapping function) for the odd slot of the subframe is not used; the same cyclic time shift as in the first slot of the subframe is used.

16.3.7 Scheduling Request (SR) Transmission on PUCCH (Format 1)

The structure of the SR PUCCH format 1 is the same as that of the ACK/NACK PUCCH format 1a/1b explained in Section 16.3.5, where a cyclic time shift of the base RS sequence is modulated with time-domain orthogonal block spreading. The SR uses simple On–Off

Table 16.8: Multiplexing of ACK/NACK (format 1/1a/1b) and CSI (format 2/2a/2b) from different UEs in the same (mixed) PUCCH RB by using different sets of cyclic time shifts.

Cyclic shift index	Cyclic shift index allocation
0	Format 1/1a/1b (HARQ ACK/NACK, SR) cyclic shifts
1	
2	
⋮	
$N_{cs}^{(1)}$	
$N_{cs}^{(1)} + 1$	Guard cyclic shift
$N_{cs}^{(1)} + 2$	Format 2/2a/2b (CSI) cyclic shifts
⋮	
10	
11	Guard cyclic shift

keying, with the UE transmitting an SR using the modulation symbol $d(0) = +1$ (i.e. the same constellation point as is used for DTX/NACK or DTX/(NACK,NACK) for ACK/NACK transmission using PUCCH format 1a/1b – see Figure 16.10) to request a PUSCH resource (positive SR transmission), and transmitting nothing when it does not request to be scheduled (negative SR).

Since the HARQ ACK/NACK structure is reused for the SR, different PUCCH resource indices (i.e. different cyclic time shift/orthogonal code combinations) in the same PUCCH region can be assigned for SR (Format 1) or HARQ ACK/NACK (Format 1a/1b) from different UEs. This results in orthogonal multiplexing of SR and HARQ ACK/NACK in the same PUCCH region. The PUCCH resource index to be used by a UE for SR transmission, $m_{\text{PUCCH,SRI}}^{(1)}$, is configured by UE-specific higher-layer signalling.

If a UE needs to transmit a positive SR in the same subframe as a scheduled CSI transmission, the CSI is dropped and only the SR is transmitted, in order to maintain the low CM of the transmit signal. In the case of SRS coinciding with an ACK/NACK or positive SR in the same subframe, the UE drops its SRS transmission if the parameter 'ackNackSRS-SimultaneousTransmission' is set to 'FALSE' and transmits the SRS otherwise (see [6, Section 8.2]).

If an SR and ACK/NACK happen to coincide in the same subframe, the UE transmits the ACK/NACK on the assigned SR PUCCH resource for a positive SR and transmits ACK/NACK on its assigned ACK/NACK PUCCH resource in case of a negative SR (see [6, Section 7.3]).

The expected behaviour in some other cases of different types of UCI and SRS coinciding in the same subframe can be found in [17].

16.4 Multiplexing of Control Signalling and UL-SCH Data on PUSCH

When UCI is to be transmitted in a subframe in which the UE has been allocated transmission resources for the PUSCH, the UCI is multiplexed together with the UL-SCH data prior to DFT spreading, in order to preserve the low CM single-carrier property; the PUCCH is never transmitted in the same subframe as the PUSCH in Releases 8 and 9. The multiplexing of CQI/PMI, HARQ ACK/NACK, and RI with the PUSCH data symbols onto uplink resource elements (REs) is shown in Figure 16.15.

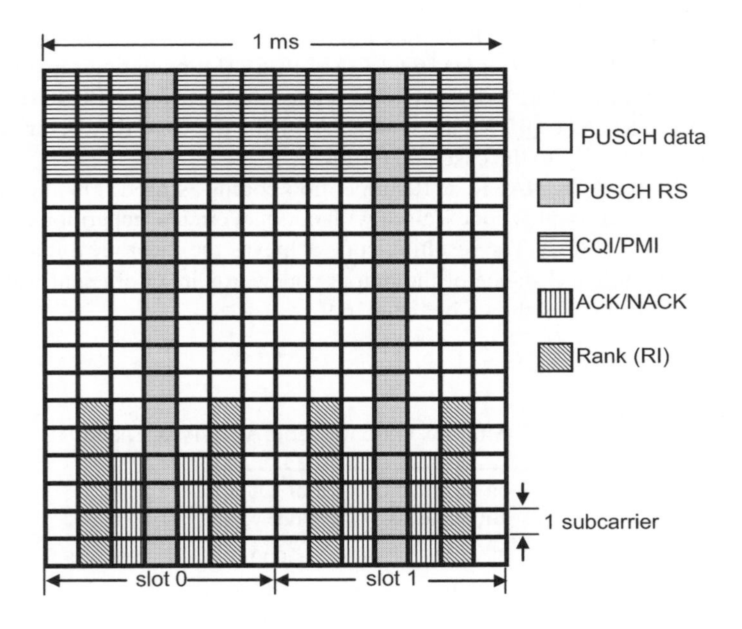

Figure 16.15: Multiplexing of control signalling with UL-SCH data.

The number of REs used for each of CQI/PMI, ACK/NACK and RI is based on the MCS assigned for PUSCH and an offset parameter, $\Delta^{\text{CQI}}_{\text{offset}}$, $\Delta^{\text{HARQ-ACK}}_{\text{offset}}$ or $\Delta^{\text{RI}}_{\text{offset}}$, which is semi-statically configured by higher-layer signalling (see [2, Section 5.2.2.6]). This allows different code rates to be used for the different types of UCI. PUSCH data and UCI are never mapped to the same RE. UCI is mapped in such a way that it is present in both slots of the subframe. Since the eNodeB has prior knowledge of UCI transmission, it can easily demultiplex the UCI and data.

As shown in Figure 16.15, CQI/PMI resources are placed at the beginning of the UL-SCH data resources and mapped sequentially to all SC-FDMA symbols on one subcarrier before continuing on the next subcarrier. The UL-SCH data is rate-matched (see Section 10.3.2.4) around the CQI/PMI data. The same modulation order as UL-SCH data on PUSCH is used for CQI/PMI. For small CQI and/or PMI report sizes up to 11 bits, a (32, k) block code, similar to the one used for PUCCH, is used, with optional circular repetition of encoded data (see [2, Section 5.2.2.6.4]); no Cyclic Redundancy Check (CRC) is applied. For large CSI reports (> 11 bits), an 8-bit CRC is attached and channel coding and rate matching is performed using the tail-biting convolutional code as described in Chapter 10.

The HARQ ACK/NACK resources are mapped to SC-FDMA symbols by puncturing the UL-SCH PUSCH data. Positions next to the RS are used, so as to benefit from the best possible channel estimation. The maximum amount of resource for HARQ ACK/NACK is 4 SC-FDMA symbols.

The coded RI symbols are placed next to the HARQ ACK/NACK symbol positions irrespective of whether ACK/NACK is actually present in a given subframe. The modulation of the 1- or 2-bit ACK/NACK or RI is such that the Euclidean distance of the modulation symbols carrying ACK/NACK and RI is maximized (see [2, Section 5.2.2.6]). The outermost constellation points of the higher-order 16/64-QAM PUSCH modulations are used, resulting in increased transmit power for ACK/NACK/RI relative to the average PUSCH data power.

The coding of the RI and CQI/PMI are separate, with the UL-SCH data being rate-matched around the RI REs similarly to the case of CQI/PMI.

In the case of 1-bit ACK/NACK or RI, repetition coding is used. For the case of 2-bit ACK/NACK/RI, a (3, 2) simplex code is used with optional circular repetition of the encoded data (see [2, Section 5.2.2.6]). The resulting code achieves the theoretical maximum values of the minimum Hamming distance of the output codewords in an efficient way. The (3, 2) simplex codeword mapping is shown in Table 16.9.

Table 16.9: (3, 2) Simplex code for 2-bit ACK/NACK and RI.

2-bit Information Bit Sequence	3-bit Output Codeword
00	000
01	011
10	101
11	110

Control signalling (using QPSK modulation) can also be scheduled to be transmitted on PUSCH without UL-SCH data. The control signalling (CQI/PMI, RI, and/or HARQ ACK/NACK) is multiplexed and scrambled prior to DFT spreading, in order to preserve the low CM single-carrier property. The multiplexing of HARQ ACK/NACK and RI with the CQI/PMI QPSK symbols onto uplink REs is similar to that shown in Figure 16.15. HARQ ACK/NACK is mapped to SC-FDMA symbols next to the RS, by puncturing the CQI data and RI symbols, irrespective of whether ACK/NACK is actually present in a given subframe. The number of REs used for each of ACK/NACK and RI is based on a reference MCS for CQI/PMI and offset parameters, $\Delta_{\text{offset}}^{\text{CQI}}$, $\Delta_{\text{offset}}^{\text{HARQ-ACK}}$ or $\Delta_{\text{offset}}^{\text{RI}}$. The reference CQI/PMI MCS is computed from the CSI payload size and resource allocation. The channel coding and rate matching of the control signalling without UL-SCH data is the same as that of multiplexing control with UL-SCH data as described above.

16.5 ACK/NACK Repetition

It is possible for the network to configure a UE to repeat each ACK/NACK transmission in multiple successive subframes. This is particularly beneficial for ensuring the reliability of the ACK/NACK signalling from power-limited UEs at the cell edge, especially in large cell deployments.

The number of subframes over which each ACK/NACK transmission is repeated is signalled by the UE-specific parameter 'ackNackRepetition'.

16.6 Multiple-Antenna Techniques

In LTE Releases 8 and 9, simultaneous transmissions from multiple-transmit antennas of a single UE are not supported. Only a single power-amplifier is assumed to be available at the UE. However, LTE does support closed-loop antenna selection transmit diversity in the uplink from UEs which have multiple transmit antennas.

LTE is also designed to support uplink SDMA (or 'Virtual MU-MIMO'), which is discussed in more detail in Section 16.6.2.

More advanced multiple-antenna techniques, including closed-loop spatial multiplexing for SU-MIMO and transmit diversity for control signalling are introduced in Release 10 for LTE-Advanced, as explained in Section 29.4.

16.6.1 Closed-Loop Switched Antenna Diversity

Uplink closed-loop antenna selection (for up to two transmit antennas) is supported as an optional UE capability in LTE and configured by higher layers (see [18, Section 4.3.4.1]).

If a UE signals that it supports uplink antenna selection, the eNodeB may take this capability into consideration when configuring and scheduling the UE.[7]

16.6.1.1 UE Antenna Selection Indication for PUSCH

When closed-loop antenna selection is enabled, the eNodeB indicates which antenna should be used for the PUSCH by implicitly coding this information in the uplink scheduling grant (Downlink Control Information (DCI) Format 0[8] – see Section 9.3): the 16 CRC parity bits are scrambled (modulo-2 addition) by one of two antenna selection masks [19], $\langle 0, 0, 0, 0, 0, 0, 0, 0, 0, 0, 0, 0, 0, 0, 0, 1\rangle$ for the UE's first transmit antenna and $\langle 0, 0, 0, 0, 0, 0, 0, 0, 0, 0, 0, 0, 0, 0, 0, 1\rangle$ for the second antenna.[9] The antenna selection mask is applied in addition to the UE-ID masking which indicates for which UE the scheduling grant is intended. This implicit encoding avoids the use of an explicit antenna selection bit which would result in an increased overhead for UEs not supporting (or not configured) for transmit antenna selection.

[7] Alternatively, the eNodeB may permit the UE to use open-loop antenna selection, in which case the UE is free to determine which antenna to transmit from. This may be based on uplink–downlink channel reciprocity, for example in the case of TDD operation (see Section 23.5.2.5.)

[8] Note that closed-loop antenna selection is not supported in conjunction with DCI Format 4 for SU-MIMO transmission in Release 10.

[9] This is applicable only for FDD and half-duplex FDD.

The minimum Hamming distance between the antenna selection masks is only 1 rather than the maximum possible Hamming distance of 16; since the CRC is masked by both the antenna selection indicator and the 16-bit UE-ID, the minimum Hamming distance between the correct UE-ID/antenna selection mask and the nearest erroneous UE-ID/antenna selection mask is 1 for any antenna selection mask. Out of the possible $2^{16} - 1$ incorrect masks, the vast majority ($2^{16} - 2$) result in the misidentification of the UE-ID, such that the performance is similar regardless of the Hamming distance between antenna selection masks. The primary advantage of using the chosen masks is the ease of implementation due to simpler half-space identification, as the eNodeB can allocate UE-IDs with a fixed Most Significant Bit (e.g. MSB set to '0', or, equivalently, UE-IDs from 0 to $2^{15} - 1$). The UE-ID can be detected directly from the 15 least significant bits of the decoded mask without needing to use the transmitted antenna selection mask (bit 16).

The UE behaviour for adaptive/non-adaptive HARQ retransmissions when configured for antenna selection is as follows [19]:

- **Adaptive HARQ.** The antenna indicator (via CRC masking) is always sent in the uplink grant to indicate which antenna to use. For example, for a high-Doppler UE with adaptive HARQ, the eNodeB might instruct the UE to alternate between the transmit antennas or, alternatively, to select the primary antenna. In typical UE implementations, a transmit antenna gain imbalance of 3 to 6 dB between the secondary and primary antennas is not uncommon.
- **Non-adaptive HARQ.** The UE behaviour is unspecified as to which antenna to use. Thus, for low-Doppler conditions, the UE could use the same antenna as that signalled in the uplink grant, while at high-Doppler the UE could hop between antennas or just select the primary antenna. For large numbers of retransmissions with non-adaptive HARQ, the antenna indicated in the uplink grant may not be the best and it is better to let the UE select the antenna to use. If the eNodeB wishes to instruct the UE to use a specific antenna for the retransmissions, it can use adaptive HARQ.

16.6.1.2 Antenna Selection for SRS

If the eNodeB enables a UE's closed-loop antenna selection capability, the SRS transmissions then alternate between the transmit antennas in successive configured SRS transmission subframes, irrespective of whether frequency hopping is enabled or disabled, except when the UE is configured for a single one-shot SRS transmission (see [6, Section 8.2]).

16.6.2 Multi-User 'Virtual' MIMO or SDMA

Uplink MU-MIMO consists of multiple UEs transmitting on the same set of RBs, each using a single transmit antenna. From the point of view of an individual UE, such a mode of operation is hardly visible, being predominantly a matter for the eNodeB to handle in terms of scheduling and uplink reception.

However, in order to support uplink MU-MIMO, LTE specifically provides orthogonal DM-RS using different cyclic time shifts (see Section 15.2.2) to enable the eNodeB to derive independent channel estimates for the uplink from each UE.

A cell can assign up to eight different cyclic time shifts using the 3-bit PUSCH cyclic time shift offset in the uplink scheduling grant. As a maximum of eight cyclic time shifts can be assigned, SDMA of up to eight UEs can be supported in a cell. SDMA between cells (i.e. uplink inter-cell cooperation) is supported in LTE by assigning the same base sequence-groups and/or RS hopping patterns to the different cells as explained in Section 15.3.

16.7 Summary

The main uplink physical channels are the PUSCH for data transmission and the PUCCH for control signalling.

The PUSCH supports resource allocation for both frequency-selective scheduling and frequency-diverse transmissions, the latter being by means of intra- and/or inter-subframe frequency hopping. In Release 10, dual-cluster resource allocations on the PUSCH are also supported, as explained in Section 28.3.6.2.

Control signalling (consisting of ACK/NACK, CQI/PMI and RI) is carried by the PUCCH when no PUSCH resources have been allocated. The PUCCH is deliberately mapped to resource blocks near the edge of the system bandwidth, in order to reduce out-of-band emissions caused by data transmissions on the inner RBs, as well as maximizing flexibility for PUSCH scheduling in the central part of the band.

The control signalling from multiple UEs is multiplexed via orthogonal coding by using cyclic time shifts and/or time-domain block spreading.

Simple multiple antenna techniques in the uplink were already incorporated in LTE Releases 8 and 9, in particular through the support of closed-loop switched antenna diversity and SDMA. These techniques are cost-effective for a UE implementation, as they do not assume simultaneous transmissions from multiple UE antennas. SU-MIMO is introduced in Release 10, as explained in Section 29.4.

References[10]

[1] Motorola, 'R1-072671: Uplink Channel Interleaving', www.3gpp.org, 3GPP TSG RAN WG1, meeting 49bis, Orlando, USA, June 2007.

[2] 3GPP Technical Specification 36.212, 'Evolved Universal Terrestrial Radio Access (E-UTRA); Multiplexing and Channel Coding', www.3gpp.org.

[3] 3GPP Technical Specification 36.211, 'Evolved Universal Terrestrial Radio Access (E-UTRA); Physical Channels and Modulation', www.3gpp.org.

[4] B. Classon, P. Sartori, V. Nangia, X. Zhuang and K. Baum, 'Multi-dimensional Adaptation and Multi-User Scheduling Techniques for Wireless OFDM Systems' in *Proc. IEEE International Conference on Communications*, Anchorage, Alaska, May 2003.

[5] Motorola, 'R1-071340: Considerations and Recommendations for UL Sounding RS', www.3gpp.org, 3GPP TSG RAN WG1, meeting 48bis, St Julian's, Malta, March 2007.

[6] 3GPP Technical Specification 36.213, 'Evolved Universal Terrestrial Radio Access (E-UTRA); Physical Layer Procedures', www.3gpp.org.

[7] Motorola, 'R1-073756: Benefit of Non-Persistent UL Sounding for Frequency Hopping PUSCH', www.3gpp.org, 3GPP TSG RAN WG1, meeting 50, Athens, Greece, August 2007.

[10] All web sites confirmed 1st March 2011.

[8] A. Ghosh, R. Ratasuk, W. Xiao, B. Classon, V. Nangia, R. Love, D. Schwent and D. Wilson, 'Uplink Control Channel Design for 3GPP LTE' in *Proc. IEEE International Symposium on Personal, Indoor and Mobile Radio Communications*, Athens, Greece, September 2007.

[9] Motorola, 'R1-070084: Coexistance Simulation Results for 5 MHz E-UTRA → UTRA FDD Uplink with Revised Simulation Assumptions', www.3gpp.org, 3GPP TSG RAN WG4, meeting 42, St Louis,USA, February 2007.

[10] S. Zhou, G. B. Giannakis and Martret C. L., 'Chip-Interleaved Block-Spread Code Division Multiple Access'. *IEEE Trans. on Communications*, Vol. 50, pp. 235–248, February 2002.

[11] Texas Instruments, 'R1-080190: Embedding ACK/NAK in CQI Reference Signals and Receiver Structures', www.3gpp.org, 3GPP TSG RAN WG1, meeting 51bis, Sevilla, Spain, January 2008.

[12] Samsung, 'R1-073564: Selection of Orthogonal Cover and Cyclic Shift for High Speed UL ACK Channels', www.3gpp.org, 3GPP TSG RAN WG1, meeting 50, Athens, Greece, August 2007.

[13] Samsung, Panasonic, Nokia, Nokia Siemens Networks and Texas Instruments, 'R1-080035: Joint Proposal on Uplink ACK/NACK Channelization', www.3gpp.org, 3GPP TSG RAN WG1, meeting 51bis, Sevilla, Spain, January 2008.

[14] Motorola, 'R1-062072: Uplink Reference Signal Multiplexing Structures for E-UTRA', www.3gpp.org, 3GPP TSG RAN WG1, meeting 46, Tallinn, Estonia, August 2006.

[15] Panasonic, Samsung and ETRI, 'R1-080983: Way forward on the Cyclic Shift Hopping for PUCCH', www.3gpp.org, 3GPP TSG RAN WG1, meeting 52, Sorrento, Italy, February 2008.

[16] Nokia Siemens Networks, Nokia and Texas Instruments, 'R1-080931: ACK/NACK channelization for PRBs containing both ACK/NACK and CQI', www.3gpp.org, 3GPP TSG RAN WG1, meeting 52, Sorrento, Italy, February 2008.

[17] Alcatel-Lucent Shanghai Bell, Alcatel-Lucent, LG Electronics, Panasonic and Qualcomm, 'R1-104135: Confirmation of UE behaviour in case of simultaneously-triggered SR, SRS and CQI', www.3gpp.org, 3GPP TSG RAN WG1, meeting 61bis, Dresden, Germany, June 2010.

[18] 3GPP Technical Specification 36.306, 'Evolved Universal Terrestrial Radio Access (E-UTRA); User Equipment (UE) Radio Access Capabilities', www.3gpp.org.

[19] Motorola, Mitsubishi Electric and Nortel, 'R1-081928: Way Forward on Indication of UE Antenna Selection for PUSCH', www.3gpp.org, 3GPP TSG RAN WG1, meeting 53, Kansas City, USA, May 2008.

[20] Motorola, 'R1-082109: UE Transmit Antenna Selection', www.3gpp.org, 3GPP TSG RAN WG1, meeting 53, Kansas City, USA, May 2008.

17

Random Access

Pierre Bertrand and Jing Jiang

17.1 Introduction

An LTE User Equipment (UE) can only be scheduled for uplink transmission if its uplink transmission timing is synchronized. The LTE Random Access CHannel (RACH) therefore plays a key role as an interface between non-synchronized UEs and the orthogonal transmission scheme of the LTE uplink radio access. In this chapter, the main roles of the LTE RACH are elaborated, together with its differences from the Wideband Code Division Multiple Access (WCDMA) RACH. The rationale for the design of the LTE Physical Random Access CHannel (PRACH) is explained, and some possible implementation options are discussed for both the UE and the eNodeB.

17.2 Random Access Usage and Requirements in LTE

In WCDMA, the RACH is primarily used for initial network access and short message transmission. LTE likewise uses the RACH for initial network access, but in LTE the RACH cannot carry any user data, which is exclusively sent on the Physical Uplink Shared CHannel (PUSCH). Instead, the LTE RACH is used to achieve uplink time synchronization for a UE which either has not yet acquired, or has lost, its uplink synchronization. Once uplink synchronization is achieved for a UE, the eNodeB can schedule orthogonal uplink transmission resources for it. Relevant scenarios in which the RACH is used are therefore:

(1) A UE in RRC_CONNECTED state, but not uplink-synchronized, needing to send new uplink data or control information (e.g. an event-triggered measurement report);

LTE – The UMTS Long Term Evolution: From Theory to Practice, Second Edition.
Stefania Sesia, Issam Toufik and Matthew Baker.

(2) A UE in RRC_CONNECTED state, but not uplink-synchronized, needing to receive new downlink data, and therefore to transmit corresponding ACKnowledgement/Negative ACKnowledgement (ACK/NACK) in the uplink;

(3) A UE in RRC_CONNECTED state, handing over from its current serving cell to a target cell;

(4) For positioning purposes in RRC_CONNECTED state, when timing advance is needed for UE positioning (See Section 19.4);

(5) A transition from RRC_IDLE state to RRC_CONNECTED, for example for initial access or tracking area updates;

(6) Recovering from radio link failure.

One additional exceptional case is that an uplink-synchronized UE is allowed to use the RACH to send a Scheduling Request (SR) if it does not have any other uplink resource allocated in which to send the SR. These roles require the LTE RACH to be designed for low latency, as well as good detection probability at low Signal-to-Noise Ratio (SNR) (for cell edge UEs undergoing handover) in order to guarantee similar coverage to that of the PUSCH and Physical Uplink Control CHannel (PUCCH).[1]

A successful RACH attempt should allow subsequent UE transmissions to be inserted among the scheduled synchronized transmissions of other UEs. This sets the required timing estimation accuracy which must be achievable from the RACH, and hence the required RACH transmission bandwidth: due to the Cyclic Prefix (CP) of the uplink transmissions, the LTE RACH only needs to allow for round-trip delay estimation (instead of the timing of individual channel taps), and this therefore reduces the required RACH bandwidth compared to WCDMA.

This is beneficial in minimizing the overhead of the RACH, which is another key consideration. Unlike in WCDMA, the RACH should be able to be fitted into the orthogonal time-frequency structure of the uplink, so that an eNodeB which wants to avoid interference between the RACH and scheduled PUSCH/PUCCH transmissions can do so. It is also important that the RACH is designed so as to minimize interference generated to adjacent scheduled PUSCH/PUCCH transmissions.

17.3 Random Access Procedure

The LTE random access procedure comes in two forms, allowing access to be either *contention-based* (implying an inherent risk of collision) or *contention-free*.

A UE initiates a contention-based random access procedure for all use-cases listed in Section 17.2. In this procedure, a random access preamble signature is randomly chosen by the UE, which may result in more than one UE simultaneously transmitting the same signature, leading to a need for a subsequent contention resolution process.

For the use-cases (2) (new downlink data) and (3) (handover) the eNodeB has the option of preventing contention occurring by allocating a dedicated signature to a UE, resulting in contention-free access. This is faster than contention-based access – a particularly important factor for the case of handover, which is time-critical.

[1] See Chapter 26 for an analysis of the coverage of the PUSCH and PUCCH.

Unlike in WCDMA, a fixed number (64) of preamble signatures is available in each LTE cell, and the operation of the two types of RACH procedure depends on a partitioning of these signatures between those for contention-based access and those reserved for allocation to specific UEs on a contention-free basis.

The two procedures are outlined in the following sections.

17.3.1 Contention-Based Random Access Procedure

The contention-based procedure consists of four-steps as shown in Figure 17.1:

- Step 1: Preamble transmission;
- Step 2: Random access response;
- Step 3: Layer 2 / Layer 3 (L2/L3) message;
- Step 4: Contention resolution message.

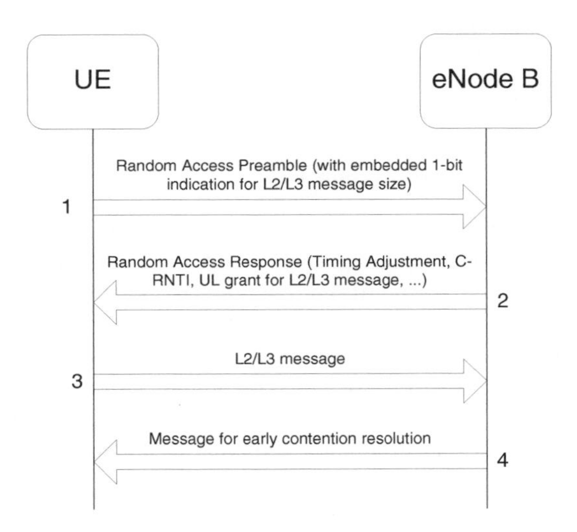

Figure 17.1: Contention-based random access procedure. Reproduced by permission of © 3GPP.

17.3.1.1 Step 1: Preamble Transmission

The UE selects one of the $64 - N_{cf}$ available PRACH contention-based signatures, where N_{cf} is the number of signatures reserved by the eNodeB for contention-free RACH. The set of contention-based signatures is further subdivided into two subgroups, so that the choice of signature can carry one bit of information relating to the amount of transmission resource needed to transmit the message at Step 3. The broadcast system information indicates which signatures are in each of the two subgroups (each subgroup corresponding to one value of the one bit of information), as well as the meaning of each subgroup. The UE selects a signature from the subgroup corresponding to the size of transmission resource needed for the appropriate RACH use case (some use cases require only a few bits to be transmitted at Step 3, so choosing the small message size avoids allocating unnecessary uplink resources). In selecting the appropriate resource size to indicate, the UE takes into account the current downlink path-loss and the required transmission power for the Step 3 message, in order to avoid being granted resources in Step 3 for a message size that would need a transmission power exceeding that which the UE's maximum power would allow. The transmission power required for the Step 3 message is calculated based on some parameters broadcast by the eNodeB, in order that the network has some flexibility to adapt the maximum size for the Step 3 message. The eNodeB can control the number of signatures in each subgroup according to the observed loads in each group.

The initial preamble transmission power setting is based on an open-loop estimation with full compensation for the path-loss. This is designed to ensure that the received power of the preambles is independent of the path-loss. The UE estimates the path-loss by averaging measurements of the downlink Reference Signal Received Power (RSRP). The eNodeB may also configure an additional power offset, depending for example on the desired received Signal to Interference plus Noise Ratio (SINR), the measured uplink interference and noise level in the time-frequency slots allocated to RACH preambles, and possibly also on the preamble format (see Section 17.4.2.2).

17.3.1.2 Step 2: Random Access Response

The Random Access Response (RAR) is sent by the eNodeB on the Physical Downlink Shared CHannel (PDSCH), and addressed with an ID, the Random Access Radio Network Temporary Identifier (RA-RNTI), identifying the time-frequency slot in which the preamble was detected. If multiple UEs had collided by selecting the same signature in the same preamble time-frequency resource, they would each receive the RAR.

The RAR conveys the identity of the detected preamble, a timing alignment instruction to synchronize subsequent uplink transmissions from the UE, an initial uplink resource grant for transmission of the Step 3 message, and an assignment of a temporary Cell Radio Network Temporary Identifier (C-RNTI) (which may or may not be made permanent as a result of the next step – contention resolution). The RAR message can also include a 'backoff indicator' which the eNodeB can set to instruct the UE to back off for a period of time before retrying a random access attempt.

The UE expects to receive the RAR within a time window, of which the start and end are configured by the eNodeB and broadcast as part of the cell-specific system information. The earliest subframe allowed by the specifications occurs 2 ms after the end of the preamble subframe, as illustrated in Figure 17.2. However, a typical delay (measured from the end of the preamble subframe to the beginning of the first subframe of RAR window) is more likely to be 4 ms. Figure 17.2 shows the RAR consisting of the step 2 message (on PDSCH) together with its downlink transmission resource allocation message 'G' (on the Physical Downlink Control CHannel (PDCCH) – see Section 9.3.5).

If the UE does not receive a RAR within the configured time window, it selects a signature again (as in Step 1) and transmits another preamble. The minimum delay for the transmission of another preamble after the end of the RAR window is 3 ms. If the UE receives the PDCCH signalling the downlink resource used for the RAR but cannot satisfactorily decode the RAR

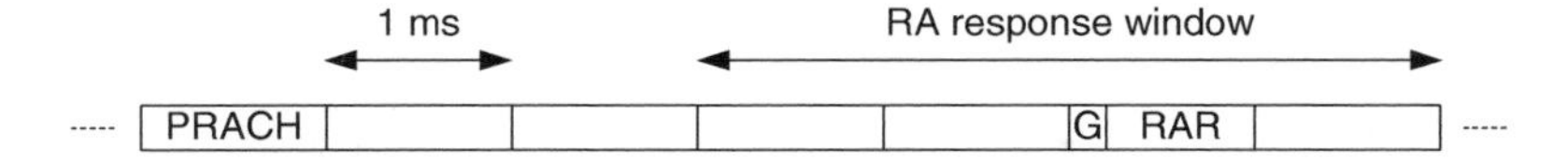

Figure 17.2: Timing of the Random Access Response (RAR) window.

message itself, the minimum delay is increased to 4 ms, to allow for the time taken by the UE in attempting to decode the RAR.

The eNodeB may configure *preamble power ramping* so that the transmission power for each transmitted preamble is increased by a fixed step. However, since the random access preambles in LTE are normally orthogonal to other uplink transmissions, it is less critical than it was in WCDMA to ensure that the initial preamble power is kept low to control interference. Therefore, the proportion of random access attempts which succeed at the first preamble transmission is likely to be higher than in WCDMA, and the need for power ramping is likely to be reduced.

17.3.1.3 Step 3: Layer 2/Layer 3 (L2/L3) Message

This message is the first scheduled uplink transmission on the PUSCH and makes use of Hybrid Automatic Repeat reQuest (HARQ). It conveys the actual random access procedure message, such as an RRC connection request, tracking area update, or scheduling request, but no Non-Access Stratum (NAS) message. It is addressed to the temporary C-RNTI allocated in the RAR at Step 2 and carries either the C-RNTI if the UE already has one (RRC_CONNECTED UEs) or an initial UE identity (the SAE[2] Temporary Mobile Subscriber Identity (S-TMSI) or a random number). In case of a preamble collision having occurred at Step 1, the colliding UEs will receive the same temporary C-RNTI through the RAR and will also collide in the same uplink time-frequency resources when transmitting their L2/L3 message. This may result in such interference that no colliding UE can be decoded, and the UEs restart the random access procedure after reaching the maximum number of HARQ retransmissions. However, if one UE is successfully decoded, the contention remains unresolved for the other UEs. The following downlink message (in Step 4) allows a quick resolution of this contention.

If the UE successfully receives the RAR, the UE minimum processing delay before message 3 transmission is 5 ms minus the round-trip propagation time. This is shown in Figure 17.3 for the case of the largest supported cell size of 100 km.

17.3.1.4 Step 4: Contention Resolution Message

The contention resolution message uses HARQ. It is addressed to the C-RNTI (if indicated in the L2/L3 message) or to the temporary C-RNTI, and, in the latter case, echoes the UE identity contained in the L2/L3 message. In case of a collision followed by successful decoding of the L2/L3 message, the HARQ feedback is transmitted only by the UE which detects its own UE identity (or C-RNTI); other UEs understand there was a collision, transmit no HARQ feedback, and can quickly exit the current random access procedure and

[2]System Architecture Evolution.

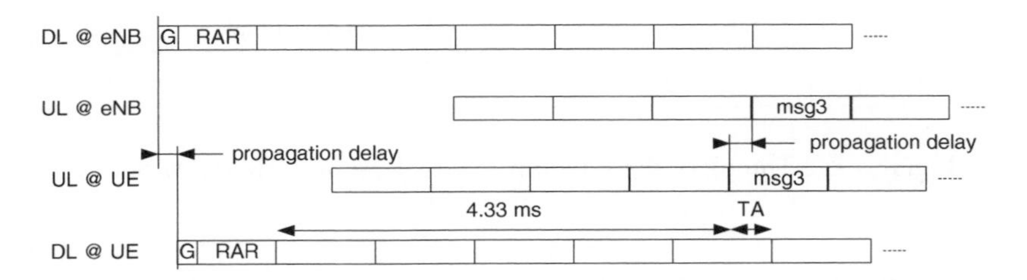

Figure 17.3: Timing of the message 3 transmission.

start another one. The UE's behaviour upon reception of the contention resolution message therefore has three possibilities:

- The UE correctly decodes the message and detects its own identity: it sends back a positive ACKnowledgement, 'ACK'.
- The UE correctly decodes the message and discovers that it contains another UE's identity (contention resolution): it sends nothing back (Discontinuous Transmission, 'DTX').
- The UE fails to decode the message or misses the DL grant: it sends nothing back ('DTX').

17.3.2 Contention-Free Random Access Procedure

The slightly unpredictable latency of the random access procedure can be circumvented for some use cases where low latency is required, such as handover and resumption of downlink traffic for a UE, by allocating a dedicated signature to the UE on a per-need basis. In this case the procedure is simplified as shown in Figure 17.4. The procedure terminates with the RAR.

17.4 Physical Random Access Channel Design

The random access preamble part of the random access procedure is mapped at the physical layer onto the PRACH. The design of the preamble is crucial to the success of the RACH procedure, and therefore the next section focuses on the details of the PRACH design.

17.4.1 Multiplexing of PRACH with PUSCH and PUCCH

The PRACH is time- and frequency-multiplexed with PUSCH and PUCCH as illustrated in Figure 17.5. PRACH time-frequency resources are semi-statically allocated within the PUSCH region, and repeat periodically. The possibility of scheduling PUSCH transmissions within PRACH slots is left to the eNodeB's discretion.

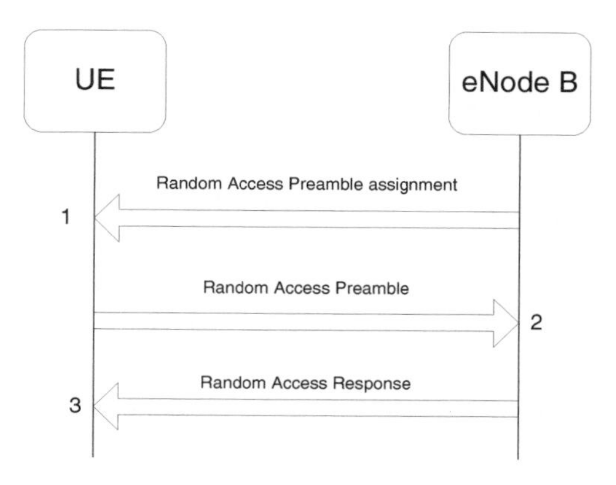

Figure 17.4: Contention-free random access procedure. Reproduced by permission of © 3GPP.

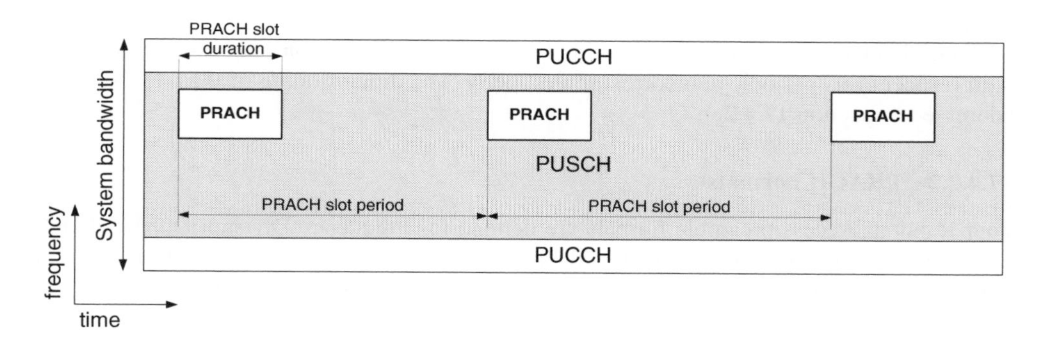

Figure 17.5: PRACH multiplexing with PUSCH and PUCCH.

17.4.2 The PRACH Structure

17.4.2.1 DFT-S-OFDM PRACH Preamble Symbol

Similarly to WCDMA, the LTE PRACH preamble consists of a complex sequence. However, it differs from the WCDMA preamble in that it is also an OFDM symbol following the DFT-S-OFDM structure of the LTE uplink, built with a CP, thus allowing for an efficient frequency-domain receiver at the eNodeB. As shown in Figure 17.6, the end of the sequence is appended at the start of the preamble, thus allowing a periodic correlation at the PRACH receiver (as opposed to a less efficient aperiodic correlation) [1].

The UE aligns the start of the random access preamble with the start of the corresponding uplink subframe at the UE assuming a timing advance of zero (see Section 18.2), and the preamble length is shorter than the PRACH slot in order to provide room for a Guard Time (GT) to absorb the propagation delay. Figure 17.6 shows two preambles at the eNodeB received with different timings depending on the propagation delay: as for a conventional

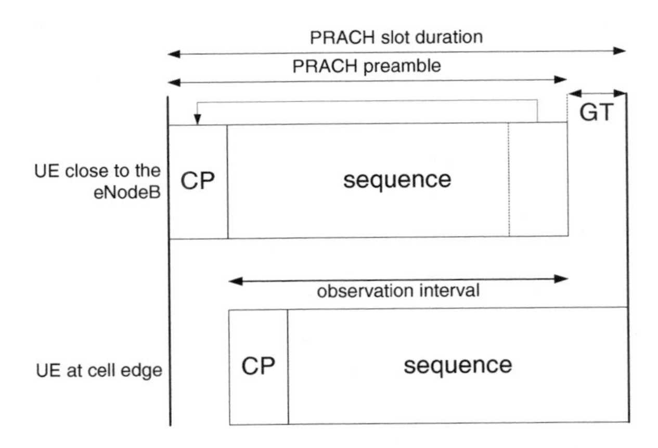

Figure 17.6: PRACH preamble received at the eNodeB.

OFDM symbol, a single observation interval can be used regardless of the UE's delay, within which periodic correlation is possible.

As further elaborated in Section 17.4.3, the LTE PRACH preamble sequence is optimized with respect to its periodic autocorrelation property. The dimensioning of the CP and GT is addressed in Section 17.4.2.4.

17.4.2.2 PRACH Formats

Four Random Access preamble formats are defined for Frequency Division Duplex (FDD) operation [2]. Each format is defined by the durations of the sequence and its CP, as listed in Table 17.1. The format configured in a cell is broadcast in the System Information.

Table 17.1: Random access preamble formats. Reproduced by permission of © 3GPP.

Preamble format	T_{CP} (μs)	T_{SEQ} (μs)	Typical usage
0	103.13	800	Normal 1 ms random access burst with 800 μs preamble sequence, for small to medium cells (up to ∼14 km)
1	684.38	800	2 ms random access burst with 800 μs preamble sequence, for large cells (up to ∼77 km) without a link budget problem
2	203.13	1600	2 ms random access burst with 1600 μs preamble sequence, for medium cells (up to ∼29 km) supporting low data rates
3	684.38	1600	3 ms random access burst with 1600 μs preamble sequence, for very large cells (up to ∼100 km)

The rationale behind these choices of sequence duration, CP and GT, are discussed below.

17.4.2.3 Sequence Duration

The sequence duration, T_{SEQ}, is driven by the following factors:

- Trade-off between sequence length and overhead: a single sequence must be as long as possible to maximize the number of orthogonal preambles (see Section 7.2.1), while still fitting within a single subframe in order to keep the PRACH overhead small in most deployments;
- Compatibility with the maximum expected round-trip delay;
- Compatibility between PRACH and PUSCH subcarrier spacings;
- Coverage performance.

Maximum round-trip time. The lower bound for T_{SEQ} must allow for unambiguous round-trip time estimation for a UE located at the edge of the largest expected cell (i.e. 100 km radius), including the maximum delay spread expected in such large cells, namely 16.67 μs. Hence

$$T_{SEQ} \geq \frac{200 \cdot 10^3}{3 \cdot 10^8} + 16.67 \cdot 10^{-6} = 683.33\ \mu s \qquad (17.1)$$

Subcarrier spacing compatibility. Further constraints on T_{SEQ} are given by the Single-Carrier Frequency Division Multiple Access (SC-FDMA – see Chapter 14) signal generation principle (see Section 17.5), such that the size of the DFT and IDFT[3], N_{DFT}, must be an integer number:

$$N_{DFT} = f_s T_{SEQ} = k, \quad k \in \mathbb{N} \qquad (17.2)$$

where f_s is the system sampling rate (e.g. 30.72 MHz). Additionally, it is desirable to minimize the orthogonality loss in the frequency domain between the preamble subcarriers and the subcarriers of the surrounding uplink data transmissions. This is achieved if the PUSCH data symbol subcarrier spacing Δf is an integer multiple of the PRACH subcarrier spacing Δf_{RA}:

$$\Delta f_{RA} = \frac{f_s}{N_{DFT}} = \frac{1}{T_{SEQ}} = \frac{1}{kT_{SYM}} = \frac{1}{k}\Delta_f, \quad k \in \mathbb{N} \qquad (17.3)$$

where $T_{SYM} = 66.67$ μs is the uplink subframe symbol duration. In other words, the preamble duration must be an integer multiple of the uplink subframe symbol duration:

$$T_{SEQ} = kT_{SYM} = \frac{k}{\Delta f}, \quad k \in \mathbb{N} \qquad (17.4)$$

An additional benefit of this property is the possibility to reuse the FFT/IFFT[4] components from the SC-FDMA signal processing for the PUSCH to implement the large DFT/IDFT blocks involved in the PRACH transmitter and receiver (see Section 17.5).

[3] Discrete Fourier Transform and Inverse Discrete Fourier Transform.

[4] Fast Fourier Transform / Inverse Fast Fourier Transform.

Coverage performance. In general a longer sequence gives better coverage, but better coverage requires a longer CP and GT in order to absorb the corresponding round-trip delay (Figure 17.6). The required CP and GT lengths for PRACH format 0, for example, can therefore be estimated from the maximum round-trip delay achievable by a preamble sequence which can fit into a 1 ms subframe.

Under a noise-limited scenario, as is typical of a low density, medium to large suburban or rural cell, coverage performance can be estimated from a link budget calculation. Under the assumption of the Okumura-Hata empirical model of distance-dependent path-loss $L(r)$ [3,4] (where r is the cell radius in km), the PRACH signal power P_{RA} received at the eNodeB baseband input can be computed as follows:

$$P_{\mathrm{RA}}(r) = P_{\max} + G_a - L(r) - \mathrm{LF} - P_L \text{ (dB)} \tag{17.5}$$

where the parameters are listed in Table 17.2 (mainly from [5]).

Table 17.2: Link budget parameters for analysis of PRACH preamble coverage.

Parameter	Value
Carrier frequency (f)	2000 MHz
eNodeB antenna height (h_b)	30 m / 60 m
UE antenna height (h_m)	1.5 m
UE transmitter EIRP[a] ($P_{\max}$)	24 dBm (250 mW)
eNodeB Receiver Antenna Gain (including cable loss) (G_a)	14 dBi
Receiver noise figure (N_f)	5.0 dB
Thermal noise density (N_0)	−174 dBm/Hz
Percentage of the area covered by buildings (α)	10%
Required E_p/N_0 (eNodeB with 2 Rx antenna diversity)	18 dB (six-path Typical Urban channel model)
Penetration loss (P_L)	0 dB
Log-normal fade margin (LF)	0 dB

[a]Equivalent Isotropic Radiated Power.

The required PRACH preamble sequence duration T_{SEQ} is then derived from the required preamble sequence energy to thermal noise ratio E_p/N_0 to meet a target missed detection and false alarm probability, as follows:

$$T_{\mathrm{SEQ}} = \frac{N_0 N_f}{P_{\mathrm{RA}}(r)} \frac{E_p}{N_0} \tag{17.6}$$

where N_0 is the thermal noise power density (in mW/Hz) and N_f is the receiver noise figure (in linear scale).

Assuming that $E_p/N_0 = 18$ dB is required to meet missed detection and false alarm probabilities of 10^{-2} and 10^{-3} respectively (see Section 17.4.3.3), Figure 17.7 plots the coverage performance of the PRACH sequence as a function of the sequence length T_{SEQ}.

It can be observed from Figure 17.7 that the potential coverage performance of a 1 ms PRACH preamble is in the region of 14 km. As a consequence, the required CP and GT

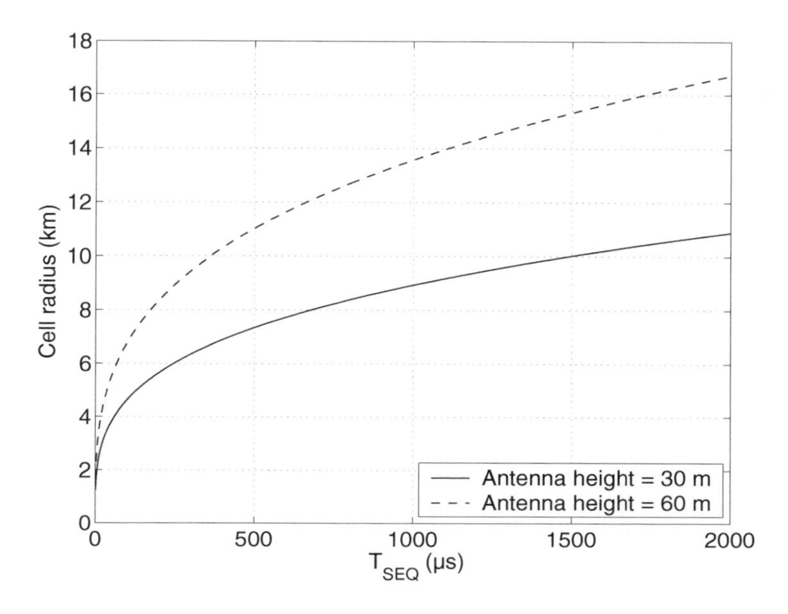

Figure 17.7: PRACH coverage performance versus sequence duration.

lengths are approximately $(2 \cdot 14000)/(3 \cdot 10^8) = 93.3$ μs, so that the upper bound for T_{SEQ} is given by

$$T_{\mathrm{SEQ}} \leq 1000 - 2 \cdot 93.33 = 813 \text{ μs} \tag{17.7}$$

Therefore, the longest sequence simultaneously satisfying Equations (17.1), (17.4) and (17.7) is $T_{\mathrm{SEQ}} = 800$ μs, as used for preamble formats 0 and 1. The resulting PRACH subcarrier spacing is $\Delta f_{\mathrm{RA}} = 1/T_{\mathrm{SEQ}} = 1.25$ kHz.

The 1600 μs preamble sequence of formats 2 and 3 is implemented by repeating the baseline 800 μs preamble sequence. These formats can provide up to 3 dB link budget improvement, which is useful in large cells and/or to balance PUSCH/PUCCH and PRACH coverage at low data rates.

17.4.2.4 CP and GT Duration

Having chosen T_{SEQ}, the CP and GT dimensioning can be specified more precisely.

For formats 0 and 2, the CP is dimensioned to maximize the coverage, given a maximum delay spread d: $T_{\mathrm{CP}} = (1000 - 800)/2 + d/2$ μs, with $d \approx 5.2$ μs (corresponding to the longest normal CP of a PUSCH SC-FDMA symbol). The maximum delay spread is used as a guard period at the end of CP, thus providing protection against multipath interference even for the cell-edge UEs.

In addition, for a cell-edge UE, the delay spread energy at the end of the preamble is replicated at the end of the CP (see Figure 17.8) and is therefore within the observation interval. Consequently, there is no need to include the maximum delay spread in the GT dimensioning. Hence, instead of locating the sequence in the centre of the PRACH slot, it is shifted later by half the maximum delay spread, allowing the maximum Round-Trip Delay

(RTD) to be increased by the same amount. Note that, as for a regular OFDM symbol, the residual delay spread at the end of the preamble from a cell-edge UE spills over into the next subframe, but this is taken care of by the CP at the start of the next subframe to avoid any Inter-Symbol Interference (ISI).

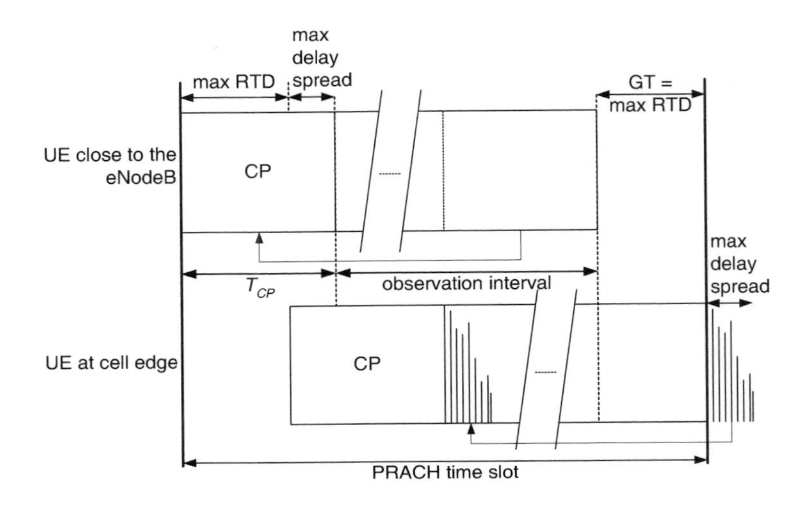

Figure 17.8: PRACH CP/GT dimensioning for formats 0 and 2.

For formats 1 and 3, the CP is dimensioned to address the maximum cell range in LTE, 100 km, with a maximum delay spread of $d \approx 16.67$ μs. In practice, format 1 is expected to be used with a 3-subframe PRACH slot; the available GT in 2 subframes can only address a 77 km cell range. It was chosen to use the same CP length for both format 1 and format 3 for implementation simplicity. Of course, handling larger cell sizes than 100 km with suboptimal CP dimensioning is still possible and is left to implementation.

Table 17.3 shows the resulting cell radius and delay spread ranges associated with the four PRACH preamble formats. The CP lengths are designed to be integer multiples of the assumed system sampling period for LTE, $T_S = 1/30.72$ μs.

Table 17.3: Field durations and achievable cell radius of the PRACH preamble formats.

Preamble format	Number of allocated subframes	CP duration in μs	CP duration as multiple of T_S	GT duration in μs	GT duration as multiple of T_S	Max. delay spread (μs)	Max. cell radius (km)
0	1	103.13	3168	96.88	2976	6.25	14.53
1	2	684.38	21024	515.63	15840	16.67	77.34
2	2	203.13	6240	196.88	6048	6.25	29.53
3	3	684.38	21024	715.63	21984	16.67	100.16

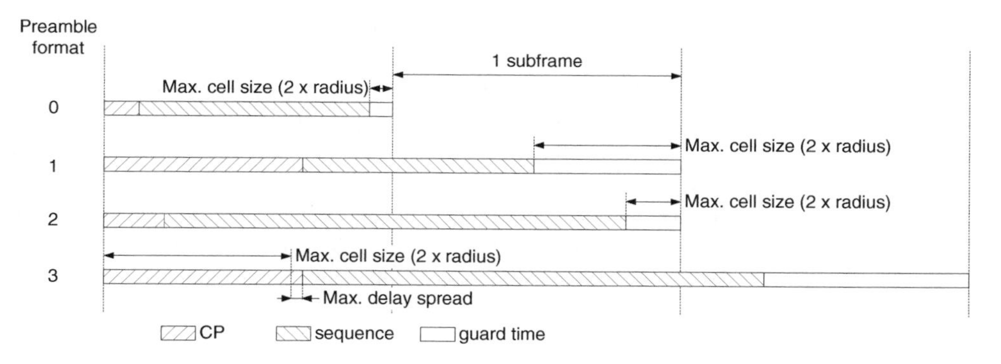

Figure 17.9: PRACH preamble formats and cell size dimensioning.

These are also illustrated in Figure 17.9.

17.4.2.5 PRACH Resource Configurations

The PRACH slots shown in Figure 17.5 can be configured to occur in up to 16 different layouts, or *resource configurations*. Depending on the RACH load, one or more PRACH resources may need to be allocated per PRACH slot period. The eNodeB has to process the PRACH very quickly so that message 2 of the RACH procedure can be sent within the required window, as shown in Figure 17.2. In case of more than one PRACH resource per PRACH period it is generally preferable to multiplex the PRACH resources in time (Figure 17.10 – right) rather than in frequency (Figure 17.10 – left). This helps to avoid processing peaks at the eNodeB.

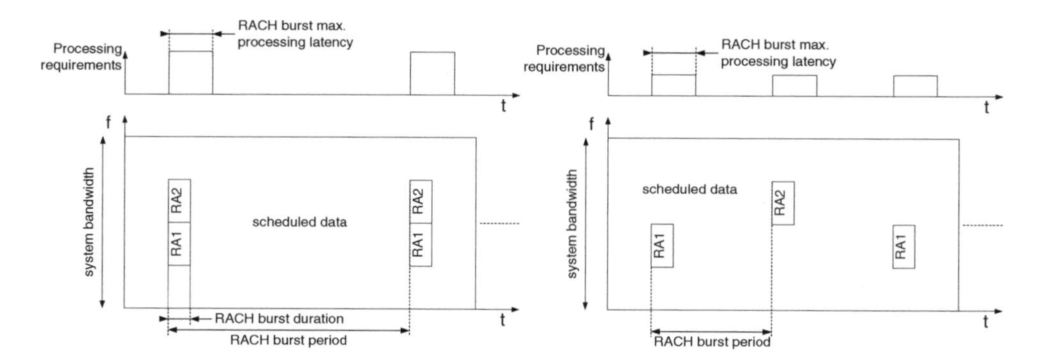

Figure 17.10: Processing peaks of the random access preamble receiver.

Extending this principle, the available slot configurations are designed to facilitate a PRACH receiver which may be used for multiple cells of an eNodeB, assuming a periodic pattern with period 10 ms or 20 ms.

Assuming an operating collision probability per UE, $p_{\text{coll}}^{\text{UE}} = 1\%$, one PRACH time-frequency resource (with 64 signatures) per 10 ms per 5 MHz can handle an offered load $G = -64 \ln(1 - p_{\text{coll}}^{\text{UE}}) = 0.6432$ average PRACH attempts, which translates into 128 attempts per second in 10 MHz. This is expected to be a typical PRACH load in LTE.

Assuming a typical PRACH load, example usages and cell allocations of the 16 available resource configurations are shown in Figure 17.11 for different system bandwidths and different numbers of cells per eNodeB. Resource configurations 0–2 and 15 use a 20 ms PRACH period, which can be desirable for small bandwidths (e.g. 1.4 MHz) in order to reduce the PRACH overhead at the price of higher waiting times.

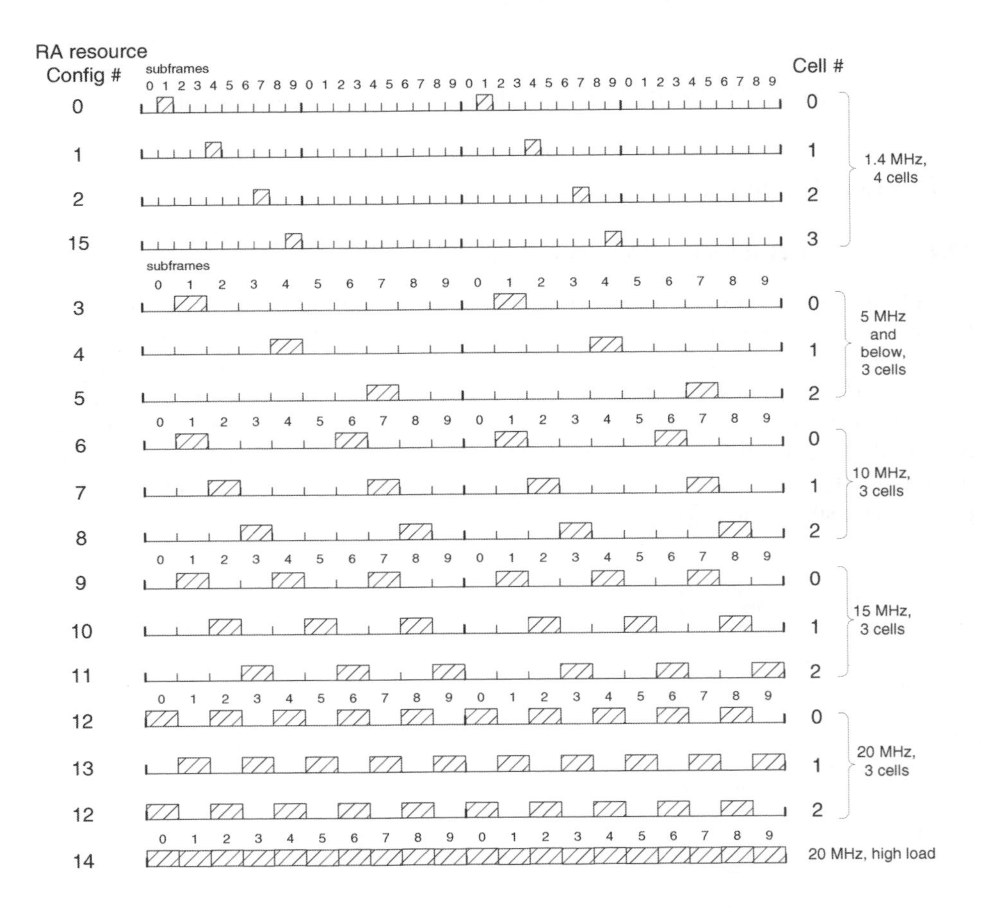

Figure 17.11: Random Access resource configurations.

As can be observed from Figure 17.11, in a three-cell scenario, time collision of PRACH resources can always be avoided except for the 20 MHz case, where collisions can be minimized to two PRACH resources occurring in the same subframe. It should also be noted that in a scenario with six cells per eNodeB, at most two PRACH resources will occur in the same subframe for bandwidths below 20 MHz.

The variety of configurations provided therefore enables efficient dimensioning of a multicell PRACH receiver.

17.4.3 Preamble Sequence Theory and Design

As noted above, 64 PRACH signatures are available in LTE, compared to only 16 in WCDMA. This can not only reduce the collision probability, but also allow for 1 bit of information to be carried by the preamble and some signatures to be reserved for contention-free access (see Section 17.3.2). Therefore, the LTE PRACH preamble called for an improved sequence design with respect to WCDMA. While Pseudo-Noise (PN) based sequences were used in WCDMA, in LTE prime-length Zadoff–Chu (ZC) [6, 7] sequences have been chosen (see Chapter 7 for an overview of the properties of ZC sequences). These sequences enable improved PRACH preamble detection performance. In particular:

- The power delay profile is built from periodic instead of aperiodic correlation;
- The intra-cell interference between different preambles received in the same PRACH resource is reduced;
- Intra-cell interference is optimized with respect to cell size: the smaller the cell size, the larger the number of orthogonal signatures and the better the detection performance;
- The eNodeB complexity is reduced;
- The support for high-speed UEs is improved.

The 800 μs LTE PRACH sequence is built by cyclicly-shifting a ZC sequence of prime-length N_{ZC}, defined as

$$x_u(n) = \exp\left[-j\frac{\pi un(n+1)}{N_{ZC}}\right], \quad 0 \leq n \leq N_{ZC} - 1 \tag{17.8}$$

where u is the ZC sequence index and the sequence length $N_{ZC} = 839$ for FDD.

The reasons that led to this design choice are elaborated in the next sections.

17.4.3.1 Preamble Bandwidth

In order to ease the frequency multiplexing of the PRACH and the PUSCH resource allocations, a PRACH slot must be allocated a bandwidth BW_{PRACH} equal to an integer multiple of Resource Blocks (RBs), i.e. an integer multiple of 180 kHz.

For simplicity, BW_{PRACH} in LTE is constant for all system bandwidths; it is chosen to optimize both the detection performance and the timing estimation accuracy. The latter drives the lower bound of the PRACH bandwidth. Indeed, a minimum bandwidth of ~1 MHz is necessary to provide a one-shot accuracy of about ±0.5 μs, which is an acceptable timing accuracy for PUCCH/PUSCH transmissions.

Regarding the detection performance, one would intuitively expect that the higher the bandwidth, the better the detection performance, due to the diversity gain. However, it is important to make the comparison using a constant signal energy to noise ratio, E_p/N_0, resulting from accumulation (or despreading) over the same preamble duration, and the same false alarm probability, p_{fa}, for all bandwidths. The latter requires the detection threshold to be adjusted with respect to the search window size, which increases with the bandwidth. Indeed, it will become clear from the discussion in Section 17.5.2.3 that the larger the search window, the higher p_{fa}. In other words, the larger the bandwidth the higher the threshold relative to the noise floor, given a false alarm target p_{fa_target} and cell size L; equivalently, the larger the cell size, the higher the threshold relative to the noise floor, given a target p_{fa_target} and bandwidth. As a result, under the above conditions, a smaller bandwidth will perform better than a large bandwidth in a single-path static AWGN channel, given that no diversity improvement is to be expected from such a channel.

Figure 17.12 shows simulation results for the TU-6 fading channel,[5] comparing detection performance of preamble bandwidths BW_{PRACH} of 6, 12, 25 and 50 RBs. For each bandwidth, the sequence length is set to the largest prime number smaller than $1/BW_{PRACH}$, the false alarm rate is set to $p_{fa\_target} = 0.1\%$, the cell radius is 0.7 km and the receiver searches for 64 signatures constructed from 64 cyclic shifts of one root ZC sequence.

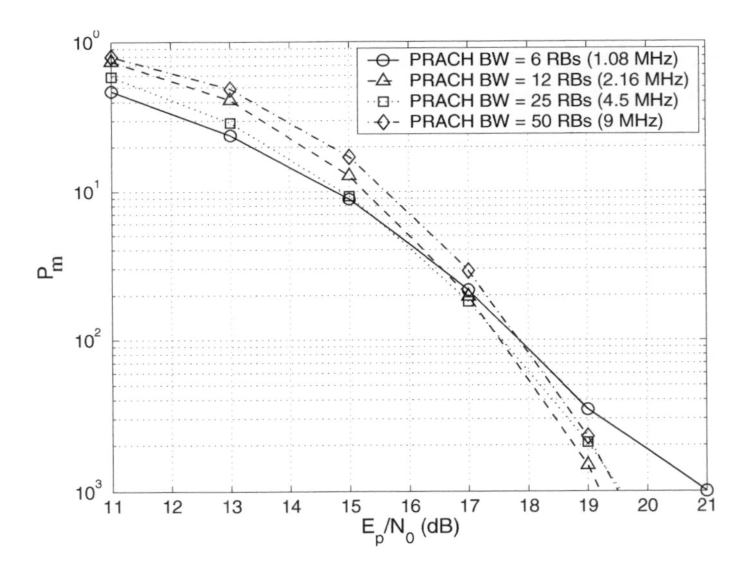

Figure 17.12: PRACH missed detection performance comparison for different BW_{PRACH} of 6, 12, 25 and 50 RBs.

We can observe that the best detection performance is achieved by preambles of 6 RBs and 12 RBs for low and high SNRs respectively. The 25-RB preamble has the overall best performance considering the whole SNR range. Thus the diversity gain of large bandwidths only compensates the increased detection threshold in the high SNR region corresponding to misdetection performances in the range of 10^{-3} and below. At a typical 10^{-2} detection probability target, the 6-RB allocation only has 0.5 dB degradation with respect to the best case.

Therefore, a PRACH allocation of 6 RBs provides a good trade-off between PRACH overhead, detection performance and timing estimation accuracy. Note that for the smallest

[5]The six-path Typical Urban (TU) channel model [8].

system bandwidth (1.4 MHz, 6 RBs) the PRACH overlaps with the PUCCH; it is left to the eNodeB implementation whether to implement scheduling restrictions during PRACH slots to avoid collisions, or to let PRACH collide with PUCCH and handle the resulting interference.

Finally, the exact preamble transmission bandwidth is adjusted to isolate PRACH slots from surrounding PUSCH/PUCCH allocations through guard bands, as elaborated in the following section.

17.4.3.2 Sequence Length

The sequence length design should address the following requirements:

- Maximize the number of ZC sequences with optimal cross-correlation properties;
- Minimize the interference to/from the surrounding scheduled data on the PUSCH.

The former requirement is guaranteed by choosing a prime-length sequence. For the latter, since data and preamble OFDM symbols are neither aligned nor have the same durations, strict orthogonality cannot be achieved. At least, fixing the preamble duration to an integer multiple of the PUSCH symbol provides some compatibility between preamble and PUSCH subcarriers. However, with the 800 μs duration, the corresponding sequence length would be 864, which does not meet the prime number requirement. Therefore, shortening the preamble to a prime length slightly increases the interference between PUSCH and PRACH by slightly decreasing the preamble sampling rate.

The interference from PUSCH to PRACH is further amplified by the fact that the operating E_s/N_0 of PUSCH (where E_s is the PUSCH symbol energy) is much greater than that of the PRACH (typically as much as 24 dB greater if we assume 13 dB E_s/N_0 for 16QAM PUSCH, while the equivalent ratio for the PRACH would be −11 dB assuming $E_p/N_0 = 18$ dB and adjusting by $-10\log_{10}(864)$ to account for the sequence length). This is illustrated in Figure 17.13 showing the missed detection rate (P_m) with and without data interference adjacent to the PRACH. The simulations assume a TU-6 channel, two receive antennas at the eNodeB and 15 km/h UE speed. The PRACH shows about 1 dB performance loss at $P_m = 1\%$.

The PRACH uses *guard bands* to avoid the data interference at preamble edges. A cautious design of preamble sequence length not only retains a high inherent processing gain, but also allows avoidance of strong data interference. In addition, the loss of spectral efficiency (by reservation of guard subcarriers) can also be well controlled at a fine granularity ($\Delta f_{\text{RA}} =$ 1.25 kHz). Figure 17.13 shows the missed detection rate for a cell radius of 0.68 km, for various preamble sequence lengths with and without 16QAM data interference.

In the absence of interference, there is no significant performance difference between sequences of similar prime length. In the presence of interference, it can be seen that reducing the sequence length below 839 gives no further improvement in detection rate. No effect is observed on the false alarm rate.

Therefore the sequence length of 839 is selected for LTE PRACH, corresponding to 69.91 PUSCH subcarriers in each SC-FDMA symbol, and offers 72 − 69.91 = 2.09 PUSCH subcarriers protection, which is very close to one PUSCH subcarrier protection on each side of the preamble. This is illustrated in Figure 17.14; note that the preamble is positioned

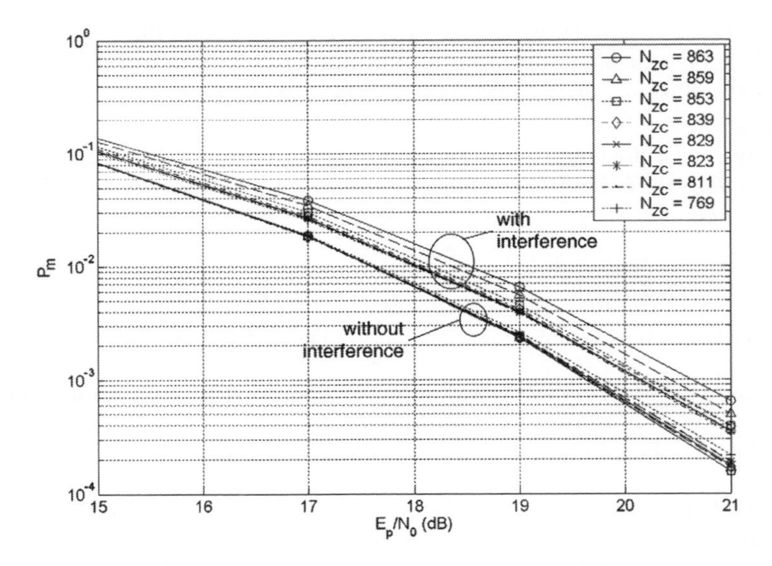

Figure 17.13: Missed detection rates of PRACH preamble with and without 16QAM interferer for different sequence lengths (cell radius of 0.68 km).

centrally in the block of 864 available PRACH subcarriers, with 12.5 null subcarriers on each side.

Finally, the PRACH preamble signal $s(t)$ can therefore be defined as follows [2]:

$$s(t) = \beta_{\text{PRACH}} \sum_{k=0}^{N_{\text{ZC}}-1} \sum_{n=0}^{N_{\text{ZC}}-1} x_{u,v}(n) \cdot \exp\left[-j\frac{2\pi nk}{N_{\text{ZC}}}\right]$$
$$\times \exp[j2\pi[k + \varphi + K(k_0 + \tfrac{1}{2})]\Delta f_{\text{RA}}(t - T_{\text{CP}})] \tag{17.9}$$

where $0 \leq t < T_{\text{SEQ}} + T_{\text{CP}}$, β_{PRACH} is an amplitude scaling factor and $k_0 = n_{\text{PRB}}^{\text{RA}} N_{\text{SC}}^{\text{RB}} - N_{\text{RB}}^{\text{UL}} N_{\text{SC}}^{\text{RB}}/2$. The location in the frequency domain is controlled by the parameter $n_{\text{PRB}}^{\text{RA}}$, expressed as an RB number configured by higher layers and fulfilling $0 \leq n_{\text{PRB}}^{\text{RA}} \leq N_{\text{RB}}^{\text{UL}} - 6$. The factor $K = \Delta f / \Delta f_{\text{RA}}$ accounts for the ratio of subcarrier spacings between the PUSCH and PRACH. The variable φ (equal to 7 for LTE FDD) defines a fixed offset determining the frequency-domain location of the random access preamble within the RBs. $N_{\text{RB}}^{\text{UL}}$ is the uplink system bandwidth (in RBs) and $N_{\text{SC}}^{\text{RB}}$ is the number of subcarriers per RB, i.e. 12.

17.4.3.3 Cyclic Shift Dimensioning (N_{CS}) for Normal Cells

Sequences obtained from cyclic shifts of *different* ZC sequences are not orthogonal (see Section 7.2.1). Therefore, orthogonal sequences obtained by cyclically shifting a single root sequence should be favoured over non-orthogonal sequences; additional ZC root sequences should be used only when the required number of sequences (64) cannot be generated by cyclic shifts of a single root sequence. The cyclic shift dimensioning is therefore very important in the RACH design.

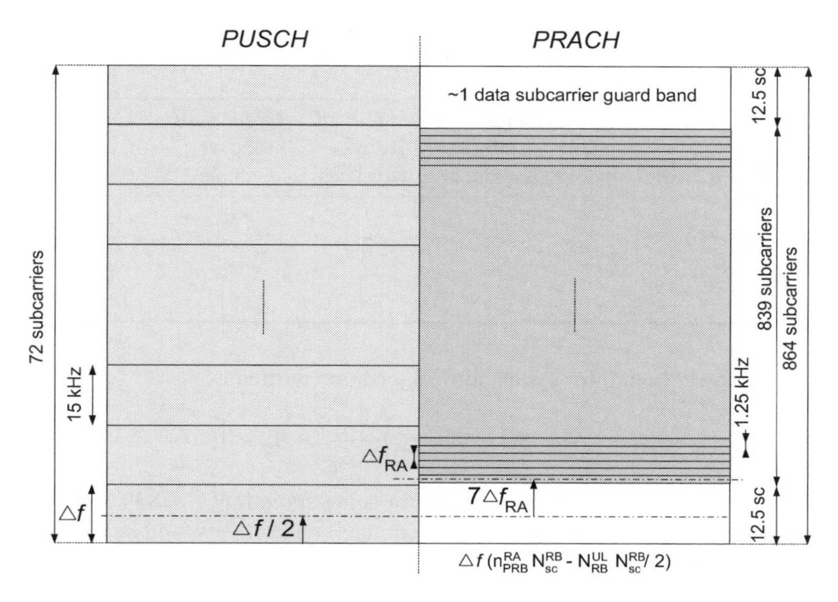

Figure 17.14: PRACH preamble mapping onto allocated subcarriers.

The cyclic shift offset N_{CS} is dimensioned so that the Zero Correlation Zone (ZCZ) (see Section 7.2.1) of the sequences guarantees the orthogonality of the PRACH sequences regardless of the delay spread and time uncertainty of the UEs. The minimum value of N_{CS} should therefore be the smallest integer number of sequence sample periods that is greater than the maximum delay spread and time uncertainty of an uplink non-synchronized UE, plus some additional guard samples provisioned for the spill-over of the pulse shaping filter envelope present in the PRACH receiver (Figure 17.15).

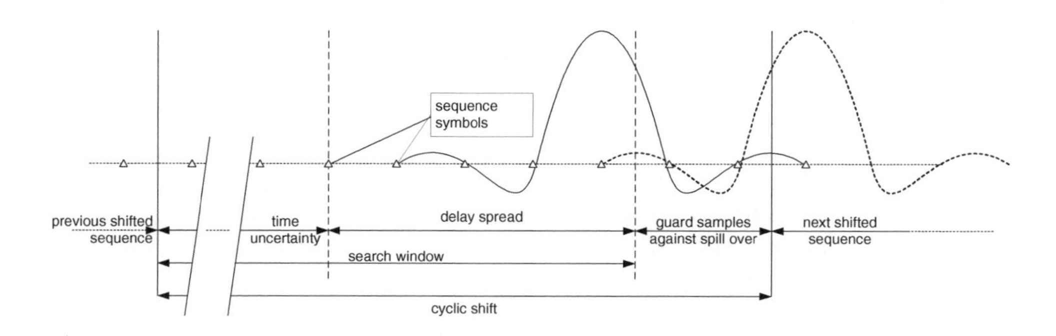

Figure 17.15: Cyclic shift dimensioning.

Table 17.4: Cell scenarios with different cyclic shift increments.

Cell scenario	Number of cyclic shifts per ZC sequence	Number of ZC root sequences	Cyclic shift size N_{CS} (samples)	Cell radius (km)
1	64	1	13	0.7
2	32	2	26	2.5
3	18	4	46	5
4	9	8	93	12

The resulting lower bound for cyclic shift N_{CS} can be written as

$$N_{CS} \geq \left\lceil \left(\frac{20}{3}r - \tau_{ds}\right)\frac{N_{ZC}}{T_{SEQ}} \right\rceil + n_g \tag{17.10}$$

where r is the cell size (km), τ_{ds} is the maximum delay spread, $N_{ZC} = 839$ and T_{SEQ} are the PRACH sequence length and duration (measured in μs) respectively, and n_g is the number of additional guard samples due to the receiver pulse shaping filter.

The delay spread can generally be assumed to be constant for a given environment. However, the larger the cell, the larger the cyclic shift required to generate orthogonal sequences, and consequently, the larger the number of ZC root sequences necessary to provide the 64 required preambles.

The relationship between cell size and the required number of ZC root sequences allows for some system optimization. In general, the eNodeB should configure N_{CS} independently in each cell, because the expected inter-cell interference and load (user density) increases as cell size decreases; therefore smaller cells need more protection from co-preamble interference than larger cells.

Some practical examples of this optimization are given in Table 17.4, showing four cell scenarios resulting from different N_{CS} values configured by the eNodeB. For each scenario, the total number of sequences is 64, but resulting from different combinations of the number of root sequences and cyclic shifts.

N_{CS} set design. Given the sequence length of 839, allowing full flexibility in signalling N_{CS} would lead to broadcasting a 10-bit parameter, which is over-dimensioning. As a result, in LTE the allowed values of N_{CS} are quantized to a predefined set of just 16 configurations. The 16 allowed values of N_{CS} were chosen so that the number of orthogonal preambles is as close as possible to what could be obtained if there were no restrictions on the value of N_{CS} [9]. This is illustrated in Figure 17.16 (left), where the cell radii are derived assuming a delay spread of 5.2 μs and 2 guard samples n_g for the pulse shaping filter.

The effect of the quantization is shown in Figure 17.16 (right), which plots the probability p_2 that two UEs randomly select two preambles on the same root sequence, as a function of the cell radius, for both the quantized N_{CS} set and an ideal unquantized set. The larger p_2, the better the detection performance. The figure also shows an ideal unquantized set. It can be seen that the performance loss due to the quantization is negligible.

Figure 17.17 illustrates the range of N_{CS} values and their usage with the various preamble formats. Note that this set of N_{CS} values is designed for use in low-speed cells. LTE also

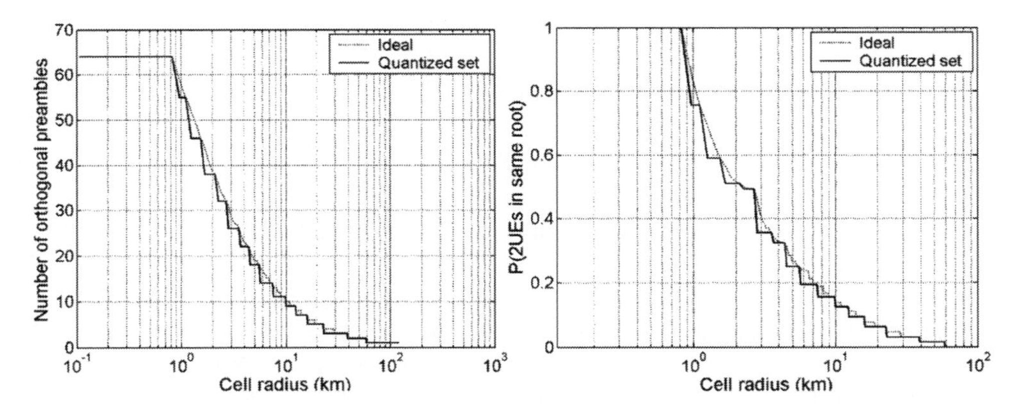

Figure 17.16: Number of orthogonal preambles (left) and probability that two UEs select two orthogonal preambles (right).

provides a second N_{CS} set specially designed for high-speed cells, as elaborated in the following sections.

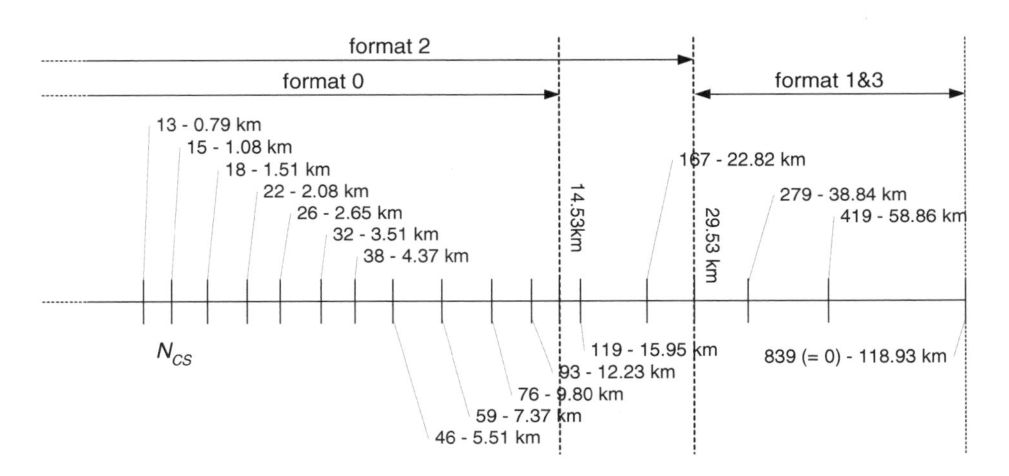

Figure 17.17: N_{CS} values and usage with the various preamble formats (low speed cells).

17.4.3.4 Cyclic Shift (N_{CS}) Restriction for High-Speed Cells

The support of 64 RACH preambles as described above assumes little or no frequency shifting due to Doppler spread, in the presence of which ZC sequences lose their zero auto-correlation property. In the presence of a frequency offset δf, it can be shown that the PRACH

ZC sequence in Equation (17.8) is distorted as follows:

$$
\begin{aligned}
x_u(n, \delta f) &= \exp\left[-j\pi \frac{u(n-1/u)(n-1/u+1)}{N_{\mathrm{ZC}}}\right] \\
&\quad \times \exp\left[j2\pi \frac{n}{N_{\mathrm{ZC}}}(\delta f\, T_{\mathrm{SEQ}} - 1)\right] \exp\left[-\frac{j\pi}{N_{\mathrm{ZC}}} \frac{u-1}{u}\right] \\
&= x_u(n - 1/u) \exp\left[j2\pi \frac{n}{N_{\mathrm{ZC}}}(\delta f\, T_{\mathrm{SEQ}} - 1)\right] e^{j\Phi_u}
\end{aligned} \tag{17.11}
$$

A similar expression can be written for the opposite frequency offset.

As can be observed, frequency offsets as large as one PRACH subcarrier ($\delta f = \pm\Delta f_{\mathrm{RA}} = \pm 1/T_{\mathrm{SEQ}} = \pm 1.25$ kHz) result in cyclic shifts $d_u = (\pm 1/u) \bmod N_{\mathrm{ZC}}$ on the ZC sequence $x_u(n)$. (Note that $u \cdot d_u \bmod N_{\mathrm{ZC}} = \pm 1$.) This frequency offset δf can be due to the accumulated frequency uncertainties at both UE transmitter and eNodeB receiver and the Doppler shift resulting from the UE motion in a Line of Sight (LOS) radio propagation condition.

Figure 17.18 illustrates the impact of the cyclic shift distortion on the received Power Delay Profile (PDP): it creates false alarm peaks whose relative amplitude to the correct peak depends on the frequency offset. The solution adopted in LTE to address this issue is referred to as 'cyclic shift restriction' and consists of 'masking' some cyclic shift positions in the ZC root sequence. This makes it possible to retain an acceptable false alarm rate, while also combining the PDPs of the three uncertainty windows, thus also maintaining a high detection performance even for very high-speed UEs.

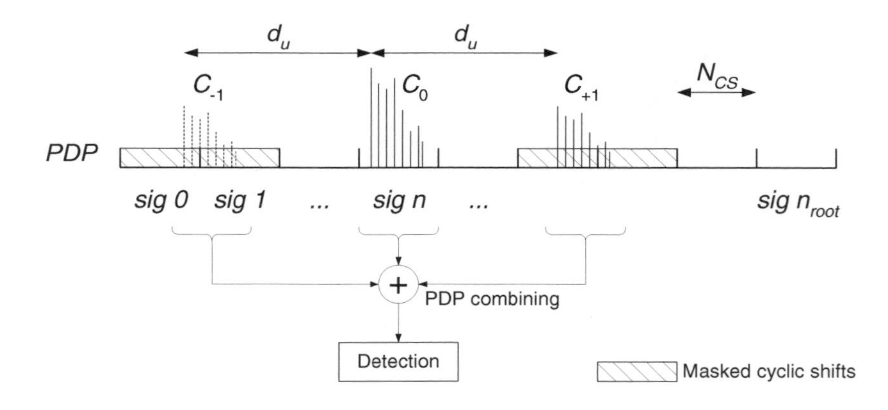

Figure 17.18: Side peaks in PDP due to frequency offset.

It should be noted that at $|\delta f| = \Delta f_{\mathrm{RA}}$, the preamble peak completely disappears at the desired location (as per Equation (17.11)). However, the false image peak begins to appear even with $|\delta f| < \Delta f_{\mathrm{RA}}$. Another impact of the side peaks is that they restrict the possible cyclic shift range so as to prevent from side peaks from falling into the cyclic shift region (see Figure 17.19). This restriction on N_{CS} is captured by Equations (17.12) and (17.13) and is important for the design of the high-speed N_{CS} set (explained in Section 17.4.3.5) and the order in which the ZC sequences are used (explained in Section 17.4.3.6):

$$N_{CS} \leq d < (N_{ZC} - N_{CS})/2 \tag{17.12}$$

where

$$d = \begin{cases} d_u, & 0 \leq d_u < N_{ZC}/2 \\ N_{ZC} - d_u, & d_u \geq N_{ZC}/2 \end{cases} \tag{17.13}$$

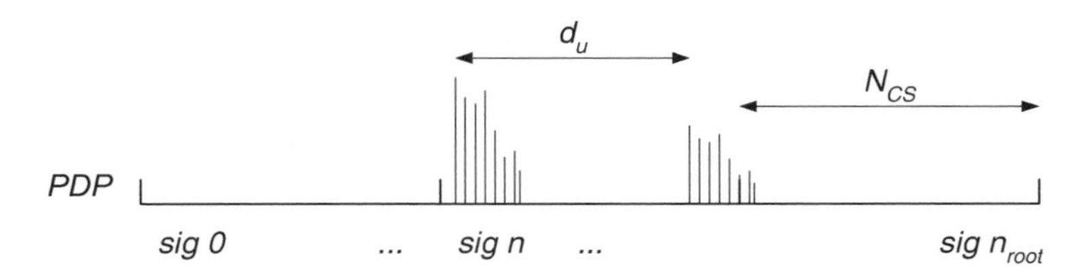

Figure 17.19: Side peaks within the signature search window.

We use C_{-1} and C_{+1} to denote the two wrong cyclic shift windows arising from the frequency offset, while C_0 denotes the correct cyclic shift window (Figure 17.18). The cyclic restriction rule must be such that the two wrong cyclic shift windows C_{-1} and C_{+1} of a cyclicly-shifted ZC sequence overlap none of C_0, C_{-1} or C_{+1} of other cyclicly-shifted ZC sequences, nor the correct cyclic shift window C_0 of the same cyclicly-shifted ZC sequence, nor each other [10]. Finally, the restricted set of cyclic shifts is obtained such that the minimum difference between two cyclic shifts is still N_{CS} but the cyclic shifts are not necessarily multiples of N_{CS}.

It is interesting to check the speed limit beyond which it is worth considering a cell to be a high-speed cell. This is done by assessing the performance degradation of the PRACH at the system-level as a function of the UE speed when no cyclic shift restriction is applied.

For this analysis, we model the RACH access attempts of multiple concurrent UEs with a Poisson arrival rate. A preamble detection is considered to be correct if the timing estimation is within 2 μs. A target E_p/N_0 of 18 dB is used for the first preamble transmission, with a power ramping step of 1 dB for subsequent retransmissions. The cell radius is random between 0.5 and 12 km, with either AWGN or a six-path TU channel, and a 2 GHz carrier frequency. eNodeB and UE frequency errors are modelled randomly within ±0.05 ppm. The access failure rate is the measure of the number of times a UE unsuccessfully re-tries access attempts (up to a maximum of three retransmissions), weighted by the total number of new access attempts.

Figure 17.20 shows the access failure rate performance for both channel types as a function of the UE speed, for various offered loads G. It can be observed that under fading conditions, the RACH failure rates experience some degradation with the UE speed (which translates into Doppler spread), but remains within acceptable performance even at 350 km/h. For the AWGN channel (where the UE speed translates into Doppler shift) the RACH failure rate stays below 10^{-2} up to UE speeds in the range 150 to 200 km/h. However, at 250 km/h and above, the throughput collapses. Without the cyclic shift restrictions the upper bound for useful performance is around 150–200 km/h.

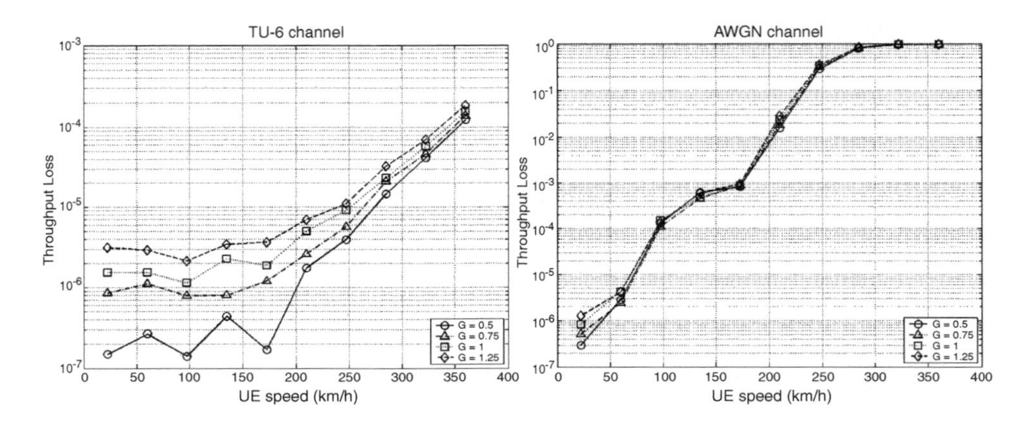

Figure 17.20: Random access failure rate as a function of UE speed.

17.4.3.5 Cyclic Shift Configuration for High-Speed Cells

The cyclic shift dimensioning for high-speed cells in general follows the same principle as for normal cells, namely maximizing the sequence reuse when group quantization is applied to cyclic shift values. However, for high-speed cells, the cyclic shift restriction needs to be considered when deriving the sequence reuse factor with a specific cyclic shift value. Note that there is no extra signalling cost to support an additional set of cyclic shift configurations for high-speed cells since the one signalling bit which indicates a 'high-speed cell configuration' serves this purpose.

The N_{CS} values for high-speed cells are shown in Figure 17.21 for the number of available and used preambles, with both consecutive and non-consecutive (quantized) cyclic shift values. The number of available preambles assumes no cyclic shift restriction at all, as in low-speed cells. It should be noted that with the cyclic shift restriction above, the largest usable high-speed cyclic shift value among all root sequences is 279 (from Equation (17.12)). As is further elaborated in the next section, only the preambles with Cubic Metric (CM) (see Section 21.3.3) below 1.2 dB are considered in Figure 17.21.

Since for small N_{CS} values the sequence usage is not so tight with a generally high sequence reuse factor, a way to simplify design, while still achieving a high reuse factor, is to reuse the small N_{CS} values for normal cells. In Figure 17.21, N_{CS} values up to 46 are from the normal cyclic shift values, corresponding to a cell radius up to 5.8 km. At the high end, the value of 237 rather than 242 is chosen to support a minimum of two high-speed cells when all the 838 sequences are used. The maximum supportable high-speed cell radius is approximately 33 km, providing sufficient coverage for preamble formats 0 and 2.

17.4.3.6 Sequence Ordering

A UE using the contention-based random access procedure described in Section 17.3.1 needs to know which sequences are available to select from. As explained in Section 17.4.3.3, the full set of 64 sequences may require the use of several ZC root sequences, the identity of each of which must be broadcast in the cell. Given the existence of 838 root sequences, in LTE the

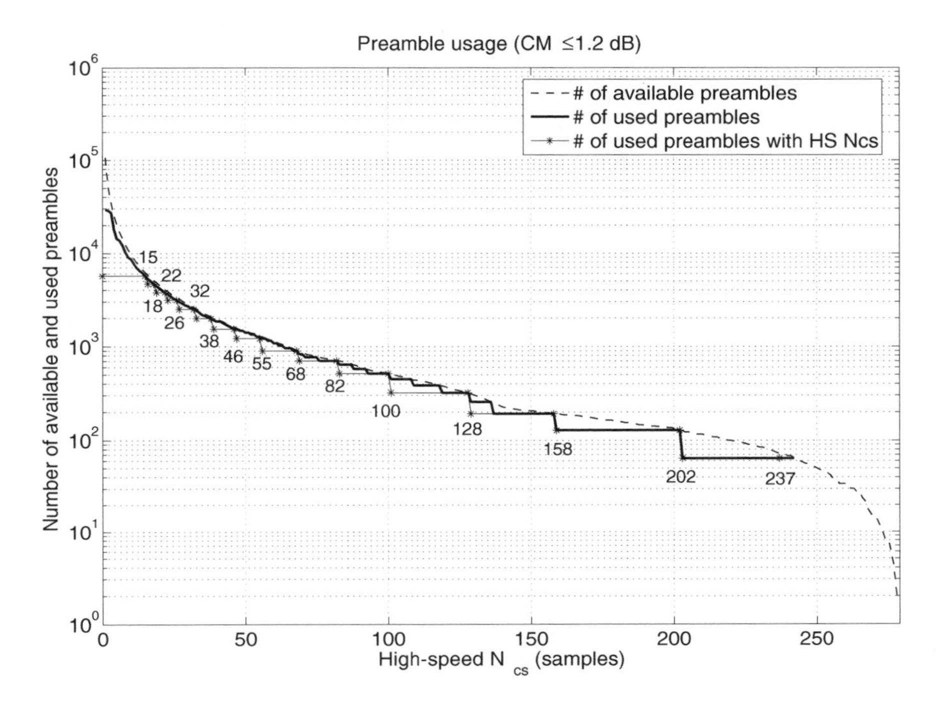

Figure 17.21: Number of available and used preambles in low CM group.

signalling is streamlined by broadcasting only the index of the *first* root sequence in a cell, and the UE derives the other preamble signatures from it given a predefined ordering of all the sequences.

Two factors are taken into account for the root sequence ordering, namely the CM [11, 12] of the sequence, and the maximum supportable cell size for high-speed cells (or equivalently the maximum supported cyclic shift). Since CM has a direct impact on cell coverage, the first step in ordering the root sequences is to divide the 838 sequences into a low CM group and a high CM group, using the CM of Quadrature Phase Shift Keying (QPSK) (1.2 dB) as a threshold. The low CM group would be used first in sequence planning (and also for high-speed cells) since it is more favourable for coverage.

Then, within each CM group, the root sequences are classified into subgroups based on their maximum supportable cell radius, to facilitate sequence planning including high-speed cells. Specifically, a sequence subgroup g is the set of all root sequences with their maximum allowed cyclic shifts ($S_{\max}$) derived from Equation (17.12) lying between two consecutive high-speed N_{CS} values according to

$$N_{\text{CS}}(g) \leq S_{\max} < N_{\text{CS}}(g+1), \quad \text{for } g = 0, 1, \ldots, G-2, \quad \text{and}$$
$$S_{\max} \geq N_{\text{CS}}(G-1) \tag{17.14}$$

for G cyclic shift values, with the set of N_{CS} values being those for high-speed cells. Sequences in each subgroup are ordered according to their CM values.

Figure 17.22 shows the CM and maximum allowed cyclic shift values at high speed for the resulting root sequence index ordering used in LTE. This ordering arrangement reflects a continuous CM transition across subgroups and groups, which ensures that consecutive sequences always have close CM values when allocated to a cell. Thus, consistent cell coverage and preamble detection can be achieved in one cell.

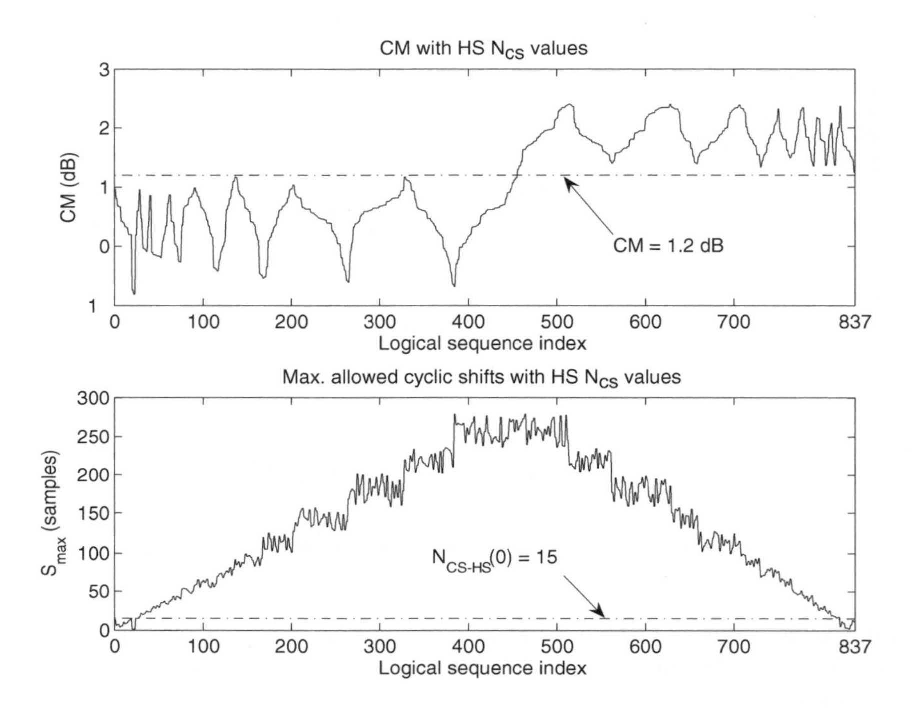

Figure 17.22: CM and maximum allowed cyclic shift of reordered group sequences.

The LTE specifications define the mapping from root sequence index u to a reordered index in a table [2], an enhanced extract of which is given in Figure 17.23. The first 16 subgroups are from the low CM group, and the last 16 from the high CM group. Figure 17.23 shows the corresponding high-speed N_{CS} value for each subgroup. Note that sequences with S_{max} less than 15 cannot be used by any high-speed cells, but they can be used by any normal cell which requires no more than 24 root sequences from this group for a total of 64 preambles. Note also that the ordering of the physical ZC root sequence indices is pairwise, since root sequence indices u and $N_{ZC} - u$ have the same CM and S_{max} values. This helps to simplify the PRACH receiver, as elaborated in Section 17.5.2.2.

17.5 PRACH Implementation

This section provides some general principles for practical implementation of the PRACH function.

CM group	Sub-group no.	N_{CS} (High-Speed)	Logical index (i.e. re-ordered)	Physical root sequence index u (in increasing order of the corresponding logical index number)
Low	0	-	0~23	129, 710, 140, 699, 120, 719, 210, 629, 168, 671, 84, 755, 105, 734, 93, 746, 70, 769, 60, 779, 2, 837, 1, 838
	1	15	24~29	56, 783, 112, 727, 148, 691
	2	18	30~35	80, 759, 42, 797, 40, 799
	3	22	36~41	35, 804, 73, 766, 146, 693
	4	26	42~51	31, 808, 28, 811, 30, 809, 27, 812, 29, 810
	...	...	...	...
	15	23 7	384~455	3, 836, 19, 820, 22, 817, 41, 798, 38, 801, 44, 795, 52, 787, 45, 794, 63, 776, 67, 772, 72 767, 76, 763, 94, 745, 102, 737, 90, 749, 109, 730, 165, 674, 111, 728, 209, 630, 204, 635, 117, 722, 188, 651, 159, 680, 198, 641, 113, 726, 183, 656, 180, 659, 177, 662, 196, 643, 155, 684, 214, 625, 126, 713, 131, 708, 219, 620, 222, 617, 226, 613
High	16	23 7	456~513	230, 609, 232, 607, 262, 577, 252, 587, 418, 421, 416, 423, 413, 426, 411, 428, 376, 463, 395, 444, 283, 556, 285, 554, 379, 460, 390, 449, 363, 476, 384, 455, 388, 451, 386, 453, 361, 478, 387, 452, 360, 479, 310, 529, 354, 485, 328, 511, 315, 524, 337, 502, 349, 490, 335, 504, 324, 515
	...	...	...	...
	29	18	810~815	309, 530, 265, 574, 233, 606
	30	15	816~819	367, 472, 296, 543
	31	-	820~837	336, 503, 305, 534, 373, 466, 280, 559, 279, 560, 419, 420, 240, 599, 258, 581, 229, 610

Figure 17.23: Example of mapping from logical index to physical root sequence index.

17.5.1 UE Transmitter

The PRACH preamble can be generated at the system sampling rate, by means of a large IDFT as illustrated in Figure 17.24.

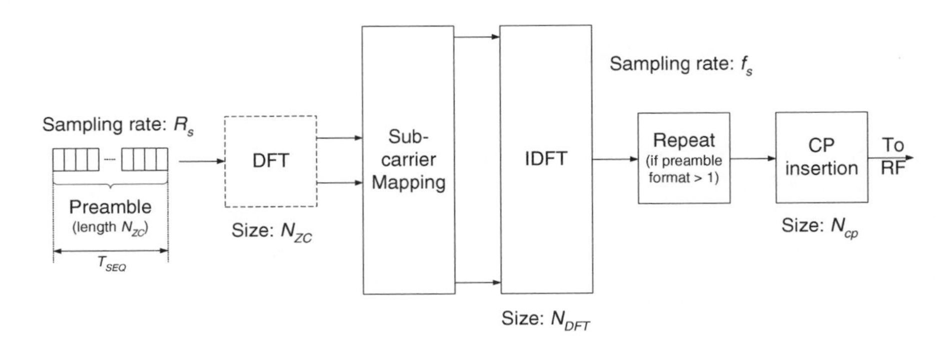

Figure 17.24: Functional structure of PRACH preamble transmitter.

Note that the DFT block in Figure 17.24 is optional as the sequence can be mapped directly in the frequency domain at the IDFT input, as explained in Section 7.2.1. The cyclic shift can be implemented either in the time domain after the IDFT, or in the frequency domain before the IDFT through a phase shift. For all possible system sampling rates, both CP and sequence durations correspond to an integer number of samples.

The method of Figure 17.24 does not require any time-domain filtering at baseband, but leads to large IDFT sizes (up to 24 576 for a 20 MHz spectrum allocation), which are cumbersome to implement in practice.

Therefore, another option for generating the preamble consists of using a smaller IDFT, actually an IFFT, and shifting the preamble to the required frequency location through time-domain upsampling and filtering (hybrid frequency/time-domain generation, shown in Figure 17.25). Given that the preamble sequence length is 839, the smallest IFFT size that can be used is 1024, resulting in a sampling frequency $f_{\mathrm{IFFT}} = 1.28$ Msps. Both the CP and sequence durations have been designed to provide an integer number of samples at this sampling rate. The CP can be inserted before the upsampling and time-domain frequency shift, so as to minimize the intermediate storage requirements.

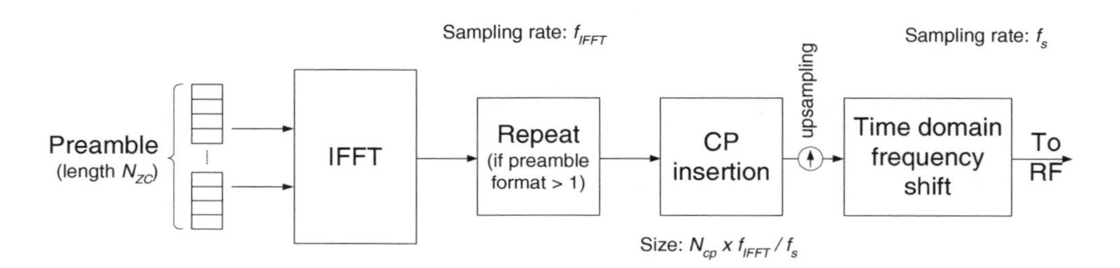

Figure 17.25: Hybrid frequency/time domain PRACH generation.

17.5.2 eNodeB PRACH Receiver

17.5.2.1 Front-End

In the same way as for the preamble transmitter, a choice can be made for the PRACH receiver at the eNodeB between full frequency-domain and hybrid time/frequency-domain approaches. As illustrated in Figure 17.26, the common parts to both approaches are the CP removal, which always occurs at the front-end at the system sampling rate f_{s}, the PDP computation and signature detection. The approaches differ only in the computation of the frequency tones carrying the PRACH signal(s).

The full frequency-domain method computes, from the 800 μs worth of received input samples during the observation interval (Figure 17.6), the full range of frequency tones used for UL transmission given the system bandwidth. As a result, the PRACH tones are directly extracted from the set of UL tones, which does not require any frequency shift or time-domain filtering but involves a large DFT computation. Note that even though $N_{\mathrm{DFT}} = n \cdot 2^m$, thus allowing fast and efficient DFT computation algorithms inherited from the building-block construction approach [13], the DFT computation cannot start until the complete sequence is stored in memory, which increases delay.

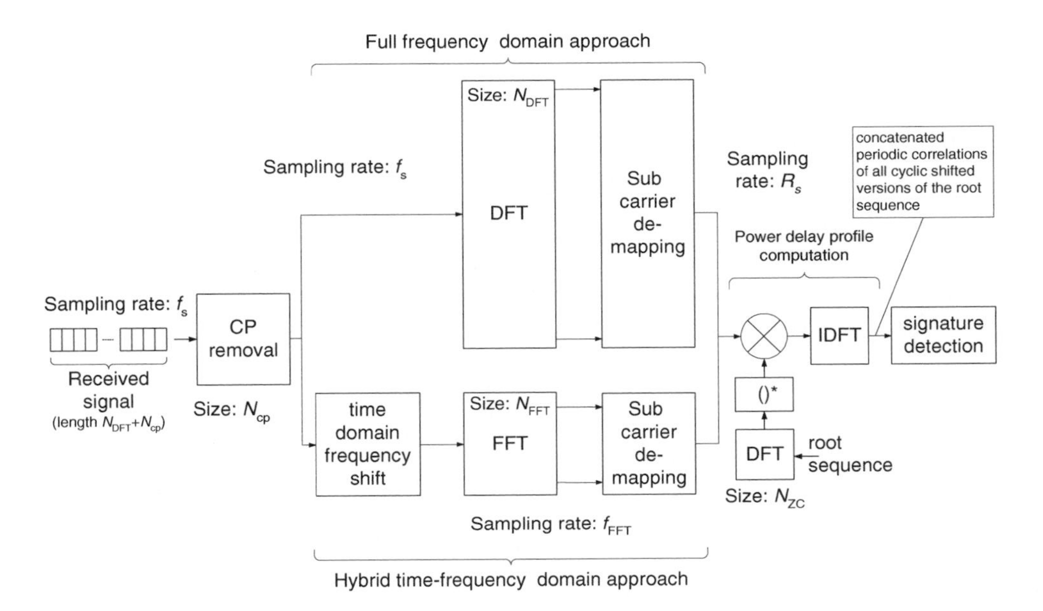

Figure 17.26: PRACH receiver options.

On the other hand, the hybrid time/frequency-domain method first extracts the relevant PRACH signal through a time-domain frequency shift with down-sampling and anti-aliasing filtering. There follows a small-size DFT (preferably an FFT), computing the set of frequency tones centered on the PRACH tones, which can then be extracted. The down-sampling ratio and corresponding anti-aliasing filter are chosen to deliver a number of PRACH time samples suitable for an FFT or simple DFT computation at a sampling rate which is an integer fraction of the system sampling rate. Unlike the full frequency-domain approach, the hybrid time/frequency-domain computation can start as soon as the first samples have been received, which helps to reduce latency.

17.5.2.2 Power Delay Profile (PDP) Computation

The LTE PRACH receiver can benefit from the PRACH format and Constant Amplitude Zero AutoCorrelation (CAZAC) properties as described in the earlier sections by computing the PRACH power delay profile through a frequency-domain periodic correlation. The PDP of the received sequence is given by

$$\mathrm{PDP}(l) = |z_u(l)|^2 = \left| \sum_{n=0}^{N_{ZC}-1} y(n)x_u^*[(n+l)_{N_{ZC}}] \right|^2 \tag{17.15}$$

where $z_u(l)$ is the discrete periodic correlation function at lag l of the received sequence $y(n)$ and the reference searched ZC sequence $x_u(n)$ of length N_{ZC}, and where $(\cdot)^*$ denotes the complex conjugate. Given the periodic convolution of the complex sequences $y(n)$ and $x_u(n)$

defined as

$$[y(n) * x_u(n)](l) = \sum_{n=0}^{N_{ZC}-1} y(n)x_u[(l-n)_{N_{ZC}}], \tag{17.16}$$

$z_u(l)$ can be expressed as follows:

$$z_u(l) = [y(n) * x_u^*(-n)](l) \tag{17.17}$$

Let $X_u(k) = R_{X_u}(k) + jI_{X_u}(k)$, $Y_u(k) = R_Y(k) + jI_Y(k)$ and $Z_u(k)$ be the DFT coefficients of the time-domain ZC sequence $x_u(n)$, the received baseband samples $y(n)$, and the discrete periodic correlation function $z_u(n)$ respectively. Using the properties of the DFT, $z_u(n)$ can be efficiently computed in the frequency domain as

$$\begin{cases} Z_u(k) = Y(k)X_u^*(k) & \text{for } k = 0, \ldots, N_{ZC} - 1 \\ z_u(n) = \text{IDFT}\{Z_u(k)\}_n & \text{for } n = 0, \ldots, N_{ZC} - 1 \end{cases} \tag{17.18}$$

The PDP computation is illustrated in Figure 17.27.

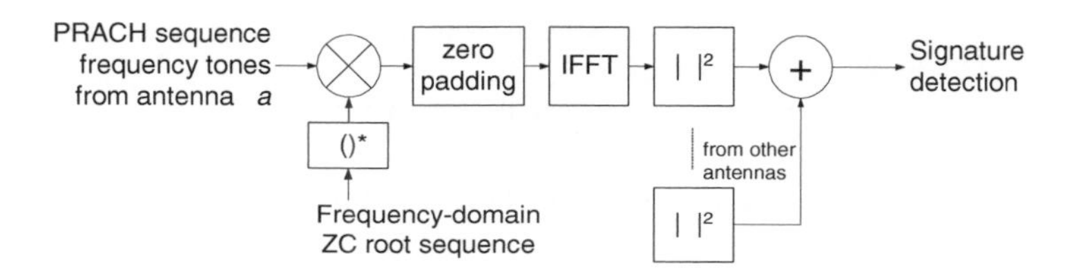

Figure 17.27: PDP computation per root sequence.

The zero padding aims at providing the desired oversampling factor and/or adjusting the resulting number of samples to a convenient IFFT size. Note that for high-speed cells, additional non-coherent combining over three timing uncertainty windows can be performed for each receive antenna, as shown in Figure 17.18.

In addition, the pairwise sequence indexing (Section 17.4.3.6) allows further efficient 'paired' matched filtering.

Let $W_u(k)$ be defined as

$$W_u(k) = Y(k)X_u(k) \quad \text{for } k = 0, \ldots, N_{ZC} - 1 \tag{17.19}$$

The element-wise multiplications $Z_u(k)$ and $W_u(k)$ can be computed jointly [14], where the partial products $R_Y(k)R_{X_u}(k)$, $I_Y(k)I_{X_u}(k)$, $I_Y(k)R_{X_u}(k)$ and $I_{X_u}(k)R_Y(k)$ only need to be computed once for both $Z_u(k)$ and $W_u(k)$ frequency-domain matched filters.

Now let $w_u(l)$ be defined as

$$w_u(l) = \text{IDFT}\{W_u(k)\}_l \quad \text{for } l = 0, \ldots, N_{ZC} - 1 \tag{17.20}$$

By substituting Equation (17.19) into Equation (17.20) and using the property $\text{DFT}\{x_{N_{ZC}-u}(n-1)\}_k = X_u^*(k)$, we can obtain [15]

$$w_u(l) = \sum_{n=0}^{N_{ZC}-1} y(n)x_{N_{ZC}-u}(l+n-1) = [y(n) * x_{N_{ZC}-u}(-n)](l-1) \tag{17.21}$$

As a result, the IDFTs of the joint computation of the frequency-domain matched filters $Z_u(k)$ and $W_u(k)$ provide the periodic correlations of time-domain ZC sequences $x_u(n)$ and $x_{N_{ZC}-u}(n)$, the latter being shifted by one sequence sample.

17.5.2.3 Signature Detection

The fact that different PRACH signatures are generated from cyclic shifts of a common root sequence means that the frequency-domain computation of the PDP of a root sequence provides in one shot the concatenated PDPs of all signatures derived from the same root sequence.

Therefore, the signature detection process consists of searching, within each ZCZ defined by each cyclic shift, the PDP peaks above a detection threshold over a search window corresponding to the cell size. Figure 17.28 shows the basic functions of the signature detector.

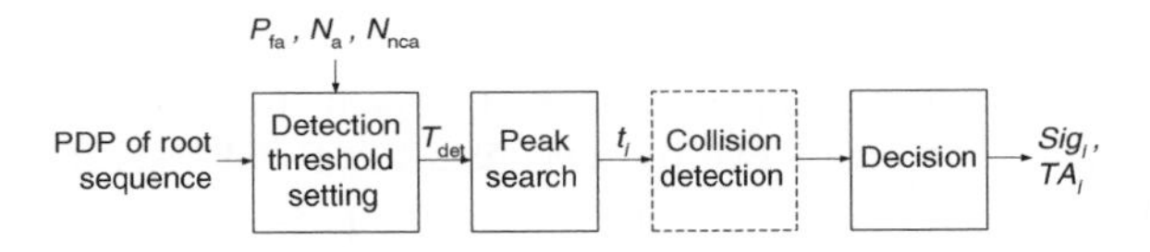

Figure 17.28: Signature detection per root sequence.

Detection threshold setting. The target false alarm probability $p_{fa}(T_{det})$ drives the setting of the detection threshold T_{det}.

Under the assumption that the L samples in the uncertainty window are uncorrelated[6] Gaussian noise with variance σ_n^2 in the absence of preamble transmission, the complex sample sequence $z_a^m(\tau)$ received from antenna a (delayed to reflect a targeted time offset τ of the search window, and despread over a coherent accumulation length (in samples) N_{ca} against the reference code sequence) is a complex Gaussian random variable with variance $\sigma_{n,ca}^2 = N_{ca}\sigma_n^2$. In practice, N_{ca} is the size of the IFFT in Figure 17.27. The non-coherent accumulation $z_{nca}(\tau)$ is modelled as follows:

$$z_{nca}(\tau) = \sum_{a=1}^{N_a} \sum_{m=0}^{N_{nca}-1} |z_a^m(\tau)|^2 \tag{17.22}$$

where N_a is the number of antennas and N_{nca} is the number of additional non-coherent accumulations (e.g. in case of sequence repetition).

$z_{nca}(\tau)$ follows a central chi-square distribution with $2N = 2N_a \cdot N_{nca}$ degrees of freedom, with mean (defining the noise floor) $\gamma_n = N\sigma_{n,ca}^2$ and Cumulative Distribution Function (CDF) $F(T_{det}) = 1 - p_{fa}(T_{det})^L$. It is worth noticing that instead of the absolute threshold we can

[6]The assumption of no correlation between samples holds true in practice up to an oversampling factor of 2.

consider the threshold T_r relative to the noise floor γ_n as follows:

$$T_r = \frac{T_{\text{det}}}{\gamma_n} = \frac{T_{\text{det}}}{N_a \cdot N_{\text{nca}} N_{\text{ca}} \sigma_n^2} \tag{17.23}$$

This removes the dependency of $F(T_r)$ on the noise variance [16]:

$$F(T_r) = 1 - e^{-N_a \cdot N_{\text{nca}} T_r} \sum_{k=0}^{N_a \cdot N_{\text{nca}} - 1} \frac{1}{k!} (N_a \cdot N_{\text{nca}} T_r)^k \tag{17.24}$$

As a result, the relative detection threshold can be precomputed and stored.

Noise floor estimation. For the PDP arising from the transmissions of each root sequence, the noise floor can be estimated as follows:

$$\gamma_n = \frac{1}{N_s} \sum_{i=0, z_{\text{nca}}(\tau_i) < T_{\text{det\_ini}}}^{L-1} z_{\text{nca}}(\tau_i) \tag{17.25}$$

where the summation is over all samples less than the absolute noise floor threshold $T_{\text{det\_ini}}$ and N_s is the number of such samples. In a real system implementation, the number of additions can be made a power of two by repeating some additions if needed. The initial absolute threshold $T_{\text{det\_ini}}$ is computed using an initial noise floor estimated by averaging across all search window samples.

Collision detection. In any cell, the eNodeB can be made aware of the maximum expected delay spread. As a result, whenever the cell size is more than twice the distance corresponding to the maximum delay spread, the eNodeB may in some circumstances be able to differentiate the PRACH transmissions of two UEs if they appear distinctly apart in the PDP. This is illustrated in Figure 17.29, where the upper PDP reflects a small cell, where collision detection is never possible, while the lower PDP represents a larger cell where it may sometimes be possible to detect two distinct preambles within the same ZCZ. If an eNodeB detects a collision, it would not send any random access response, and the colliding UEs would each randomly reselect their signatures and retransmit.

17.5.2.4 Timing Estimation

The primary role of the PRACH preamble is to enable the eNodeB to estimate a UE's transmission timing. Figure 17.30 shows the CDF of the typical performance achievable for the timing estimation. From Figure 17.30 one can observe that the timing of 95% of UEs can be estimated to within 0.5 μs, and more than 98% within 1 μs. These results are obtained assuming a very simple timing advance estimation algorithm, using only the earliest detected peak of a detected signature. No collision detection algorithm is implemented here. The IFFT size is 2048 and the system sampling rate 7.68 MHz, giving an oversampling rate of 2.44.

17.5.2.5 Channel Quality Estimation

For each detected signature, the relative frequency-domain channel quality of the transmitting UE can be estimated from the received preamble. This allows the eNodeB to schedule the L2/L3 message (message 3) in a frequency-selective manner within the PRACH bandwidth.

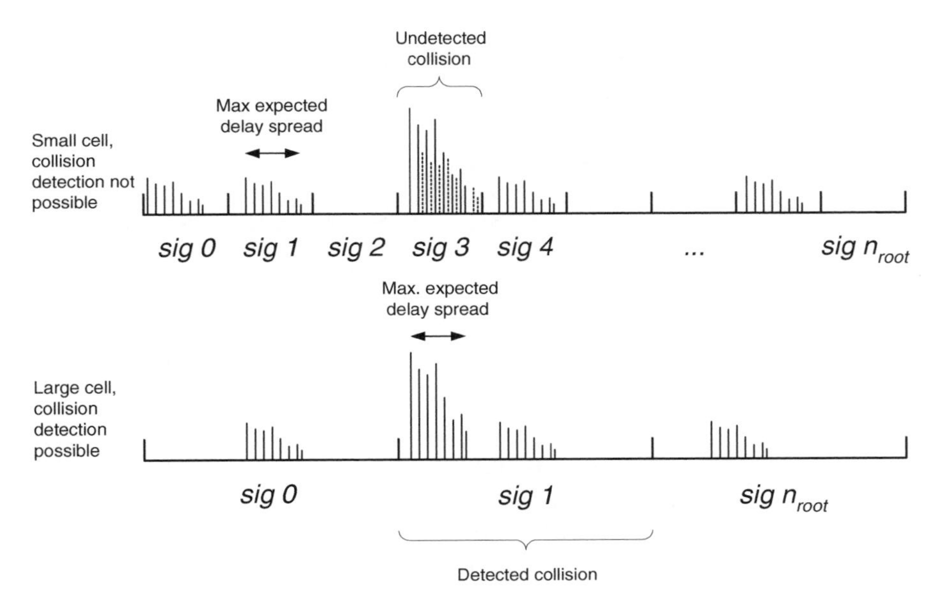

Figure 17.29: Collision detection.

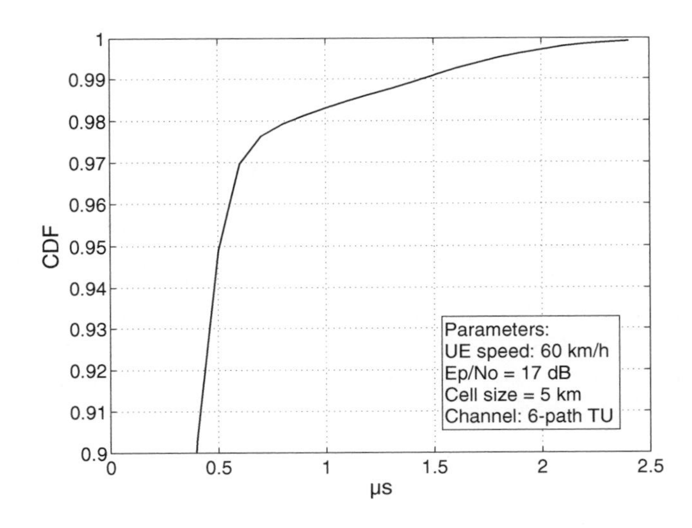

Figure 17.30: CDF of timing estimation error from the PRACH preamble.

Figure 17.31 shows the BLock Error Rate (BLER) performance of the L2/L3 message of the RACH procedure when frequency-selectively scheduled or randomly scheduled, assuming a typical 10 ms delay between the PRACH preamble and the L2/L3 message. A Least Squares (LS) filter is used for the frequency-domain interpolation, and a single RB is assumed for the size of the L2/L3 message.

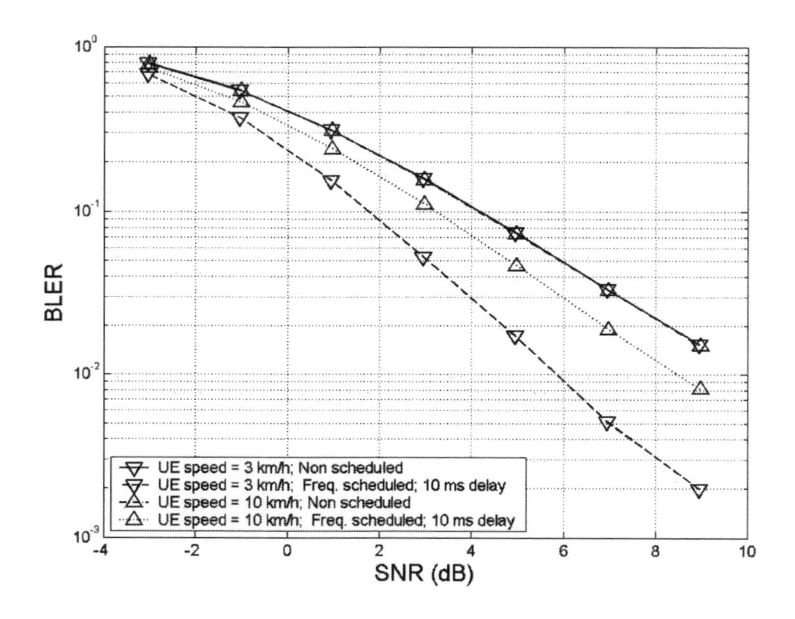

Figure 17.31: BLER performance of post-preamble scheduled data.

It can be seen that the performance of a frequency-selectively scheduled L2/L3 message at 10% BLER can be more than 2 dB better than 'blind scheduling' at 3 km/h, and 0.5 dB at 10 km/h.

17.6 Time Division Duplex (TDD) PRACH

As discussed in Chapter 23, one design principle of LTE is to maximize the commonality of FDD and TDD transmission modes. With this in mind, the random access preamble formats 0 to 3 are supported in both FDD and TDD operation. In addition, a short preamble format, 'format 4', is supported for TDD operation. Format 4 is designed to fit into the short uplink special field known as UpPTS[7] for small cells (see Section 23.4). Unlike in FDD where there can be at most one PRACH opportunity in one subframe, TDD allows frequency multiplexing of up to six PRACH opportunities in one subframe, to compensate for the smaller number of uplink subframes.

17.6.1 Preamble Format 4

A ZC sequence of length 139 is used for preamble format 4. The preamble starts 157 μs before the end of the UpPTS field at the UE; it has a T_{CP} of 14.6 μs and T_{SEQ} is 133 μs.

Unlike preamble formats 0 to 3, a restricted preamble set for high-speed cells is not necessary for preamble format 4, which uses a 7.5 kHz subcarrier spacing. With a random access duration of two OFDM symbols (157 μs), the preamble format 4 is mainly used for

[7] Uplink Pilot TimeSlot.

small cells with a cell radius less than 1.5 km, and where cyclic shift restrictions for high UE velocities (see Section 17.4.3.4) are not needed. Therefore, considering that Layer 2 always sees 64 preambles and a sequence length of 139, a smaller set of cyclic shift configurations can be used, as shown in Table 17.5.

Table 17.5: Cyclic shift configuration for preamble format 4. Reproduced by permission of © 3GPP.

N_{CS} configuration	N_{CS} value	Required number of ZC root sequences per cell
0	2	1
1	4	2
2	6	3
3	8	4
4	10	5
5	12	6
6	15	8

Unlike for preamble formats 0 to 3, the root ZC sequence index for preamble format 4 follows the natural pairwise ordering of the physical ZC sequences, with no special restrictions related to the CM or high-speed scenarios. The sequence mapping for preamble format 4 can be formulated as

$$u = \left((-1)^{v}\left\lfloor \frac{v}{2} + 1 \right\rfloor\right) \bmod N_{ZC} \tag{17.26}$$

from logical sequence index v to physical sequence index u, therefore avoiding the need to prestore a mapping table.

17.7 Concluding Remarks

In this chapter, the detailed design choices of the LTE PRACH have been explained based on theoretical derivations and performance evaluations. In particular, it can be seen how the PRACH preamble addresses the high performance targets of LTE, such as high user density, very large cells, very high speed, low latency and a plurality of use cases, while fitting with minimum overhead within the uplink SC-FDMA transmission scheme.

Many of these aspects benefit from the choice of ZC sequences for the PRACH preamble sequences in place of the PN sequences used in earlier systems. The properties of these sequences enable substantial numbers of orthogonal preambles to be transmitted simultaneously.

Considerable flexibility exists in the selection of the PRACH slot formats and cyclic shifts of the ZC sequences to enable the LTE PRACH to be dimensioned appropriately for different cell radii and loadings.

Some options for the implementation are available, by which the complexity of the PRACH transmitter and receiver can be minimized without sacrificing the performance.

References[8]

[1] A. V. Oppenheim and R. W. Schafer, *Discrete-Time Signal Processing*. Englewood Cliffs, NJ: Prentice Hall, 1999.

[2] 3GPP Technical Specification 36.211, 'Evolved Universal Terrestrial Radio Access (E-UTRA); Physical Channels and Modulation', www.3gpp.org.

[3] M. Shafi, S. Ogose and T. Hattori, *Wireless Communications in the 21st Century*. New York: Wiley Inter-Science, 2002.

[4] Panasonic and NTT DoCoMo, 'R1-062175: Random Access Burst Design for E-UTRA', www.3gpp.org, 3GPP TSG RAN WG1, meeting 46, Tallinn, Estonia, August 2006.

[5] 3GPP Technical Report 25.814 'Physical layer aspects for E-UTRA (Release 7)', www.3gpp.org.

[6] R. L. Frank, S. A. Zadoff and R. Heimiller, 'Phase Shift Pulse Codes with Good Periodic Correlation Properties'. *IRE IEEE Trans. on Information Theory*, Vol. 7, pp. 254–257, October 1961.

[7] D. C. Chu, 'Polyphase Codes with Good Periodic Correlation Properties'. *IEEE Trans. on Information Theory*, Vol. 18, pp. 531–532, July 1972.

[8] ETSI EN 300 910, 'Radio Transmission and Reception (Release 1999),' www.etsi.org.

[9] Huawei, 'R1-072325: Multiple Values of Cyclic Shift Increment NCS', www.3gpp.org, 3GPP TSG RAN WG1, meeting 49, Kobe, Japan, May 2007.

[10] Panasonic and NTT DoCoMo, 'R1-073624: Limitation of RACH Sequence Allocation for High Mobility Cell', www.3gpp.org, 3GPP TSG RAN WG1, meeting 50, Athens, Greece, August 2007.

[11] Motorola, 'R1-040642: Comparison of PAR and Cubic Metric for Power De-rating', www.3gpp.org, 3GPP TSG RAN WG1, meeting 37, Montreal, Canada, May 2004.

[12] Motorola, 'R1-060023: Cubic Metric in 3GPP-LTE', www.3gpp.org, 3GPP TSG RAN WG1, LTE adhoc meeting, Helsinki, Finland, January 2006.

[13] W. W. Smith and J. M. Smith, *Handbook of Real-time Fast Fourier Transforms*. New York: Wiley Inter-Science, 1995.

[14] Panasonic, 'R1-071517: RACH Sequence Allocation for Efficient Matched Filter Implementation', www.3gpp.org, 3GPP TSG RAN WG1, meeting 48bis, St Julians, Malta, March 2007.

[15] Huawei, 'R1-071409: Efficient Matched Filters for Paired Root Zadoff–Chu Sequences,' www.3gpp.org, 3GPP TSG RAN WG1, meeting 48bis, St Julians, Malta, March 2007.

[16] J. G. Proakis, *Digital Communications*. New York: McGraw–Hill, 1995.

[8] All web sites confirmed 1st March 2011.

18

Uplink Transmission Procedures

Matthew Baker

18.1 Introduction

In this chapter two procedures are explained which are fundamental to the efficient operation of the LTE uplink: *Timing control* is essential for the orthogonal uplink intra-cell multiple access scheme, while *power control* is important for maintaining Quality-of-Service (QoS), ensuring acceptable User Equipment (UE) battery life and controlling inter-cell interference.

18.2 Uplink Timing Control

18.2.1 Overview

As explained in Chapters 14 and 16, a key feature of the uplink transmission scheme in LTE is that it is designed for orthogonal multiple access in time and frequency between the different UEs. This is fundamentally different from WCDMA, in which the uplink transmissions are non-orthogonal; in WCDMA, from the point of view of the multiple access, there is therefore no need to arrange for the uplink signals from different UEs to be received with any particular timing at the NodeB. The dominant consideration for the uplink transmission timing in WCDMA is the operation of the power control loop, which was designed (in most cases) for a loop delay of just one timeslot (0.666 ms). This is achieved by setting the uplink transmission timing as close as possible to a fixed offset relative to the received downlink timing, without taking into account any propagation delays. Propagation delays in uplink and downlink are absorbed at the NodeB, by means of reducing the time spent measuring the Signal-to-Interference Ratio (SIR) to derive the next power control command.

LTE – The UMTS Long Term Evolution: From Theory to Practice, Second Edition.
Stefania Sesia, Issam Toufik and Matthew Baker.

For LTE, uplink orthogonality is maintained by ensuring that the transmissions from different UEs in a cell are time-aligned at the receiver of the eNodeB. This avoids intra-cell interference occurring, both between UEs assigned to transmit in consecutive subframes and between UEs transmitting on adjacent subcarriers.

Time alignment of the uplink transmissions is achieved by applying a *timing advance* at the UE transmitter, relative to the received downlink timing. The main role of this is to counteract differing propagation delays between different UEs, as shown in Figure 18.1. A similar approach is used in GSM.

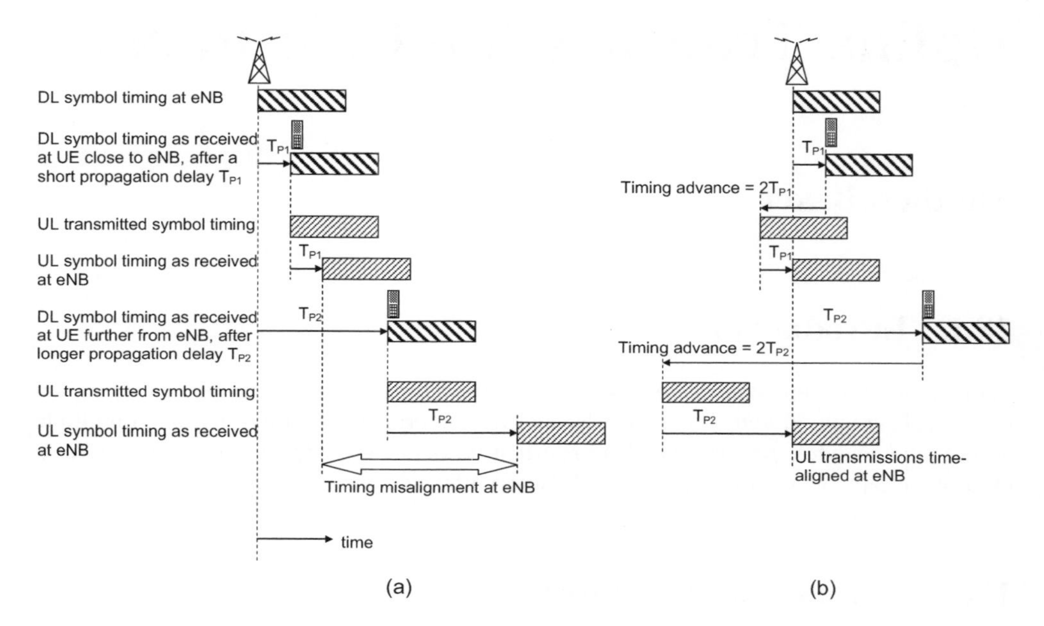

Figure 18.1: Time alignment of uplink transmissions: (a) without timing advance; (b) with timing advance.

18.2.2 Timing Advance Procedure

18.2.2.1 Initial Timing Advance

After a UE has first synchronized its receiver to the downlink transmissions received from the eNodeB (see Section 7.2), the initial timing advance is set by means of the random access procedure described in Section 17.3. This involves the UE transmitting a random access preamble from which the eNodeB estimates the uplink timing and responds with an 11-bit initial timing advance command contained within the Random Access Response (RAR) message. This allows the timing advance to be configured by the eNodeB with a granularity of 0.52 μs from 0 up to a maximum of 0.67 ms,[1] corresponding to a cell radius

[1] This is equal to $16T_s$, where $T_s = 1/30.72$ μs is the basic unit of time introduced in Section 5.4.1.

of 100 km.[2] The timing advance was limited to this range in order to avoid further restricting the processing time available at the UE between receiving the downlink signal and having to make a corresponding uplink transmission (see Figures 10.13 and 10.14). In any case, a cell range of 100 km is sufficient for most practical scenarios, and is far beyond what could be achieved with the early versions of GSM, in which the range of the timing advance restricted the cell range to about 35 km. Support of cell sizes even larger than 100 km in LTE is left to the eNodeB implementation to handle.

The granularity of 0.52 μs enables the uplink transmission timing to be set with an accuracy well within the length of the uplink CP (the smallest value of which is 4.7 μs). This granularity is also significantly finer than the length of a cyclic shift of the uplink reference signals (see Chapter 15). Simulations have shown [1, 2] that timing misalignment of up to at least 1 μs does not cause significant degradation in system performance due to increased interference. Thus the granularity of 0.52 μs is sufficiently fine to allow for additional timing errors arising from the uplink timing estimation in the eNodeB and the accuracy with which the UE sets its initial transmission timing – the latter being required to be better than 0.39 μs[3] (except if the downlink system bandwidth is 1.4 MHz, in which case the UE initial transmission timing accuracy is relaxed to 0.79 μs) (see [3, Section 7.1]).

18.2.2.2 Timing Advance Updates

After the initial timing advance has been established for each UE, it will then need to be updated from time to time to counteract changes in the arrival time of the uplink signals at the eNodeB. Such changes may arise from:

- The movement of a UE, causing the propagation delay to change at a rate dependent on the velocity of the UE relative to the eNodeB; at 500 km/h (the highest speed considered for LTE), the round-trip propagation delay would change by a maximum of 0.93 μs/s;
- Abrupt changes in propagation delay due to existing propagation paths disappearing and new ones arising; such changes typically occur most frequently in dense urban environments as the UEs move around the corners of buildings;
- Oscillator drift in the UE, where the accumulation of small frequency errors over time may result in timing errors; the frequency accuracy of the oscillator in an LTE UE is required to be better than 0.1 ppm (see [4, Section 6.5.1]), which would result in a maximum accumulated timing error of 0.1 μs/s;
- Doppler shift arising from the movement of the UE, especially in Line-Of-Sight (LOS) propagation conditions, resulting in an additional frequency offset of the uplink signals received at the eNodeB.[4]

The updates of the timing advance to counteract these effects are performed by a closed-loop mechanism whereby the eNodeB measures the received uplink timing and issues timing

[2]In theory it would also be possible to support small negative timing advances (i.e. a timing delay) up to the duration of the Cyclic Prefix (CP), for UEs very close to the eNodeB, without causing loss of uplink time-domain orthogonality. However, this is not supported in LTE.

[3]This is equal to $12T_s$.

[4]In non-LOS conditions, this becomes a Doppler spread, where the error is typically a zero-mean random variable.

advance update commands to instruct the UE to adjust its transmission timing accordingly, relative to its current transmission timing.[5]

In deriving the timing advance update commands, the eNodeB may measure any uplink signal which is useful. This may include the Sounding Reference Signals (SRSs), DeModulation Reference Signals (DM-RS), Channel Quality Indicator (CQI), ACKnowledgements/Negative ACKnowledgements (ACK/NACKs) sent in response to downlink data, or the uplink data transmissions themselves. In general, wider-bandwidth uplink signals enable a more accurate timing estimate to be made, although this is not, in itself, likely to be a sufficient reason to configure all UEs to transmit wideband SRS very frequently. The benefit of highly accurate timing estimation has to be traded off against the uplink overhead from such signals. In addition, cell-edge UEs are power-limited and therefore also bandwidth-limited for a given uplink Signal-to-Interference-plus-Noise Ratio (SINR); in such cases, the timing estimation accuracy of narrower-bandwidth uplink signals can be improved through averaging multiple measurements over time and interpolating the resulting power delay profile. The details of the uplink timing measurements at the eNodeB are not specified but left to the implementation.

A timing advance update command received at the UE is applied at the beginning of the uplink subframe which begins 4–5 ms after the end of the downlink subframe in which the command is received (depending on the propagation delay), as shown in Figure 18.2. For a TDD or half-duplex FDD system configuration, the new uplink transmission timing takes effect at the start of the first uplink transmission after this point. In the case of an increase in the timing advance relative to the previous transmission, the first part of the subframe in which the new timing is applied is skipped.

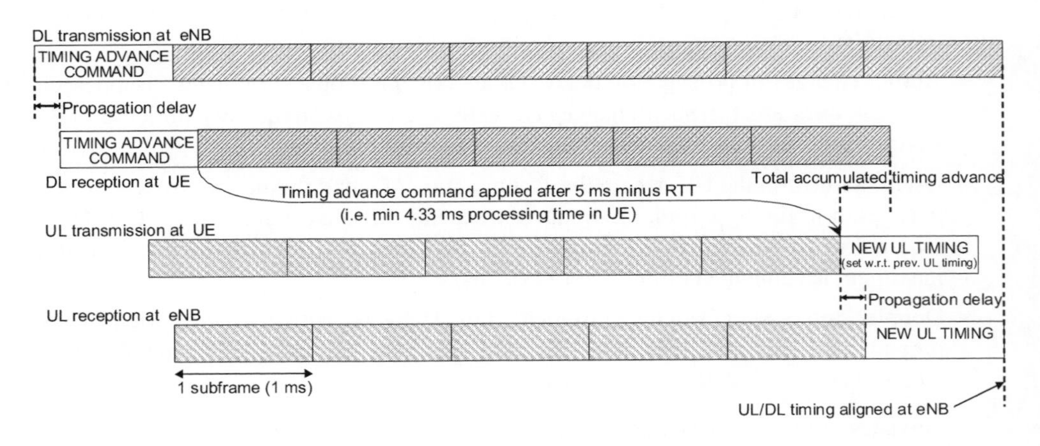

Figure 18.2: Application of timing advance update commands.

[5]Transmission timing adjustments arising from timing advance update commands are always made relative to the latest uplink timing. The initial transmission timing is set relative to the received downlink timing; further, if no timing advance update commands are received, the UE is required autonomously to adjust its uplink transmission timing to track changes in the received downlink timing, as specified in [3, Section 7.1.2].

The timing advance update commands are generated at the Medium Access Control (MAC) layer in the eNodeB and transmitted to the UE as MAC control elements which may be multiplexed together with data on the Physical Downlink Shared CHannel (PDSCH). Like the initial timing advance command in the response to the Random Access CHannel (RACH) preamble, the update commands have a granularity of 0.52 μs; in the case of the update commands, the UE is required to implement them with an accuracy of ±0.13 μs. The range of the update commands is ±16.6 μs, allowing a step change in uplink timing equivalent to the length of the extended CP (i.e. 16.67 μs). They would typically not be sent more frequently than about 2 Hz. In practice, fast updates are unlikely to be necessary, as even for a UE moving at 500 km/h the change in round-trip time is not more than 0.93 μs/s. The eNodeB must balance the overhead of sending regular timing advance update commands to all the UEs in the cell against a UE's ability to transmit quickly when data arrives in its transmit buffer. The eNodeB therefore configures a timer for each UE, which the UE restarts each time a timing advance update command is received; if the UE does not receive another timing advance update command before the timer expires, it must then consider its uplink to have lost synchronization.[6] In such a case, in order to avoid the risk of generating interference to uplink transmissions from other UEs, the UE is not permitted to make another uplink transmission of any sort without first transmitting a random access preamble to reinitialize the uplink timing.

One further use of timing advance is to create a switching time between uplink reception at the eNodeB and downlink transmission for TDD and half-duplex FDD operation. This switching time can be generated by applying an additional timing advance offset to the uplink transmissions, to increase the amount of timing advance beyond what is required to compensate for the round-trip propagation delay. Typically a switching time of up to 20 μs may be needed. This is discussed in more detail in Section 23.4.1.

18.3 Power Control

18.3.1 Overview

Uplink transmitter power control in a mobile communication system serves an important purpose: it balances the need for sufficient transmitted energy per bit to maintain the link quality corresponding to the required Quality-of-Service (QoS), against the needs to minimize interference to other users of the system and to maximize the battery life of the mobile terminal.

In achieving this purpose, the power control has to adapt to the characteristics of the radio propagation channel, including path-loss, shadowing and fast fading, as well as overcoming interference from other users – both within the same cell and in neighbouring cells.

The requirements for uplink interference management in LTE are quite different from those for WCDMA. In WCDMA, the uplink is basically non-orthogonal,[7] and the primary source of interference which has to be managed is intra-cell interference between different users in the same cell. Uplink users in WCDMA share the same time-frequency resources,

[6] Note that loss of uplink synchronization is possible without leaving RRC_CONNECTED state (see Chapter 3).

[7] The later releases of WCDMA do, however, introduce a greater element of orthogonality into the uplink transmissions, by means of lower spreading factors and greater use of time-division multiplexing of different users in HSDPA and HSUPA.

and they generate an interference rise above thermal noise at the NodeB receiver; this is known as 'Rise over Thermal' (RoT), and it has to be carefully controlled and shared between users. The primary mechanism for increasing the uplink data rate for a given user in WCDMA is to reduce the spreading factor and increase the transmission power accordingly, consuming a larger proportion of the total available RoT in the cell.

By contrast, in LTE the uplink is basically orthogonal by design, and intra-cell interference management is consequently less critical than in WCDMA. The primary mechanisms for varying the uplink data rate in LTE are varying the transmitted bandwidth and varying the Modulation and Coding Scheme (MCS), while the transmitted Power Spectral Density (PSD) could typically remain approximately constant for a given MCS.

Moreover, in WCDMA the power control [5] was primarily designed with continuous transmission in mind for circuit-switched services, while in LTE fast scheduling of different UEs is applied at 1 ms intervals. This is reflected in the fact that power control in WCDMA is periodic with a loop delay of 0.67 ms and a normal power step of ±1 dB, while LTE allows for larger power steps (which do not have to be periodic), with a minimum loop delay of about 5 ms (see Figure 18.5).

With these considerations in mind, the power control scheme provided in LTE employs a combination of open-loop and closed-loop control. This in theory requires less feedback than a purely closed-loop scheme, as the closed-loop feedback is only needed to compensate for cases when the UE's own estimate of the required power setting is not satisfactory.

A typical mode of operation for power control in LTE involves setting a coarse operating point for the transmission PSD[8] by open-loop means, based on path-loss estimation. This would give a suitable PSD for a reference MCS in the prevailing path-loss and shadowing conditions. Faster adaptation can then be applied around the open-loop operating point by closed-loop power control. This can control interference and fine-tune the power setting to suit the channel conditions (including fast fading). However, due to the orthogonal nature of the LTE uplink, the LTE closed-loop power control does not need to be as fast as in WCDMA – in LTE it would typically be expected to operate at no more than a few hundred Hertz.

Meanwhile, the fastest and most frequent adaptation of the uplink transmissions is by means of the uplink scheduling grants, which vary the transmitted bandwidth (and accordingly the total transmitted power), together with setting the MCS, in order to reach the desired transmitted data rate.

With this combination of mechanisms, the power control scheme in LTE in practice provides support for more than one mode of operation. It can be seen as a 'toolkit' from which different power control strategies can be selected and used depending, for example, on the deployment scenario or system loading.

18.3.2 Detailed Power Control Behaviour

Detailed power control formulae are specified in LTE for the Physical Uplink Shared CHannel (PUSCH), Physical Uplink Control CHannel (PUCCH) and the Sounding Reference Signals (SRSs) [6]. The formula for each of these uplink signals follows the same basic principles; though they appear complex, in all cases they can be considered as a summation of two main terms: a basic open-loop operating point derived from static or semi-static parameters

---

[8]In LTE, the PSD is set as a power per Resource Block (RB); if multiple RBs are transmitted by a UE in a subframe the power per RB is the same for all RBs.

signalled by the eNodeB, and a dynamic offset updated from subframe to subframe:

Power per resource block = basic open-loop operating point + dynamic offset .

18.3.2.1 Basic Open-Loop Operating Point

The basic open-loop operating point for the transmit power per Resource Block (RB) depends on a number of factors including the inter-cell interference and cell load. It can be further broken down into two components:

- a semi-static base level, P_0, comprising a nominal power level that is common for all UEs in the cell (measured in dBm per RB) and a UE-specific offset;
- an open-loop path-loss compensation component.

Different base levels can be configured for PUSCH data transmissions depending on the scheduling mode: those which are dynamically scheduled (i.e. using Physical Downlink Control CHannel (PDCCH) signalling) and those which use Semi-Persistent Scheduling (SPS) (see Section 4.4.2.1). This in principle allows different BLER (BLock Error Rate) operating points to be used for dynamically scheduled and SPS transmissions. One possible use for different BLER operating points is to achieve a lower probability of retransmission for SPS transmissions, hence avoiding the PDCCH signalling overhead associated with dynamically scheduled retransmissions; this is consistent with using SPS for delivery of services such as VoIP with minimal signalling overhead.

The UE-specific offset component of the base level P_0 enables the eNodeB to correct for systematic offsets in a UE's transmission power setting, for example arising from errors in path-loss estimation or in absolute output power setting.

The path-loss compensation component is based on the UE's estimate of the downlink path-loss, which can be derived from the UE's measurement of Reference Signal Received Power (RSRP) (see Section 22.3.1.1) and the known transmission power of the downlink reference signals, which is broadcast by the eNodeB. In order to obtain a reasonable indication of the uplink path-loss, the UE should filter the downlink path-loss estimate with a suitable time-window to remove the effect of fast fading but not shadowing. Typical filter lengths are between 100 and 500 ms for effective operation.

For the PUSCH and SRS, the degree to which the uplink PSD is adapted to compensate for the path-loss can be set by the eNodeB, on a scale from 'no compensation' to 'full compensation'. This is an important feature of power control in LTE and is known as *fractional power control*; it is configured by means of a fractional path-loss compensation factor, referred to as α.

In principle, the combination of the base level P_0 and the path-loss compensation component together allow the eNodeB to configure the degree to which the UE responds to the path-loss. At one extreme, the eNodeB could configure the base level to the lowest level (−126 dBm) and rely entirely on the UE's path-loss measurement to raise the power towards the cell edge; alternatively, the eNodeB can set the base level to a higher value, possibly in conjunction with only partial path-loss compensation.

Disregarding the UE-specific offset, the range of the base level P_0 for the PUSCH (−126 dBm to +24 dBm per RB) is designed to cover the full range of target SINR values for different degrees of path-loss compensation, transmission bandwidths and interference levels. For example, the highest value of P_0, +24 dBm per RB, corresponds to the maximum likely transmission power of an LTE UE and would typically only be used if path-loss compensation was not being used at all. The lowest value of P_0 for the PUSCH, −126 dBm, is relevant to a case when full path-loss compensation is used and the uplink transmission and reception conditions are optimal: for example, taking a single RB transmission, with a target SINR at the eNodeB of −5 dB (around the lowest useful SINR), interference-free reception and a 0 dB noise figure for the eNodeB receiver (see Section 21.4.4.2), then the required value of P_0 is the thermal noise level in one RB (180 kHz) minus 5 dB, which gives $P_0 = -126$ dBm.

In general, the maximum path-loss that can be compensated (either by P_0 or by the path-loss compensation component) depends on the required SINR and the transmission bandwidth. Some examples are shown in Figure 18.3, for typical ranges of SINR from −5 dB to +30 dB, interference rise above thermal noise from 0 dB to +30 dB, and transmission bandwidth from one RB to the maximum LTE system bandwidth of 110 RBs (19.8 MHz). Note that this assumes full path-loss compensation and ignores the dynamic offset.

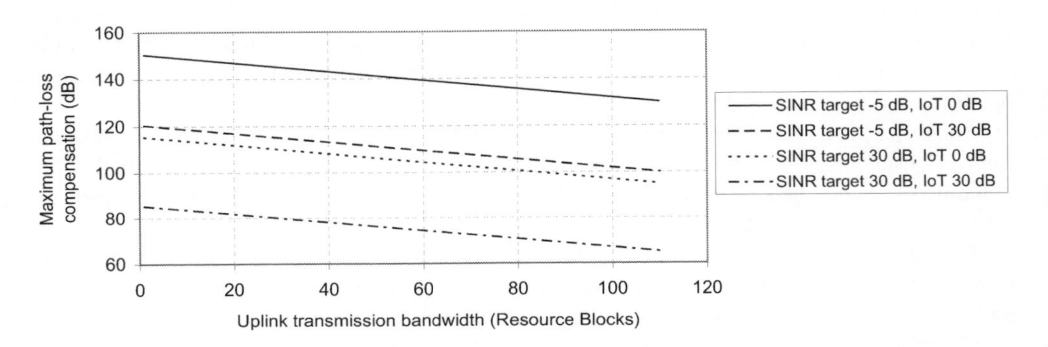

Figure 18.3: Maximum path-loss compensation in typical scenarios for a 23 dBm UE.

The fractional path-loss compensation factor α can be seen as a tool to trade off the fairness of the uplink scheduling against the total cell capacity. Full path-loss compensation maximizes fairness for cell-edge UEs. However, when considering multiple cells together as a system, the use of partial path-loss compensation can increase the total system capacity in the uplink, as less resources are spent ensuring the success of transmissions from cell-edge UEs and less inter-cell interference is caused to neighbouring cells. Path-loss compensation factors around 0.7–0.8 typically give a close-to-maximal uplink system capacity (typically around 15–25% greater than can be achieved with full path-loss compensation) without causing significant degradation to the cell-edge data rate.

The principle of fractional power control is illustrated in Figure 18.4. The target received PSD for a given MCS is reduced as the path-loss increases, so that cell-edge UEs cause less inter-cell interference.

Inter-cell interference is of particular concern for UEs located near the edge of a cell, as they may disrupt the uplink transmissions in neighbouring cells. LTE consequently provides an interference coordination mechanism whereby a frequency-dependent ‘Overload Indicator’ (OI) may be signalled directly between eNodeBs to warn a neighbouring eNodeB of high uplink interference levels in specific RBs. In response to this, the neighbouring eNodeB may reduce the permitted power per RB of the UEs which are scheduled in the corresponding RBs

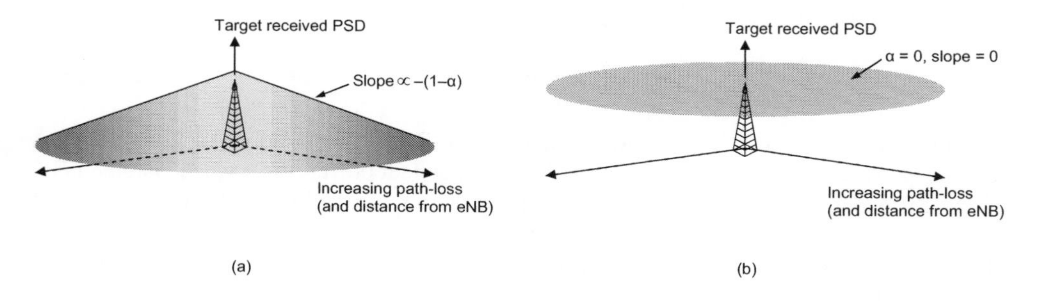

Figure 18.4: (a) Fractional and (b) non-fractional power control.

in its cell(s). It is also possible for eNodeBs to cooperate to avoid scheduling cell-edge UEs in neighbouring cells to transmit in the same RBs. This is discussed further in Section 12.5.

In summary, the basic operating point for the transmit power per RB can be expressed as:

$$\text{Basic operating point} = P_0 + \alpha \cdot \text{PL} \tag{18.1}$$

where α is the fractional path-loss compensation factor which allows the trade-off between total uplink capacity and cell-edge data rate, and PL is the downlink path-loss estimate.

For the low-rate PUCCH (carrying ACK/NACK and CQI signalling), full path-loss compensation is always applied, as the PUCCH transmissions from different users are code-division-multiplexed. Full path-loss compensation facilitates good control of the interference between the different users, and hence helps to maximize the number of users which can be accommodated simultaneously on the PUCCH. A different base level P_0 is also provided for the PUCCH compared to that used for the PUSCH, assuming a reference PUCCH format.

For the SRS, an additional semi-static offset relative to the PUSCH power operating point may be configured by RRC signalling.

18.3.2.2 Dynamic Offset

The dynamic offset part of the power per RB can also be broken down into two components:

- A component dependent on the MCS;
- Explicit Transmitter Power Control (TPC) commands.

MCS-dependent component. The MCS-dependent component (referred to in the LTE specifications as Δ_{TF}, where TF stands for 'Transport Format') allows the transmitted power per RB to be adapted according to the transmitted information data rate. Ideally, the transmission power required for a given information data rate should follow the fundamental capacity limit, such that

$$R_N = \log_2(1 + \text{SNR}) \tag{18.2}$$

where R_N is the normalized information data rate per unit bandwidth and can be calculated as the number of information bits per Resource Element (RE) in the RB, denoted here BPRE (Bits Per RE), and SNR is the Signal-to-Noise Ratio. Practical limitations of the system and receiver can be modelled with a scaling factor k (> 1):

$$\text{BPRE} = \frac{1}{k}\log_2(1 + \text{SNR}) \tag{18.3}$$

It follows that the transmission power required per RB is proportional to $2^{k\cdot\text{BPRE}} - 1$. A suitable value for k is taken as 1.25 for the MCS-dependent power offset when enabled [7].

The MCS-dependent component of the transmit power setting can act like a power control command, as the MCS is under the direct control of the eNodeB scheduler: by changing the MCS which the UE has permission to transmit, the eNodeB can quickly apply an indirect adjustment to the UE's transmit PSD via the MCS-dependent component of the transmit power setting. This may be done to take into account the instantaneous buffer status, available power headroom and QoS requirements of the UE.

The MCS-dependent component can also be used to allow an element of frequency-dependent power control, for example in cases where explicit power control commands (discussed in more detail below) are not transmitted frequently and are therefore following only the wideband fading characteristics; for example, by scheduling a low-rate MCS when the UE is granted permission to transmit in a particular part of the band, the eNodeB can dictate a low transmission power in those RBs.

Another use for the MCS-dependent component is in cases where the number of uplink RBs allocated to a UE in a subframe is not matched to the desired data rate and SIR. One example is to enable the transmit power to be reduced if the amount of data to be transmitted is less than the rate supported by the radio channel in a single RB.

The MCS-dependent component for the PUSCH can be set to zero if it is not needed, for example if fast Adaptive Modulation and Coding (AMC) is used instead.[9]

Transport-format-dependent power control is also particularly relevant for the PUCCH, as the PUCCH bandwidth for a UE does not vary depending on the amount of information to be transmitted in a given subframe (ranging from a single bit for a scheduling request or ACK/NACK, to 22 bits for combined dual-codeword ACK/NACK and CQI together – see Section 16.3). For the PUCCH, the magnitude of the power offset for each combination of control information can be adjusted semi-statically by the eNodeB, in order to set a suitable error-rate operating point for each PUCCH format (see Table 16.2). This is analogous to the different power offsets which may be set in HSDPA[10] for ACK/NACK and CQI signalling according to the error rate desired by the network.

UE-specific power control commands. The other component of the dynamic offset is the UE-specific TPC commands. These can operate in two different modes: *accumulative* TPC commands (available for PUSCH, PUCCH and SRS) and *absolute* TPC commands (available for PUSCH only). For the PUSCH, the switch between these two modes is configured semi-statically for each UE by RRC signalling – i.e. the mode cannot be changed dynamically.

With the accumulative TPC commands, each TPC command signals a power step relative to the previous level. This is the default mode and is particularly well-suited to fine-tuning of the transmission power, and to situations where a UE receives power control commands in groups of successive subframes. This mode is similar to the closed-loop power control

[9]In Release 10, the MCS-dependent component cannot be used in conjunction with PUSCH transmission mode 2 for multiple codeword uplink Single UserŰMIMO transmission (see Section 29.4.1).

[10]High Speed Downlink Packet Access.

operation in WCDMA, except that the exact values of the power steps are different: in LTE, two sets of power step values are provided: $\{-1, +1\}$ dB and $\{-1, 0, +1, +3\}$ dB (compared to the sets $\{-1, +1\}$ dB and $\{-2, +2\}$ dB in WCDMA). Which of these two sets of power steps is used is determined by the format of the TPC commands and RRC configuration. The maximum size of power step that can be made using accumulative TPC commands is therefore $+3/-1$ dB, but the range over which the power can be adjusted relative to the semi-static operating point is unlimited (except for the maximum and minimum power limits according to the UE power class – see Section 21.3.1.2). Larger power steps can be achieved by combining an accumulative TPC command with an MCS-dependent power step, by changing the MCS. The provision of one set of power step values containing a 0 dB step size enables the transmit power to be kept constant if needed (i.e. without necessarily having to change the transmission power every time a scheduling grant is sent). This is useful, for example, in scenarios where the interference is not expected to vary significantly over time.

By contrast, the transmit power setting that results from an absolute TPC command is independent of the sequence of TPC commands that may have been received previously; the transmit power setting depends only on the most recently received absolute TPC command, which independently signals a power offset relative to the semi-static operating point.[11] The set of offsets which can be signalled by absolute TPC commands is $\{-4, -1, +1, +4\}$ dB. Thus the absolute power control mode can only control the power within a range of ± 4 dB from the semi-static operating point, but a relatively large power step can be triggered by a single command (up to ± 8 dB). This mode is therefore suited to scenarios where the scheduling of the UE's uplink transmissions may be intermittent; an absolute TPC command enables the UE's transmission power to be adjusted to a suitable level in a single step after each transmission gap. Absolute TPC commands can also be useful for dynamic frequency-domain inter-cell interference coordination.

The timing of the closed-loop TPC commands is illustrated in Figure 18.5. The transmission power change resulting from a TPC command usually takes effect at the fourth uplink subframe after the TPC command is received.[12] Transmission power changes have to be completed within 20 μs of the relevant subframe boundary (see [4, Section 6.3.4]).

18.3.2.3 Total Transmit Power Setting

Finally, for the PUSCH and SRS, the total transmit power of the UE in each subframe is scaled up linearly from the power level derived from the semi-static operating point and dynamic offset, according to the number of RBs actually scheduled for transmission from the UE in the subframe.

Thus the overall power control equation is as follows:

$$\text{UE transmit power} = \underbrace{P_0 + \alpha \cdot \text{PL}}_{\substack{\text{basic open-loop} \\ \text{operating point}}} + \underbrace{\Delta_{\text{TF}} + f(\Delta_{\text{TPC}})}_{\text{dynamic offset}} + \underbrace{10 \log_{10} M}_{\text{bandwidth factor}}$$

where Δ_{TPC} denotes a TPC command, $f(\cdot)$ represents accumulation in the case of accumulative TPC commands, and M is the number of allocated RBs.

[11]The absolute TPC mode can be seen as a low-overhead way to adjust the UE-specific offset in the base level component of the semi-static operating point.

[12]For TDD, the execution of the power changes may occur later, depending on the availability of uplink subframes; full details can be found in [6, Section 5.1].

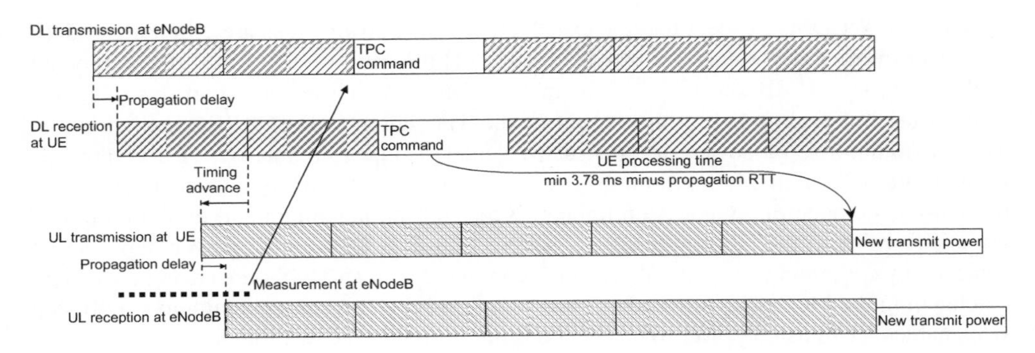

Figure 18.5: Timing of the uplink power control loop.

This overall power control formula allows the UE's transmit power to be controlled with a granularity of 1 dB within the range set −40 dBm to +23 dBm (corresponding to a maximum transmission power of 0.2 W). The maximum transmission power of a UE may, however, be subject to additional restrictions:

- The maximum transmission power of the UEs in a cell may be restricted to a lower level by RRC signalling, for example in a hospital scenario;
- In some configurations, reductions in maximum output power may be applied in order to satisfy emissions requirements (see Section 21.3.1.2;
- For UEs which support aggregation of multiple uplink carriers (from Release 10 onwards), the transmission power of an individual carrier or uplink physical channel may have to be scaled down according to defined rules in order to satisfy the total output power constraints (see Section 28.3.5).

The accuracy with which a UE is required to be able to set its total transmission power depends on the length of time since the last uplink transmission (greater or less than 20 ms) and the size of the required change in transmission power; details can be found in Section 21.3.1.2 and in [4, Section 6.3.5].

18.3.2.4 Transmission of TPC Commands

TPC commands for the dynamic offset part of the power control are sent to the UE in messages on the PDCCH. The UE is required to check for a TPC command in every subframe unless it is specifically configured in Discontinuous Reception (DRX – see Section 4.4.1.1). However, unlike in WCDMA, the TPC commands in LTE are not necessarily periodic.

One method by which TPC commands are transmitted to the UEs is in the uplink resource scheduling grant messages for each specific UE. This is logical as it results in all the applicable information for an uplink transmission (set of RBs, transport format, and power setting) being included in a single message.

Additionally, individual accumulative TPC commands for multiple UEs can be jointly coded into a special PDCCH message dedicated to power control (PDCCH Formats 3 and 3A – see Section 9.3.5.1). Such grouped TPC commands may be useful for controlling the power of uplink SPS transmissions (see Section 4.4.2.1), SRSs, or non-adaptive PUSCH retransmissions (see Section 4.4.1.1). Furthermore, for the PUCCH only, TPC commands can be sent in downlink resource assignment messages on the PDCCH. Both these methods for TPC command transmission enable the power control loop to track changes in channel conditions even when the UE is not scheduled for uplink data transmission, and they can therefore be seen as an alternative to the use of absolute TPC commands. The LTE specifications do not allow jointly coded TPC commands on the PDCCH to be used if the UE is configured in the absolute power control mode.

Due to the structure of the PDCCH signalling (see Section 9.3.5), in all cases the TPC commands are protected by a CRC (Cyclic Redundancy Check); this means that they should be considerably more reliable than in WCDMA. The only likely source of error in LTE would be the UE's failure to detect a PDCCH message, which should typically have a probability around 1% (compared to a typical power control error rate of 4–10% in WCDMA).

The eNodeB can use a number of techniques to determine how to command each UE to adjust its transmit power. One method will be the received SIR, based, for example, on measurements of the SRS and uplink demodulation RSs; in addition the BLER experienced on the decoding of uplink data packets may be used.

The eNodeB may also take into account interference coordination with neighbouring cells, for example if it has received an OI indicating that interference from a UE is causing a problem in a neighbouring cell. Note, however, that although eNodeBs may signal OIs to each other, an eNodeB receiving an OI cannot know for certain whether the overload situation is caused by a UE in its cell or not; it can only infer that, if the received OI relates to a group of RBs where it has scheduled a cell-edge UE, then it is possible that the interference arises from its cell and it should therefore react. Further details of interference coordination are explained in Section 12.5.

18.3.3 UE Power Headroom Reporting

In order to assist the eNodeB to schedule the uplink transmission resources to different UEs in an appropriate way, it is important that the UE can report its available power headroom to the eNodeB. The eNodeB can use the Power Headroom Reports (PHRs) to determine how much more uplink bandwidth per subframe a UE is capable of using. This can help to avoid allocating uplink transmission resources to UEs which are unable to use them; as the uplink is basically orthogonal in LTE, no other UE would be able to use such resources, so system capacity would be wasted.

The range of the PHR is from +40 dB to −23 dB. The negative part of the range enables the UE to signal to the eNodeB the extent to which it has received an uplink resource grant which would require more transmission power than the UE has available. This would enable the eNodeB to reduce the size (i.e. the number of RBs in the frequency domain) of a subsequent grant, thus freeing up transmission resources to allocate to other UEs.

A PHR can only be sent in subframes in which a UE has an uplink transmission grant; the report relates to the subframe in which it is sent. The PHR is therefore a prediction rather than a direct measurement; the UE cannot directly measure its actual transmission power headroom for the subframe in which the report is transmitted. It, therefore, relies on reasonably accurate calibration of the UE's power amplifier output, especially at high output powers when reliable knowledge of the headroom is more critical to system performance.

A number of criteria are defined to trigger a PHR. These include:

- A significant change in estimated path-loss since the last PHR;
- More than a configured time elapse since the previous PHR (controlled by the 'PHR prohibit timer');
- More than a configured number of TPC commands implemented by the UE.

The eNodeB can configure parameters to control each of these triggers depending on, for example, the system loading and the requirements of its scheduling algorithm.

In Release 10, some additional aspects are included in the PHRs; these are explained in Section 28.3.5.

18.3.4 Summary of Uplink Power Control Strategies

In summary, a variety of degrees of freedom are available for uplink power control in LTE. Not every parameter will be actively used in every network deployment, but each deployment will select a mode of power control appropriate to the scenario or scheduling strategy. The use of fractional power control facilitates an appropriate trade-off between fairness and system capacity.

One typical mode of operation would be to set the semi-static operating point (via P_0 and the fractional path-loss compensation factor α) to achieve at least the required SINR at the eNodeB for the required QoS for each UE, compensating for path-loss and wideband shadowing. Further control for interference management and rate adaptation can be exercised by means of frequency-domain scheduling and bandwidth adaptation – these being degrees of freedom for power management which were not available in WCDMA. Bandwidth adaptation may also be used in conjunction with changing the MCS to set different BLER operating points for different HARQ processes.

Finally, dynamic transmission power offsets can be used to give a finer degree of control, by means of the MCS-dependent offsets and the closed-loop corrections using the explicit TPC commands.

References[13]

[1] Nokia, 'R1-063377: UL Timing Control Accuracy and Update Rate', www.3gpp.org, 3GPP TSG RAN WG1, meeting 47, Riga, Latvia, November 2006.

[2] Texas Instruments, 'R1-072841: Simulation of Uplink Timing Error Impact on PUSCH', www.3gpp.org, 3GPP TSG RAN WGI, meeting 49bis, Orlando, USA, June 2007.

[3] 3GPP Technical Specification 36.133, 'Evolved Universal Terrestrial Radio Access (E-UTRA); Requirements for Support of Radio Resource Management', www.3gpp.org.

[4] 3GPP Technical Specification 36.101, 'Evolved Universal Terrestrial Radio Access (E-UTRA); User Equipment (UE) radio transmission and reception', www.3gpp.org.

[5] M. P. J. Baker and T. J. Moulsley, 'Power Control in UMTS Release 99' in *Proc. First IEE Int. Conf. on 3G Communications*, March 2000.

[6] 3GPP Technical Specification 36.213, 'Evolved Universal Terrestrial Radio Access (E-UTRA); Physical Layer Procedures (Release 8)', www.3gpp.org.

[7] Ericsson, 'R1-080881: Range and Representation of Delta_MCS', www.3gpp.org, 3GPP TSG RAN WG1, meeting 52, Sorrento, Italy, February 2008.

[13] All web sites confirmed 1st March 2011.

Part IV

Practical Deployment Aspects

19

User Equipment Positioning

Karri Ranta-aho and Zukang Shen

19.1 Introduction

The ability both to locate an object and to communicate with it is a combination that enables a wide range of location-based services – from navigator-like map services to location-based advertising to tracking children, cars or even convicted criminals. This provides a natural motivation for mobile phones to have positioning capabilities. Another strong motivation is a requirement from the Federal Communications Commission (FCC) of the USA that emergency calls, whether fixed or mobile, can be located with a high degree of accuracy.

The most straightforward method by which a mobile phone can know its location is to use a GPS[1] receiver operating completely independently of the cellular network. Alternatively, the cellular network itself can establish the approximate location of a connected mobile phone using knowledge of the coverage area of each cell.

The first version of LTE (Release 8) does not provide any direct protocol support for locating the User Equipment (UE). However, a Release 8 LTE UE can nonetheless be located by means of Assisted Global Navigation Satellite System (A-GNSS) and Enhanced Cell-ID-based techniques (see Sections 19.2 and 19.4 respectively) in conjunction with a general-purpose positioning protocol known as Secure User Plane Location (SUPL), defined by the Open Mobile Alliance (OMA). SUPL operates as a service in the application layer and requires only a normal User Plane (UP) connection between a server in the network (known in OMA as a SUPL Location Platform (SLP) and in LTE as an Evolved Serving Mobile Location Centre (E-SMLC)) and the SUPL client application in the UE. These techniques can provide sufficient average positioning accuracy for many commercial applications, but even when used together they may not meet the emergency call positioning requirements of the FCC with sufficiently high probability in all deployment scenarios.

[1]Global Positioning System: www.gps.gov.

LTE – The UMTS Long Term Evolution: From Theory to Practice, Second Edition.
Stefania Sesia, Issam Toufik and Matthew Baker.

The FCC requires all mobile network operators in the USA to comply with the 'E911 Phase II' requirements by January 2013 [1]. For UE-based techniques such as GPS:

- 67% of emergency calls need to be located within 50 m;
- 80% of emergency calls need to be located within 150 m (rising to 90% within 6 years).

GPS can easily meet the accuracy aspect of the requirements, but it cannot provide the required availability due to the satellite signals being blocked in indoor and urban environments. Therefore other techniques are needed in addition.

In GERAN[2] and UTRAN[3] deployments, cell-ID-based location techniques have been successfully used in conjunction with GPS. However, cell-ID-based techniques are limited in their accuracy by the density of deployed cell sites, and therefore the second release of the LTE specifications (Release 9) introduced support for Observed Time Difference Of Arrival (OTDOA) positioning. OTDOA was already defined for UTRAN and a similar technique has been successfully used in CDMA2000 systems.

In addition to support for UP-connection-based positioning (which is transparent to the Radio Access Network (RAN)), support for control plane techniques was also introduced in LTE Release 9 to provide a solution that is more robust against network congestion in emergency scenarios, providing a direct link between the location server and the network nodes.

Table 19.1 summarizes the UE positioning techniques supported in LTE Release 9 [2]. The only positioning technique that can operate in a purely UE-based mode is A-GNSS positioning; for all other techniques the location determination takes place in a location server in the network. eNodeB involvement is not needed for the A-GNSS or OTDOA methods, with the positioning procedure operating directly between the UE and the location server. For Enhanced Cell-ID, the location server may request additional measurement reports from the eNodeB, such as the Angle of Arrival (AoA) of the received signal. These techniques are explained in detail in the following sections.

Table 19.1: UE positioning techniques supported in LTE Release 9.

	UE-based	UE-assisted, E-SMLC-based	eNodeB-assisted
A-GNSS	Yes	Yes	No
OTDOA	No	Yes	No
Enhanced Cell-ID	No	Yes	Yes

[2]GSM/EDGE Radio Access Network.

[3]Universal Terrestrial Radio Access Network.

19.2 Assisted Global Navigation Satellite System (A-GNSS) Positioning

A-GNSS-based positioning refers in general to any satellite-based positioning system, such as GPS, Galileo[4] or GLONASS,[5] in conjunction with assistance provided over a terrestrial network to improve the sensitivity and/or speed of detection of the satellites, as illustrated in Figure 19.1.

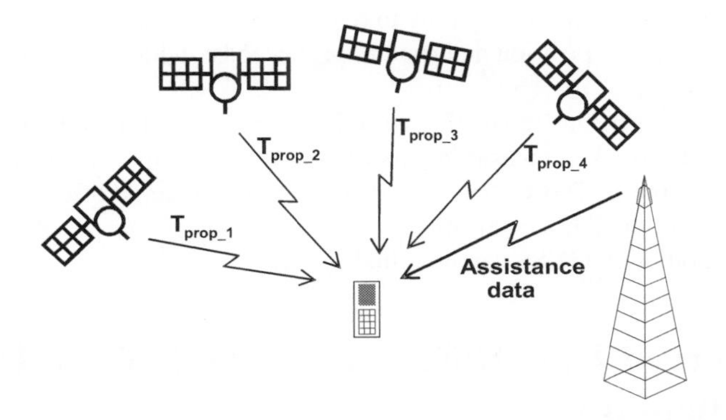

Figure 19.1: Basic principle of A-GNSS.

Basic GNSS-based positioning relies on accurate knowledge of the locations of the satellites and the transmission times of their signals. With some simplification it can be said that a GNSS receiver measures the exact time at which it receives the signal of each satellite it can detect. Using this information, it is possible to calculate the location of the UE. Since GNSS receivers typically do not have high-accuracy atomic clocks, the problem to be solved is four-dimensional (x, y, z and time), and at least four satellites have to be detected for a position estimate.

A GNSS receiver typically has a 'Time To First Fix' (TTFF)[6] that is too long for most practical uses of mobile terminal positioning. First the receiver must detect the signals transmitted by at least four satellites and decode a message modulated on each satellite's signal providing details of its position in orbit. Only then does the GNSS receiver have sufficient information to perform the measurements and calculate its position. Stand-alone GPS navigation devices typically solve this problem by remaining powered on continually, so that although the first fix may take some time, subsequently the position is always known provided that the satellite signals remain sufficiently strong. However, this would be impractical in a mobile phone as the GNSS receiver would quickly drain the battery even if the location information was rarely needed.

[4]The GNSS of the European Union: www.esa.int/esaNA/galileo.html.

[5]GLObal'naya NAvigatsionnaya Sputnikovaya Sistema, the Russian GNSS: www.federalspace.ru.

[6]TTFF describes the time required for a GPS device to acquire the necessary satellite signals and calculate an accurate position.

In order to reduce the TTFF, the terrestrial cellular network can provide assistance data to UEs equipped with GNSS receivers. The simplest such assistance data comprises the messages the satellites broadcast to enable the receiver to calculate their orbital positions at any given time. Receiving this information over the fast cellular network rather than the slow satellite link (50 bps for GPS) directly reduces the TTFF. In addition, the GNSS receiver sensitivity and/or detection speed can be improved by reducing the size of the required satellite signal search window by providing fine time assistance and reference location data to the UE. The more accurate the UE's knowledge of location and time, the smaller the search window it needs for the satellites' signals.

Note that the assistance data serves only to reduce the time taken to determine the position and the probability that the position determination succeeds; it does not improve the accuracy of the location estimate.

A-GNSS can also operate in a UE-assisted mode, where rather than determining its location itself the UE makes measurements of the timing of the received satellite signals and reports them to the E-SMLC. The E-SMLC may then take additional information into account in calculating the UE's position.

More information on A-GNSS can be found in [3].

19.3 Observed Time Difference Of Arrival (OTDOA) Positioning

The principle behind OTDOA positioning is similar to GPS, but for OTDOA the signals measured by the UE are the terrestrial downlink transmissions from the eNodeBs. The location determination is typically distributed between the UE and the network. Unlike GPS positioning, the UE does not acquire an accurate reference time, but the position estimate is based on the received time difference of at least two pairs of cells.

OTDOA in LTE is based on the UE measuring the time difference observed (by the UE receiver) between the Reference Signals (RS – see Section 8.2) of a neighbour cell and those of the serving cell; this is known as a Reference Signal Time Difference (RSTD) measurement (see Figure 19.2). In order to be able to calculate the UE's location, the network needs to know accurately the locations of the eNodeB transmit antennas and the transmission timing of each cell. The fact that the eNodeBs are not orbiting the earth at a high velocity makes the location determination mathematics somewhat simpler than for GPS. An efficient mathematical method for calculating the UE location based on OTDOA measurements is described in [4].

OTDOA positioning does not suffer from long TTFF like GPS, but the requirement that the eNodeB transmission timings are accurately known is not necessarily straightforward if the cellular network is asynchronous. Assuming, however, that these basic requirements can be fulfilled, the key factor governing the achievable performance of a cellular OTDOA system is whether the signals to be measured can be detected by the UE sufficiently fast and with sufficiently high probability.

The requirement that the UE can detect cells of at least three different eNodeBs with a high probability does not sit well with the fact that LTE was designed primarily to provide high-speed data services with good spectral efficiency, for which reason a frequency reuse factor of one is typically used. This can render the Release 8 synchronization signals and RSs of

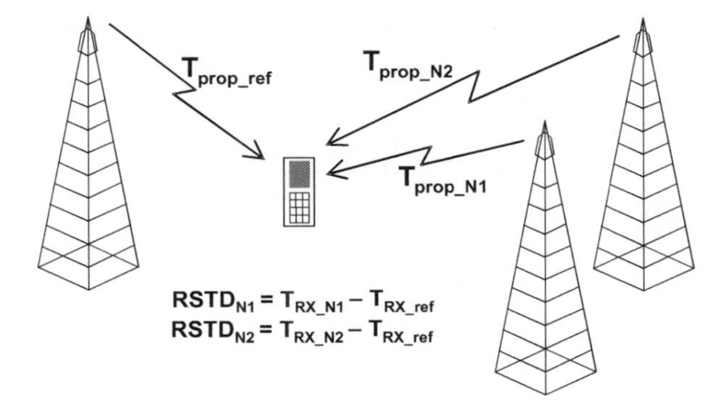

Figure 19.2: Basic principle of OTDOA in LTE.

distant neighbour cells undetectable to the UE;[7] on many occasions this would prevent the RSTD measurements required by OTDOA from being made if the available signals were only those defined in Release 8 of the LTE specifications.

Figure 19.3 shows that in an interference-limited hexagonal cell deployment the probability that the SINR of the cell-specific RS of the third-best cell (i.e. the second-best neighbour cell) is above −6 dB is about 70%, which clearly does not give a sufficient level of availability. Figure 19.3 also shows that, even in a 500 m inter-site distance deployment with no interference from user data traffic, the 5-percentile for the SINR of the cell-specific RS of the third best cell is about −13 dB (i.e. for the 95% availability requirement of the FCC E911 Phase II requirements, an SINR of only −13 dB can be assured). Thus it is obvious that even if the UE were enabled to skip the detection of the synchronization signals by providing network assistance information (indicating, for example, the potential neighbour cells for which to search), OTDOA would not work sufficiently well if it were based on the Release 8 cell-specific RSs alone, especially if an interference-free environment could not be guaranteed.

For this purpose, special Positioning Reference Signals (PRSs) are introduced in Release 9 of the LTE specifications, as detailed in the following section.

19.3.1 Positioning Reference Signals (PRS)

In order to increase the probability that the UE can detect sufficient neighbour cells, and therefore achieve good OTDOA positioning reliability, special 'positioning subframes' are designated in LTE Release 9. Positioning subframes are designed to aid the 'hearability' of neighbour cells by reducing the interference and increasing the RS energy: typically they do not carry any Physical Downlink Shared CHannel (PDSCH) data, but provide PRSs in addition to the Release 8 cell-specific RSs.

[7]A neighbour cell is only considered detectable if the Signal-to-Interference-plus-Noise Ratio (SINR) of the synchronization signals is at least −6 dB.

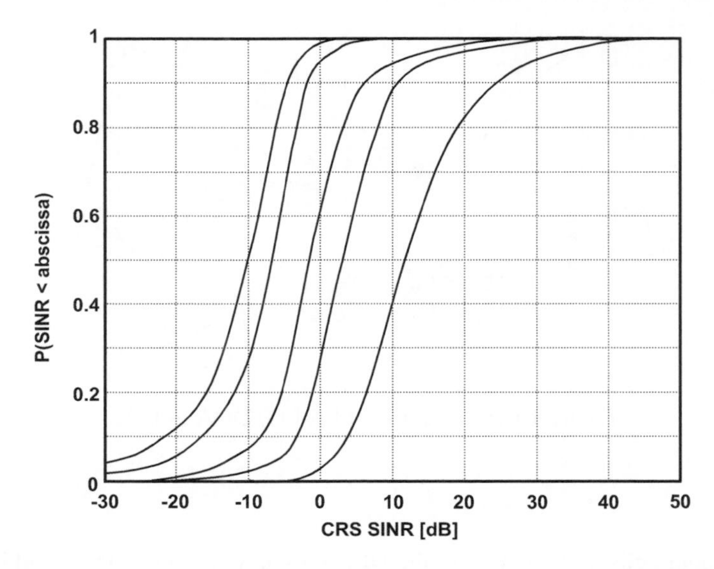

Figure 19.3: SINR of the Cell-specific RS (CRS) for the best 5 cells, inter-site distance = 500 m (unloaded system) [5]. Reproduced by permission of © Motorola.

Figure 19.4 shows the arrangement of the PRSs in a Resource Block (RB).[8] The PRS pattern is designed so that it never overlaps with the Physical Downlink Control CHannel (PDCCH), nor with the cell-specific RS of any other antenna port. Overlap between PRS patterns in neighbouring cells can be avoided by means of a cell-specific frequency shift of a number of subcarriers given by the Physical Cell ID (PCI) modulo 6, allowing six different non-overlapping frequency shifts. The PRS sequence is constructed in the same way as the cell-specific RS (see Section 8.2.1).

The PRSs are designed to provide more RS energy and a larger reuse factor (namely 6) than is available with the Release 8 cell-specific RS. In synchronous network deployments, even if the interference from data transmissions is eliminated by providing subframes free of PDSCH transmission, the cell-specific RS of cells with two or more antenna ports have a reuse factor of only 3, with the result that they interfere with each other, resulting in reduced hearability of the neighbour cells.

The positioning subframes occur in groups of consecutive downlink subframes known as 'positioning occasions', consisting of between one and six positioning subframes (see [6, Section 6.10.4.3]). In TDD deployments, uplink subframes and special subframes cannot contain PRSs.

In order for the UEs to benefit fully from the reduced inter-cell interference towards the PRS of other cells during positioning subframes, the positioning subframes should overlap (at least partially) between the serving cell and all the neighbour cells to be detected, and they need to occur sufficiently frequently for accurate RSTD measurements to be made. The

[8]The eNodeB antenna port used for transmission of the PRS is defined as antenna port number six and is used for PRS transmission only.

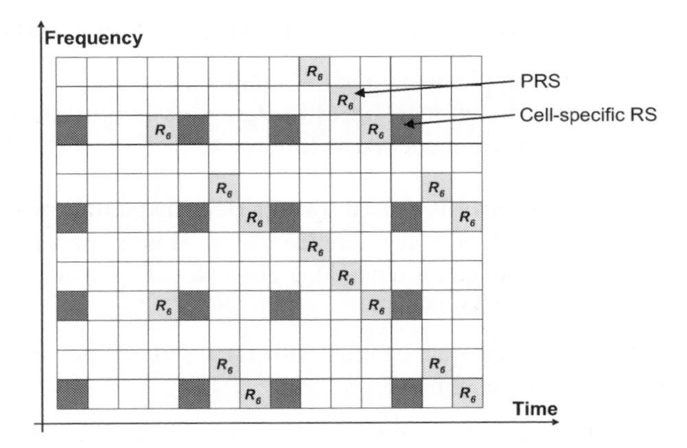

Figure 19.4: PRS arrangement (R_6) in a cell with two cell-specific transmit antenna ports. Reproduced by permission of © 3GPP.

best overlap can be achieved in a fully synchronized network, but even partial overlap can be shown to be beneficial.

In some cases a network may further reduce the inter-cell interference by muting the PRS transmissions themselves from time to time, which can be beneficial if the PRS patterns in neighbouring cells overlap. If this is the case, signalling is provided to inform the UEs as to the subframes in which the PRSs are muted. For simplicity, muting can only occur in all or none of the subframes in a given positioning occasion. The signalled pattern of muted positioning occasions is given by a bitmap and repeats periodically; it is applicable in both the serving cell and the neighbour cells. If PRS muting is not applied in a positioning occasion, the UEs can assume that the PRS power is the same in all PRS subframes in the positioning occasion.

In any case, the network should assist the UE in its neighbour-cell search by providing the relative transmission time differences of each neighbour cell relative to the serving cell (in a synchronous network these would always be zero) and the search window length in terms of the maximum propagation delay difference that occurs in the deployment environment. Such assistance can significantly reduce the UE's search space and increase the detection sensitivity.

It should also be noted that OTDOA measurements based on neighbour cell PRS require the UE to have a priori knowledge of the PRS sequences for which to search. As the location servers know the exact geographical locations of the transmit antennas, they can construct a list of candidate neighbour cells for each cell without traditional neighbour list planning. The network can then provide this information to the UEs.

The requirements for the time within which a UE must be capable of reporting an RSTD measurement after receiving OTDOA assistance information can be found in [7].

19.3.2 OTDOA Performance and Practical Considerations

In practical network and UE implementations, a number of sources of error can have an impact on the final accuracy of the location estimates. This is illustrated in Figure 19.5, where the main sources of error are modelled as follows:

- eNodeB synchronization error is caused by imperfect signal transmission timing from the antenna compared to a perfect reference clock. This is modelled by a Gaussian distribution with 100 ns standard deviation. The eNodeB transmit antenna locations are assumed to be perfectly known.
- The UE RSTD measurement quantization error is caused by the UE receiver sampling rate which is assumed to be 32 ns.
- Multipath propagation introduces error to the desired Line-Of-Sight (LOS) propagation delay estimation. In Figure 19.5, the UE takes the centre of gravity of the received multipath signal as the time of reception, and the error is the difference between that estimate and the actual LOS propagation delay.
- Timing offset estimation error and UE frequency instability directly affect the error in the RSTD measurements themselves. The magnitudes of these errors are a function of the SINR.

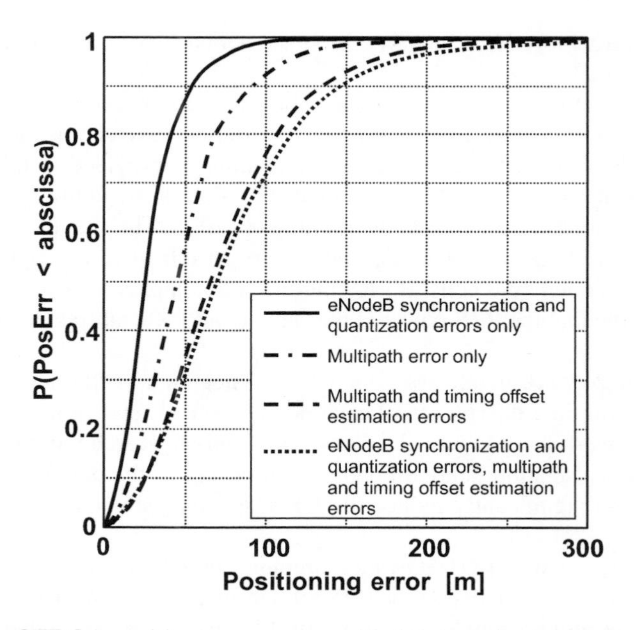

Figure 19.5: OTDOA positioning accuracy with 4 detected neighbouring cells, inter-site distance = 1732 m, Extended Typical Urban channel model [8].

19.4 Cell-ID-based Positioning

Positioning based on Cell-ID (CID) uses geographical knowledge of a UE's serving cell. To improve the accuracy, measurements made by the UE and/or the eNodeB can be utilized in addition.

19.4.1 Basic CID Positioning

Basic CID positioning estimates the location of a UE using only the coordinates of its serving eNodeB or cell. Knowledge of the serving eNodeB or cell can be obtained by paging, tracking area updates or other methods. Typically, basic CID positioning provides only coarse estimation of the UE location, with accuracy of roughly the same order as the cell radius.

19.4.2 Enhanced CID Positioning using Round Trip Time and UE Receive Level Measurements

Enhanced CID positioning uses additional information beyond the identity of the eNodeB or cell that is serving the UE. The distance of a UE from its serving eNodeB or cell can be estimated from the Round Trip Time (RTT).

When timing advance is operating (see Section 18.2), Figure 19.6 illustrates the reception and transmission timings at the eNodeB and UE. Two measurements are defined in LTE Release 9 by which an eNodeB can indicate the RTT to the E-SMLC. These are referred to as the 'Timing Advance Type 1' and 'Timing Advance Type 2' measurements [9]. Alternatively, the E-SMLC or SLP may obtain the timing advance setting directly from the UE.

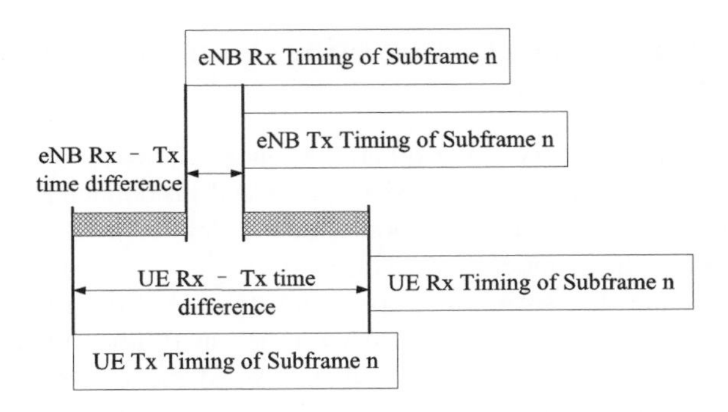

Figure 19.6: RTT estimation in the presence of timing advance.

The Type 1 measurement is the time difference defined as the sum of the receive–transmit time difference at the eNodeB (either a positive or a negative value) and the receive–transmit time difference at the UE (a positive value, due to the operation of timing advance). The result is represented by the shaded duration in Figure 19.6, corresponding to the RTT. The UE reports its receive–transmit time difference to its serving eNodeB, and the eNodeB calculates its own receive–transmit time difference. The Type 1 method can only be used for UEs which support the reporting of this information – i.e. UEs conforming to Release 9 of the LTE specifications or later.

The Type 2 measurement is simply defined as the receive–transmit time difference at the eNodeB, measured in a radio frame containing a Physical Random Access CHannel (PRACH – see Section 17.4) transmission from the UE whose location is to be ascertained. For the PRACH, the transmit timing of the UE should be aligned with the UE's receive timing. The Type 2 measurement is equally applicable to Release 8 UEs as to later UEs.

The distance of the UE from the eNodeB or cell is estimated as $d = c \times \mathrm{RTT}/2$, where c is the speed of light. The accuracy of the Type 1 measurement is mainly determined by the receive timing accuracy of the eNodeB and UE, which is typically of the order of 0.3 μs, corresponding to a distance of around 45 m. The accuracy of the Type 2 measurement depends on the PRACH detection accuracy, which is typically of the order of 1–2 μs depending on the multipath characteristics of the radio channel (see [10]). Thus the Type 1 measurement can potentially provide better positioning accuracy.

In addition to the receive–transmit time difference measurement, the UE may be configured to perform Reference Signal Received Power (RSRP) and Reference Signal Received Quality (RSRQ) (see Sections 22.3.1.1 and 22.3.1.2 respectively) measurements for the serving cell and for neighbouring cells. These measurements can be used to help enhance CID-based positioning by enabling an RSRP/RSRQ map to be compiled throughout the network coverage area. The estimate of a UE's location can then be enhanced by comparing its reported RSRP/RSRQ with the RSRP/RSRQ map. The usefulness of this information depends on the accuracy of the UE's RSRP/RSRQ reports and on the accuracy of the RSRP/RSRQ map.

19.4.3 Enhanced CID Positioning using Round Trip Time and Angle of Arrival

Although the RTT can be used to estimate distance, it does not provide directional information for the UE location. Angle of Arrival (AoA) [9] is defined as the estimated angle of a UE with respect to a reference direction which is defined as geographical North; the value of AoA is positive in an anticlockwise direction.

The eNodeB estimates the AoA using uplink transmissions from the UE. The eNodeB antenna configuration is a key factor for AoA estimation. With a linear array of equally spaced antenna elements, the received signals at any two adjacent elements are phase-rotated by a fixed amount θ. The value of θ is a function of the AoA as well as the antenna element spacing and carrier frequency. The latter parameters are known by the eNodeB, and therefore the AoA can be estimated. In general, any uplink signal from the UE can be used to estimate the AoA, but typically a known signal such as the Sounding Reference Signals (SRSs) or DeModulation Reference Signals (DM-RS) would be used (see Sections 15.6 and 15.5 respectively). With knowledge of both the AoA and the RTT, a UE's position can be estimated [11].

Figure 19.7 shows an example of estimated UE positions with AoA+RTT and GPS for a cellular deployment in Guangzhou, China, where positioning based on AoA and RTT estimation is used in a TD-SCDMA network. Typically, AoA+RTT based positioning can provide an accuracy of 150 m and 50 m in Non-LOS and LOS environments respectively [11, 12].

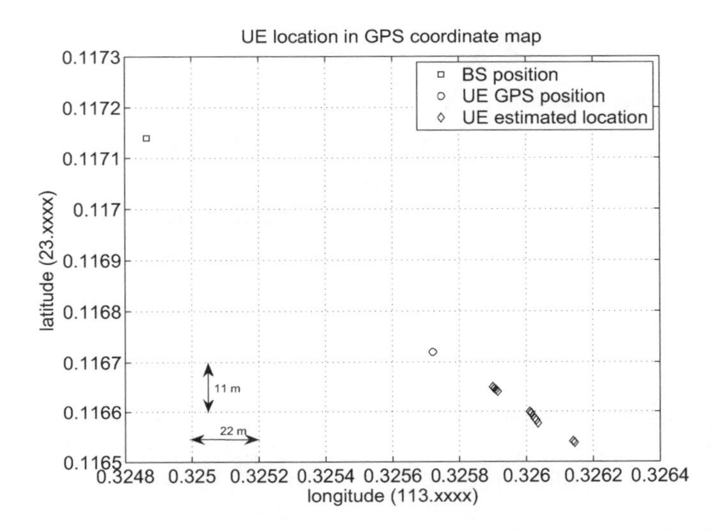

Figure 19.7: AoA+RTT positioning results compared to GPS in a TD-SCDMA deployment in Guangzhou, China.

19.5 LTE Positioning Protocols

Figure 19.8 illustrates the signalling connections for the location servers for LTE UE positioning. For the Control Plane (CP), the E-SMLC is shown, while for the User Plane (UP) the SUPL Location Platform (SLP) is shown. The E-SMLC–UE communication is tunnelled through the MME and eNodeB in the CP protocols and the E-SMLC–eNodeB communication is achieved by terminating the tunnel in the eNodeB. The SLP interacts directly with the UE client using normal UP data transport and the same protocol as is used for the E-SMLC–UE communication. It should be noted that the E-SMLC and SLP are logical entities and may therefore be implemented in a single physical server.

Figure 19.9 shows the protocol stack used to carry the LTE Positioning Protocol (LPP) [13] between the UE and the E-SMLC. LPP is a point-to-point protocol, and naturally the E-SMLC can have a number of parallel connections to different UEs. The LPP messages can be divided into four functional groups:

- UE positioning capability information transfer to the E-SMLC;
- Positioning assistance data delivery from the E-SMLC to the UE;
- Location information transfer;
- Session management (for procedure abortion and error handling).

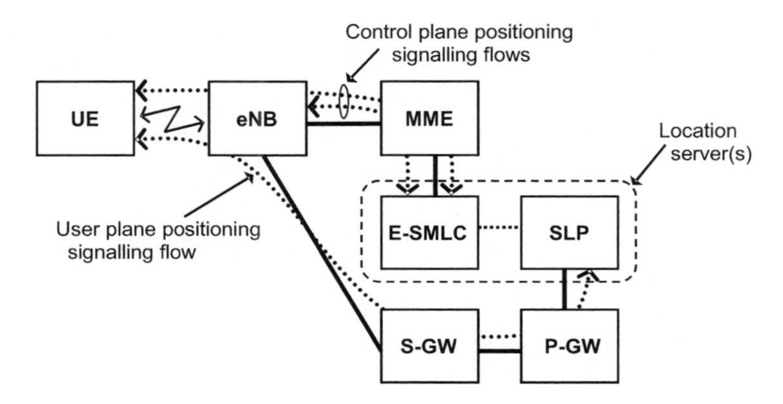

Figure 19.8: CP and UP architecture for positioning.

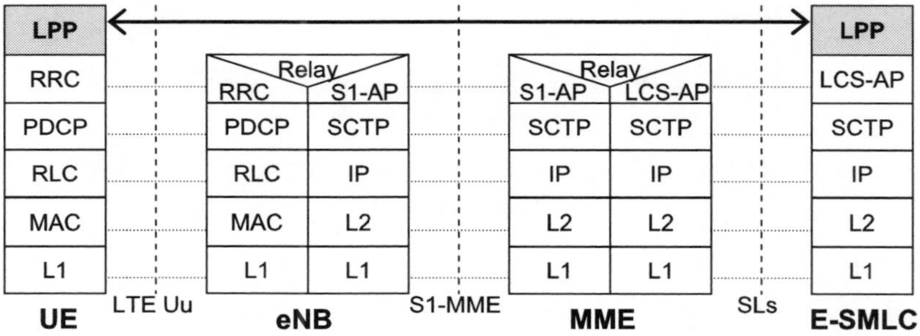

UE	eNB		MME		E-SMLC
LPP					LPP
RRC	RRC (Relay)	S1-AP	S1-AP (Relay)	LCS-AP	LCS-AP
PDCP	PDCP	SCTP	SCTP	SCTP	SCTP
RLC	RLC	IP	IP	IP	IP
MAC	MAC	L2	L2	L2	L2
L1	L1	L1	L1	L1	L1

Interfaces: LTE Uu (UE–eNB), S1-MME (eNB–MME), SLs (MME–E-SMLC)

Figure 19.9: The LPP stack between the E-SMLC and the UE.
Reproduced by permission of © 3GPP.

The underlying protocols and the nodes between the eNodeB and the MME act solely as transport and do not take part in the LPP procedures. This means that LPP is a stand-alone protocol that can use any UE-server delivery mechanism. In particular, SUPL can use the LPP as is and can use normal UP data transport to carry LPP instead of the CP protocols. This is especially beneficial in deployments where both CP and UP positioning solutions are deployed together. The same server can support both architectures and directly use the same protocol for positioning. Similarly, UEs supporting both solutions need to implement only one positioning protocol.

In addition to supporting the 3GPP-defined positioning mechanisms for LTE, LPP includes an 'External Protocol Data Unit (EPDU)', which provides a transparent message exchange mechanism between the UE and the server for the support of additional positioning techniques that may be defined in the future.

In addition to the LPP protocol used in E-SMLC–UE communication, the LTE Positioning Protocol A (LPPa) [14] is defined for E-SMLC–eNodeB communication. Figure 19.10 shows the LPPa protocol stack. LPPa is used in Enhanced CID positioning if eNodeB measurements such as AoA are used to aid the location determination. LPPa can also be used for the E-SMLC to request details of the eNodeB configuration for OTDOA, which is needed when constructing the OTDOA assistance data to be delivered to the UE. Examples of these parameters are the cell IDs and transmission frequencies, the PRS configuration and the cell timings. Like LPP, LPPa is also transparent to the MME and is tunnelled through it.

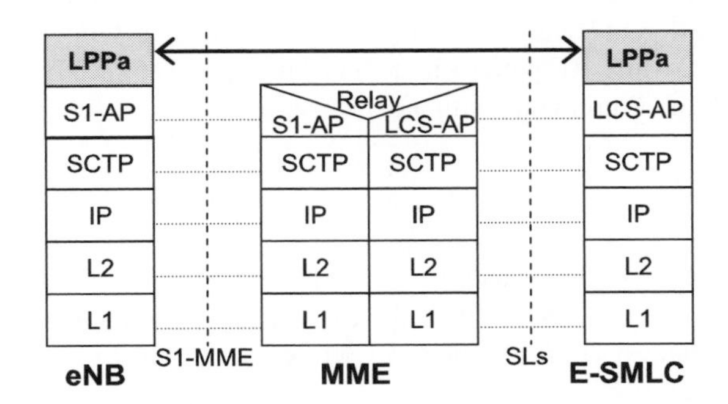

Figure 19.10: The LPPa protocol stack between the E-SMLC and the eNodeB. Reproduced by permission of © 3GPP.

19.6 Summary and Future Techniques

Whatever the motivation for deploying the capability to locate a UE, be it regulatory or commercial, the LTE Release 9 specifications provide suitable methods. Different techniques inherently have varying degrees of accuracy, availability and cost of deployment, so a careful case-by-case assesment is needed to identify which techniques should be deployed in a given environment.

A-GNSS is assumed to be the primary positioning technique for LTE UEs, providing very good positioning accuracy and reasonable availability. However, satellite-based techniques alone are not sufficient to provide a location estimate in all cases, due to satellite hearability problems in dense urban and indoor situations.

OTDOA provides a useful alternative positioning technique to A-GNSS in environments where the satellite signals cannot be received but a sufficiently dense deployment of eNodeBs exists to provide a good probability of OTDOA yielding a location estimate. The probability of obtaining a location estimate by OTDOA can be increased by means of Positioning Reference Signals. OTDOA accuracy depends on the deployment density of the eNodeBs and the accuracy with which the antenna locations and their transmission timings are known by the network.

Cell-ID-based techniques give a reasonably good estimate of the UE location, especially in urban environments with dense eNodeB deployments, and can further benefit from UE measurements of the radio environment as well as eNodeB AoA measurements if suitable antenna arrays are deployed. The smaller the cells, the more accurate the positioning estimate. A key benefit of Cell-ID-based techniques is the fact that if the UE is able to connect to a network then the technique is guaranteed to provide a location estimate.

The UE-to-server positioning communications can take place either via the UP (as with other normal client–server application-layer communications), or via the CP in a similar fashion to RRC, NAS or S1AP signalling.

Further development of LTE in Release 11 is expected to add additional positioning techniques to the range already available, namely Uplink Time Difference of Arrival (UTDOA) and RF pattern matching. UTDOA uses eNodeB measurements of uplink signal reception timings at different cells to estimate the location of a UE without any specific involvement of the UE, while RF pattern matching aims to locate a UE by comparing any available radio channel measurements with a database.

References[9]

[1] Federal Communications Commission Rules, Section 20.18(h), www.fcc.gov/oet/info/rules.

[2] 3GPP Technical Specification 36.305, 'Evolved Universal Terrestrial Radio Access Network (E-UTRAN); Stage 2 functional specification of User Equipment (UE) positioning in E-UTRAN (Release 9)', www.3gpp.org.

[3] A. Kupper, *Location-Based Services: Fundamentals and Operation*. John Wiley & Sons Ltd/Inc., 2005.

[4] Y. Chan and K. Ho, 'A simple and efficient estimator for hyperbolic location'. *IEEE Transactions on Signal Processing*, Vol. 42, pp. 1905–1915, August 1994.

[5] Motorola, 'R1-091336: Study on hearability of reference signals in LTE positioning support', www.3gpp.org, 3GPP TSG RAN WG1, meeting 56bis, Seoul, South Korea, March 2009.

[6] 3GPP Technical Specification 36.211, 'Evolved Universal Terrestrial Radio Access (E-UTRA); Physical Channels and Modulation (Release 9)', www.3gpp.org.

[7] 3GPP Technical Specification 36.133, 'Evolved Universal Terrestrial Radio Access (E-UTRA); Requirements for support of radio resource management (Release 9)', www.3gpp.org.

[8] Alcatel-Lucent, 'R1-092307: Analysis of UE Subframe Timing Offset Measurement Sensitivity to OTDOA Performance', www.3gpp.org, 3GPP TSG RAN WG1, meeting 57bis, Los Angeles, USA, June 2009.

[9] 3GPP Technical Specification 36.214, 'Evolved Universal Terrestrial Radio Access (E-UTRA); Physical layer - Measurements (Release 9)', www.3gpp.org.

[10] 3GPP Technical Specification 36.104, 'Evolved Universal Terrestrial Radio Access Network (E-UTRAN); Base Station (BS) radio transmission and reception (Release 9)', www.3gpp.org.

[11] CATT, 'R1-090936: UE Positioning Based on AoA+TA for LTE Rel-9', www.3gpp.org, 3GPP TSG RAN WG1, meeting 56, Athens, Greece, February 2009.

[12] RITT, CATT, 'R1-091979: Further performance evaluation of UE positioning based on AoA+TA', www.3gpp.org, 3GPP TSG RAN WG1, meeting 57, San Francisco, USA, May 2009.

[13] 3GPP Technical Specification 36.355, 'Evolved Universal Terrestrial Radio Access (E-UTRA); LTE Positioning Protocol (LPP) (Release 9)', www.3gpp.org.

[14] 3GPP Technical Specification 36.455, 'Evolved Universal Terrestrial Radio Access (E-UTRA); LTE Positioning Protocol A (LPPa) (Release 9)', www.3gpp.org.

[9] All web sites confirmed 1st March 2011.

20

The Radio Propagation Environment

Juha Ylitalo and Tommi Jämsä

20.1 Introduction

Realistic modelling of propagation characteristics is essential for two main reasons in relation to the LTE system. Firstly, the link- and system-level performance of LTE can be evaluated accurately only when the radio channel models are realistic. In particular, the spatial characteristics of the channel models have a significant effect on the performance of multi-antenna systems. Secondly, the model used for radio propagation plays an important role in the network planning phase of LTE deployment.

Different environments such as rural, suburban, urban, and indoor set different requirements for network planning, antenna configuration, and the preferred spatial transmission mode. Moreover, the actual deployment scenarios for LTE cover numerous special cases ranging from mountainous rural surroundings to dense urban and outdoor-to-indoor situations. As an example of the range of different propagation environments of relevance to high-bandwidth Multiple-Input Multiple-Output (MIMO) systems such as LTE, the IST-WINNER[1] project [1] developed wideband radio channel models for 12 propagation environments.

The propagation characteristics are in a large part affected by the carrier frequency. In November 2007 the International Telecommunication Union (ITU) World Radiocommunication Conference (WRC-2007) allocated new frequency bands for International Mobile Telecommunications (IMT) radio access systems between 450 MHz and 3.6 GHz. Thus depending on the deployed carrier frequency, the LTE radio channel characteristics will vary

[1] Information Society Technologies – Wireless world INitiative NEw Radio.

LTE – The UMTS Long Term Evolution: From Theory to Practice, Second Edition.
Stefania Sesia, Issam Toufik and Matthew Baker.

significantly even within a particular type of propagation environment. The carrier frequency used also has an impact on the deployment of MIMO for LTE, due to the fact that the size of the antenna arrangement depends strongly on the carrier wavelength.

As discussed in Chapter 11, the ability to use a variety of MIMO techniques is an important feature of LTE. While conventional wireless systems are designed to counteract multipath fading, MIMO approaches such as those used in LTE are able to take advantage of multipath scattering to increase capacity [2, 3]. Therefore, it is important to create realistic standardized MIMO radio channel models for the evaluation of the performance of LTE and its future enhancements.

It is especially important to model accurately the correlation between the signals of different antenna branches, since this dictates, to a large extent, the preferred spatial transmission mode and its performance in practice. The spatial correlation properties of the MIMO radio channel model define the ultimate limit of the theoretical channel capacity. The applied channel model has to reflect all the instantaneous space-time-frequency characteristics which affect the configuration of diversity, beamforming, and spatial multiplexing techniques. The model also has to support all the antenna designs that are of practical interest.

In the following sections we discuss first the 3GPP Single-Input Single-Output (SISO) and Single-Input Multiple-Output (SIMO) channel models which include only a single antenna at the transmitter and one or two antennas at the receiver respectively. This is followed by a detailed analysis and comparison of the characteristics and modelling principles of a MIMO radio channel, focusing in particular on the models used in 3GPP, up to and including the latest models defined by ITU-R for IMT-Advanced systems.

20.2 SISO and SIMO Channel Models

In practice, realistic modelling of the radio channel propagation characteristics requires extensive measurement campaigns with appropriate carrier frequencies and bandwidths, in the radio environments which are identified to be the most relevant for deployment. The research projects COST207, COST231, COST259, CODIT (UMTS Code Division Testbed), and ATDMA (Advanced TDMA Mobile Access) created extensive wideband measurement datasets for SISO and SIMO channel modelling in the 1980s and 1990s [4–8]. The corresponding channel models form a basis for the ITU models which were largely applied in the development of the third generation wireless communication systems.

Figure 20.1 illustrates a multipath propagation scenario in which the UE has an approximately omnidirectional antenna. The transmitted signal traverses three paths with different delays. As explained in Chapter 5, in wideband communications the delay spread of the propagation paths is larger than the symbol period and the receiver observes the multipath components separately. In the frequency domain this corresponds to frequency-selective fading since the coherence bandwidth of the radio channel is smaller than the signal bandwidth. The coherence bandwidth is proportional to the inverse of the root mean square (r.m.s.) delay spread which can be calculated from the Power Delay Profile (PDP) of the radio channel (see Section 8.2.1). The PDP also defines the *maximum excess delay*, which is a measure of the largest delay difference between the propagation paths. This measure is important when considering the length of the Cyclic Prefix (CP) in LTE, as discussed in Section 5.4.

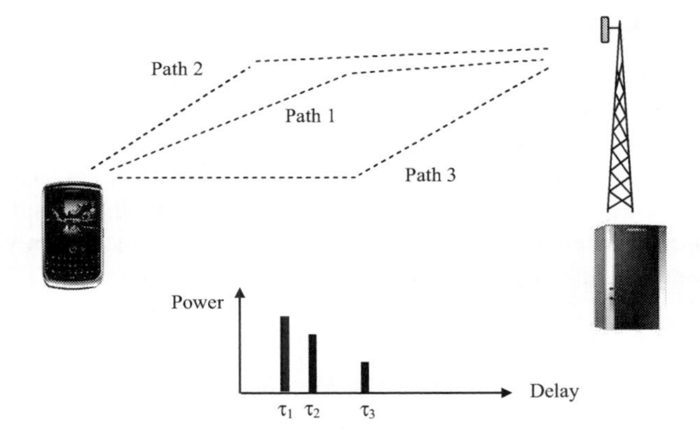

Figure 20.1: Multipath propagation and PDP.

Three main phenomena can be identified which affect the received signal properties:

- Propagation path-loss;
- Shadow (slow, or 'large-scale') fading;
- Multipath (fast, or 'small-scale') fading.

The propagation loss determines the average signal level at the receiver. Typical radio access systems can cope with a path-loss up to around 150 dB. The received signal level varies relatively slowly due to shadowing effects from large obstructions such as buildings and hills. Finally, the multipath characteristics define the frequency, time, and space correlations. As several multipath signal components with approximately the same propagation delay add up incoherently at the receiver antenna, the composite signal level can vary by up to 30–40 dB as the User Equipment (UE) moves. This variation can be very fast since the distance between two consecutive fades can be of the order of half wavelength.

20.2.1 ITU Channel Model

ITU channel models were used in developing the Third Generation 'IMT-2000' family[2] of radio access systems [9]. The main user scenarios covered indoor office, outdoor-to-indoor, pedestrian and vehicular radio environments. The key parameters to describe each propagation model include time delay spread and its statistical variability, path-loss and shadow fading characteristics, multipath fading characteristics, and operating radio frequency. Each environment is defined for two cases which have different probabilities of occurrence: a smaller delay spread case and a larger delay spread case. The numbers of paths are as for the 3GPP Channel Models introduced in the following section.

[2]The initial 'IMT-2000' systems of the IMT family were proposed with the year 2000 in view.

20.2.2 3GPP Channel Model

The 3GPP specifications for UMTS made use of propagation conditions largely based on the ITU models. As an example, Table 20.1 shows the propagation conditions used for performance evaluation in different multipath fading environments [10]. All taps use the classical Doppler spectrum [11, 12] (see Section 8.3.1), which assumes that the propagation paths at a receiver antenna are uniformly distributed over 360° in azimuth. The fading of the signals at different antennas is assumed to be independent, which is only appropriate for SISO and some antenna diversity cases. MIMO was not introduced in UMTS until Release 7.

Table 20.1: Propagation conditions for multipath fading in the ITU models.

	ITU Pedestrian A		ITU Pedestrian B		ITU Vehicular A	
Tap number	Relative delay (ns)	Relative mean power (dB)	Relative delay (ns)	Relative mean power (dB)	Relative delay (ns)	Relative mean power (dB)
1	0	0	0	0	0	0
2	110	−9.7	200	−0.9	310	−1.0
3	190	−19.2	800	−4.9	710	−9.0
4	410	−22.8	1200	−8.0	1090	−10.0
5			2300	−7.8	1730	−15.0
6			3700	−23.9	2510	−20.0

20.2.3 Extended ITU Models

The evaluation of LTE techniques demands channel models with increased bandwidth compared to UMTS models, to reflect the fact that the characteristics of the radio channel frequency response are connected to the delay resolution of the receiver. In 3GPP the 20 MHz LTE channel models were based on a synthesis of existing models such as the ITU and 3GPP models. Specifically the six ITU models covering an excess (maximum) delay spread from 35 ns to 4000 ns were chosen as a starting point, together with the Typical Urban (TU) model from GSM[3] which has a maximum excess delay of 5000 ns. In this way, extended wideband models with low, medium, and large delay spread values could be identified. The low delay spread gives an Extended Pedestrian A (EPA) model which is employed in an urban environment with fairly small cell sizes (or even up to about 2 km in suburban environments with low delay spread), while the medium and large delay spreads give an Extended Vehicular A (EVA) model and Extended TU (ETU) model respectively. The ETU model has a large maximum excess delay of 5000 ns, which in fact is not very typical in urban environments but it applies to some extreme urban, suburban, and rural cases which occur seldom but which are important in evaluating LTE performance in the most challenging environments. Table 20.2 shows the r.m.s. delay spread values for the three extended models [13].

It was also decided that the extended channel models are applied with low, medium, and high Doppler shifts, namely 5 Hz, 70 Hz and 300 Hz, which at a 2.5 GHz carrier frequency

[3]Global System for Mobile Communications.

Table 20.2: r.m.s. delay spread for the extended ITU models.

Category	Channel model	Acronym	r.m.s. delay spread (ns)
Low delay spread	Extended Pedestrian A	EPA	43
Medium delay spread	Extended Vehicular A model	EVA	357
High delay spread	Extended Typical Urban model	ETU	991

correspond roughly to mobile velocities of 2, 30 and 130 km/h respectively. Combinations which are likely to be used are EPA 5 Hz, EVA 5 Hz, EVA 70 Hz and ETU 70 Hz. The classical Doppler spectrum is again assumed. The tapped delay line propagation conditions for the LTE performance evaluation are hence summarized in Table 20.3.

Table 20.3: Power delay profiles of extended ITU models.

	EPA model		EVA model		ETU model	
Tap number	Excess tap delay (ns)	Relative power (dB)	Excess tap delay (ns)	Relative power (dB)	Excess tap delay (ns)	Relative power (dB)
1	0	0.0	[0	0.0	0	−1.0
2	30	−1.0	30	−1.5	50	−1.0
3	70	−2.0	150	−1.4	120	−1.0
4	80	−3.0	310	−3.6	200	0.0
5	110	−8.0	370	−0.6	230	0.0
6	190	−17.2	710	−9.1	500	0.0
7	410	−20.8	1090	−7.0	1600	−3.0
8			1730	−12.0	2300	−5.0
9			2510	−16.9	5000	−7.0

LTE conformance tests also include propagation conditions for two high-speed train scenarios. These are based on 300 and 350 km/h, with different Doppler shift trajectories, in non-fading propagation channels [14].

20.3 MIMO Channel Models

As already mentioned, the performance of MIMO systems depends strongly on the underlying propagation conditions. First of all, the Signal-to-Interference-plus-Noise Ratio (SINR) at the MIMO receiver determines the ultimate gains of spatial multiplexing. The SINR is a direct function of the path-loss, so that a terminal far from the base station tends to have a relatively small SINR compared to a terminal close to the base station. A second important factor is the correlation between signals at different antennas. Fading decorrelation is facilitated by the presence of a large number of multipath components at both the transmitter and the receiver, as typically experienced in a Non-Line-Of-Sight (NLOS) situation. The antenna separation also has a strong impact on the spatial correlation. The largest MIMO gains are obtained in scenarios with large SINR and low spatial correlation (where the channel matrix has a high rank), which in practice may rarely occur. For a detailed analysis of the effect of spatial correlation on MIMO system performance, the interested reader is referred to [2, 3] and references therein.

There are two commonly applied types of MIMO channel model:

- Correlation matrix based channel models;
- Geometry-based channel models.

Models of both types are used for LTE. The 3GPP Spatial Channel Model (SCM) and Spatial Channel Model – Extension (SCME) (see Sections 20.3.1 and 20.3.2 respectively) are geometry-based stochastic models, whereas the extended ITU models (EPA, EVA, ETU) are correlation matrix based models. In Section 20.3.3 we then discuss the geometry-based IST-WINNER channel model to give an insight into the state of the art in channel modelling. It should be noted that it is possible to create a correlation matrix based channel model from a geometry-based one, but not vice versa.

20.3.1 SCM Channel Model

In order to model MIMO channel characteristics realistically, originally for the evaluation of different MIMO schemes for High Speed Downlink Packet Access (HSDPA), 3GPP developed jointly with 3GPP2 a geometry-based SCM [15]. The SCM includes simple tapped-delay line models for calibration purposes and a geometry-based stochastic model for system-level simulations. The time-delay properties of the SCM calibration model are the same as in the 3GPP SISO/SIMO models discussed in Section 20.2.2. The spatial characteristics of the MIMO channel are defined by the Angular Spread (AS) and directional distribution of the propagation paths at both the base station and the UE.

System simulations typically consist of multiple cells/sectors, multiple base stations and multiple UEs. Performance metrics such as throughput and delay are gathered over a large number of simulation runs, called 'drops', which may consist of a predefined number of radio frames. During a drop, the large-scale parameters are fixed but the channel undergoes fast fading according to the motion of the terminals. The UEs may feed back channel state information about the instantaneous radio channel conditions and the base station can schedule its transmissions accordingly. Figure 20.2 defines the geometrical framework for the spatial parameters.

The SCM includes three main propagation environments: urban microcell, urban macrocell, and suburban macrocell. Additionally, it is possible to modify the basic scenarios by applying a LOS component in an urban micro case, and adding a far-scattering cluster or an urban canyon in an urban macro case.

The stochastic geometry-based model of the SCM enables realistic modelling of spatial correlation at both the base station and the UE. It is based on the ITU SISO models described in Section 20.2.1, so the number of propagation paths in each environment is six. However, the delay and angular spreads, as well as the directions of the scattering clusters, are random variables with normal or log-normal distribution. The elevation (i.e. vertical) domain is not modelled. Table 20.4 shows the main SCM parameters.

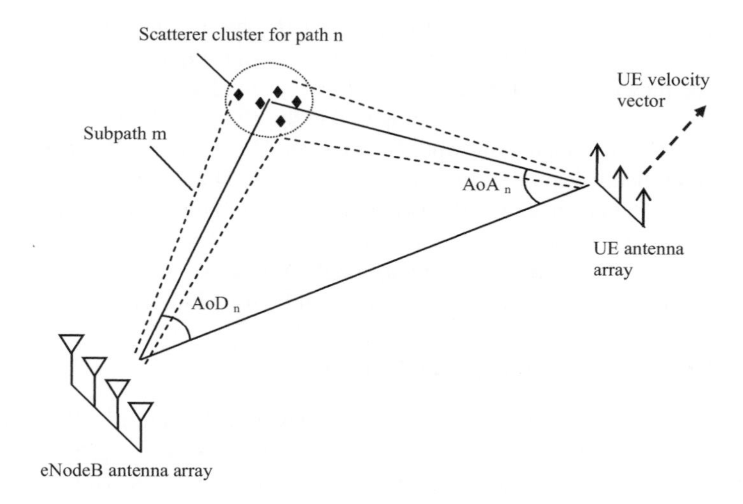

Figure 20.2: Geometry for generating spatial parameters for base station and UE.

Table 20.4: Main parameters for SCM.

Parameter	Suburban macro	Urban macro	Urban micro
Number of paths	6	6	6
Number of subpaths per path	20	20	20
Mean AS at base station	5°	8°, 15°	19°
Per-path AS at base station (fixed)	2°	2°	5°
Mean AS at UE	68°	68°	68°
Per-path AS at UE (fixed)	35°	35°	35°
Mean total r.m.s. delay spread	0.17 μs	0.65 μs	0.25 μs
Standard deviation for log-normal shadowing	8 dB	8 dB	NLOS: 10 dB; LOS: 4 dB
Path-loss model (dB, d = distance in metres from base station to UE)	$31.5 + 35 \log_{10}(d)$	$34.5 + 35 \log_{10}(d)$	NLOS: $34.53 + 38 \log_{10}(d)$; LOS: $30.18 + 26 \log_{10}(d)$

The SCM model is designed so that different antenna constellations can be applied. For example, the base station antenna spacing and beam pattern can be varied, including beam patterns for 3-sector and 6-sector cells. Typical values for the antenna spacing at the base station vary from 0.5 to 10 carrier wavelengths. At the base station the per-path angular spread is rather small (<5°) in all three environments, which is typical of antennas well above rooftop level. The mean angular spread is also relatively small, the largest value being 19° in the urban micro scenario, which is small compared to the beamwidth of a 4-antenna array. At the UE, the per-path and mean angular spreads are large (35°), which reflects the fact that the UE antennas are embedded in the scattering environment. The angular spread is composed of 6×20 subpaths. The per-path angular spread at both the base station and the UE follows a Laplacian distribution, which is generated by giving the 20 subpaths an equal power and fixed azimuth directions with respect to the nominal direction of the corresponding path.

A Matlab® implementation of the SCM model can be found in [1].

20.3.2 SCM-Extension Channel Model

An extension to the 3GPP/3GPP2 SCM model was developed in the IST-WINNER project. Known as the SCME (SCM-Extension) model, it is described in [16]. The extension was designed to remain backward-compatible, simple and consistent with the initial SCM concept. The main development is a broadening of the channel model bandwidth from 5 MHz to 100 MHz.

The bandwidth extension is carried out by introducing so-called midpaths which define the intra-cluster delay spread (i.e. the delay spread within a cluster of paths in a similar direction). The midpaths have fixed delay and power offsets in order to keep the SCME model backward-compatible with SCM. Therefore, the low-pass-filtered SCME impulse response corresponds closely to the respective SCM impulse response. As a result of the bandwidth extension the number of delay taps increases from six in SCM to 18 or 24, depending on the scenario. Table 20.5 shows the midpath delays and powers for the SCME propagation scenarios. Frequency-dependent factors are also added to the path-loss formulae, to extend the frequency range from 2 GHz to 5 GHz.

Table 20.5: SCME midpath powers and delays.

Scenario	Urban macro, suburban macro		Urban micro	
No. of midpaths per path	3		4	
	Midpath power	Relative delay (ns)	Midpath power	Relative delay (ns)
Midpath 1	10/20	0	6/20	0
Midpath 2	6/20	7	6/20	5.8
Midpath 3	4/20	26.5	4/20	13.5
Midpath 4	—	—	4/20	27.6

Another important contribution of the SCME model is the introduction of fixed spatio-temporal Tapped Delay Line (TDL) models, referred to as 'Clustered Delay Line (CDL) models'. The model parameters include the arrival and departure angles in addition to the traditional power and delay, thus covering all the MIMO propagation channel parameters except the polarization information. The TDL models are intended for the calibration of simulators. The spatio-temporal TDL models of the SCM extension resemble closely the SCM system model; they can actually be viewed as a realization of the system model which is optimized for low correlation in the frequency domain.

The SCME model has a number of optional features which can be applied depending on simulation purposes. In addition to the SCM 'drop' concept, the SCME model introduces optional drifting of delays and arrival/departure angles of the propagation paths. In the original SCM all the propagation parameters remain fixed during a drop and the only variation is the fast fading caused by the Doppler effects. Time evolution of the delay taps may be used for example for evaluating receiver synchronization algorithms. Drifting of the path arrival/departure angles is targeted to the testing of beamforming algorithms. Time evolution of shadow fading is also an optional feature; since the correlation of the shadow fading at two locations is related to the distance between them, this is modelled by an exponentially-shaped spatial auto-correlation function applied to the shadow fading such that the correlation of the shadow fading reduces exponentially with distance. Finally, the SCME model includes parameterization of the LOS condition for all the macrocellular scenarios.

A Matlab® implementation of the SCM model can be found in [1].

20.3.3 WINNER Model

The WINNER radio channel models were developed in the two phases of the IST-WINNER project [1]. Since the initial SCME model was not adequate for more advanced simulations, new measurement-based models were developed.

The models are ray-based double-directional multilink models which are antenna-independent, scalable and capable of creating arbitrary geometry-based MIMO channel models. Statistical distributions and channel parameters extracted from measurements in any propagation scenarios can be fitted to the models.

The latest model, known as the Phase II model, forms the basis for the ITU IMT-Advanced models introduced in Section 20.3.6.1. It extends the frequency range to cover frequencies from 2 to 6 GHz and covers 13 propagation scenarios including indoor, outdoor-to-indoor (and vice versa), urban micro- and macrocell and corresponding difficult urban scenarios, suburban and rural macrocell, feeder links, and moving networks. It is possible to vary the number of antennas, the antenna configurations, the geometry and the antenna beam pattern without changing the basic propagation model. This method enables the use of the same channel data in different link- and system-level simulations. The details of the model are described in [17]. The models also include CDL models for system calibration.

Path-losses have been specified for all the scenarios, divided into two subscenarios according to propagation conditions (LOS/NLOS) when applicable. The general structure of the path-loss is given by

$$PL = A\log_{10}(d) + B + C\log_{10}\left(\frac{f_c}{5}\right) + X \tag{20.1}$$

where d is the distance in metres and f_c is the carrier frequency in GHz. For each (sub)scenario, a set of constants A, B and C are defined in [17], as is the additional term X, for special cases. X is an environment-dependent factor, for example modelling additional propagation attenuation due to walls or floors (e.g. if the transmitter and receiver are located on different floors). A Matlab® implementation of the WINNER model can be found in [1].

20.3.4 LTE Evaluation Model

Channel models have been used in the design of LTE for many purposes, including evaluation of UE and eNodeB performance requirements, Radio Resource Management (RRM) requirements (see Chapter 22), system concepts and RF system scenarios (see Chapters 21 and 23). Conformance tests of the UE and eNodeB are discussed in Section 20.4.1.

Fixed TDL models of the SCME were proposed for both link- and system-level simulations in [18]. Two simplification steps from the SCM approach were taken. Firstly, the fixed TDL models, which were originally defined for calibration simulations only, were planned to be applied also in system-level simulations. The stochastic drop concept of SCM was discarded. As a slight modification to the SCME TDL models, the delays were quantized to the resolution which corresponds to fourfold over-sampling of the LTE bandwidth. Secondly, the MIMO characteristics were described by correlation matrices instead of directional propagation parameters such as the angles of arrival/departure and the polarization characteristics. For this purpose, antenna array characteristics were defined for the eNodeB and UE for different scenarios. The LTE channel model scenarios and antenna configurations are listed in Table 20.6 where λ is the carrier wavelength.

Table 20.6: Scenarios of LTE performance evaluation model.

Name	Propagation scenario	eNodeB arrangement	UE arrangement
SCM-A	Suburban Macro	3-sector, 0.5λ spacing	Handset, talk position
SCM-B	Urban Macro (low spread)	6-sector, 0.5λ spacing	Handset, data position
SCM-C	Urban Macro (high spread)	3-sector, 4λ spacing	Laptop
SCM-D	Urban Micro	6-sector, 4λ spacing	Laptop

In the eNodeB, a cross-polarized antenna configuration is used with four antenna elements. Two dual-polarized slanted +45° and −45° elements are spatially separated by a distance d, as shown in Figure 20.3. The polarization of each element is for simplicity assumed to be unchanged over all departure angles. The antenna element patterns are identical to those proposed for the link calibration model of SCM and can be either 3-sector or 6-sector. The radiation pattern as a function of angle, $A(\theta)$, is as follows:

$$A(\theta) = -\min\left[12\left(\frac{\theta}{\theta_{3\text{ dB}}}\right)^2, A_m\right] \quad \text{where } -180^\circ \leq \theta \leq 180^\circ \tag{20.2}$$

where, for a 3-sector arrangement, the 3 dB beamwidth $\theta_{3\text{ dB}} = 70^\circ$, $A_m = 20$ dB and the maximum gain is 14 dBi,[4] while for a 6-sector arrangement $\theta_{3\text{ dB}} = 35^\circ$, $A_m = 23$ dB and the maximum gain is 17 dBi.

Two types of mobile antenna scenarios are assumed: a small handset with two orthogonally-polarized antennas (see Figure 20.4), and a laptop with two spatially separated dual-polarized antennas (see Figure 20.5). It is assumed that the polarizations are purely horizontal and vertical in all directions when the antennas are in the nominal position. In the 'talk position'

[4]This indicates the maximum forward gain of the antenna relative to the gain of a hypothetical omnidirectional antenna.

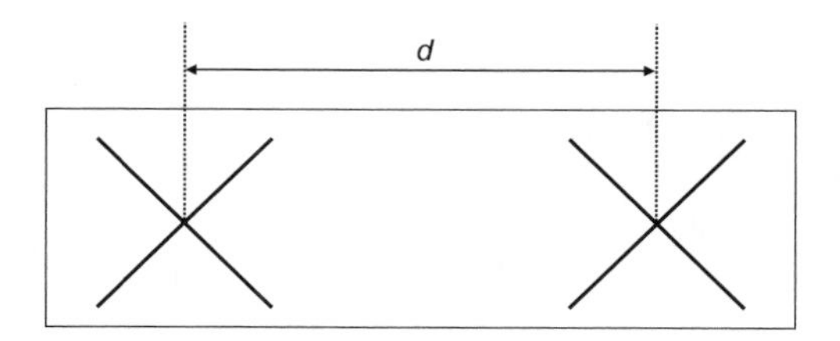

Figure 20.3: Base station antenna configuration.

case, the lobe is in the horizontal direction and the handset is rotated 60° (with the polarizations also being rotated). In the data position, the mobile is tilted 45° such that the radiation lobe has its maximum slanted downwards. The azimuthal orientations of the mobile antennas are selected such that the angle of arrival of the first tap occurs at +45° in all scenarios. The antenna element patterns are given by the same function as the eNodeB patterns, with the following parameters:

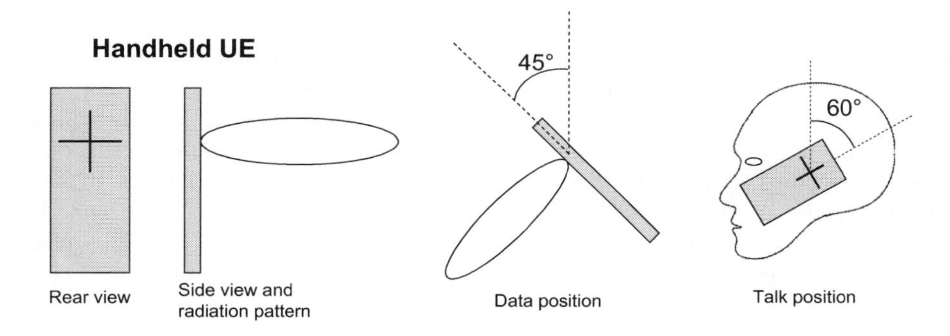

Figure 20.4: Handset antenna configuration.

- Handheld, talk position: $\theta_{3\text{ dB}} = 120°$, $A_m = 15$ dB, maximum gain = 3 dBi (vertical), 0 dBi (horizontal);
- Handheld, data position: $\theta_{3\text{ dB}} = 120°$, $A_m = 5$ dB, maximum gain = 3 dBi (vertical), 0 dBi (horizontal);
- Laptop: $\theta_{3\text{ dB}} = 90°$, $A_m = 10$ dB, maximum gain = 7 dBi, spatial separation 2λ.

Spatial correlations of the LTE performance evaluation model are calculated based on the antenna characteristics and the path arrival/departure angles defined in the SCME model. Polarization covariance matrices are determined based on the antenna polarizations and a cross polarization discrimination value of 8 dB. The Kronecker assumption is applied – i.e. fully separable transmitter and receiver power azimuth spectra. Finally, 8×8 correlation matrices can be determined for the SCM-A and -B scenarios, and 16×16 correlation matrices for the SCM-C and -D scenarios (see Table 20.6).

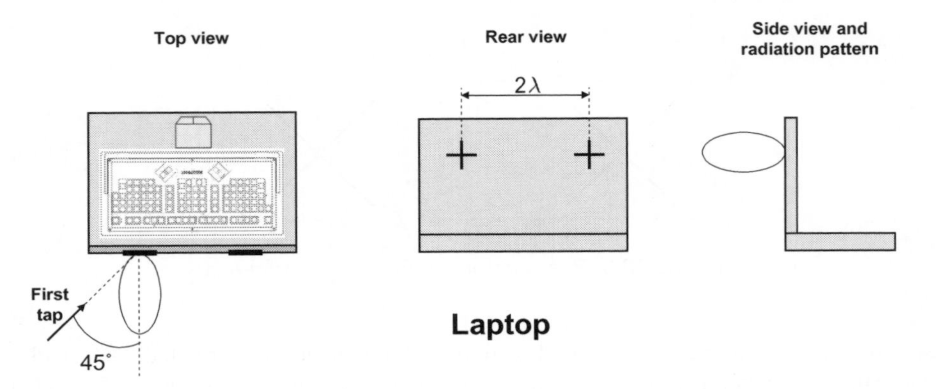

Figure 20.5: Laptop antenna configuration.

20.3.5 Extended ITU Models with Spatial Correlation

The extended ITU models introduced in Section 20.2.3 (EPA, EVA, ETU) were developed for MIMO conformance tests in a simple way by allocating the same correlation matrix to all the multipath components [20]. Three degrees of spatial correlation between different antenna signals are defined: high, medium and low. The low correlation case actually represents a case where the antennas are fully uncorrelated. The high and medium correlations are defined by specific correlation matrices based on practical antenna constellations. In the high correlation case the eNodeB and UE antenna arrays are co-polarized and have small inter-antenna distances of 1.5 and 0.5 wavelengths respectively. Medium correlation is achieved by means of orthogonally polarized antennas. These cases are illustrated in Figure 20.6.

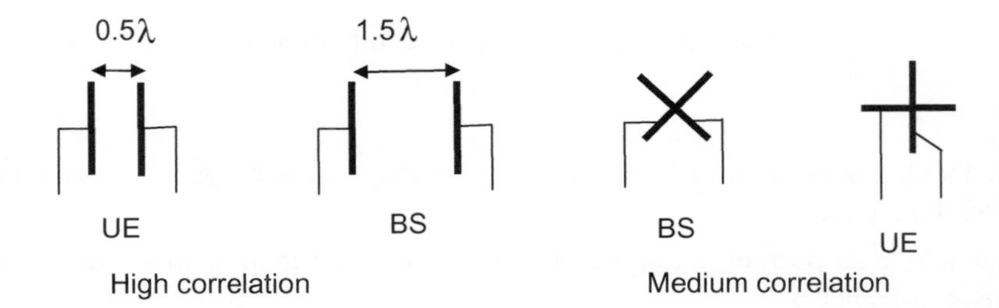

Figure 20.6: Antenna configurations for medium and high spatial correlation for extended ITU models.

In the 2×2 MIMO case the base station and UE correlation matrices are defined respectively as follows:

$$\mathbf{R}_{\mathrm{BS}} = \begin{pmatrix} 1 & \alpha \\ \alpha^* & 1 \end{pmatrix}, \quad \mathbf{R}_{\mathrm{UE}} = \begin{pmatrix} 1 & \beta \\ \beta^* & 1 \end{pmatrix} \tag{20.3}$$

and accordingly the spatial correlation is defined as

$$\mathbf{R}_{\text{spat}} = \mathbf{R}_{\text{BS}} \otimes \mathbf{R}_{\text{UE}} = \begin{pmatrix} 1 & \alpha \\ \alpha^* & 1 \end{pmatrix} \otimes \begin{pmatrix} 1 & \beta \\ \beta^* & 1 \end{pmatrix} \tag{20.4}$$

where $\otimes$ is the Kronecker product. The spatial correlation in the different cases can be represented as shown in Table 20.7.

Table 20.7: Spatial correlation with extended ITU channel models.

Low correlation		Medium correlation		High correlation	
α	β	α	β	α	β
0	0	0.3	0.9	0.9	0.9

Further, by applying the polarization matrices for the medium and high correlation cases, the 4×4 spatial correlation matrices can be obtained for the 2×2 MIMO case [21]. Similar spatial channel models for the 4×2 and 4×4 MIMO cases are also specified. The above correlation models are relatively simple. They apply reasonably well to basic MIMO concepts but may not be adequate for advanced beamforming/MIMO concepts due to the fact that all the delay taps have the same correlation properties. In real environments each delay tap will have different directions and angular spreads which define the per-path spatial correlation properties.

20.3.6 ITU Channel Models for IMT-Advanced

In November 2008, ITU-R approved radio channel models for the evaluation of the performance of the candidate technologies for IMT-Advanced, including LTE-Advanced [22, 23].

20.3.6.1 Test Environments

ITU-R has defined four representative test environments which are designed to span sufficiently the range of possible radio environments [22]:

- Indoor: a model for isolated indoor cells (e.g. in offices or hotspots), focusing on stationary and pedestrian users, with high user throughput or high user density.
- Microcellular: an urban model for high traffic loads, focusing on high densities of pedestrian and slow vehicular users. This scenario also includes outdoor-to-indoor coverage with the base station antennas situated below rooftop level.
- Base coverage urban: a macrocellular model providing continuous coverage for users with pedestrian to fast vehicular mobility. The base station antennas are above rooftop level.
- High speed: an environment of large rural macrocells with users in high speed vehicles and trains.

The IMT-Advanced channel modelling approach includes a primary module and an extension module.

The primary module is based on the WINNER Phase II channel model, and it covers the essential parameters needed for evaluation of the IMT-Advanced candidate radio interface technologies. The primary module includes path-loss models, a generic geometry-based stochastic channel model, and optional CDL models. The CDL models are used for calibration purposes only, under which all the radio channel parameters except for the phase angles of the propagation path signals are fixed. The extension module is based on the Time-Spatial Propagation (TSP) model described in [24].

The ITU-R IMT-Advanced channel model is a geometry-based stochastic model which is targeted at multi-cell, multi-user system simulations. Similarly to the 3GPP/3GPP2 SCM it does not explicitly specify the positions of the scatterers, but instead, models the directions of the propagation paths. It allows separate definition of the antenna constellation and radiation patterns. The channel parameters for individual snapshots of the radio channel are determined stochastically, based on statistical distributions derived from channel measurements. Channel realizations are generated by summing contributions of individual propagation paths with specific small scale parameters like delay, power, Angle-of-Arrival (AoA) and Angle-of-Departure (AoD). The IMT-Advanced channel model is based on a similar drop concept as used in the SCM model. Some of the main parameters of the IMT-Advanced channel model are listed in Table 20.8.

Table 20.8: Some main parameters of the IMT-Advanced channel models [22].

Test Environment	Indoor	Microcellular	Base coverage urban	High speed
Deployment scenario	Indoor Hotspot ('InH')	Urban Micro ('UMi')	Urban Macro ('UMa')	Rural Macro ('RMa')
Network layout	Indoor floor	Hexagonal grid	Hexagonal grid	Hexagonal grid
Site-to-site distance	60 m	200 m	500 m	1732 m
Carrier frequency	3.4 GHz	2.5 GHz	2 GHz	800 MHz
BS antenna height	6 m, ceiling-mounted	10 m, below rooftop	25 m, above rooftop	35 m, above rooftop
BS antenna gain	0 dBi	17 dBi	17 dBi	17 dBi
No of BS antennas	up to 8	up to 8	up to 8	up to 8
BS Tx power	21 dBm/ 20 Mhz 24 dBm/ 40 Mhz	41 dBm/ 10 Mhz 44 dBm/ 20 Mhz	46 dBm/ 10 Mhz 49 dBm/ 20 Mhz	46 dBm/ 10 Mhz 49 dBm/ 20 Mhz
No of UE antennas	up to 2	up to 2	up to 2	up to 2
UE antenna gain	0 dBi	0 dBi	0 dBi	0 dBi
UE Tx power	21 dBm	24 dBm	24 dBm	24 dBm
UE velocity	3 km/h	3 km/h	30 km/h	120 km/h
Inter-site interference modelling	Explicitly modelled	Explicitly modelled	Explicitly modelled	Explicitly modelled

20.3.6.2 Primary Module

The primary channel models are specified in general in the frequency range from 2 GHz to 6 GHz. However, the Rural Macro cell (RMa) model extends down to 450 MHz. The path-loss depends strongly on whether LOS conditions apply or not; Figure 20.7 shows the LOS probabilities in the different propagation scenarios.

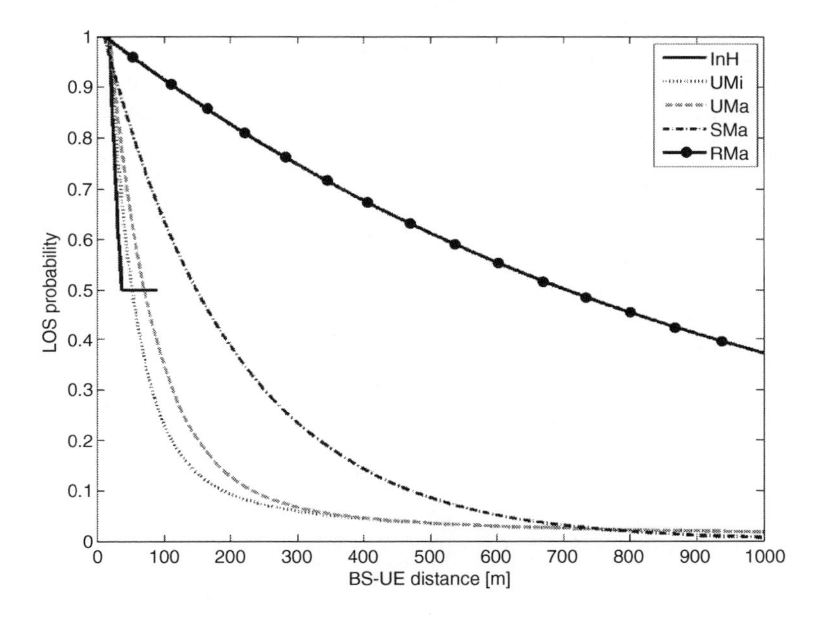

Figure 20.7: LOS probabilities for the ITU deployment scenarios.

The detailed parameters of the path-loss and shadow fading for the primary module in LOS and non-LOS cases for each deployment scenario can be found in [22].

The long-term shadow fading (in dB) follows a log-normal distribution around the mean path-loss. The normalized autocorrelation function computed at distance Δx can be approximated by an exponential function as follows

$$R(\Delta x) = \exp(-\Delta x / d_{corr})$$

where the correlation length d_{corr} depends on the environment.

In addition to the test environments described above, the primary module also includes an optional suburban model.

Table 20.9 summarizes the multi-antenna radio channel parameters in the different test environments in LOS and NLOS conditions.

20.3.6.3 Extension Module

The optional extension module can be applied for specific macrocell cases where the radio channel parameters defined in the primary module do not accurately describe the actual case

Table 20.9: Radio channel parameters for the ITU IMT-Advanced test environments (LOS and NLOS) [22]

Scenarios		**InH**		**UMi**			**UMa**		**RMa**	
		LOS	**NLOS**	**LOS**	**NLOS**	**Outdoor-to-Indoor**	**LOS**	**NLOS**	**LOS**	**NLOS**
Delay spread (DS) $\log_{10}$(s)	μ	-7.70	-7.41	-7.19	-6.89	-6.62	-7.03	-6.44	-7.49	-7.43
	σ	0.18	0.14	0.40	0.54	0.32	0.66	0.39	0.55	0.48
AoD spread (ASD) $\log_{10}$(degrees)	μ	1.60	1.62	1.20	1.41	1.25	1.15	1.41	0.90	0.95
	σ	0.18	0.25	0.43	0.17	0.42	0.28	0.28	0.38	0.45
AoA spread (ASA) $\log_{10}$(degrees)	μ	1.62	1.77	1.75	1.84	1.76	1.81	1.87	1.52	1.52
	σ	0.22	0.16	0.19	0.15	0.16	0.20	0.11	0.24	0.13
Shadow fading (SF) (dB)	σ	3	4	3	4	7	4	6	4	8
K-factor (*K*) (dB)	μ	7	N/A	9	N/A	N/A	9	N/A	7	N/A
	σ	4	N/A	5	N/A	N/A	3.5	N/A	4	N/A
Cross-correlations	ASD vs DS	0.6	0.4	0.5	0	0.4	0.4	0.4	0	-0.4
	ASA vs DS	0.8	0	0.8	0.4	0.4	0.8	0.6	0	0
	ASA vs SF	-0.5	-0.4	-0.4	-0.4	0	-0.5	0	0	0
	ASD vs SF	-0.4	0	-0.5	0	0.2	-0.5	-0.6	0	0.6
	DS vs SF	-0.8	-0.5	-0.4	-0.7	-0.5	-0.4	-0.4	-0.5	-0.5
	ASD vs ASA	0.4	0	0.4	0	0	0	0.4	0	0
	ASD vs *K*	0	N/A	-0.2	N/A	N/A	0	N/A	0	N/A
	ASA vs *K*	0	N/A	-0.3	N/A	N/A	-0.2	N/A	0	N/A
	DS vs *K*	-0.5	N/A	-0.7	N/A	N/A	-0.4	N/A	0	N/A
	SF vs *K*	0.5	N/A	0.5	N/A	N/A	0	N/A	0	N/A
Delay distribution		Exp	Exp	Exp	Exp	Exp	Exp	Exp	Exp	Exp
AoD and AoA distribution		Laplacian		Wrapped Gaussian			Wrapped Gaussian		Wrapped Gaussian	
Delay scaling parameter r_τ		3.6	3	3.2	3	2.2	2.5	2.3	3.8	1.7
cross-polarization ratio (XPR) (dB)	μ	11	10	9	8.0	9	8	7	12	7
Number of clusters		15	19	12	19	12	12	20	11	10
Number of rays per cluster		20	20	20	20	20	20	20	20	20
Cluster ASD		5	5	3	10	5	5	2	2	2
Cluster ASA		8	11	17	22	8	11	15	3	3
Per cluster shadowing standard deviation ζ (dB)		6	3	3	3	4	3	3	3	3
Correlation distance (m)	DS	8	5	7	10	10	30	40	50	36
	ASD	7	3	8	10	11	18	50	25	30
	ASA	5	3	8	9	17	15	50	35	40
	SF	10	6	10	13	7	37	50	37	120
	K	4	N/A	15	N/A	N/A	12	N/A	40	N/A

[1] AoA: angle-of-arrival, AoD: angle-of-departure, std: standard deviation, DS: rms delay spread, ASA: rms azimuth spread of arrival angles, ASD: rms azimuth spread of departure angles, SF: shadow fading, and *K*: Ricean *K*-factor.

(e.g. with respect to the BS antenna height, street width, city structure, etc). The extension module is intended to cover those cases beyond the evaluations of the IMT Advanced candidate radio interface technologies [22].

20.3.7 Comparison of MIMO Channel Models

The SCM, SCME, and WINNER channel models are compared in [19]. Table 20.10 compares the main features of the SCM, 3GPP LTE-evaluation, SCME, WINNER, and IMT-Advanced channel models, and Table 20.11 shows the numerical values of some of the key parameters.

Table 20.10: Feature comparison of SCM, LTE-evaluation, SCME, WINNER, and IMT-A models.

	SCM	LTE-eval	SCME	WINNER	IMT-A
Bandwidth > 20 MHz	No	Yes	Yes	Yes	Yes
Indoor scenarios	No	No	No	Yes	Yes
Outdoor-to-indoor scenarios	No	No	No	Yes	Yes
AoA / AoD elevation	No	No	No	Yes	No
Intra-cluster delay spread	No	Yes	Yes	Yes	Yes
Cross-correlation between large-scale parameters	No	No	No	Yes	Yes
Time evolution of model parameters	No	No	Yes*	Yes	No

**Continuous time evolution.

Table 20.11: Numerical comparison of SCM, LTE, SCME, WINNER, and IMT-A models.

Parameter	Unit	SCM	LTE-eval	SCME	WINNER	IMT-A
Max. bandwidth	MHz	5	20*	100*	100**	100**
Frequency range	GHz	2	N/A	2 Ű 5	2 – 6	2 – 6***
No. of scenarios		3	4	3	12	5
No. of clusters		6	6	6	8 – 20	10 – 20
No. of midpaths per cluster		1	3	3 – 4	1 – 3	1 – 3
No. of subpaths per cluster		20	N/A	20	20	20
No. of taps		6	18	18 – 24	12 – 24	14 – 24
Base station angle spread	deg	5 – 19	5 – 18	5 – 18	3 – 58	6 – 42
UE angle spread	deg	68	62 – 68	62 – 68	16 – 55	30 – 74
Delay spread	ns	160 – 650	231 – 841	231 – 841	16 – 630	20 – 365
Shadow fading standard deviation	dB	4 – 10	N/A	4 – 10	3 – 8	3 – 8

* Artificial extension from 5 MHz bandwidth.
** Based on 100 MHz measurements.
*** Rural model applies from 450 MHz to 6 GHz.

The LTE evaluation channel models are simplified models which have fixed delay and angle spreads. In that sense they apply well to link-level calibration models and also to conformance testing. However, it could be argued that they are too simplified for the most thorough of system-level performance evaluations. The SCM, SCME and WINNER models provide random delay and angle spreads for different users and therefore apply well to system level (multicell, multi-user) simulations. Thus the delay spread can take quite different values during different drops for a single user or during parallel drops to different users – this increases the variability and dynamics of the radio channels in multi-user simulations. From the MIMO modelling point of view, the LTE and SCM/SCME/WINNER/IMT-A models differ significantly in the way the spatial correlation is defined. In SCM/SCME/WINNER/IMT-A the spatial correlation is defined by the nominal direction and the angular geometry of the subpaths of each delay tap, while the LTE evaluation model generates deterministic correlation values for different clusters. Another significant difference is that the SCM/SCME/WINNER/IMT-A models specify the path-loss and shadow fading for each of the user environments, which are important in system-level performance evaluation.

20.4 Radio Channel Implementation for Conformance Testing

Radio channel phenomena can be mimicked by software (simulation) or by hardware (emulation). Software simulation enables easy programmability and low cost, and is typically utilized for link- and system-level performance evaluations. On the other hand, hardware emulation may be used to test real hardware implementations, to verify and validate products, and to speed up testing. It typically offers more realistic testing.

20.4.1 Performance and Conformance Testing

Performance testing is a general term for all testing where some performance metrics such as power, voltage, sensitivity, Bit Error Rate (BER), BLock Error Rate (BLER), throughput or handover success rate are measured. Test instruments such as signal generators and radio channel emulators need to be calibrated in order to ensure the accuracy of the measurements. The standardized channel models are useful to compare different products and technologies.

The purpose of conformance testing is not to secure optimal product operation in the field, but just to validate that the various products in the network conform to specified requirements and do not cause unexpected problems. A conformance test typically results in either a 'pass' or a 'fail' and is often performed by an external organization, such as a certified conformance test laboratory. Traditionally, the models used to conduct conformance testing have been the simple tapped-delay line models in order to allow their direct implementation in hardware-based radio channel emulators. However, more recent hardware-based channel emulators implement much more realistic channel models. For the LTE conformance test, the extended ITU channel models EPA, EVA and ETU are used.

20.4.2 Future Testing Challenges

The modelling of the radio channel is one of the key challenges for accurate performance measurement and testing. A direct cable connection (conductive testing) to transfer the

RF energy between the eNodeB and the UE avoids the uncertainty of the radio channel and allows measurements within well-defined and repeatable parameters. However, as UE antennas become more highly integrated, such testing becomes more difficult since the antenna connectors are no longer available. Conductive testing using RF connectors also has the drawback that it separates the antenna testing from the overall RF characteristics of the UEs which can lead to unrealistic results; moreover, the test connectors and cables themselves may affect the RF performance. Therefore, interest in Over-The-Air (OTA) testing, which eliminates the need for cable connections, is increasing.

Studies have taken place in 3GPP to examine measurements of radiated performance for MIMO and multi-antenna reception for HSPA and LTE terminals [25]. The particular focus is on methodologies based on anechoic RF chambers and on reverberation chambers. Anechoic chambers are free from reverberations, and multipath propagation is created in a controlled manner with multiple antennas and a radio channel emulator. By contrast, reverberation chambers create random echoes to mimic multipath propagation; this can favour specific MIMO terminals due to the unrealistically rich scattering environment with short delay spread.

20.5 Concluding Remarks

It is important to specify realistic propagation conditions for the evaluation of LTE performance. Accurate modelling of the spatial characteristics of the MIMO radio channel is needed in order to determine the best-performing multi-antenna transmission schemes, and to evaluate the LTE system performance in different radio environments.

3GPP specifications define advanced geometry-based radio channel models (SCM, SCME) which fulfil most of the sophisticated demands for system performance testing of LTE. SCME, WINNER and IMT-Advanced channel models provide additional features for further evaluations and simulations of LTE. The area of MIMO propagation modelling will continue to be important as the LTE conformance tests are developed, and as new versions of LTE with even larger numbers of antennas are standardized in the future.

References[5]

[1] Commission of the European Communities, 'IST-WINNER Project', www.ist-winner.org.

[2] G. J. Foschini, 'Layered Space-time Architecture for Wireless Communication in a Fading Environment when Using Multi-element Antennas'. *Bell Labs Tech. Journal*, Vol. 1, pp. 41–59, 1996.

[3] E. Telatar, 'Capacity of multi-antenna Gaussian channels'. *European Trans. on Telecommunications*, Vol. 10, pp. 585–595, November 1999.

[4] Commission of the European Communities, 'Digital Land Mobile Radio Communications-COST 207 Final Report', 1989.

[5] Commission of the European Communities, 'Digital Mobile Radio towards Future Generation Systems-COST 231 Final Report', 1993.

[6] Commission of the European Communities, 'Wireless Flexible Personalised Communications-COST 259 Final Report', 2001.

[5] All web sites confirmed 1st March 2011.

[7] W. R. Braun and U Dersch, 'A Physical Mobile Radio Channel Model'. *IEEE Trans. on Vehicular Technology*, Vol. 40, pp. 472–482, 1991.

[8] Commission of the European Communities, 'RACE-II advanced TDMA mobile access project: An Approach for UMTS'. Springer, Berlin/Heidelberg, 1994.

[9] ITU-R M.1225 International Telecommunication Union, 'Guidelines for evaluation of radio transmission technologies for IMT-2000', 1997.

[10] 3GPP Technical Specification 25.101, 'UE Radio Transmission and Reception (FDD)', www.3gpp.org.

[11] W. C. Jakes, *Microwave Mobile Communications*. New York: John Wiley & Sons Ltd/Inc, 1974.

[12] R. H. Clarke, 'A Statistical Theory of Mobile-radio Reception'. *Bell Syst. Tech. Journal.*, pp. 957–1000, July 1968.

[13] Ericsson, Nokia, Motorola, and Rohde & Schwarz, 'R4-070572: Proposal for LTE Channel Models', www.3gpp.org, 3GPP TSG RAN WG4, meeting 43, Kobe, Japan, May 2007.

[14] 3GPP Technical Specification 36.141, 'Base Station (BS) conformance testing ', www.3gpp.org.

[15] 3GPP Technical Specification 25.996, 'Spatial Channel Model for Multiple Input Multiple Output (MIMO) Simulations', www.3gpp.org.

[16] D. S. Baum, J. Salo, G. Del Galdo, M. Milojevic, P. Kyösti and J. Hansen, 'An Interim Channel Model for Beyond-3G Systems' in *Proc. IEEE Vehicular Technology Conference* (Stockholm, Sweden), May 2005.

[17] Commission of the European Communities, IST-WINNER II project deliverable D1.1.2 'WINNER II Channel Models', www.ist-winner.org.

[18] Ericsson, Elektrobit, Nokia, Motorola, and Siemens, 'R4-060334: LTE Channel Models and simulations', www.3gpp.org, 3GPP TSG RAN WG4, meeting 38, Denver, USA, February 2006.

[19] M. Narandzic, C. Schneider, R. Thomä, T. Jämsä, P. Kyösti and X. Zhao, 'Comparison of SCM, SCME, and WINNER Channel Models', in *Proc. IEEE Vehicular Technology Conference* (Dublin, Ireland), April 2007.

[20] Agilent Technologies, 'R4-071318: LTE MIMO Correlation Matrices', August 2007.

[21] 3GPP Technical Specification 36.101, 'User Equipment (UE) radio transmission and reception', www.3gpp.org.

[22] ITU-R Report M.2135, 'Guidelines for evaluation of radio interface technologies for IMT-Advanced', www.itu.int/itu-r.

[23] 3GPP Technical Report 36.814, 'Further Advancements for E-UTRA Physical Layer Aspects', www.3gpp.org.

[24] ITU-R Recommendation P.1816 'The prediction of the time and the spatial profile for broadband land mobile services using UHF and SHF bands', www.itu.int/

[25] Vodafone, 'RP-0900352: Measurement of radiated performance for MIMO and multi-antenna reception for HSPA and LTE terminals', www.3gpp.org, 3GPP TSG RAN, meeting #43, Biarritz, France, March 2009.

21

Radio Frequency Aspects

Moray Rumney, Takaharu Nakamura, Stefania Sesia, Tony Sayers and Adrian Payne

21.1 Introduction

Radio Frequency (RF) signal processing constitutes the final interface between the baseband techniques described in earlier chapters and the transmission medium – the air. The RF processing presents its own unique challenges for practical transceiver design, arising in particular from the fact that the air interface is a shared resource between multiple RF carriers, each with their own assigned portion of spectrum. The RF transmitter must be designed in such a way as not only to generate a clean signal within the assigned spectrum portion, but also to keep Inter-Carrier Interference (ICI) within acceptable levels. The receiver likewise must reliably demodulate the wanted signal, in order to avoid requiring excessive energy to be transmitted, whilst also rejecting interference from neighbouring carriers. Performance requirements for these RF aspects aim to ensure that equipment authorized to operate on an LTE carrier meets certain minimum standards.

In general, the performance requirements for LTE transceivers are intended not to be significantly more complex than for UMTS[1] in terms of implementation and testing. Consequently, many of the RF requirements for LTE are derived from those already defined for UMTS.

There are, however, a number of key differences between LTE and UMTS which affect the RF complexity and performance.

One such difference is the use of a variable channel bandwidth in LTE, up to a maximum of 20 MHz. Even the lowest category of LTE User Equipment (UE) is required to support all the bandwidths specified for the bands in which the UE is designed to operate, which in general includes 1.4, 3.0, 5.0, 10.0, 15.0 and 20.0 MHz. This

[1] Universal Mobile Telecommunications System.

LTE – The UMTS Long Term Evolution: From Theory to Practice, Second Edition.
Stefania Sesia, Issam Toufik and Matthew Baker.

requires a set of RF requirements to be defined for each bandwidth, in contrast to the Frequency Division Duplex (FDD) mode of UMTS which only supports a single channel bandwidth of 5 MHz. For LTE, the variable bandwidth represents a new challenge for the RF design and testing. For example, a transceiver for a constant-bandwidth radio system can potentially employ fixed filters at a number of points in the signal processing. Such filters are designed to pass a signal with known characteristics and reject particular frequencies. If, however the bandwidth of the transceiver is variable over a wide range, it is clear that fixed filters cannot be used. Frequencies which must be passed in 20 MHz operation may need to be rejected in the narrower bandwidth modes. This implies that LTE transceivers must be more adaptable than those of previous systems, while also being cleaner in transmission and having better selectivity in reception.

A second difference between UMTS and LTE is that in LTE it is assumed that the UE has at least two receive antennas. This means that multiple RF signal paths are needed at all times.

Thirdly, LTE is even more adaptable than UMTS in terms of the range of data rates it supports in order to suit different Signal-to-Interference-plus-Noise Ratio (SINR) () conditions (e.g. from 4×2 Multiple-Input Multiple-Output (MIMO) 64QAM in high SINR conditions to Single-Input Single-Output (SISO) QPSK in low SINR conditions (see Section 10.2)). Together with the ability to vary the instantaneous bandwidth to a given user, this implies a large number of modes of operation and flexibility in signal-handling capabilities. The reception of the maximum data rate requires high SINR at the highest bandwidth, which is particularly challenging for the analogue-to-digital converter in the receiver.

Fourthly, the LTE signal structure itself alters which specific RF aspects are the most critical. As already discussed extensively in Chapters 5 and 14, the use of Orthogonal Frequency Division Multiplexing (OFDM) in the downlink and Single Carrier Frequency Division Multiple Access (SC-FDMA) in the uplink provides an inherent robustness against multipath propagation, which means that amplitude and phase distortions from receiver and transmitter filters are not as critical as for UMTS's single-carrier Wideband Code Division Multiple Access (WCDMA) modulation. On the other hand, OFDM requires better frequency synchronization and is more sensitive to phase noise as shown in Chapter 5 and further discussed in Section 21.5.

It is also worth noting that, for WCDMA, the RF requirement specifications are separate for FDD and TDD (Time Division Duplex) [1–4], while in LTE the commonality between FDD and TDD is such that they can be handled in the same specifications [5,6]. Nevertheless, a more detailed discussion of aspects specific to TDD operation is given in Chapter 23.

Some further useful background on the transmitter and receiver RF characteristics can be found in [7] and [8] for the LTE UE and eNodeB respectively.

This chapter seeks to describe the impact of the LTE physical layer radio requirements on the implementation complexities of an LTE transceiver, compared to the well-known UMTS system. The relevant spectrum bands for LTE are introduced in Section 21.2. Radio requirements for the transmitter and receiver are discussed in Sections 21.3 and 21.4 respectively, and some of the implementation challenges are highlighted. Both the UE and eNodeB are addressed, although with the greater emphasis on the UE side. This is followed by a discussion of the RF impairments for both transmitter and receiver in Section 21.5, including mathematical models of the most common impairments to help elucidate the impact of RF imperfections on the overall radio behaviour.

21.2 Frequency Bands and Arrangements

LTE and UMTS are both defined for a wide range of different frequency bands, in each of which one or more independent carriers may be operated. Tables 21.1 and 21.2 give details of the frequency bands for FDD and TDD operation respectively. For FDD, the duplex separation is not actually defined, but typically the uplink and downlink carriers in a pair are in a similar position in their respective bands so that the duplex separation is usually approximately $F_{\mathrm{DL\,low}}$–$F_{\mathrm{UL\,low}}$ as shown in Table 21.1.

Table 21.1: UMTS and LTE frequency bands for FDD.

Band Number	Uplink (MHz)	Downlink (MHz)	Band Gap (MHz)	Duplex Separation (MHz)	UMTS Usage	LTE Usage
	$F_{\mathrm{UL\,low}}$–$F_{\mathrm{UL\,high}}$	$F_{\mathrm{DL\,low}}$–$F_{\mathrm{DL\,high}}$				
1	1920–1980	2110–2170	130	190	Y	Y
2	1850–1910	1930–1990	20	80	Y	Y
3	1710–1785	1805–1880	20	95	Y	Y
4	1710–1755	2110–2155	355	400	Y	Y
5	824–849	869–894	20	45	Y	Y
6*	830–840	875–885	35	45	Y	Y
7	2500–2570	2620–2690	50	120	Y	Y
8	880–915	925–960	10	45	Y	Y
9	1749.9–1784.9	1844.9–1879.9	60	95	Y	Y
10	1710–1770	2110–2170	340	400	Y	Y
11	1427.9–1447.9	1475.9–1495.9	28	48	Y	Y
12	698–716	728–746	12	30	Y	Y
13	777–787	746–756	21	31	Y	Y
14	788–798	758–768	20	30	Y	Y
17	704–716	734–746	18	30	N	Y
18**	815–830	860–875	30	45	N	Y
19**	830–845	875–890	30	45	Y	Y
20**	832–862	791–821	11	41	Y	Y
21**	1447.9–1462.9	1495.9–1510.9	33	48	Y	Y
23***	2000–2020	2180–2200	160	180	N	Y
24***	1626.5–1660.5	1525–1559	-135.5	-101.5	N	Y
25***	1850–1915	1930–1995	15	80	Y	Y
26****	814–849	859–894	10	45	Y	Y

* This band was defined in the context of Release 8; it is replaced by Band 19 for later releases (Release 9 and 10). Only legacy terminals would use band 6.

** These bands were specified in the timeframe of Release 9, although all bands are release-independent and can be implemented by UEs conforming to any release.

*** These bands were specified in the timeframe of Release 10, although all bands are release-independent and can be implemented by UEs conforming to any release.

**** This band was under consideration at the time of going to press.

Table 21.2: UMTS and LTE frequency bands for TDD.

Band	F_{low}–F_{high} (MHz)	UMTS	LTE
33	1900–1920	Y	Y
34	2010–2025	Y	Y
35	1850–1910	Y	Y
36	1930–1990	Y	Y
37	1910–1930	Y	Y
38	2570–2620	Y	Y
39	1880–1920	N	Y
40	2300–2400	Y	Y
41*	2496–2690	N	Y
42*	3400–3600	N	Y
43*	3600–3800	N	Y

* These bands were specified in the timeframe of Release 10, although all bands are release-independent and can be implemented by UEs conforming to any release.

In Table 21.1 band 15 and 16 are not shown; these bands were introduced by the European Telecommunication Standard Institute (ETSI) and are applicable to Europe only.

A typical UE would support a certain subset of these bands depending on the desired market, since supporting all would be challenging for the transceiver, in particular for the front-end components such as Power Amplifiers (PAs), filters and duplexers. The set of frequency bands chosen defines the capability of the UE to switch bands, roam between national operators and roam internationally.

Each frequency band is regulated to allow operation in only a certain set of channel bandwidths, as shown in Table 21.3. Some frequency bands do not allow operation in the narrow bandwidth modes below 5 MHz, while others do not allow operation in the wider bandwidths, generally 15 MHz or above. '—' denotes that the channel bandwidth is not supported in the specific band; 'X' indicates the channel bandwidth is too wide to be supported in the band.

In LTE-Advanced, new requirements are defined for combinations of carriers in the same and different bands for carrier aggregation. These combinations are discussed in Section 28.4.3.

Due to LTE's flexibility in the frequency domain, two new concepts are defined: 'transmission bandwidth configuration', which defines the maximum number of Resource Blocks (RBs) for any channel bandwidth, and 'transmission bandwidth' for a specific transmission, which depends on the scheduled resource allocation and can be less than or equal to the transmission bandwidth configuration as shown in Figure 21.1.

Moreover it is important to distinguish between the transmission bandwidth configuration, generally given in RBs, and the channel bandwidth, which is expressed in MHz as shown in Table 21.4.

Table 21.3: Number of supported non-overlapping channels in each frequency band and bandwidth.

LTE band	Downlink bandwidth	Channel bandwidth (MHz)					
		1.4	3	5	10	15	20
1	60	—	—	12	6	4	3
2	60	42	20	12	6	4*	3*
3	75	53	25	15	7	5*	3*
4	45	32	15	9	4	3	2
5	25	17	8	5	2*	—	—
6	10	—	—	2	1*	X	X
7	70	—	—	14	7	4	3*
8	35	25	11	7	3*	—	—
9	35	—	—	7	3	2*	1*
10	60	—	—	12	6	4	3
11	25	—	—	4	2*	1*	1*
12	18	12	6	3*	1*	—	X
13	10	—	—	2*	1*	X	X
14	10	—	—	2*	1*	X	X
...							
17	12	—	—	2*	1*	X	X
18	15	—	—	3	1*	1*	X
19	15	—	—	3	1*	1*	X
20	30	—	—	6	3*	2*	1
21	15	—	—	3	1*	1*	X
23	20	14	6	4	2	—	—
24	34	—	—	6	3	—	—
25	65	46	21	13	6	4*	3*
26	35	25	11	7	3*	2*	—
...							
33	20	—	—	4	2	1	1
34	15	—	—	3	1	1	X
35	60	42	20	12	6	4	3
36	60	42	20	12	6	4	3
37	20	—	—	4	2	1	1
38	50	—	—	10	5	3	2
39	40	—	—	8	4	3	2
40	100	—	—	20	10	6	5
41	194	—	—	38	19	12	9
42	200	—	—	40	20	13	10
43	200	—	—	40	20	13	10

* A relaxation of the specified UE receiver sensitivity requirement is allowed.

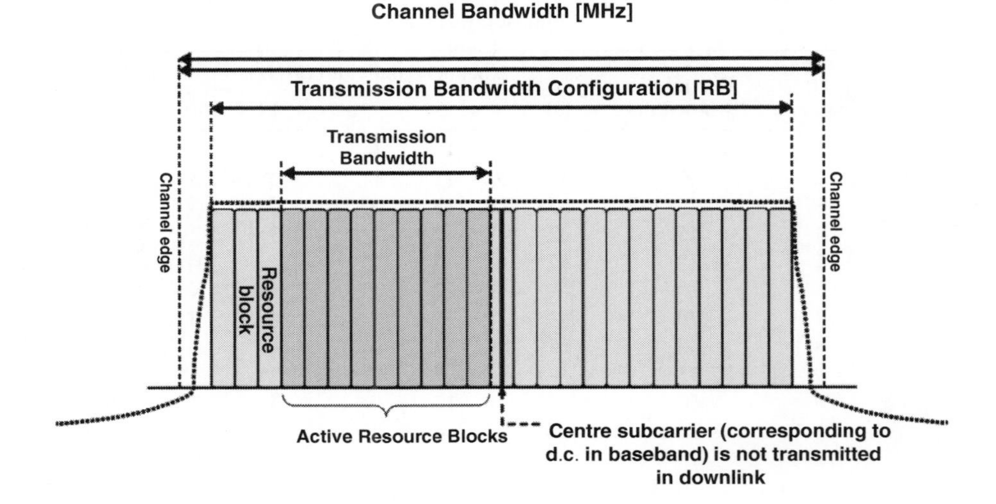

Figure 21.1: Definition of transmission bandwidth configuration and transmission bandwidth. Reproduced by permission of © 3GPP.

Table 21.4: Transmission bandwidth configuration for each channel bandwidth.

Channel bandwidth	MHz	1.4	3	5	10	15	20
Transmission bandwidth configuration	N_{RB}	6	15	25	50	75	100
	MHz	1.06	2.7	4.5	9	13.5	18

21.3 Transmitter RF Requirements

This section explains the implications of the LTE RF performance specification for the design of its transmitters.

Each transmitter must satisfy two categories of requirements: those relating to the power level and quality of the intended transmissions, and those prescribing the level of unwanted emissions which can be tolerated. The latter is usually the more challenging for the radio designer.

21.3.1 Requirements for the Intended Transmissions

21.3.1.1 Signal Quality: Error Vector Magnitude (EVM)

The quality of the transmitted radio signal has to fulfil certain requirements. The main parameter used to measure this quality is the Error Vector Magnitude (EVM), which is a measure of the distortion introduced by the RF imperfections of practical implementations.[2] It is defined as the magnitude of the difference between a theoretical reference signal (i.e. the signal defined by the physical layer specification equations) and the actual transmitted signal (normalized by the intended signal magnitude). The EVM sets the maximum possible SNR of a radio link in the absence of any noise, interference, propagation loss and other distortions introduced by the radio channel; it therefore serves to determine the maximum useful modulation order and code rate.

The EVM measures the quality of the transmitted signal across all the allocated RBs. The measurement duration is one slot for the UE (uplink) and one subframe for eNodeB (downlink) and takes into account all the symbols belonging to the modulation scheme under test. Figures 21.2 and 21.3 show the EVM measurement points in the downlink and uplink transmission chains respectively [5, 6].

The EVM measurement is taken after an equalizer in the test equipment, which carries out per-subcarrier channel correction [5, 6]. The equalizer is used in order to obtain a measurement which realistically shows what a receiver might experience. It is intended to reflect the fact that the equalizer in the receiver is capable of correcting some of the impairments of the transmitted signal to some extent. A zero-forcing equalizer is used for EVM measurement;[3] however, real receiver implementations may use different equalization techniques, and therefore the measured EVM may not exactly correspond to the actual signal quality experienced by all receivers.

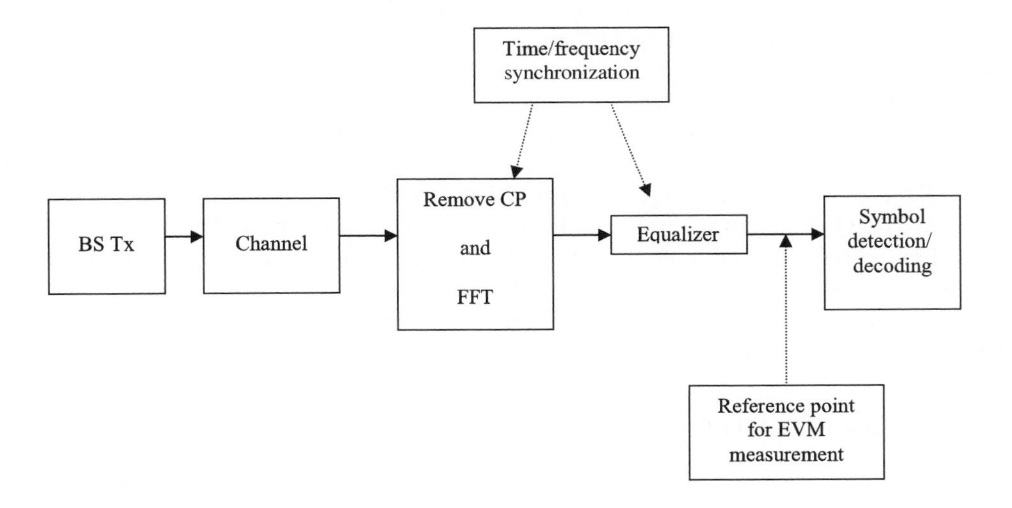

Figure 21.2: Measurement points for the EVM for the downlink signal.

[2]Note that distortion may also be introduced in the process of balancing in-channel and out-of-channel performance, especially in the eNodeB.

[3]Details of how to compute the coefficients of the equalizer can be found in [5, 6].

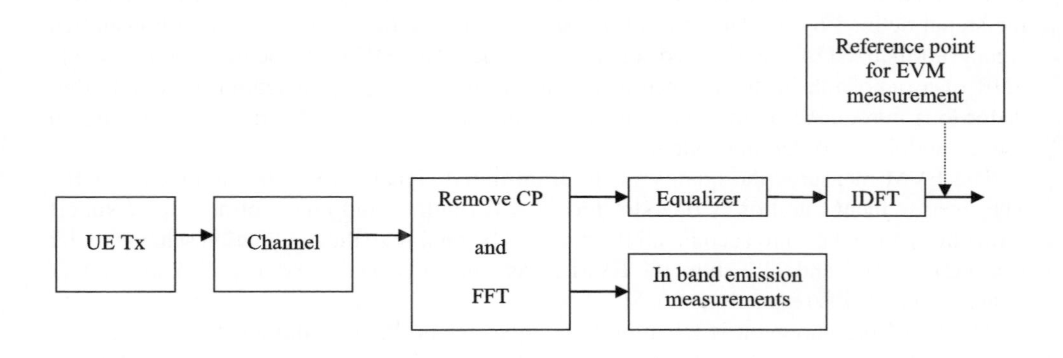

Figure 21.3: Measurement points for the EVM for the uplink signal.

It should be noted that before measuring the EVM, time and frequency synchronization must be carried out. The test equipment then computes the EVM for two extreme values of sample timing difference between the FFT processing window and the nominal timing of the ideal signal; these two extreme values correspond to the beginning and end of the window, the length of which is expressed as a percentage of the Cyclic Prefix (CP) length. Finally the measurement is averaged over 20 slots for UE (uplink) and 10 subframes for eNodeB (downlink) to reduce the impact of noise.

In LTE the EVM is required to be less than 17.5% for QPSK, 12.5% for 16QAM and 8% for 64QAM (64QAM being applicable to the downlink only) [5, 6]. These EVM values are designed to correspond to no more than a 5% loss in average and cell-edge throughputs in typical deployment scenarios. At the link level, EVM is equivalent to an SNR loss.

A description of the main transmitter impairments which generally give rise to non-zero EVM is given in Section 21.5.1.

21.3.1.2 Transmit Output Power

The transmitted output power directly influences the inter-cell interference experienced by neighbouring cells using the same channel, as well as the magnitude of unwanted emissions outside the transmission band. This affects the ability of the LTE system to maximize spectral efficiency, and it is therefore important that the transmitters can set their output power accurately.

For the eNodeB, the maximum output power must remain within ±2 dB of the rated power, (P_{RAT}) declared by the manufacturer. There are three classes of eNodeB (see Section 24.4): the Home eNodeB with $P_{\rm RAT} \leq 20$ dBm, the Local Area eNodeB with $P_{\rm RAT} \leq 24$ dBm and the Wide Area eNodeB for which no upper limit has been defined. In addition, the dynamic range in the frequency domain (computed as the difference between the power of a given

Resource Element (RE) and the average RE power) must not exceed specified limits [5, 6] depending on the modulation order, in order to avoid saturating the UE receivers.[4]

For the UE, a requirement is defined for only one power class, known as 'power class 3' for which the maximum output power is 23 dBm. Like the eNodeB, the UE must satisfy this requirement within a range of ±2 dB.

The UE maximum output power requirements may be modified by a number of factors [5]:

- **Maximum Power Reduction (MPR).** The purpose of MPR is to allow the UE, in some demanding configurations, to lower its maximum output power in order to meet the general requirements on signal quality and Out-Of-Band (OOB) emissions. Note that the MPR is an allowance and the UE does not have to use it. Table 21.5 shows the allowed values of MPR as a function of the modulation scheme, the channel bandwidth and the transmission bandwidth (number of transmitted RBs).

Table 21.5: Maximum power reduction for power class 3.

Modulation	Channel bandwidth (MHz)/ transmission bandwidth (RB)						MPR (dB)
	1.4	3	5	10	15	20	
QPSK	>5	>4	>8	>12	>16	>18	≤ 1
16QAM	≤ 5	≤ 4	≤ 8	≤ 12	≤ 16	≤ 18	≤ 1
16QAM	>5	>4	>8	>12	>16	>18	≤ 2

- **Additional MPR (A-MPR).** The eNodeB may inform the UE of the possibility of further lowering its maximum power by signalling an A-MPR. The need for A-MPR occurs with certain combinations of E-UTRA bands, channel bandwidths and transmission bandwidths for which the UE must meet additional (more stringent) requirements for spectrum emission mask and spurious emissions (see Section 21.3.2). As with MPR, the A-MPR is an allowance, not a requirement, and it applies in addition to MPR. Regardless of whether the UE makes use of the allowed MPR and A-MPR, the additional requirements for spectrum emission mask and spurious emissions that are signalled by the network always apply. The reason for the complex set of conditions for the relaxation is the expected intermodulation products which may fall into adjacent bands which have different levels of sensitivity (e.g. public safety bands).
- **ΔT_C.** ΔT_C is a 1.5 dB reduction in the lower limit of the maximum output power range when the signal is within 4 MHz of the channel edge.
- **Power Management MPR (P-MPR).** Introduced in Release 10, the P-MPR allows a UE to reduce its maximum output power when other constraints are present. In particular, multi-RAT[5] terminals may have to limit the LTE transmission power if

[4]For example, a Physical Downlink Control CHannel (PDCCH) RE power can range between −6 dB and +4 dB around its average.

[5]Radio Access Technology.

transmissions on another RAT are taking place simultaneously. Such power restrictions may arise, for example, from regulations on Specific Absorption Rate (SAR) of radio energy into a user's body or from out-of-band emission requirements that may be affected by the inter-modulation products of the simultaneous radio transmissions. The P-MPR is not aggregated with MPR or A-MPR, since any reduction in a UE's maximum output power for the latter factors helps to satisfy the requirements that would have necessitated P-MPR.

Taking these factors into account, the UE has to configure its nominal maximum power P_{CMAX} (i.e. the highest power at which the UE will attempt to transmit) within the following upper and lower limits:

$$P_{\text{CMAX, L}} = \min\left(P_{\text{EMAX}} - \Delta T_C, P_{\text{PowerClass}} - \max\left(\text{MPR} + \text{A-MPR}, \text{P-MPR}\right) - \Delta T_C\right)$$
$$P_{\text{CMAX, H}} = \min\left(P_{\text{EMAX}}, P_{\text{PowerClass}}\right) \quad (21.1)$$

where $P_{\text{PowerClass}}$ is the original power class of the UE and P_{EMAX} is a maximum power that may be signalled by the network.

Note that when carrier aggregation is configured (for UEs of Release 10 and beyond), P_{CMAX} becomes a component-carrier-specific nominal maximum power, $P_{\text{CMAX,c}}$, as explained in Section 28.3.5.

Finally, the actual transmitted power is allowed to vary over a wider range due to uncertainties in the transmit chain. The added tolerance is a function of P_{CMAX} which varies from 2 dB at high powers to 7 dB for powers below 8 dBm.

In summary, the concept of maximum power for the UE is a complex and dynamic function of many variables designed so that the needs of specific operating conditions can be met without over-specifying the transmitter design. Details of the maximum power specification can be found in [5].

21.3.1.3 Output Power Dynamics

LTE, like UMTS and HSPA, needs to ensure user orthogonality in the time domain. In WCDMA, the requirements for this were relatively straightforward being based on so-called 'on–off masks' which define the allowable transmit power during the 'off' and 'on' periods of transmission. For LTE, similar general requirements exist for the eNodeB in both FDD and TDD [6]. However, for the LTE UE the requirements are considerably more complex than those for WCDMA due to the characteristics of the SC-FDMA scheme used for the uplink transmissions. Figure 21.4 shows the general 'on/off time mask' which applies whenever the UE is required to switch on or off.[6] Note that although the requirement is given in terms of a 'mask', it actually applies to the average power during the 'on' and 'off' periods [5].

In UMTS the transient period (20 μs) is centred on the slot boundary, while in LTE the transient period is in general shifted forwards into the next subframe. Hence, the first few SC-FDMA samples are vulnerable to being corrupted by insufficient transmit power or inter-UE interference whereas the samples at the end of the subframe are protected. For the Physical Random Access CHannel (PRACH) the transient periods are located outside the preamble in order to protect the whole PRACH preamble. The same applies to Sounding Reference

[6]This is also applicable for Discontinuous Transmission (DTX) measurement gaps – see Chapter 22.

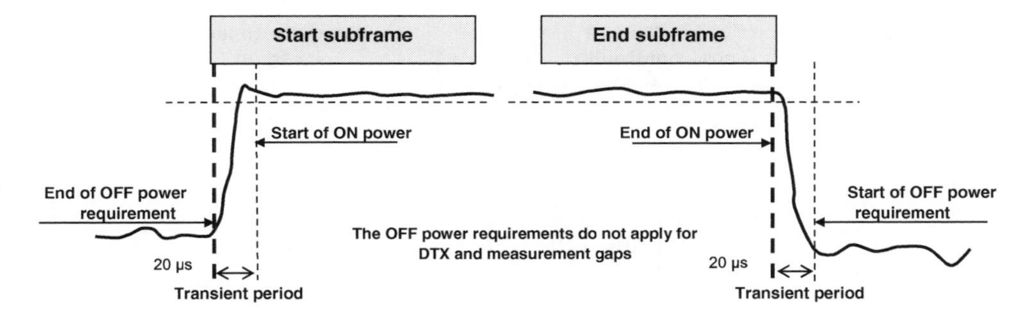

Figure 21.4: General on–off time mask for the UE. Reproduced by permission of © 3GPP.

Signals (SRSs) (see Section 15.6) which need to be protected to ensure the eNodeB can exploit them entirely[7] to carry out reliable uplink channel sounding without impairments arising from transient periods. For TDD, where two SRSs can be transmitted on adjacent symbols in the UpPTS field, the transient periods are located between the two SRSs.

The general time masks apply in the case of transitions into and out of the off power state. A general output power dynamic requirement also exists for continuous transmission at slot boundaries where frequency hopping occurs and at subframe boundaries for either frequency hopping or power changes. In all these cases, the transient periods are symmetric about the slot or subframe boundaries. Transient power masks also apply in the following cases:

- Transition from PUCCH/PUSCH to SRS with DTX after SRS;
- Transition from PUCCH/PUSCH to SRS to PUCCH/PUSCH;
- Transition from DTX to SRS to PUSCH/PUCCH;
- Transition from PUCCH/PUSCH to DTX in an SRS symbol to PUCCH/PUSCH.

21.3.2 Requirements for Unwanted Emissions

Ideally, the radio should transmit nothing at all outside its designated transmission channel. However, in practice this is very far from being the case.

Outside the intended channel bandwidth, the LTE specifications define two separate kinds of unwanted emission: *Out-Of-Band (OOB) emissions* and *spurious emissions*. These are shown schematically in Figure 21.5.

Out-of-band emissions are those which fall in a band close to the intended transmission, while spurious emissions can be at any other frequency. The precise boundary between the OOB range and the spurious range is different for different aspects of the LTE specifications.

LTE defines requirements for both types of unwanted emission, with those for spurious emissions being the more stringent. As mentioned in Section 21.3.1.2, the specified requirements must be fulfilled when indicated by network signalling or under particular conditions. This means that, unlike the relatively fixed RF requirements of earlier systems,

[7] Note that the SRS is only 66.67 μs in duration.

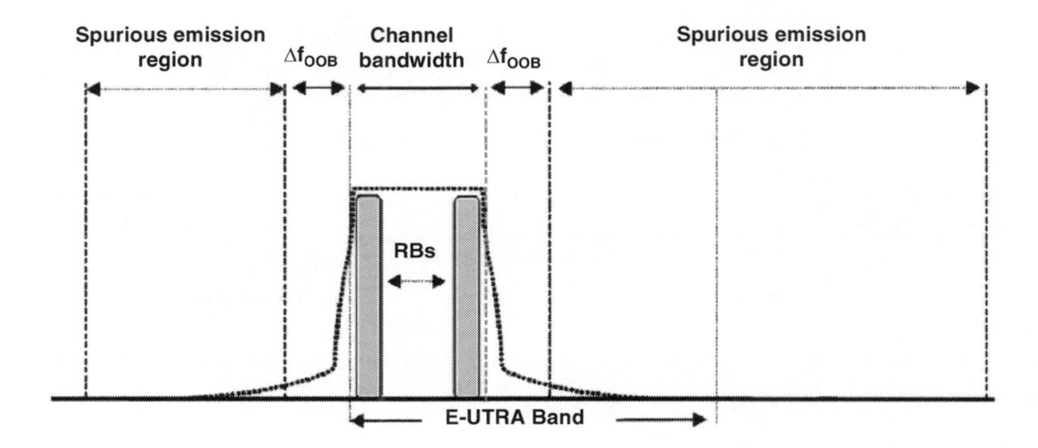

Figure 21.5: Transmitter spectrum. Reproduced by permission of © 3GPP.

the emission requirements may change in different scenarios. If the transmitter cannot satisfy a particular requirement, it can either switch the transmitter off, or adapt the transmitter characteristics (for example by reducing the transmitter power or using software-defined radio techniques) to alter the performance in the required manner. This is explained in more detail in Section 21.3.3.

21.3.2.1 Out-of-Band Emissions

Since OOB emissions occur close to the wanted transmission, increasing the power level of the wanted transmission will usually increase the level of the unwanted emissions. Conversely, reducing the transmitted power is usually an effective method for reducing the OOB emissions, thus providing one possible method for responding to network-signalled requirements as discussed above.

OOB emissions may be an almost inevitable by-product of the modulation process itself, and are also often caused by non-linearities in PAs. A summary of the main practical transmission impairments which can cause OOB emissions is given in Section 21.5.1.

In previous fixed-bandwidth radio systems (including UMTS), the OOB emissions requirements have been defined with respect to the centre frequency of the transmission. As the bandwidth of the LTE system is variable, it is more convenient to define OOB requirements with respect to the edge of the channel bandwidth rather than the centre of the channel.

In LTE, OOB emissions are defined by means of Spectrum Emission Masks (SEMs) and Adjacent Channel Leakage Ratio (ACLR) requirements. The SEM has a much narrower reference bandwidth than the ACLR, which is a stricter requirement.

Spectrum Emission Mask

The Spectrum Emission Mask (SEM) is a mask defined for out-of-channel emissions relative to the in-channel power. The spectrum emission mask of the UE applies to frequencies within Δf_{OOB} (MHz) of the edge of the assigned LTE channel bandwidth, as shown diagrammatically in Figure 21.5. Figure 21.6 shows the basic SEM requirement for a UE transmitter.

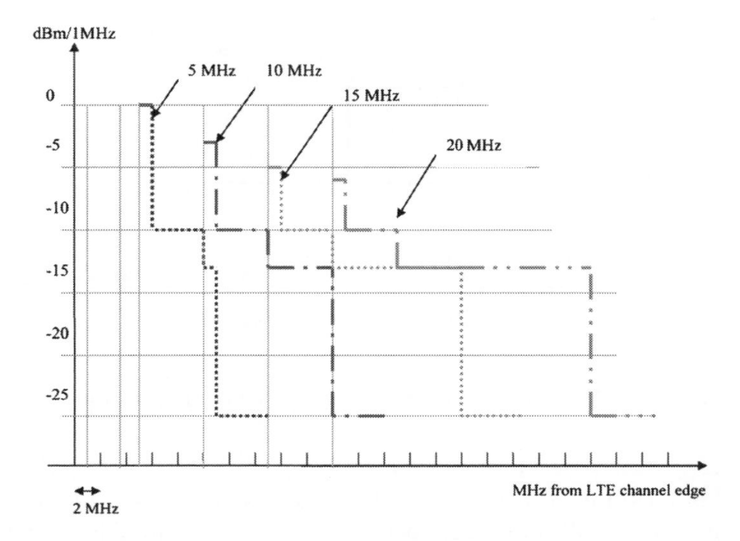

Figure 21.6: Spectrum Emission Mask (SEM) for a UE transmitter, for each channel bandwidth.

LTE also provides a set of 'Additional SEMs' (A-SEMs) which the network may instruct the UE to use in particular deployment scenarios.[8] Such an instruction may, for example, be signalled on handover to a new cell [5].

For the eNodeB, the SEM is defined in a similar manner; an example is given in Figure 21.7. It should again be noted that this represents only the basic requirement, and there are many conditions and additional cases to be taken into account in different circumstances [6].

Adjacent Channel Leakage Ratio

The second method used to measure OOB emissions in LTE is the ACLR. While the SEM measures the performance of the transmitter, the ACLR measures the power which actually leaks into certain specific nearby radio channels and thus estimates how much a neighbouring radio receiver will be affected by the OOB emissions from the transmitter.

The ACLR is defined as the ratio of the filtered mean power centred on the assigned channel frequency to the filtered mean power centred on an adjacent channel frequency.

The LTE specifications not only set ACLR requirements for adjacent 20 MHz LTE channels, but also for UMTS channels which may be located in two adjacent 5 MHz channels (i.e. a total of 10 MHz).

[8]These are Network Signals 03, 04, 06 and 07.

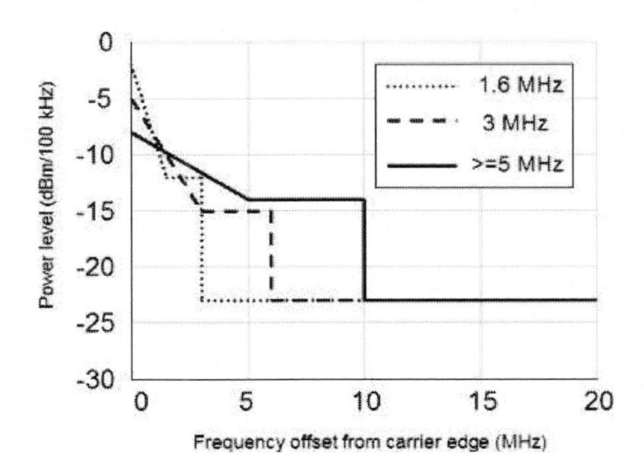

Figure 21.7: Spectrum emission mask for eNodeB transmitter.

The ACLR of a UE for an adjacent LTE channel is required to be >30 dB for the whole 20 MHz Bandwidth, >33 dB for the adjacent UMTS channel deployed in the closest 5 MHz of the OOB portion of the spectrum and >36 dB for the adjacent UMTS channel deployed in the next 5 MHz portion of the OOB spectrum [5].

21.3.2.2 Spurious Emissions

Spurious emissions occur well outside the bandwidth necessary for transmission[9] and may arise from a large variety of non-ideal effects including harmonic emissions and intermodulation products.[10] The magnitude of the spurious emissions may or may not vary with transmitter power.

The basic spurious emissions requirements for an LTE UE are given in Table 21.6.

Table 21.6: Spurious emissions limits.

Frequency range	Maximum level (dBm)	Measurement bandwidth
9 kHz ≤ 150 kHz	−36	1 kHz
150 kHz ≤ 30 MHz	−36	10 kHz
30 MHz ≤ 1000 MHz	−36	100 kHz
1 GHz ≤ 12.75 GHz	−30	1 MHz

[9]The spurious emission limits apply for frequency ranges that are more than Δf_{OOB} (MHz) from the edge of the channel bandwidth – see [5, Section 6.6.3.1].

[10]Intermodulation products are unwanted frequencies generated whenever two or more tones are present in a non-linear device, such as an amplifier or mixer. The possible combinations of generated frequencies can be described by a power series.

These basic requirements are more stringent than the OOB requirements, but are not usually difficult to meet. The main challenge for the transmitter designer lies in some 'additional spurious emissions' requirements which are specified for certain frequency bands. The additional spurious emissions requirements exist to protect the receiver on the UE, colocated receivers, or nearby receivers from the transmission. In the case of FDD radios, the receiver and transmitter operate at the same time, and the receiver relies on the duplex spacing to separate the weak received signal from its strong transmitted signal (see Section 21.4.2 for further details on transmitted signal leakage in the receiver band). If the transmitter emits anything significant on its own receiver frequency, then reception will be blocked or degraded. This results in the additional spurious emissions requirements being very stringent. In addition, some further additional spurious emissions requirements may be signalled by the network if needed for specific deployment scenarios.

21.3.3 Power Amplifier Considerations

The ACLR is typically directly related to the operating point of the power amplifier. In general, leakage into adjacent channels increases sharply as the PA is driven into its non-linear operating region at the highest output power levels, due in particular to the intermodulation products. Consequently it is important that the peak output power of the UE does not cause the PA to enter too far into this region. On the other hand, most PAs are designed to operate efficiently only in a small region at the top of the linear operating region. This corresponds to the *rated* output power of the PA, below which the efficiency falls sharply. Since high efficiency is crucial to ensuring a long battery life for the UE, it is therefore also desirable to keep the PA operating as close as possible to the top of the linear operating region.

If the ACLR and spectrum mask requirements cannot be met, typically the UE output power must be reduced to bring the leakage to acceptable levels. This can be achieved without undue loss of efficiency by reducing the peak output power of the PA – a process known as *de-rating*. As discussed in Section 21.3.1.2, the LTE specifications provide an MPR for this purpose. The amount of de-rating required is highly dependent on the Peak to Average Power Ratio[11] (PAPR) and bandwidth of the transmitted signal. For example, for any given channel bandwidth and PA, transmissions with a larger occupied bandwidth create more OOB emissions, resulting in larger adjacent channel leakage than transmissions with lower occupied bandwidth; this is shown in Figure 21.8, in which WCDMA and LTE occupied bandwidths and adjacent channel leakage are compared in a 5 MHz channel.

The increase in OOB emissions from the larger occupied bandwidth of the LTE signal is mainly due to increased adjacent channel occupancy by 3rd and 5th order InterModulation products (called 3rd and 5th order IM plateau in Figure 21.8).

For an LTE uplink transmitted signal, certain combinations of RB allocations and modulation schemes create more OOB emissions than others. For QPSK and 16QAM the number of RBs which can be allocated at the edge of the channel for a given amount of power de-rating is shown in Figure 21.9. For applications such as VoIP, which typically use low-bandwidth transmissions with QPSK modulation, no power de-rating is required from the normal rated maximum UE output power (see Section 21.3.1.2). This helps to ensure

[11]The concept of peak to average power ratio is discussed later in this section and in Section 5.2.2.

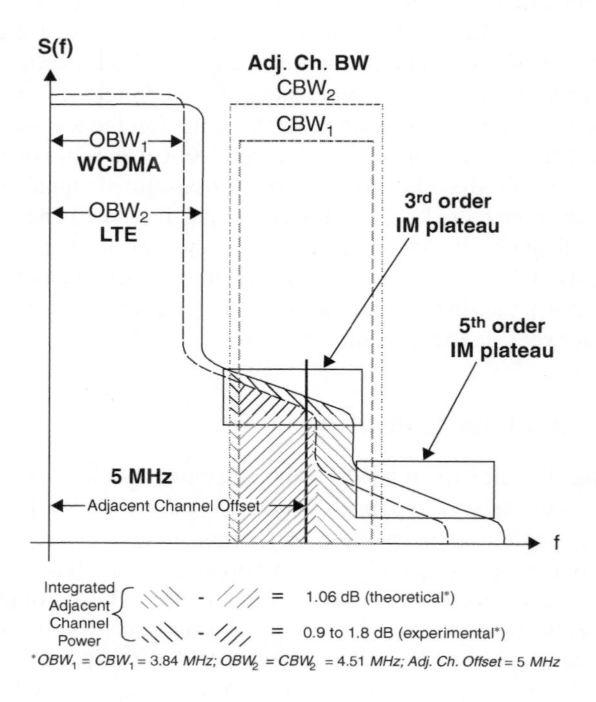

Figure 21.8: Adjacent channel power increase (2–3 dB) for 4.51 versus 3.84 MHz Occupied BandWidth (OBW) [9]. Reproduced by permission of © 2006 Motorola.

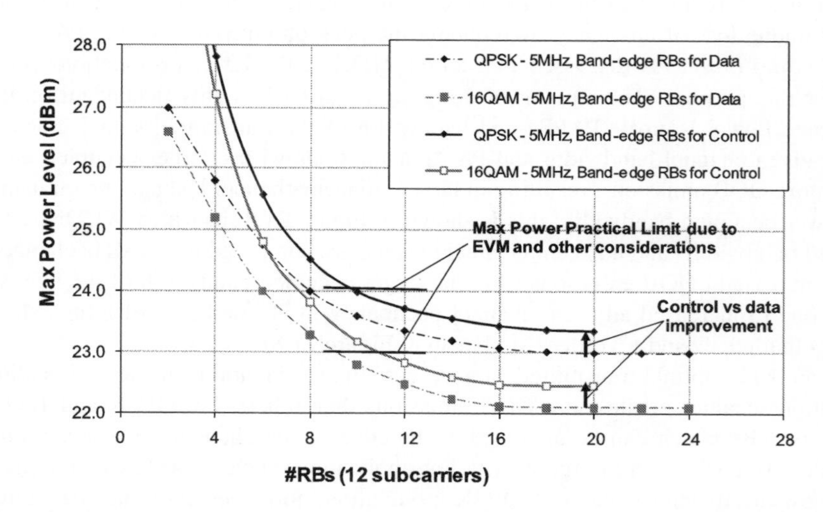

Figure 21.9: Maximum power level versus the number of RBs which can be allocated. Reproduced by permission of © 2006 Motorola.

wide-area coverage for such applications, since the full rated output power of the PA can be used to counteract path-loss at the cell edge.

Moreover, the structure of the LTE uplink signal, with control information being usually situated at the channel edge and high-bandwidth data transmissions toward the middle of the band (see Section 16.3), helps to improve OOB emissions and reduce ACLR.

Taking the above considerations into account, the Total Power De-rating (TPD) required to meet a given ACLR requirement can be broken down into an element corresponding to the occupied bandwidth (as a proportion of the channel bandwidth), termed here the Occupied Bandwidth Power De-rating (OBPD), and an element corresponding to the waveform of the transmitted signal, termed here the Waveform Power De-rating (WPD).

OBPD can be approximately expressed as a ratio of the transmitted Occupied Bandwidth (OBW) and the Reference Occupied Bandwidth ($\mathrm{OBW_{ref}}$) [10]:

$$\mathrm{OBPD} \propto 10 \log_{10}\left(\frac{\mathrm{OBW}}{\mathrm{OBW_{ref}}}\right) \tag{21.2}$$

The other element of the required TPD, the WPD, arises in a large part from the dynamic range of the signal. This is usually quantified in terms of its *Peak to Average Power Ratio* (PAPR) and *Cubic Metric* (CM). These measures are typically used as indications of how much PA power headroom is required (i.e. how far from the rated power the PA generally has to operate) to avoid entering the non-linear region of operation. A high PAPR means that on average the operating point of the PA has to be lower in order to avoid the non-linear region and achieve the required ACLR. This results in a reduction in efficiency. A rule of thumb is that for each 1 dB increase in required power amplifier headroom, a corresponding 10 to 15% increase in PA current drain occurs, leading to a corresponding reduction in battery life. De-rating the PA is therefore usually a preferable strategy (as opposed to increasing the power headroom), since it achieves the required reduction in adjacent channel leakage by reducing the total maximum output power (including that of the wanted signal), while allowing the PA still to operate partly in the non-linear region. De-rating therefore enables a UE to meet the OOB emission requirements without a loss of efficiency, at the expense of a loss in coverage due to the reduced power in the wanted channel.

The CM of a waveform can be of interest because it characterizes the effects of the 3rd order (cubic) non-linearity of a PA on the waveform of interest relative to a reference waveform, in terms of the power de-rating needed to achieve the same ACLR as that achieved by the reference waveform at the PA's rated maximum power level. As discussed in Chapter 14, one criterion which was instrumental in selecting SC-FDMA instead of OFDMA for the LTE uplink was the low inherent CM of the SC-FDMA waveform compared to OFDMA, as shown in Figure 21.10, thus enabling more efficient operation for a given ACLR and maximum total output power.

However, it should be noted that the CM does not predict the spectral location of the non-linear distortion, and it is possible that two waveforms with different RB allocations have the same CM but one creates worse out-of-band emissions than the other; therefore, CM has not been defined in the LTE specifications.

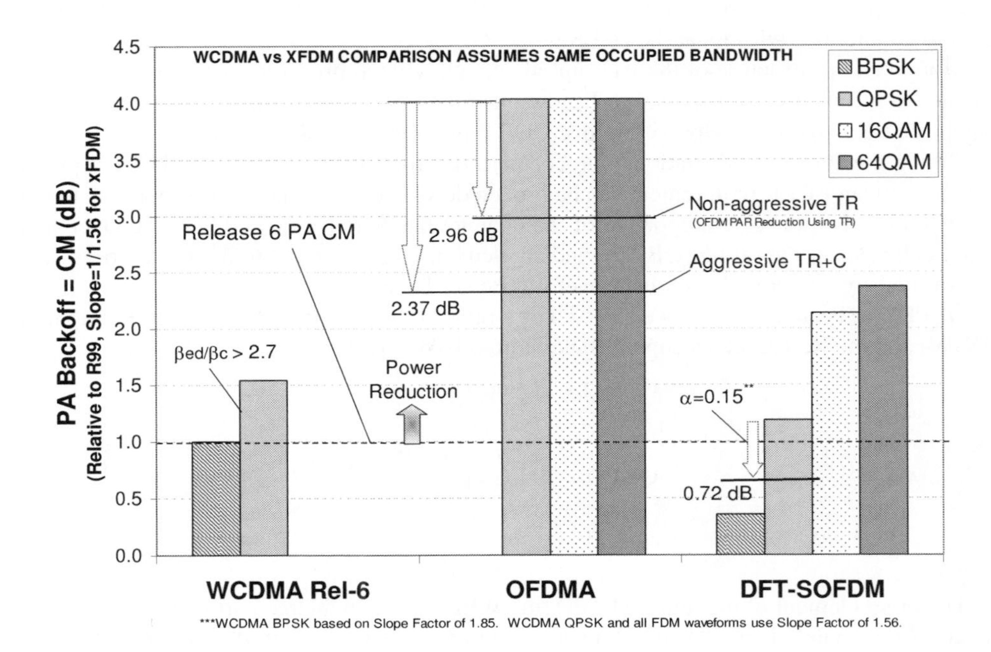

Figure 21.10: Cubic Metric (CM) for different waveforms.
Reproduced by permission of © 2006 Motorola.

21.4 Receiver RF Requirements

Like the transmitter, the LTE receiver requirements are based largely on those of UMTS, with many of the RF requirements being identical or similar. The general intention is for the implementation of an LTE receiver not to be significantly more complex than UMTS, so as to reduce redesign effort. An underlying assumption is that there is no need for tightening the coexistence specifications, as the basic cellular scenarios remain similar.

The main differences between the LTE and UMTS RF receiver requirements arise from the variable channel bandwidth and the new multiple access schemes.

The emphasis in this section is on the UE receiver requirements for FDD operation.

21.4.1 Receiver General Requirements

The receiver requirements are based on a number of key assumptions for testing purposes:

- The receiver has integrated antennas with a 0 dBi gain.
- The receiver has two antenna ports. Requirements for four ports may be added in the future.
- Test signals of equal power level are applied to each port, with Maximum Ratio Combining (MRC) being used to combine the signals. It is assumed that the signals

come from independent Additive White Gaussian Noise (AWGN) channels, so that signal addition gives a 3 dB diversity gain.

Although LTE supports a variety of Modulation and Coding Schemes (MCSs) (see Chapter 10), the RF receiver specifications are defined for just two MCSs (referred to as 'reference channels'), near the extremes of the available range, in order to reduce the number of type approval tests which have to be performed. The 'low SNR' reference channel uses QPSK with a code rate of 1/3, while the 'high SNR' reference channel uses 64QAM with a code rate of 3/4. For each of these reference channels, the SINR requirements at which 95% throughput should be achieved are specified.

21.4.2 Transmit Signal Leakage

When an LTE receiver is tested in full-duplex FDD mode, the transmitter must also be operating so that signal leakage from the transmitter is taken into account. This does not apply to half-duplex FDD or to TDD operation, which are discussed in detail in Chapter 23.

An FDD receiver's exposure to the transmitted signal from the same equipment is largely due to insufficient isolation within the duplexer. The power of the leakage of the transmitted signal is proportional to the transmitted power.

Not only can the power of the fundamental components of the transmitted signal interfere with the receiver, but also the OOB phase noise of the transmitted signal can fall into the receive band. In LTE the spurious emissions from the UE transmitter in its own receive band are required to be −47 dBm or lower, measured in a 1 MHz bandwidth. This corresponds to −107 dBm/Hz. The maximum transmit power for a mobile device is +23 dBm (power class 3), so the spurious emissions requirement is −130 dBc/Hz. UMTS has a tougher requirement of −60 dBm spurious emissions in 3.84 MHz bandwidth, which is −125.8 dBm/Hz or −149.8 dBc/Hz for Class 3 +24 dBm output power. However, for UMTS the toughest requirement is in one specific receive band ('Band 8') which requires −153 dBc/Hz (when considering the maximum output power). The transmitter may therefore be designed to achieve this more broadly, including its own receive band.

For both UMTS and LTE, receiver sensitivity is measured with the transmitter operating at the full power for its class. In order to allow the transmitter to degrade the receiver sensitivity by no more than 0.5 dB, the transmitter noise needs to be 9 dB below the Noise Floor (NF) at the antenna (NF −9 dB = −183 dBm/Hz). If we assume that the spurious emissions of the transmitter in its receive band are just at the limit of the specifications, it can be shown that the duplex isolation needs to be at least 78 dB for LTE and 59 dB for UMTS.[12]

However, making duplexers smaller and cheaper is often achieved at the expense of isolation, and typical duplexers might provide just 45 to 55 dB isolation. Potentially, therefore, the transmit signal could reduce receiver sensitivity substantially. This is not acceptable, so within the receive band the LTE transmitter has to achieve spurious emissions about 20 dB lower than the transmitter specifications require.

The transmit power at the input to the Low Noise Amplifier (LNA) is equal to the transmit power at the antenna plus the duplexer loss from transmit port to antenna (about 2 dB) minus the duplexer transmit-to-receiver isolation (assume 50 dB). Therefore, at the maximum LTE

[12]The duplex isolation is obtained as NF − 9 dB − 2 dB − SE, where NF is the noise floor (−174 dBm/Hz), 9 dB comes from the degradation of the receiver sensitivity, 2 dB comes from the loss from transmit port to antenna and SE is the Spurious Emission requirement.

UE output power of 23 dBm, the mean transmitter power leaking back to the LNA, $P_{\text{Tx\_leak}}$ is

$$P_{\text{Tx\_leak}} \simeq 23 + 2 - 50 = -25 \text{ dBm}. \tag{21.3}$$

The equivalent figure for UMTS would be 1 dB higher.

However, most of the receiver specifications are defined with the transmitter power at 4 dB below maximum, so then the transmit leakage could be −29 dBm. This value is for the average transmit power, but the transmitted signal contains amplitude modulation and therefore the peak signal will be higher. This scenario is illustrated in Figure 21.11 which shows a simplified block diagram of the transceiver with direct-conversion architectures for both receiver and transmitter.

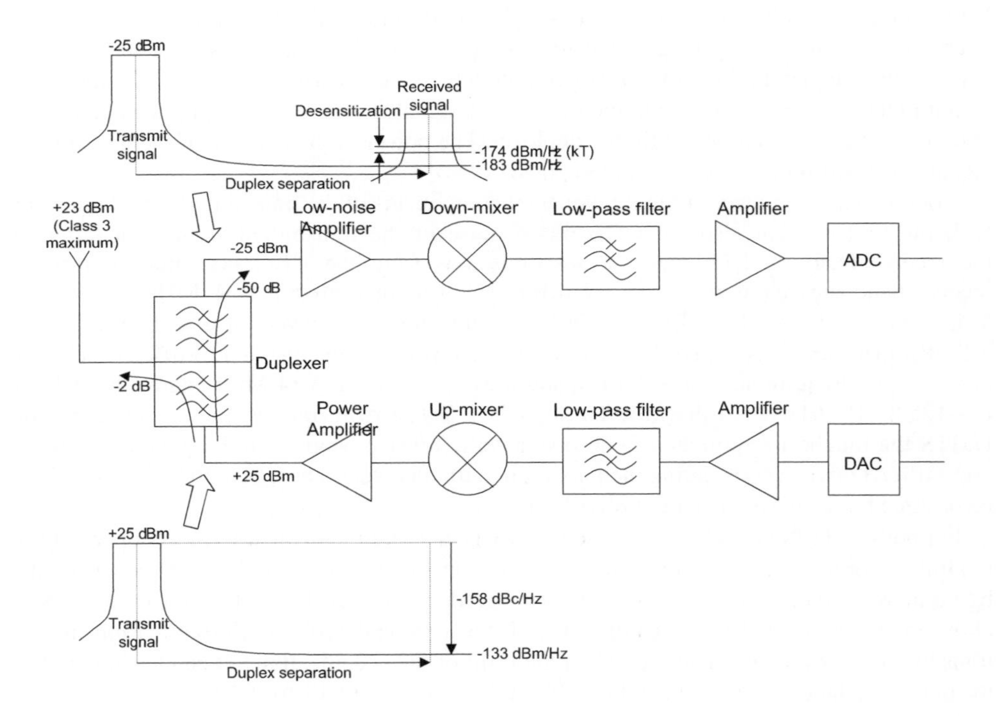

Figure 21.11: Transmit leakage problem in LTE FDD UE transceiver.

The transmit leakage problem presents a particularly challenging requirement if applied to RF bands with a small duplex separation. In FDD, when performing a selectivity or blocking test with a single interfering signal in one of the adjacent channels, the receiver is also exposed to its own transmitted signal as a second interferer, which, in the worst scenario, can cause cross-modulation to occur, in which the interferers mix together to generate in-band intermodulation distortion products.

21.4.3 Maximum Input Level

The maximum input level is the maximum mean received signal strength, measured at each antenna port, at which there is sufficient SINR for a specified modulation scheme to meet a minimum throughput requirement. Some wireless standards also specify a higher level at which it is guaranteed that no damage will be done to the receiver, but this is not covered in the 3GPP specifications. Although in UMTS the maximum input level is defined for the high and low SINR reference channels, in LTE it is only specified for the high SINR reference channel, on the assumption that the high SINR reference channel is about 4 dB more demanding due to the higher PAPR of the signal. The downlink requirement is for a maximum average input level of −25 dBm at any channel bandwidth with full RB allocation. For FDD operation, the requirement must be met when the transmitter is set at 4 dB below the maximum output power. The corresponding UMTS requirement is the same, applied to the combination of all the channels rather than full RB allocation.

The peak maximum input level is higher than the maximum average input level by an amount corresponding to the PAPR of the signal. PAPR is a quantity which increases the longer the duration of the measurement. If PAPR is measured on a bit-by-bit basis and plotted as a complementary cumulative probability distribution, then for the LTE downlink it reaches a value of around 11 dB for QPSK and 11.5 dB for 64QAM, over a window of 10^6 bits. In OFDMA, the PAPR is dominated by the multicarrier nature of the signal, and therefore does not vary much with modulation. The UMTS downlink signal consists of a combination of CDMA codes which also leads to a relatively high PAPR of around 8 dB.

The downlink peak maximum input level is therefore about −13.5 dBm for LTE and −17 dBm for UMTS.

As explained above, the maximum input level requirement must be satisfied with the transmitter operating at 4 dB below the maximum output power. For a UMTS UE, this can result in a mean transmit signal leakage into the receiver of −28 dBm,[13] while the typical uplink PAPR of 3.4 dB results in a peak transmit signal leakage of of −24.6 dBm. For LTE, the mean leakage from the uplink transmitter can be −29 dBm, resulting in a peak transmitter leakage of up to about −21.5 dBm for QPSK (or −20 dBm for 64QAM[14]). The peak value of the total received power is the combination of the received input signal and the leakage from the transmitter and thus reaches a peak of about −12 dBm for LTE and −15.2 dBm for UMTS.

Under the hypothesis considered above, the transmitter leakage raises the total received signal power by less than 2 dB.

Some operating margin is needed between this total peak received signal power and the non-linear operating region of the amplifiers in the receiver; typically this margin is chosen relative to the 1 dB compression point of the amplifiers, at which the output power is 1 dB below the expected linear gain.

[13]The mean transmit leakage is obtained from Equation (21.3) as the maximum output power (24 dBm for UMTS or 23 dB for LTE) minus 4 dB because the transmitter power is set at 4 dB below maximum, plus the duplexer loss from transmit port to antenna (about 2 dB) minus the isolation between transmitter and receiver (assume 50 dB).

[14]Note that only the highest category of LTE UE supports 64QAM transmission in the uplink.

21.4.4 Small Signal Requirements

21.4.4.1 SINR Requirements for Adaptive Modulation and Coding

The LTE specifications will define requirements for the demodulation error rate of the different modulation and coding schemes. An extra Implementation Margin (IM) is included to account for the difference in SINR requirement between theory and practicable implementation. This can include degradation of the signal due to any (digital) processing of the signal before the demodulator (such as filtering and re-sampling) and the use of a non-ideal demodulator, as well as the diversity gain being less than 3 dB. A 2.5 dB IM has been defined for QPSK with a 1/3-rate code in low SINR conditions; the IM will be higher for other modes. In general, 2.5 dB is a reasonable implementation margin for all QPSK modes, while 3 dB and 4 dB could be expected for 16QAM and 64QAM respectively.

Typical assumptions for the SINR values for different modulation and coding schemes are given in Table 21.7.

Table 21.7: Downlink SINR requirements.

System	Modulation	Code rate	SINR (dB)	IM (dB)	SINR+IM (dB)
LTE UE	QPSK	1/8	−5.1	2.5	−2.6
		1/5	−2.9		−0.4
		1/4	−1.7		0.8
		1/3	−1		1.5
		1/2	2		4.5
		2/3	4.3		6.8
		3/4	5.5		8.0
		4/5	6.2		8.7
	16QAM	1/2	7.9	3	10.9
		2/3	11.3		14.3
		3/4	12.2		15.2
		4/5	12.8		15.8
	64QAM	2/3	15.3	4	19.3
		3/4	17.5		21.5
		4/5	18.6		22.6
UMTS UE	QPSK	1/3	1.2	2	3.2

21.4.4.2 Thermal Noise and Receiver Noise Figure

In the LTE specifications the thermal noise density, kT, is defined to be −174 dBm/Hz where k is Boltzmann's constant (1.380662×10^{-23}) and T is the temperature of the receiver (assumed to be 15°C). No account is taken of the small variations in temperature over normal operating conditions (typically + 15° to +35°C) or extreme operating conditions (−10° to +55°C).

kTB represents the thermal noise level in a specified noise bandwidth B, where $B = N_{RB} \times$ 180 kHz in LTE, N_{RB} is the number of RBs and 180 kHz is the bandwidth of one RB.

The receiver Noise Figure (NF) is a measure of the SINR degradation caused by components in the RF signal chain. This includes the antenna filter losses, the noise introduced by the analogue part of the receiver, the degradation of the signal due to imperfections of the analogue part of the receiver (such as I/Q imbalance), the noise introduced by the Analogue to Digital Converter (ADC) and any other noise sources. The NF is the ratio of actual output noise to that which would remain if the receiver itself did not introduce noise.

LTE defines an NF requirement of 9 dB for the UE, the same as UMTS. This is somewhat higher than the NF of a state-of-the-art receiver, which would be in the region of 5–6 dB, with typically about 2.5 dB antenna filter insertion loss and an NF for the receiver integrated circuit of 3 dB or less. Thus, a practical 3–4 dB margin is allowed. The eNodeB requirement is for an NF of 5 dB.

21.4.4.3 Reference Sensitivity

The reference sensitivity level is the minimum mean received signal strength applied to both antenna ports at which there is sufficient SINR for the specified modulation scheme to meet a minimum throughput requirement of 95% of the maximum possible. It is measured with the transmitter operating at full power.

REFerence SENSitivity (REFSENS) is a range of values that can be calculated using the formula:

$$\text{REFSENS} = kTB + \text{NF} + \text{SINR} + \text{IM} - 3 \ \ \text{(dBm)}$$

where kTB is the thermal noise level introduced above, in units of dBm, in the specified bandwidth (B), NF is the prescribed maximum noise figure for the receiver, 'SINR' is the SINR requirement for the chosen modulation and coding scheme, IM is the implementation margin and −3 dB represents the diversity gain.

For UMTS, the NF, SINR and IM are not specified separately, but the total is 12 dB. Typical REFSENS values can be calculated using the assumptions outlined above for a few example cases, as in Table 21.8. The REFSENS requirements can be found in [5] Table 7.3.1-1.

REFSENS is plotted for an LTE UE over the complete range of bandwidths and likely modulation and coding schemes in Figure 21.12.

Effect of receiver sensitivity on coverage

Using the path-loss equations described in Section 20.2, and knowing the downlink transmit power, the effect of receiver sensitivity on the practical operating coverage can be estimated. An example is shown in Figure 21.13, for the downlink with an eNodeB power of +46 dBm for a 5 MHz channel, ~900 MHz RF frequency (Band 5) and REFSENS of -100 dB and -80 dB for QPSK rate 1/3 and 64QAM rate 3/4 respectively. The corresponding throughput versus range for different MCSs is illustrated in Figure 21.14.

Table 21.8: Example of REFSENS computation for band 1

System	Modulation	Channel BW (MHz)	*kTB* (dBm)	NF (dB)	SINR (dB)	IM (dB)	REFSENS (dBm)
LTE UE	QPSK 1/3	5	−107.5	9	−1	2.5	−100
	QPSK 1/3	20	−101.4	9	−1	2.5	−94
	64QAM* 3/4	5	−107.5	9	17.5	4	−80
	64QAM* 3/4	20	−101.4	9	17.5	4	−74
LTE BS	QPSK 1/3	5	−107.5	5	1.5	2.5	−101.5
UMTS UE	QPSK 1/3	3.84	−108.2	9	1.2–21.1 (21.1 dB spreading gain)	2.5	−117

* Note that the REFSENS is specified only for QPSK modulation.

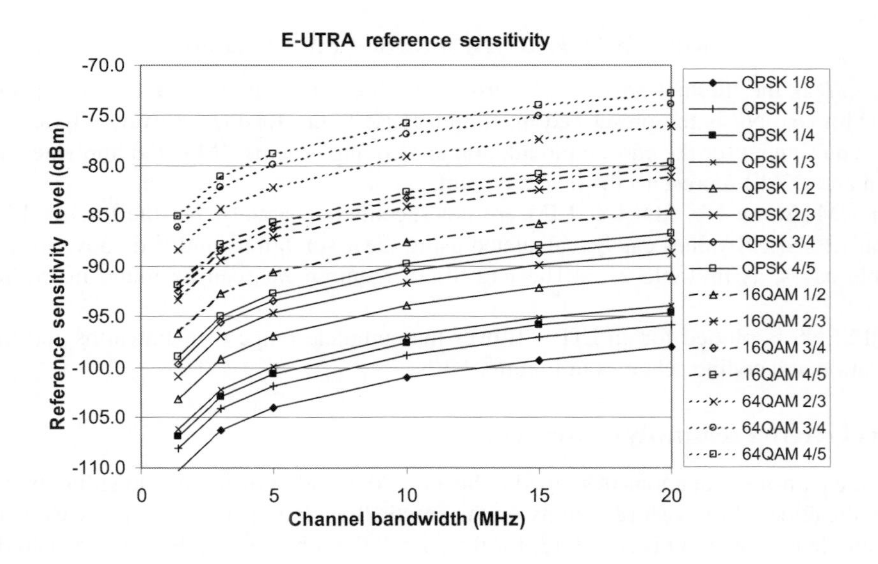

Figure 21.12: LTE UE reference sensitivity.

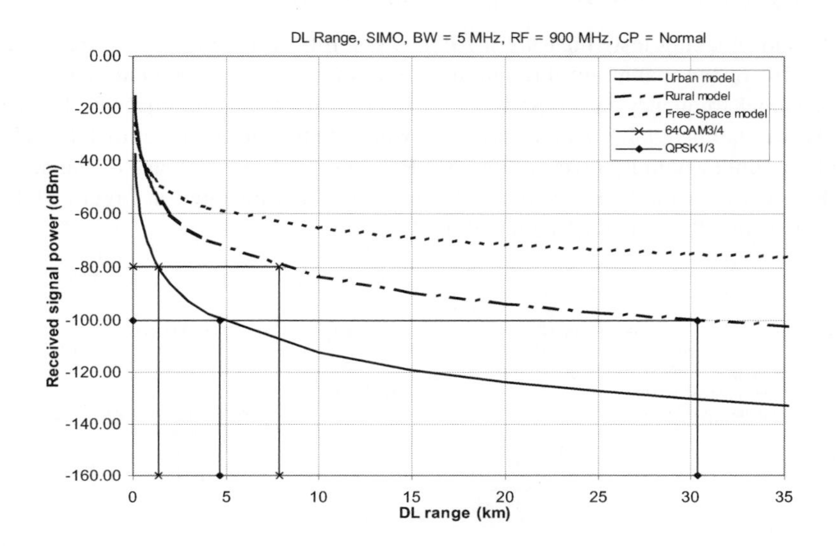

Figure 21.13: Downlink coverage taking receiver sensitivity into account.

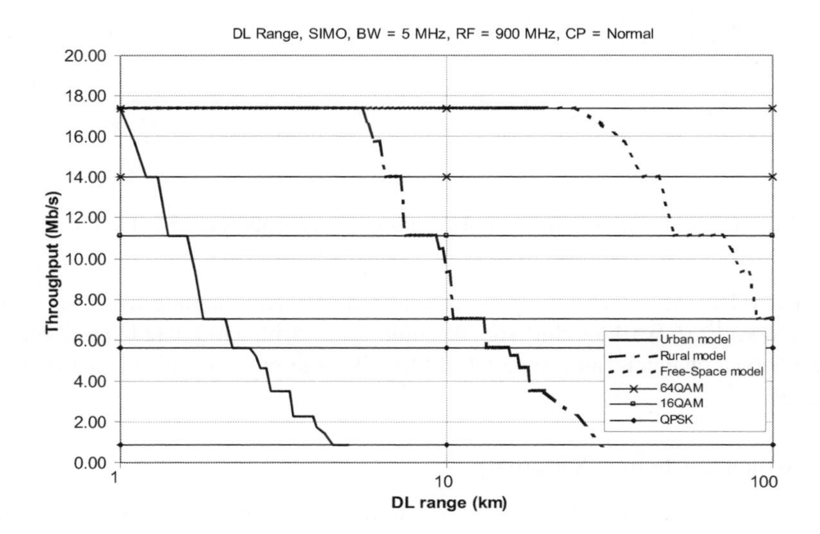

Figure 21.14: Throughput versus coverage.

21.4.5 Selectivity and Blocking Specifications

Selectivity and blocking tests measure a receiver's ability to receive the wanted signal (to achieve ≥ 95% of the maximum throughput of the reference measurement channels) at its assigned channel frequency in the presence of interfering signals in adjacent channels and beyond. As usual, this requirement must be met when the transmitter is set to 4 dB below the supported maximum output power. A low SINR is assumed. A summary of the selectivity and blocking tests for the LTE UE in a 5 MHz channel is illustrated in Figure 21.15. The requirements generally scale with bandwidth.

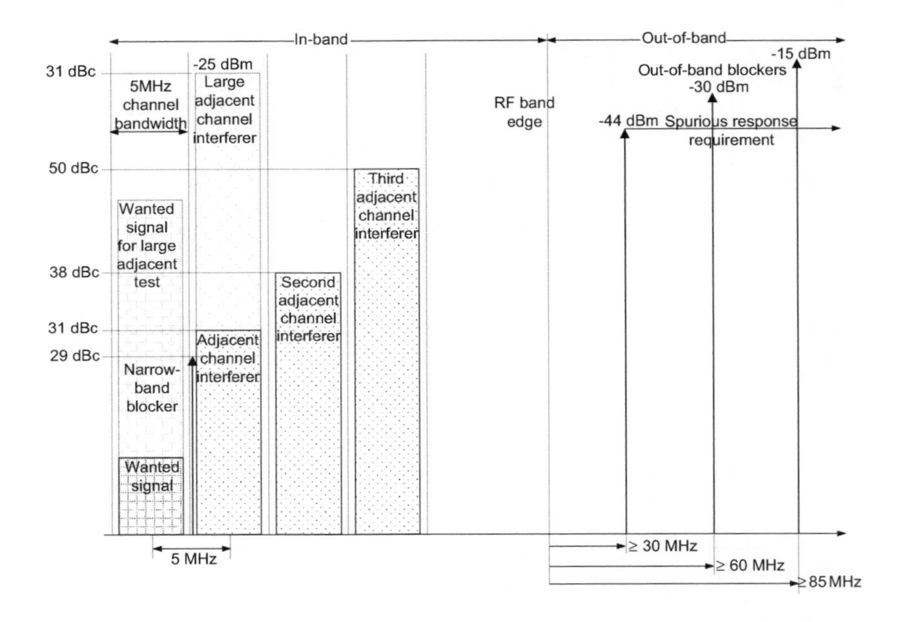

Figure 21.15: Selectivity and blocking requirements for a 5 MHz UE.

As with UMTS, the selectivity and blocking requirements include the case of a close Continuous-Wave[15] (CW) blocking signal, some cases with modulated interferers in the first three adjacent channels, and some OOB blocker requirements with a spurious response allowance [5].

21.4.5.1 Adjacent Channel Selectivity

Adjacent Channel Selectivity (ACS) is a measure of a receiver's ability to receive a wanted signal at its assigned channel frequency in the presence of an adjacent channel interfering signal at a given frequency offset from the centre frequency of the assigned channel, without the interfering signal causing a degradation of the receiver performance beyond a specified

[15] A 'continuous wave' signal is an unmodulated tone.

limit. ACS is predominantly defined by the ratio of the receive filter attenuation on the assigned channel frequency to the receive filter attenuation on the adjacent channel.

For LTE, ACS is defined following the same principles as UMTS and requiring similar performance capability up to a 10 MHz bandwidth, but is more relaxed for 15 and 20 MHz bandwidths. The LTE ACS is defined for each bandwidth using a modulated LTE signal as the interferer, and is only defined for low SINR conditions.

In order to check the ability of the receiver to handle the full required dynamic range, the ACS requirement is specified for two cases – a small adjacent channel interferer power and a large adjacent channel interferer power. These cases are explained in more detail below.

ACS with a small adjacent interferer

In this ACS case, the wanted signal is, like in UMTS, 14 dB above REFSENS (given in Table 21.8) and therefore takes a different absolute level for each bandwidth. The Carrier-to-Interference Ratio (C/I) is set at −31.5 dB for bandwidths up to 10 MHz and at -28.5 dB and -25.5 dB for 15 and 20 MHz respectively; the ACS is quoted as being 33 dB including a 2.5 dB implementation margin for bandwidths up to 10 MHz and relaxed to 30 dB and 27 dB for 15 and 20 MHz respectively.

Up to a bandwidth of 5 MHz, the bandwidth of the interferer is the same as the bandwidth of the wanted signal. Above 5 MHz, the bandwidth of the interferer stays at 5 MHz, which means that the RF test equipment does not need to be able to generate a wide bandwidth interferer. For bandwidths above 5 MHz the interferer is positioned at the near edge of the channel. The consequence of this is that the interferer power is concentrated at the edge of the adjacent channel at which the filters used in the receiver have least selectivity. The digital filters will have a sharp cut-off because they are designed for OFDM, but the analogue filters in the RF front-end of the receiver could have much less attenuation at the near edge of the channel, which will push up the dynamic range at the ADC. To compensate, the ACS is relaxed by 3 dB and 6 dB for the 15 MHz and 20 MHz modes respectively as mentioned above.

Level diagrams for ACS requirements for two bandwidths (5 and 20 MHz) are shown in Figure 21.16.

The gap between the edge of the wanted signal and the edge of the interferer is a function of the used bandwidth of both signals. Given that both use a full allocation of RBs, this gap can be calculated as 1.25 MHz for 20 MHz bandwidth reducing to a minimum of 300 kHz for 3 MHz bandwidth. Below this bandwidth the channel usage reduces, so the gap increases. Note that in percentage terms, a 1.25 MHz gap adjacent to a 20 MHz channel is actually smaller than a 300 kHz gap adjacent to a 3 MHz channel (6.25% compared to 10%). The channel filters should be designed to scale with bandwidth, so it is this percentage ratio which determines the filter roll-off requirements. Comparing the 20 and 5 MHz modes we see that the percentage has roughly doubled but the specification is relaxed by 6 dB, so the filter roll-off requirements should be similar.

ACS with a large adjacent interferer

The large adjacent interferer requirement uses an adjacent channel interferer power of −25 dBm. Similarly to the case of the small adjacent interferer, the SIRC/I is fixed at −31.5 dB for bandwidths of 10 MHz and below (i.e. the wanted signal power −56.5 dBm)

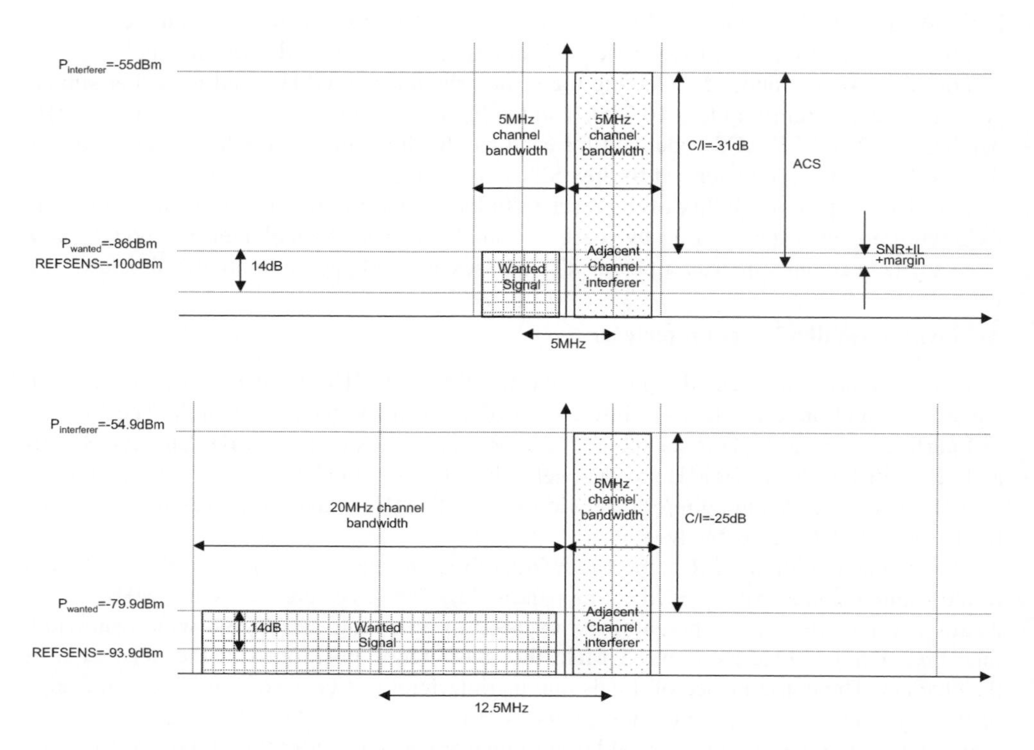

Figure 21.16: Adjacent channel selectivity illustrated for Case 1 [5].

and −28.5 dB and −25.5 dB for 15 and 20 MHz respectively (i.e. the wanted signal power is −53.5 dBm and −51.5 dBm). The margin between the wanted signal power and REFSENS varies for each bandwidth in the range 41.0 to 50.2 dB, which is well above the noise floor and therefore not of much significance (see [5, Section 7.5, ACS Case 2]).

In practice, the toughest test for the receiver is likely to be somewhere between the small interferer and large interferer ACS tests. Assuming that the dynamic range of the receiver is limited, there is a point at which the interferer power first becomes high enough to require that the front-end gain needs to be reduced. This gain reduction will degrade the noise figure of the receiver at a point at which the wanted signal power is not very high. The gain control algorithm used by the receiver must be well planned to avoid such problems.

Adjacent Channel Interference Ratio (ACIR)

The ACS and ACLR (see Section 21.3.2.1) together give the Adjacent Channel Interference Ratio (ACIR). The ACIR is the ratio of the total power transmitted from a source to the total interference power affecting a victim receiver, resulting from both transmitter and receiver imperfections.

It follows that

$$\text{ACIR} \cong \frac{1}{\dfrac{1}{\text{ACLR}} + \dfrac{1}{\text{ACS}}}$$

ACLR and ACS have been extensively used for coexistence studies.

21.4.5.2 Narrowband Blocking (in Adjacent Channel)

The blocking characteristic is a measure of the receiver's ability to receive a wanted signal (to achieve $\geq 95\%$ of the maximum throughput of the reference measurement channels) at its assigned channel frequency in the presence of an unmodulated unwanted interferer on frequencies other than those of the spurious response or the adjacent channels, without this unwanted input signal causing a degradation of the performance of the receiver beyond a specified limit. The blocking performance applies at all frequencies except those at which a spurious response occurs.

The narrowband blocking specification is a severe test of the receiver's ability to reject 3rd order intermodulation products resulting from cross-modulation of the transmitter leakage which appears around the narrowband blocker. The frequency of the unwanted cross-modulation product depends only on the narrowband blocker frequency and not on the frequency of the transmitter, or any other modulated blocker.

The LTE 'narrowband' blocking performance requirement uses a CW interferer very close to the wanted signal at an offset less than the nominal channel spacing. For such small offsets nearly half of the transmitter leakage will appear in-band.

The CW blocker is positioned at approximately 200 kHz from the near edge of the adjacent channel. For example, for a 5 MHz bandwidth the offset is ~2.7 MHz, which is 450 kHz from the edge of the wanted signal. The offset of the blocker from the edge of the wanted signal reduces with bandwidth, reaching just 350 kHz for the 3 MHz channel bandwidth. Below 3 MHz bandwidth the channel occupancy reduces, which compensates for the reduced gap.

For this test, for bandwidths of 10 MHz and below, the power of the wanted signal is set at the REFSENS level plus a bandwidth-specific offset (22, 18, 16, 13, 14 and 16 dB for 1.4, 3, 5,10, 15 and 20 MHz respectively – see [5, Section 7.6.3],); the blocker power is set to −55 dBm independently from the bandwidth. Compared to the ACS test, the gap between the wanted and interfering signals is between 30 and 50 kHz less, which makes the narrowband blocking test a little more demanding than the ACS test.

In UMTS the wanted signal power is 13 dB higher than the REFSENS and the interferer power is set to −56 dBm. The most important difference between the UMTS narrowband blocking specification and the LTE case is that the blocker for UMTS is actually a GMSK[16] modulated signal, not CW. A GMSK signal is a little narrower than QPSK (although clearly not as narrow as a CW signal), and the modulation has a constant envelope so there is no PAPR variation which could increase non-linear distortion. On balance, it can be concluded that the UMTS and LTE narrowband blocking specifications are similarly demanding.

21.4.5.3 Non-Adjacent Channel Selectivity (In-Band Blocking)

Non-adjacent Channel Selectivity (NACS) is a measure of the receiver's ability to receive a wanted signal at its assigned channel frequency in the presence of unwanted interfering

[16]Gaussian Minimum-Shift Keying, as used in GSM.

signals falling into the receive band beyond the adjacent channel or at less than 15 MHz offset from the edge of the receive band. The interfering signals are modulated and occupy the same bandwidths as specified for ACS. The LTE specifications refer to this test as 'in-band blocking', although, unlike the other blocking tests, NACS does not use CW signals.

There are two requirements to be met, the first with an interferer of −56 dBm[17] in the second adjacent channel or further (referred to as Case 1), the second with an interferer of −44 dBm[18] in the third adjacent channel or any larger frequency offset up to 15 MHz out of band (referred to as Case 2). Furthermore, a specific requirement which applies to assigned UE channel bandwidth of 5 MHz and for Band 17 (referred to as Case 3) is also provided [5].

Unlike the ACS tests, NACS does not need to be repeated at higher signal levels, so it does not test dynamic range to the same extent. However, the wanted signal level is much lower, at just 6 dB above REFSENS for bandwidths up to 10 MHz, and 7 dB and 9 dB above REFSENS for 15 and 20 MHz bandwidths respectively. Consequently the C/I ratios are much lower, falling for example to −50 dB for the third adjacent channel.[19] The total filtering requirement will therefore be of the order of −60 dB at three times the bandwidth.

A summary of the NACS requirements for LTE and UMTS UEs is shown in Tables 21.9 and 21.10. For UMTS, the wanted signal power includes 21 dB spreading gain.

Table 21.9: Non-adjacent (N ± 2) channel selectivity.

System	LTE						UMTS
Bandwidth of wanted signal (MHz)	1.4	3	5	10	15	20	3.84
Own signal power above REFSENS (dB)	6	6	6	6	7	9	3
Power of interferer (dBm)	−56						
Frequency offset of interferer (MHz)	±2.8	±6	±10	±12.5	±15	±17.5	±10
Bandwidth of interferer (MHz)	1.4	3	5	5	5	5	3.84

21.4.5.4 Out-of-Band Blocking

The LTE OOB blocking tests measure the receiver's ability to receive a wanted signal at its assigned channel frequency in the presence of unwanted interfering signals falling outside the receive band, at 15 MHz or more offset from the edge of the band.

The wanted signal power is at the same level as for the in-band blocking test (at 6 dB above REFSENS for bandwidths of up to 10 MHz, and relaxed to 7 or 9 dB above REFSENS for bandwidths of 15 MHz or 20 MHz respectively).

[17] Same as for UMTS.

[18] Same as for UMTS.

[19] This example is for a REFSENS of −100 dBm for Band 1.

Table 21.10: Non-adjacent (N ± 3) channel selectivity.

System	LTE						UMTS
Bandwidth of wanted signal (MHz)	1.4	3	5	10	15	20	3.84
Own signal power above REFSENS (dB)	6	6	6	6	7	9	3
Power of interferer (dBm)	−44						
Frequency offset of interferer (MHz)	±4.2	±9	±15	±17.5	±20	±22.5	±15
Bandwidth of interferer (MHz)	1.4	3	5	5	5	5	3.84

The blocker is a CW signal with a power of −44 dBm at 15 to 60 MHz offset, −30 dBm from 60 to 85 MHz offset and −15 dBm from 85 to 12750 MHz offset.[20] These values are all the same as UMTS.

The actual offset of the blocker is the specified offset from the band edge plus half the wanted signal bandwidth plus the RF guard band, which is 2.4 MHz. Hence, for example, for a 5 MHz bandwidth, the −44 dBm blocker is at a minimum offset of 15 + 2.5 + 2.4 = 19.9 MHz, which is close to four times the bandwidth.

As in UMTS, there is an additional requirement for RF bands 2 (DCS1800), 5 (GSM850), 12 and 17, comprising a −15 dBm blocker coming from the transmit band, which is at an offset of just 20 MHz for bands 2 and 5, 12 MHz for band 12 and 18 MHz for band 17, see Table 21.1. This latter requirement is clearly far tougher than all the other blocking specifications.

A summary of the selectivity and blocking requirements is shown in Figure 21.17. This includes the specified 2 or 3 dB margin added to the C/I requirements.

When normalizing all frequency offsets to the wanted signal bandwidth, it can be seen that the 20 MHz LTE bandwidth requires the most severe filter frequency response relative to its bandwidth (and also to the digital sampling frequency). Additionally, the LTE 5 MHz selectivity requirement is relatively tougher than UMTS.

21.4.5.5 Spurious Response Specifications

Frequencies for which the throughput does not meet the requirements of the OOB blocking test are called spurious response frequencies. Spurious responses occur at specific frequencies at which an interfering signal mixes with the fundamental or harmonic of the receiver local oscillator and produces an unwanted baseband frequency component. Spurious responses are measured by recording when the OOB blocking test is not passed.

At the spurious response frequencies, the LTE receiver must still achieve the required throughput with a −44 dBm CW blocker and a wanted signal level set to the power level specified in the OOB blocking test.

[20]This blocker test is defined for UEs supporting all bands a part from 2, 5, 12 and 17.

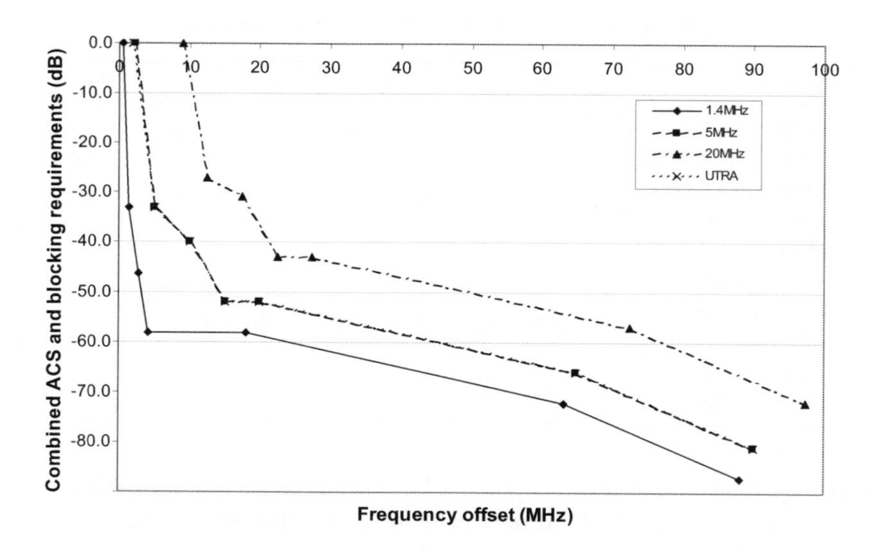

Figure 21.17: Selectivity and blocking requirements.

21.4.6 Spurious Emissions

The spurious emissions power is the power of emissions generated or amplified in a receiver which appear at the antenna connector. For LTE, limits are set covering the range 30 MHz to 12.75 GHz. The LTE specification is the same as UMTS, namely not more than −57 dBm between 30 MHz and 1 GHz (measured in 100 kHz bandwidth), and −47 dBm from 1 GHz to 12.75 GHz (measured in 1 MHz bandwidth); these requirements correspond to −107 dBm/Hz across the full range.

For UMTS, some additional spurious emissions requirements are specified for the cellular bands. Generally the requirement is −60 dBm measured in 100 kHz bandwidth (−110 dBm/Hz), but there are some more demanding requirements, of which the toughest is for the 935–960 MHz band to protect the GSM downlink receive band, i.e. −79 dBm measured in 100 kHz (−129 dBm/Hz).

Spurious emissions are caused by power from the local oscillator or the amplified received signal leaking back to the antenna. The required isolation from the amplified received signal can be estimated as follows. The maximum received power is −25 dBm, and the minimum received bandwidth 1.08 MHz; thus at the maximum input signal density the power in the 100 kHz measurement bandwidth is −35.3 dBm. The isolation required to prevent the amplified received signal from breaking the spurious emission limit will depend on the amplifier gain, which will vary with the signal and is likely to be at its minimum at the maximum input signal level. Assuming a 20 dB amplifier gain, then the isolation would need to be about 64 dB to reach the −79 dBm spurious emission limit. This requirement could be tougher than for UMTS because spreading enables the signal levels to be lower. However, the toughest isolation requirement is likely to be for the local oscillator signal which drives the mixer, since it typically operates at a higher power, for example around −4 dBm. The worst case isolation requirement for the local oscillator is thus about 75 dB, higher than the received signal isolation requirement.

21.4.7 Intermodulation Requirements

21.4.7.1 Intermodulation Distortion

As mentioned in Section 21.3.2.2, when two or more tones are present in a non-linear device, such as an amplifier in a receiver, intermodulation products are created. A power series describes all of the possible combinations of generated frequencies. Intermodulation products generated from mixing with a tone outside the wanted band may themselves fall into the wanted band. The problem is more severe for in-band interfering tones, since in that case the RF filter provides no attenuation.

As a device is driven further into its non-linear region, the amplitudes of the intermodulation products increase, while the power of the original tones at the output decreases. If the device was not limited in output power, then the powers of the intermodulation products would increase to the level of the original tones.

Intermodulation response rejection is a measure of the capability of the receiver to receive a wanted signal on its assigned channel frequency in the presence of two or more interfering signals, where the frequencies of the interfering signals relative to the wanted signal are such that the InterModulation Distortion (IMD) products fall into the wanted signal band.

The strongest interference is caused by third-order intermodulation products. The standard procedure for measuring the third-order intermodulation performance of a receiver is with two CW interfering tones offset from the wanted carrier frequency, with the offset of one tone being twice that of the other. Positioning one tone with twice the offset of the other results in the IMD product which occurs at the difference frequency falling into the wanted band. Such a test scenario is defined for UMTS, where it is referred to as a *narrowband intermodulation requirement*. UMTS also includes a *wideband intermodulation* test, with interferers at larger offsets, where one of the interferers is a modulated UMTS signal. This is designed to assess the effect of IMD products arising from UMTS transmissions in other channels.

For LTE, a narrowband intermodulation requirement could be defined following the same principles as in UMTS, using two CW signals as interferers. However, as a narrow bandwidth IMD product falling on top of the wanted signal would only degrade the performance of few subcarriers, the impact on throughput would be minimal. Hence, this is not defined in the current specification.

For the LTE wideband intermodulation performance requirement, the modulated interferer is defined to have the same bandwidth as the ACS test, denoted here 'BW_{int2}'.[21] The CW interferer is at offset $BW_{channel} + 1.5 \times BW_{int2}$, where $BW_{channel}$ is the channel bandwidth; thus for the 5 MHz bandwidth case the CW interferer is at a 10 MHz offset and the modulated interferer at 20 MHz offset, twice the offset of the CW interferer, as for UMTS. The interferer powers, are both −46 dBm, the same as for UMTS. The wanted signal is at a variable level between 6 and 12 dB above REFSENS depending on the channel bandwidth. This variation

[21] BW_{int2} is equal to the channel bandwidth for the 1.4, 3 and 5 MHz cases, and it is limited to 5 MHz for higher channel bandwidths.

of the wanted signal level is designed to keep the wanted signal at an absolute value close to −94 dBm and thus ensure a consistent IMD requirement for each bandwidth. For the 5 MHz bandwidth case, the wanted signal is 6 dB above REFSENS, 3 dB higher than UMTS (because of the 3 dB diversity gain).

For bandwidths of 10 MHz and above, the modulated interferer is always 5 MHz. The consequence of this is that the power of the interferer is spread over a narrower bandwidth than the wanted signal, so the interfering power on each subcarrier is higher than the wanted power of each subcarrier by the ratio of the bandwidths. This increases the IMD requirement: for example, for the 20 MHz bandwidth, a value of $10 \times \log_{10}(\mathrm{BW}_{\mathrm{channel}}/\mathrm{BW}_{\mathrm{int2}}) = 6$ dB should be taken into account when calculating IMD requirements.

The usual way to quantify the IMD performance of a receiver is to determine the *third-order intercept point*, or IP3. IP3 is a theoretical point where the amplitudes of the intermodulation tones are equal to the amplitude of the fundamental tones. The third-order IMD products, IM3, increase as the cube of the power of the input tones (i.e. at 60 dB/decade, three times the rate of the fundamental). Therefore, at some power level the distortion products will overtake the fundamental signal; the point at which the curves of the fundamental signal and the IM3 intersect on a log-log scale is the IP3. The corresponding input power level is known as Input IP3 (IIP3), and the output power when this occurs is the Output IP3 (OIP3) point, which is higher than IIP3 by an amount equal to the gain of the receiver. The IIP3 can be computed as follows:

$$\mathrm{IIP3} = (3P_{\mathrm{in}} - P_{\mathrm{IMD3,in}})/2 = \mathrm{OIP3} - G \qquad (21.4)$$

where P_{in} is the power of a single tone at the input of the system, $P_{\mathrm{IMD3,in}}$ the input-referred third-order intermodulation distortion product and G the passband gain (all in dB).

It is not possible to measure IP3 directly, because by the time the non-linear amplifier reached this point it would be heavily overloaded. Instead the amplifier is measured at a lower input tone power and IP3 is found by extrapolation.

The wide-band intermodulation test can be used to derive an equivalent IIP3 requirement. The IM3 products fall in-band and add to the existing receiver noise. The maximum power of the IM3 products that can be tolerated, referred back to the input (i.e. so that the gain of the receiver is factored out), $P_{\mathrm{IMD3,in}}$, is such that, when combined with the existing noise, it reaches a level that is equal to the maximum tolerable noise floor. This noise floor is below the wanted signal power by the SINR requirement plus implementation margin. Thus, if the wanted signal is 3 dB above REFSENS the IM3 products and existing noise can be the same, but as the wanted signal level is further raised above the REFSENS level the existing noise floor will be less significant and the IM3 products a little higher.

As an example, we can calculate the IIP3 requirement for a 15 MHz bandwidth UE supporting band 1. The SINR requirement plus implementation margin is assumed to be 1.5 dB. The wanted signal is 7 dB above REFSENS (i.e. at −88.2 dBm), so it can be calculated that the IM3 products must be 0.97 dB[22] below the maximum tolerable noise floor. The factor of three difference in bandwidth must be accounted for by a factor of 4.77 dB.[23] Thus, $P_{\mathrm{IMD3,in}} = -88.2 - 1.5 - 0.97 - 4.77 = -95.4$ dBm and IIP3 =

[22]The IM3 noise plus the thermal noise combines to use up to 7 dB margin, $0.97 = -10 \cdot \log_{10}(1 - 10^{-7/10})$, therefore $10 \cdot \log_{10}(10^{-7/10} + 10^{-0.97/10}) = 0$.

[23]This comes from the fact that the modulated interferer always has a 5 MHz bandwidth when the channel bandwidth is 10 MHz or more.

$[(3 \times (-46)) + 95.4]/2 = -21.3$ dBm.[24] IIP3 can be calculated for the other bandwidths and reaches a maximum of −20.6 dBm for 5 or 10 MHz bandwidth.

The IIP3 requirement for UMTS can be calculated in a similar manner.

21.4.7.2 Out-of-Band Intermodulation Distortion

As explained above, the LTE transmit signal may leak into the receiver. If a blocker is also present, IM3 products may result. If the blocker is half-way between the transmit and receive bands (i.e. between 15 and 200 MHz from the wanted band, depending on the duplex spacing) then the IM3 products will fall into the wanted band. For Band 5, the duplex spacing is 45 MHz, so a blocker at a 22.5 MHz offset, which can have a power of −44 dBm, would be the worst case. For the OOB blocking tests, the wanted signal is 3 dB above REFSENS. Therefore $P_{\mathrm{IMD3,in}}$ can be calculated as described above (equal to −101.5 dBm for the 5 MHz bandwidth). If we assume the transmitter leakage is −29 dBm and the blocker is at −44 dBm, then the average input power is −31.9 dBm. Thus the OOB IP3 requirement is $\mathrm{IIP3} = [(3 \times (-31.9)) + 101.5]/2 = +2.9$ dBm and the maximum IIP3 is +6 dBm for the 1.4 MHz bandwidth.

For Band 1 the duplex spacing is 190 MHz, so a blocker at 95 MHz, which can be at −15 dBm, would cause IMD. If the RF filter gives less than 29 dB attenuation of the blocker, then this band would present an even tougher OOB IIP3 requirement.

A similar calculation for UMTS for Band 5, assuming +20 dBm transmit signal leaking into the receiver at −28 dBm and a −44 dBm blocker, would yield a $P_{\mathrm{IMD3,in}}$ requirement of −98 dBm and an OOB IIP3 of +2.7 dBm.

Since the receiver's IP3 is dominated by that of the mixer, the problem of cross-modulation can be reduced for the most demanding bands by using a band-pass filter between the LNA and the mixer, so that most of the cross-modulation occurs in the LNA.

21.4.8 Dynamic Range

The input dynamic range is a key factor affecting the cost of the receiver. The larger the dynamic range, the larger must be the linear operating region of the receiver components. There are many ways to analyse the dynamic range and resulting signal handling requirements for a receiver, with widely differing results.

The simplest definition of the dynamic range of a receiver is the ratio of the maximum and minimum signal levels required to maintain a specified throughput or error rate. For UMTS the maximum and minimum wanted signal powers are −44 and −117 dBm respectively, giving a dynamic range of 73 dB. For an LTE UE, the maximum input is −25 dBm (assumed to be applicable to any modulation), while the minimum signal level for a 5 MHz bandwidth is −100 dBm (minimum REFSENS value across the different bands), giving a maximum dynamic range of 75 dB.

Dynamic range can also be specified as the ratio between the maximum signal and the noise floor. Even more usefully, the maximum signal level could include a margin to allow for the variation of the peak signal power above its average (PAPR). Such a measure of dynamic range gives an indication of the total signal handling requirements of the receiver in the absence of gain control. In Section 21.4.3, the total peak received signal power was

[24] −46 dBm is the interferer power as specified in [5].

derived, which includes the PAPR and the contribution from the transmitter leakage; the values were −12.0 dBm for LTE and −15.2 dBm for UMTS. For the LTE 5 MHz bandwidth the noise floor at the antenna is −107.5 dBm for UMTS it is −108.2 dBm. Thus this dynamic range estimate is −12.0 − (−107.5) = 94.9 dB for LTE and −15.2 − (−108.2) = 92.9 dB for UMTS (note that for UMTS the REFSENS, −117dBm, is below the noise floor).

In practice, gain control for large signals is normally used in a receiver to reduce the dynamic range. The minimum dynamic range that the receiver should handle linearly, measured at the antenna, is the maximum SINR requirement, plus implementation loss, plus NF, plus a margin for PAPR (assumed to be 11.5 dB). For LTE this gives 17.5 + 4 + 9 + 11.5 = 42 dB, while for UMTS using QPSK it is only 12.2 + 8 = 20.2 dB.

However, this does not take account of any interferers which could be present, nor leakage from the transmitter. At REFSENS, for an FDD LTE UE, there could be transmitter leakage at −25 dBm (peak about −16 dBm); the difference between the peak transmitter leakage and the noise floor is 91.5 dB for LTE and 83.6 dB for UMTS, although the frequency offset is large so the analogue filters will reduce the interference.

In conclusion, is clear that the higher SINR and higher PAPR requirements of LTE present a higher signal-handling dynamic range requirement than for UMTS, but the linearity requirements due to interferers are similar.

21.5 RF Impairments

The RF parts of the transmitter and receiver are comprised of non-ideal components which can have a strong impact on the ultimate demodulation performance. Different access technologies have different sensitivities to these RF non-idealities. For example, OFDM-based systems are particularly sensitive to any distortion which may remove the orthogonality between the subcarriers, resulting in Inter-Carrier Interference (ICI), as discussed in Section 5.2.3. The RF impairments can have a non-negligible impact on Bit Error Rate (BER), as shown for example in [11, 12] and references therein, where some compensation algorithms are also discussed.

The goal of this section is to show how the most common RF impairments introduce errors in an OFDM signal. First a simplified model of the impairments is discussed for the purpose of establishing SINR limitations for LTE performance assessment. This is followed in Section 21.5.2 by a mathematical framework for analysing the sensitivity of the LTE to particular impairments.

21.5.1 Transmitter RF Impairments

A generic model of the typical RF impairments of a transmitter is given in Figure 21.18. It comprises the following components:[25]

- **Digital to Analogue Converter (DAC):** a source of quantization noise – here we assume a uniform linear quantizer;
- **Up-sampling function:** up-samples the OFDMA/SC-FDMA sample rate process by a factor (not shown in diagram);

[25]It is not implied that a real implementation would necessarily use exactly the components specified; the aim here is to give a representative view of the typical impairments.

- **Baseband equivalent filter:** equivalent to the concatenation of linear filtering components in the transmit path, including digital and analogue elements;
- **Quadrature error component:** corresponding to the loss of I/Q orthogonality in the frequency conversion process;
- **d.c. offset:** arising from, for example, direct conversion Local Oscillator (LO) leakage effects;
- **Phase noise:** corresponding to the LO phase noise process;
- **Frequency error:** due to LO long-term frequency error;
- **InterModulation Distortion (IMD):** attributable to, for example, PA non-linearity.

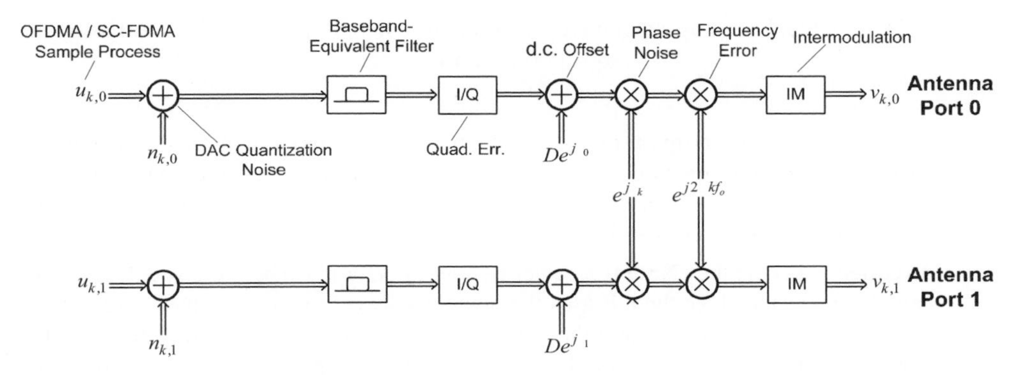

Figure 21.18: Model of multi-antenna transmitter impairments. Reproduced by permission of © 2006 Motorola.

A representative model of the RF impairments in the receiver, shown in Figure 21.19 for two receiver antennas, comprises the following conceptual components:

- **Adjacent Channel Interference (ACI):** important when establishing deployment and coexistence guidelines and for estimating total system spectral efficiency. The effect of ACI may be neglected, however, in an initial LTE system performance analysis by assuming that Co-Channel Interference (CCI) is the dominant interference-limiting effect.
- **Antenna Gain Imbalance (AGI):** in practical multiantenna UE designs, the antenna gains can be affected by multiple factors including user grip pattern, orientation and proximity to the user's body.
- **InterModulation (IM):** receiver non-linearity in general, is a critical consideration when operating at the upper limits of the receiver's dynamic range and in the presence of strong adjacent channel interference.
- **Thermal Noise:** applicable noise figures for LTE UE and eNodeB devices are specified in [13]. One further practical consideration is the potential for non-uniform noise figures applicable to each LTE antenna port.

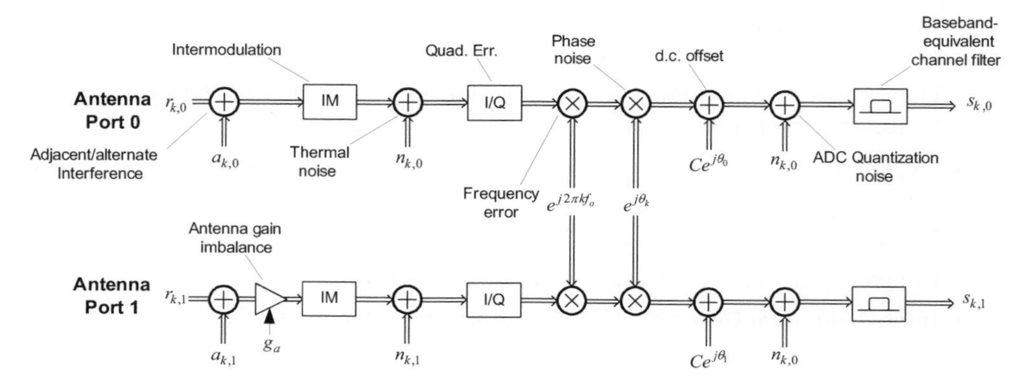

Figure 21.19: Model of multi-antenna receiver impairments. Reproduced by permission of © 2006 Motorola.

- **Quadrature error component:** as with the transmitter, this element models the loss of quadrature in the frequency conversion process. As an initial assumption, quadrature error may be neglected in eNodeB receivers, but is an essential element in direct conversion UE receiver modelling.
- **Frequency error:** the eNodeB receiver frequency error attributed to eNodeB LO error may be neglected since the UE uses the downlink waveform as a frequency reference. Clearly, in some circumstances there can be a significant frequency shift between the downlink signal received by the UE and the resulting uplink signal observed by the eNodeB.
- **Phase noise:** this corresponds to the eNodeB and UE LO phase noise process.
- **d.c. offset:** as for the transmitter model, this can arise due to LO leakage effects.
- **Analogue to Digital Converter (ADC):** similarly to the transmitter, this can be modelled as a quantization noise source.
- **Baseband equivalent filter:** equivalent to the concatenation of linear filtering components in the receive path, including digital and analogue elements.
- **Down-sampling function (not shown):** decimates the OFDMA/SC-FDMA sample rate process by a factor.

Often, a simple additive white Gaussian model can be considered to simplify the analysis of the impact of the RF impairments on link performance (see, for example, [11]). In fact the distortion generated by any non-linear device present for example in the downlink transmitter[26] can be modelled by using Bussgang's theorem [14] as follows:

$$\widetilde{\mathbf{x}} = \alpha\mathbf{x} + \mathbf{d} \tag{21.5}$$

where $\widetilde{\mathbf{x}}$ is the OFDM signal vector distorted by non-linearities, $\mathbf{x}$ is the column vector of the ideal OFDM symbol, $\mathbf{d}$ is the vector of the equivalent interference term due to distortions

[26]The analysis here considers the downlink OFDM transmitter, but it can equally well be applied to the uplink transmitter.

which are uncorrelated with the signal $\mathbf{x}$ and α is a complex gain factor accounting for the attenuation and phase rotation.

The parameter α can be derived as

$$\alpha = \frac{\mathbb{E}[\widetilde{\mathbf{x}}^{\mathrm{H}}\mathbf{x}]}{\mathbb{E}[\mathbf{x}^{\mathrm{H}}\mathbf{x}]} \tag{21.6}$$

where $\mathbb{E}[\cdot]$ is the expectation operator.

Assuming, for the sake of simplicity, an AWGN[27] propagation channel, the distorted signal $\widetilde{\mathbf{x}}$ generated by the RF transmitter is further corrupted by complex white circular Gaussian noise $\mathbf{n} \sim \mathcal{N}(0, \sigma_n^2\mathbf{I})$ such that the signal received by the OFDM receiver at the output of the FFT can be expressed as

$$\mathbf{R} = \alpha\mathbf{F}\mathbf{x} + \mathbf{F}\mathbf{d} + \mathbf{F}\mathbf{n}$$

Assuming coherent detection in the OFDM receiver,[28] the phase rotation induced by the scaling factor α has no effect as it will be ideally compensated in the receiver's channel estimation process. However, in both coherent and non-coherent OFDM systems, the impact of the magnitude of α translates into a power penalty for the useful signal. The reader is referred to [14] for further details.

Assuming FFT processing of large blocks and applying the central limit theorem, the frequency-domain distortion term $\mathbf{D} = \mathbf{F}\mathbf{d}$ can be closely approximated by complex white circular Gaussian noise $\mathbf{D} \sim \mathcal{N}(0, \sigma_d^2\mathbf{I})$, and the SINR of the distorted received OFDM signal can be simply expressed as

$$\mathrm{SINR_d} = \frac{|\alpha|^2\mathrm{tr}\{\mathbf{C_{XX}}\}}{\mathrm{tr}\{\mathbf{C_{DD}}\} + \sigma_{\mathbf{n}}^2} = \frac{|\alpha|^2}{\sigma_{\mathbf{d}}^2 + \sigma_{\mathbf{n}}^2} \tag{21.7}$$

where $\mathbf{C_{XX}} = \mathbf{F}\mathbf{C_{xx}}\mathbf{F}^{\mathrm{H}}$ and $\mathbf{C_{DD}} = \mathbf{F}\mathbf{C_{dd}}\mathbf{F}^{\mathrm{H}}$, assuming that the signal power $\mathbf{X}$ is normalized.

This model can be used in determining the impact of RF impairments on the LTE performance, with the additive distortion term being modelled as a transmitter EVM source.

21.5.2 Model of the Main RF Impairments

In order to gain insight into the impact of typical impairments, not only in terms of the impact on a single-user's communication link but also in terms of their overall effect on system-level spectral efficiency, it is instructive to develop a more precise analytical model. In particular, three sources of errors are included here:

- Phase noise;
- Modulator image interference (amplitude and phase imbalance);
- Non-linear distortion and intermodulation products which fall into adjacent RBs.

[27]Additive White Gaussian Noise. Note that the reasoning can be straightforwardly extended to the case of a frequency-selective convolutive channel and an OFDM system with CP.

[28]Coherent detection means that the phase of the received signal is known to the receiver a priori, for example from Reference Signals (RSs) as discussed in Chapter 8.

The model here is developed considering an OFDM signal (downlink) but it can be easily generalized for an SC-FDMA signal (uplink). The OFDM signal can be written as the sum of closely-spaced tones which do not interfere with each other due to their orthogonality. In particular the discrete-time signal for symbol ℓ is written as

$$x_\ell[n] = \sum_{k=-N/2}^{N/2-1} S_{k,\ell} \exp\left[j2\pi k\Delta f \frac{n}{N}\right] \tag{21.8}$$

where N is the number of subcarriers, $S_{k,\ell}$ is the constellation symbol sent on the kth subcarrier for the ℓth OFDM symbol and n represents the component within the ℓth OFDM symbol, $n \in [0, N-1]$. In the following, we can drop the dependency from the variable ℓ without loss of generality.

21.5.2.1 Amplitude and Phase Imbalance

Let us consider the continuous time version of the signal $x(t)$ modulated to frequency f_c, i.e.

$$r(t) = \text{Re}\{x(t)e^{j2\pi f_c t}\} = x(t)e^{j2\pi f_c t} + x^\star(t)e^{-j2\pi f_c t} \tag{21.9}$$

With a non-ideal I/Q (de-)modulator, the recovered signal can be written as follows:

$$\begin{aligned} y_{\text{IQ}}(t) &= \text{LP}\{r(t)(\cos(2\pi f_c t) - j\beta \sin(2\pi f_c t + \phi))\} \\ &\overset{(1)}{=} \text{LP}\{(x(t)e^{j2\pi f_c t} + x^\star(t)e^{-j2\pi f_c t})(\gamma_1 e^{-j2\pi f_c t} + \gamma_2 e^{j2\pi f_c t})\} \\ &\overset{(2)}{=} x(t)\gamma_1 + x^\star(t)\gamma_2 \end{aligned} \tag{21.10}$$

where LP means Low-Pass filtering, (1) is because of Equation (21.9) with $\gamma_1 = (1 + \beta e^{-j\phi})/2$ and $\gamma_2 = (1 - \beta e^{-j\phi})/2$ and (2) is obtained by low-pass filtering the signal. By substituting Equation (21.8) into (21.10), after a variable change it follows that

$$y_{\text{IQ}}(t) = \sum_{k=-N/2}^{N/2-1} \left(\frac{1+\beta e^{-j\phi}}{2} S_{k,\ell} + \frac{1-\beta e^{-j\phi}}{2} S^\star_{-k,\ell}\right) \exp\left[j2\pi k\Delta f \frac{t}{N}\right] \tag{21.11}$$

By letting $\beta = 1 + q$ and $Q(\beta, \phi) = (q - j\phi - qj\phi)/2$, using the approximation that $\cos(\phi) = 1$ and $\sin(\phi) = \phi$ for small angle ϕ, it follows that

$$\begin{aligned} y_{\text{IQ}}(t) &= \sum_{k=-N/2}^{N/2-1} S_{k,\ell} \exp\left[j2\pi k\Delta f \frac{t}{N}\right] \\ &\quad + Q(\beta, \phi) \sum_{k=-N/2}^{N/2-1} (S_{k,\ell} - S^\star_{-k,\ell}) \exp\left[j2\pi k\Delta f \frac{t}{N}\right] \end{aligned} \tag{21.12}$$

Equation (21.12) shows clearly that the I/Q imbalance creates two different error terms: the first is the self-interference created by the same signal at the same subcarrier frequency, while the second is the signal at the frequency mirror-image subcarrier. The ideal case corresponds to $q = 0$ and $\phi = 0$. Note also that the amplitude imbalance and the phase imbalance create the same effect, i.e. only the factor $Q(\beta, \phi)$ changes, in case of only amplitude imbalance being present ($\phi = 0$) $Q(\beta, \phi) = q/2$, and when only phase imbalance is present the multiplicative coefficient becomes $Q(\beta, \phi) = j\phi/2$.

For high-order modulation (increasing number of states in the constellation) the error term acts as noise and it spreads the constellation points as shown in [15].

In general the amplitude and phase imbalance can vary depending on the frequency, on a subcarrier basis. This happens when there is a timing offset between the in-phase and quadrature signal paths.

21.5.2.2 Phase Noise

When the LO frequency at the transmitter is not matched to the LO at the receiver, the frequency difference implies a shift of the received signal spectrum at the baseband. In OFDM, this creates a misalignment between the bins of the FFT and the peaks of the sinc pulses of the received signal. This results in a loss of orthogonality between the subcarriers and a leakage between them. Each subcarrier interferes with every other (although the effect is strongest on adjacent subcarriers) and, as there are many subcarriers, this is a random process equivalent to Gaussian noise. Thus, a LO frequency offset lowers the SINR of the receiver. An LTE receiver will need to track and compensate for this LO offset and quickly reduce it to substantially less than the 15 kHz subcarrier spacing. This is a tougher requirement than for UMTS. The phase noise impairment has been widely studied both for a single antenna [15–17] and for the MIMO case [18], especially in the context of WiMAX.

The ideal baseband transmitted signal, neglecting the additive white Gaussian noise, is given in Equation (21.8). The received baseband signal, in the presence of only phase noise can be written as

$$y_\theta(t) = \sum_{k=-N/2}^{N/2-1} S_{k,\ell} \exp\left[j2\pi k\Delta f \frac{t}{N}\right] e^{j\theta(t)} \tag{21.13}$$

where θ is the phase noise. In [11] it is shown that the single-sideband phase noise power follows a Lorentzian spectrum [11, 19]:

$$L(f) = \frac{2}{\pi\Delta f_{3\text{dB}}} \frac{1}{1 + [2f/(\Delta f_{3\text{dB}})]^2} \tag{21.14}$$

where $\Delta f_{3\text{dB}}$ is the two-sided 3 dB bandwidth of phase noise. The power spectrum given in Equation (21.14) can be considered as an approximation to practical oscillator spectra.

After the FFT, the symbol transmitted on the mth subcarrier in the ℓth OFDM symbol can be written as

$$z_{m,\ell} = S_{m,\ell} \frac{1}{N} \sum_{n=-N/2}^{N/2-1} e^{j\theta(n)} + \frac{1}{N} \sum_{k=-N/2,k\neq m}^{N/2-1} S_{k,\ell} \sum_{n=-N/2}^{N/2-1} \exp\left[j2\pi\Delta f \frac{n}{N}(k-m)\right] e^{j\theta(n)} \tag{21.15}$$

By defining $C(k) = (1/N) \sum_{n=-N/2}^{N/2-1} e^{j2\pi\Delta f nk/N + j\theta(n)}$, as in [16], Equation (21.15) can be rewritten as

$$z_{m,\ell} = S_{m,\ell}C(0) + \sum_{k=-N/2,k\neq m}^{N/2-1} S_{k,\ell}C(k-m) \tag{21.16}$$

Equation (21.16) shows that the effect of phase noise is twofold: the useful symbol transmitted on the mth subcarrier is scaled by a coefficient $C(0)$ which depends on the phase noise realization on the ℓth OFDM symbol, but it is independent of the subcarrier index – i.e. all the subcarriers are rotated by the same quantity $C(0)$. This is referred to as the Common Phase Rotation (CPR). This term can be estimated from the RSs and removed. In LTE, four symbols per subframe contain RS (see Section 8.2), so that in theory low frequency phase noise up to about 2 kHz can be compensated. However common loop bandwidths of PLLs[29] are in general in the order of 10 to 100 kHz, so that a major part of the phase noise energy is outside the region which can be compensated. The second term causes Inter-Carrier Interference (ICI), the amount of which is given by the coefficient $C(k - m)$ for subcarrier k interfering on subcarrier m. The ICI due to phase noise creates a fuzzy constellation [15]. In [18] the above equations are generalized for the MIMO case, showing that phase noise has similar effects in that case.

Time-domain offsets

Any timing synchronization error results in a misalignment between the samples contained in the OFDM symbol and the samples actually processed by the FFT. If the timing error is large, some samples could be from the wrong symbol, which would cause a serious error. It is more likely that the timing error will be just a few samples and the presence of the cyclic prefix should give enough margin to prevent samples from the wrong symbol being used. Assuming this is the case, the shift in time is equivalent to a linearly-increasing phase rotation of the constellation points.

The use of higher modulation schemes and wider channel bandwidths mean that timing synchronization needs to be performed more accurately for LTE than for UMTS.

Another source of timing error is a sampling frequency error. This can occur when the baseband sampling rate of the receiver is offset from that of the transmitter, or if there is jitter of the receiver's clock. When the sampling rate is too high this contracts the spectrum of the signal; similarly, slow sampling widens the spectrum. Either of these effects misaligns the FFT bin locations, resulting in a loss of orthogonality and hence reduced SINR. A sampling rate converter, perhaps driven by an error estimate, could correct for a systematic sampling rate error or sampling rate drift. Sampling jitter should be kept low by the choice of a clean crystal reference oscillator.

Group delay distortion

Filter group delay and amplitude ripple variation create deviation from the wanted impulse response of the receiver and cause inter-symbol interference. Unfortunately analogue filters cannot have both flat group delay and flat amplitude response, so a compromise is needed. To achieve this the analogue channel filter is usually slightly larger than the channel to allow reduced in-band distortion at the expense of out-of-band attenuation and in-band noise suppression.

One of the inherent advantages of multicarrier OFDM is that it is quite resilient to group delay variation and amplitude ripple, more so than the single-carrier QPSK modulation used by UMTS.

[29] Phase-Locked Loop.

Additionally, the OFDM symbol rate is relatively low and thus there are greater margins for delays introduced by large filters. Overall there is more freedom with LTE to design the receiver analogue and digital filters to achieve high selectivity.

Reciprocal mixing

An unwanted phase offset of the received signal may include a large fixed (or slowly varying), phase offset, which can be corrected by the equalizer. However, the LO also introduces small random variations and jitter of the frequency and phase, which manifests itself as shaped phase noise, tending to reduce with offset from the carrier frequency. This cannot be corrected by the equalizer. Crystal oscillators have low levels of phase noise, but synthesizers are not so clean, especially if they are PLL-based.

Each of the elements of a frequency synthesizer produces noise which contributes to the overall noise appearing at the output. The noise within the PLL loop bandwidth arises from the phase detector and the reference, whereas outside the loop bandwidth the VCO[30] is the main noise source. The phase noise tends to reduce from the edge of the PLL loop bandwidth until eventually it reaches a flat noise floor.

A serious problem for receivers caused by phase noise is reciprocal mixing. Reciprocal mixing occurs when the phase noise on the receiver LO mixes with a strong interfering signal to produce a signal which falls inside the pass-band of the receiver. Intermediate-Frequency (IF) filters, if used, do not give any rejection of such signals; instead the problem must be solved by keeping the LO phase noise at a sufficiently low level. The various interference requirements (see Sections 21.4.6 to 21.4.7.2) together create an overall requirement for the receiver LO phase noise. The interferer which gets mixed onto the received signal needs to be weaker by a margin of the SINR requirement plus the implementation margin, plus a further margin of about 10 dB if it is to have no impact on the received signal. Therefore, the LO noise, L, needs to satisfy the following requirement:

$$L \leq \text{SINR} + \text{IM} - C/I + 10 + 10 \cdot \log_{10}(B) \ \text{(dBc/Hz)}$$

21.5.2.3 Non-Linear Distortion and Intermodulation Products

IMD results from non-linearities of the PAs. The high PAPR (see Section 5.2.2) associated with multicarrier signals is one of the principal challenges in the implementation of OFDM systems. It requires linear operation of the PA over a large dynamic range, and this imposes a considerable implementation cost and reduced efficiency. The linearity and efficiency of a power amplifier are mutually exclusive specifications.

As already discussed, practical power amplifiers have a non-linear response. Numerous models exist, a selection of which can be found in [20]. Here we describe a simple polynomial memoryless model, by which the output signal can be written as

$$y_{\text{PA}}(t) = a_1 y(t) + a_2 y^2(t) + a_3 y^3(t) \tag{21.17}$$

[30] Voltage-Controlled Oscillator.

where a_1, a_2, a_3 are independent coefficients which can be found by measurement. From Equation (21.8), (21.17) can be written as[31]

$$\begin{aligned} y_{\mathrm{PA}}(t) &= a_1 \sum_{k=-N/2}^{N/2-1} S_{k,\ell} \exp\left[j2\pi k\Delta f \frac{t}{N}\right] \\ &+ a_2 \sum_{k=-N/2}^{N/2-1} \sum_{p=-N/2}^{N/2-1} S_{k,\ell} S_{p,\ell} \exp\left[j2\pi(k+p)\Delta f \frac{t}{N}\right] \\ &+ a_3 \sum_{k=-N/2}^{N/2-1} \sum_{p=-N/2}^{N/2-1} \sum_{v=-N/2}^{N/2-1} S_{k,\ell} S_{p,\ell} S_{v,\ell} \exp\left[j2\pi(k+p+v)\Delta f \frac{t}{N}\right] \end{aligned} \quad (21.18)$$

The non-linear response of the PA creates ICI. The intermodulation products contribute to a noise-like cloud surrounding each constellation point.[32] For higher-order modulation in particular (such as 64QAM), these constellation clouds contribute to an increase in error rate for each subcarrier. Thus, in an OFDM modem design, linearity must be carefully controlled.

21.6 Summary

In this chapter, we have reviewed the RF requirements affecting the practical implementation of LTE equipment, especially the UE. We have seen how the effects of typical RF impairments of the transmitter and receiver can be analysed, including the use of a mathematical model for quantitative analysis.

The LTE RF requirements can be compared to those of UMTS. In many aspects, the RF requirements of LTE are similarly demanding as for UMTS, including for example narrowband blocking of the receiver and self-interference in case of full-duplex FDD operation. However, the LTE specifications are more challenging for other aspects, such as

- The channel bandwidth is variable over a wide range;
- At least two receive antennas are expected for receive diversity, and only the lowest UE category does not have to support reception of MIMO spatial multiplexing;
- There is a large variety of modulation and coding schemes, with some of them requiring a high SNR;
- OFDM is more sensitive to phase noise.

On the other hand, LTE is more robust than UMTS against amplitude and phase distortions from receiver and transmitter filters.

[31] This can be derived by considering ideal (de-)modulator components; thus, the signal in Equation 21.9 can be recovered by a simple low-pass filtering.

[32] Note that the interference is correlated with the symbol transmitted on subcarrier k, and hence strictly it cannot be considered as white noise.

References[33]

[1] 3GPP Technical Specification 25.101, 'User Equipment (UE) Radio Transmission and Reception (FDD)', www.3gpp.org.

[2] 3GPP Technical Specification 25.104, 'Base Station (BS) Radio Transmission and Reception (FDD)', www.3gpp.org.

[3] 3GPP Technical Specification 25.102, 'User Equipment (UE) Radio Transmission and Reception (TDD)', www.3gpp.org.

[4] 3GPP Technical Specification 25.105, 'Base Station (BS) Radio Transmission and Reception (TDD)', www.3gpp.org.

[5] 3GPP Technical Specification 36.101, 'Evolved Universal Terrestrial Radio Access (E-UTRA); User Equipment (UE) Radio Transmission and Reception', www.3gpp.org.

[6] 3GPP Technical Specification 36.104, 'Evolved Universal Terrestrial Radio Access (E-UTRA); Base Station (BS) Radio Transmission and Reception', www.3gpp.org.

[7] 3GPP Technical Report 36.803, 'Evolved Universal Terrestrial Radio Access (E-UTRA); User Equipment (UE) Radio Transmission and Reception', www.3gpp.org.

[8] 3GPP Technical Report 36.804, 'Evolved Universal Terrestrial Radio Access (E-UTRA); Base Station (BS) Radio Transmission and Reception', www.3gpp.org.

[9] Motorola, 'R1-060144: UE Power Management for E-UTRA', www.3gpp.org, 3GPP TSG RAN WG1, LTE Ad-hoc, Helsinki, Finland, January 2006.

[10] Motorola, 'R1-060023: Cubic Metric in 3GPP-LTE', www.3gpp.org, 3GPP TSG RAN WG1, LTE Ad-hoc, Helsinki, Finland, January 2006.

[11] B. E. Priyanto, T. B. Sorensen, O. K. Jensen, T. Larsem, T. Kolding and P. Mogensen, 'Assessing and Modeling the Effect of RF Impairments on UTRA LTE Uplink Performance', in *Proc. IEEE Vehicular Technology Conference*, Baltimore, MD, USA, September 2007.

[12] M. Moonen and D. Tandur, 'Compensation of RF Impairments in MIMO OFDM Systems' in *Proc. IEEE International Conference on Acoustics, Speech and Signal Processing*, Las Vegas, NV, USA, April 2008.

[13] 3GPP Technical Report 25.814, 'Physical Layer Aspects for Evolved UTRA, (Release 7)', www.3gpp.org.

[14] J. Bussgang, 'Crosscorrelation Function of Amplitude Distorted Gaussian Signals', Technical Report 216, Research Laboratory of Electronics, Massachusetts Institute of Technology, Cambridge, Massachusetts, USA, March 1952.

[15] Agilent Technologies, 'R4-061276: Effects of Physical Impairments on OFDM Signals', www.3gpp.org, 3GPP TSG RAN WG4, meeting 33, Riga, Latvia, November 2006.

[16] F. Munier, T. Eriksson and A. Svensson, 'Receiver Algorithms for OFDM Systems in Phase Noise and AWGN' in *Proc. IEEE International Symposium on Personal, Indoor and Mobile Radio Communications*, Barcelona, Spain, September 2004.

[17] S. Wu and Y. Bar-Ness, 'Performance Analysis on the Effect of Phase Noise in OFDM Systems', in *Proc. IEEE International Symposium on Spread Spectrum Techniques and Applications*, Prague, Czech Republic, 2002.

[18] T. C. W. Schenk, T. Xiao-Jiao, P. F. M. Smulders and E. R. Fledderus, 'On the Influence of Phase Noise Induced ICI in MIMO OFDM Systems'. *IEEE Communications Letters*, August 2005.

[19] E. Costa and S. Pupolin, 'M-QAM-OFDM System Performance in the Presence of a Nonlinear Amplifier and Phase Noise'. *IEEE Trans. on Communications*, pp. 462–472, March 2002.

[33] All web sites confirmed 1st March 2011.

[20] S. C. Cripps, *RF Power Amplifiers for Wireless Communications*. Artech House, Norwood, MA, USA, 1999.

22

Radio Resource Management

Muhammad Kazmi

22.1 Introduction

Radio Resource Management (RRM) encompasses a wide range of techniques and procedures, including power control, scheduling, cell search, cell reselection, handover, radio link or connection monitoring, and connection establishment and re-establishment. Advanced features like interference management, location services, Self-Optimizing Networks[1] (SON) and some network planning methods make use of RRM-related techniques based on radio related measurements made by the User Equipment (UE) or eNodeB. In this chapter, we address the RRM techniques and reporting mechanisms that support UE mobility in the LTE network (E-UTRAN), including cell search, radio measurements, cell reselection, handover and radio link monitoring. We focus here on the performance requirement aspects, while the procedures themselves are described in Chapters 3, 4, 7, 12, 17 and 18.

The RRM-related actions undertaken by the UE can be broadly divided into those relevant in the RRC_IDLE state and those relevant in the RRC_CONNECTED state, as illustrated in Figures 22.1 and 22.2 respectively.

RRM in E-UTRAN is designed to handle the challenges posed by the fundamental characteristics of the LTE system, including:

- The packet-oriented transmission of LTE, realized by fast time- and frequency-domain scheduling, may lead to large and swift interference fluctuations. This may affect the accuracy of the signal quality estimates required for mobility decisions in certain scenarios. Time-domain filtering is used to help smooth out the interference variations.
- The provision of a wide range of Discontinuous Reception (DRX) cycle lengths, including periods up to 2.5 seconds, yields substantial benefits in terms of UE power saving. However, such DRX cycles also mean that it is not feasible to enforce the same mobility performance in all cases.

[1]Details on user equipment positioning and self optimization of the network are described in Chapters 19 and 25 respectively.

LTE – The UMTS Long Term Evolution: From Theory to Practice, Second Edition.
Stefania Sesia, Issam Toufik and Matthew Baker.

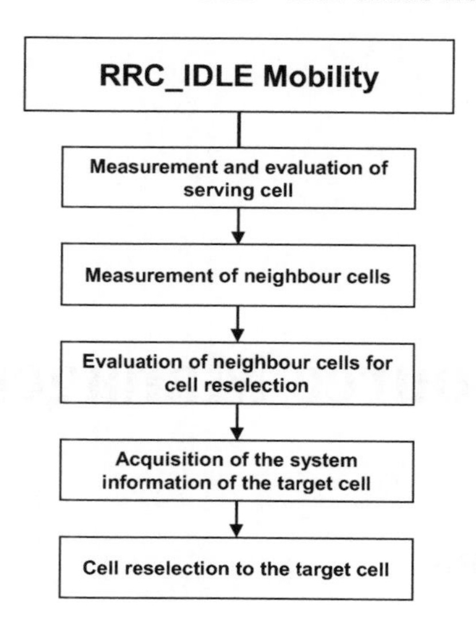

Figure 22.1: UE actions related to RRM Procedures in RRC_IDLE state.

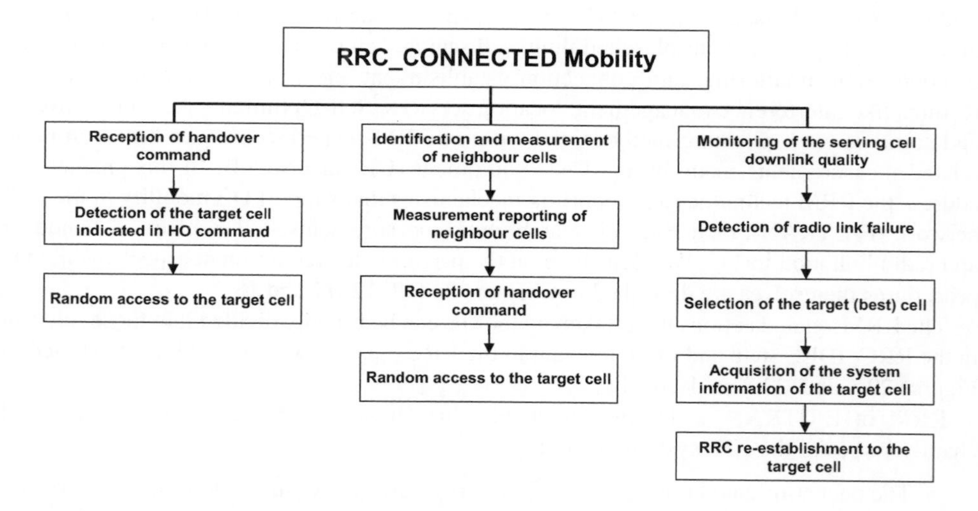

Figure 22.2: UE actions related to RRM Procedures in RRC_CONNECTED state.

- LTE is designed to support seamless mobility not only within itself (both intra- and inter-frequency) but also with the legacy 3GPP Radio Access Technologies (RATs) (e.g. GERAN[2] and UTRAN[3]) and with certain non-3GPP RATs (e.g. CDMA2000

[2]GSM EDGE Radio Access Network.
[3]Universal Terrestrial Radio Access Network.

1xRTT and HRPD[4]). Mobility between these different RATs requires the UE to detect and measure cells on the target technologies as requested by the network, each with its own different channel structure.

- LTE is designed to support a wide range of cell sizes (ranging from a few tens of metres to tens of kilometres) and deployment/propagation scenarios, yet the RRM techniques are required to be as generic as possible and to exhibit consistent mobility performance.
- Despite the differences between the Frequency Division Duplex (FDD) and Time Division Duplex (TDD) frame structures in LTE (see Section 6.2), and the fact that in FDD deployments the eNodeBs may be either synchronized or unsynchronized, mobility requirements are expected to be the same (or at least similar) in order to minimize UE implementation complexity and simplify network planning.
- The low overall latency requirements in LTE place constraints on the time taken to make and report measurements and to perform handover. Additionally, to meet the challenges of certain specific mobility procedures, enhanced requirements with substantially lower latency are also specified.
- The time-multiplexing of Multimedia Broadcast/Multicast Service (MBMS) and unicast data on a single LTE carrier is attractive in that it obviates the need for a separate carrier frequency to offer mobile broadcast services. However, it reduces the opportunities for the UEs to perform downlink cell measurements to support mobility; nevertheless, the RRM performance should remain consistent.

22.2 Cell Search Performance

Cell search is one of the most fundamental aspects of mobility. As explained in Chapter 7, it enables the UE to acquire the carrier frequency, timing and cell identity of cells. In LTE the cell search performance requirements relate to the case of neighbour cell search under the assumption that the UE has already acquired carrier frequency synchronization; the requirements are only applicable in RRC_CONNECTED.

22.2.1 Cell Search within E-UTRAN

The objective of cell search within E-UTRAN is to identify one of the 504 unique Physical Cell Identities (PCIs) (see Chapter 7 and [1, Section 6.11]). The cell search algorithm is not specified and is left for UE implementation; however, typically, the UE performs cell search in a hierarchical manner (see Section 7.2).

An important characteristic of the cell identification requirements is that the same requirements are applicable in a wide range of propagation conditions and for both FDD (with or without synchronization of the eNodeBs) and TDD (where synchronization of the eNodeBs can be assumed).

The requirements are specified in terms of the maximum permissible cell identification delay, which includes the time taken for Reference Signal Received Power (RSRP) or Reference Signal Received Quality (RSRQ) physical layer measurements (see Sections 22.3.1.1 and 22.3.1.2 respectively).

[4]High Rate Packet Data.

22.2.1.1 E-UTRAN Intra-frequency Cell Search

In case of intra-frequency cell search, the UE identifies E-UTRA cells on the same carrier frequency as that of the serving cell. The time required to detect a cell depends upon a number of factors, most notably the received quality of the synchronization signals, the received level of the Reference Signals (RSs) and the time available for performing the search. The latter factor stems from the fact that the available time for intra-frequency measurements may be reduced by measurement gaps for inter-frequency or inter-RAT measurements as explained in Section 22.2.1.2. The cell search delay also depends upon the configured DRX cycle period.

If no DRX is configured, and for DRX cycles up to 40 ms,[5] the UE is required to detect an E-UTRA FDD or TDD intra-frequency target cell within 800 ms if no inter-frequency measurement gaps are configured, provided that the target cell's received synchronization signal quality $\hat{E}_s/I_{ot}$ (defined as the energy per Resource Element (RE) of the synchronization signals divided by the total received energy of noise and interference on the same RE) is at least −6 dB. This is the 'minimum', or worst case, requirement.

The cell search delay can be shorter if the received signal quality is higher than the minimum cell detection threshold. The performance in some typical deployment conditions is illustrated in Figures 22.3 and 22.4. Here, scenarios covering both synchronized and unsynchronized eNodeBs are analysed, as summarized in Table 22.1. ETU5 (Extended Typical Urban with UE speed 5 km/h), ETU300 (UE speed 300 km/h) and EPA5 (Extended Pedestrian A with UE speed 5 km/h) propagation models are used,[6] and two receive antennas are assumed at the UE.[7] Further details of the modelled scenarios can be found in [2].

Table 22.1: Cell identification test parameters.

	Unit	Cell1	Cell2	Cell3 (target cell)
Relative delay of 1st path for synchronized case	ms	0	0	Half CP length
Relative delay of 1st path for unsynchronized case	ms	0	1.5	3
SNR	dB	5.18	0.29	−0.75 (worst case)
PSS for case of different PSS		PSS1	PSS2	PSS3
PSS for case of same PSS		PSS1	PSS2	PSS1

The cell search performance is measured in terms of the 90-percentile cell identification delay, i.e. the maximum time required to detect the target cell 90% of the time.

Various scenarios are analysed to examine the impact on the detection performance of different combinations of PSS[8] and SSS[9] sequences as indicated in Table 22.2. More detailed performance results can be found in [3].

[5] This is designed to ensure robust mobility performance for delay-sensitive services like Voice over IP (VoIP), which typically requires short DRX cycles.

[6] Further details of these propagation models are given in Chapter 20.

[7] No margin is included for non-ideal UE receiver implementation or reporting delay for the RSRP measurement to the network.

[8] Primary Synchronization Signal.

[9] Secondary Synchronization Signal.

Table 22.2: Cell identification test scenarios.

Test case (synch, asynch eNodeBs)	Cell3 (Target)		Cell1 (Interference)		Cell2 (Interference)	
1,5	PSS3	SSS3a, SSS3b	PSS1	SSS1a, SSS1b	PSS2	SSS2a, SSS2b
2,6	PSS1	SSS3a, SSS3b	PSS1	SSS1a, SSS1b	PSS2	SSS2a, SSS2b
3,7	PSS1	SSS1a, SSS3b	PSS1	SSS1a, SSS1b	PSS2	SSS2a, SSS2b
4,8	PSS3	SSS1a, SSS1b	PSS1	SSS1a, SSS1b	PSS2	SSS2a, SSS2b

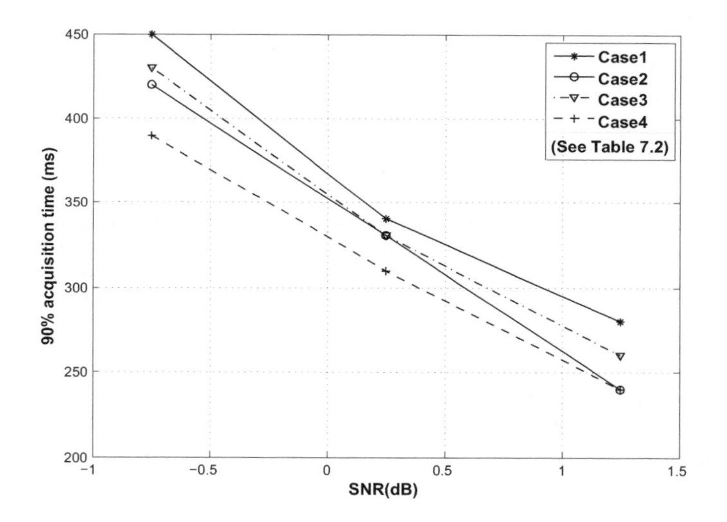

Figure 22.3: Cell search performance with synchronized eNodeBs. Reproduced by permission of © NXP Semiconductors.

When measurement gaps are configured for inter-frequency or inter-RAT measurements the UE has less opportunity to detect a cell. Hence, in this case the cell identification delay may be larger than the baseline 800 ms delay, with the actual value depending upon the periodicity of the gaps. Furthermore, for DRX cycles larger than 40 ms, the cell identification delay increases in proportion to the length of the DRX cycle, allowing UE to save battery power.

22.2.1.2 E-UTRAN Inter-frequency Cell Search

In the case of inter-frequency cell search, the UE identifies E-UTRA cells operating on carrier frequencies other than that of the serving cell (and possibly also in different frequency bands and/or with different duplex modes). Inter-frequency measurements, including cell identification, are performed during periodic measurement gaps unless the UE has more than one receiver. Two possible gap patterns can be configured by the network, each with a gap length of 6 ms: in gap pattern #0, the gap occurs every 40 ms, while in gap pattern #1 the

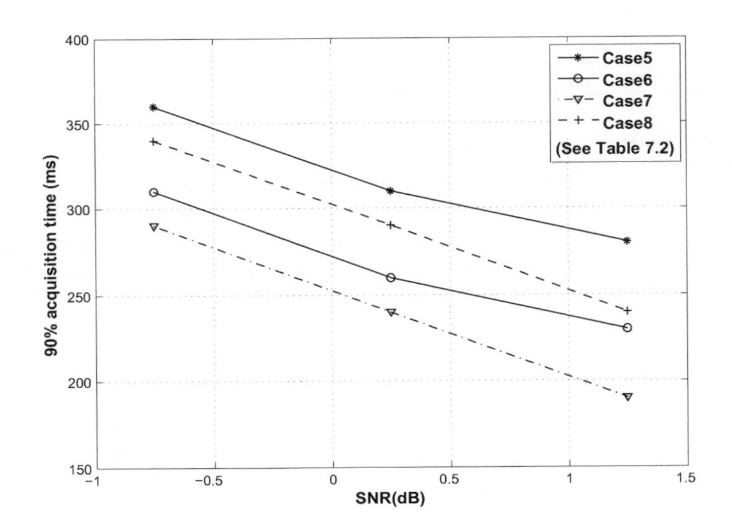

Figure 22.4: Cell search performance with unsynchronized eNodeBs.
Reproduced by permission of © NXP Semiconductors.

gap occurs every 80 ms, as shown in Figure 22.5. There is an obvious trade-off between these two gap patterns: the former yields a shorter cell identification delay but a greater interruption in data transmission and reception. Only one gap pattern can be configured at a time for measuring all frequency layers (both inter-frequency and inter-RAT).

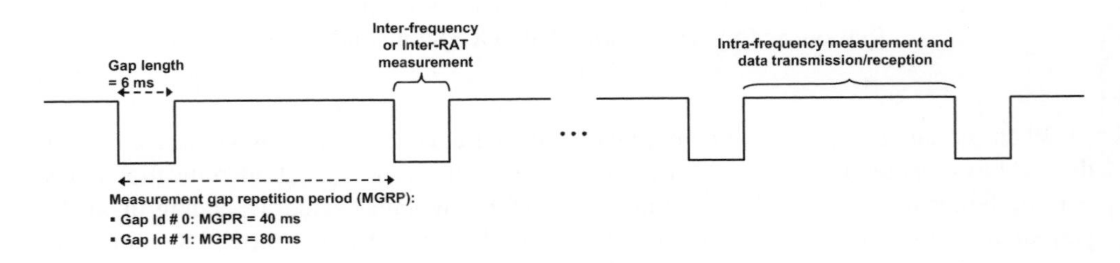

Figure 22.5: Measurement gap patterns for inter-frequency and inter-RAT cell search and measurements.

If no DRX is used, or if the DRX cycle length is less than or equal to 160 ms, the UE is required to identify an E-UTRA FDD or TDD inter-frequency cell within 3.84 s, provided the received synchronization signal quality is at least −4 dB (assuming gap pattern #0). As for intra-frequency cell search, for DRX cycles larger than 160 ms the cell identification delay increases proportionally.

22.2.2 E-UTRAN to E-UTRAN Cell Global Identifier Reporting Requirements

In addition to the basic PCI cell search requirements specified in LTE Release 8, E-UTRAN Cell Global Identifier[10] (ECGI) reporting requirements were introduced in Release 9 for both FDD and TDD. The requirements are the same for FDD and TDD for their respective frequency bands, and therefore the description in the subsequent sections is generic for both.

The UE is required to identify and report the ECGI of a target E-UTRA cell upon request from the serving eNodeB. The ECGI measurement report can be used by the eNodeB for various features. For instance, it can assist the eNodeB to establish the neighbour cell relations automatically (see Chapter 25.2) when a new neighbour cell is added or an existing one is removed [5]. The ECGI measurement report can also be used by the serving eNodeB for performing inbound mobility to a Home eNodeB[11] (HeNB) in RRC_CONNECTED state (see Chapter 24 for more details on HeNBs). Due to the small size of HeNB cells, it is not unlikely that a PCI may be used more than once among the HeNBs located within the coverage area of a macro-eNodeB. Hence an important objective of the ECGI report is to enable the serving eNodeB to identify unambiguously the target HeNB.

The acquisition of ECGI requires the UE first to read the Master Information Block (MIB), which is transmitted on Physical Broadcast CHannel (PBCH) with a periodicity of 40 ms (see Section 9.2.1). This is followed by the reading of the System Information Block type 1 (SIB1), which contains the ECGI and is transmitted with a periodicity of 80 ms on the DownLink Shared CHannel (DL-SCH) (see Section 3.2.2). The UE is not required to receive or transmit in the serving cell while acquiring the ECGI of a target cell; a UE is permitted to create gaps autonomously in downlink reception and uplink transmission to read the MIB and SIB1 of a target cell.

HeNBs can be deployed on either a shared carrier or a dedicated carrier. Similarly the SON Automatic Neighbouring Relation (ANR) is applicable to both intra- and inter-frequency cells, and therefore EGCI requirements are specified for both intra-frequency and inter-frequency ECGI reporting.

22.2.2.1 Intra-frequency E-UTRAN Cell Global Identifier Reporting

The intra-frequency ECGI reporting requirement is specified in terms of the maximum time allowed for the UE to identify the ECGI of a new intra-frequency target E-UTRA cell whose synchronization signal quality $\hat{\mathrm{E}}_s/\mathrm{I}_{\mathrm{ot}}$ (as defined in Section 22.2.1.1) is at least -6 dB; the identification delay is not allowed to exceed 150 ms with 90% confidence, excluding the procedure delay of the ECGI request message. The same ECGI identification delay requirements are applicable with and without DRX.

The initial autonomous gap created by the UE for correcting the frequency and for decoding the first MIB block of the target cell comprises 9 subframes. Each subsequent autonomous gap for decoding the remaining MIB and SIB1 blocks to read the whole of the target cell's ECGI typically consists of 4 subframes. The maximum aggregated duration of autonomous gaps created by the UE is not allowed to exceed 90 subframes when identifying

[10]The E-UTRAN Cell Global Identifier is a globally unique identifier broadcast by the cell. It is composed of 3-bytes of Public Land Mobile Network (PLMN) Identity and 28 bits to identify the cell within that PLMN – see [4, Section 6.3.4], and Chapter 25.

[11]Inbound mobility to Home eNodeB refers to the case of a UE moving into a Home eNodeB cell.

an intra-frequency cell's ECGI; this is designed to ensure that certain minimum serving cell reception and transmission performance is maintained.

22.2.2.2 Inter-frequency E-UTRAN Cell Global Identifier Reporting

The inter-frequency ECGI requirements (i.e. ECGI identification delay and serving cell reception/transmission performance) are the same as those specified for the intra-frequency requirement, except that the synchronization signal quality under which the requirement should be satisfied is relaxed to -4 dB. An autonomous gap created by the UE for decoding the initial MIB block of the target inter-frequency cell comprises 9 subframes, and each of the subsequent gaps typically consists of 4 subframes. The maximum aggregated duration of autonomous gaps created by the UE is likewise not allowed to exceed 90 subframes when identifying an inter-frequency cell's ECGI. This ensures that certain minimum serving cell reception and transmission performance is maintained while reading the inter-frequency target cell ECGI.

22.2.3 E-UTRAN to UTRAN Cell Search

The E-UTRAN to UTRAN cell search procedure allows the UE to identify a target FDD or TDD UTRA cell. The E-UTRAN-UTRAN cell search requirements apply until an explicit neighbour cell list is received by the UE from the serving E-UTRA cell. This list indicates the primary scrambling codes of up to 32 neighbour cells per UTRA FDD or TDD carrier. The signals provided by UTRA FDD and TDD differ, and therefore separate cell identification requirements are specified for each technology. In each case, the cell search algorithm is not specified, but the search is typically carried out in hierarchical manner. In UTRA FDD, where Primary and Secondary Synchronization CHannels (P-SCH and S-SCH) are provided, the UE typically identifies the target UTRA cell via the following steps:

Slot timing acquisition: The UE acquires the slot timing by performing a matched filter correlation over the Primary Synchronization Code (PSC), which is transmitted in the first 256 chips in every slot (0.666 ms). The PSC is common to all UTRA cells.

Frame timing acquisition and code group identification: The UE acquires the frame timing and the identity of the group of 8 codes to which the cell's primary scrambling code belongs by performing correlations over the Secondary Synchronization Codes (SSCs) transmitted in the initial 256 chips of every slot (i.e. at the same time as the PSC). A sequence of 15 SSCs in one 10 ms radio frame uniquely identifies the code group. In good propagation conditions the information contained within three slots is sufficient to identify both the frame timing and the code group.

Primary scrambling code identification: A given UTRA cell uses one code from the group indicated by the SSC as the scrambling code for all downlink channels, including the Primary Common PIlot CHannel (P-CPICH). The UE therefore performs a simple correlation against the known CPICH sequence (which is the same in all UMTS cells) scrambled in turn by each of the eight possible scrambling sequences in the code group, in order to determine which primary scrambling code is being used in the cell, and hence the cell identity.

In order to report an identified cell, the UE needs to measure the P-CPICH E_c/N_0 (where E_c is the CPICH energy per chip and N_0 the noise power spectral density) or Received Signal Code Power (RSCP) over the 'physical layer measurement period'.

The same measurement gap patterns are used for E-UTRA inter-frequency and inter-RAT measurements. In both DRX and non-DRX scenarios, a target UTRA FDD cell is considered detectable if the CPICH E_c/I_0 is at least -20dB and the synchronization channel SCH_E_c/I_0 is at least -17dB, where I_0 is the power spectral density of the total input signal at the UE antenna connector and SCH_E_c is the transmit energy per chip of the SCH, including the downlink signal transmitted by the serving cell [6]. However, in terms of reporting delay, separate sets of UTRA cell identification requirements are specified for the cases with and without DRX.

If no DRX is used, or for DRX cycles up to 40 ms, the UE is required to identify a detectable UTRA FDD cell within $2.4 * N_{\text{freq}}$ seconds and $4.8 * N_{\text{freq}}$ seconds using gap patterns #0 and #1 respectively (where N_{freq} is the number of carrier frequencies to be monitored). For DRX cycles larger than 40 ms, the UTRA cell search delay increases proportionally and is expressed in terms of a number of DRX cycles. For example in case of DRX cycle length of 1.28 seconds, the maximum permissible cell search delay is $25.6 * N_{\text{freq}}$ seconds or $20 *$ DRX cycle length $* N_{\text{freq}}$; where the factor of 20 incorporates the required UTRA FDD cell search occasions and CPICH measurement samples.

22.2.4 E-UTRAN to GSM Cell Search

In GSM the Broadcast Control CHannel (BCCH) frequency reuse factor is greater than one. Therefore in principle the GSM BCCH Received Signal Strength Indicator (RSSI) measurement, which is measured on the BCCH carrier (see Section 22.3.3) can identify a GSM cell, since a different RSSI measurement can be reported for each carrier. However, if small cluster sizes are used with BCCH frequency reuse, the GSM carrier RSSI alone cannot always reliably identify the correct GSM cell: for example, the network may not be able to distinguish RSSI measurement reports for GSM cells whose BCCHs use the same carrier. This may lead to calls being dropped, especially in deployment scenarios with tight reuse. Therefore, the UE is also required to decode the Base Station Identity Code (BSIC)[12]. In E-UTRAN, GSM measurements are always reported together with the verified BSIC [4], in contrast to the legacy GSM system, where measurements can be reported with or without a verified BSIC. The BSIC verification comprises initial BSIC identification and BSIC re-confirmation. These are summarized below; A more extensive description of the GSM common channels is provided in [7].

Initial BSIC identification comprises the BSIC decoding for a GSM cell for the first time when the UE does not have any knowledge about the relative timing between the E-UTRA and GSM cells. The UE is required to decode the BSIC of the 8 strongest BCCH carriers

[12]The BSIC allows a UE to distinguish two different GSM cells which share the same beacon frequency. It is a 6-bit field composed of two 3- bit fields: the Base station Colour Code (BCC) and the Network Colour Code (NCC). The BCC is used to identify the Training Sequence Code (TSC) to be used when reading the BCCH. The NCC is used to differentiate between operators utilizing the same frequencies, e.g. on an international border when both operators have been allocated the same frequency or frequencies.

of the GSM cells that are provided in the GSM inter-RAT cell information list[13] [4]. The UE decodes the BSIC of the BCCH carriers in order of decreasing RSSI. Through the BSIC decoding the UE acquires the initial timing information of these GSM cells.

BSIC re-confirmation consists of the BSIC decoding of a GSM cell after the initial BSIC identification has been performed. The UE updates its stored values of the timing of up to 8 identified GSM cells every time the BSIC is decoded, in order to compensate for cell timing drift relative to the serving cell to which the UE is locked.

BSIC verification requirements are applicable down to the reference sensitivity level defined in [8]. Initial BSIC identification and BSIC re-confirmation requirements are expressed in terms of the time required to decode the BSIC for the first time, $T_{\text{identity,GSM}}$ and the time required to re-confirm the initially identified BSIC $T_{\text{re-confirm,GSM}}$ under the above constraints, which depend on the total number of carrier frequency layers of all RATs since the same gap pattern is shared by all RATs.

22.2.5 Enhanced Inter-RAT Measurement Requirements

As mentioned earlier, the measurement requirements are generally minimum requirements for worst-case scenarios. In order to support Circuit-Switched Fall Back (CSFB)[14] scenarios in E-UTRAN, which require shorter overall delay, enhanced measurement requirements for UTRAN FDD and GSM were introduced in LTE Release 9.

For both UTRA FDD and GSM, the enhanced measurement requirements are specified under the assumption that CSFB from a serving E-UTRA cell to a target UTRA FDD or GSM cell is used when the serving cell and the target cell have overlapping coverage. Under these conditions the received signal quality of the target cell is typically higher than the minimum levels specified for normal cell search.

In the case of UTRA FDD, the enhanced UTRA FDD cell identification requirements are therefore specified assuming that both received CPICH E_c/I_0 and synchronization channel SCH_E_c/I_0 levels from the UTRA FDD cell are at least -15 dB. At these levels when no DRX is used or when DRX $\leq$ 40 ms, the UTRA FDD cell search delay is about 1 s (assuming gap pattern #0). Hence, compared to the minimum requirements, the enhanced UTRA FDD measurement reporting delay of an unknown UTRA FDD cell is reduced by a factor of 2.

In the case of GSM, the enhanced BSIC verification requirements are specified assuming that the target cell is received at about 10 dB above the reference sensitivity level or reference interference levels (as specified in [8]). Under this condition and without DRX or with DRX $\leq$ 40 ms, the UE can complete initial BSIC decoding in about 1.3 s, assuming gap pattern #0 is used only for monitoring GSM measurements.

[13]When camped on an LTE cell, the UE is provided with a Neighbour Cell List (NCL) containing at least 32 GSM carrier numbers (i.e. Absolute Radio Frequency Channel Numbers (ARFCNs)) indicating the frequencies of neighbouring cells, and optionally an associated BSIC for each GSM carrier in the NCL.

[14]CSFB enables voice and other circuit-switched services to be provided to UEs served by E-UTRAN by reusing the existing circuit switched infrastructure such as GERAN or UTRAN [9] – see Section 2.4.2.1.

22.3 Mobility Measurements

In order to support mobility within E-UTRAN and between E-UTRAN and other RATs (UTRAN FDD and TDD, GERAN and CDMA2000), a number of radio-related UE measurements are specified. All the E-UTRAN and inter-RAT measurements described in the following sections are reported by the UE to the serving eNodeB only in RRC_CONNECTED state. In RRC_IDLE, the measurements are not reported but may be used autonomously by the UE for cell reselection. Therefore although a particular measurement may be used in both RRC_CONNECTED and RRC_IDLE states, the corresponding requirements only apply to RRC_CONNECTED.

22.3.1 E-UTRAN Measurements

For mobility within E-UTRAN and from other RATs to E-UTRAN, two UE measurements are defined: RSRP and RSRQ [10].

22.3.1.1 Reference Signal Received Power (RSRP)

RSRP is measured by the UE over the cell-specific Reference Signals (RSs) within the measurement bandwidth over a measurement period. RSRP is a type of signal strength measurement and is indicative of the cell coverage. It is defined as the linear average over the power contributions (in Watts) of the REs that carry cell-specific RSs within the considered measurement frequency bandwidth. Normally the RSs transmitted on the first antenna port are used for RSRP determination, but the RSs on the second antenna port can also be used if the UE can reliably determine that they are being transmitted. If receive diversity is in use by the UE, the combined RSRP must be at least as large as the RSRP of any of the individual diversity branches. It is applicable in both RRC_IDLE and RRC_CONNECTED states and is used for cell reselection and handover within E-UTRAN (intra-frequency and inter-frequency) and to E-UTRAN from any of UTRAN FDD or TDD, GSM, CDMA2000 1xRTT or HRPD.

RSRP is therefore considered to be the most important measurement quantity for E-UTRAN.

An accuracy requirement is defined for the RSRP measurement (both intra- and inter-frequency); this is applicable for received signal quality $\hat{E}_s/I_{ot}$ as low as -6 dB. A measurement bandwidth equivalent to the central 6 Resource Blocks (RBs) is assumed. In the time domain the physical layer measurement periods when no DRX is used (and for short DRX cycles) are 200 ms and 480 ms for intra-frequency and inter-frequency RSRP respectively. The inter-frequency measurement naturally takes longer as it can only be performed during the measurement gaps; furthermore, the inter-frequency RSRP physical layer measurement period increases linearly with the number of carrier frequencies that has to be measured during the gaps, N_{freq}. The physical layer measurement period also increases in proportion to the DRX cycle for DRX cycles larger than 40 ms for intra-frequency RSRP and larger than 80 ms for inter-frequency RSRP. The measurement sampling rate is not specified but is left to the UE implementation.

The UE is required to be able to measure RSRP from at least 8 identified intra-frequency cells over the physical layer measurement period.[15] Similarly, the UE is also required to measure RSRP from at least 4 identified inter-frequency cells per inter-frequency carrier for up to 3 carriers (i.e. a total of 12 inter-frequency cells).

22.3.1.2 Reference Signal Received Quality (RSRQ)

RSRQ is the ratio of RSRP to RSSI for an E-UTRA carrier. The RSSI part of RSRQ is the total received power including interference from all sources, including serving and non-serving cells, adjacent channel interference and thermal noise. Unlike RSRP, it is measured on all REs in the OFDM symbols containing RSs for antenna port 0 (not just the REs containing RSs themselves), within the measurement bandwidth.[16]

This interference component of RSRQ enables the UE to quantify the received signal quality considering both signal strength and interference, which may vary with the UE's location in the cell [11].

In the first release of LTE (Release 8), RSRQ was applicable only in RRC_CONNECTED state. It is therefore used for handover within E-UTRAN, and from other RATs to E-UTRAN.

However, in order to prevent outages caused by high interference situations, in Release 9 RSRQ was also introduced for RRC_IDLE state; this gives the network the option to configure the UE to use RSRQ as a metric for performing cell reselection, at least in the cases of cell reselection within E-UTRAN, from UTRAN FDD to E-UTRAN and from GSM to E-UTRAN.

Both intra- and inter-frequency RSRQ measurement requirements are specified in [12]. The RSRQ and RSRP together have been shown to be particularly beneficial for performing inter-frequency quality-based handover [13]. RSRQ is inherently a relative quantity which to some extent eliminates absolute measurement errors and leads to better accuracy than is possible for RSRP. Like RSRP, the RSRQ accuracy requirements are applicable for $\hat{\mathrm{E}}_s/\mathrm{I}_{\mathrm{ot}}$ down to −6 dB (based on a measurement bandwidth equivalent to the central 6 RBs). The time domain physical layer measurement periods for RSRP are equivalent to those of RSRP for the corresponding DRX cases. The measurement sampling rate is also UE implementation dependent.

The UE is also required to measure RSRQ from the same number of intra- and inter-frequency cells as for RSRP.

22.3.2 UTRAN Measurements

The UTRAN measurements can be categorized into those specific to UTRAN FDD and those for E-UTRAN TDD. For mobility to UTRAN FDD, CPICH[17]RSCP, CPICH E_c/N_0[18] and RSSI measurements are defined [10]. The CPICH measurements are the most important for

[15]When measurement gaps are used, and depending upon the gap periodicity, the number of measured cells may be lower than 8 due to the fact that the measurement gaps reduce the time available for the UE to make intra-frequency measurements.

[16]Note that in Release 10 the RSRQ can be configured to be measured on all OFDM symbols if mechanisms for time-domain inter-cell interference coordination are configured – see Section 31.2.4.2.

[17]Common PIlot CHannel.

[18]Energy per chip divided by the noise power spectral density.

mobility between E-UTRAN and UTRAN FDD systems. For mobility to UTRAN TDD P-CCPCH[19] RSCP and RSSI measurements are used.

All these UTRAN measurements can be used in both RRC_IDLE and RRC_CONNECTED states. However, the network has the freedom to configure specific measurements in different mobility scenarios. Therefore, all the measurement requirements (such as accuracies) are applicable only in RRC_CONNECTED state.

22.3.2.1 UTRAN FDD CPICH RSCP

UTRAN FDD CPICH RSCP is measured on the Primary-CPICH of the target UTRAN cell and is used for both cell reselection and handover from E-UTRAN. For inter-RAT mobility, the CPICH RSCP absolute accuracy requirement should be fulfilled over the physical layer measurement period, whose length depends upon the gap pattern and DRX cycle (e.g. 480 ms for gap pattern #0 with DRX cycle not greater than 40 ms). The UE is required to measure the CPICHs from 6 identified UTRA FDD cells.

22.3.2.2 UTRAN FDD CPICH E_c/N_0

The UTRAN FDD CPICH E_c/N_0 is the ratio of the CPICH RSCP to the UTRA carrier RSSI. CPICH E_c/N_0 can be used for both cell reselection and handover.

The measurement periods for CPICH E_c/N_0 with and without DRX are the same as for CPICH RSCP.

22.3.2.3 UTRAN FDD Carrier RSSI

The UTRAN FDD carrier RSSI is the total received wideband power, including thermal noise and noise generated in the receiver, within the bandwidth defined by the receiver pulse shaping filter. In practice this measurement quantity is rarely used in RRC_IDLE, and its usage is mainly for performing RRC_CONNECTED handovers between E-UTRAN and UTRAN FDD. The measurement periods are the same as those described in Section 22.3.2.1.

22.3.2.4 UTRAN TDD P-CCPCH RSCP

The UTRAN TDD P-CCPCH RSCP is the received power of the P-CCPCH of a UTRA TDD neighbour cell. It is considered to be the most fundamental measurement quantity for all mobility scenarios between E-UTRAN and UTRAN TDD. It is therefore used for performing both cell reselection and handover to UTRA TDD. The P-CCPCH RSCP physical layer measurement period depends on the configured gap periodicity and the DRX cycle in the same way as for UTRAN FDD.

22.3.2.5 UTRAN TDD carrier RSSI

The UTRAN TDD carrier RSSI is the total received wideband power, including thermal noise and noise generated in the receiver, within the bandwidth defined by the receiver pulse shaping filter within a specified timeslot. In principle it can be used for both cell reselection

[19]Primary Common Control Physical CHannel.

and handover from E-UTRAN to UTRAN TDD. The measurement periods of UTRAN TDD carrier RSSI with and without DRX are the same as for the P-CCPCH RSCP measurement.

22.3.3 GSM Measurements: GSM Carrier RSSI

For mobility from E-UTRAN to GSM in RRC_IDLE mode (i.e. cell reselection) and RRC_CONNECTED mode (i.e. handover) the GSM carrier RSSI is used.

The GSM carrier RSSI is measured on a GSM BCCH carrier [10]. In accordance with the GSM core specification [14] the UE is required to take at least 3 GSM carrier RSSI samples per GSM carrier as evenly spaced as possible during the physical layer measurement period. The GSM carrier RSSI physical layer measurement period without DRX and for DRX cycles up to 80 ms is 480 ms regardless of the gap pattern, and increases linearly with the number of carrier frequencies, N_{freq}, monitored during the measurement gaps. For DRX cycles larger than 80 ms, the measurement period increases with the length of the DRX cycle.

22.3.4 CDMA2000 Measurements

For mobility from E-UTRAN to CDMA2000 systems the CDMA2000 1xRTT pilot strength and CDMA2000 High Rate Packet Data (HRPD) Pilot Strength measurement quantities are used in RRC_IDLE and RRC_CONNECTED states.

22.3.4.1 CDMA2000 1xRTT Pilot Strength

CDMA2000 1xRTT Pilot Strength is the strength of the received pilot signals as defined in [15, Section 2.6.6.2.2]. The required measurement accuracy, specified in [16, Section 3.2.4], is to be fulfilled over the measurement periods of about $1.1 * N_{\text{freq}}$ seconds and $2.1 * N_{\text{freq}}$ seconds for gap patterns #0 and #1 respectively. It is used for cell reselection and handover from E-UTRAN to CDMA2000 1xRTT. A CDMA20000 1xRTT-capable UE is required to monitor CDMA2000 1xRTT Pilot Strength of CDMA2000 1xRTT cells for up to five CDMA2000 1xRTT carriers.

22.3.4.2 CDMA2000 HRPD Pilot Strength

CDMA2000 HRPD pilot strength is the strength of the received pilot signals as defined in [17, Section 8.7.6.1.2.3]. It is used for cell reselection and handover from E-UTRAN to CDMA2000 HRPD. An HRPD-capable UE is required to monitor CDMA2000 HRPD Pilot Strength of HRPD cells for up to five HRPD carriers.

22.4 UE Measurement Reporting Mechanisms and Requirements

As explained in Section 22.3, in LTE the UE reports measurements to the serving eNodeB only in RRC_CONNECTED state using the Dedicated Control CHannel (DCCH). The measurement reporting mechanism can be periodic, event-triggered or event-triggered and periodic.

These reporting mechanisms are applicable both with and without DRX. The interval between periodic reports is configurable by the eNodeB and ranges from 120 ms to 3600 s [4]. The events which can trigger reporting by the UE are described in Section 3.2.5.1.

The total number of measurement reporting criteria which can be evaluated by the UE in parallel is limited to 21. This includes 9 E-UTRA intra-frequency reporting criteria, 7 E-UTRA inter-frequency reporting criteria and 5 inter-RAT reporting criteria. The actual number and types of reporting criteria are configured by the network and can be based on any of the three possible reporting mechanisms mentioned above.

The event triggered reporting delay requirements, which also apply to the first report in event-triggered periodic reporting, are explained in the following sections for different E-UTRA and inter-RAT cases.

22.4.1 E-UTRAN Event Triggered Reporting Requirements

The E-UTRAN event-triggered reporting consists of intra- and inter-frequency event reporting. Five such events (A1-A5) and their corresponding triggering thresholds are configured in the UE by the serving eNodeB as explained in section 3.2.5.2. When an event is triggered, the UE reports the event to the eNodeB, which may take an appropriate mobility decision.

For both intra-frequency and inter-frequency event-triggered measurements, the reporting delay[20] is less than the cell search delay (see Sections 22.2.1.1 and 22.2.1.2) for the corresponding cases. Due to signal variations and user mobility, the measured quality of a cell may vary between detectable and undetectable threshold levels.

To cater for this typical scenario, requirements are specified for both intra- and inter-frequency event reporting as follows: Regardless of the number of times the cell quality varies between the detectable and undetectable levels, whenever an event (i.e. any of the intra- or inter-frequency events A1-A5) is triggered the UE is required to send the event-triggered measurement report within a duration less than the relevant physical layer measurement period without layer 3 filtering (e.g. within 200 ms RSRP/RSRQ for the intra-frequency non-DRX case), provided certain conditions are met. When layer 3 filtering is used, a reporting delay longer than the physical layer measurement period is expected. For an intra-frequency event reporting, these conditions require that the cell which has been detectable over the intra-frequency cell search delay, does not become undetectable for more than 5 s. It is also required that the timing to the intra-frequency cell does not change by more than $\pm 50T_s$.[21] In case of inter-frequency event reporting the corresponding condition requires that the timing to the inter-frequency cell does not change by more than $\pm 50T_s$ during periods when the measurement gaps are not available (i.e. between successive measurement gaps). This is because in the absence of measurement gaps the UE cannot appropriately track the timing of an inter-frequency cell.

22.4.2 Inter-RAT Event-Triggered Reporting

Event-triggered reporting requirements are specified for E-UTRAN to UTRAN FDD, E-UTRAN to UTRAN TDD and E-UTRAN to GSM inter-RAT scenarios. Two inter-RAT

[20]Here we refer to the delay for the transmitted measurement reports without layer 3 filtering being applied. The behaviour of the Layer 3 filters is standardized and their configuration provided by RRC signalling. See [4, Section 5.5.3.2] for more details.

[21]T_s is the basic time unit and is equal to $1/(15000 \times 2048)$ s.

events (B1 and B2) and their corresponding triggering thresholds are also configured in the UE by the serving eNodeB as explained in section Section 3.2.5.2. When one of these events is triggered, the UE reports it to the eNodeB, allowing it to take an appropriate inter-RAT mobility decision.

E-UTRAN to UTRAN event-triggered reporting. The UTRA FDD and TDD event-triggered measurement reporting delays without Layer 3 filtering are less than the corresponding cell search delays (see Section 22.2.3). However, if a UTRA FDD cell, which was previously detectable over a period equal to the UTRA FDD cell search delay, enters or leaves the reporting range, then the event-triggered measurement reporting delay is less than the UTRA FDD physical layer measurement period without layer 3 filtering provided that the timing to that UTRA FDD cell has not changed more than ± 32 chips when the measurement gap has not been available. Note also that the inter-RAT physical layer measurement period is typically a few times shorter than the corresponding inter-RAT cell search delay.

E-UTRAN to GSM event-triggered reporting. The event-triggered measurement reporting delay for a GSM cell with BSIC verified without layer 3 filtering is less than twice the GSM measurement period (see Section 22.3.3).

22.5 Mobility Performance

The mobility procedures comprising of cell selection and cell reselection in RRC_IDLE state and handover in RRC_CONNECTED state are described in Chapter 3. Here we focus on the performance requirements for these mobility procedures.

22.5.1 Mobility Performance in RRC_IDLE State

The performance requirements for mobility in RRC_IDLE state aim to ensure that a UE camps on a cell which guarantees good paging reception, that substantial UE battery power saving is achieved and that the interruption in paging reception during cell reselection is minimized.

All intra-frequency, inter-frequency and inter-RAT cell reselection requirements have the following characteristics:

- Requirements are specified for a selected set of typical DRX cycles: 0.32 s, 0.64 s, 1.28 s and 2.56 s;
- Cell reselection involves detection of new neighbour cells (both E-UTRA and inter-RAT), and measurement of those cells and of previously detected neighbour cells;
- Measurement and evaluation of cells are carried out at specific rates, which depend upon the DRX cycle in use;
- Cell reselection decisions are autonomously taken by the UE but are governed by pre-defined standardized rules, network control parameters and performance requirements;
- No performance requirements are specified for E-UTRA or inter-RAT cell identification, RSRP/RSRQ or inter-RAT measurements in RRC_IDLE, since in RRC_IDLE the UE is not required to report any event or measurement to the network.

The key elements of cell reselection in E-UTRAN are described in the following subsections.

22.5.1.1 Measurement and Evaluation of Serving Cell

The UE is required to measure both RSRP and RSRQ of the serving cell in order to evaluate the serving cell selection criterion (the 'S-criterion' – see Section 3.3) at least once every DRX cycle.

Failure of the S-criterion may imply that the UE is on the verge of losing the serving cell and it is therefore important that the UE identifies a potential new serving cell. Therefore if the S-criterion for the serving cell is not met over a certain number of consecutive DRX cycles (depending on the DRX cycle length in use), the UE initiates measurements of all the neighbour cells (i.e. over all frequency layers indicated by the serving cell) regardless of the measurement rules. For example in case of a 1.28 s DRX cycle, the UE starts measuring all the neighbour cells if the S-criterion for the serving cell is not met for 2 consecutive DRX cycles (i.e. for 2.56 s).

22.5.1.2 Intra-frequency Cell Reselection

Cell reselection to a neighbour cell on the same frequency is governed by a set of requirements for measurement and evaluation of intra-frequency cells.

Measurement of intra-frequency cells. Unlike the measurement of the serving cell, the neighbour cell measurements may not be performed every DRX cycle. The UE initiates the measurement of intra-frequency neighbour cells when the serving cell's RSRP or RSRQ fall below their respective thresholds.

However, the cell ranking for cell reselection is only based on RSRP. The UE is required to detect and measure the RSRP of neighbour cells whose received quality is above the following thresholds, without an explicit intra-frequency neighbour cell list:

- Synchronization signal $\hat{E}_s/I_{ot} \geq -4$ dB
- RSRP and synchronization signal received power ≥ respective thresholds, which depend on the frequency band e.g. -124 dBm for Band 1.

The RSRP of the identified intra-frequency cells is measured once every $T_{measure,\ E\text{-}UTRA\_intra}$, as shown in Table 22.3.

Table 22.3: Measurement, detection and evaluation rates for intra-frequency cells.

DRX cycle length (s)	$T_{detect,EUTRAN\_intra}$ (s) (number of DRX cycles)	$T_{measure,EUTRAN\_intra}$ (s) (number of DRX cycles)	$T_{evaluate,EUTRAN\_intra}$ (s) (number of DRX cycles)
0.32	11.52 (36)	1.28 (4)	5.12 (16)
0.64	17.92 (28)	1.28 (2)	5.12 (8)
1.28	32(25)	1.28 (1)	6.4 (5)
2.56	58.88 (23)	2.56 (1)	7.68 (3)

Evaluation of intra-frequency cells. The measured intra-frequency cells are evaluated for possible cell reselection based on cell ranking (see Section 3.3.4.3). .

If an intra-frequency cell is detectable but not yet detected then the UE is required to evaluate whether it meets the reselection criteria based on ranking within $T_{detect,EUTRAN\_Intra}$ (see Table 22.3) when $T_{reselection} = 0$. On the other hand if the cell is already detected then the UE is required to evaluate whether this intra-frequency cell meets the reselection criteria based on ranking within $T_{evaluate,EUTRAN\_Intra}$ (see Table 22.3) when $T_{reselection} = 0$, provided that the target intra-frequency cell is ranked at least 3 dB above the serving cell. If the timer $T_{reselection}$ has a non-zero value and the target intra-frequency cell is found to be better ranked than the serving cell over the $T_{reselection}$ time, then that intra-frequency cell is selected.

22.5.1.3 Inter-frequency Cell Reselection

Similarly to the reselection of intra-frequency cells, requirements are also defined for measurement and evaluation of inter-frequency cells.

Measurement of inter-frequency cells. An inter-frequency layer may have a lower, equal or higher priority than that of the serving frequency layer. The UE is required to detect and measure the relevant measurement quantity under the same conditions as are applicable for intra-frequency cells.

If the serving cell's RSRP and RSRQ are above their respective thresholds the UE searches higher-priority inter-frequency layers at least once every $T_{higher\_priority\_search}$:

$$T_{higher\_priority\_search} = (60 * N_{layers}) \text{ seconds} \quad (22.1)$$

where N_{layers} is the total number of configured higher-priority inter-frequency and inter-RAT frequency layers. The relevant measurement quantities should be measured at least every $T_{measure, E\text{-}UTRAN\_Inter}$ as defined in Table 22.4.

If the serving cell's RSRP and RSRQ become equal to or fall below their respective thresholds, the UE searches and measures all inter-frequency cells regardless of their priority. The relevant measurement quantity of the identified inter-frequency cells is measured every $K_{carrier} * T_{measure, E\text{-}UTRA\_Inter}$ seconds, as shown in Table 22.3, where $K_{carrier}$ is the number of inter-frequency carriers configured by the serving cell.

Table 22.4: Measurement and Evaluation of Inter-frequency Cells.

DRX cycle length (s)	$T_{detect,EUTRAN\_Inter}$ (s) (number of DRX cycles)	$T_{measure,EUTRAN\_Inter}$ (s) (number of DRX cycles)	$T_{evaluate,EUTRAN\_Inter}$ (s) (number of DRX cycles)
0.32	11.52 (36)	1.28 (4)	5.12 (16)
0.64	17.92 (28)	1.28 (2)	5.12 (8)
1.28	32 (25)	1.28 (1)	6.4 (5)
2.56	58.88 (23)	2.56 (1)	7.68 (3)

Evaluation of inter-frequency cells. The evaluation of inter-frequency cells for possible cell reselection is based on cell ranking for layers of equal priority and on absolute priorities for layers of different priority. Only RSRP is allowed for cell ranking, while the network can configure either RSRP or RSRQ for the evaluation of cells with unequal priorities [18].

If an inter-frequency cell of lower or equal priority is detectable but not yet detected then the UE is required to evaluate whether it meets the reselection criteria (see Sections 3.3.4.2 and 3.3.4.3) within $K_{\text{carrier}} * T_{\text{detect,EUTRAN\_Inter}}$ seconds (see Table 22.4) provided that the cell reselection criteria can be met by a certain margin (5 dB for cell ranking and 6 dB for absolute priority based reselection) and the timer value $T_{\text{reselection}}$ is zero.

However, if an inter-frequency cell is already detected then the UE is required to evaluate whether it meets the reselection criteria within a shorter duration, $K_{\text{carrier}} * T_{\text{evaluate,EUTRAN\_Inter}}$ seconds (see Table 22.4) provided the same two conditions above are satisfied.

If the timer $T_{\text{reselection}}$ has a non-zero value and the target inter-frequency cell is found to be better ranked than the serving cell over the time $T_{\text{reselection}}$, then that inter-frequency cell is reselected.

22.5.1.4 Inter-RAT Cell Reselection

Inter-RAT cell reselection covers reselection to UTRAN FDD, UTRAN TDD, GSM, CDMA2000 1xRTT and CDMA2000 HRPD.

As for inter- and intra-frequency reselection, cell reselection to an inter-RAT neighbour cell is also governed by the measurement and evaluation of the inter-RAT cells.

Measurement of inter-RAT cells. An inter-RAT frequency layer may have either lower or higher priority than that of the serving frequency layer. If the quality of the serving cell is above the threshold 'Snonintrasearch'[22] the UE searches higher priority inter-RAT frequency layers at least once every $T_{\text{higher\_priority\_search}}$ according to Equation (22.1). The detected higher-priority inter-RAT cells of RAT_j are to be measured at least every $T_{\text{measure,RAT,j}}$ as defined in Table 22.5.

If the serving cell quality becomes equal to or falls below the Snonintrasearch threshold the UE searches and measures all inter-RAT cells of RAT_j (i.e. cells of higher, lower or equal priority frequency layers of RAT_j). The relevant measurement quantity of the identified inter-RAT cells is measured every $N_{\text{RAT},j} * T_{\text{measure, RAT,j}}$, as illustrated in Table 22.5, where $N_{\text{RAT},j}$ is the number of frequency layers of RAT_j configured by the serving cell (for GSM, $N_{\text{RAT}} = 1$). The relevant inter-RAT measurement quantities in RRC_IDLE states are described in Section 22.3.

Evaluation of inter-RAT cells. Evaluation of inter-RAT cells for possible cell reselection is based only on priority (see Section 3.3.4.2).

If an inter-RAT cell is detectable but not yet detected then the UE is required to evaluate whether it meets the reselection criteria within $N_{\text{RAT},j} * T_{\text{detect, RAT, j}}$ (see Table 22.5) provided that the cell reselection criteria can be met by a 6 dB margin (in the case of UTRA) and the timer value $T_{\text{reselection}} = 0$.

[22]See Section 3.3.4.1; 'Snonintrasearch' is defined in [18, Section 5.2.4.7].

Table 22.5: Measurement and Evaluation of Inter-RAT Cells.

DRX cycle length (s)	apply to UTRA FDD and TDD	$T_{detect,\,RAT,j}$ (s); $T_{measure,RAT,j}$ (s) (number of DRX cycles); apply to all RATs	$T_{evaluate,\,RAT,j}$ (s) (number of DRX cycles); apply to UTRA FDD, TDD and CDMA2000
0.32	30	5.12 (16)	15.36 (48)
0.64	30	5.12 (8)	15.36 (24)
1.28	30		6.4 (5) 19.2 (15)
2.56	60	7.68 (3)	23.04 (9)

For an inter-RAT cell that has already been detected, the UE is required to evaluate that it meets the reselection criteria within a shorter duration, $N_{\mathrm{RAT,\,j}} * T_{\mathrm{evaluate,\,RAT,\,j}}$ (see Table 22.4) provided that the cell reselection criteria of the respective RAT are satisfied.

22.5.1.5 Paging Interruption during Cell Reselection

During intra-frequency, inter-frequency or inter-RAT cell reselection procedures, the UE monitors the serving cell for paging messages until the UE is able to monitor the paging channels of the target cell. In order to complete the cell reselection successfully and camp on the new cell the UE has to acquire the relevant system information of the target cell. Therefore, during this period paging interruption might occur as the UE is not required simultaneously to receive paging and acquire the system information of the target cell.

The interruption in paging reception should not exceed $T_{\mathrm{Interrupt}} = T_{\mathrm{SI\text{-}RAT}} + 50(\mathrm{ms})$, where $T_{\mathrm{SI\text{-}RAT}}$ is the time required for acquiring all the relevant system information of the target RAT.

22.5.2 Mobility Performance in RRC_CONNECTED State

In E-UTRAN, only hard handovers are possible (see Section 3.2.3.4) resulting in a delay including a short interruption. In order to limit the length of the interruption, requirements are specified for various E-UTRAN handover scenarios as described in the following subsections. These requirements are expressed in terms of handover delay, which is the sum of the RRC procedure delay and the interruption time. This principle applies whether the target cell is known to the UE (referred to as *non-blind handover*), or unknown (*blind handover*). The interruption time is defined as the time from the end of the last subframe containing the handover command on the Physical Downlink Shared CHannel (PDSCH) from the serving cell and the moment the UE starts transmission on the relevant uplink physical channel in the target cell. Figure 22.6 illustrates the radio interface signalling of the handover procedure to a known target.

As explained in Section 3.2.3.4, the network does not require measurement reports from target cells for performing a blind handover. This is particularly useful in case of multiple frequency layers, which can only be monitored using the same gap pattern (i.e. either pattern #0 or pattern #1). In such a scenario the network may request the UE to perform measurements on only a subset of the frequency layers while relying on blind handovers for

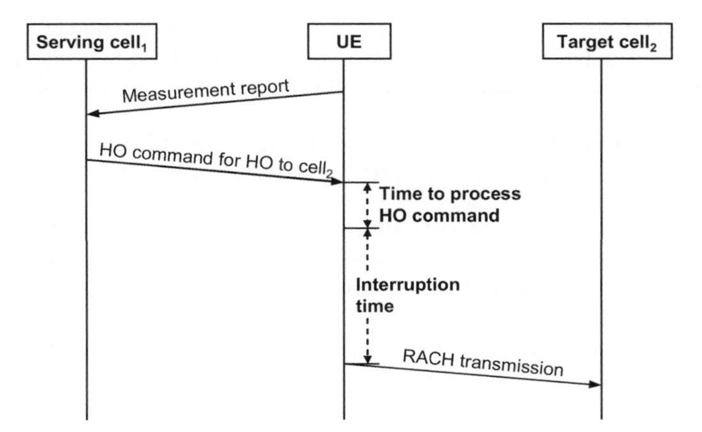

Figure 22.6: Handover to a known target cell.

the remaining ones. Handover to an unknown target cell results in a longer interruption time since the UE has to detect the target cell prior to accessing it. The process of handover to an unknown target cell is shown in Figure 22.7.

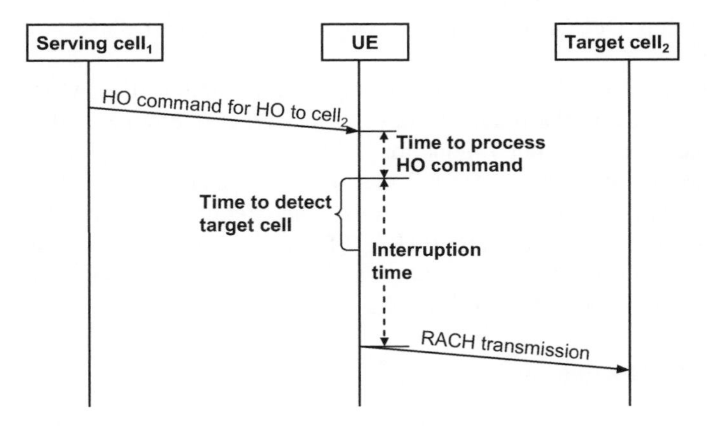

Figure 22.7: Handover to an unknown target cell.

22.5.2.1 E-UTRAN to E-UTRAN Handover

An E-UTRAN to E-UTRAN handover is completed when the UE starts transmission on the Physical Random Access CHannel (PRACH) in the target E-UTRA cell. The interruption time is expressed as $T_{interrupt} = T_{search} + T_{IU} + 20$ (ms), where:

- T_{search} is the cell search delay, which is 0 for known target cells and 80 ms for unknown target cells. The target cell is considered to be known if the cell search requirements were met in the last 5 s; otherwise the cell is considered to be unknown.
- T_{IU} is the timing uncertainty due to the timing of the PRACH occasions, and thus depends upon the PRACH configuration (see Section 17.4.2.5).

As an example, suppose the PRACH is allowed once in the middle of every frame (i.e. in every 4th subframe). If the target cell is known, the total intra-frequency or inter-frequency handover delay is 50 ms including 15 ms of RRC procedure delay (see [4, Section 11.2]). For the same PRACH configuration, in case of blind handover, the total handover delay is 130 ms including the 15 ms RRC procedure delay. In this case the cell search is much shorter than the usual cell search requirements (see Section 22.2.1) due to the fact that it is assumed that the target cell is sufficiently strong to be detected by the UE on the first correlation attempt.

22.5.2.2 Handover to Other 3GPP RATs

E-UTRAN to UTRAN handover

UTRA FDD or TDD cells can be known or unknown for the purposes of handover. The target cell is known if it has been measured by the UE during the last 5 s; otherwise it is considered unknown. An E-UTRAN-UTRAN FDD handover is completed when the UE starts transmission on the uplink Dedicated Physical Control CHannel (DPCCH) in the target cell. An E-UTRAN-UTRAN TDD handover is completed when the UE starts transmission of either the uplink DPCCH or an uplink synchronization code ('SYNC-UL') in the target cell.

Depending upon the choice of parameters, the interruption time when the target cell is known is typically in the order of 100–150 ms. When the target cell is unknown, the interruption increases by about 100 ms for UTRA FDD and 160 ms for UTRA TDD. The RRC procedure delay is 50 ms (see [12, Sections 5.3.1 and 5.3.2]) as this requirement is based on the UTRAN requirement [19]. Hence the overall handover delay can be of the order of 150–200 ms when the target UTRA (FDD or TDD) cell is known and 250–300 ms when the target cell is unknown. The exact handover delay can be derived from the relevant expressions in [12, Sections 5.3.1 and 5.3.2].

E-UTRAN to GSM handover

Handovers from the serving eNodeB to both known and unknown target GSM cells are supported in E-UTRAN. A handover to GSM is completed when the UE starts uplink transmission in the target GSM cell.

The total handover delays to known and unknown target GSM cells are 90 ms and 190 ms respectively, which includes 50 ms delay for processing the handover command (see [12, Section 5.3.3.1]).

22.5.2.3 Handover to CDMA2000 1xRTT or HRPD

Non-3GPP RAT handovers defined from E-UTRAN are for CDMA2000 1xRTT and HRPD. Handovers to both known and unknown target CDMA2000 1xRTT or HRPD cells are possible. In both of these CDMA2000 technologies, a cell is considered known if it has been measured by the UE during the last 5 s. When performing handover to CDMA2000 1xRTT or HRPD the interruption time is influenced primarily by the following factors :

- Uncertainty due to changing the timing from the old E-UTRA serving cell to the new CDMA2000 1xRTT or HRPD cell (the delay can be up to one frame, i.e. 20 ms for CDMA2000 1xRTT and 26.7 ms for HRPD;
- The number of known and unknown target cells, which is signalled to the UE in the E-UTRAN System Information Block 8 (SIB8);
- The search window sizes for searching the known and unknown target cells; this is expressed in CDMA2000 1xRTT chips and signalled to the UE in SIB8.

The RRC procedure delay for processing the handover command is 130 ms for CDMA2000 1xRTT and 50 ms for HRPD. In order to derive the handover delay for different combinations of parameters, the reader is referred to [12, Section 5.4.2]. As a typical example the handover delay can be about 200 ms for CDMA2000 1xRTT and 130 ms for HRPD, assuming one target cell (regardless of whether it is known or unknown) and a typical window size of 60 chips.

22.6 RRC Connection Mobility Control Performance

In LTE the establishment and maintenance of the RRC connection is governed by the two main control plane functions: RRC Connection Re-establishment and Random Access.

22.6.1 RRC Connection Re-establishment

RRC connection re-establishment is initiated when a UE in RRC_CONNECTED state loses its RRC connection (e.g. due to radio link failure, handover failure or RRC connection reconfiguration failure), as specified in [4, Section 5.3.7.3]. The UE tries to re-establish the RRC connection with the strongest E-UTRA cell. Successful RRC re-establishment means that the UE is able to send the 'RRCConnectionReestablishmentRequest' message within $\mathrm{T}_{\text{re-establish\_delay}}$, including the delay in acquiring the uplink grant for sending the message and $\mathrm{T}_{\text{UE-re-establish\_delay}}$ which is defined as the delay from the moment when the UE detects the need for RRC re-establishment until it transmits a random access signal to the target cell:

$$\mathrm{T}_{\text{UE-re-establish\_delay}} = 50\text{ms} + N_{\text{freq}} * \mathrm{T}_{\text{search}} + \mathrm{T}_{\text{SI}} + \mathrm{T}_{\text{PRACH}} \tag{22.2}$$

where N_{freq} is the total number of E-UTRA carrier frequencies available for RRC re-establishment,$\mathrm{T}_{\text{search}}$ is the target cell search delay and depends upon whether the target cell is known or unknown to the UE, T_{SI} is the time required to read the target cell System Information (SI) and $\mathrm{T}_{\text{PRACH}}$ is the delay due to random access.

22.6.2 Random Access

In LTE the random access procedure serves several purposes as described in Sections 17.2 and 17.3. Requirements are specified for both contention-based and contention-free random access. The primary objectives of the requirements are to ensure the correct UE behavior when performing random access, and that the UE transmit power accuracy and transmit timing error when sending random access are within suitable limits. These requirements include:

- If the UE does not receive a random access response matching its PRACH preamble identity within the random access response window, it should retransmit a preamble, up to and not exceeding the maximum number of preamble transmissions configured by the eNodeB.
- The transmit power accuracy of the UE's PRACH preamble transmissions should fulfil the requirements defined in [20, Sections 6.3.5.1 and 6.3.5.2].
- The required transmit timing accuracy of the UE's PRACH preamble transmissions depends upon the transmission bandwidth, since at lower bandwidth the UE typically uses lower sampling rate. The transmit timing error due to all PRACH transmissions (initial and subsequent preamble transmissions) should be within $\pm\ 24T_s$ and $\pm\ 12T_s$ for system bandwidths equal to 1.4 MHz and $\geq$ 3 MHz respectively, as specified in [12, Section 7.1.2].

22.7 Radio Link Monitoring Performance

The purpose of the radio link monitoring function in the UE is to monitor the downlink radio link quality of the serving cell in RRC_CONNECTED state and is based on the cell-specific RSs (see Section 8.2.1). This in turn enables the UE when in RRC_CONNECTED state to determine whether it is *in-sync* or *out-of-sync* with respect to its serving cell (see [21, Section 4.2.1]).

In case of a certain number of consecutive out-of-sync indications (called 'N310'), the UE starts a network-configured radio link failure timer 'T310'. The timer is stopped if a number 'N311' of consecutive in-sync indications are reported by the UE's physical layer. Both the out-of-sync and in-sync counters (N310 and N311) are configurable by the network. Upon expiry of the timer T310, Radio Link Failure (RLF) occurs. As a consequence the UE turns off its transmitter to avoid interference and is required to re-establish the RRC connection within $T_{\text{UE-re-establish\_delay}}$ as explained in Section 22.6.1. The various actions pertaining to radio link monitoring and the subsequent RRC re-establishment to the target cell are shown in Figure 22.8.

22.7.1 In-sync and Out-of-sync Thresholds

The UE's estimate of the downlink radio link quality is compared with out-of-sync and in-sync thresholds, Q_{out} and Q_{in}, for the purpose of radio link monitoring. These thresholds are expressed in terms of the BLock Error Rate (BLER) of a hypothetical Physical Downlink Control Channel (PDCCH) transmission from the serving cell (see [12, Section 7.6] for details). Specifically, Q_{out} corresponds to a 10% BLER while Q_{in} corresponds to a 2% BLER. The same threshold levels are applicable with and without DRX.

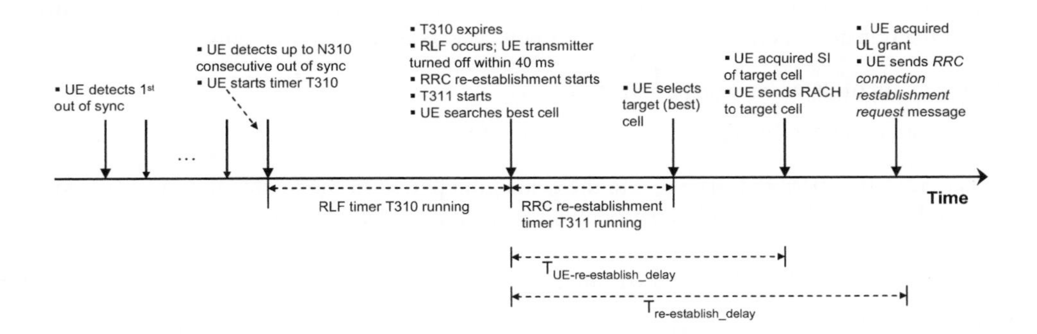

Figure 22.8: Radio link monitoring of the serving cell followed by RRC re-establishment to the target cell.

The mapping between the cell specific RS based downlink quality and the hypothetical PDCCH BLER is up to the UE implementation. However, the performance is verified by conformance tests defined for various environments [22].

22.7.2 Requirements without DRX

When no DRX is configured, out-of-sync occurs when the downlink radio link quality estimated over the last 200 ms period becomes worse than the threshold Q_{out}. Similarly without DRX the in-sync occurs when the downlink radio link quality estimated over the last 100 ms period becomes better than the threshold Q_{in}. Upon detection of out-of-sync, the UE initiates the evaluation of in-sync. The occurrences of out-of-sync and in-sync are reported internally by the UE's physical layer to its higher layers, which in turn may apply layer 3 (i.e. higher layer) filtering for the evaluation of RLF.

22.7.3 Requirements with DRX

When DRX is in use, in order to enable sufficient UE power saving the out-of-sync and in-sync evaluation periods are extended and depend upon the configured DRX cycle length. The UE starts in-sync evaluation whenever out-of-sync occurs. Therefore the same period ($\text{T}_{\text{Evaluate}\_Q\text{out\_DRX}}$) is used for the evaluation of out-of-sync and in-sync. However, upon starting the RLF timer (T310) until its expiry, the in-sync evaluation period is shortened to 100 ms, which is the same as without DRX. If the timer T310 is stopped due to N311 consecutive in-sync indications, the UE performs in-sync evaluation according to the DRX-based period ($\text{T}_{\text{Evaluate}\_Q\text{out\_DRX}}$).

22.7.4 Requirements during Transitions

In LTE a transition phase is caused by switching between DRX and non-DRX operation or switching between short and long DRX or vice versa. These scenarios can occur often, and therefore the UE behaviour is specifically defined for the evaluation of the radio link quality during the transition period. There are two main aspects of the requirements:

- Length of the transition period, T_P;
- Evaluation period, $T_{Evaluate\_Transition}$, during the transition.

The transition period T_P is equal to the evaluation period of the mode after the transition. During this phase the evaluation period $T_{Evaluate\_Transition}$ is defined as follows:

$$T_{Evaluate\_Transition} \geq \min(T_{Evaluate\_mode1}, T_{Evaluate\_mode2}) \tag{22.3}$$

where $T_{Evaluate\_mode1}$ and $T_{Evaluate\_mode2}$ correspond to the evaluation periods of the first and the second mode respectively.

Equation (22.3) applies to both in-sync and out-of-sync evaluations. After the transition period, the UE uses an evaluation period corresponding to the second mode. The evaluation periods during the transition from short to long DRX and vice versa are illustrated in Figure 22.9.

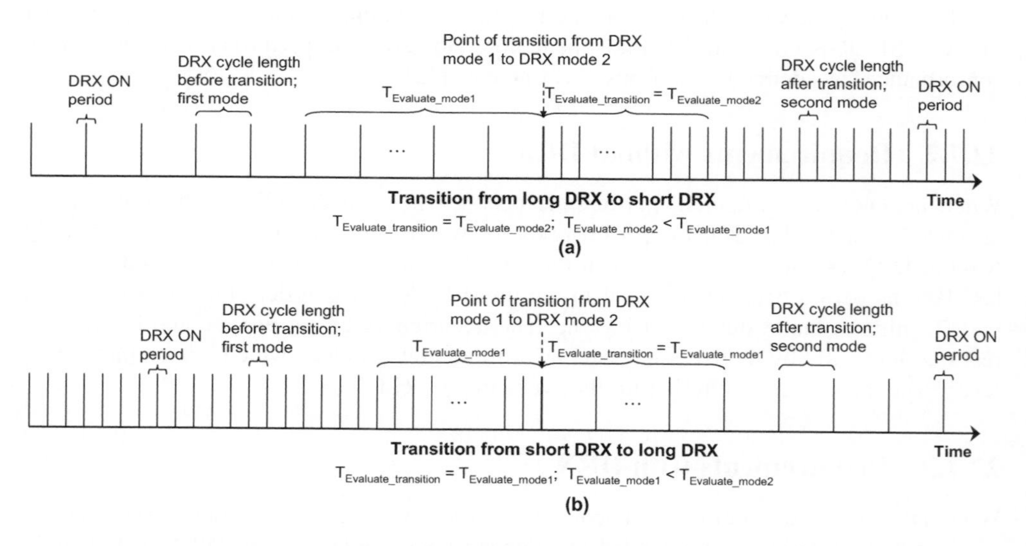

Figure 22.9: Radio link monitoring evaluation period (a) during transition from long DRX to short DRX, and (b) during transition from short DRX to long DRX.

22.8 Concluding Remarks

In this chapter we have explained the main aspects of the LTE RRM performance requirements. These requirements play a key role in ensuring that the UE meets the desired mobility performance in a wide range of practical scenarios envisaged for LTE. More specifically, by virtue of these requirements, robust mobility in both RRC_IDLE and RRC_CONNECTED states is ensured, both within E-UTRAN and between E-UTRAN and other RATs including UTRAN FDD/TDD, GSM and CDMA2000 technologies.

References[23]

[1] 3GPP Technical Specification 36.211, 'Evolved Universal Terrestrial Radio Access (E-UTRA); Physical Channels and Modulation', www.3gpp.org.

[2] Texas Instruments, NXP, Motorola, Ericsson, and Nokia, 'R4-072215: Simulation Assumptions for Intra-frequency Cell Identification', 3GPP TSG RAN WG4, meeting 45, Jeju, Korea, November 2007, www.3gpp.org.

[3] NXP, 'R4-080691: LTE Cell Identification Performance in Multi-cell Environment', www.3gpp.org 3GPP TSG RAN WG4, meeting 46bis, Shenzen, China, February 2008.

[4] 3GPP Technical Specification 36.331, ' Evolved Universal Terrestrial Radio Access (E-UTRA); Radio Resource Control (RRC); Protocol specification', www.3gpp.org.

[5] 3GPP Technical Specification 36.300, 'Evolved Universal Terrestrial Radio Access (E-UTRA) and Evolved Universal Terrestrial Radio Access Network (E-UTRAN); Overall description; Stage 2', www.3gpp.org.

[6] 3GPP Technical Specification 25.101, 'User Equipment (UE) Radio Transmission and Reception (FDD)', www.3gpp.org.

[7] M. Mouly and M.-B. Pautet, *The GSM System for Mobile Communications*. Cell & Sys, 1992.

[8] 3GPP Technical Specification 45.005, ' Radio transmission and reception', www.3gpp.org.

[9] 3GPP Technical Specification 23.272, 'Circuit Switched (CS) fallback in Evolved Packet System (EPS); Stage 2', www.3gpp.org.

[10] 3GPP Technical Specification 36.214, 'Evolved Universal Terrestrial Radio Access (E-UTRA); Physical Layer Measurements', www.3gpp.org.

[11] Ericsson, 'R1-073041: Reference Signal Received Quality, RSRQ, Measurement', www.3gpp.org, 3GPP TSG RAN WG1, meeting 49bis, Orlando, USA, June 2007.

[12] 3GPP Technical Specification 36.133, 'Evolved Universal Terrestrial Radio Access (E-UTRA); Requirements for Support of Radio Resource Management', www.3gpp.org.

[13] M. Kazmi et al., 'Evaluation of Inter-Frequency Quality Handover Criteria in E-UTRAN', IEEE 69th Vehicular Technology Conference, Spring 2009.

[14] 3GPP Technical Specification 45.008, 'Radio subsystem link control', www.3gpp.org.

[15] 3GPP2 Technical Specification C.S0005-D, 'Upper Layer (Layer 3) Signaling Specification for cdma2000 Spread Spectrum Systems', www.3gpp2.org.

[16] 3GPP2 Technical Specification C.S0011-C, 'Recommended Minimum Performance Standards for cdma2000 Spread Spectrum Mobile Stations', www.3gpp2.org.

[17] 3GPP2 Technical Specification C.S0024-A, 'cdma2000 High Rate Packet Data Air Interface Specification', www.3gpp2.org.

[18] 3GPP Technical Specification 36.304, 'Evolved Universal Terrestrial Radio Access (E-UTRA); User Equipment (UE) Procedures in Idle Mode', www.3gpp.org.

[19] 3GPP Technical Specification 25.331, 'Radio Resource Control (RRC) Protocol Specification', www.3gpp.org.

[20] 3GPP Technical Specification 36.101, 'Evolved Universal Terrestrial Radio Access (E-UTRA); User Equipment (UE) Radio Transmission and Reception', www.3gpp.org.

[23] All web sites confirmed 1st March 2011.

[21] 3GPP Technical Specification 36.213, 'Evolved Universal Terrestrial Radio Access (E-UTRA); Physical Layer Procedures', www.3gpp.org.

[22] 3GPP Technical Specification 36.521-3, 'Evolved Universal Terrestrial Radio Access (E-UTRA); User Equipment (UE) conformance specification; Radio Transmission and Reception; Part 3: Radio Resource Management Conformance Testing', www.3gpp.org.

23

Paired and Unpaired Spectrum

Nicholas Anderson

23.1 Introduction

Expansion of consumer demand for cellular communications, as well as for the multitude of other wireless systems and applications, places a corresponding strain on the basic physical resource needed to support them: spectrum. The suitability of spectral resources to a particular application is governed by a range of inter-related factors of a technological, commercial or regulatory nature. Technological considerations influencing the choice of frequency band include propagation characteristics, antenna size and separation, the viability of Radio Frequency (RF) circuitry, and design implications resulting from the need to coexist with systems operating in neighbouring spectrum without causing (or suffering from) undue interference. These considerations determine in part the commercial viability of a system. Signal range and spectral efficiency govern coverage and capacity, and hence determine the required number of base station sites and the capital outlay, whilst terminal costs and form factor affect the acceptability of products in the marketplace.

From a regulatory perspective, in spite of significant coordination at an international level through the International Telecommunications Union (ITU), regional variations in governmental policy regarding spectrum and technology are inescapable due to the differing needs of each region and the different historical factors which have shaped their present-day spectrum allocations. Thus, the availability of globally harmonized spectrum for a particular technology cannot be guaranteed. Furthermore, in many regions, a tendency towards technology neutrality within viable spectrum assignments (see, for example, [1]) is beginning to liberalize the traditional one-to-one mappings between technologies and their addressable spectrum and is hence opening up new markets.

LTE – The UMTS Long Term Evolution: From Theory to Practice, Second Edition.
Stefania Sesia, Issam Toufik and Matthew Baker.

Against this background, a key design goal of LTE was to enable deployment in a diverse range of spectrum environments in terms of bandwidth, uplink-downlink duplex spacing[1] and uplink-downlink asymmetries. By supporting paired spectrum allocations (separate uplink and downlink carriers) in addition to stand-alone unpaired allocations (uplink and downlink operating on the same carrier frequency), wastage of valuable spectrum can be avoided.

It is also the case that traffic is increasingly data-centric and often asymmetric. In unpaired spectrum, asymmetry may be provided through the use of unequal duty cycles in the time domain for uplink and downlink. In paired spectrum, asymmetry is also possible via the deployment of unequal bandwidths for uplink and downlink.

Considering the nature of the regulatory environment, an increase in demand and competition for spectrum and the trends of mobile traffic and service usage, it is advantageous for the LTE system to be designed such that it may be flexibly adapted to diverse spectral assignments including both paired and unpaired bands. The market addressable by the technology is thereby increased.

23.2 Duplex Modes

The term 'duplex' refers to bidirectional communication between two devices, as distinct from unidirectional communication which is referred to as 'simplex'. In the bidirectional case, transmissions over the link in each direction may take place at the same time ('full duplex') or at mutually exclusive times ('half-duplex').

It should be noted that a communication link being half-duplex need not imply that only half-duplex user services are supported (such as 'push-to-talk' voice applications). Full-duplex services such as a normal telephone conversation may of course be carried over half-duplex communication systems. The key factor differentiating the duplex nature of the service from that of the underlying communication link is the timescale over which the communication direction is cycled in relation to the service duplex timescale: if the link direction is cycled at a sufficiently high rate then the duplex nature of the link can be concealed when viewed from the perspective of the user application.

In the case of a full-duplex transceiver, the frequency domain is used to separate the inbound and outbound communications; that is, different carrier frequencies are employed for each link direction. This is referred to as Frequency Division Duplex (FDD). As discussed in Section 21.4.2, the ability of a full-duplex transceiver to transmit and receive at the same time instant is enabled via the use of a duplexer – a tuned filter network able to provide a high degree of isolation between the inbound and outbound signals on the different carrier frequencies (often sharing the same antenna). However, these filter networks incur some signal attenuation. For the receiver, this attenuation occurs before the low-noise amplifier in the signal path and hence contributes directly to the receiver noise figure and degrades its sensitivity. For the transmitter, the duplexer follows the high power amplification stage, requiring either a higher-powered amplification device to overcome the loss, or tolerance of the corresponding reduction in communication range.

Conversely, in the case of a half-duplex transceiver, the time domain provides the necessary separation between the inbound and outbound communications. When the same

[1] 'Duplex spacing' is the term used to describe the size of the frequency separation between uplink and downlink carriers.

carrier frequency is used for each link direction, the system is said to be purely Time Division Duplex (TDD) and incorporates half-duplex transceivers at each end of the radio link. Alternatively different carrier frequencies may be used, in which case the system is often known as 'Half-Duplex FDD' (HD-FDD). The popular GSM system utilizes HD-FDD, with uplink and downlink communications taking place for a particular user not only on distinct carrier frequencies but also at different times. HD-FDD can therefore be seen as a hybrid combination of FDD and TDD. By scheduling users (or sets of users) at mutually exclusive times, full occupancy of the transmission resources may be achieved without requiring simultaneous transmission and reception at each mobile terminal. Thus for HD-FDD, the base station is full-duplex whilst the terminals are half-duplex.

The three duplex modes outlined above are represented diagrammatically in Figure 23.1.

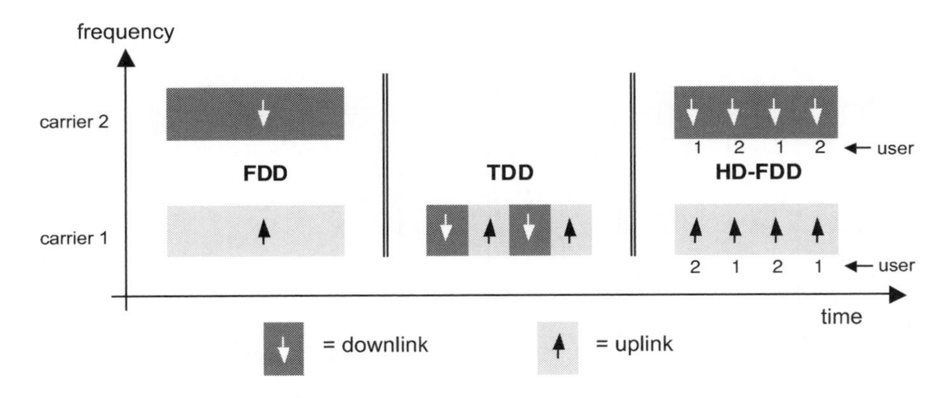

Figure 23.1: FDD, TDD and HD-FDD duplex modes.

Both HD-FDD and TDD carry the advantage that terminals may be developed without the need for duplexers, thus simplifying the design. HD-FDD operation is also potentially useful for paired bands in which the duplex spacing is relatively narrow (e.g. less than a few times the system bandwidth) since it avoids the need for a technically challenging duplexer design in the terminal and can thereby help in reducing its cost. At the base station there is more scope to implement such technically challenging designs due to less onerous size and cost constraints and because base station products are usually targeted towards a specific frequency band.

The spectrum bands defined for LTE are listed in Section 21.2. The paired bands are all currently symmetrically dimensioned between uplink and downlink, although this may change in the future.

23.3 Interference Issues in Unpaired Spectrum

As discussed in Chapter 21, the spectral emissions from an eNodeB or from a UE are unfortunately not strictly band-limited within the desired carrier bandwidth. This is due to practical limitations on filter technology and on the linearity of amplification.

In addition, receivers suffer from non-ideal suppression of signals falling outside the desired reception band. There is thus the potential for interference between transmitter and receiver. This is shown in Figure 23.2, where all sources of attenuation of the signal between transmitter and receiver (due to phenomena such as propagation, antenna radiation patterns and cable losses) are denoted by a single 'coupling loss'. In the region marked 'A', energy from the transmitter falls directly within the desired passband of the receiver, while in region 'B' the non-ideal characteristic of the receiver collects energy from inside the passband of the transmitter. In region 'C' transmitted energy falling outside both the transmit and receive passbands is collected by the non-ideal receiver characteristic. The aggregated effects of regions A, B and C represent the total unwanted energy from the interfering transmitter that is captured by the receiver.[2]

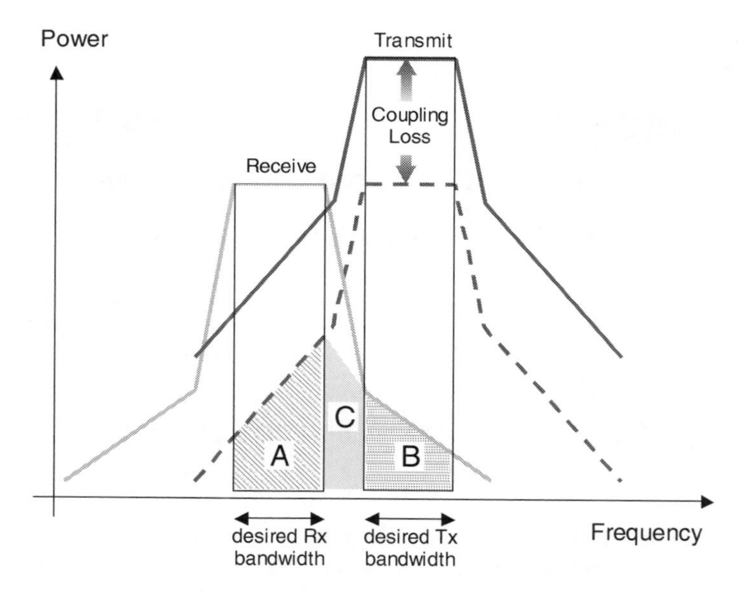

Figure 23.2: Frequency-domain interference between transmitter and receiver.

Interference arising from regions A and B is controlled by performance requirements imposed on the transmitting and receiving devices respectively, while in region C both transmitter performance and receiver performance have an impact. The interference effects are generally small for large frequency separations (as in the majority of FDD cases), and when there is a large coupling loss between the interfering transmitter and a 'victim' receiver, as is the case when there is a large physical separation between the two.

Naturally, the interference effects are most pronounced when the transmitter and receiver operate on adjacent carrier frequencies (excepting the co-channel case) and with low coupling loss (e.g. due to small spatial separation). These worst-case adjacent-channel scenarios form the basis upon which key requirements for transmitter and receiver are generally set.

[2]The limits on these emissions according to the LTE specification are explained in Chapter 21.

The overall 'leakage' (whether at the transmit side or the receive side) from a transmission on one carrier into a receiver operating on an adjacent carrier is described by the *Adjacent Channel Interference Ratio* (ACIR), which is derived from the transmitter's *Adjacent Channel Leakage Ratio* (ACLR) and the receiver's *Adjacent Channel Selectivity* (ACS) as defined in Section 21.3.2.1. It is worth noting that it is often the case that the ACIR is dominated by either the ACS or the ACLR. For example, it is technologically difficult to design a mobile transmitter with an ACLR that approaches or exceeds the ACS of a base station receiver.

As is discussed in the following section, the presence of imperfect ACIR has implications for the deployment of systems at a boundary between unpaired and paired spectrum allocations, and also for unsynchronized systems operating in closely spaced unpaired allocations.

23.3.1 Adjacent Carrier Interference Scenarios

For an FDD cellular system, adjacent channel frequency separation of an interfering transmitter and a victim receiver naturally implies that the interferer and victim are of differing equipment types (i.e. one is a mobile terminal whilst the other is a base station). Transmitter–receiver interference between one User Equipment (UE) and another, or between one eNodeB and another is avoided by virtue of the duplex spacing.

The same is also generally true in TDD systems if they are time-synchronized so that overlap between uplink and downlink transmission periods is avoided. However, when synchronization is not or cannot be provided, or when TDD systems operate on carriers adjacent to an FDD system, the possibility arises for interferer and victim to be of the same device type. Figure 23.3 depicts a relatively common scenario in which an unpaired spectral allocation is located in a region between an FDD downlink band and an FDD uplink band (as is the case for the 2.5–2.6 GHz UMTS extension band, for example).

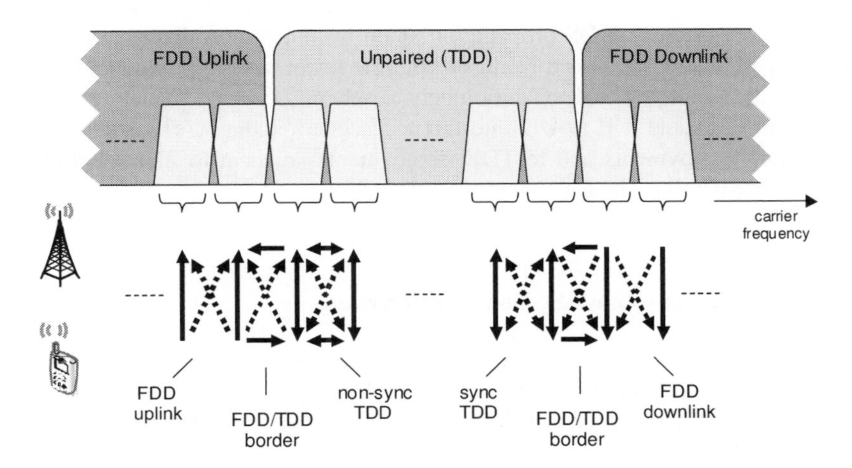

Figure 23.3: Possibilities for Adjacent Carrier Interference (ACI) between FDD and TDD systems.

The vertical arrows in Figure 23.3 represent the desired communication between the base station and the mobile terminal, unidirectional on a per-carrier basis for FDD and bi-directional for TDD. The diagonal dotted lines represent base-station-to-mobile and mobile-to-base-station adjacent channel interference that results from imperfect ACIR.

Two TDD scenarios are shown, synchronized and non-synchronized – the latter encompassing any general possibility for partial or full uplink-downlink overlap in time between two systems. For the synchronized case, the interference scenarios between the base station and mobile terminal (and vice versa) are the same as for the corresponding FDD-to-FDD adjacent channel cases. However, for the non-synchronized case additional interference paths exist between TDD mobiles and between TDD base stations, represented by the horizontal bidirectional arrows. At the FDD/TDD border regions, these same 'horizontal' interference paths exist but are unidirectional in nature.

There are many facets of a deployment which affect the severity of these various interference paths. For example, the locations of the interfering transmitter and victim receiver, as well as the characteristics of the propagation between them, clearly influence the overall coupling that exists. Macrocellular deployments typically use base station transmit antennas mounted on masts located above roof-top level, thereby resulting in an increased likelihood of Line-Of-Sight (LOS) propagation between base stations, with a correspondingly low path-loss exponent. The common use of macrocell base station antennas with vertical directivity and hence high gain can further worsen this situation. These aspects are less problematic for microcellular and dense-urban deployments in which LOS propagation between base stations is less likely due to their antennas being located below rooftop level.

Coupling between mobile terminals is often mitigated by the surrounding local clutter, and due to the lower antenna gain in the terminals. One typical scenario is depicted in Figure 23.4 in which non-co-located macro base stations have some potential to exhibit stronger mutual coupling than between the mobiles which they respectively serve. However, the figure also shows that it is not always possible to rely on local clutter to provide the necessary isolation between terminals, due to the fact that when terminals are closely spaced (for example, in the same office or café), LOS propagation again becomes more likely and the potential for interference is increased – as a result of both the lower path-loss exponent between the terminals and the small physical separation between them.

The base-to-base and UE-to-UE interference scenarios that are particular to unsynchronized TDD deployments and to TDD deployments adjacent to FDD deployments are reviewed in more detail in the following two subsections.

23.3.1.1 Base-Station to Base-Station Interference

Base stations of relevance to a particular base-to-base interference scenario may be either co-located (i.e. antennas mounted at the same cell-site) or non-co-located. Nevertheless, base-to-base interference is generally deterministic. This is because the locations of the base stations are fixed, and furthermore the link adaptation strategy typically employed for the LTE downlink usually results in all available transmit power being used to maximize the throughput of the link. It is therefore reasonable to analyse the interference assuming full transmit power from each base station.

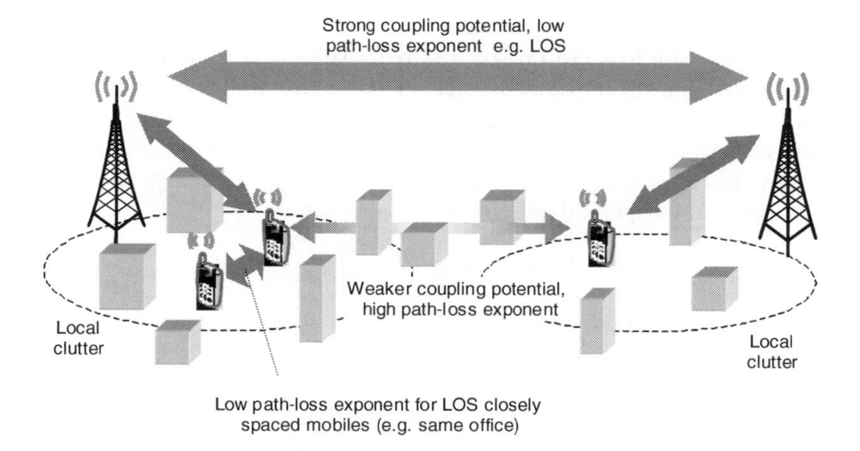

Figure 23.4: Typical RF interference scenario for a TDD system.

In general, systems operating on adjacent carriers may use differing deployment topologies and cell sizes, giving rise to varying distances between the base stations on the adjacent carriers. We therefore consider the worst-case base-to-base distance only, and assume that at some point in the network the transmit and receive antenna patterns are aligned to provide maximum gain at this worst-case distance. The co-channel Power Spectral Density (PSD) of the received interference at the victim base station antenna connector can then be written as:

$$\mathrm{PSD}_{\mathrm{Rx}} = \mathrm{PSD}_{\mathrm{Tx}} + 2G_{\mathrm{BS}} - \rho_{\mathrm{BS\text{-}BS}}(x_0) \tag{23.1}$$

where $\mathrm{PSD}_{\mathrm{Tx}}$ is the transmitted PSD at each interfering base station antenna connector, G_{BS} is the antenna gain at each base site, $\rho_{\mathrm{BS\text{-}BS}}(x)$ is the path-loss between base sites as a function of distance x in metres and x_0 is the worst-case (smallest) distance between base sites of different carriers.

In order for the inter-system interference to have only a minor effect, one can assume that the PSD of the interference after benefiting from any available ACIR should be of the order of $\mathrm{PSD}_{\mathrm{N}} + \mathrm{NF}_{\mathrm{BS}} - 6$ dB or less if it is to produce no more than a 1 dB desensitization of the base station receiver (where $\mathrm{NF}_{\mathrm{BS}}$ denotes the base station noise figure and $\mathrm{PSD}_{\mathrm{N}}$ is the PSD of thermal noise, e.g. -174 dBm/Hz at typical temperatures):

$$(\mathrm{PSD}_{\mathrm{Rx}} - \mathrm{ACIR}) \leq (\mathrm{PSD}_{\mathrm{N}} + \mathrm{NF}_{\mathrm{BS}} - 6)\ \mathrm{dB} \tag{23.2}$$

Note that in the case of an SC-FDMA[3] uplink victim receiver, consideration needs to be paid not only to the ACIR averaged over the system bandwidth, but, more challengingly, to the localized frequency resource blocks located closest to the interfering carrier, especially if the important uplink control signalling on the Physical Uplink Control CHannel (PUCCH) is to be protected (see Section 16.3.1). In general, however, the ACIR requirement varies directly with $\mathrm{PSD}_{\mathrm{Tx}}$:

$$\mathrm{ACIR} \geq \mathrm{PSD}_{\mathrm{Tx}} + 2G_{\mathrm{BS}} - \rho_{\mathrm{BS\text{-}BS}}(x_0) - \mathrm{PSD}_{\mathrm{N}} - \mathrm{NF}_{\mathrm{BS}} + 6\ \mathrm{dB} \tag{23.3}$$

[3]Single Carrier Frequency Division Multiple Access.

In order to arrive at an ACIR requirement, we must therefore know the transmitted PSD and the intervening path-loss. To do so, it is reasonable to assume that the base station transmit power capabilities are dimensioned in order that each of the two systems are interference-limited on the downlink (at least this can apply for small- and medium-sized cells without exceeding the eNodeB output power capabilities). We therefore assume here that the spectral density of the downlink signal for 95% of the total area is γ_{DL} dB larger than the spectral density of the thermal noise in the UE receivers. Thus

$$\text{PSD}_{\text{Tx}} = \text{PSD}_{\text{N}} + \text{NF}_{\text{UE}} + \gamma_{\text{DL}} + \rho'_{\text{BS-UE}} - G_{\text{BS}} \tag{23.4}$$

where NF_{UE} is the noise figure of the UE receiver, γ_{DL} is the received downlink signal-to-noise ratio in dB which is exceeded at 95% of the UE receivers, and $\rho'_{\text{BS-UE}}$ is the 95-percentile path-loss between UEs and their serving eNodeBs in the interfering network.

For ease of representation, an empirical approximation to $\rho'_{\text{BS-UE}}$ is applied here specific to this particular example deployment: let σ be the standard deviation of the log-normal shadow fading between eNodeB and UE and L_{b} represent the additional building penetration loss (assuming indoor coverage); then with $\rho_{\text{BS-UE}}(x)$ denoting the path-loss between base stations and UEs separated by distance x metres, we can write

$$\rho'_{\text{BS-UE}} \approx \rho_{\text{BS-UE}}(0.58s) + 0.7\sigma + L_{\text{b}} \tag{23.5}$$

Substituting Equations (23.5) and (23.4) into (23.3) we obtain an approximate expression for the necessary ACIR to maintain an acceptable adjacent channel interference level at a victim base station:

$$\text{ACIR} \geq \text{NF}_{\text{UE}} - \text{NF}_{\text{BS}} + \gamma_{\text{DL}} + \rho_{\text{BS-UE}}(0.58s) + 0.7\sigma + L_{\text{b}} + G_{\text{BS}} - \rho_{\text{BS-BS}}(x_0) + 6\text{ dB} \tag{23.6}$$

This is plotted in Figure 23.5 as a function of the smallest inter-carrier eNodeB–eNodeB separation x_0 for several selected values of the inter-site spacing s between co-channel base-stations on the interfering carrier under the following assumptions:

- uniform base station deployment on the interfering carrier;
- free-space propagation between eNodeBs $\rho_{\text{BS-BS}}(x)$;
- $\rho_{\text{BS-UE}}(x) = 128.1 + 37.6$ dB, from [2];
- $\gamma_{\text{DL}} = 6$ dB, $\text{NF}_{\text{UE}} = 9$ dB, $\text{NF}_{\text{BS}} = 5$ dB, $G_{\text{BS}} = 14$ dBi, $L_{\text{b}} = 20$ dB.

The transmit PSD of the eNodeB (P_{eNB}) for 95% coverage is also listed in the legend for each Inter-Site Distance (ISD).

For a given ISD, the required ACIR naturally decreases as the worst-case separation between interfering eNodeBs is increased. Notice, however, that the base-to-base problem worsens significantly as the ISD in the interfering network increases, due to the fact that the path-loss exponent from eNodeB to UE is higher than the path-loss exponent between eNodeBs. The transmit power needed by the interfering eNodeB to reach UEs at its cell edge increases at a faster rate than can be compensated by the path-loss to a victim eNodeB receiver at the same cell-edge location. It should be remembered, however, that this is representative of a macrocellular scenario with eNodeB antennas mounted above rooftop level; for the smaller cell sizes (characteristic of microcells) the eNodeB-to-eNodeB situation

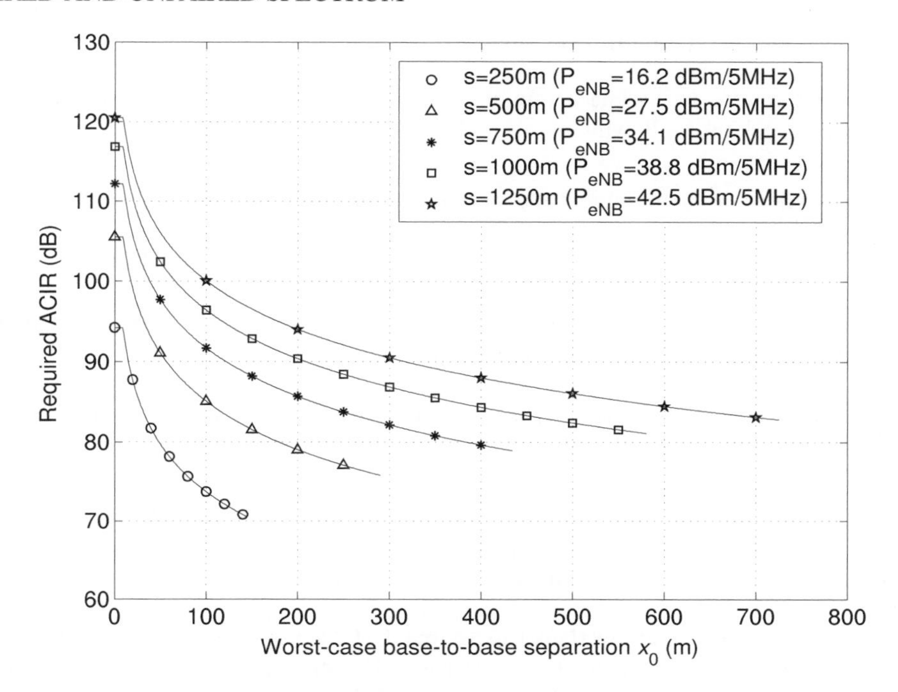

Figure 23.5: Required ACIR for 1 dB desensitization in an eNodeB-to-eNodeB interference scenario.

will be significantly improved by the higher propagation exponent between eNodeBs whose antennas are mounted below rooftop level.

For co-sited eNodeBs (i.e. very small x_0), a Minimum Coupling Loss (MCL) value of 30 dB has been used to replace $2G_{\text{BS}} - \rho_{\text{BS-BS}}(x_0)$ in Equation (23.1), resulting in

$$\text{ACIR}_{\text{co-siting}} \geq \text{PSD}_{\text{Tx}} - \text{MCL} - \text{PSD}_{\text{N}} - \text{NF}_{\text{BS}} + 6\ \text{dB} \tag{23.7}$$

For wide area base stations, the ACIR required for co-siting can rise towards a challenging 120 dB or so. This problem, however, is not new or specific to LTE and has been encountered previously for FDD/TDD coexistence in WCDMA. Some practical solutions to this problem have been documented in [3], in which RF bandpass cavity resonator filters were used to improve greatly the ACLR and ACS of base station transmitters and receivers respectively either side of a TDD/FDD boundary. These significantly exceed the standardized minimum requirements which were not intended to cope unaided with the case of co-sited base stations.

Similar techniques also apply to LTE yet remain significantly challenging. With careful design, however, adjacent channel deployment of FDD and TDD LTE base stations, or of two non-synchronized LTE TDD base stations, should be feasible, even for co-sited arrangements, provided that appropriate measures are adopted in both the interfering and victim base stations.

23.3.1.2 Mobile-to-Mobile Interference

The UE-to-UE interference scenario requires a more probabilistic approach than for base-to-base interference for the following reasons:

- the locations of the interferers and victims are variable and dynamic;
- the physical resources assigned by the scheduler to the UEs are variable;
- the transmit powers of the interfering UEs are a function of their channel conditions and of the power control policy implemented;
- the received levels of the wanted base station signals at the victim UEs are also variable as a function of the UEs' channel conditions.

It is difficult, therefore, to formulate a definitive analysis of UE-to-UE interference for LTE. Nonetheless, a basic analysis is presented here together with some discussion of the attributes of the LTE system which have some bearing on the magnitude of the interference effects.

We consider two similar overlaid tri-sectored LTE deployments on adjacent 5 MHz carriers. One deployment contains the interfering UEs while the other contains the victim UEs. Both carriers have regions in which uplink transmissions overlap in time with downlink transmissions. The base stations of the deployments are either co-located or maximally spaced non-co-located. Cells in the interfering network each schedule groups of four contiguous uplink Resource Blocks (RBs) to a number of randomly selected UEs, resulting in six simultaneously scheduled interferers per subframe.

The impacts of the scheduled interferers' transmissions on a UE receiver in the victim network are calculated for the case in which a randomly selected victim UE is scheduled a downlink transmission resource in the one RB next to the band edge separating the two carriers. The impact caused by the interfering adjacent-carrier UEs is analysed in terms of the mean percentage reduction in victim UE downlink throughput, R_{loss}, caused by the presence of the interfering UEs.

As assumed in [2], the ACIR increases by an additional 13 dB for localized SC-FDMA interferer transmissions located anywhere other than the four RBs next to the band edge of the interferer network. Additionally, for the purposes of this analysis an uplink power control strategy is employed whereby the transmit PSD of each UE is set such that it is not received at any co-channel non-serving eNodeB receiver any higher than 6 dB above the eNodeB receiver's thermal noise floor.

The transmit PSD of the eNodeB in the victim network is set via Equations (23.4) and (23.5) in the same manner as for the eNodeB-to-eNodeB analysis, such that the downlink is in an interference-limited region of operation (but is not excessively 'over-powered'). The same path-loss model between eNodeBs and UEs is assumed. The path-loss between UEs is assumed to be given by Equation (23.8) (with a carrier frequency f_c of 2000 MHz), based upon a simple two-slope microcellular model from [4] with break point at $x_b = 45$ m to reflect the likelihood of free-space-like propagation (with exponent 2) for low separation distances (i.e. $x \leq x_b$), and increased attenuation exponent $z = 6.7$ due to local clutter at higher distances ($x > x_b$).

$$\rho_{\text{UE-UE}}(x) = \begin{cases} -27.56 + 10\log_{10}(f_c^2 x^2)\ \text{dB} & \text{for } x \leq x_b \\ -27.56 + 10\log_{10}\left(\dfrac{f_c^2 x^z}{x_b^{z-2}}\right)\ \text{dB} & \text{for } x > x_b \end{cases} \tag{23.8}$$

A microcellular model is considered applicable to UE-to-UE interference as it reflects the case where both the transmitting and receiving antennas are below rooftop level. Other system parameters assumed for this analysis are generally in line with those of [4] for macrocell simulation. With these assumptions, and for $L_b = 20$ dB, the results of Figure 23.6 are obtained, displaying the relationship between the band-edge ACIR and the throughput loss R_{loss} for various values of ISD s.

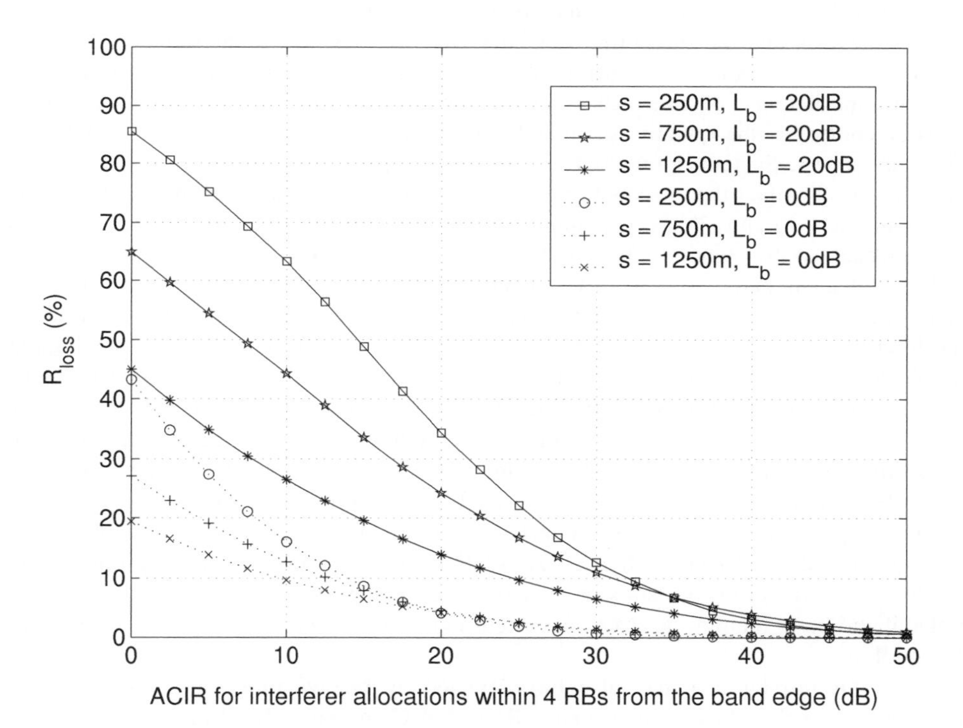

Figure 23.6: Throughput reduction due to UE-to-UE interference.

The results in Figure 23.6 are given for the case of co-located base stations. Those for the non-co-located case are very similar showing only a small further degradation for low ACIR values; the absence of a significant difference is a result of the uplink power control strategy employed as described above (whereby the mobile transmit power is correlated with the strongest non-serving cell path-loss rather than with the path-loss of the serving cell).

The LTE specifications are based upon a 30 dB ACLR which would provide an ACIR of 28 dB assuming an ACS of 33 dB. In this case, it can be seen from Figure 23.6 that the worst-case throughput loss for the band-edge downlink resource block would be between 7% and 16% depending on cell size.

This analysis is, however, rather sensitive to certain assumptions, especially the value of the in-building penetration loss L_b. When the penetration loss is increased, the serving eNodeB may instruct the UE to increase its transmit power by the same amount in order to maintain its received Signal-to-Interference-plus-Noise Ratio (SINR) at the serving eNodeB (subject to maximum UE output power constraints) without causing additional interference to non-serving eNodeBs which are also protected by the same building penetration loss. However, the path-loss to a worst-case nearby victim UE (e.g. with LOS propagation between the UEs) is not affected by the increased building penetration loss. Thus, increased building penetration loss can have the effect of increasing worst-case UE-to-UE interference levels. This effect is clearly evident from the fact that the curves of Figure 23.6 for the case of $L_b = 20$ dB show a greater R_{loss} than the dotted curves for $L_b = 0$ dB.

This suggests that the susceptibility of the system to UE-to-UE interference is lowered considerably in an outdoor scenario. In these cases, the system's need to control the uplink inter-cell interference between UEs and neighbouring eNodeBs in a frequency-reuse-1 network constrains the quantity of interference power that is injected by those UEs into adjacent carriers. System throughput loss is then minimal (~2%) for commonly expected ACIR levels. The fact that a reduction in penetration loss mitigates UE-to-UE interference also points towards possible cell-planning solutions to alleviate the problem, for example using picocells or home base stations rather than macrocells to provide in-building coverage.

The statistical nature of UE-to-UE interference is also of relevance when assessing its impact. LTE allows for randomization of the allocated radio resources for both the interferer and the victim in both the time and frequency domains. Uplink frequency hopping is able to provide the necessary randomization in frequency, and in the time domain a degree of randomization can be provided by different resource scheduling strategies, as well as the possibility for differing retransmission delays due to the fact that the downlink retransmissions to the victim UE are dynamically scheduled rather than synchronous. The use of Hybrid Automatic Repeat reQuest (HARQ) also provides robustness against those instantaneous events in which high interference levels are experienced. Thus, in the case of frequency-adjacent LTE systems, UE-to-UE interference may be heavily randomized. Its effects can therefore be 'smoothed' and shared amongst all users of the system on a probabilistic basis, helping to avoid persistent effects on specific pairs of users with close RF coupling.

UE-to-UE interference may also be alleviated by receive processing at the UE. As discussed in Section 26.2.6, Interference Rejection Combining (IRC) receivers [5] can be used to maximize the received SINR, taking into account the instantaneous direction of arrival of the wanted and interfering signals. Forms of the IRC receiver that make use of averaged correlation (e.g. in time or in frequency) of the received signals across antennas (e.g. [6]) are particularly applicable to the adjacent-channel UE-to-UE interference scenario in which explicit channel estimation of interferers is likely to be impractical. A nearby closely coupled UE on an adjacent channel would typically present a single dominant interference source, enabling an IRC receiver to provide a gain when the instantaneous interference-to-thermal-noise ratio is relatively high. Thus, one could anticipate that use of the IRC receiver may provide some additional robustness against UE-to-UE interference. Figure 23.7 confirms that this is the case (here shown for a 500 m ISD). This figure shows the benefits that an IRC receiver can deliver compared to a Maximum Ratio Combining (MRC) receiver in terms of mean user throughput in the presence of UE-to-UE interference from an adjacent carrier. The figure also shows that an IRC receiver provides benefits even in the absence of adjacent-carrier UE-to-UE interference (due to the removal of interference from co-channel eNodeBs).

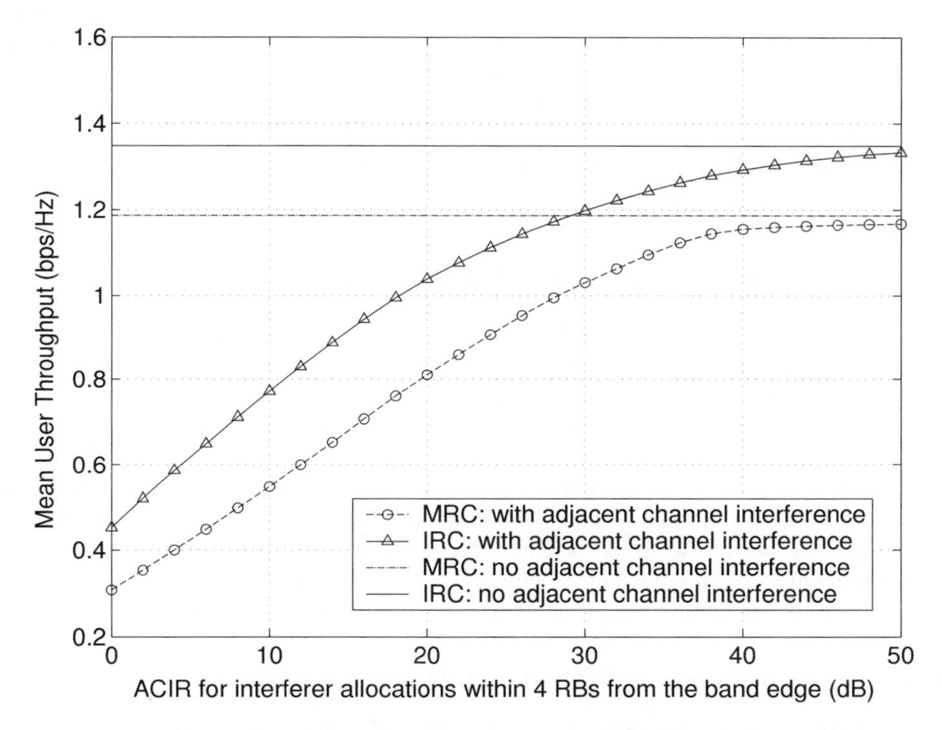

Figure 23.7: Comparison of IRC and MRC receivers with and without adjacent channel UE–UE interference.

More generally, Figure 23.7 reveals that for the scenario considered, a system using ideal IRC receivers in the presence of UE-to-UE interference with 29 dB ACIR could perform as well as a system using Maximum Ratio Combining (MRC) receivers with no UE-to-UE interference at all. Thus, the detrimental effects of UE-to-UE interference may be mitigated via a combination of moderate ACIR together with IRC receivers.

23.3.2 Summary of Interference Scenarios

The preceding sections have provided some discussion of the issues facing the deployment of LTE TDD on carriers adjacent either to other non-synchronized TDD systems, or to FDD downlink or uplink carriers. In these deployment scenarios, one must consider not only eNodeB-to-UE and UE-to-eNodeB interference but also the nature and severity of eNodeB-to-eNodeB and UE-to-UE interference.

In the case of eNodeB-to-eNodeB interference, the discussions of Section 23.3.1.1 have shown that very stringent ACLR and ACS are needed, especially in the case of co-siting. Nonetheless, co-siting would appear to be technically feasible, for example by means of cavity-based low-loss RF filtering solutions at the base stations.

In the case of UE-to-UE interference, system-level analysis suggests that by using appropriate radio resource management on the uplink, together with interference randomization and possibly also interference suppression at the receiver, the overall system throughput loss and user experience in the presence of the adjacent carrier interference are likely to remain acceptable for many deployment scenarios.

Although it is generally desirable to enable alignment of uplink–downlink switching points between the LTE TDD system and LTE or legacy TDD systems on an adjacent carrier, it should be remembered that this cannot always be relied upon.

23.4 Half-Duplex System Design Aspects

For half-duplex (including TDD and HD-FDD) operation, the restriction that a UE may not transmit and receive at the same time instant has consequences for the physical layer (and to a lesser extent, higher-layer) design of LTE.

In order to enable the exploitation of economies of scale, LTE has generally followed a design principle in which differences between the duplex modes are introduced only where necessary for correct system operation, or where they offer a significant performance advantage when used for a particular duplex mode. With careful design, only a relatively small set of attributes need to be modified. Nonetheless, these have the potential to alter certain behaviours and structures of the physical layer, primarily in terms of frame structure, HARQ operation and control signalling, which are discussed in the following sections.

23.4.1 Accommodation of Transmit–Receive Switching

For TDD systems, switching between transmit and receive functions occurs on the transition from uplink to downlink (for the UE) and on the transition from downlink to uplink (for the eNodeB). For half-duplex FDD systems, switching only occurs at the UE, as the eNodeB is assumed to be full-duplex.

In order to preserve the frequency-domain orthogonality of the LTE uplink multiple access scheme, propagation delays between an eNodeB and the UEs under its control are compensated by means of timing advance as explained in Section 18.2.

At a half-duplex UE, the timing-advanced uplink transmission cannot be allowed to overlap with reception of any preceding downlink. For TDD, to prevent the overlap, a transmission gap or 'guard period' between transmission and reception at the eNodeB is created (T_{G1}) to accommodate the greatest possible timing advance and any required switching delay (including power amplifier ramp-up or ramp-down to avoid excessive wideband emissions). A further guard period (T_{G2}) is also required at the TDD eNodeB transition between uplink and downlink to cater for switching and power ramping delays only (this being independent of the propagation delay or timing advance). These are illustrated in Figure 23.8.

Four switching times are therefore of relevance in the case of TDD operation. These correspond to the transmit-to-receive and receive-to-transmit delays at the UE (denoted

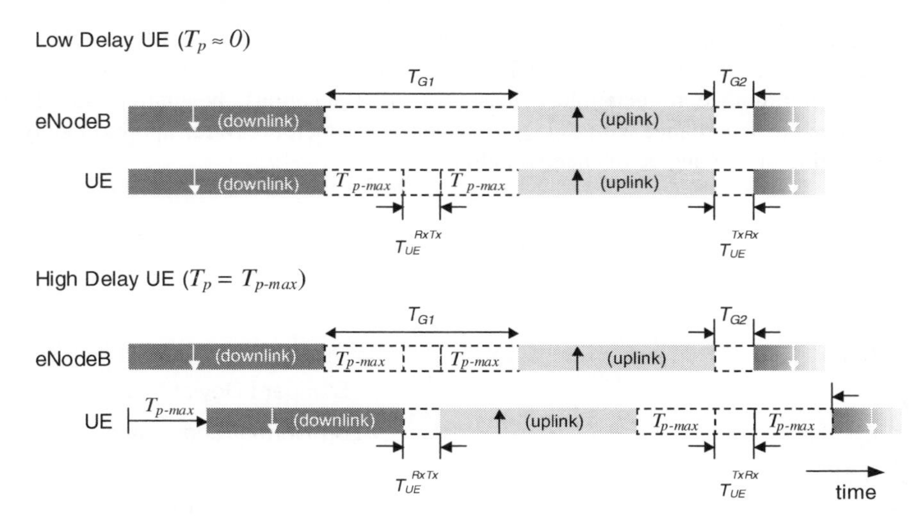

Figure 23.8: TDD signal timings in the presence of uplink timing advance.

as time intervals $T_{\text{UE}}^{\text{TxRx}}$ and $T_{\text{UE}}^{\text{RxTx}}$ respectively) and likewise at the eNodeB (denoted $T_{\text{eNB}}^{\text{TxRx}}$ and $T_{\text{eNB}}^{\text{RxTx}}$). Figure 23.8 depicts two cases corresponding to the two extremes of propagation delay T_{P} within a TDD cell ($T_{\text{P}} = 0$ for a UE physically close to the eNodeB and $T_{\text{P}} = T_{\text{P\_max}}$ for a UE at the border of a cell, where $T_{\text{P\_max}} = d_{\text{max}}/c$ corresponds to the maximum one-way propagation delay supported by the cell, occurring at distance d_{max}). Note that the switching delays are exaggerated for diagrammatical clarity and that UE switching delays are assumed to be longer than those at the eNodeB.

It is apparent from Figure 23.8 that the time available at the UE for downlink to uplink transition is a function of the propagation delay T_{P} (most stringent for the case of high delay) whereas the time available at the eNodeB for the same transition is constant and equal to T_{G1}:

$$T_{G1} = 2T_{\text{P\_max}} + T_{\text{UE}}^{\text{RxTx}} \tag{23.9}$$

The time interval T_{G2} at the eNodeB is independent of the propagation delay. To support the case for which $T_{\text{P}} \rightarrow 0$ (i.e. a UE close to the eNodeB), T_{G2} needs to be dimensioned such that

$$T_{G2} = \max(T_{\text{UE}}^{\text{TxRx}}, T_{\text{eNB}}^{\text{RxTx}}) \tag{23.10}$$

In the case of HD-FDD, $T_{\text{eNB}}^{\text{RxTx}} = 0$, so T_{G2} is determined only by the time $T_{\text{UE}}^{\text{TxRx}}$. In order to support the case of low T_{P}, the (full duplex) eNodeB must still allow sufficient time for this UE switching delay if the uplink and downlink subframes surrounding the switching point are both active for a particular user. Hence in practice the uplink frame timing at the eNodeB should be advanced for the whole cell by an amount $T_{\text{UE}}^{\text{TxRx}}$ relative to the downlink frame timing at the eNodeB even for a full-duplex eNodeB if it supports HD-FDD UEs in the cell. Full duplex FDD UEs communicating with the same eNodeB will likewise need to have their timing advanced to maintain uplink orthogonality with the HD-FDD UEs.

It is important to note that d_{max} may be significantly larger than the notional cell radius r_0 (i.e. half the ISD), due to propagation effects such as shadow fading. This effect is shown for one example of a tri-sectored deployment with frequency reuse factor 1 in Figure 23.9 (the shadow fading is assumed to be log-normal with standard deviation σ). It can be observed that in order to accommodate, for example, at least 98% of UE locations, the guard period should be dimensioned in accordance with $d_{\max} \geq \gamma r_0$ where γ is a factor between approximately 1.5 and 3 depending on the degree of shadow fading.

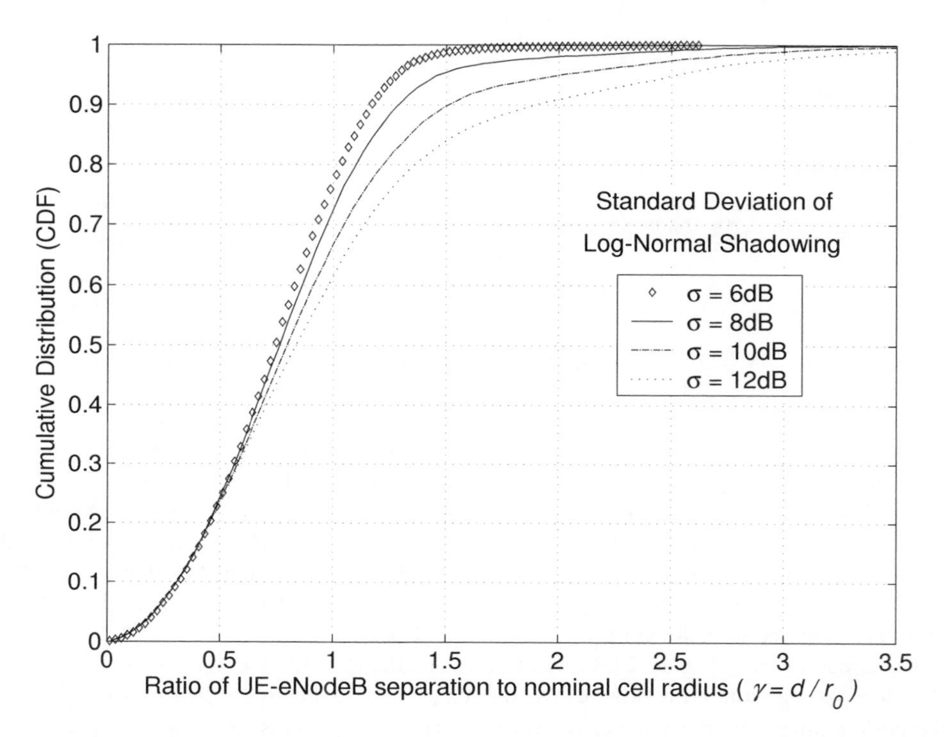

Figure 23.9: Cumulative distribution of serving cell propagation delays relative to nominal cell radius.

Overall, the total guard time T_G at a TDD eNodeB per uplink–downlink cycle is equal to the sum of T_{G1} and T_{G2}, as given by Equation (23.11). In the LTE specifications this is represented by a single guard period amalgamating both parts (with the uplink subframe timing advanced by an amount T_{G2} at the eNodeB with respect to the downlink timing). This is only a matter of representation, however, and the end result is essentially identical to the presence of the two separate guard periods:

$$T_G = T_{G1} + T_{G2} = 2T_{\text{P\_max}} + T_{\text{UE}}^{\text{RxTx}} + \max(T_{\text{UE}}^{\text{TxRx}}, T_{\text{eNB}}^{\text{RxTx}}) \tag{23.11}$$

By assuming reasonable values for the UE and eNodeB switching times (typically of the order of 10 to 20 μs) the length of this amalgamated guard period can be dimensioned (in multiples of the OFDM[4] symbol duration) for a particular deployment. Thus although

[4]Orthogonal Frequency Division Multiplexing.

guard periods represent an undesirable overhead for TDD, by allowing for a flexible and configurable guard period duration, systems can be tailored to the topology of the deployment whilst minimizing the spectral efficiency loss.

The LTE specifications [7] support a set of guard period durations ranging (non-contiguously) from 1 to 10 OFDM symbols for the normal CP (or from 1 to 8 OFDM symbols for the extended CP). A duration of 1 OFDM symbol should be sufficient for many of the anticipated cellular deployments of LTE (up to around 2 km nominal cell radius for $\gamma = 2$), whereas at the other end of the scale, guard period durations of the order of 700 μs support one-way propagation-path delays of the order of 100 km.

The guard period in LTE TDD is located within a mixed uplink/downlink subframe (known as a 'special subframe') as shown in Figure 6.2 and further discussed in Section 23.4.2.

23.4.2 Coexistence between Dissimilar Systems

There are scenarios in which an LTE TDD system could need to coexist with other (non-LTE) radio access technologies within the same frequency bands. As mentioned in Section 23.3.2, the ability to time-align the uplink and downlink transmissions between neighbouring systems can be used to avoid base-to-base and UE-to-UE interference paths and hence to alleviate the dependency on the other coexistence measures mentioned above. This also applies to the case of dissimilar neighbouring systems.

As shown in Figure 6.2, the LTE TDD system has a 10 ms radio frame supporting either one pair of switching points per 5 ms, or one pair of switching points per 10 ms, which we denote here as '5 ms switching' and '10 ms switching' respectively.

The provision for both 5 ms and 10 ms switching options in LTE TDD was introduced to enable switching point alignment between LTE and the UTRA TDD modes. Ideally, for maximum flexibility of alignment, the location of the downlink-to-uplink transition within the 5 ms or 10 ms cycle would be fully adjustable with symbol-level granularity, although certain practical considerations must also be taken into account. The overall intention is first to align the location of the uplink-to-downlink transitions between the dissimilar systems by means of a frame-timing offset, and then to adjust the position of the LTE downlink-to-uplink transition such that it is approximately aligned with that of the other system.

The TDD variant of Mobile WiMAX (based upon the IEEE 802.16e amendment) also utilizes 5 ms switching, and can itself accommodate a variable position of the switching point with symbol-level granularity (i.e. with adjustments of approximately ±102.8 μs). Hence switching point alignment between TDD WiMAX and TDD LTE is possible with sufficient resolution.

The presence of a switching point that is adjustable at the OFDM symbol level results in a mixed subframe, known as the 'special subframe', potentially containing both downlink and uplink regions (referred to as 'DwPTS' and 'UpPTS' respectively[5]) as shown in Figure 23.10 (see also Figure 6.2).

The special subframe is the most significant difference between the FDD and TDD physical layers in LTE, giving rise to a number of details which must be considered in the design. Uplink and downlink transmission durations in this irregular subframe are effectively

[5]DwPTS and UpPTS is terminology inherited from Time-Division Synchronous Code Division Multiple Access (TD-SCDMA) where the terms denoted Downlink Pilot Time Slot and Uplink Pilot Time Slot respectively. This, however, does not well reflect the usage of the fields bearing the same names in LTE.

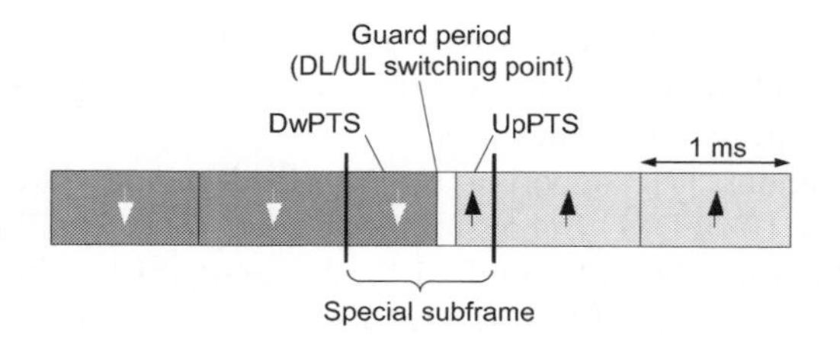

Figure 23.10: Special subframe for downlink–uplink switching in TDD operation.

reduced compared to a normal subframe, implying that less forward error correction redundancy can be employed for a given transport block size. Alternatively, the transport block size itself can be reduced, although this smaller transport block size must then also be used for HARQ retransmissions which may occur in a regular 1 ms subframe. In general, the use of HARQ with incremental redundancy, as described in Section 10.3.2.5, limits the impact of the fact that the special subframe has less downlink transmission resource than a normal subframe.

For the downlink control signalling on the Physical Downlink Control CHannel (PDCCH), the special subframe does not have a significant effect, as the PDCCH is anyway contained within the first few OFDM symbols of a subframe. Hence PDCCH transmission in the DwPTS region is possible, with the exception that it is constrained in this case to a duration of a maximum of two OFDM symbols (instead of up to three in normal subframes, or even four in cases of narrow system bandwidths, as explained in Section 9.3).

The length of the UpPTS field in the mixed subframe is constrained in the LTE specifications to support only lengths of one and two SC-FDMA symbols. This field does not therefore support uplink data transmission and is instead used only for a shortened random access preamble (suitable only for small cells – see Section 17.6) and for transmission of uplink Sounding Reference Signals (SRSs) – see Section 15.6. One downside of the absence of support for UpPTS lengths other than one or two symbols is that it restricts the set of possible configurations for switching point alignment with other TDD systems. Nonetheless, Figure 23.11 shows some examples of uplink-downlink alignment that are possible between LTE TDD and other TDD systems.

23.4.3 HARQ and Control Signalling for TDD Operation

As explained in Section 10.3.2.5, transmission of downlink or uplink data with HARQ requires that an ACKnowledgement ACK or Negative ACK) be sent in the opposite direction to inform the transmitting side of the success or failure of the packet reception.

In the case of FDD operation, acknowledgement indicators related to data transmission in a subframe k are transmitted in the opposite direction during subframe $k+4$, such that a one-to-one synchronous mapping exists between the instant at which a transport block is transmitted and its corresponding acknowledgement. However, in the case of TDD operation, subframes are designated on a cell-specific basis as uplink or downlink (with the exception of the mixed subframe), thereby constraining the times at which resource grants, data transmissions, acknowledgements and retransmissions can be sent in their respective

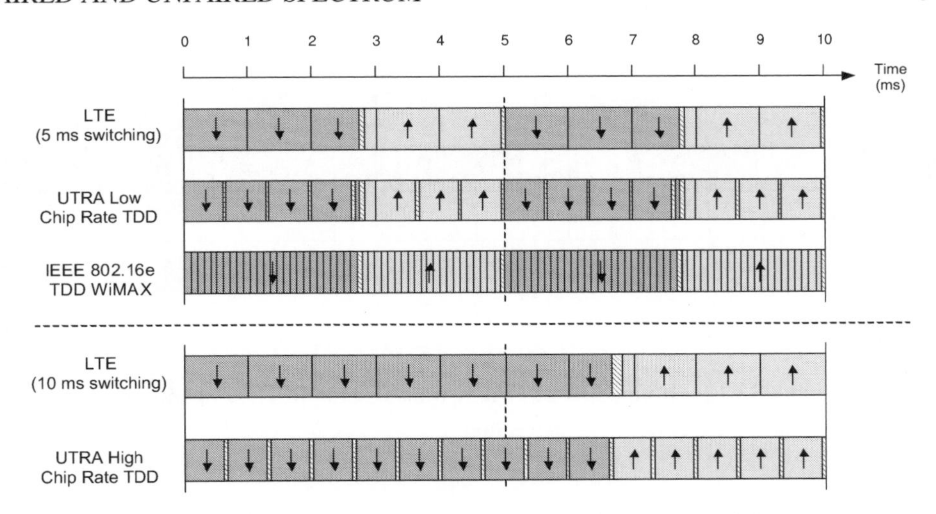

Figure 23.11: Examples of switching point alignment between LTE and non-LTE TDD systems.

directions. The synchronous scheme used for FDD cannot therefore be directly reused for TDD operation.

For TDD in asymmetric uplink-downlink cases, it is inevitable that for one of the link directions there are insufficient subframes to support a synchronous one-to-one mapping between transmitted data and its acknowledgement. The LTE design for TDD therefore supports grouped ACK/NACK transmission to carry multiple acknowledgements within one subframe.

For uplink HARQ, the sending (in one downlink subframe) of multiple acknowledgements on the Physical Hybrid ARQ Indicator CHannel (PHICH) is not problematic since, when viewed from the eNodeB, this is not significantly different from the case in which single acknowledgements are sent to multiple simultaneous UEs. However, for downlink HARQ, if the asymmetry is downlink-biased, the uplink control signalling (PUCCH) formats of FDD are insufficient to carry the additional ACK/NACK information. This situation is evident from the example of Figure 23.12 where only 4 uplink sub-frames per radio frame are available to carry acknowledgements for the 6 subframes that may contain downlink data transport blocks. In this particular example, this is solved by means of uplink subframe numbers 2 and 7 each carrying acknowledgement data corresponding to two downlink subframes (numbers 5 and 6, and 0 and 1 respectively). Each of the TDD subframe configurations in LTE (shown in Figure 6.2) has its own such mapping predefined between downlink and uplink subframes for HARQ purposes (see [8, Section 7.3]), with the mapping being designed to achieve a balance between minimization of acknowledgement delay and an even distribution of ACK/NACKs across the available uplink subframes.

Two mechanisms are provided for grouping the acknowledgement information carried in the uplink in TDD operation, termed 'ACK/NACK bundling' and 'ACK/NACK multiplexing'. Selection between these mechanisms is by higher-layer (Radio Resource Control (RRC)) configuration.

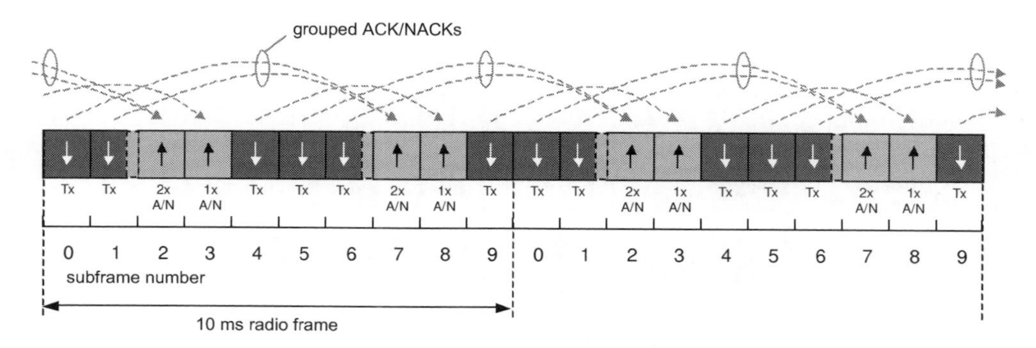

Figure 23.12: Grouped ACK/NACK transmission in TDD operation.

ACK/NACK bundling reuses the same 1- and 2-bit Physical Uplink Control CHannel (PUCCH) formats ('1a' and '1b' – see Section 16.3) which are used for FDD. For each downlink codeword (up to two if downlink spatial multiplexing is used), only a single acknowledgement indicator is derived by performing a logical 'AND' operation of the acknowledgements across the group of downlink subframes associated with that uplink subframe; this indicates whether zero or more than zero transport blocks in the group were received in error.

For ACK/NACK multiplexing, a separate acknowledgement indicator is returned for each of the associated downlink subframes. However, to limit the amount of signalling information that this would generate, acknowledgements from multiple codewords on different spatial layers within a subframe are first condensed into a single acknowledgement, again by means of a logical 'AND' operation. This is known as 'spatial ACK/NACK bundling'. For the more extreme asymmetries, however, there remains a need to transmit more than two bits of ACK/NACK information in one uplink subframe. This is achieved using the normal 1- and 2-bit PUCCH formats augmented with a code selection scheme whereby the PUCCH code selected by the UE conveys the surplus information to the eNodeB (see [8, Section 10.1].

A disadvantage of these lossy compression schemes for grouped acknowledgements is that the eNodeB does not know exactly which transport block(s) failed in decoding. In the event of a NACK, all transport blocks in the same group must be resent, increasing retransmission overheads and reducing link throughput. A more subtle impact is that the average HARQ round trip time (and hence latency) can be increased due to the fact that some blocks cannot be acknowledged until the remainder of the group have been received.

A further complication arises because the PDCCH control signalling is not 100% reliable and there is some possibility that the UE will miss some downlink resource assignments. This would introduce the possibility of HARQ protocol errors, including the erroneous transmission of ACK in the case when one or more downlink assignments were missed in a group of subframes. In order to help avoid this problem, a 'Downlink Assignment Index' (DAI) is included in the PDCCH to communicate to the UE the number of subframes in a group that actually contain a downlink transmission. In the case of ACK/NACK bundling, this helps the UE to detect missed downlink assignments and avoid returning ACK if one or more downlink assignments were missed, while in the case of ACK/NACK multiplexing the DAI helps the UE to determine how many bits of ACK/NACK information should be returned. Nevertheless, this mechanism alone cannot safeguard against all possible error cases.

In Release 10, where even larger numbers of ACK/NACK bits may need to be transmitted in a single subframe due to carrier aggregation, new PUCCH mechanisms are provided, as explained in Section 28.3.2.

The presence of link asymmetry in TDD generally increases the overall HARQ round trip times (compared to the 8 ms of FDD as discussed in Section 10.3.2.5) as a result of the additional 'waiting times' needed for the appropriate link direction to become available for transmission of an acknowledgement, grant or retransmission. This, together with the degree of asymmetry, also means that the number of HARQ processes in TDD varies depending on the TDD uplink-downlink subframe configuration: in the downlink, the number of HARQ processes varies from 4 for configuration 0 to 15 for configuration 5, while in the uplink it varies from 1 for configuration 5 to 7 for configuration 0 – see [8, Tables 7-1 and 8-1].

The signalling of the granted uplink transmission resources also requires some specific attention for TDD. For FDD, the downlink subframe in which an uplink grant is sent implicitly also signals the specific uplink subframe that has been assigned (located four subframes later), whereas for TDD this relationship cannot always hold due to the various uplink-downlink configurations. An alternative linkage is therefore formulated for each specific uplink-downlink configuration to associate each uplink subframe with one preceding downlink subframe that controls it (maintaining the same four-subframe spacing as FDD wherever possible). For configuration 0 (shown in Figure 6.2), there are more uplink subframes per 10 ms than the number of subframes available for PDCCH, and here an additional 'UL Index' field in the uplink grant message is used to indicate the uplink subframe to which the grant relates. In addition, one value of the 'UL Index' field can be used to enable a single PDCCH message to grant uplink resources in two uplink subframes (see [8, Section 8.0]).[6]

23.4.4 Half-Duplex FDD (HD-FDD) Physical Layer Operation

In principle, one of two paths could be taken for the physical layer design of a HD-FDD system:

- a derivative of the FDD system, in which the scheduling is arranged such that the terminals are not required to simultaneously transmit and receive;
- a derivative of the TDD system, in which the uplink and downlink happen to reside on different carrier frequencies.

The scheme selected for HD-FDD operation in LTE is in accordance with the FDD derivative. This has helped to maintain the desirable commonality with FDD and has reduced the overall complexity of the solution. Unless otherwise informed, a UE in the FDD-derived HD-FDD scheme has no a priori knowledge of the uplink-downlink transmission pattern. Instead the UE checks any subframe which has not otherwise been pre-assigned to uplink transmission for the presence of PDCCH control signalling addressed towards it. The eNodeB is of course aware of any uplink transmission grants it has sent using the PDCCH, and does

[6]Note, however, that a UE is not required to receive more than one uplink grant message on the PDCCH in a single downlink subframe.

not therefore expect the UE to be able to receive downlink transmissions in the corresponding uplink subframe(s).

In this scheme, the fixed one-to-one association between downlink and uplink subframes (arising from the timing relationships between resource grants, data transmission and HARQ acknowledgements) is retained as in normal FDD operation. It is worth noting, however, that this approach results in the situation that, for a given user, no more than 50% of subframes may be used for any one link direction for a given UE.

Nonetheless, the perceived impacts to a user in anything other than an unloaded system are likely to remain small. The instantaneous (i.e. in one subframe) peak data rate is not affected and, in any normally loaded or even partially loaded system, the eNodeB is in any case unable to dedicate its full time resources to a user for any lasting period due to its need to service other users. Furthermore, at least for the downlink, scheduling strategies may be adapted to increase the amount of frequency resource allocated to a HD-FDD UE at each scheduling instant in order to alleviate reliance on frequent transmissions in the time domain. For the uplink this is also possible except for situations in which the UE is transmission-power-limited. In such cases, the application of wider bandwidths may not allow for increased instantaneous data rates and therefore is also unlikely to allow for a reduction in the fraction of time-domain resources needed to achieve a targeted aggregate uplink rate.

In order to allow sufficient time for downlink-to-uplink switching in HD-FDD operation, the UE is not expected to be able to receive the last symbol(s) of a downlink subframe that precedes an uplink subframe in which the UE is active. The length of the switching period is very similar to that previously given in Equation (23.11), with the exception that $T_{\mathrm{eNB}}^{\mathrm{RxTx}} = 0$ and the cell-specific maximum propagation delay value $T_{\mathrm{P\_max}}$ may be replaced with a value T_{P} that is applicable to the particular UE in question. Thus for half-duplex FDD:

$$T_G = T_{G1} + T_{G2} = 2T_{\mathrm{P}} + T_{\mathrm{UE}}^{\mathrm{RxTx}} + T_{\mathrm{UE}}^{\mathrm{TxRx}} \tag{23.12}$$

During this time, the eNodeB may or may not continue to transmit the remainder of the data towards the UE. In order to simplify the system and to minimize the differences from full-duplex FDD the transport channel coding structure continues to output the same number of channel-coded bits as for a normal-length subframe. Correct reception of the data in spite of the UE's lack of reception during the switching period is therefore reliant on the presence of sufficient forward error correction. For larger switching times the effects on performance obviously become more pronounced due to the increasing code rate, and in these instances the scheduler could reduce the transport block size mapped to the downlink subframe containing the switching period.

Due to the fact that the eNodeB is full-duplex, cell-specific Reference Signals (RSs), common signalling and any other data directed towards other UEs may also continue to be transmitted by the eNodeB throughout a particular UE's switching period and uplink subframes; that is to say, the switching periods and designated uplink subframes are UE-specific in the case of HD-FDD, rather than cell-specific as is the case for TDD.

23.5 Reciprocity

Reciprocity is a general phenomenon encountered in wave propagation over linear media. The details of the underlying theory are not reiterated here, and the interested reader is referred to the comprehensive treatment which can be found, for example in [9–12]. We cover here only the practical implications for cellular radio communication systems, and more specifically those for LTE TDD.

The efficiency of an antenna in converting between electrical and electromagnetic energy is equal in both directions at a given frequency. The result is that the directional sensitivity of an antenna (when acting as a receiver) is also equal to its far-field radiated pattern (when acting as a transmitter).

For a simple source and a simple receiver in a linear medium separated by sufficient distance such that one does not affect the load of the other, the reciprocity theorem states that the locations of the source and receiver may be interchanged yet the transfer function between them remains unchanged. Therefore when two transceivers A and B each utilize the same antenna for transmission and reception, the frequency-domain transfer function $H(f)$ (or equivalently the time-domain impulse response $h(t)$) observed at B when A is transmitting will be the same as the transfer function observed at A when B is transmitting. This is illustrated in Figure 23.13.

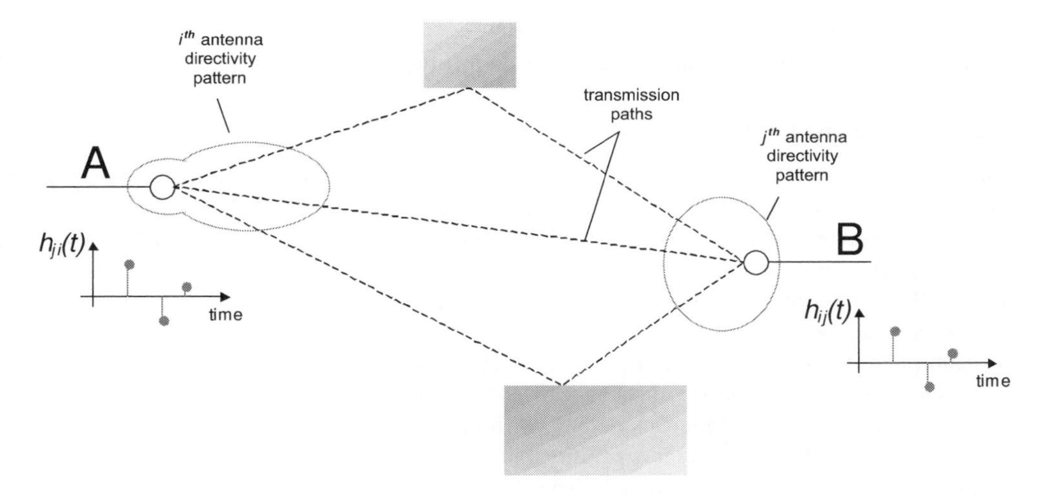

Figure 23.13: Reciprocity between a pair of transceivers.

This holds for any pair of antennas in each other's far field, and hence is easily generalized to the MIMO case in which each transceiver has multiple antennas ($i = 1, \dots, I$ for transceiver A and $j = 1, \dots, J$ for transceiver B):

$$H_{ij}(f) = H_{ji}(f) \tag{23.13}$$

Considering for simplicity a particular frequency k, and denoting the MIMO channel matrix between A and B at frequency k as $H_{k,\mathrm{AB}}$ where $H_{k,\mathrm{AB}}(i, j) = H_{ij}(k)$, it is evident that

$$H_{k,\mathrm{AB}} = H_{k,\mathrm{BA}}^{\mathrm{T}} \tag{23.14}$$

In order to exploit reciprocity in a radio communications link, the set of frequencies used for transmission and reception at each transceiver must usually overlap, as is the case for TDD systems. Short-term reciprocity cannot be exploited for FDD systems unless the nature of each constituent channel $H_{ij}(f)$ is frequency-invariant over a range of frequencies spanning both distinct transmit and receive bands, which is rarely the case. However, long-term reciprocity of the channel covariance matrix can sometimes still be exploited, as discussed in Section 23.5.2.7.

23.5.1 Conditions for Reciprocity

Reciprocity can be used to control the characteristics of an outbound transmission in anticipation of the channel response that the transmission will experience. In these instances, knowledge of the outbound channel is inferred at the transmitter from measurement of the inbound channel responses. For such techniques to be successful, however, one must first consider the following:

- the topology or configuration of the MIMO antenna system;
- the rate of change in the mobile radio channel;
- calibration of the transmitters and receivers involved.

23.5.1.1 Antenna Configuration

Equation (23.14) holds for any number of antennas at each end of the radio link, yet it may be the case that not every antenna element is used for both the transmit and receive functions at a given transceiver. For example, implementation of a transmit chain (and associated power amplifier) per receive antenna element may be feasible at an eNodeB whereas it may not be so in a UE for reasons of form-factor, cost and power consumption.

In these cases, knowledge of the Channel State Information (CSI) of the outbound channel is incomplete if obtained only using inbound RSs. Additional measures would then need to be adopted to determine the missing CSI components. These could include the use of supplementary and explicit feedback of the CSI from the remote receiver for any of its receive antennas that do not have transmit capability. Alternatively, the inbound MIMO channel may be sounded using RSs sent for example from only one transmit antenna at a time. This can then allow for a UE with only one transmit chain and power amplifier to sound the channel from all of its receive antennas by means of a time-controlled switch.

23.5.1.2 Time Variations in the Radio Channel

Mobile radio channels are of course time-varying in nature, as discussed in Chapter 8. For TDD systems, uplink and downlink transmissions occur at different times, thereby violating the reciprocity principle in a time-varying channel. However, the channels can still be considered as approximately reciprocal when viewed over a time span short enough to preclude any appreciable change. The duration between the instant at which the CSI is measured and the instant at which it is applied thus becomes of paramount importance.[7]

[7] Note that this applies also to systems using closed-loop feedback signalling, and not only to those attempting to exploit uplink and downlink channel reciprocity. It is simply the case that for control schemes reliant upon reciprocity the CSI is measured at the output of the inbound link rather than at the output of the outbound link.

The time over which a channel's impulse response $h(t)$ remains relatively constant is termed its *coherence time* and is defined as the delay δt at which the squared magnitude of its autocorrelation first drops below a certain fraction of its value at zero delay (a fraction of 50% is often used to define the coherence time).

Clearly in order to benefit from channel reciprocity, the delay between the time at which the observation of the inbound channel is made and the time at which this knowledge is applied to the outbound signal, must be shorter than the coherence time of the channel. Typical relationships between mobile speed and coherence time (in this instance for a carrier frequency of 2 GHz) are shown in Figure 23.14.

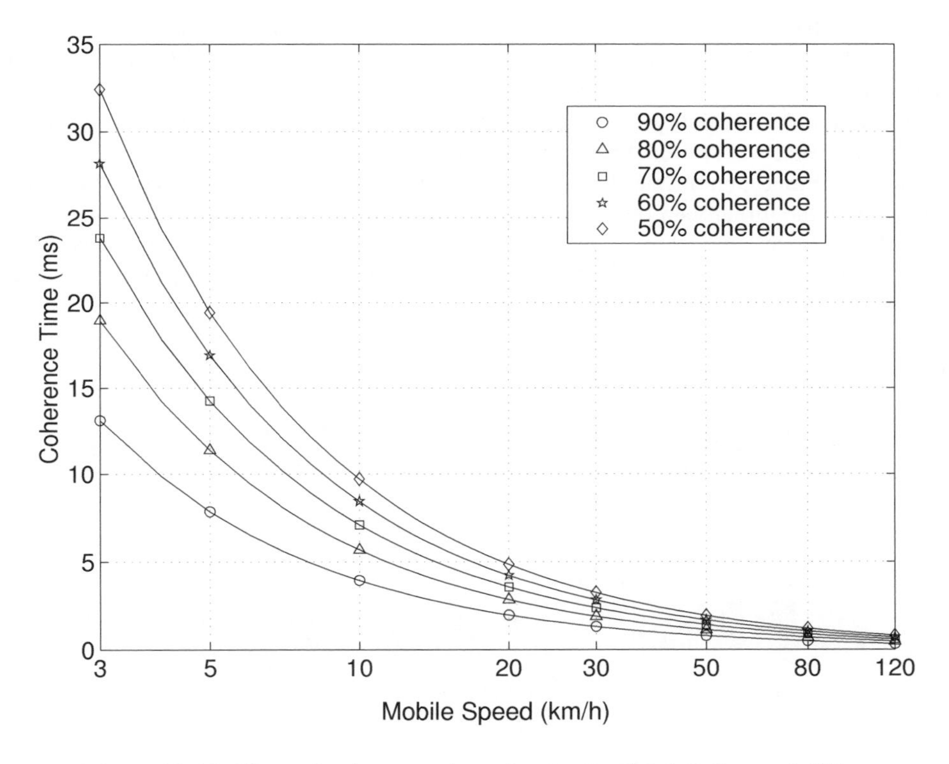

Figure 23.14: Channel coherence times for flat Rayleigh fading at 2 GHz.

From a practical perspective, the processes to be executed within the channel coherence time include receiving the necessary RSs, estimating the CSI, computing the required action and encoding and processing the outbound signals. With present-day technology this turnaround time is unlikely to be much lower than 1 ms. Even if it were, the LTE TDD frame structure would need to support uplink–downlink transitions at this rate to ensure continual updating of the CSI on the inbound link (interspersed by transmission on the outbound link); however, due to the minimum switching period of 5 ms, this is not possible. The CSI cannot be applied earlier than the next occurrence of a subframe in the return direction. It may also be the case that a particular UE will not be scheduled in the first available subframe of the return direction, potentially further increasing the delay between measurement of the CSI and its application.

The RSs commonly used for exploiting reciprocity in LTE are the cell-specific RSs in the downlink[8] and the Sounding RSs (SRSs) in the uplink. The cell-specific downlink RSs are present in each downlink subframe as shown in Section 8.2.1, while the SRSs are located either in the short UpPTS field of the mixed subframe (see Figure 23.10) or on one symbol of a normal uplink subframe where configured.

Under the assumption that one would prefer to use UpPTS for sounding in TDD, then with a 5 ms switching cycle and a balanced DL/UL configuration (TDD configuration 1 in Figure 6.2), the shortest possible turnaround times to exploit reciprocity for the two downlink subframes and the DwPTS would be 2 ms, 3 ms and 4 ms respectively, as shown in Figure 23.15(a). These correspond to mobile speeds of up to around 24 km/h at a 50% coherence level at 2 GHz.

For the uplink, the ability of the system to take advantage of reciprocity is not limited by the occurrence interval of the wideband cell-specific RSs (as they are present in every downlink subframe), but primarily by the length of the contiguous uplink period within the frame and by the turnaround processing time at the UE. Figure 23.15(b) gives one example in which the processing delay at the UE is assumed to be of the order of 2 ms, resulting in similar turnaround latencies to the downlink case. With faster processing at the UE, these could be reduced a little further.

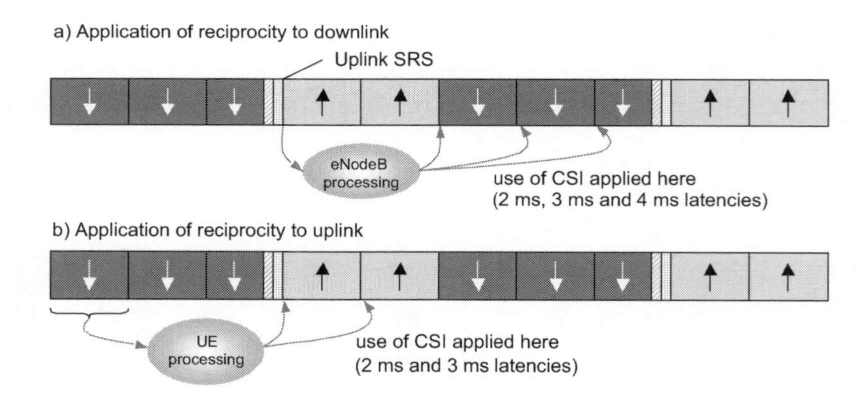

Figure 23.15: Example turnaround times for exploitation of reciprocity in LTE TDD: (a) application of reciprocity to downlink; (b) application of reciprocity to uplink.

23.5.1.3 Calibration

Under conditions of channel reciprocity, the forward and reverse responses between the antennas at two transceivers A and B are the same, although this does not mean that the overall responses between the baseband processors at those transceivers are also equivalent

[8] In Release 10, the CSI-RS (see Section 29.1.2) may also be used.

in both directions. The responses of the individual transmitters and receivers themselves must also be taken into account, as shown for a 2×2 system transmitting from baseband A to B in Figure 23.16.

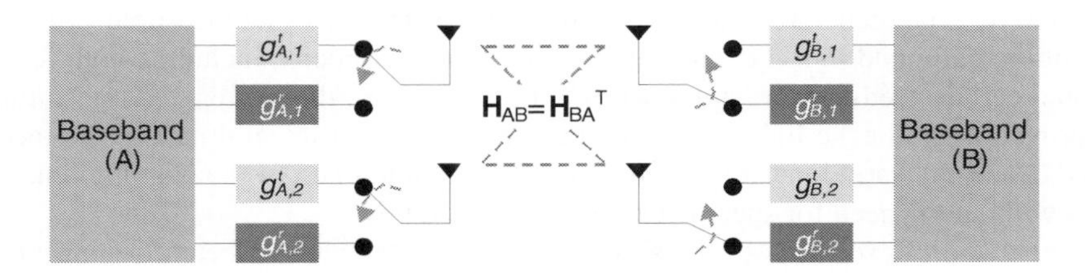

Figure 23.16: Transceiver elements in a reciprocal channel.

At a given frequency the responses of the multiple transceiver elements may be represented together in matrix form (assuming here that variations within the desired system bandwidth are small). Without coupling effects between elements, these are simply diagonal matrices containing the per-element complex gains, such that for I antennas at A and J antennas at B:

$$G_{\text{Tx,A}} = \begin{pmatrix} g^t_{\text{A},1} & \cdots & 0 \\ \vdots & \ddots & \vdots \\ 0 & \cdots & g^t_{\text{A},I} \end{pmatrix}, \; G_{\text{Rx,B}} = \begin{pmatrix} g^r_{\text{B},1} & \cdots & 0 \\ \vdots & \ddots & \vdots \\ 0 & \cdots & g^r_{\text{B},J} \end{pmatrix} \tag{23.15}$$

A similar formulation may be applied to $G_{\text{Tx,B}}$ and $G_{\text{Rx,A}}$ in the reverse direction assuming each antenna is capable of both transmit and receive. The composite transfer functions between A and B (and vice versa) at a given frequency then become

$$Z_{\text{AB}} = G_{\text{Tx,A}} H_{\text{AB}} G_{\text{Rx,B}} \quad \text{and} \quad Z_{\text{BA}} = G_{\text{Tx,B}} H_{\text{BA}} G_{\text{Rx,A}} \tag{23.16}$$

It is clear that Z_{AB} and Z_{BA} are not necessarily equal. Nonetheless, if the time between the measurements $\hat{Z}_{\text{AB}}$ and $\hat{Z}_{\text{BA}}$ of Z_{AB} and Z_{BA} respectively is sufficiently smaller than the coherence time, then using Equation (23.14) it can be deduced that (for invertible G matrices):

$$\hat{Z}_{\text{AB}} = G_{\text{Tx,A}} (G^{\text{T}}_{\text{Rx,A}})^{-1} \hat{Z}^{\text{T}}_{\text{BA}} (G^{\text{T}}_{\text{Tx,B}})^{-1} G_{\text{Rx,B}} \tag{23.17}$$

The general goal therefore of a system wishing to exploit reciprocity is to infer $\hat{Z}_{\text{AB}}$ from the observation $\hat{Z}_{\text{BA}}$ using knowledge of the relevant transceiver gains in G.

There are numerous methods by which this can, in theory, be achieved. One such method is via an explicit signalling exchange between transceivers of the measured composite channel responses ($\hat{Z}_{\text{AB}}$ and $\hat{Z}_{\text{BA}}$), as observed within the coherence time of the channel. By analysing the relation between these, an active transmitter may then derive the necessary transfer function needed to correctly infer the composite outbound channel from future measurements of the composite inbound channels [13]. Note that the update rate of such feedback to track variations in the transceiver responses should be relatively slow – otherwise a pure feedback-based approach that is not reliant on reciprocity becomes a more natural choice. Fortunately, the variations in transceiver responses are predominantly driven by temperature changes, and therefore a low update rate is usually possible (although other factors can also arise, such as interaction between the antenna and nearby objects such as human hands at the UE side). In theory, self-calibration methods are also possible whereby the transceivers make use of accurate reference transmitters or receivers, or of auxiliary calibration signals.

Calibration is not generally required for link control systems using measured CSI from the output of the outbound link (i.e. those not reliant upon reciprocity, including both feedback-based and certain feed-forward approaches[9]). In these cases, the aggregated transmitter and receiver responses for the link under control already form an integral part of the measured CSI and hence any variations are accommodated as a natural part of the ongoing link control process without the need for specific calibration procedures.

One noteworthy exception to the above, however, is the case of beamforming in which the ability of an array to steer energy towards (or collect energy from) a given direction is impaired if the phase between the elements is not adequately controlled. In this instance calibration would generally only be required at the transceiver attached to the array (such as at the eNodeB); an opposing transceiver without an array (such as a mobile terminal) would not require calibration.

In general therefore, it can be said that systems attempting to utilize channel reciprocity require calibration of their constituent transceivers in order to restore the overall reciprocity between them at baseband. Calibration is also required for antenna arrays reliant upon directionality, such as transmit and receive beamforming, although this applies irrespective of the presence of channel reciprocity.

23.5.2 Applications of Reciprocity

For the purposes of this discussion, we define the 'outbound link' as the link which is being actively controlled in response to the channel.

Active control techniques may be generally classified as feed-forward (open-loop) or feed-back (closed-loop). Note that both may involve feedback of information from the receiving side to the transmitting side, the distinction lying in the particular nature of this information. Receivers in closed-loop feedback systems return information regarding how well the transmitting side is performing in relation to a target. Conversely, feed-forward transmitters do not know how well their transmissions are performing, and simply use whatever information is available (via fed-back information or otherwise) to try to pre-empt one or more aspects of the radio channel.

There also exists a class of passive open-loop systems in which the transmitting side takes no pre-emptive action, and exploitation of the channel is confined solely to the receiving side. Figure 23.17 shows this classification of link control methods.

For active feed-forward control systems, the information required by the transmitter often takes the form of CSI which supplies information regarding the response between transmitter and receiver. It is really only this class of system for which reciprocity may be used – the outbound link CSI being inferred from the inbound link CSI given that the conditions discussed in Section 23.5.1 are met and that inbound RSs covering the desired range of frequencies are available for this purpose.

[9]Feed-back and feed-forward control mechanisms are discussed further in Section 23.5.2.

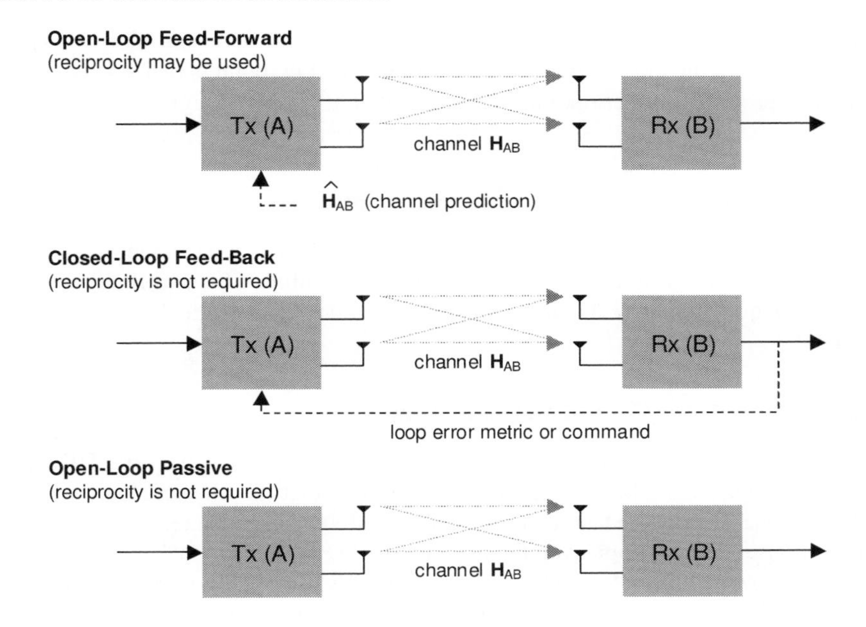

Figure 23.17: Feed-forward, feed-back and open-loop exploitation of the radio channel.

However, for wireless communication, it is often not only the channel between the controlling transmitter and the receiver that is of interest; information relating to interference caused by other transmitters is also relevant and cannot be obtained via reciprocity.

Some link control methods make use of only amplitude information from the channel (e.g. open-loop uplink power control and frequency-domain scheduling), whereas others, such as MIMO precoding, also need phase information. Even in a TDD deployment, it cannot always be assumed that reciprocity can be freely exploited to achieve this. In this regard, we discuss below some link control examples of relevance in a TDD LTE system.

23.5.2.1 Downlink Frequency-Domain Scheduling

The eNodeB scheduler can exploit the frequency-selective nature of the channel by attempting to schedule downlink data to a user using frequency resources that are currently experiencing higher than average SINR (see Chapter 12). Although reciprocity allows for the frequency response of the downlink radio channel to be inferred via the observed response of the uplink channel, it does not provide information concerning the downlink interference and noise levels observed at the UE. Hence knowledge of the channel provides only partial knowledge of the SINR.

23.5.2.2 Uplink Frequency-Domain Scheduling

As for the downlink case, this is also an interference-responsive mechanism. Contrary to the downlink, however, the eNodeB in this instance does have some knowledge of the interference and noise levels affecting the uplink transmission at its own receiver. Together with the required CSI of the uplink channel gleaned from the SRS transmissions, this provides the eNodeB with all the information required. However, at no point is reciprocity required in order for this process to function, due to the fact that all involved signals exist only in one link direction.

23.5.2.3 Downlink Power Control and Link Adaptation

This function requires interference feedback from the UE if it is to operate effectively. Reciprocity can be used by the eNodeB to infer the downlink CSI, and thereby to provide partial information, but pure feedback-based techniques tend to be better suited to this application.

23.5.2.4 Uplink Power Control and Link Adaptation

In theory, reciprocity could be used in the feed-forward sense to adjust the UE's transmit power and coding/modulation. In this instance, the fact that the interference levels may be controlled by the eNodeB means that the received SINR can be better correlated with the channel response, thereby improving the usefulness of reciprocity.

For uplink power control, the LTE system only makes use of long-term reciprocity via the downlink path-loss estimation (see Section 18.3.2.1). For uplink link adapation, autonomous selection of the modulation and coding rate by the UE is deliberately not permitted, in order to enable the eNodeB to exercise full control of the uplink and its optimization.

23.5.2.5 Phase-Passive MIMO

Phase-passive MIMO schemes are those in which the transmitter does not attempt to adapt to the phase of the channel and responds only to amplitude information. Two techniques in LTE that fall into this category are open-loop downlink MIMO (i.e. transmission mode 3 not using Precoding Matrix Indicator (PMI) feedback – see Section 9.2.2.1) with per-stream Modulation and Coding Scheme (MCS) selection, and switched transmit antenna diversity for the uplink. The former essentially comprises parallel instances of downlink rate control and hence the discussion of Section 23.5.2.3 is applicable.

LTE supports switched antenna diversity for the uplink in one of two configurable modes. In the first mode, the eNodeB selects the transmit antenna that the UE should use as described in Section 16.6.1 and signals this decision to the UE. This mode does not require reciprocity in order to function. In the second mode, the UE is allowed to select the transmit antenna for uplink transmission, and in TDD operation this can therefore be based on reciprocity using for example the cell-specific downlink RSs to provide the UE with the necessary information on which to base its decision.

23.5.2.6 Short-Term MIMO Precoding

As discussed in Chapter 11, precoding refers to the application of a set of antenna weights per transmitted spatial layer (the precoding matrix) which, when applied to a forthcoming transmission, interact beneficially with the radio channel in terms of improved reception or separation of the layers at the intended receiver(s). Both feed-back and feed-forward techniques are possible in terms of precoding matrix selection strategy. Feed-back approaches (using PMI in LTE) do not rely on reciprocity, only on the feedback loop delay being shorter than the channel coherence time. For feed-forward approaches, the necessary CSI can, in theory, be obtained in one of two ways:

1. using 'Direct Channel Feed-Back' (DCFB), in which the CSI observed by the receiver is explicitly signalled back to the transmitter;
2. using channel sounding RSs (e.g. SRS on uplink) to sound the link inbound to the precoding transmitter.

DCFB obviates the need for reciprocity, whereas the RS-assisted method is reliant upon it. LTE does not support DCFB, and hence reciprocity and TDD are prerequisites for feed-forward-based short-term precoding, whereas the feedback-based (PMI) scheme is applicable to all duplex modes.

As described in Chapter 11, explicit control signalling is required for the PMI-based scheme, whereas for a reciprocity-based scheme the role of control signalling is undertaken by UE-specific RSs in the downlink (see Section 8.2.2) and SRS in the uplink.

A reciprocity-based scheme carries the advantage that the applied precoding matrix is not constrained to a finite set of possibilities[10] (a codebook). However, the support for UE-specific RSs was limited in the first release of LTE (Release 8) to allow only a single spatial layer,[11] thereby precluding the use of spatial multiplexing using non-codebook-based precoding. In Release 9, the UE-specific RS design was extended to support two layers (see Sections 8.2.3, 11.2.2.3) thereby also enabling dual-layer short-term precoding using techniques based on reciprocity. In Release 10, the use of UE-specific RSs is further extended to up to eight layers, as explained in Section 29.1.1.

23.5.2.7 Long-Term MIMO Precoding (Beamforming)

At a beamforming transmitter, energy is focused in the directions which are more commonly observed in the inbound channel than others, assuming that a similar angular distribution also applies in the mean sense to the outbound channel. Beamforming is reliant, therefore, only on longer-term statistics of the radio channel and does not require short-term correlation between the inbound and outbound channels in order to function. As such, this scheme is applicable to both FDD and TDD; short-term channel reciprocity in the sense described above is not required.

23.5.3 Summary of Reciprocity Considerations

Exploitation of channel reciprocity for LTE TDD is possible in certain areas. These include its application to uplink transmit antenna selection (using only amplitude information) and to non-codebook-based downlink precoding (also using phase information). Support for the latter has been greatly improved since the first release of LTE via the introduction of multi-layer support for UE-specific RSs. The benefits arising from the use of reciprocity are generally dependent upon the intended deployment scenario, and in some cases pure feedback-based approaches are sufficient.

---

[10]Referred to as non-codebook-based precoding, this can potentially improve performance by removing mismatch between the best-fit precoding matrix in the codebook and the precoding matrix optimal for the channel.

[11]The initially intended application of UE-specific RSs was to support single-layer beamforming.

In order to realize a true state of reciprocity between a base station and a UE, the feed-forward loop delay must be within the coherence time of the channel. Calibration of the transceivers is also required (notably when exploiting phase information), and the antenna elements involved must generally be shared by both transmit and receive functions.

References[12]

[1] Ofcom, 'Spectrum Framework Review: Implementation Plan', Ofcom consultation document, www.ofcom.org.uk, January 2005.

[2] 3GPP Technical Report 36.942, 'Evolved Universal Terrestrial Radio Access (E-UTRA); Radio Frequency (RF) System Scenarios (Release 8)', www.3gpp.org.

[3] T. Wilkinson and P. Howard, 'The Practical Realities of UTRA TDD and FDD Co-Existence and their Impact on the Future Spectrum', in *Proc. 15th IEEE International Symposium Personal and Indoor Mobile Radio Communications (PIMRC 2004)* (Barcelona, Spain), September 2004.

[4] 3GPP Technical Report 25.814, 'Physical Layer Aspects for Evolved Universal Terrestrial Radio Access (UTRA)', www.3gpp.org.

[5] J. H. Winters, 'Optimum Combining in Digital Mobile Radio with Cochannel Interference'. *IEEE Journal on Selected Areas in Communications*, Vol. 2, pp. 528–539, July 1984.

[6] E. G. Larsson, 'Model-Averaged Interference Rejection Combining'. *IEEE Trans. on Communications*, Vol. 55, pp. 271–274, February 2007.

[7] 3GPP Technical Specification 36.211, 'Evolved Universal Terrestrial Radio Access (E-UTRA); Physical Channels and Modulation', www.3gpp.org.

[8] 3GPP Technical Specification 36.213, 'Evolved Universal Terrestrial Radio Access (E-UTRA); Physical Layer Procedures', www.3gpp.org.

[9] J. W. S. Rayleigh, *The Theory of Sound* [Macmillan & Co., 1877 and 1878]. Reprinted by Dover, New York, 1945.

[10] H. A. Lorentz, 'Het theorema van Poynting over de energie in het electronmagnetisch veld en een paar algemeene stellingen over de voortplanting van licht'. Verhandelingen en bijdragen uitgegeven door de Afdeeling Natuurkunde, Koninklijke Nederlandse Akademie van Wetenschappen te Amsterdam, Vol. 4, pp. 176–187, 1895–1896.

[11] J. R. Carson, 'Reciprocal Theorems in Radio Communication' in *Proc. Institute of Radio Engineers*, June 1929.

[12] S. Ballantine, 'Reciprocity in Electromagnetic, Mechanical, Acoustical, and Interconnected Systems', in *Proc. Institute of Radio Engineers*, June 1929.

[13] M. Guillaud, D. Slock and R. Knopp, 'A Practical Method for Wireless Channel Reciprocity Exploitation through Relative Calibration', in *Proc. Eighth International Symposium on Signal Processing and Its Applications*, August 2005.

[12] All web sites confirmed 1st March 2011.

24

Picocells, Femtocells and Home eNodeBs

Philippe Godin and Nick Whinnett

24.1 Introduction

The use of small cells is becoming increasingly important due to their ability to provide increased system capacity compared to a homogeneous network of macrocells.[1] Small cells can generally be characterized as either *picocells* (also referred to as hotzone cells), controlled by a pico eNodeB, or *femtocells*, controlled by a Home eNB (HeNB).

The definitions of picocells and femtocells are somewhat variable, but typically the main differentiating features can be summarized as follows:

- A pico eNodeB usually controls multiple small cells which are planned by the network operator in a similar way as the macrocells. Picocells are often mounted at low elevation and operate at lower power than the macrocells. A HeNB controls only one cell and is deployed by the customer (the registered owner), usually without planning.
- Femtocells are typically Closed Subscriber Group (CSG) cells accessible only to a limited group of users; in contrast, picocells are usually open to all users (Open Subscriber Group (OSG)) but may offer preferential treatment to some users, e.g. to the staff of a particular establishment.
- The transmission power of femtocells is lower, being designed typically to cover a house or apartment. Picocells usually operate with a higher transmission power to cover an enterprise, mall or other hotzone, or simply to extend macrocellular coverage.

[1] A combined network of macrocells and small cells is often known as a 'heterogeneous network'.

LTE – The UMTS Long Term Evolution: From Theory to Practice, Second Edition.
Stefania Sesia, Issam Toufik and Matthew Baker.

- Femtocells do not necessarily have the same network interfaces as macro eNodeBs, whereas pico eNodeBs follow the same logical architecture principles as macro eNodeBs as described in Chapter 2.

Small cells may be deployed on the same carrier as the macrocells or on a dedicated carrier. If a dedicated carrier is used it would typically be at a higher frequency to deliver high data rates within a limited area without causing excessive interference. If the small cells share a carrier with the macrocells, careful attention needs to be given to the interference that may arise between the small cells and the macrocells. This is discussed in Section 24.3.

This chapter first addresses the architectural aspects specific to femtocells/HeNBs. Then the potential interference issues are analysed and some possible interference management schemes are discussed. Finally some of the Radio Frequency (RF) challenges of both HeNBs and pico eNodeBs are discussed.

24.2 Home eNodeB Architecture

The main architectural aspects of HeNBs are based on the general concepts applicable to all types of LTE cell as described in Chapter 2. The following sections describe the aspects specific to HeNBs.

24.2.1 Architecture Overview

The architecture of a network involving HeNBs has been designed to be flexible and scalable with respect to the number of HeNBs. In particular, if the network contains a large number of HeNBs, a HeNB Gateway (HeNB GTW) can optionally be deployed to manage the HeNBs from the perspective of the Evolved Packet Core (EPC). The capacity of a HeNB GTW can be up to several tens of thousands of HeNBs. A HeNB GTW can be deployed on the Control Plane (HeNB GTW CP) only, or it may also include a User Plane (HeNB GTW UP) part. The logical architecture is shown in Figure 24.1.

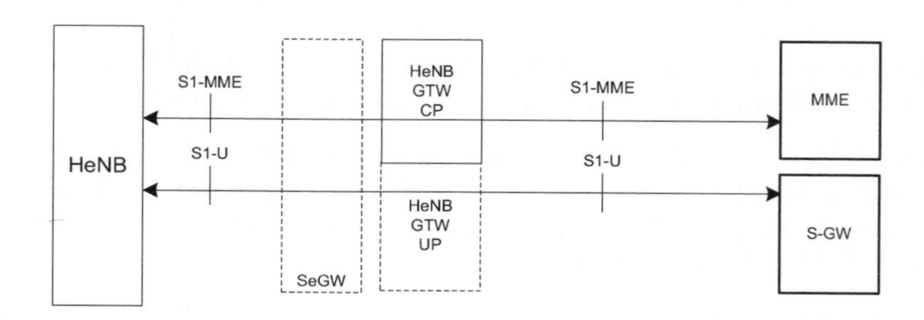

Figure 24.1: E-UTRAN HeNB logical architecture. Reproduced by permission of © 3GPP.

When a HeNB GTW is deployed, it serves as a concentrator for the CP S1-MME interface (see Section 2.3). Each HeNB therefore has only one Stream Control Transmission Protocol

(SCTP, see [1] and Section 2.5.1) association towards the HeNB GTW and each Mobility Management Entity (MME) likewise has only one SCTP association towards the HeNB GTW. This significantly reduces the overall number of SCTP associations needed in a pool area.[2] For example if the number of MME nodes is M in the pool area and the number of HeNBs is N, then only $N + M$ SCTP associations will be necessary instead of NM (assuming only one HeNB GTW is used in the pool area).

The S1-U interface may also optionally be terminated in the HeNB GTW UP. In this case the UP connection for one E-UTRAN Radio Access Bearer (E-RAB)[3] of one UE consists of two GPRS Tunnelling Protocol (GTP) tunnels instead of one – i.e. one from the Serving-GW (S-GW) to the HeNB GTW UP and one from the HeNB GTW UP down to the HeNB. This option may, for example, be used to avoid the SŰGW having to manage paths to tens of thousands of HeNBs. Figure 24.1 also shows the Security Gateway (SeGW) which is optional and is not necessarily part of the HeNB GTW.

The HeNB GTW is assumed to be *transparent* for the S1 interface – i.e. the HeNB sees the HeNB GTW exactly as if it were an MME and conversely the MME sees the HeNB GTW as a regular eNodeB. In terms of protocol this also means that the same S1-MME interface (and the same CP protocol, S1 Application Protocol – S1-AP) is used between the HeNB GTW and the EPC, between the HeNB and the HeNB GTW, between the HeNB and the EPC, and between the eNB and the EPC.

Despite its transparency, the HeNB GTW when deployed ensures the essential function of the S1-flex[4] in a pool area on behalf of the HeNBs. This means that a particular HeNB only sees one MME (in fact the HeNB GTW) and does not need to implement the S1-flex, while the HeNB GTW is in charge of selecting, on a per-UE basis when a UE attaches[5] to the network, which one of the MMEs in the pool will handle the UE's requests. This also means that when a HeNB is connected to an HeNB GTW, it will not simultaneously connect to another HeNB GTW or another MME. However, when no HeNB GTW is deployed, the HeNB can still use the S1-flex like any regular eNodeB.

24.2.2 Functionalities

When a HeNB is directly connected to the EPC, it supports exactly the same functions as a normal eNodeB. However, an HeNB is limited to the support of a single cell and it may support an X2 interface only from Release 10 onwards. When the HeNB is connected via the intermediate HeNB GTW it cannot support the Non-Access Stratum (NAS) node selection function used in S1-flex for the reasons explained above, but it does support some additional functions such as initial discovery of a suitable HeNB GTW.

HeNBs can in principle operate in any of three different access modes:

- **Closed access mode**: The HeNB belongs to one or more specific Closed Subscriber Groups (CSGs) identified by CSG Identifiers (CSG IDs). Only UEs which have the corresponding CSG ID included in their CSG subscription list are allowed access. This is the most common mode for HeNBs.

[2] The pool area is the common area served by a set of MME/S-GWs – see Section 2.2.2.

[3] See Section 2.4.

[4] See Section 2.2.2.

[5] See Section 2.2.1.1.

- **Hybrid access mode**: Like for closed access mode, the HeNB belongs to a particular CSG. It provides service to all UEs but gives preferential treatment to UEs which include the corresponding CSG ID in their CSG subscription list; such UEs are called 'CSG members'.
- **Open access mode**: The HeNB behaves as a regular eNodeB.

When a HeNB operates in closed or hybrid access modes, during incoming handovers it has to check the validity of the CSG ID which has been used by the MME for the Access Control or Membership Verification of the user.

The HeNB GTW hosts two main functions:

- It transparently relays the UE-associated S1-AP messages between the MME and the HeNB which serves the UE. All protocol functions associated with a UE-dedicated S1 procedure resides within the HeNB and the MME only.
- It runs the non-UE associated S1-AP procedures between it and the HeNB on one side and between it and the EPC on the other side, and interfaces between them.

In addition, the HeNB GTW can perform some optional functions such as terminating the S1-U interface (see Sections 24.2.1 and 2.2.2) and hence relaying the user plane data between the HeNB and the S-GW, and paging optimization based on a list of CSG IDs provided by the MME in the PAGING message (avoiding paging a UE in closed access mode HeNB cells whose CSG ID is not in the list).

The MME performs Access Control for UEs accessing or handing over to a closed access HeNB, and Membership Verification in the case of hybrid access HeNBs – i.e. the MME checks that the CSG ID of the closed or hybrid HeNB is in the list of allowed CSG IDs for the UE; for a closed access HeNB, the MME rejects the UE if this check is unsuccessful while for a hybrid access HeNB it indicates to the HeNB whether the user is a 'CSG member' or not in order to allow any special treatment to be triggered. The MME also controls the routing of handover messages towards HeNBs, either using dedicated HeNB addresses or, if the HeNB is connected through an HeNB GTW, using a 'Tracking Area Identity' (TAI) (see [2, Section 19.4.2.3]). The MME may also optionally support paging optimization.

24.2.3 Mobility

Several kinds of mobility are supported via the MME regardless of whether the HeNB is connected via a HeNB GTW:

- Mobility from HeNB to an eNodeB, i.e. outbound mobility, supported from Release 8 onwards;
- Mobility from eNodeB to HeNB, i.e. inbound mobility from macrocell, via S1 handover, supported from Release 9 onwards;
- Mobility from HeNB to HeNB, i.e. inbound mobility from an HeNB cell, via S1 handover, supported from Release 9 onwards;
- Optimized mobility from HeNB to HeNB, i.e. inbound mobility from an HeNB cell via X2 handover, supported from Release 10 onwards.

The HeNB GTW should preferably connect to the EPC in such a way that inbound and outbound mobility to cells served by the HeNB GTW do not necessarily require inter-MME handovers. An example of the general call flow for an inbound mobility via S1 handover is shown in Figure 24.2. The messages may pass through the HeNB GTW (if present) without any functional change.

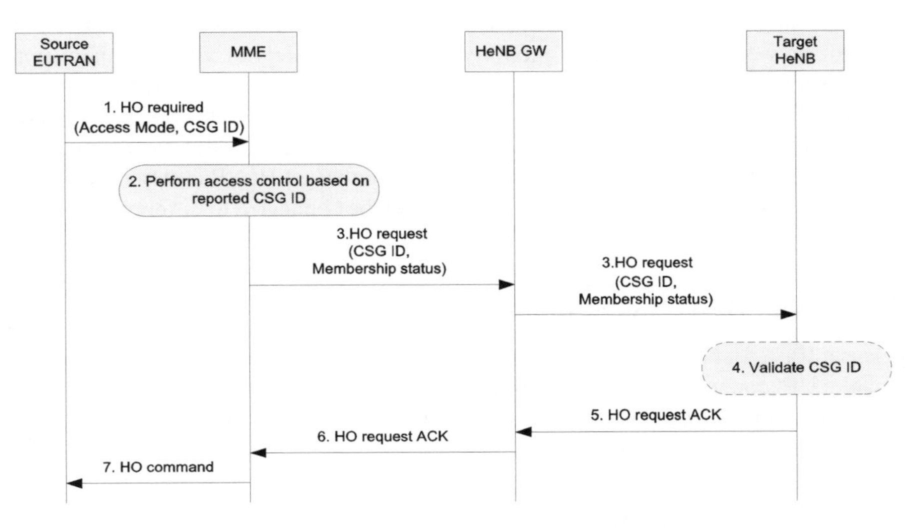

Figure 24.2: A typical call flow for mobility towards a HeNB. Reproduced by permission of © 3GPP.

The source node for the handover indicates to the MME the ECGI,[6] CSG ID and access mode of the target HeNB. These elements are generally reported by UEs which are able to read the System Information (SI) of the target cell, but may be otherwise inferred by the source node if it can resolve any Physical Cell Identifier (PCI) confusion. The MME performs the Access Control or Membership Verification based on the received information and the CSG subscription data of the UE. If the UE is not a member of the CSG and the target HeNB is in closed access mode, the MME ends the handover procedure. Otherwise the MME sends the 'HANDOVER REQUEST' message (see Section 2.6) to the target HeNB including the value of the CSG ID. If the target HeNB is configured for hybrid access, the MME also indicates in the 'HANDOVER REQUEST' message whether the UE is a CSG member or not. The target HeNB further checks whether the received CSG ID is correct and allocates the appropriate resources. The HeNB sends back the 'HANDOVER REQUEST ACKNOWLEDGE' message to complete the handover.

If the received CSG ID is found to be incorrect by the target HeNB, the incoming handover will fail if the HeNB is in closed access mode, while in the case of hybrid access mode the HeNB will accept the incoming handover but report back to the MME (in the 'HANDOVER REQUEST ACKNOWLEDGE' message) the valid CSG ID so that the MME can apply accurate charging.

[6]E-UTRAN Cell Global Identifier. See Section 22.2.2 for reporting requirements.

In Release 10, mobility between two HeNBs does not necessarily need to use S1 handover via the MME; it can instead use X2 handover directly. This optimization reduces backhaul traffic and delay, even though in practice this X2 connection may still need to be routed via a central SeGW (if present) if the environment is not secure.

X2 handover between two HeNBs is only allowed for cases where the MME does not need to perform access control – i.e. when the source and target HeNBs are in closed or hybrid access mode and have the same CSG ID, or when the target HeNB is in open access mode.

The call flow for this optimized mobility between two HeNBs via X2 handover is the same as the general call flow for an X2 handover between two eNodeBs (see Section 2.6.3.1), even though the PATH SWITCH REQUEST/ACKNOWLEDGE messages may pass (transparently) through the HeNB GTW if present. This optimized mobility works regardless of the type of S1 connectivity employed by the source and target HeNBs (they may be directly connected to an MME or to a HeNB GW, or even each be connected to different HeNB GWs).

24.2.4 Local IP Access Support

In Release 10, a HeNB may operate in Local IP Access (LIPA) mode. LIPA enables an IP-capable UE connected via a HeNB to access other IP-capable devices in the same residential/enterprise IP network without the user plane traversing the mobile operator's network, as shown in Figure 24.3.

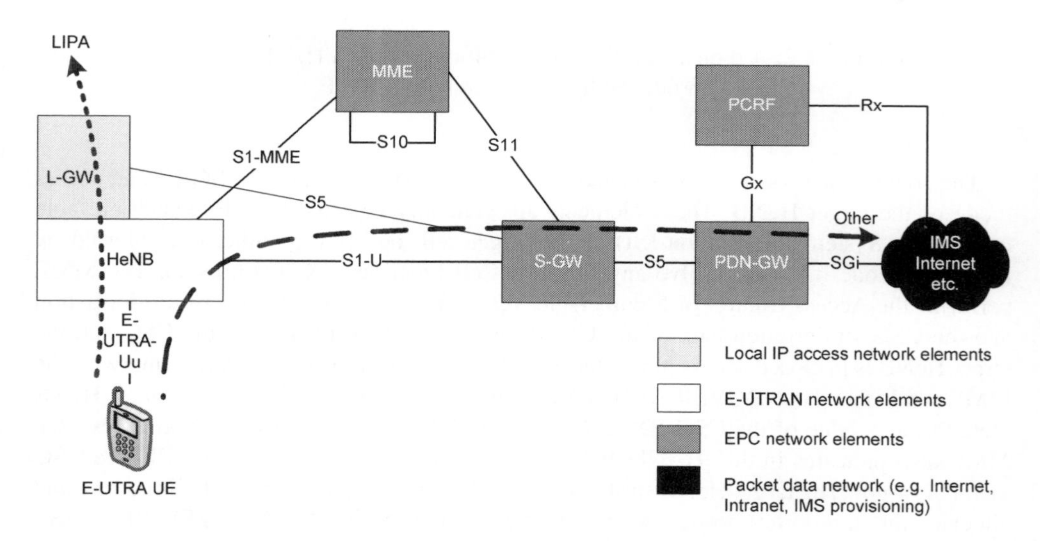

Figure 24.3: HeNB operating in LIPA mode in Release 10. Reproduced by permission of © 3GPP.

When a HeNB operates in LIPA mode, it supports an S5 interface towards the S-GW and an SGi interface towards the IP network to direct the LIPA user plane traffic. The HeNB may reuse the S1-interface IP address for the S5 interface in order to be able to reuse the S1

IPSec tunnel, or it may use a different IP address which would result in the establishment of a different IPSec tunnel. A LIPA connection is always released on outbound handover.

In order to be able to operate in LIPA mode, HeNBs support the following additional functions, regardless of the presence of a HeNB GTW:

- Transfer of the co-located LIPA GateWay (L-GW) IP address of the HeNB via the S1-MME interface to the EPC on every idle–active transition and on every uplink NAS PDU transfer in connected mode;
- Transfer over the S5 and S1-MME interfaces of the co-located L-GW uplink Tunnelling End IDs (TEIDs, see Section 2.3.1) used for the LIPA bearers for correlation purposes between the co-located L-GW function and the HeNB;
- Support of basic Packet Data Network GateWay (P-GW) functions in the co-located L-GW, such as support of the SGi interface corresponding to LIPA;
- Additional support of first packet sending, buffering of subsequent packets and internal direct L-GW–HeNB user path management;
- Support of a restricted set of S5 procedures relevant to LIPA.

24.3 Interference Management for Femtocell Deployment

Interference management is a key issue for heterogeneous network deployments of macrocells and small cells in LTE. The problem is most acute for femtocells operating in closed access mode on the same carrier frequency as the macrocells. Interference is typically more manageable in systems with picocells or other open- or hybrid-access small cells because UEs which are causing or suffering from interference can be handed over freely between the macro and small cells. For example, if a Macro UE (MUE) (i.e. a UE connected to a macrocell) is near the edge of the macrocell and is also close to a CSG femtocell, then it will be transmitting at high power and may interfere with the uplink of the small cell. If the MUE can be handed over to the small cell, the interference to the small cell would be eliminated. Similarly in the downlink, a MUE near the macrocell edge might suffer interference from a nearby small cell, which would be eliminated if the MUE could be handed over to the femtocell. The most severe interference occurs when the small cells and macrocells are deployed on the same carrier frequency; otherwise, the interference depends upon the Spectrum Emission Mask (SEM), and, in the case of adjacent channel operation, on the Adjacent Channel Leakage Ratio (ACLR) and Adjacent Channel Selectivity (ACS).

Therefore in this section we focus on the case of co-channel deployments of femtocells and macrocells. Section 31.2 addresses co-channel deployments of picocells and macrocells.

Femtocells are usually installed by the consumer in an ad-hoc fashion, rather than being part of a planned deployment. They are therefore designed to be self-configuring, in that they are required to sense their environment (e.g. to detect and measure neighbouring macrocells and femtocells) and adapt their operation accordingly (see, for example, Section 24.4.1 regarding adjacent channel protection. If the femtocell can choose its carrier frequency, then interference may be controlled by the appropriate carrier selection (for example based on measured Reference Signal Received Power (RSRP) and cell reselection priority information).

Note that co-channel interference between femtocells situated close to each other may also occur.

24.3.1 Interference Scenarios

Here we outline the possible interference scenarios of relevance./footnoteNote that these scenarios apply for both FDD and TDD systems; in the case of TDD it is assumed that synchronization of uplink/downlink switching points is achieved between the macrocell and small cells.)

Downlink transmissions from the femtocell suffer interference from transmissions from a macrocell as shown by interference path A in Figure 24.4.

Macro to Femto, Downlink. UEs connected to femtocells (known as Femto UEs or FUEs) are more susceptible to this interference when they (and their associated HeNB) are closer to the macrocell, since the transmit power of the macrocell is much higher than that of the femtocell and thus the received interference power will be higher. FUEs are also more susceptible when located far from the serving HeNB, especially if they are outside the house or apartment the femtocell is designed to cover.

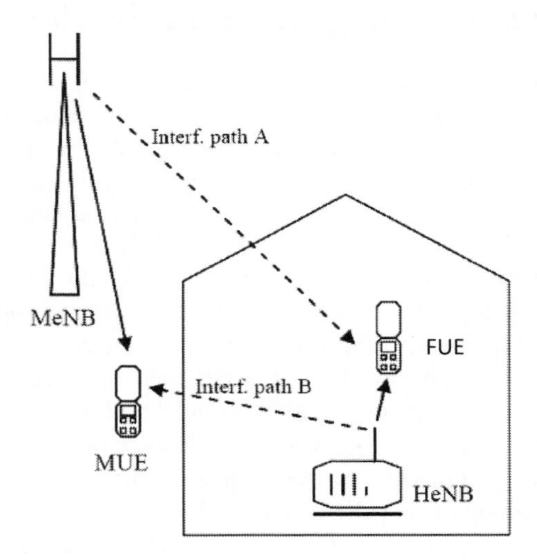

Figure 24.4: Macro/femto downlink interference scenarios.

Potential interference mitigation approaches for this scenario may include:

- Control channel protection (Physical Downlink Control CHannel (PDCCH), synchronization signals (PSS/SSS), Physical Broadcast CHannel (PBCH)) by arranging control channels to be orthogonal in time and/or frequency, e.g. by applying a subframe boundary offset in the femtocells relative to the macrocells.

- Data channel protection (PDSCH). If partial frequency re-use (see Section 12.5) is employed in the macrocell then a HeNB can schedule data on the RBs with low transmission power from the macrocell (e.g. RBs being used for cell centre UEs by the macrocell). The HeNB could ascertain the frequency partition information of the MeNB by various means, e.g. by configuration or by monitoring the macrocell transmissions.

Femto to Macro, Downlink. Downlink transmissions from a macrocell suffer interference from transmissions from a HeNB as shown by interference path B in Figure 24.4. This can cause a 'deadzone' around a HeNB within which a MUE is unable to receive transmissions from the macrocell. Such deadzones are larger for HeNBs near the edge of the macrocell, where the signal received from the macrocell is weakest, or for MUEs located indoors in the coverage of a CSG HeNB. Potential interference mitigation approaches for this scenario may include:

- Enabling hybrid or open access if possible. In the case of hybrid access, the power settings of the HeNB could be adapted differently to the closed access case, taking into account the total system performance (macro + hybrid cell) and the resources consumed by 'visiting' non-CSG UEs.

- Downlink power setting. The HeNB can limit the maximum downlink power (or power per RB) according to its environment. In the case of co-channel deployment of macrocells and femtocells with closed access, there are a number of possible ways this could be achieved [3], such as setting the power to achieve a trade-off between coverage and interference, based on the estimated path-loss between the HeNB and the victim MUEs, and the coverage requirements of the femtocell. Such a solution could also involve detecting the presence of nearby victim MUEs and correspondingly reducing the downlink transmission power; this could be done by detecting uplink transmissions at the HeNB, or by means of measurement reports from victim MUEs to the serving MeNB if it is possible to then signal this information to the HeNB.

- Time-domain coordination making use of Almost Blank Subframes (ABS), a concept introduced in Release 10 (see Section 31.2.2). ABSs contain only certain essential transmissions, leading to a reduction in interference to victim UEs. Typically, an aggressor HeNB would set up a pattern of ABSs resulting in reduced interference to victim MUEs. In Release 10 there is no X2 interface between HeNBs and macro eNodeBs, and therefore ABS patterns at a HeNB would need to be configured either by Operation and Maintenance (O&M) or autonomously by the HeNB; for example, an ABS pattern could be set up at an HeNB to protect subframes containing PSS/SSS/PBCH and paging occasions at a macro eNodeB, assuming the HeNB has obtained time synchronization with the macro eNodeB. Additional mitigation of the residual interference due to the essential transmissions is possible e.g. by arranging the frequency positions of CRS to be different in victim and aggressor cells.

System simulation results [3] are given in Table 24.1 showing the effect of introducing HeNBs, setting the HeNB power appropriately and enabling hybrid access. The system bandwidth is 5 MHz with a co-channel deployment of macrocells (with 1 km inter-site distance) and HeNBs, with 22 UEs per macrocell, 12 of which are FUEs.

Table 24.1: Effect of various small cell configurations on system throughput.

	Outage Probability (SNR< −6 dB)	Worst 20% UE throughput (kbps)	Median throughput (kbps)
No HeNB	12.7%	35	150
CSG HeNB, Fixed Power 8 dBm	18.9%	100	5600
CSG HeNB, Power Control (−10 , +10 dBm)	9.8%	250	3300
Hybrid HeNB, Fixed Power 8 dBm	2%	900	5100
Hybrid HeNB, Power Control (−10 , +10 dBm)	3%	400	3400

Macro to Femto, Uplink. Uplink transmissions from an FUE suffer interference from transmissions from a MUE as shown by interference path C in Figure 24.5.

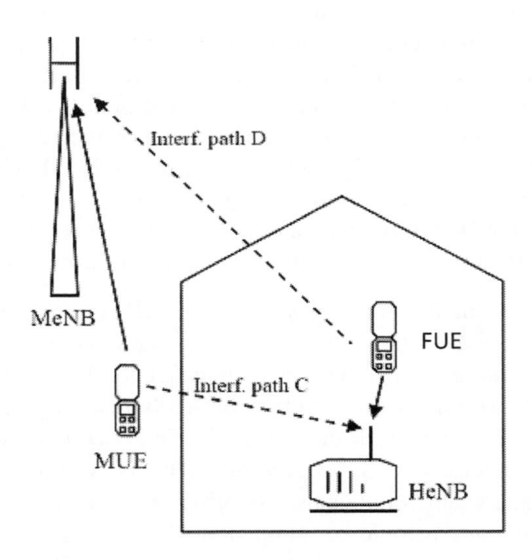

Figure 24.5: Macro/femto uplink interference scenarios.

Femtocells are more susceptible to this interference when they are located close to the edge of the macrocell edge since MUEs in the vicinity of the HeNB will be transmitting at higher power in this case. System simulations [3] have suggested that this scenario does not create a significant interference impact because in most cases MUEs cannot operate very close to the HeNB, and therefore the FUEs will typically be closer to the HeNB than the MUEs are. Nevertheless, the interference may be severe if the MUE and HeNB are both situated indoors; in this case the following potential interference mitigation approaches may be relevant:

- Uplink power control: the HeNB can control the power of its FUEs to overcome the interference from neighbouring MUEs;
- Control channel protection (PUCCH): if the PUCCH resources are over-provisioned at the HeNB then the RBs used for PUCCH at the HeNB can be moved away from the band edges (see Section 16.3.1) such that they do not overlap with the RBs used for PUCCH by the macrocell.

Femto to Macro, Uplink. Uplink transmissions from an MUE suffer interference from transmissions from an FUE as shown by interference path D in Figure 24.5. The impact of this interference increases as the density of femtocells within the macrocell coverage area increases. The severity of the impact also depends on whether the FUEs are operating outdoors (in which case the FUEs will tend to be transmitting at higher power and the path-loss to the macrocell will be lower). Potential interference mitigation approaches for this scenario include:

- Uplink power control: the HeNB can control the power of its FUEs to limit the interference to neighbouring macrocells. This can, for example, be based on the estimated path-loss between the FUEs and the macrocells or the estimated path-loss between the FUEs and the HeNB [3].
- Control channel protection (PUCCH): as described above for mitigating macro-to-femto uplink interference.

Femto to Femto, Downlink Downlink transmissions from one femtocell suffer interference from transmissions from another femtocell as shown by interference paths E and F in Figure 24.6. This figure assumes an apartment scenario, with just two apartments for simplicity; however, it should be borne in mind that in practice interference can occur between multiple apartments on the same and different floors.

Potential interference mitigation approaches for this scenario include setting up ABS patterns and/or frequency re-use schemes whereby each HeNB determines its neighbours (from measurements made at the HeNB or FUE) and the associated path-losses. This neighbour information is then used to construct orthogonal patterns of RBs or ABSs to be used by neighboring HeNBs. Both distributed approaches (with or without information exchange between HeNBs [4,5]) and centralized approaches (e.g. at the HeNB GW [3]) have been proposed for constructing the orthogonal sets. Direct X2 connectivity between HeNBs was added inRelease 10 for the support of mobility, and this also allows coordination of ABS patterns between HeNBs.

Femto to Femto, Uplink. Uplink transmissions from one FUE suffer interference from transmissions from another FUE connected to another HeNB as shown by interference paths G and H in Figure 24.6. Uplink power control is one potential interference mitigation technique for this scenario.

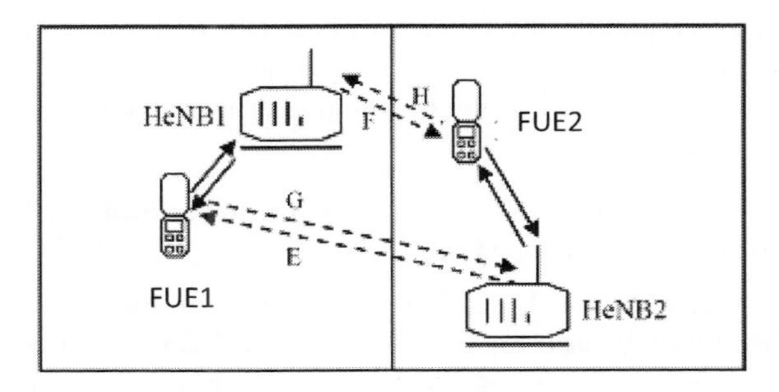

Figure 24.6: Femto/femto interference scenarios.

24.3.2 Network Listen Mode

Many of the interference mitigation techniques discussed above require the HeNB to make measurements of surrounding macrocells and femtocells. HeNBs are therefore commonly designed to have a Network Listen Mode (NLM) of operation which involves making measurements and decoding system information from neighbouring eNodeBs. This may be done at initial system setup and periodically thereafter. Examples of measurements made in NLM include:

- Uplink interference power, to infer the presence of nearby MUEs;
- Determination of cell IDs and CSG status/ID, by decoding the SI of the neighbour cells;
- Co-channel RSRP and RS transmission power to estimate the path-loss to neighbour cells, which in turn may be used for uplink and downlink power control at the HeNB;
- Reference Signal Received Quality (RSRQ) which can be used together with RSRP to determine the reliability of coverage of a macrocell, to help determine a suitable power setting for a femtocell operating in hybrid access mode.

In addition, TDD femtocells may obtain time synchronization from neighbouring macrocells.

24.4 RF Requirements for Small Cells

The unique characteristics of small cells, and in particular the potential interference scenarios, give rise to specific Radio Frequency (RF) challenges and requirements. These are explained in this section and compared to those defined for macrocells (see Chapter 21). In Release 8 of the LTE specifications, RF requirements were defined only for the Wide Area eNodeB class which covers macrocell applications, while Release 9 saw the introduction of the Local Area eNodeB class (covering picocell applications) and the Home eNodeB class (covering femtocell applications).

A major factor influencing the RF requirements for pico and Home eNodeBs is the lower minimum separation distance (and hence lower coupling loss) between UEs and the pico or Home eNodeBs when compared to the Wide Area eNodeB case. As explained in

Section 24.1, picocells tend to be small cells planned by the network operator, with the pico eNodeBs mounted at low elevation (e.g. interior walls), in contrast to femtocells for which the Home eNodeBs are typically self-installed in a home environment by the end user. Both picocells and femtocells will have lower Minimum Coupling Loss (MCL) to UEs than macrocells. Typically an MCL of 70 dB is assumed for macrocells compared to 45 dB for picocells and femtocells [6, 7].

Other factors influencing the differences from the Wide Area eNodeB requirements include lower transmit power, degraded receiver sensitivity (reflecting the lower cost), and for Home eNodeBs, the lack of a planned deployment. For these reasons in many cases the Home eNodeB class generally has the most stringent RF requirements, with the Local Area class requirements being intermediate between those of the Wide Area and the Home classes.

24.4.1 Transmitter Specifications

A comparison of transmission requirements between Home, Local Area and Wide Area eNodeBs is provided in Table 24.2.

Table 24.2: Summary of transmitter requirements.

Requirement	Wide Area	Local Area	Home
Maximum Output Power[a] (MOP) (dBm)	46	24	20
Power adjustment for adjacent channel	No	No	Yes
Adjacent Channel Leakage Ratio (ACLR) lower limit (dBm)	−15	−32	−50
Spurious emissions (dBm/100 kHz) [b]	−62	−62	−71
Frequency Error (ppm)	0.05	0.10	0.25

[a] Summed over transmit antennas.

[b] For co-existence with UTRA or E-UTRA downlinks in other bands.

Maximum Output Power (MOP). For the Local Area class, the MOP was selected to provide acceptable performance loss to overlaid Wide Area macrocellular systems on the same or adjacent carriers, and also to ensure safety from electromagnetic radiation [8, 9]. For the Home eNodeB class, the adjacent channel selectivity of UEs was also taken into account, such that UEs are still able to operate when a HeNB at 45 dB MCL is transmitting in a channel adjacent to that being received by the UE [7].

An HeNB is required to detect if the downlink adjacent channel is being used by another operator for either UTRA or E-UTRA, and to set its MOP according to the measured power of the adjacent channel base station and the total measured noise plus interference in its channel, I_{oh} [6]. This aims to reduce the impact of adjacent channel leakage on neighbouring operators. For a UTRA victim system the requirements are the same as for a UMTS Home NodeB (HNB) [10]. For an LTE victim system, different MOP levels can be set depending on the value of the cell-specific Reference Signal (CRS) Received Power (RSRP) per Resource Element (RE) of the adjacent-channel eNodeB, $\hat{E}_{c,\mathrm{CRS}}$ [6]. If $\hat{E}_{c,\mathrm{CRS}} < -127$ dBm, then there are unlikely to be UEs connected to this adjacent channel eNodeB in the vicinity of the

HeNB, since the signal is weak, so no power reduction by the HeNB is required. If $\hat{E}_{c,\text{CRS}}$ is less than a threshold, then it is considered that the power measurement on the adjacent channel is unreliable due to adjacent channel interference from the uplink channel being used by the HeNB, and the MOP is set to an intermediate value of 10 dBm. Otherwise, the MOP is set between 8 and 20 dBm depending on the value of $\hat{E}_{c,\text{CRS}}$ [6].

Adjacent Channel Leakage Ratio. For all eNodeB classes the ACLR is defined by a relative value (45 dBc), subject to an absolute lower limit [6]. For the HeNB class this lower limit is lower than Wide Area eNodeB class due to the lower MCL [11].

Spurious emissions. Spurious emissions limits are defined for co-existence with HeNBs operating in other frequency bands. In deriving the HeNB requirement it was assumed that the HeNB is in an adjacent apartment to a neighbouring HNB or HeNB operating in a different band, and that an MCL of 47 dB applies in this case. Moreover, a 0.8 dB desensitization criterion was assumed [12]. Spurious emission requirements for the protection of the receiver band (Home and Local Area FDD classes), and for co-located base stations (Local area class) are all relaxed by 8 dB, corresponding to the relaxation in the sensitivity requirement relative to the Wide area eNodeB.

Frequency error. Table 24.2 also shows the requirements for frequency error. The total error seen by a UE is the sum of the errors due to frequency error at the eNodeB and Doppler shift due to mobility. For Home and Local Area eNodeBs the mobility is assumed to be restricted to 30 km/h and 50 km/h respectively, compared to a maximum speed of 350 km/h for Wide Area eNodeBs. Therefore for the same total error, the component arising from frequency error can be larger for the Home and Local Area eNodeB classes. In [12] it is shown that 0.25 ppm is a sufficient accuracy considering handover measurement performance, demodulation performance and maintenance of timing synchronization for TDD HeNBs. For the Local Area eNodeB class the frequency error is the same as that already supported in GSM and UTRA Local Area base stations, i.e. 0.1 ppm [13].

Spectrum Emission Mask (SEM). The SEM for the Home and Local Area eNodeB classes differs from the Wide Area class in both the absolute value and the level relative to the carrier power. For the Home class, the mask at high frequency offsets is a function of the transmitted power (summed over all antennas), subject to a minimum of 2 dBm and a maximum of 20 dBm. This is shown in Figure 24.7 for different system bandwidths.

24.4.2 Receiver Specifications

This section introduces the receiver specifications for the Local Area and Home eNodeB classes. Table 24.3 summarizes the wanted and interfering signal levels for these classes for the case of a 10 MHz LTE system.

Reference sensitivity and noise figure. For Local Area and Home eNodeBs the maximum path-loss between a served UE and the eNodeB is considerably less than for a Wide Area eNodeB. In addition, lower implementation cost is important, especially for HeNBs. For these reasons the assumed noise figure for Local Area and Home eNodeBs is 13 dB, 8 dB higher than for Wide Area eNodeBs.

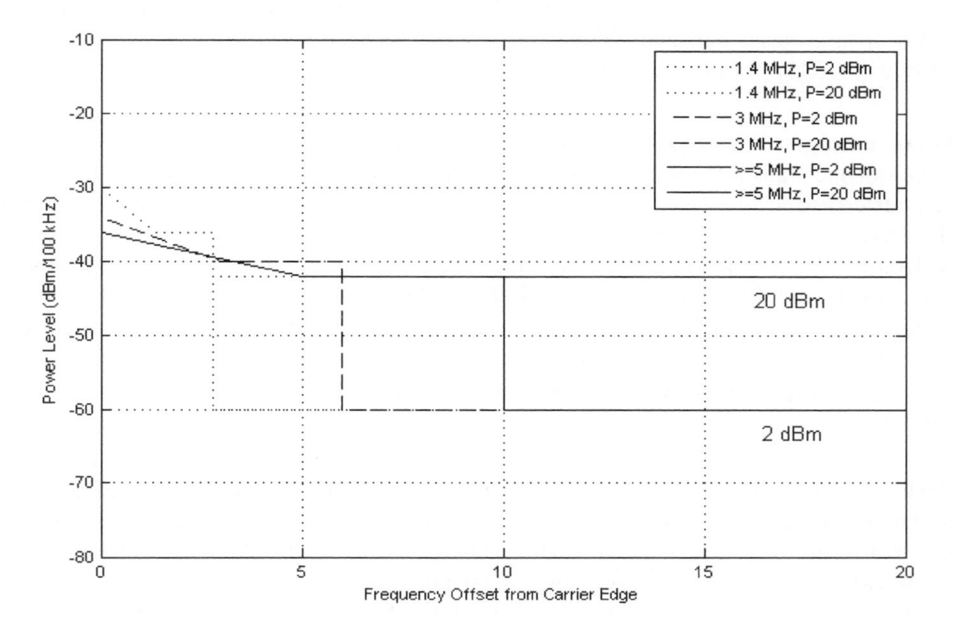

Figure 24.7: Spectrum Emission Mask for Home eNodeB class.

Table 24.3: Wanted and interfering signal power level requirements for 10 MHz system bandwidth.

	Wide Area		Local Area		Home Area	
	Wanted Signal (dBm)	Inter-ference (dBm)	Wanted Signal (dBm)	Inter-ference (dBm)	Wanted Signal (dBm)	Inter-ference (dBm)
Reference sensitivity	−101.5	–	−93.5	–	−93.5	–
Dynamic range	−70.2	−79.5	−62.2	−71.5	−25.7	−35.0
In-channel selectivity	−98.5	−77.0	−90.5	−69.0	−90.5	−69.0
Narrowband blocking	−95.5	−49.0	−87.5	−41.0	−79.5	−33.0
ACS	−95.5	−52.0	−87.5	−44.0	−71.5	−28.0
Receiver IM	−95.5	−52.0	−87.5	−44.0	−79.5	−36.0
Out-of-band blocking	−95.5	−15.0	−87.5	−15.0	−79.5	−15.0
In-band blocking	−95.5	−43.0	−87.5	−35.0	−79.5	−27.0
Co-located blocking*	−95.5	+16.0	−87.5	−6.0	––	––

* For the Local Area class these requirements are adopted in the case of co-location with UTRA or E-UTRA (however, the interfering signal level is slightly different in the case of co-location with GSM/DCS – see [6, Table 7.6.2.1-2]).

This increase of the noise figure has an impact on several other receiver requirements. Specifically, both the wanted signal level and interfering signal level are increased by 8 dB for the in-channel selectivity (for both Local Area and Home classes), in-band blocking (for the Local Area class), narrowband blocking (for the Local Area class) and Adjacent Channel Selectivity (ACS) (for the Local area class).

Dynamic range. The required dynamic range in an eNodeB depends not only on the ability to control the transmit power of the uplink transmissions from served UEs, but also on the interference levels seen from UEs in neighbouring cells, in particular from co-channel UEs served by the macrocellular network which could be transmitting with high power. For the Home eNodeB class operating in closed access mode, the fact that the deployment is unplanned means that such co-channel MUEs could come close to the HeNB. However, as discussed in Section 24.3.1, in such cases the HeNB will create a 'deadzone' around itself due to the interference caused to MUEs in the downlink. As a result of this deadzone, a coupling loss higher than the MCL of 45 dB was assumed when deriving the dynamic range requirement.

Blocking, Narrowband Blocking, Adjacent Channel Selectivity (ACS) and Receiver Intermodulation. The eNodeB blocking requirements consist of three components as shown in Figure 24.8. The first component is Out-Of-Band (OOB) blocking, which is defined over a wide frequency range,[7] excluding the operating band.

The second component is the in-band blocking, for which the interfering signal is an LTE signal at a specified frequency separation from the wanted signal. For the HeNB class, the interfering signal level is increased relative to the other classes to take account of the unplanned nature of HeNB deployments. This is also the case for the narrowband blocking and ACS requirements. For these cases, system simulations were used to define the interference powers [12].

The third component is the co-located blocking, which is defined for co-located (e)NodeBs operating in other frequency bands. No requirements are currently defined for the HeNB class. For the Local Area class, requirements are defined which assume the same 30 dB coupling loss as for the Wide Area class and a maximum transmit power of 24 dBm from interfering Local Area E-UTRA eNodeBs.

Due to the lower probability of two large interfering signals being present simultaneously, the power level of the interfering signals for the inter-modulation requirement is lower compared to the blocking requirement [12].

24.4.3 Demodulation Performance Requirements

For the Home eNodeB class, the maximum served UE speed is assumed to be 30 km/h which corresponds to a maximum Doppler of about 70 Hz at a 2.5 GHz carrier frequency. For the Local Area class, the maximum served UE speed is assumed to be 50 km/h. Furthermore, due to the small cell size in the indoor environment, for both these classes the multipath delays tend to be significantly smaller than for the Wide Area eNodeB class. Therefore only a subset of the Wide Area demodulation performance requirements are specified, consisting of the

[7] For most operating bands, it is defined for a frequency range of 20 MHz either side of the operating band.

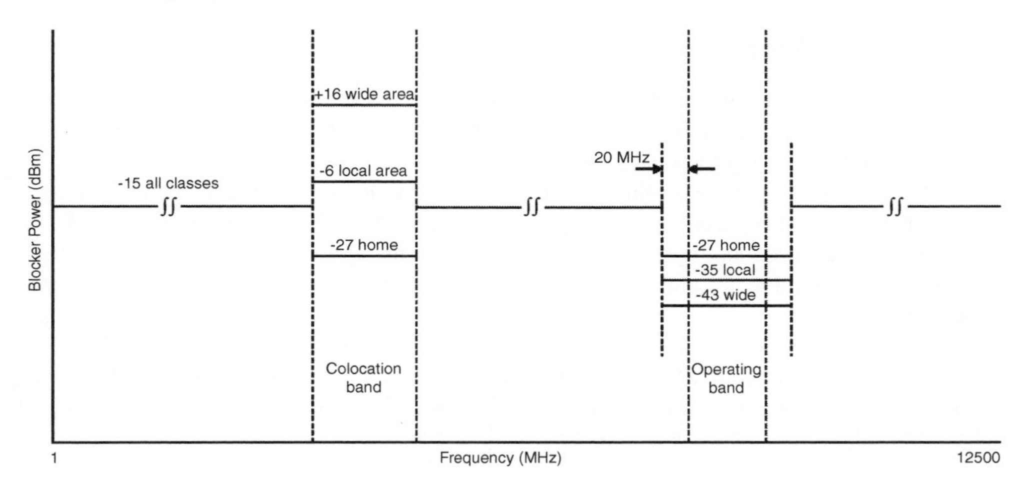

Figure 24.8: Blocking requirements.

EPA and EVA propagation models[8] with a maximum Doppler frequency no larger than 70 Hz. In these cases the performance requirements are the same as for the Wide Area eNodeB class. In addition for the Home class, new requirements are defined for the decoding of Hybrid Automatic Repeat reQest ACKnowledgements (HARQ-ACKs) and channel quality information feedback.

24.4.4 Time Synchronization for TDD Operation

Time synchronization in TDD systems is an important consideration for the avoidance of interference between uplink and downlink transmissions on neighbouring eNodeBs (see Section 23.3). For the HeNBs this is particularly important in view of the potential interference scenarios and unplanned deployment. Therefore a mandatory accuracy requirement is specified for HeNBs, even though, in common with all eNodeBs, no specific method is mandated. The difference in radio frame start timing, measured at the transmit antenna connectors, between the HeNB and any other HeNB or eNodeB which has overlapping coverage [12] is required to be less than 3 μs, except in cases where timing is obtained by monitoring another eNodeB which is more than 500 m away. In this case the requirement is relaxed in line with the additional one-way propagation delay beyond 500 m.

The requirement is designed to ensure that the combination of synchronization error, propagation delay to a victim, and multipath delay spread remains less than the smallest Cyclic Prefix (CP) length, taking into account that multipath delay spreads tend to be small in femtocells; if this condition is satisfied, interference at the uplink/downlink switching points within the TDD radio frame will be avoided.

[8]Extended Pedestrian A and Extended Vehicular A – see Chapter 20.

24.5 Summary

LTE Release 9 saw the introduction of new classes of base stations: Home eNodeBs or femtocells to cover homes and apartments, usually on a closed subscriber group basis, and pico eNodeBs to cover hotzones and other local areas.

While the basic architectural principles remain the same as for LTE macro eNodeBs, some new features were introduced specifically to handle the challenges posed by small cells – most notably the introduction of the Home eNodeB Gateway to address the needs of very large densities of femtocells, and, in Release 10, direct X2-interface handover between Home eNodeBs.

Heterogeneous deployments of small cells and macrocells bring new problems in terms of interference; some potential interference management schemes are described in this Chapter.

Finally, in the last part of the chapter, some of the RF challenges and requirements are discussed for both pico and Home eNodeBs.

References[9]

[1] Request for Comments 4960 The Internet Engineering Task Force (IETF), Network Working Group, 'Stream Control Transmission Protocol', www.ietf.org.

[2] 3GPP Technical Specification 23.003, 'Numbering, addressing and identification (Release 9)', www.3gpp.org.

[3] 3GPP Technical Report 36.921, 'FDD Home eNodeB (HeNB) RF requirements analysis (Release 9)', www.3gpp.org.

[4] Qualcomm, 'R4-091906: Frequency reuse results with full buffer', 3GPP TSG RAN WG4, meeting 51, San Francisco, USA, May 2009.

[5] CMCC, 'R4-092872: Downlink interference coordination between HeNBs', 3GPP TSG RAN WG4, meeting 52, Shenzhen, China, August 2009.

[6] 3GPP Technical Specification 36.104, 'Evolved Universal Terrestrial Radio Access (E-UTRA); Base Station (BS) Radio Transmission and Reception (Release 9)', www.3gpp.org.

[7] CATT, 'R4-094074: Home eNode B Maximum output power', 3GPP TSG RAN WG4, meeting 52bis, Miyazaki, Japan, October 2009.

[8] CATT, 'R4-092852: Proposal of maximum output power for Pico eNB', 3GPP TSG RAN WG4, meeting 52, Shenzhen, China, August 2009.

[9] Huawei, 'R4-092810: LTE Pico NodeB maximum output power', 3GPP TSG RAN WG4, meeting 52, Shenzhen, China, August 2009.

[10] 3GPP Technical Specification 25.104, 'Base Station (BS) Radio Transmission and Reception', www.3gpp.org.

[11] CMCC, 'R4-091789: Analysis of absolute ACLR1 requirements for LTE TDD HeNB', 3GPP TSG RAN WG4, meeting 51, San Francisco, USA, May 2009.

[12] 3GPP Technical Report 36.922, 'TDD Home eNodeB (HeNB) RF requirements analysis (Release 9)', www.3gpp.org.

[13] CATT and Huawei, 'R4-092809: TP on frequency error for Pico eNodeB', 3GPP TSG RAN WG4, meeting 52, Shenzhen, China, August 2009.

[9] All web sites confirmed 1st March 2011.

25

Self-Optimizing Networks

Philippe Godin

25.1 Introduction

The provision of Self-Optimizing Network (SON) functions is one of the key differentiators of LTE compared to previous generations of cellular systems such as UMTS and GSM. Self-optimization of the network is a tool to derive the best performance in a cost-effective manner, especially in changing radio environments. It allows the network operator to automate key aspects of the network configuration processes, and thus reduces the need for centralized planning and human intervention. For these reasons, this feature has been given a high priority and was a cornerstone around which the LTE radio, S1 and X2 procedures were designed. This makes SON functionality particularly efficient in the LTE system. The involvement of the User Equipment (UE) in the SON functionality of LTE is another key contributor to its success.

This chapter starts with an explanation of the Automatic Neighbour Relation (ANR) Function, the functionality for self-configuration of the eNodeB and MME[1] and the automatic Physical Cell Identity (PCI) configuration as natively implemented within the basic S1 and X2 interface procedures.[2] These three SON functions, which were included in Release 8, are of particular relevance for the initial deployment of an LTE network.

The chapter then explores other SON functions developed in Release 9 which are designed to optimize deployed LTE networks, such as the Mobility Load Balancing (MBL), Mobility Robustness Optimization (MRO) and Random Access CHannel (RACH) optimization functions. The chapter also includes the latest Release 10 SON enhancements which further optimize advanced LTE networks. Finally, as SON is a continuously evolving area, the

[1] Mobility Management Entity.

[2] For details about the S1 interface see Section 2.5 and [1]; for X2 see Section 2.6 and [2].

LTE – The UMTS Long Term Evolution: From Theory to Practice, Second Edition.
Stefania Sesia, Issam Toufik and Matthew Baker.

chapter concludes with examples of other new SON functions which are envisioned to complement the SON family in the near future.

Further details of the specified SON techniques can be found in [3, 4].

25.2 Automatic Neighbour Relation Function (ANRF)

The ANR Function (ANRF) is an example of a SON function which exploits both the design of the LTE radio interface and the UE to enable the network to optimize itself. The purpose of the ANRF is to relieve the network operator from the burden of manually managing relations between neighbouring cells.

25.2.1 Intra-LTE ANRF

The ANRF relies on the cells broadcasting their globally unique identity, termed the E-UTRAN Cell Global Identifier (ECGI). The ECGI is composed of 3 bytes carrying the Public Land Mobile Network (PLMN) ID and 28 bits to identify the cell within that PLMN. The function involves User Equipments (UEs), when requested by their serving eNodeB, reading and reporting the ECGI broadcast by a neighbouring cell that has been detected previously by that UE or another UE.

When an eNodeB receives from a UE a Physical Cell Identity (PCI) of a neighbour cell as part of a normal measurement report, and the eNodeB does not recognize the PCI, the eNodeB can instruct the UE to execute a new dedicated reporting procedure which uses the newly discovered PCI as a parameter. Through this procedure, the UE reads and reports to the requesting eNodeB some system information of the detected neighbouring cell, including the ECGI, the Tracking Area Code (TAC) and all available PLMN IDs. An example of this procedure is illustrated in Figure 25.1.

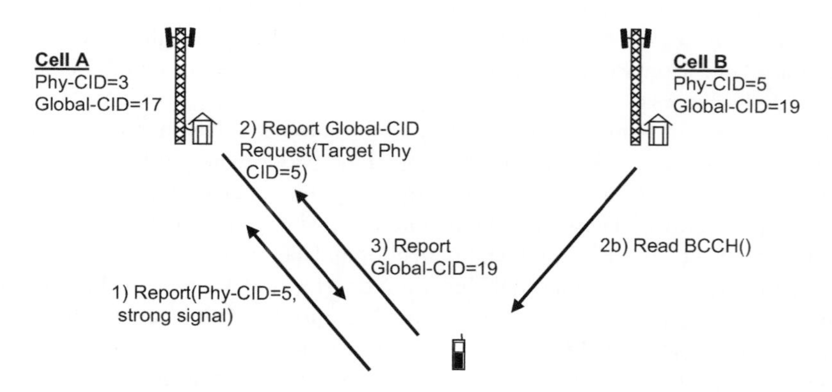

Figure 25.1: The intra-LTE ANR Function. Reproduced by permission of © 3GPP.

By means of the ANRF, each eNodeB can thus automatically populate a Neighbour Relation Table (NRT) for each cell it controls, containing all the Neighbour cell Relations (NRs) of the cell. An existing NR is defined as a unidirectional cell-to-cell relation from one source cell controlled by the eNodeB to a target cell for which the eNodeB controlling the source cell:

- knows the ECGI and PCI;
- has an entry in the NRT for the source cell that identifies the target cell;
- has the attributes (detailed in the following subsection) in this NRT entry defined either by Operation and Maintenance (O&M) or set to default values.

25.2.2 Automatic Neighbour Relation Table

From the discussion above, it can be seen that the ANRF is a cell-related function managed by the eNodeB. In the NRT, each NR can have a number of attributes, including:

- **No Remove flag:** The eNodeB shall not remove the NR from the NRT if this flag is set. This is typically used if the NR is certain, for example because it has been configured by O&M.
- **No Handover flag:** The NR shall not be used by the eNodeB for handover purposes if this flag is checked. This may, for example, be used if handovers are not useful from the source cell to the target cell, but the NR is nevertheless useful for the purpose of exchanging interference information.
- **No X2 flag:** The NR shall not use an X2 interface to initiate procedures towards the eNodeB that controls the target cell. This may, for example, be due to the source and target cell belonging to different PLMNs.

The ANRF incorporates three main functions for the management of NRs in the NRT, as shown in Figure 25.2:

- The 'NRT management function' which manages the table, including modifying NR attributes;
- The 'neighbour detection function' which finds new neighbour cells and adds them to the NRT;
- The 'neighbour removal function' which removes outdated NRs, for example after expiry of a timer.

The ANRF also allows the NRTs to be managed by O&M. O&M functionality can add or delete NRs or change the attributes of an NR. Conversely, O&M is also informed about changes in an NRT.

25.2.3 Inter-RAT or Inter-Frequency ANRF

The ANRF also exists between LTE and other Radio Access Technologies (RATs) or towards other frequencies. For inter-RAT and inter-frequency ANR, each cell is assigned an inter-frequency search list containing all frequencies that should be searched. The target RATs for inter-RAT ANRF are UTRAN[3], GERAN[4] and CDMA2000. The inter-RAT/inter-frequency ANR function can be divided into five steps:

[3]UMTS Terrestrial Radio Access Network.

[4]GSM EDGE Radio Access Network.

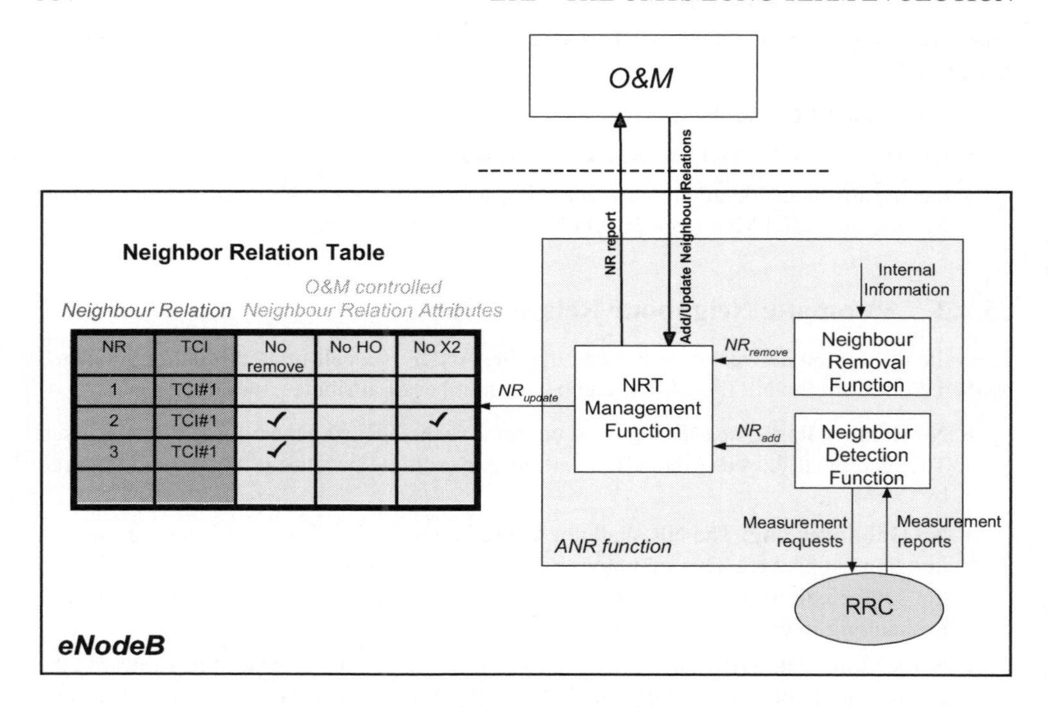

Figure 25.2: The ANR function and the ANR table. Reproduced by permission of © 3GPP.

1. In connected state, the eNodeB instructs a UE to search for neighbour cells in the target RATs/frequencies (it may need to schedule appropriate idle periods for this).
2. The UE reports the PCIs of the detected cells in the target RATs/frequencies (note that each RAT has its own specific format of PCI).
3. The eNodeB instructs the UE, using the newly discovered PCI as a parameter, to read the key RAT-specific cell identification parameters from the broadcast channel of the cell (e.g. the CGI and Routing Area Code (RAC) for GERAN; CGI, Local Area Code (LAC) and RAC for UTRAN; CGI for CDMA2000).
4. The UE reports these key RAT-specific cell identification parameters to the eNodeB of the serving cell.
5. the eNodeB updates its inter-RAT/inter-frequency NRT. The eNodeB can then make use of the NR, for example when triggering subsequent handovers.

25.3 Self-Configuration of eNodeB and MME

Self-configuration of the eNodeB/MME is a SON function which is implemented in the basic S1 and X2 interface procedures [1, 2].

25.3.1 Self-Configuration of eNodeB/MME over S1

With the native support of the S1-flex function in LTE (see Section 2.5), an eNodeB must set up an S1 interface towards each MME of the pool area to which it belongs. The list of MME nodes of the pool area together with an initial corresponding remote Internet Protocol (IP) address can be directly configured in the eNodeB at deployment (although other means may also be used). Once the eNodeB has initiated a Stream Control Transmission Protocol (SCTP, see Section 2.5.1.1) association with each MME of the pool area using that IP address, they can exchange, via the 'S1 Setup procedure' (see Section 2.5), some application-level configuration data which is essential for the system operation. This automatic configuration process thus saves some manual configuration effort for the network operator, together with the associated risk of human error.

Examples of such application-level configuration data exchanged between the eNodeB and the MMEs include Tracking Area (TA) identities,[5] lists of PLMNs of different operators who may be sharing the network so that all the PLMN IDs can be broadcast over the air for their respective UEs, and Closed Subscriber Group (CSG) IDs to allow auto-configuration of the Home eNodeB (HeNB) gateway when it connects to thousands of HeNBs.[6] Once all the data to be broadcast over the radio interface have been configured within each and every eNodeB, they are sent automatically to all the relevant MME nodes of the pool area via the S1 Setup procedure.[7]

The eNodeB can later update the configuration data it had previously sent to the MME by sending an 'eNB CONFIGURATION UPDATE' message. In this case it only sends the updated configuration data, and the data that is not included is interpreted by the MME as being unchanged. Conversely, the MME can also send updates of its data to the eNodeBs by means of an 'MME CONFIGURATION UPDATE'. The updated configuration data is assumed to be stored in both the eNodeB and the MME for the duration of the SCTP association or until any further update occurs.

25.3.2 Self-Configuration of IP address and X2 interface

Similarly to the S1 interface, self-configuration of IP addresses and of the X2 interface is implemented in the basic X2 and S1 interface procedures. The X2 interface may be established between one eNodeB and one of its neighbour eNodeBs when they need to exchange load, interference or handover related information (see Section 2.6). The automatic initialization of the X2 interface consists of three steps:

1. The eNodeB identifies a suitable neighbour;
2. The eNodeB retrieves a suitable IP address for this neighbour if not already available and sets up an SCTP association with it;
3. The two eNodeBs exchange configuration data.

[5]TAs correspond to the zones in which UEs are paged, and their mapping to eNodeBs must remain consistent between the E-UTRAN and the Evolved Packet Core (EPC).

[6]This enables the paging optimization feature in the HeNB gateway. Further details can be found in Chapter 24.

[7]Note that in case of CSG, the S1 Setup messages are exchanged via the Home eNodeB gateway to the MME (see Section 24.2).

The first step can be achieved either by configuration or by using the ANRF described in Section 25.2. If the ANRF is used, this step basically consists of the eNodeB being made aware of the ECGI and Tracking Area Identity (TAI) of the detected neighbour.

For the second step, the eNodeB needs to know the IP address of the neighbour in order to set up an SCTP association. This IP address may again be either configured or retrieved via the network (the latter being used if the ANRF is used during the first step). Auto-configuration of the IP address is achieved by the requesting eNodeB sending over the S1 interface a dedicated 'eNB CONFIGURATION TRANSFER' message that includes both routing information (such as the ECGI of the detected target cell) and the nature of the information that is requested – in this case an IP address for the purpose of X2 initiation. The requesting eNodeB also includes its own ECGI to be used for routing back the answer.

If the receiving eNodeB agrees, it returns one or more IP addresses which can be used for the establishment of an X2 interface. When this procedure is complete, the requesting eNodeB can set up the SCTP association with its neighbour by sending an SCTP INIT message. In Release 10, in order to protect the eNodeBs from malicious SCTP INIT requests from unauthorized parties, this auto-configuration process has been enhanced to enable the use of an Access Control List (ACL)[8] of authorized source IP addresses in the receiving eNodeB. For this purpose, the 'eNB CONFIGURATION TRANSFER' message has been enhanced with the possibility for the requesting eNodeB to include one or several IP addresses that the receiving eNodeB can store in its ACL. Thus, whenever a further SCTP INIT message is received to set up an X2 interface, the receiving eNodeB can first check that the source IP address corresponds to one notified earlier. The protocol also allows the requesting eNodeB to provide IP addresses for an IPSec[9] transport endpoint for scenarios where IPSec is expected to be used (e.g. routing via a security gateway). Finally, the requesting eNodeB can also include the IP addresses that it intends to use for the data-forwarding GPRS Tunnelling Protocol (GTP) tunnels that it will later establish. This would also allow for checking of the user plane traffic at the receiving eNodeB. The reciprocal behaviour is also supported: the receiving eNodeB may similarly provide in the 'eNB CONFIGURATION TRANSFER' message to the requesting eNodeB all the IP addresses it intends to use for its control plane SCTP endpoint, user plane GTP endpoints and/or IPSec endpoint.

Once an SCTP association exists between these two neighbour eNodeBs, the third step can be started, i.e. exchanging configuration data. This consists of application-level data similar to the data exchanged during the self-configuration of the S1 interface (see Section 25.3.1). In this case, the 'X2 Setup' procedure is used for the exchange. For example, an eNodeB can report, via the 'X2 SETUP REQUEST' message to a neighbour eNodeB, information about each cell it manages, such as the cell's PCI, frequency band, TAI and/or associated PLMNs. More detailed radio parameters can also be included, such as the cyclic prefix length (see Section 5.4.1), the transmission bandwidth or the uplink-downlink subframe configuration (see Section 6.2) for Time Division Duplex (TDD) cells.

An eNodeB can also exchange the list of pool areas to which it belongs with a neighbour eNodeB. The neighbour eNodeB can thus automatically learn if it shares a pool area in common and therefore whether it will need to use the S1 or the X2 handover procedure

[8]A receiving network node where ACL functionality is applied may only accept connections from other peer network nodes once the source addresses of the sending network node are known in the receiving node.

[9]IP Security – a collection of protocols and algorithms for IP security, including key management.

to transfer UEs. Indeed, if the eNodeBs do not share a pool area in common, the MME associated with the UE must be relocated on handover and the S1 handover has to be used (see Sections 2.5.6 and 2.6.3 respectively).

25.4 Automatic Configuration of Physical Cell Identity

The application-level configuration data exchanged during the X2 setup procedure is also the core of another SON feature: automatic self-configuration and self-optimization of the Physical Cell Identities (PCIs). This helps the eNodeBs to select PCIs that avoid collisions and hence cell confusion. Cell confusion arises when two neighbouring cells broadcast the same PCI so that a UE cannot discriminate between the two cells when it reports measurements. As a consequence the serving eNodeB of that UE cannot determine which one of these two cells should be the handover target for the UE. Increased inter-cell interference may also arise. The possibility for cell confusion stems from the fact that only about five hundred PCI values are available. PCI collision can be avoided by careful configuration on the part of the network operator, i.e. by selecting the PCIs allocated to each cell so that they are unique within clusters of adjacent neighbouring cells; however, this is a laborious operation, and moreover there remains an ongoing risk of further PCI collisions due to the start up of new eNodeBs as the network is densified.

The SON solution to this problem relies on the exchange of PCI values between neighbour eNodeBs during the X2 Setup procedure. In both the 'X2 SETUP REQUEST' and the 'X2 SETUP RESPONSE' messages, an eNodeB can include the list of PCI values used not only by its own cells but also by the 'direct neighbours' of its own cells. A direct neighbour of a cell is defined as any cell controlled by an eNodeB that is a neighbour of the eNodeB controlling the first cell (even if that cell has not yet been reported by any UE). This exchange of direct neighbour PCI values over the X2 interface enables an eNodeB to become quickly aware of the set of PCI values that are being used in the cluster to which it belongs. In particular the eNodeB can easily identify any collision in this cluster and can decide to change the PCI of one of its cells if needed; if it does so, it can signal the change to its neighbours in an 'eNB CONFIGURATION UPDATE' message. However, the exact PCI change algorithm supported by eNodeBs is not standardized and remains up to vendor implementations. An example of self configuration of PCIs is illustrated in Figure 25.3.

Via O&M, the network operator can use a variety of degrees of self-configuration at start up of the network: the operator could assign no PCI values and let each eNodeB select PCIs fully autonomously, or alternatively a range of possible PCI values can be assigned in order to assist the convergence of the self-configuration algorithms.

25.5 Mobility Load Balancing Optimization

Release 9 incorporates SON load balancing functionality, the objective of which is to counteract local traffic load imbalance between neighbouring cells with the aim of improving the overall system capacity and reducing congestion. The feature functions by first detecting any traffic imbalance and then applying solutions such as adjusting the cell reselection/handover parameters (such as handover thresholds). These parameters can be autonomously changed

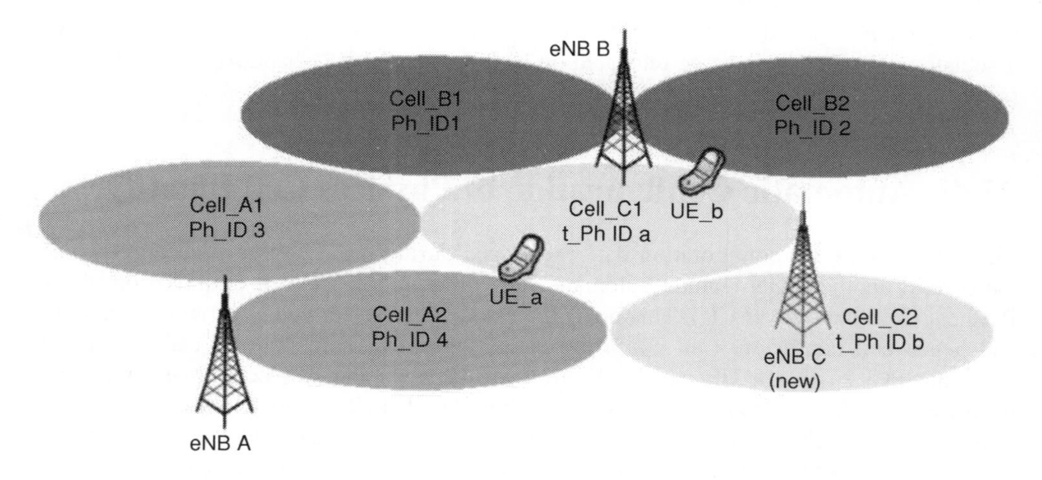

Figure 25.3: Illustration of PCI allocation. Reproduced by permission of © 3GPP.

and directly communicated between neighbouring cells by means of the X2 Parameter Negotiation procedure as explained in Section 25.5.2.

25.5.1 Intra-LTE Load Exchange

In order to detect an imbalance, it is first necessary to exchange load information between neighbouring eNodeBs over X2 for comparison. A client-server mechanism is used for this purpose: a requesting eNodeB (client) sends a 'RESOURCE STATUS REQUEST' message to request a load report from some of its neighbours. The 'RESOURCE STATUS REQUEST' message can simultaneously request multiple types of load report and may also be directed at multiple cells of the receiving eNodeB (server). The neighbours that receive the request report the requested load information over the X2 interface via the 'RESOURCE STATUS RESPONSE/UPDATE' message. The reporting of the load is periodic with period indicated in the 'RESOURCE STATUS REQUEST' message.

The reported information can indicate any of four different types of cell load information:

- Current usage of Physical Resource Blocks (PRBs), possibly partitioned into real-time and non-real-time traffic;
- Current hardware load;
- Current S1 transport load;
- Available composite load.

The first three measurements represent a global view of the current load situation in the node that reports them. The 'available composite load' indicator represents the amount of overall resources that the reporting node is ready to accept. 'Composite' means that the reporting node takes into account multiple internal resource criteria via a proprietary evaluation to build up its report. The 'available' characteristic is interesting as an estimate of the amount of non-GBR[10] traffic that can be handed over into the cells controlled by the eNodeB. For example, in a case when one UE was using all PRBs, the 'current usage of PRBs' would typically indicate that all PRBs were fully utilized whereas those PRBs would in fact still be available to be shared by the traffic of a second UE. Two formats may be used by the operator for the 'available composite load' indicator:

- A simple percentage of the total E-UTRAN resources available (i.e. of the total cell uplink or downlink bandwidth known from the X2 Setup procedure);
- A percentage weighted according to a cell capacity class value.[11]

25.5.2 Intra-LTE Handover Parameter Optimization

As a result of the exchange of load information and detection of local load imbalance, eNodeBs may take immediate action such as deciding to handover some UEs to cells which are less loaded. However, longer-term actions can also be taken to combat the imbalance more efficiently. For example when an imbalance is detected between two specific cells it may be desirable to shift the handover trigger threshold by, for example, 0.5 dB. By so doing, UEs served by an overloaded cell may find that a less-loaded neighbour cell becomes a relatively attractive handover target, thus allowing some UEs from the overloaded cell to be served reliably by the less-loaded cell. In order to avoid a ping-pong effect, it is often desirable that the lightly loaded cell shift its corresponding threshold by a similar amount as the overloaded cell but in the opposite direction. For this to happen, the lightly loaded cell must be made aware of the change applied by the overloaded cell, or alternatively the overloaded cell could first request the less-loaded cell to change its handover threshold and wait for a positive response before initiating the change.

This is the object of the 'Handover Parameter Negotiation' procedure. If the two cells in question belong to the same eNodeB, this negotiation happens within the eNodeB node itself via a proprietary algorithm; otherwise, the negotiation procedure is conducted over X2 interface via the 'Mobility Settings Change' procedure. This procedure enables one eNodeB to send a 'MOBILITY CHANGE REQUEST' message to another, to indicate the handover trigger parameter shift that the first eNodeB sees as necessary for one of the cells controlled by the receiving eNodeB. The same message can optionally indicate if the first eNodeB has already performed any configuration change of this parameter for one of its own cells. If the second eNodeB accepts the proposed handover trigger parameter modification and is able to complete it, then it replies with a 'MOBILITY CHANGE ACKNOWLEDGE' message; otherwise it responds with a 'MOBILITY CHANGE FAILURE' message which can include an 'allowed parameter modification range' if the reason for the failure was that the change proposed by the first eNodeB was outside the possible range.

[10]Guaranteed Bit-Rate – see Section 2.4.

[11]The cell capacity class value is a parameter between 0 and 100 configured by the operator to classify one particular cell capacity with respect to the other cells available in its network. The use of this parameter is mandatory when the 'available composite load' indicator is exchanged inter-RAT, e.g. towards UMTS.

25.5.3 Inter-RAT Load Exchange

In order to perform load information exchange with neighbour cells of a RAT other than LTE, an eNodeB can trigger a 'Cell Load Reporting Request/Response' procedure towards a neighbouring Radio Network Controller (RNC) (for WCDMA), Base Station Controller (BSC) (for GSM) or evolved High Speed Packet Access (HSPA) NodeB (for HSPA). This new inter-RAT procedure was purposely designed to be independent of the handover procedure in order that it can be triggered at any time by a requesting eNodeB even in the absence of any mobility event. The procedure is implemented as a generic 'SON Transfer' container carried on top of the RAN Information Management (RIM) protocol.[12] The new inter-RAT cell load reporting procedure allows an eNodeB to evaluate the load situation of GSM or UMTS/HSPA neighbour cells before triggering any load-related action such as handing traffic over to those neighbours or switching off a carrier. The new procedure may be used in all directions, i.e. from any LTE/UMTS/GSM source node to any LTE/UMTS/GSM target node of another RAT.

25.5.4 Enhanced Inter-RAT Load Exchange

One limitation of the Release 9 inter-RAT load exchange procedure described in Section 25.5.3 is that the RIM messaging which carries it passes through the core network nodes, which may become overloaded if load exchange occurs too frequently. On the other hand, variations in the radio conditions mean that it is important for the RAN nodes to have up-to-date load values from their peers in order to assess the load situation across different RATs. In order to solve these two opposing requirements, the inter-RAT load exchange procedure was enhanced in Release 10 by the introduction of two new types of reporting:

Event-triggered reporting enables a source RAN node to request a target RAN node to report only when a pre-defined cell load-level threshold is crossed in the target node. The source node signals the granularity with which it wishes to receive indications of load variation: for example, if it requests a granularity of five levels, the target RAN node will divide the cell-load scale by five, evenly distributed on a linear scale below the overload threshold and will report the target cell load each time the load changes from one reporting level to another or enters or exits the overload state. Other parameters (such as hysteresis and averaging factors) can be configured by O&M at the target nodes.

Multiple-cell reporting enables a source RAN node to request the a target RAN node to report the load of multiple cells within a single report. For example, in the case of UMTS, an eNodeB will only need to send one request to the RNC instead of individual requests for all the cells controlled by the RNC.

---

[12] RIM is a well known protocol dedicated to information exchange between two RAN nodes of different RATs. An example is given in Section 25.6.7.

25.6 Mobility Robustness Optimization

The Mobility Robustness Optimization (MRO) SON feature aims at detecting and preventing connection failures that occur as a result of mobility. These failures are of four types:

- Too-late handover, leading to a connection failure in the serving cell;
- Coverage hole, leading to a connection failure in the serving cell;
- Too-early handover, leading to a connection failure in the target cell;
- Handover to an inappropriate cell, leading to a connection failure in the wrong target cell.

Connection failure cases generally correspond to Radio Link Failures (RLFs), including handover failure. These scenarios and their respective SON solutions are described in more detail below.

25.6.1 Too-Late Handover

When a handover is executed too late, a connection failure may occur in the source cell and the UE may try to re-establish the radio link in a different cell controlled by a different eNodeB, as illustrated in Figure 25.4. In order to enable eNodeB of the source cell to improve the situation, an 'RLF INDICATION' message can be used to report the event from the eNodeB under which the UE re-establishes the connection to the original source eNodeB. The message includes the PCI of the cell in which the connection failed, the CGI of the cell where the radio link was re-established and the identity (C-RNTI[13]) that the UE had in the cell where the connection failed. The message may also include the short MAC-I bits[14] to allow verification of UE identity at the source eNodeB: the source eNodeB can recalculate these bits using the security configuration of the source cell. This mechanism makes it possible to eliminate false RLF detections that could happen due to PCI collision, for example. If a PCI collision occurs, an eNodeB would typically send the 'RLF INDICATION' message to multiple neighbour eNodeBs which all control cells with the same PCI, but the use of the short MAC-I bits enables only the source cell in which the RLF had really occurred to concern itself about the too-late handover.

25.6.2 Coverage Hole Detection

Receiving the 'RLF INDICATION' message at a source eNodeB might be caused either by a connection failure due to a too-late handover from the source to the target cell or by a coverage hole at the edge of the source cell. In order to improve the operation of the 'too-late handover' counter in the source eNodeB, it is useful to discriminate between these two scenarios. Hence, after completing the re-establishment procedure in the eNodeB where the connection is re-established, the UE may indicate to this eNodeB whether it can report the last cell measurements it has performed before the connection failed. If this is possible, this eNodeB may retrieve an 'R9 UE RLF Report' measurement and add it into the 'RLF INDICATION' message sent to the source eNodeB. By evaluating the measurements

[13]Cell-Radio Network Temporary Identifier.

[14]The 16 least-significant bits of the Message Authentication Code for Integrity (MAC-I, see Section 4.2.3).

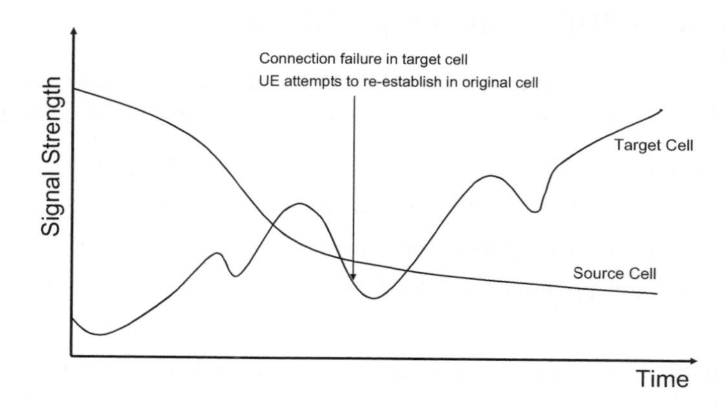

Figure 25.4: Example of signal strength fluctuation in the source and target cells during handover.

contained in this report, the source eNodeB should be able to determine if the connection failure was actually due to a coverage hole or not.

25.6.3 Too-Early Handover

When a handover is executed too early, the UE may experience a connection failure in the target cell due to fluctuations of the radio signal in cell B as illustrated in Figure 25.4, and will try to re-establish the radio link in the original source cell. Since the source eNodeB would have deleted all contexts related to that UE when the handover was completed, the re-establishment in the source cell would result in the source eNodeB sending an 'RLF INDICATION' message to the target eNodeB, exactly as if a too-late handover had taken place in the opposite direction. However, the target eNodeB can still discriminate the too-early handover from a too-late handover in the other direction if it had previously started a timer T_{early} on the incoming handover of the UE and notices that the 'RLF INDICATION' message is received from the same source cell and for the same UE before T_{early} expires. The implementation of the SON remedy for too-early handover therefore requires both a new timer to be run in the target eNodeB and a new message, 'HANDOVER REPORT', to be sent from the target eNodeB B to the source eNodeB in response to the 'RLF INDICATION' message, to signal to the source eNodeB that the source cell had a too-early handover problem towards the target cell rather than a 'too-late' handover problem.

These examples also show that implementations of SON remedies for too-late and too-early handover are best implemented in conjunction with each other as a comprehensive solution in order to avoid wrong MRO verdicts.

25.6.4 Handover to an Inappropriate Cell

Handover to an inappropriate cell shares some similarities with the too-early handover described above. Consider an example where a UE moves from a cell A of eNodeB A to a cell B of eNodeB B, but traverses a small overlap area in which a cell C of eNodeB C presents the strongest signal. This scenario is illustrated in Figure 25.5.

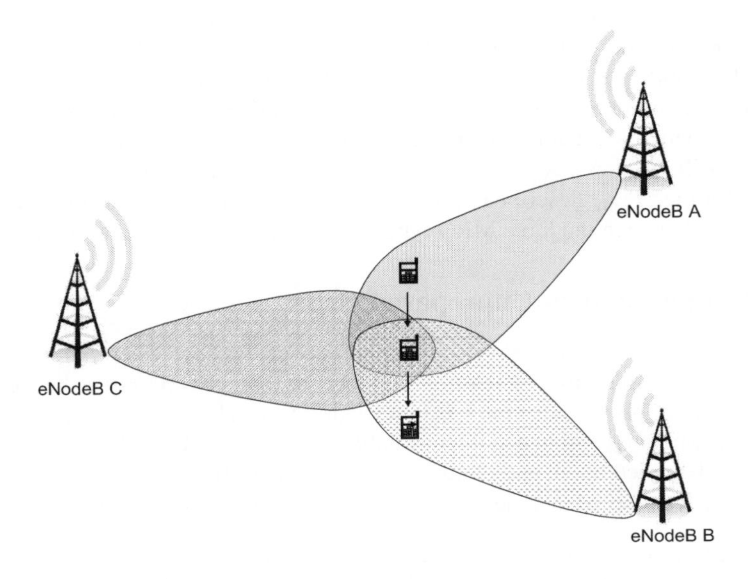

Figure 25.5: Handover to an inappropriate cell.

In this scenario, eNodeB A may be tempted to trigger the handover to cell C, but if the period for which cell C is the strongest is too short this could result in the UE experiencing a connection failure in cell C and re-establishing the radio link in the real suitable target cell B. However, the problem can again easily be reported to the originating eNodeB A: When the UE re-establishes the radio link in cell B, eNodeB B will send an 'RLF INDICATION' message to eNodeB C for that UE before the timer T_{early} expires in cell C, indicating the originating cell A. Then eNodeB C can send a 'HANDOVER REPORT' message to eNodeB A to inform it about the inappropriate handover to cell C. eNodeB A can then take appropriate corrective action if it receives multiple similar reports. The 'HANDOVER REPORT' message can also be sent from eNodeB C to eNodeB A over the X2 interface, even when both cells B and C belong to the same eNodeB and the 'RLF INDICATION' was thus internal to that eNodeB.

It can be noticed that the same 'HANDOVER REPORT' message is used to report the two MRO verdicts of too-early handover and handover to inappropriate cells, an information element 'type of handover problem' is used within the message to discriminate between these two cases.

25.6.5 MRO Verdict Improvement

As mentioned above, discrimination between a 'too-late handover' and a 'coverage hole' MRO verdict requires that the eNodeB where the UE re-establishes the connection after the connection failure wait for the completion of the re-establishment and retrieve some measurements from the UE (the 'R9 UE RLF Report') before it can send the 'RLF INDICATION' message. As a result, this message may arrive late and hence confuse the MRO verdict between too-late, too-early and wrong cell handover. In order to remove this dependency on the timing of the network delivery of the 'RLF INDICATION' message, the 'R9 UE RLF Report' was extended in Release 10 to include the 'Time elapsed since the last handover initialization until connection failure'. Since this timer runs in the UE, its value is independent of network delivery conditions and therefore more accurate. When receiving the 'RLF INDICATION' message, the source eNodeB will compare this new timer (instead of the Release 9 timer T_{early} described above) to an internal threshold and thus obtain a more accurate verdict between the three MRO types (too-late, too-early and wrong cell).

25.6.6 Handover to an Unprepared Cell

Release 10 further extends the scope of MRO to cover the case where, after a connection failure, the UE is in idle mode when it re-connects to the network, either because the RLF took too long or simply because the UE first tried to re-establish the connection on a target cell that was unprepared and hence the re-establishment failed. In such scenarios, a network context cannot be relied upon, and therefore the Release 9 MRO detection scheme cannot be used. Therefore Release 10 includes an equivalent context-less MRO solution:

1. A Release 10 UE starts a log when connection failure occurs, containing the following items:
    - The information contained in the Release 9 UE RLF report;
    - The new timer defined for the MRO verdict improvement (see Section 25.6.5), which measures the time elapsed since the last handover initialization until the connection failure;
    - The identity of the last cell that served the UE;
    - The identity of the cell which received the first (unsuccessful) re-establishment attempt;
    - The identity of the cell that served the UE at the last handover initialization.
2. The UE starts this log at connection failure detection and delivers it when it re-connects to the network from idle mode over the newly established Radio Resource Control (RRC) connection. By analysing the UE's log, the network can obtain all the elements to make a verdict between too-late handover, too-early handover, wrong-cell handover or coverage hole, despite the fact that the UE had transitioned through idle mode.

25.6.7 Unnecessary Inter-RAT Handovers

MRO was also extended in Release 10 to cover inter-RAT scenarios. The unnecessary inter-RAT handover scenario addresses the case where a UE is handed over from LTE to another RAT (e.g. UMTS or GSM) despite the fact the quality of the LTE coverage was actually good enough and would have allowed the UE to remain on LTE. This can be seen as a too-early handover to another RAT with no connection failure; it may affect the user experience if the other RAT offers lower performance and the UE cannot easily reconnect to LTE; it may also lead to a non-optimal use of network resources. It is typically due to the handover threshold for the LTE serving cell being set too high.

Such unnecessary inter-RAT handovers are detected by the target RAN node when triggered by the source eNodeB. The source eNodeB first requests the target RAN node during the inter-RAT Handover Preparation procedure to instruct the UE to continue measuring the LTE cells. It also provides the duration for which these measurements should continue and a minimum reference quality above which the target RAN node is allowed to report. It may also provide a list of selected LTE frequencies, for example to minimize the impact of compressed mode while in UMTS. Then, whenever the target RAN node receives a UE measurement above this threshold for one LTE cell for a sufficient period of time, it marks this LTE cell as to be reported. The reporting phase is performed by the target RAN node: when the measurement duration ends, the target RAN node sends a report (using a generic SON transfer container on top of the RIM protocol) to the originating LTE cell containing the results of the measurements – i.e. the list of LTE cells whose quality the UE measured to be above the configured threshold. Upon reception of the report, the originating eNodeB can take further corrective action, such as deciding if and how its parameters (e.g. the threshold to trigger inter-RAT handover) should be adjusted.

25.6.8 Potential Remedies for Identified Mobility Problems

Whenever the mechanisms described above are used in a network to detect such connection failure scenarios between cells, the source eNodeB may take some appropriate actions to combat them. For instance, in the case of repeated too-late or too-early handovers, one possible action could be to change the threshold at which the handovers are triggered. The source eNodeB can then use the same 'Mobility Settings Change' procedure as for MBL scenarios to inform the neighbour eNodeB of the change of threshold it has performed. In this case the 'MOBILITY CHANGE REQUEST' message is sent with a cause value 'handover optimization' and indicates the change (in dB) of the handover trigger parameter change performed in the source cell. For instance, in the case of repeated handover to an inappropriate cell, the source eNodeB could modify the way it builds its list of candidate target cells.

25.7 Random Access CHannel (RACH) Self-Optimization

Optimization of the RACH configuration in cells is a Release 9 SON feature that is key to optimizing the system performance of a mobile network. A poorly configured RACH may result in higher call setup and handover delays due to frequent RACH collisions, or low preamble-detection probability and limited coverage. The amount of uplink resource reserved for RACH also affects the system capacity. Therefore a network operator should carefully monitor that the RACH parameters are appropriately set, taking into account factors such as the RACH load, the uplink interference, the traffic patterns and the population under the cell coverage. The task becomes more complicated given that these factors may change dynamically. For example, if the antenna tilt is changed in a cell, it will affect the rates of call arrival and handover in this cell and the surrounding cells, and therefore the RACH load per preamble in all those cells. A change in transmission power settings or handover thresholds may have similar effects.

Whenever such a network configuration change happens, the RACH self-optimization feature should automatically make appropriate measurements of the RACH performance and usage in all the affected cells and determine any necessary updates of the RACH parameters. Some useful measurements are UE reports of the number of RACH attempts needed to obtain access, or time elapsed from the first attempt until access is finally granted. The reports may use RRC signalling or MAC[15] control elements depending on how much data is to be transmitted and on reliability and timing requirements. RACH parameters that can then be adjusted are typically the split of RACH preambles between contention-free access, contention-based access with high payload and contention-based access with low payload (see Section 17.3), the RACH back-off parameter value or the RACH transmission power ramping parameters.

In addition, the RACH self-optimization feature facilitates automatic configuration of the Physical RACH (PRACH) parameters (including the PRACH resource configuration, preamble root sequence and cyclic shift configuration – see Sections 17.4.2.5 and 17.4.3) to avoid preamble collisions with neighbouring cells. The principle of this automatic configuration is similar to the automatic PCI configuration SON feature described in Section 25.4: the PRACH configuration information is included in the 'X2 Setup' and 'eNB Configuration Update' procedures. Therefore, whenever a new eNodeB is initialized and learns about its neighbours via the ANR function, it can at the same time learn the neighbouring PRACH configurations. It can then select its own PRACH configuration to avoid conflicts with the neighbouring ones.

Whenever a conflict is identified, one of the cells should change its configuration, but the algorithm for selecting which cell should change and in what manner is not specified. The network operator can also combine PRACH self-optimization with manual configuration if necessary, but this is typically more prone to errors and more time consuming than automatic RACH optimization.

25.8 Energy Saving

Energy saving is a feature that aims at cost savings, as well as reduced environmental impact. Energy savings in the RAN typically focus on automatically adapting the capacity offered by the network to the actual traffic demand at a given point in time. Examples include radio solutions, such as not scheduling transmissions in certain subframes, and network solutions, such as switching off cells, transceivers or antennas.

Such a radio solution was introduced in Release 9 for LTE heterogeneous networks deploying LTE capacity-booster cells on top of cells that give wide-area coverage. The function enables the capacity-booster cells to be switched off when their capacity is no longer needed, and to be re-activated from a dormant state on a per-need basis.

The eNodeBs managing the capacity-booster cells operate an algorithm by which they decide autonomously whether to switch off a cell or not. The algorithm may take into account

[15]Medium Access Control.

various implementation-dependent criteria, such as some knowledge of the traffic load in the overlaid macrocell.[16]

Before switching off a capacity-booster cell, the eNodeB may initiate handover actions in order to offload any users in the cell; in this case it indicates the reason for handover to facilitate the target eNodeB in taking appropriate subsequent action, such as not selecting the switched-off cell as a target for subsequent handovers.

After switch-off, all peer eNodeBs are informed via the X2 interface by means of the eNB Configuration Update procedure, and they are then assumed to maintain the switched-off cell's configuration data while it is in the dormant state.

These peer eNodeBs may request a re-activation of the switched-off cell via the X2 interface if they need additional capacity. The algorithm that determines whether and when to request re-activation is implementation-dependent; it would typically take into account the current load condition in the macrocell(s).[17] The eNodeB managing the dormant capacity-booster cell would normally be expected to obey such a request.

After switch-on, all peer eNodeBs are again informed of the change of status of the capacity-booster cell via the X2 interface.

Importantly, the energy-saving solution described here is assumed to be under close control of the network operator, as it should have no impact on the basic coverage. Therefore it is assumed that the operator would first configure the eNodeBs that are allowed to decide to switch cells off, together with associated policies, and then configure the eNodeBs that are allowed to request re-activation of a list of dormant cells.

25.9 Emerging New SON Use Cases

Further enrichment of the LTE SON functionality is expected to include the following aspects:

- Further improvements related to mobility load balancing and mobility robustness;
- SON features for Home eNodeBs;
- SON features for Relay Nodes.

For example, further improvements of Mobility Load Balancing and Mobility Robustness in Release 11 may comprise:

- detection and prevention of intra-LTE rapid handovers – i.e. too-early handovers that do not lead to connection failures;
- detection and prevention of inter-RAT too-late handovers (which lead to connection failures);
- exchange of mobility parameters between RATs to optimize inter-RAT handover (e.g. to combat unnecessary inter-RAT handovers).

[16]The decision to switch off a cell can also be taken by O&M.

[17]Like the decision to switch off, the decision to re-activate may also be taken by O&M.

References[18]

[1] 3GPP Technical Specification 36.413, 'Evolved Universal Terrestrial Radio Access Network (E-UTRAN); S1 Application Protocol (S1AP)', www.3gpp.org.

[2] 3GPP Technical Specification 36.423, 'Evolved Universal Terrestrial Radio Access Network (E-UTRAN); X2 Application Protocol (X2AP)', www.3gpp.org.

[3] 3GPP Technical Specification 32.511, 'Automatic Neighbour Relation (ANR) management; Concepts and Requirements', www.3gpp.org.

[4] 3GPP Technical Specification 32.521, 'Self-Organizing Networks (SON) Policy Network Resource Model (NRM) Integration Reference Point IRP); Requirements', www.3gpp.org.

[18] All web sites confirmed 1st March 2011.

26

LTE System Performance

Tetsushi Abe

26.1 Introduction

The system performance requirements set for LTE Release 8 in [1] (summarized in Table 1.1) demanded substantial improvements over Release 6 of UMTS.[1] For the downlink, the average cell spectral efficiency was required to be three to four times higher than HSDPA,[2] and the cell edge spectral efficiency two to three times higher. For the uplink, average and cell edge spectral efficiencies two to three times higher than those of HSUPA[3] were required.

This chapter first summarizes the main technical features of LTE Release 8 that deliver these substantial system capacity enhancements. Evaluation results are then presented for spectral efficiency and coverage for typical LTE macrocell deployment scenarios, known as 'Case 1' and 'Case 3' [2] with inter-site distances of 500 m and 1732 m respectively, together with system performance evaluations in test environments defined by the International Telecommunication Union Radiocommunication Sector (ITU-R) [3].

26.2 Factors Contributing to LTE System Capacity

The preceding chapters have extensively explained the state-of-the-art technical features that significantly improve the system performance of LTE compared to legacy systems. The main difference in both downlink and uplink between LTE Release 8 and UMTS Release 6 (HSDPA/HSUPA) is that the LTE system provides orthogonal resource allocation in the frequency domain, which enables frequency-domain multi-user diversity gain to be exploited. In addition, the LTE downlink supports transmission with up to two or four spatial layers via

[1]Universal Mobile Telecommunications System.
[2]High Speed Downlink Packet Access.
[3]High Speed Uplink Packet Access.

LTE – The UMTS Long Term Evolution: From Theory to Practice, Second Edition.
Stefania Sesia, Issam Toufik and Matthew Baker.

multiple antennas, which enhances both the peak data rate and the cell average and cell edge spectral efficiencies. The key features are discussed in more detail in the following sections.

26.2.1 Multiple Access Techniques

Downlink OFDMA. As explained in Chapter 5, the LTE downlink is based on Orthogonal Frequency Division Multiple Access (OFDMA) which enables flexible channel-dependent multi-user resource allocation in both the frequency and time domains as illustrated in Figure 5.12. This leads to improved multi-user diversity gain. Inter-Symbol Interference (ISI) reduction by means of the Cyclic Prefix (CP) leads to a simplified receiver structure, which is well suited to Multiple-Input Multiple-Output (MIMO) transmission.

Uplink SC-FDMA. The Single-Carrier Frequency Division Multiple Access (SC-FDMA) scheme used for the LTE uplink (as explained in Chapter 14) achieves frequency-domain intra-cell orthogonality among User Equipment (UEs) while also maintaining a low Peak-to-Average Power Ratio (PAPR) which is important for maximizing data rates at the cell edge. In addition, the Sounding Reference Signals (SRSs) supported by the LTE uplink (see Section 15.6) facilitate multi-user scheduling and rate adaptation strategies to enhance spectral efficiency.

26.2.2 Frequency Reuse and Interference Management

Similarly to WCDMA,[4] LTE is designed to operate with a frequency reuse factor of one to maximize the spectral efficiency. In such a system, however, data and control channels can experience a significant level of interference from neighbour cells, which reduces the achievable spectral efficiency, especially at the cell edge. LTE therefore supports various techniques to manage and mitigate inter-cell interference.

In the downlink, these include:

- A cell-specific frequency-shift is applied to the mapping of cell-specific Reference Signals (RSs) to subcarriers to avoid inter-cell RS collisions, as described in Section 8.2.1.
- With respect to the downlink control channels, a cell-specific frequency offset is applied to the PCFICH and PHICH[5] positions, as described in Sections 9.3.3 and 9.3.4; for the Physical Downlink Control CHannel (PDCCH), interleaving provides frequency diversity and enhances the robustness against inter-cell interference (see Section 9.3.5.1).
- For the data, Inter-Cell Interference Coordination (ICIC) techniques can be applied by utilizing the Relative Narrowband Transmit Power (RNTP) messages that can be exchanged among eNodeBs over the X2 interface as explained in Section 12.5.

Uplink interference mitigation techniques in LTE include the following:

- Fractional power control (see Section 18.3.2.1) is supported to improve the throughput near the eNodeB and mitigate inter-cell interference at the cell edge. Power control

[4]Wideband Code Division Multiple Access.

[5]Physical Control Format Indicator CHannel and Physical HARQ Indicator CHannel.

can be performed jointly with frequency-domain resource allocation, whereby cell-centre UEs are allocated more Resource Blocks (RBs) to enhance the data rate, while cell-edge UEs are allocated fewer RBs for coverage extension. When devising power control strategies for interference management, it is important to control the ratio of the Interference over Thermal noise (IoT) below a target level.

- Various means are provided to avoid inter-cell RS collisions, including cyclic shift hopping, sequence-group hopping and planning, as explained in Sections 15.3 and 15.4.

- For the Physical Uplink Control CHannel (PUCCH), cell-specific symbol-level cyclic shift hopping is applied for inter-cell interference randomization, as explained in Section 16.3.3. Each PUCCH RB pair is mapped to both edges of the system bandwidth to achieve frequency diversity. In addition, the fact that the PUCCH is mapped to different RBs in the frequency domain from those of the Physical Uplink Shared CHannel (PUSCH, see Section 16.3.1) means that independent interference management techniques (such as power control) can be applied for the control and data channels.

- Uplink ICIC techniques can be applied using the Overload Indicator (OI) and High Interference Indicator (HII) messages that can be exchanged between eNodeBs over the X2 interface as explained in Section 12.5.

Scrambling is applied to the data and control channels, and to the downlink RSs, to randomize inter-cell interference.

26.2.3 Multiple Antenna Techniques

Downlink Spatial Multiplexing and Diversity. The multiple antenna schemes supported by LTE contribute much to the overall spectral efficiency gain with respect to Release 6 of UMTS. The performance of the open-loop and closed-loop spatial multiplexing modes supported in LTE is shown in Section 11.2.4.

Uplink Multi-User MIMO. In the LTE uplink, as explained in Section 16.6.2, orthogonal demodulation RSs can be assigned to multiple UEs, thus enabling the eNodeB receiver to estimate the channel of multiple UEs scheduled simultaneously to transmit in the same set of RBs.

26.2.4 Semi-Persistent Scheduling

The Semi-Persistent Scheduling (SPS) provided in LTE (see Section 4.4.2.1) can alleviate pressure on the limited downlink control channel capacity by replacing dynamic scheduling signalling with semi-static signalling. This allows a larger number of UEs to be scheduled, which is especially beneficial for services such as Voice over IP (VoIP).

26.2.5 Short Subframe Duration and Low HARQ Round Trip Time

LTE has a subframe duration of 1 ms for both uplink and downlink – shorter than the 2 ms subframe duration of UMTS. This leads to reduced latency (with a shorter HARQ[6] Round Trip Time (RTT)) and more flexible multi-user scheduling in the time domain.

26.2.6 Advanced Receivers

Advanced receivers provide an implementation method to enhance further the capacity of the LTE system. A typical example is the Linear Minimum Mean Squared Error (LMMSE) receiver with Interference Rejection Combining (IRC) [4]. Suitable for both uplink and downlink, such receivers compute the signal combining weights by exploiting statistical knowledge, such as the covariance matrix, of the inter-cell interference (unlike Maximum Ratio Combining (MRC) receivers which do not take the spatial characteristics of the interference into account). The ability of an IRC receiver to suppress interference is a function of many factors including the number and strength of the interfering signals and the number of receive antennas. For a single dominant source of interference, the mean Signal-to-Interference Ratio (SIR) gain that IRC is able to offer relative to an MRC receiver improves with the interference-to-thermal-noise ratio and not with the wanted Signal-to-Noise Ratio (SNR). However, if significant interference arrives from more sources or directions, an IRC receiver with only a small number of antennas is limited in the amount of interference suppression it can provide, especially if the multiple interference sources are received with similar powers.

Other more advanced receiver structures may also be considered, such as MMSE receivers with Successive Interference Cancellation (SIC) (particularly for downlink Single-User MIMO (SU-MIMO)) and Maximum Likelihood Detection (MLD) (see Section 11.1.3.3).

26.2.7 Layer 1 and Layer 2 Overhead

Any part of the time-frequency transmission resources that are not used directly for data transmission constitutes an overhead when considering the overall spectral efficiency. One design criterion for LTE was to minimize these overheads while achieving high system performance and flexibility. Table 26.1 summarizes the major sources of Layer 1 and Layer 2 overhead in the LTE downlink transmissions in a 10 MHz Frequency Division Duplex (FDD) deployment, as a percentage of the total transmission resources over the duration of a 10 ms radio frame.[7] The main contributors are guard bands, the OFDM CP, RSs, and control channels. It can be seen that the percentage overhead increases with the number of transmit antennas, due to the higher RS overhead for the larger number of antenna ports. The gain from MIMO therefore has to more than offset this increased overhead if it is to be worthwhile.

[6]Hybrid Automatic Retransmission reQuest.

[7]Note that transmission mode 7 (see Section 9.2.2.1) is not considered in these examples, and therefore the overhead due to UE-specific RSs is not taken into account. A worst-case downlink control channel duration of three OFDM symbols per subframe is used here, although in practice LTE supports dynamic resource allocation for the downlink control channels, so the average control channel duration would depend on the deployment scenario.

Table 26.1: Examples of percentage overhead in the LTE FDD downlink (calculated over a 10 ms radio frame for a 10 MHz system bandwidth).

Source of overhead	1 antenna port	2 antenna ports	4 antenna ports	Illustration
Guard bands (1 MHz)	10.0	10.0	10.0	Figure 21.1
OFDM CP	6.0	6.0	6.0	Figure 5.13
Cell-specific RSs	4.0	8.0	12.0	Figures 8.2 & 8.3
Control channels (3 symbols)	17.0	16.0	14.0	Figure 9.5
Synchronization signals	0.29	0.29	0.29	Figure 7.4
PBCH	0.28	0.26	0.24	Figure 9.1
Total (%)	37.6	40.6	42.6	

Table 26.2 similarly summarizes the major sources of Layer 1 and Layer 2 overhead in the LTE FDD uplink. Guard bands, CP, demodulation RSs and PUCCH constitute the main sources of uplink overhead.[8]

Table 26.2: Examples of percentage overhead in the LTE FDD uplink (calculated over a 10 ms radio frame for a 10 MHz system bandwidth).

Source of overhead	Overhead (%)	Illustration
Guard bands (1 MHz)	10.0	
SC-FDMA CP	5.9	Table 14.1
PUSCH demodulation RSs (2 symbols per subframe)	6.0	Figure 15.7
PUCCH (4 RBs)	6.7	Figure 16.2
RACH (6 RBs, 10 ms period)	1.1	Figure 17.5
SRS (48 RBs bandwidth, 10 ms period)	0.55	Table 15.1
Total (%)	30.3	

26.3 LTE Capacity Evaluation

In this section the capacity of LTE Release 8 is evaluated for various deployment scenarios. Table 26.3 shows the key parameters for the deployment scenarios designated 'Case 1' and 'Case 3' according to [2, 5].

Key parameters for the models defined by ITU-R are given in Table 20.8. Four ITU-R test environments are defined, each of which may be represented by one or more typical deployment scenarios as described in Section 20.3.6.1. The evaluations presented in this chapter focus on the Indoor Hotspot ('InH'), Urban Microcell ('UMi'), Urban Macrocell ('UMa') and Rural Macrocell ('RMa') deployment scenarios [3].

[8]The overhead due to the Random Access CHannel (RACH) depends on its configured transmission period, and that of the SRSs depends on their configured period and bandwidths.

Table 26.3: Key parameters for 3GPP Case 1 and Case 3 models.

	Case 1	Case 3
Deployment scenario	Macrocell	Macrocell
Network layout	Hexagonal grid	Hexagonal grid
Channel model	3GPP Spatial Channel Model [6]	3GPP Spatial Channel Model [6]
Inter-Site Distance (ISD)	500 m	1732 m
Carrier frequency	2 GHz	2 GHz
eNodeB antenna height	32 m	32 m
eNodeB transmit power	43 dBm (1.25 MHz, 5 MHz) 46 or 49 dBm (10 MHz, 20 MHz)	43 dBm (1.25 MHz, 5 MHz) 46 or 49 dBm (10 MHz, 20 MHz)
UE transmit power	23 or 24 dBm	23 or 24 dBm
UE speed	3 km/h	3 km/h

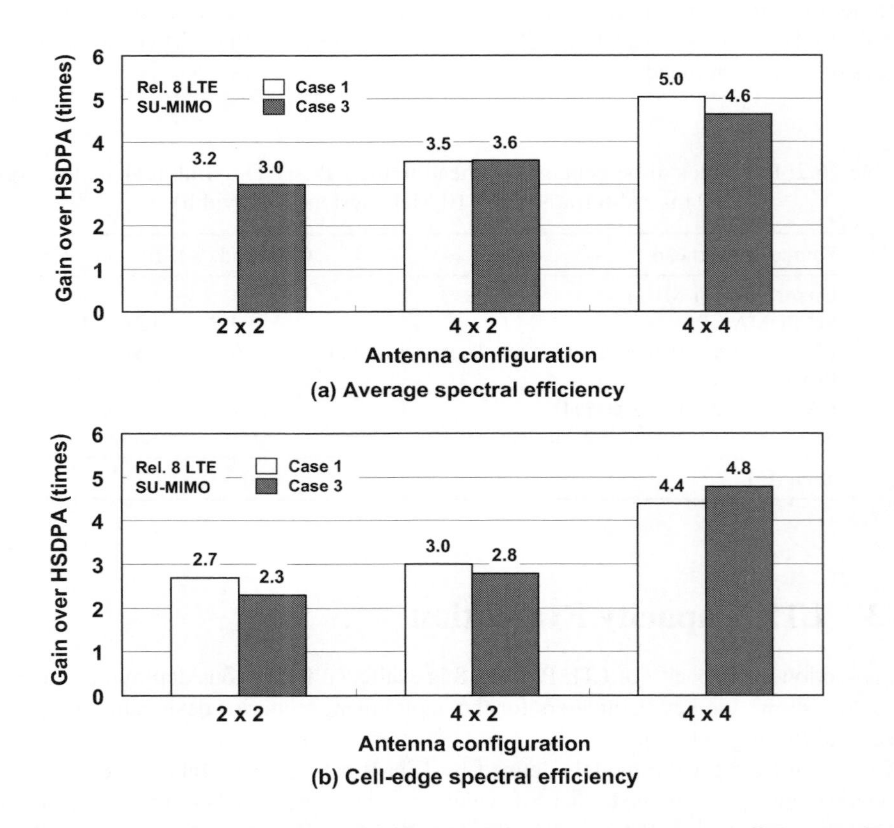

Figure 26.1: Ratio of Release 8 LTE FDD and HSDPA downlink spectral efficiencies for the Case 1 and Case 3 deployment scenarios.

26.3.1 Downlink and Uplink Spectral Efficiency

Figure 26.1 shows the average and cell-edge spectral efficiencies of the Release 8 LTE FDD downlink, as a ratio relative to Release 6 HSDPA for the 3GPP Case 1 and Case 3 deployment scenarios. Here, the cell-edge user throughput is defined as the 5-percentile user throughput. The LTE performance assumes 2×2, 4×2 and 4×4 antenna configurations with closed-loop MIMO, while 1×2 is assumed for HSDPA. Antenna separations at the eNodeB and UE are assumed to be ten wavelengths and half a wavelength respectively; eNodeB antenna tilting is not assumed.

The results show that the 2×2 configuration satisfies the LTE performance requirement of three to four times higher average spectral efficiency and two to three times higher cell-edge spectral efficiency than HSDPA. This substantial gain is mainly attributable to frequency-domain multi-user scheduling and dual-stream MIMO transmission. The additional gain from a 4×2 antenna configuration can also be seen, justifying the increased RS overhead.

Figure 26.2 shows the average and cell-edge spectral efficiencies of the Release 8 LTE FDD uplink compared to Release 6 HSUPA. For LTE a 1×2 or 1×4 antenna configuration is assumed, and 1×2 for HSUPA.

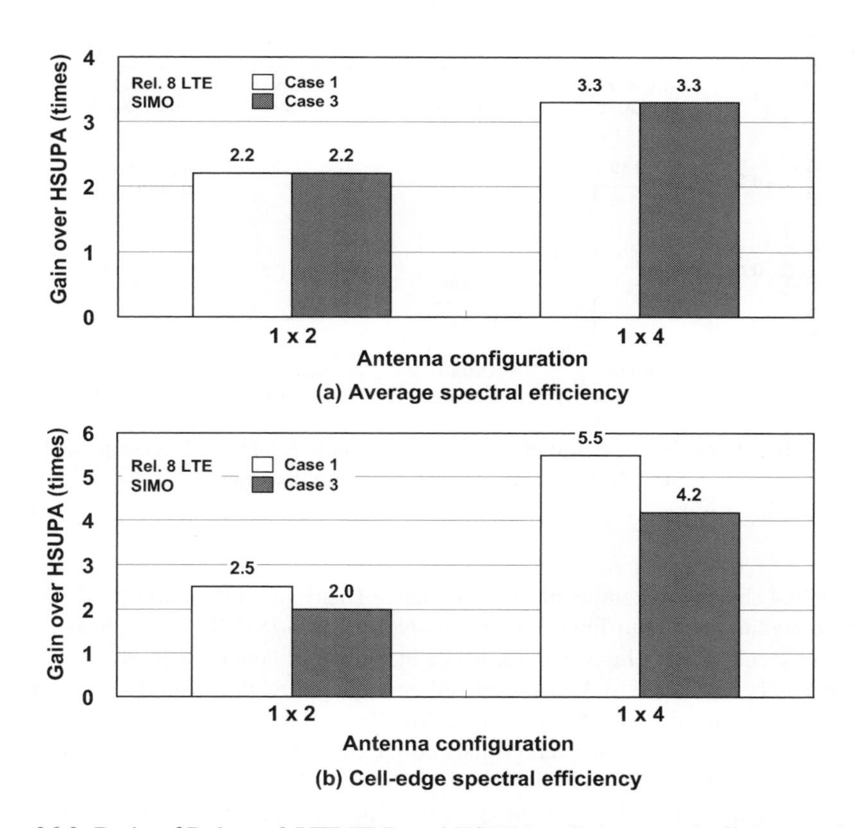

Figure 26.2: Ratio of Release 8 LTE FDD and HSUPA uplink spectral efficiencies for the Case 1 and Case 3 deployment scenarios.

The results show that the 1×2 LTE configuration satisfies the performance requirement of two to three times higher average and cell-edge spectral efficiencies than HSUPA. This is mainly attributable to intra-cell orthogonality and frequency-domain multi-user scheduling of LTE. The figures clearly show the effect of increasing the number of receive antennas.

Figure 26.3 shows the downlink average and cell-edge spectral efficiencies for Release 8 LTE FDD 4×2 downlink SU-MIMO transmission in the ITU-R deployment scenarios.[9] Vertical antenna tilting is assumed at the eNodeB, with tilt angles of 12, 12 and 6 degrees in the urban microcell, urban macrocell, and rural macrocell scenarios respectively [5]. The antenna separation is assumed to be four wavelengths at the eNodeB and half a wavelength at the UE.

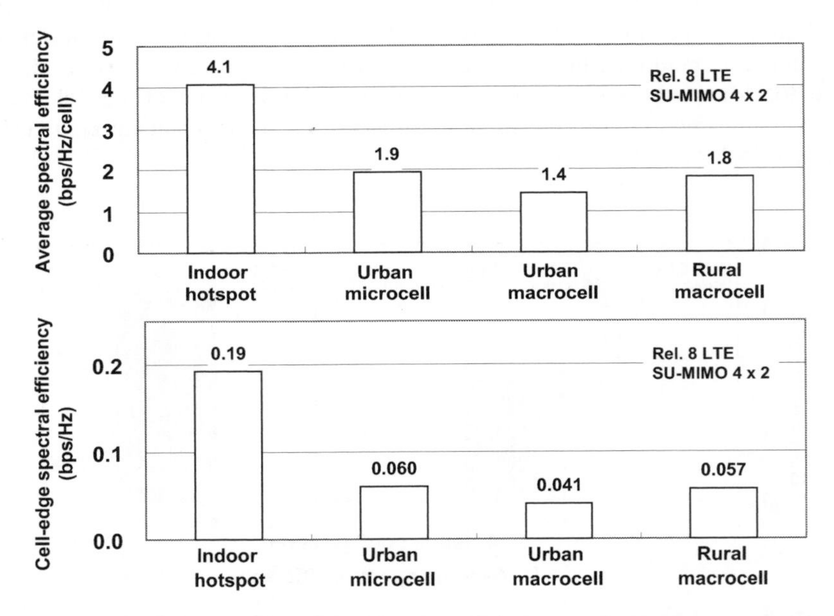

Figure 26.3: Downlink spectral efficiency of a Release 8 LTE FDD system with 4×2 SU-MIMO in the ITU-R deployment scenarios.

Figure 26.4 shows an example of the Cumulative Distribution Function (CDF) of average received downlink Signal-to-Interference-plus-Noise Ratio (SINR) for each of the ITU-R deployment scenarios. The higher performance in the indoor hotspot scenario compared with the others can be attributed to the isolated cell environment with low inter-cell interference; this is especially advantageous for high-order MIMO spatial multiplexing.

Figure 26.5 shows the uplink average and cell-edge spectral efficiencies for Release 8 LTE FDD with 1×4 Single-Input Multiple-Output (SIMO) and Multi-User MIMO (MU-MIMO) transmissions in the ITU-R deployment scenarios. These results are partially published in [7].

[9] A downlink control channel duration of three OFDM symbols is assumed. These results are partially published in [7].

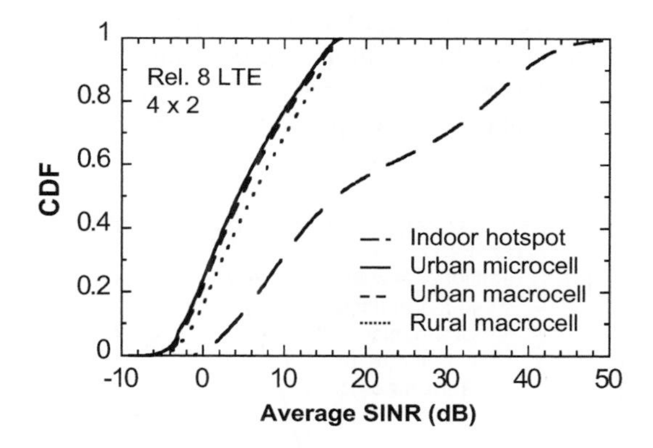

Figure 26.4: Downlink received SINR distribution for the ITU-R deployment scenarios.

The same antenna separations and eNodeB vertical antenna tilt angles are assumed as in the downlink evaluation.

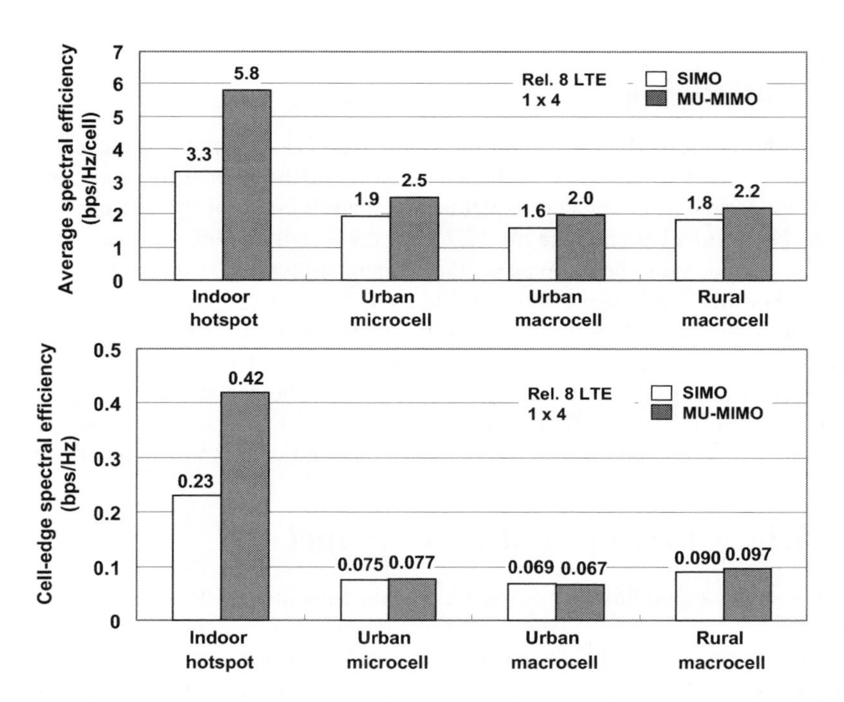

Figure 26.5: Uplink spectral efficiency of a Release 8 LTE FDD system with 1×4 SIMO and MU-MIMO in the ITU-R deployment scenarios.

Figure 26.6 shows an example of the CDF of average received uplink SINR for each deployment scenario. Similarly to the downlink, these results show that the indoor scenario provides higher throughput performance compared to the other scenarios, due to the higher SINR distribution. The potential gain of MU-MIMO can also be seen.

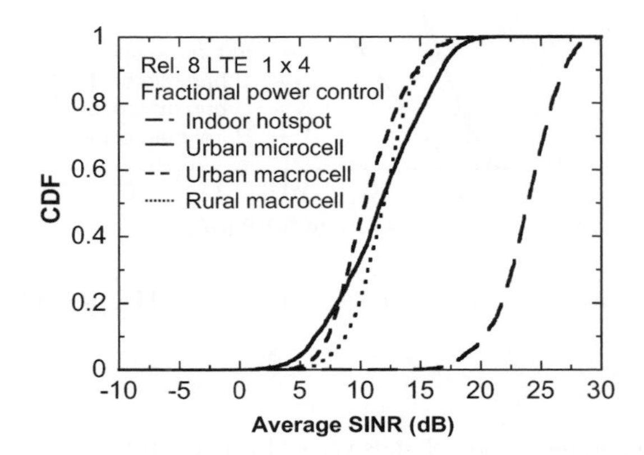

Figure 26.6: Uplink received SINR distribution for the ITU-R deployment scenarios.

26.3.2 VoIP Capacity

Figure 26.7 shows the VoIP capacity for the 3GPP and ITU-R deployment scenarios.

The VoIP capacity is typically evaluated in terms of the number of users per cell. For the ITU-R scenarios, VoIP capacity is defined as the minimum of the downlink and uplink capacities. The results here assume the 12.2 kbps codec with a 50% voice activity factor. A VoIP user is considered to be in outage if the 98-percentile radio interface latency is greater than 50 ms. For the 3GPP Case 1 and 3 evaluations, 2×2 and 1×2 antenna configurations are assumed for downlink and uplink respectively, while for the ITU-R scenarios 4×2 or 1×2 antenna configuration are assumed for the downlink and 1×4 for the uplink. The eNodeB vertical antenna tilting assumptions are the same as in the spectral efficiency evaluations above. Application of SPS to VoIP helps to avoid any limitation on the PDCCH capacity, thus enabling a high number of VoIP users to be supported in both uplink and downlink.

26.4 LTE Coverage and Link Budget

Figure 26.8 shows the uplink and downlink coverage for the 3GPP and ITU-R deployment scenarios.[10]

The PDCCH coverage in the downlink is estimated on the basis of a 44-bit Downlink Control Information (DCI) message with 8 Control Channel Elements (CCEs) (see Sections 9.3.5 and 9.3.5.1), while the PUCCH coverage in the uplink is based on a 4-bit CQI report (see

[10]The values for the ITU-R deployment scenarios are published in [7, Annex C2].

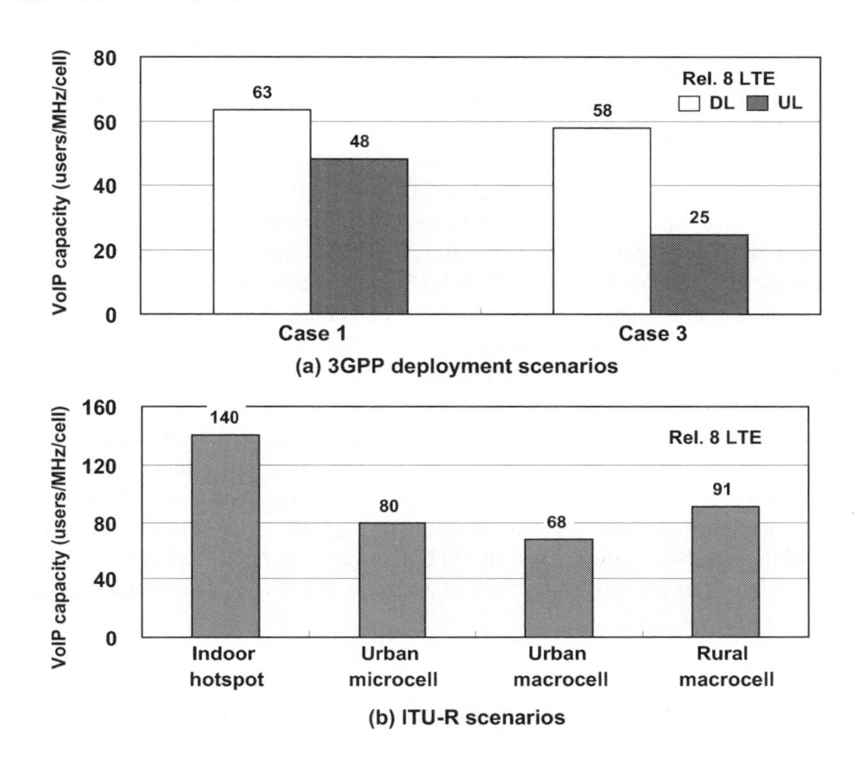

Figure 26.7: VoIP capacity of LTE Release 8 FDD.

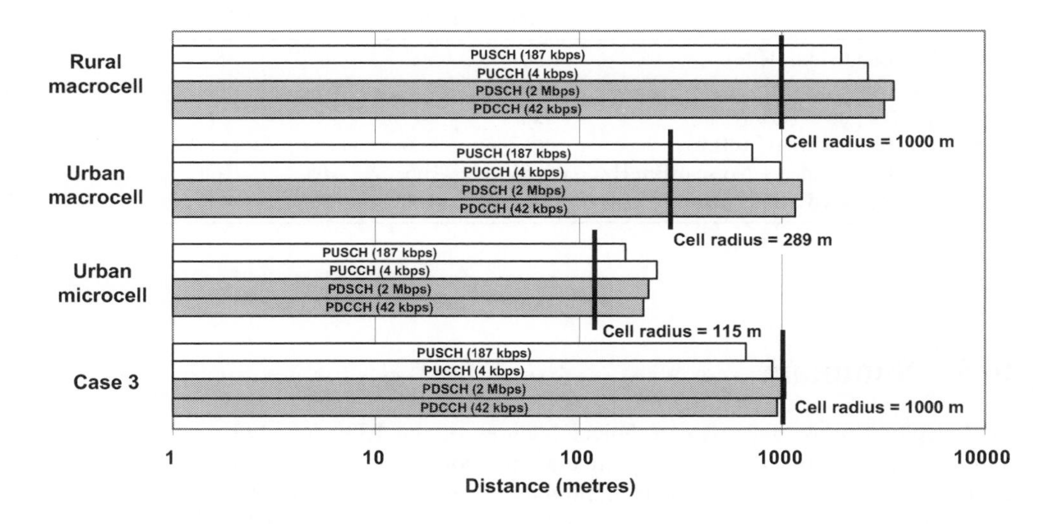

Figure 26.8: LTE coverage (10 MHz, FDD).

Sections 10.2.1 and 16.3). A 2×2 antenna configuration is assumed in the downlink and 1×4 in the uplink. The eNodeB antenna gain is 17 dBi, and the total cable loss at the transmitter and receiver is 4 dB for both uplink and downlink. Noise figures of 7 dB and 5 dB are assumed for the downlink and uplink respectively, and a receiver implementation margin of 2 dB. The thermal noise density is −174 dBm (see Section 21.4.4.2). Propagation models for each deployment scenario are discussed in Section 20.3.6.1, and full details can be found in [5]. 'Coverage' is defined here as 95% successful reception for the control channels and 90% for the data channels. The target BLock Error Rates (BLERs) are 1% for the control channels and 10% for the data channels. A 0.5 dB HARQ gain is assumed for the data channels. In the ITU-R deployment scenarios, the coverage exceeds the cell radius in each case. In 3GPP Case 3, PCFICH power boosting and narrowing the PUSCH transmission bandwidth would further improve the coverage of the PDCCH and PUSCH respectively.

Table 26.4 shows the link budget calculation in detail for Case 3. A UE transmit and receive antenna gain of 0 dBi is assumed, penetration loss of 20 dB, downlink receiver interference over thermal noise ratio of 1 dB for both data and control channels, uplink receiver interference over thermal noise ratio of 1.5 dB for data and 4.5 dB for control channels, and log-normal shadow fading margins of 10.5 dB for control and 6.7 dB for data. The same SINR threshold values as in the ITU-R rural macrocell deployment scenario are assumed. Full details of the link budget calculations for the ITU-R deployment scenarios can be found in [7, Annex C2].

Table 26.4: LTE link budget for Case 3 scenario with Non-Line-of-Sight (NLOS) propagation channel.

	Downlink		Uplink	
	PDCCH (42 kbps)	PDSCH (2 Mbps)	PUCCH (4 kbps)	PUSCH (187 kbps)
Bandwidth (RBs)	50	50	1	5
Coverage (m)	945	1053	898	668
Tx power (dBm)	46	46	24	24
Rx sensitivity* (dBm)	-98.7	-96.6	-119.8	-111.2
Interference + Noise (dBm/Hz)	-166	-166	-165	-168
Required SINR (dB)	-4.2	-1.7	-9.9	-4.8

* See Chapter 21.

26.5 Summary

This chapter reviews the main technical features of the LTE downlink and uplink that influence the system capacity. In particular, the role of the orthogonal multiple access schemes OFDMA and SC-FDMA is highlighted, together with downlink MIMO, in achieving significant capacity improvements over the Release 6 HSDPA and HSUPA.

Extensive performance evaluations for LTE have been carried out in 3GPP, and key results are presented here for cell average and cell-edge spectral efficiencies, VoIP capacity, and coverage. Considering a range of deployment scenarios defined by both 3GPP and the ITU-R, it is shown that the performance requirements set in [1] and summarized in Table 1.1 for the first version of LTE (i.e. Release 8) are satisfied.

The LTE system performance is further enhanced in Release 10 for LTE-Advanced, as explained in Chapters 27 to 32.

References[11]

[1] 3GPP Technical Report 25.913, 'Requirements for Evolved UTRA (E-UTRA) and Evolved UTRAN (E-UTRAN) (Release 7)', www.3gpp.org.

[2] 3GPP Technical Report 25.814, 'Physical Layer Aspects for Evolved Universal Terrestrial Radio Access (UTRA) (Release 7)', www.3gpp.org.

[3] ITU-R Report M.2135, 'Guidelines for Evaluation of Radio Interface Technologies for IMT-Advanced', www.itu.int/itu-r.

[4] J. H. Winters, 'Optimum Combining in Digital Mobile Radio with Cochannel Interference'. *IEEE Journal on Selected Areas in Communications*, Vol. 2, pp. 528–539, July 1984.

[5] 3GPP Technical Report 36.814, 'Further Advancements for E-UTRA Physical Layer Aspects (Release 9)', www.3gpp.org.

[6] 3GPP Technical Report 25.996, 'Spatial Channel Model for Multiple Input Multiple Output (MIMO) simulations (Release 8)', www.3gpp.org.

[7] 3GPP Technical Report 36.912, 'Feasibility Study for Further Advancements for E-UTRA (LTE-Advanced) (Release 9)', www.3gpp.org.

[11] All web sites confirmed 1st March 2011.

Part V

LTE-Advanced

27

Introduction to LTE-Advanced

Dirk Gerstenberger

27.1 Introduction and Requirements

With the completion of LTE Release 8, 3GPP started to look into ways to further evolve LTE for the future, in order to build upon the existing LTE technology and to ensure that LTE remains the leading global standard for mobile broadband.

Enhanced performance can in principle be achieved in two ways – by using more radio spectrum, and by using the available spectrum more efficiently. The International Telecommunication Union (ITU) has taken steps to ensure that more radio spectrum will be available, globally whenever possible, for systems beyond the 3rd Generation. The World Radiocommunication Conference (WRC) 2007 resulted in some new spectrum bands being earmarked for mobile services. In order to satisfy the perceived needs and to ensure that effective use is made of spectrum allocations, in March 2008 the ITU Radiocommunication Sector (ITU-R) issued a 'circular letter' [1] calling for submission of candidate Radio Interface Technologies (RITs). Successful proposals would fulfil the ITU-R's requirements for *IMT-Advanced*[1] [2].

Key features of IMT-Advanced set out in the circular letter are:

- A high degree of commonality of functionality worldwide, while retaining the flexibility to support a wide range of services and applications in a cost-efficient manner;
- Compatibility of services within IMT and with fixed networks;
- Capability of interworking with other radio access systems;
- High quality mobile services, user equipment suitable for worldwide use, user-friendly applications, services and equipment, and worldwide roaming capability;

[1] International Mobile Telecommunications - Advanced (following on from the IMT-2000 family of systems).

LTE – The UMTS Long Term Evolution: From Theory to Practice, Second Edition.
Stefania Sesia, Issam Toufik and Matthew Baker.

- Enhanced peak data rates to support advanced services and applications (100 Mbps for high mobility and 1 Gbps for low mobility were established as targets for research).

The key radio access requirements set by ITU-R for IMT-Advanced for different deployment scenarios are summarized in Table 27.1. Note that the spectral efficiency requirements in downlink and uplink are defined on a per-cell basis, and no explicit peak data rate requirements are defined.

Table 27.1: Key radio access requirements of IMT-Advanced for different deployment scenarios (see Figure 20.8 for details of the deployment scenarios).

Parameter	Downlink	Uplink
Maximum Bandwidth	At least 40 MHz	
Peak spectral efficiency (bps/Hz)	15	6.75
Average spectral efficiency (bps/Hz/Cell)	3.0 (Indoor Hotspot) 2.6 (Urban Micro) 2.2 (Urban Macro) 1.1 (Rural Macro)	2.25 (Indoor Hotspot) 1.8 (Urban Micro) 1.4 (Urban Macro) 0.7 (Rural Macro)
Cell-edge user spectral efficiency (bps/Hz)	0.1 (Indoor Hotspot) 0.075 (Urban Micro) 0.06 (Urban Macro) 0.04 (Rural Macro)	0.07 (Indoor Hotspot) 0.05 (Urban Micro) 0.03 (Urban Macro) 0.015 (Rural Macro)
VoIP capacity (user/cell/MHz)	50 (Indoor Hotspot) / 40 (Urban Micro and Urban Macro) / 30 (Rural Macro)	
User plane latency (ms)	10	
Control plane latency (ms)	100	

In response to the call for proposals from ITU-R, a workshop of 3GPP TSG RAN[2] took place in April 2008 to identify targets and potential techniques for further advancement of LTE, which led to a set of requirements being approved in May 2008 [3].

These targeted advancements became *LTE-Advanced*, specified as LTE Release 10 and beyond. LTE-Advanced is the 3GPP candidate RIT for IMT-Advanced. It is designed to meet the requirements of mobile network operators for the evolution of LTE, and to exceed the IMT-Advanced requirements.

3GPP's key radio-access targets for LTE-Advanced are outlined in Table 27.2. In addition, 3GPP set requirements on backward compatibility with earlier releases of LTE, as discussed in more detail in Section 27.3. This allows network operators to continue serving existing LTE customers while their network equipment is progressively upgraded. Requirements on spectrum deployment and flexibility, coexistence with legacy Radio Access Technologies (RATs), and complexity and service support were also defined [3].

[2]Technical Specification Group Radio Access Network – see Section 1.1.3.

Table 27.2: Key radio access targets for LTE-Advanced as set by 3GPP [4].

Parameter	Downlink	Uplink
Maximum Bandwidth	Up to 100 MHz	
Peak data rate (Mbps)	1000	500
Peak spectral efficiency (bps/Hz)	30	15
Average spectral efficiency (bps/Hz/Cell)	2.6 for 'Case 1'	2 for 'Case 1'
Cell-edge user spectral efficiency (bps/Hz)	0.09 for 'Case 1'	0.07 for 'Case 1'
VoIP capacity (user/cell/MHz)	Exceeding LTE Release 8	
User plane latency (ms)	10	
Control plane latency (ms)	50 (Idle to Active), 10 (Dormant to Active)	

3GPP's targets for LTE-Advanced were set independently from the IMT-Advanced requirements; it can be seen that some of the 3GPP targets exceed the IMT-Advanced requirements, such as the peak spectral efficiency and the control plane latency targets. This is related both to the fact that LTE Release 8 already fulfils many of the IMT-Advanced requirements (see Chapter 26), and to the fact that LTE-Advanced is not limited to LTE Release 10 but will also include new features in subsequent LTE releases. The ITU-R process called for complete descriptions of the IMT-Advanced candidates to be submitted by June 2009, with submission of the final details including a performance evaluation following by October 2009. 3GPP documented its submission in [5].

The 3GPP submission to ITU-R included an FDD[3] and a TDD[4] RIT component, which were developed with the goal of maximizing their commonality. Together, the FDD and TDD RITs comprise a 'Set of RITs' (SRIT). 3GPP's submission to ITU-R [6, 7] included detailed technology characteristics, link budget analysis and information about supported services, spectrum and technical performance.

An evaluation of LTE-Advanced was carried out by 18 companies in 3GPP, showing that LTE-Advanced completely satisfies the criteria set by the ITU-R for IMT-Advanced. The results of the evaluation are included in [7]. As a result, LTE-Advanced was accepted by the ITU as an IMT-Advanced technology in October 2010.

[3]Frequency Division Duplex.

[4]Time Division Duplex.

27.2 Overview of the Main Features of LTE-Advanced

The main components of LTE-Advanced that are added to LTE in Release 10 are:

- Carrier aggregation;
- Enhanced downlink multiple antenna transmission;
- Uplink multiple antenna transmission;
- Relaying;
- Support for heterogeneous network deployments.

Data rates of the order of 1 Gbps might theoretically be achieved using contiguous bandwidths of 40 MHz or more. However, competition for spectrum and fragmentation of the available spectrum makes it unrealistic to expect such large contiguous bandwidths in most cases. LTE-Advanced therefore makes use of *carrier aggregation* (see Chapter 28) to support such large bandwidths. This also has the advantages of limiting the cost of equipment and enabling much of the technology developed for LTE Release 8 to be reused. Each 'component carrier' within an aggregation is designed to be fundamentally similar to an LTE Release 8 carrier so that they can be configured in a backward-compatible way and used by legacy UEs if desired. Up to five component carriers with a bandwidth of up to 20 MHz each can be aggregated in LTE-Advanced to make efficient use of the available spectrum and achieve the desired total bandwidth and peak data rate. LTE-Advanced enables a variety of different arrangements of component carriers to be aggregated, including component carriers of the same or different bandwidths, adjacent or non-adjacent component carriers in the same frequency band, and component carriers in different frequency bands. The physical layer mechanisms for carrier aggregation are largely independent of the frequency location of the component carriers, but in order to minimize the number of configurations that have to be supported and avoid unnecessary terminal implementation complexity, the set of supported scenarios is carefully prioritized in 3GPP. This is discussed in more detail in Section 28.4.3.

LTE-Advanced can also make use of carrier aggregation to support deployments of *heterogeneous networks* consisting of a layer of macrocells and a layer of small cells coexisting with at least one carrier being common between them. In such a deployment, transmissions from one cell can interfere strongly with the control channels of another, thus impeding scheduling and signalling. LTE-Advanced supports *cross-carrier scheduling* (see Section 28.3.1) to enable control signalling to be transmitted on one component carrier corresponding to data transmissions on another; in this way, control channel interference between macrocells and small cells can be avoided.

Although the use of larger bandwidths by means of carrier aggregation allows higher peak data rates to be achieved, it does not increase the spectral efficiency as is required by the peak spectral efficiency targets shown in Table 27.2. LTE-Advanced therefore supports enhanced downlink MIMO transmission, by increasing the number of antennas at the eNodeB and UE, and hence the maximum number of spatial transmission layers for Single-User MIMO (SU-MIMO), from four in LTE Release 8 to eight. This may increase the multiplexing gain by a factor of two depending on the level of decorrelation between the antennas, and thus helps to achieve the spectral efficiency target of 30 bps/Hz. This is discussed in detail in Chapter 29.

Similarly to the downlink, the number of spatial layers supported in the uplink for SU-MIMO is increased from one to four in Release 10 in order to meet the peak spectral efficiency target of 15 bps/Hz. In addition, transmit diversity is introduced for the uplink control signalling.

In order to further improve the spectral efficiency, especially at the cell edge, a later release of LTE-Advanced may incorporate enhanced support for *Coordinated MultiPoint* (CoMP) schemes. CoMP transmission in the downlink entails the coordination of transmissions from multiple cells. This may take the form of coordinated scheduling to one or more UEs to reduce interference or to achieve spatial multiplexing gain by benefiting from the macro-diversity that results from the low correlation between geographically diverse base station sites. With an even higher degree of coordination, multisite beamforming approaches may be considered. Some further details can be found in Section 29.5. Release 10 supports enhanced reference signals to facilitate multicell measurements by the UEs. In the uplink, CoMP reception at different cells is already possible as part of the network implementation in LTE Release 8.

In order to support deployments of LTE in parts of the network where a wired backhaul is not available or is very expensive, *Relay Nodes* (RNs) are supported by LTE-Advanced (see Chapter 30). An LTE-Advanced RN appears to the UEs as a Release 8 cell with its own cell ID. A UE receives and transmits all its control and data signals from and to the RN, while the RN separately uses LTE-Advanced technology to transfer control and data to and from a donor cell. The main characteristics and challenges of relaying are explained in Chapter 30.

27.3 Backward Compatibility

LTE-Advanced is defined as an evolution of LTE which can also be deployed on new bands. Hence, one of the design targets for LTE-Advanced was backward compatibility between LTE Release 8 and LTE Release 10 and beyond. This is reflected in [3] by requiring that a Release 8 LTE UE can work in a Release 10 LTE-Advanced network, and that an LTE-Advanced UE can work in a Release 8 LTE network.

This is an important requirement to give operators confidence to deploy LTE and to build upon it, as it means that LTE operators upgrading their network to LTE-Advanced will be able to do so without swapping their existing UE base; Release 8 LTE UEs will be able to enjoy service continuity in an LTE-Advanced network. Backward compatibility is also of key significance for UE and network complexity, as well as for the cost of implementation and verification, since it enables implementation reuse on both the UE and the network sides, and minimization of interoperability testing.

The requirement for backward compatibility does not, in general, prevent the introduction of new features, thereby providing a degree of future-proofness for LTE. New functionality can be configured by the network on a per-UE basis without affecting legacy UEs, for example by scheduling the Physical Downlink Shared CHannel (PDSCH) in a UE-specific way. It is more difficult to introduce new cell-wide features, which have to be compatible with both Release 8 and Release 10 terminals. LTE-Advanced will therefore be visible in the 3GPP specifications simply as 'LTE Release 10 and beyond', including the base functionality of Release 8 LTE. This reflects the nature of LTE-Advanced as the further evolution of LTE.

27.4 Deployment Aspects

The existence of internationally identified common frequency bands is a key factor for significant economies of scale in the development and production of terminals [8].

A key outcome of the WRC 2007 was that a total of 136 MHz of new global spectrum was allocated for use by IMT-designated radio technologies:

- 450–470 MHz;
- 790–806 MHz;
- 2300–2400 MHz.

Other region-specific bands were also allocated:

- 790–862 MHz for ITU Region 1 (EMEA[5]) and ITU Region 3 (all other Asia Pacific);
- 698–806 MHz for ITU Region 2 (North and South America) and ITU Region 3 (nine countries, including Japan, China and India);
- 3400–3600 MHz allocated to mobile use on a primary basis for ITU Region 1 (EMEA in 82 countries), ITU Region 2 (Americas in 14 countries, except US/Canada) and Region 3.

All new bands identified by the WRC 2007 are valid generically for IMT technologies – i.e. they are not specific to IMT-2000 or IMT-Advanced only [9]. The deployment of the frequency bands for the different regions is illustrated in Figure 27.1.

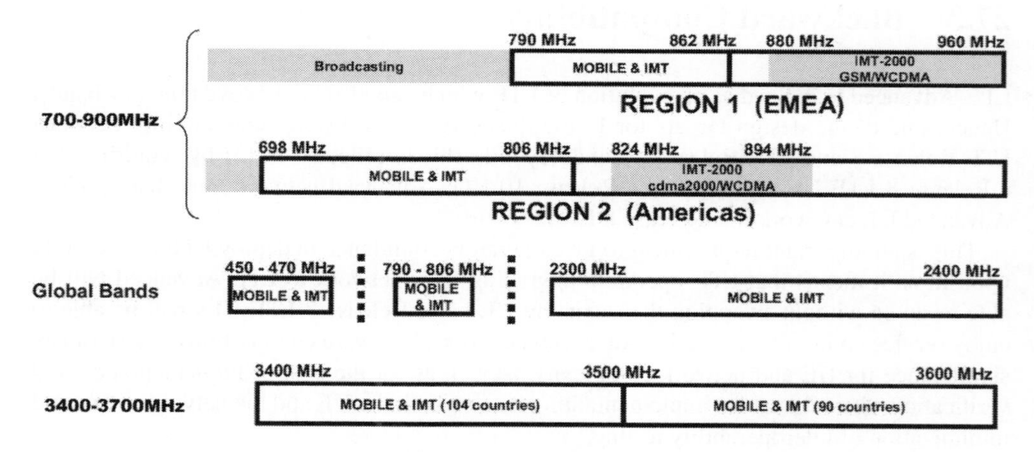

Figure 27.1: Allocation of new global spectrum resulting from WRC-07.

Carrier aggregation allows operators to gather spectrum for deployment of LTE from different parts of one band or from different bands. In order to limit the number of bands and combinations of bands to be supported by an LTE-Advanced UE, care has to be taken

[5]Europe, Middle East and Africa.

in focusing on the spectrum deployment scenarios with highest relevance for the mobile industry on a global level. 3GPP therefore prioritizes the development of requirements for different band combinations based on operators' requirements reflecting the different regions. Details of the first prioritized band combinations are given in Section 28.4.3.

27.5 UE Categories for LTE-Advanced

The definition of Release 10 UE categories builds upon the principles used in Releases 8 and 9 (see Section 1.3.4), where the number of UE categories is limited to avoid fragmentation of UE implementations and excessive variants in the market.

Three new Release 10 UE categories (6, 7 and 8) are specified [10], as shown in Table 27.3, defined in terms of their peak rates which reach about 3 Gbps in the downlink and 1.5 Gbps in the uplink. The highest UE category combines the aggregation of five 20 MHz component carriers with eight MIMO layers in the downlink and four in the uplink.

The peak data rate of categories 6 and 7 can be achieved by different means – for example, it is possible to achieve 300 Mbps either by supporting two MIMO layers together with the aggregation of 40 MHz or by four MIMO layers with a single 20 MHz carrier.[6]

Table 27.3: UE categories supported in Release 10.

	UE category		
	6	7	8
Approximate supported downlink data rate (Mbps)	300	300	3000
Approximate supported uplink data rate (Mbps)	50	100	1500
Number of downlink MIMO layers supported	2 or 4	2 or 4	8
Number of uplink MIMO layers supported	1, 2 or 4	1, 2 or 4	4
Support for 64QAM modulation in downlink	✔	✔	✔
Support for 64QAM modulation in uplink	✖	✖	✔
Relative memory requirement for downlink HARQ processing (normalized to category 1 level)	14.6	14.6	144

The Release 8 and 9 UE categories are reused and can in Release 10 support carrier aggregation; for example, a category 3 Release 10 UE may support the aggregation of two component carriers of up to 10 MHz bandwidth each. A category 6, 7 or 8 UE will also signal a category in the range 1–5, to allow backward compatibility in networks which do not yet support Release 10.

Additional UE categories are expected to be defined in the future, depending on market requirements.

[6]Note that the number of supported MIMO layers for each band combination is signalled by the UE.

References[7]

[1] Circular Letter 5/LCCE/2, 'Invitation for submission of proposals for candidate radio interface technologies for the terrestrial components of the radio interface(s) for IMT-Advanced and invitation to participate in their subsequent evaluation', Radiocommunication Bureau, International Telecommunication Union (ITU), www.itu.int.

[2] ITU-R Report M.2134, 'Requirements related to technical performance for IMT-Advanced radio interface(s)', www.itu.int/itu-r, 2008.

[3] 3GPP Technical Report 36.913, 'Requirements for Further Advancements for Evolved Universal Terrestrial Radio Access (E-UTRA) (LTE-Advanced)', www.3gpp.org.

[4] 3GPP Technical Report 25.814, 'Physical Layer Aspects for Evolved UTRA', www.3gpp.org.

[5] 3GPP Technical Report 36.912, 'Feasibility study for Further Advancements for E-UTRA (LTE-Advanced)', www.3gpp.org.

[6] 3GPP ITU-R Ad Hoc, 'RP-090736: Proposed Cover page for the October submission of "LTE Release 10 & beyond (LTE-Advanced)" ', www.3gpp.org, 3GPP TSG RAN, meeting 45, Seville, Spain, September 2009.

[7] LTE-Advanced Rapporteur, 'RP-090939: 3GPP Submission Package for IMT-Advanced', www.3gpp.org, 3GPP TSG RAN, meeting 45, Seville, Spain, September 2009.

[8] GSMA, 'The advantage of common frequency bands for mobile handset production', www.gsmworld.com.

[9] ITU World Radio Conference 2007 (WRC-07), October–November, 2007, Geneva.

[10] 3GPP Technical Specification 36.306, 'Evolved Universal Terrestrial Radio Access (E-UTRA); User Equipment (UE) Radio Access Capabilities (Release 10)', www.3gpp.org.

[7] All web sites confirmed 1st March 2011.

28

Carrier Aggregation

Juan Montojo and Jelena Damnjanovic

28.1 Introduction

As discussed in Chapter 27, LTE-Advanced aims to support peak data rates of 1 Gbps in the downlink and 500 Mbps in the uplink [1]. In order to fulfil such requirements, a transmission bandwidth of up to 100 MHz is required; however, since the availability of such large portions of contiguous spectrum is rare in practice, LTE-Advanced uses carrier aggregation of multiple Component Carriers (CCs) to achieve high-bandwidth transmission. Release 8 LTE carriers have a maximum bandwidth of 20 MHz, so LTE-Advanced supports aggregation of up to five 20 MHz CCs.

A second motivation for carrier aggregation is to facilitate efficient use of fragmented spectrum, irrespective of the peak data rate. Carrier aggregation in LTE-Advanced is designed to support aggregation of a variety of different arrangements of CCs, including CCs of the same or different bandwidths, adjacent or non-adjacent CCs in the same frequency band, and CCs in different frequency bands. Each CC can take any of the transmission bandwidths supported by LTE Release 8, namely 6, 15, 25, 50, 75 or 100 Resource Blocks (RBs), corresponding to channel bandwidths of 1.4, 3, 5, 10, 15 and 20 MHz respectively. For Frequency Division Duplex (FDD) operation, the number of aggregated carriers in uplink and downlink may be different (although Release 10 focuses on the case where the number of downlink CCs is not less than the number of uplink CCs).[1] This flexibility enables a large variety of fragmented spectrum arrangements of relevance to network operators to be supported.

A third motivation for carrier aggregation is support of *heterogeneous networks*. A heterogeneous network deployment typically consists of a layer of high-power macrocells

[1]For TDD deployment, the number of CCs and the bandwidth of each CC in uplink and downlink is expected to be the same.

LTE – The UMTS Long Term Evolution: From Theory to Practice, Second Edition.
Stefania Sesia, Issam Toufik and Matthew Baker.

and a layer of low-power small cells (e.g. picocells, Closed Subscriber Group (CSG) femtocells or relay nodes – see Chapters 24 and 30) with at least one carrier being used by both layers. In such a deployment, transmissions from one cell can interfere strongly with the control channels of another, thus impeding scheduling and signalling. Rather than simply using separate carriers for the two layers, which would result in inefficient spectrum usage, carrier aggregation enables multiple carriers to be used for a given layer, while interference can be avoided by means of *cross-carrier scheduling*. Cross-carrier scheduling allows the Physical Downlink Control Channel (PDCCH) on the CC of one serving cell to schedule transmission resources on a CC of another serving cell, as explained in detail in Section 28.3.1.

All CCs in Release 10 are designed to be *backward-compatible*. This means that it is possible to configure each CC such that it is fully accessible to Release 8 User Equipment (UEs). Therefore, essential Release 8 channels and signals such as Primary and Secondary Synchronization Signals (PSS and SSS) and System Information (SI) specific to each CC are transmitted on the respective CC. Backward-compatibility also has the advantage that the technology developed for LTE Release 8 can be reused on aggregated Release 10 CCs. From the higher-layer perspective, each CC appears as a separate cell with its own Cell ID. A UE that is configured for carrier aggregation connects to one *Primary Serving Cell* (known as the 'PCell') and up to four *Secondary Serving Cells* (known as 'SCells'). The PCell is defined as the cell that is initially configured during connection establishment; it plays an essential role with respect to security, NAS[2] mobility information, SI for configured cells, and some lower-layer functions. An SCell is a cell that may be configured after connection establishment, merely to provide additional radio resources. The term *Serving Cell* can refer to either a PCell or an SCell. The same frame structure is used in all aggregated serving cells, and, for TDD carrier aggregation, the uplink-downlink configuration (see Section 6.2) across all serving cells is the same.

The CCs corresponding to the PCell are referred to as the Downlink and Uplink Primary Component Carriers (PCCs), while the CCs corresponding to an SCell are referred to as Downlink and Uplink Secondary Component Carriers (SCCs). In a given geographic cell, all CCs that may be aggregated are assumed to be synchronized and belong to the same eNodeB. A default linkage between downlink and uplink CCs is signalled in System Information Block 2 (SIB2) on each downlink CC.

A UE's identity (C-RNTI[3]) is the same in the PCell and its configured SCells.

28.2 Protocols for Carrier Aggregation

28.2.1 Initial Acquisition, Connection Establishment and CC Management

As noted above, the PSS and SSS are transmitted on all CCs to facilitate cell search.[4] A UE establishes a connection to a cell by following the usual Release 8 and 9 procedures. After the initial security activation procedure, E-UTRAN may configure a UE supporting carrier aggregation with one or more SCells in addition to the PCell that is initially configured

[2]Non-Access Stratum.

[3]Cell Radio Network Temporary Identifier.

[4]See Chapter 7 for details of the cell search procedures defined in Release 8.

during connection establishment. The configured set of serving cells for a UE always contains one PCell and may also contain one or more SCells. The number of serving cells that can be configured depends on the aggregation capability of a UE. For each SCell, the usage of uplink resources by the UE in addition to the downlink ones is configurable – the number of downlink SCCs configured for a UE is therefore always greater than or equal to the number of uplink SCCs, and no SCell can be configured for usage of uplink resources only. From a UE viewpoint, each uplink resource belongs to only one serving cell.

The PCell provides the security inputs, NAS mobility information and SI for serving cells. A single Radio Resource Control (RRC) connection is established with the PCell, which controls all the CCs configured for a UE.

After RRC connection establishment to the PCell, reconfiguration, addition and removal of SCells can be performed by RRC. When adding a new SCell, dedicated RRC signalling is used to send all the required SI for the new SCell. While in connected mode, changes of SI for an SCell are handled by release and addition of the affected SCell, and this may be done with a single RRC reconfiguration message.

In RRC_CONNECTED state, as the radio conditions for a UE change on different CCs or the load on different CCs changes, the network may decide to change the PCell for a UE. The Release 8 signalling specifications already enable this, via the handover procedure (i.e. with security key change and the random access procedure – see Section 3.2.3.4), which is the only means by which the PCell can be changed. The detailed flow chart for PCell change is shown in Figure 28.1.

In the case of intra-LTE handover, RRC can add, remove, or reconfigure SCells for the target PCell. This enables a UE to begin immediately to use the assigned CCs after handover signalling is complete. The source PCell passes all necessary information to the target PCell (e.g. E-UTRAN Radio Access Bearer (E-RAB) attributes and RRC context). In addition, to enable SCell selection in the target PCell, the source PCell can provide a list of the best cells in decreasing order of radio quality. The target PCell decides which SCells are configured for use after handover, which may include cells other than the ones indicated by the source PCell. The UE does not autonomously release any SCell configuration at handover; as usual, control is network-based.

28.2.2 Measurements and Mobility

For the purposes of mobility, a UE sees a CC in the same way as any other carrier frequency, and a measurement object (see Section 3.2.5.1) has to be set up for each CC in order for the UE to measure it. Inter-frequency neighbour cell measurements encompass all carrier frequencies which are not configured as CCs. Release 8 measurement events (see Section 3.2.5.2) are applicable for UEs configured with carrier aggregation, and the following rules apply:

- There is at most one serving cell (PCell or SCell) per measurement identity;
- For measurement events A1 and A2, the serving cell of the event is the configured serving cell (PCell or SCell) corresponding to the measurement object (i.e. the eNodeB may configure separate events A1 and A2 for each serving cell).

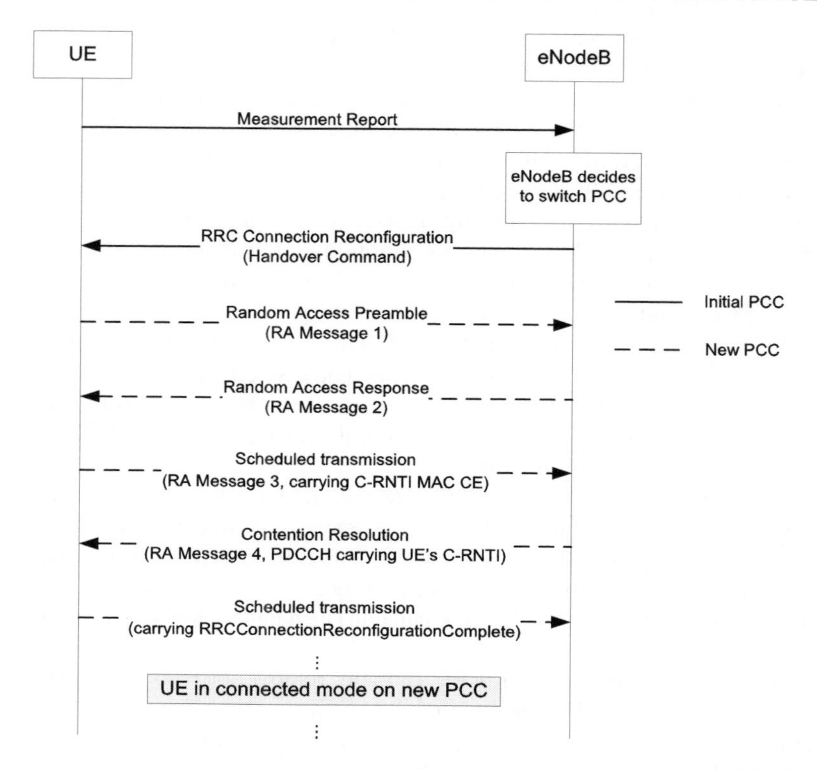

Figure 28.1: Procedure for PCell change with contention-based handover.

- For measurement events A3, A5 and B2, the serving cell used as a reference is the PCell. The measurement object linked to an A3 or A5 event can be any frequency, and, if an SCC is the target object, the corresponding SCell is included in the comparison.

In addition, a new measurement event A6 is introduced for carrier aggregation. Measurement event A6 is defined as 'intra-frequency neighbour becomes better than an offset relative to an SCell' and is intended for intra-frequency measurement events on SCCs. For this measurement, the neighbour cells on an SCC are compared to the SCell of that SCC. An example of the relationship between A3 and A6 is given in Figure 28.2.

Measurements on activated CCs can be done without measurement gaps. Measurement gaps are UE specific; UE capability signalling is used to inform the eNodeB about the need for measurement gaps independently for each supported measured band.

Measurements on all activated cells follow the Release 8 procedures and requirements (see Section 22.3). Measurement periods of deactivated SCells are configurable by RRC signalling (with a range of values from 160 ms to 1280 ms, with a default value of 320 ms). Measurement accuracy requirements are the same for all cells.

The quality threshold for cell selection (the S-criterion – see Section 3.3.3) applies to the PCell and controls all non-serving-cell measurements. In other words, when the PCell

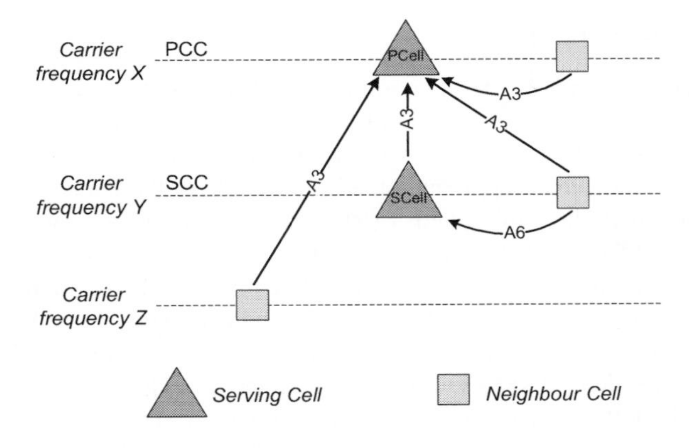

Figure 28.2: Measurement Events A3 and A6.

Reference Signal Received Power (RSRP) (after Layer 3 filtering) is higher than the S-criterion, all measurements other than those that are only on the PCell or only on an SCell can be disabled.

28.2.2.1 Radio Link Failure

Detection of a Radio Link Failure (RLF) (see Section 22.7) on the PCell triggers an RRC connection re-establishment. Triggers for RRC connection re-establishment include:

- Failure of the PCell according to the same criteria as are used for RLF detection in Release 8, based on N310/N311/T310;
- Random access failure in the PCell, as in Release 8;
- An indication from Radio Link Control (RLC) that the maximum number of retransmissions has been reached, as in LTE Release 8.

The UE does not perform radio link monitoring for downlink SCCs. The eNodeB can detect poor link quality on an SCC from CQI reports and/or existing RRM measurement reports for SCells.

28.2.2.2 Idle Mode Procedures

The same mobility procedures as defined in LTE Release 8 apply to a Release 10 UE in a network which deploys carrier aggregation. A UE in RRC_IDLE therefore always behaves as a single-carrier UE, without the possibility of having multiple aggregated CCs.

RRM requirements are defined for carrier aggregation in both idle and connected modes [2]. This is to ensure that good mobility performance is met in all cases including low- and high-mobility scenarios. The mobility performance in the different scenarios can be optimized using different network-controlled parameters for cell reselection in idle mode and for handover in active mode.

28.2.3 User Plane Protocols

From the perspective of the NAS (see Section 2.2.1), the UE is connected to the PCell, which provides the security keys at handover and the tracking area for Tracking Area Updates (TAUs). Other CCs are simply considered as additional transmission resources.

The multiple CCs of carrier aggregation are not visible to the Packet Data Convergence Protocol (PDCP) and Radio Link Control (RLC) layers, and these protocols are therefore unchanged from LTE Release 8 except to enable them to support data rates up to 1 Gbps.

At the MAC layer, each CC has its own independent Hybrid Automatic Repeat reQuest (HARQ) entity. From the perspective of the UE, the characteristics of the HARQ procedures are unchanged with respect to those defined for Release 8 (see Section 4.4). One transport block and an independent HARQ entity are scheduled per CC in the absence of spatial multiplexing, and up to two when spatial multiplexing is configured. The User Plane structures for the downlink and the uplink are shown in Figures 28.3 and 28.4 respectively.

28.2.3.1 Scheduling

It can be seen from Figure 28.3 that a single scheduler entity covers all the UEs and all their corresponding CCs. Elementary queuing theory indicates that a globally optimized scheduler will achieve better performance than a per-CC scheduler. In practice, however, independent schedulers may be utilized for each CC, depending on eNodeB implementation.

Dynamic scheduling is performed every subframe by means of grants transmitted on PDCCH. The grants may be transmitted on the same carrier as the assigned data resources or on a different carrier if cross-carrier scheduling is configured (see Section 28.3.1).

Semi-Persistent Scheduling (SPS) (see Section 4.4.2.1) can only be configured for the PCell, and only PDCCH allocations for the PCell can override an SPS resource allocation.

28.2.3.2 Random Access Procedure

As in LTE Release 8, the MAC layer is responsible for controlling the random access procedure (see Section 4.4.2.3). In the case of carrier aggregation, a UE performs the random access procedure (see Chapter 17) on the uplink CC associated with the PCell. No more than one random access procedure is ongoing at any time, irrespective of the carrier aggregation capability or configuration of the UE.

When carrier aggregation is configured, the first three steps of the contention-based random access procedure (see Section 17.3.1) occur on the PCell, while cross-carrier scheduling from the PCell (see Section 28.3.1) can be used for the contention resolution (step 4). In the non-contention-based random access procedure (see Section 17.3.2), the Random Access Preamble assignment via PDCCH of step 0, step 1 and step 2 occur on the PCell.

28.2.3.3 Discontinuous Reception Procedure

The discontinuous reception (DRX) procedures defined in Release 8 (see Section 4.4.2.5) remain applicable in Release 10. If one or more SCells are configured for a UE in addition to the PCell, the same DRX operation applies to all the serving cells. This means that the active times for PDCCH monitoring are identical across all downlink CCs.

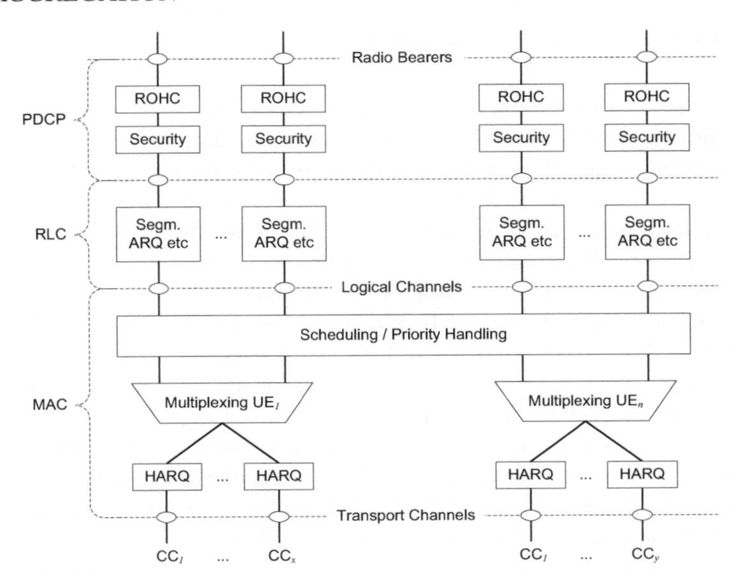

Figure 28.3: Downlink Layer 2 protocol structure for carrier aggregation. Reproduced by permission of © 3GPP.

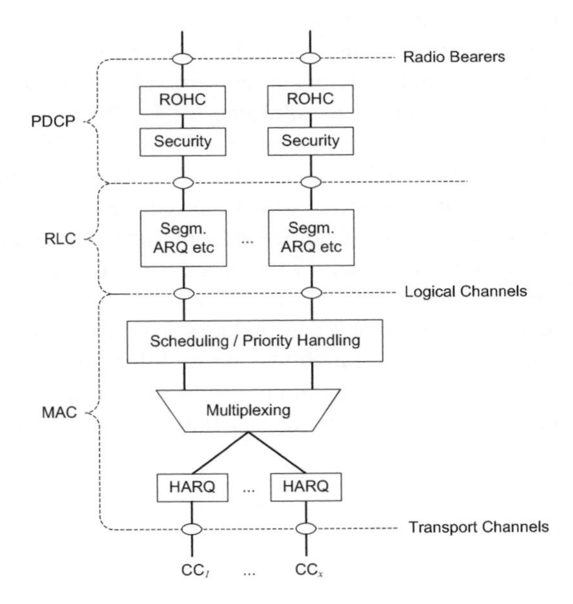

Figure 28.4: Uplink Layer 2 protocol structure for carrier aggregation. Reproduced by permission of © 3GPP.

28.2.3.4 SCell Activation and Deactivation

In addition to the DRX operation, some UE power saving may be achieved by fast activation and deactivation of individual SCells (the PCell cannot be deactivated).

When an SCell is deactivated, the UE does not have to receive data transmissions or monitor the PDCCH for that SCell. The UE is also not required to perform measurements for Channel State Information (CSI) reporting. Deactivated SCells can, however, be used as the path-loss reference for the measurements for uplink power control (see Section 28.3.5.1). It is assumed that these measurements would be less frequent while the SCell is deactivated, in order to obtain power savings at the UE. When the downlink CC of an SCell is activated or deactivated, the SIB2-linked uplink CC follows suit. Deactivation of the uplink CC includes ceasing Sounding Reference Signal (SRS) transmissions and all PUSCH transmissions (including any pending retransmissions).

Activation and deactivation of SCells is under eNodeB control. The activation and deactivation is executed by means of MAC Control Elements (see Section 4.4.2.7), which can activate or deactivate one or more SCells indicated by an 8-bit bitmap. A timer may also be used for automatic deactivation if no data or PDCCH messages are received on a CC for a certain period; this is the only case in which deactivation can be executed autonomously by the UE.[5] Even so, the duration of the timer is configured by the eNodeB and can take the value 'infinity', effectively disabling timer-based deactivation.

The timing of activation and deactivation is carefully defined in order to ensure that there is a common understanding between the eNodeB and the UE. If a MAC control element activating an SCell is received in subframe n, then the SCell has to be ready for operation in subframe $n + 8$. Hence, from subframe $n + 8$, the UE is required to monitor the PDCCH for both uplink grants and downlink assignments corresponding to the newly activated SCell. SRS transmissions can also be started in subframe $n + 8$. CSI reports are commenced in subframe $n + 8$, and CSI measurements in subframe $n + 8$ at the latest. If there is no CSI measurement available for the SCell when the UE first starts reporting CSI, the UE is expected to report the value 'out of range'. Power headroom reporting (see Section 28.3.5.3) also starts in subframe $n + 8$, and the SCell deactivation timer is started. If a MAC control element deactivating an SCell is received (or the deactivation timer expires) in subframe n, CSI reports cease from subframe $n + 8$.[6]

When an RRC reconfiguration occurs that includes mobility control information (i.e. a handover), all SCells are deactivated. If mobility control information is not included, SCells that are added to the set of serving cells are initially in the deactivated state, while any SCells that remain in the set of serving cells (either unchanged or reconfigured) do not change their activation status.

[5]Downlink SCell quality is never used to cause a UE to cease transmissions in an SCell.

[6]If the UE misses a PDCCH message, there may be a temporary misalignment between the UE's and eNodeB's understandings of the activation status of an SCell. This may affect the rate-matching (and hence decoding) of the PUSCH (see Section 16.4) due to uncertainty as to the presence of a CSI report; the eNodeB may be able to mitigate any consequent effects on uplink throughput by making another decoding attempt of the PUSCH with a different rate-matching assumption.

28.2.3.5 Buffer Status Reporting

As in Release 8, there can only be one Buffer Status Report (BSR) per transport block. However, there can be several BSRs in a subframe:

- Zero or one Regular or Periodic BSRs;
- Zero, one or more Padding BSRs of possibly different kinds, but all following the Release 8 rules (see Section 4.4.2.2).

All BSRs transmitted in a subframe reflect the buffered data that remains after all the MAC Protocol Data Units (PDUs) have been built for the CCs that are scheduled in the subframe. When more than one serving cell allows a Regular or Periodic BSR to be sent in a subframe, the UE can choose the serving cell in which the Regular or Periodic BSR is transmitted.

The amount of data that may have to be indicated by a BSR when multiple CCs are aggregated is much higher than in Release 8, due to the higher data rates supported.[7] An additional table of BSR values is therefore introduced in Release 10 to enable indication of the larger buffer sizes. The usage of the new table is controlled by RRC. All UEs which support uplink carrier aggregation or uplink MIMO must support the new BSR table.

28.2.3.6 Logical Channel Prioritization

Different CCs may provide similar QoS, and therefore the UE is allowed complete freedom in how it maps uplink data to granted resources on different CCs.

When the UE is provided with uplink grants in multiple serving cells in one subframe, the order in which the grants are processed during logical channel prioritization, and whether joint or serial processing is applied, are left up to UE implementation. A variety of approaches can fulfil the long-term Prioritized Bit Rate (PBR) (see Section 4.4.2.6) for each logical channel.

28.3 Physical Layer Aspects

At the physical layer, each transport block is mapped to a single CC of a serving cell, as shown in Figure 28.5. Even if a UE is scheduled on multiple CCs simultaneously, HARQ, modulation, coding and resource allocation, together with the corresponding signalling, are performed independently on each CC.

28.3.1 Downlink Control Signalling

Each downlink CC carries a control signalling region for the PCFICH, PDCCH and PHICH[8] at the start of each subframe, as in Release 8 (see Figure 9.5).

[7] The introduction of uplink MIMO in Release 10 (see Section 29.4) further increases the uplink data rates, with corresponding impact on the BSR.

[8] Physical Control Format Indicator Channel, Physical Downlink Control Channel and Physical Hybrid ARQ Indicator Channel.

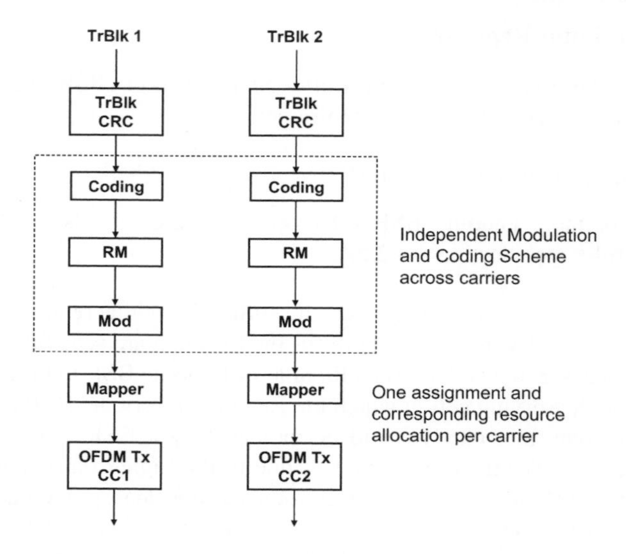

Figure 28.5: MAC to physical layer mapping for carrier aggregation.

28.3.1.1 PDCCH

As in Release 8, it is possible for a PDCCH on each downlink CC to carry downlink resource assignments applicable to the same CC, and uplink resource grants applicable to the associated uplink CC (according to the linkage indicated in SIB2).

In addition, a key feature of carrier aggregation is *cross-carrier scheduling*. This enables a PDCCH on one CC to schedule data transmissions on another CC by means of a new 3-bit *Carrier Indicator Field* (CIF) inserted at the beginning of the PDCCH messages. The rest of the Release 8 PDCCH Control Channel Element (CCE) structure, coding and message contents (as described in Section 9.3.5) is unchanged for carrier aggregation. The presence or absence of the CIF on each CC is configured semi-statically (i.e. by RRC signalling) for each UE. When configured, the CIF is only present in PDCCH messages in the UE-specific search space (see Section 9.3.5.5), not the common search space.

For data transmissions on a given CC, a UE expects to receive scheduling messages on the PDCCH on just one CC – either the same CC, or a different CC via cross-carrier scheduling; this mapping from PDCCH to PDSCH is also configured semi-statically.[9] Some example configurations are shown in Figure 28.6.

For the CC of a serving cell on which PDCCH is monitored, the UE searches for PDCCH messages at least for the same CC of the serving cell. In the example in Figure 28.6(b), the UE monitors the PDCCH on CC1 of serving cell 1 for assignments on CC1, and CC1 resources cannot be cross-scheduled from any other CC of the serving cells. The UE also searches for PDCCHs with CIF on CC1 for assignments for CC2 and CC3, without monitoring the PDCCH on CC2 or CC3 (of serving cells 2 and 3 respectively).

[9]Note that PDSCH transmissions on the PCell cannot use cross-carrier scheduling – their corresponding PDCCH messages must also be transmitted on the PCell.

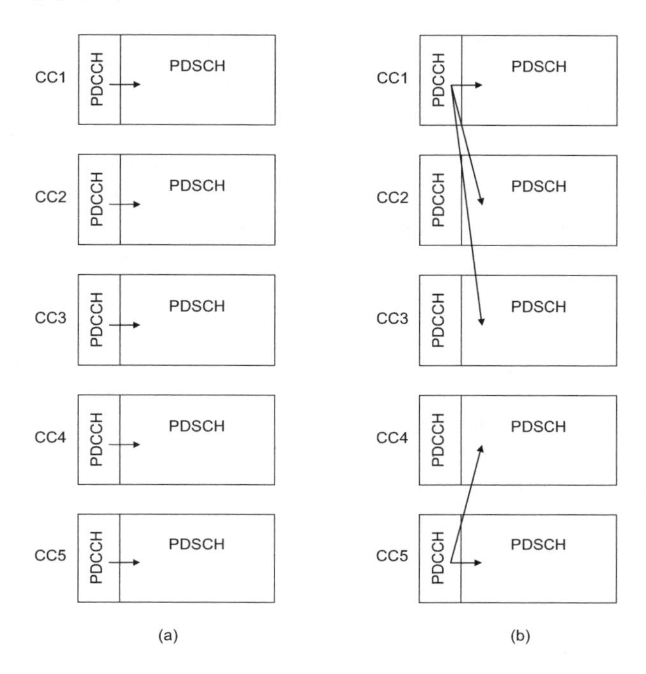

Figure 28.6: Examples of semi-statically configured mappings from PDCCH scheduling messages to CCs for data transmission:
(a) without cross-carrier scheduling; (b) with cross-carrier scheduling.

A UE configured with the CIF for a serving cell uses the CIF value from the detected PDCCH to identify the serving cell on which the corresponding PUSCH or PDSCH transmission will take place. For simplicity, the CIF value is set to be the same as the cell index.

If CIF is not configured for a UE, an uplink grant or downlink assignment received on a given serving cell corresponds to PUSCH or PDSCH transmission on the same serving cell.

In general, it is obvious that the amount of processing that the UE must perform in case of carrier aggregation is significantly larger than in Release 8. This is also true for the PDCCH decoding, where, in the worst case, the number of blind decodes the UE must perform is expected to increase linearly with the number of configured CCs. As discussed in Section 9.3.5.5, the maximum number of blind decodes in any subframe for single-carrier operation is 44 (12 in the common search space and 32 in the UE-specific search space).[10]

When carrier aggregation is configured, the maximum total number of blind decodes a UE is required to perform is 44 for the PCell, plus 32 for each active downlink SCC.[11]

For any downlink CC where the UE monitors PDCCH without the CIF being configured, the search space is the same as in Release 8.

[10]In addition, in Release 10 a further 16 blind decodes are necessary to support uplink MIMO, taking the total number of blind decodes per CC to 60.

[11]A further 16 blind decodes are needed for each uplink CC that is configured for uplink MIMO operation.

In the case of cross-carrier scheduling, the total search space size (in terms of number of CCEs) is extended beyond the Release 8 size. For a given UE, the UE-specific search spaces located in the control region of a CC are individually defined per aggregation level for each PDSCH/PUSCH CC linked to that CC for control signalling. UE-specific search spaces corresponding to different CCs in a given control region are shared if the Downlink Control Information (DCI) format size is the same between the CCs. For any downlink CC with CIF where the UE monitors the PDCCH, a UE-specific search space $S_k^{(L)}$ in subframe k for the PDSCH/PUSCH CC c at aggregation level L={1, 2, 4, 8} is defined by a set of PDCCH candidates. The CCEs corresponding to PDCCH candidate m of the search space $S_k^{(L)}$ are given by:

$$S_k^{(L)} = L \cdot \left\{\left(Y_k + m + M^{(L)} \cdot c\right) mod \lfloor N_{CCE,k}/L \rfloor\right\} + i. \tag{28.1}$$

where Y_k is the output of the UE-specific subframe-to-subframe search space hopping sequence (see Section 9.3.5.5 and [3, Section 9.1.1]), $i = 0, \ldots, L - 1$, and $m = 0, \ldots, M^{(L)} - 1$, $M^{(L)}$ is the number of PDCCH candidates to monitor in the given search space. $N_{CCE,k}$ is the total number of CCEs in the control region of subframe k.

This UE-specific search space design is shown in Figure 28.7.

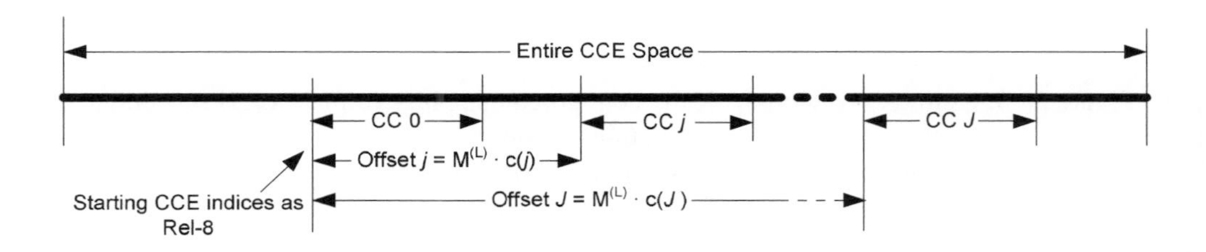

Figure 28.7: UE-specific search space for multiple CCs.

For the common search space, the term $M^{(L)} \cdot c$ in Equation (28.1) is set to zero. The same applies if a UE is not configured with CIF.

Cross-Carrier Scheduling in Heterogeneous Networks

The main motivation for cross-carrier scheduling in LTE-Advanced is to provide support for Inter-Cell Interference Coordination (ICIC) for the PDCCH in heterogeneous network deployments, as mentioned in Section 28.1. Figure 28.8 shows a typical heterogeneous network scenario where macrocells and small cells share two downlink CCs, denoted CC1 and CC2. The small cells use both CCs at low transmit power, and the macrocells use CC1 at high power and CC2 at reduced power. The macrocells' transmissions on CC1 would cause high interference to the small cells, and therefore it is beneficial for the small cells to be able to use PDCCH messages on CC2 to perform cross-carrier scheduling for data transmissions on CC1. To facilitate this, the macrocells can refrain from transmitting PDCCHs on CC2 (or transmit only with low power), instead using CC1 to schedule data transmissions on both CC1 and CC2, with cross-carrier scheduling for the latter.[12] This effectively provides ICIC for the PDCCH, while the Release 8 ICIC mechanisms may be utilized for PDSCH data.[13]

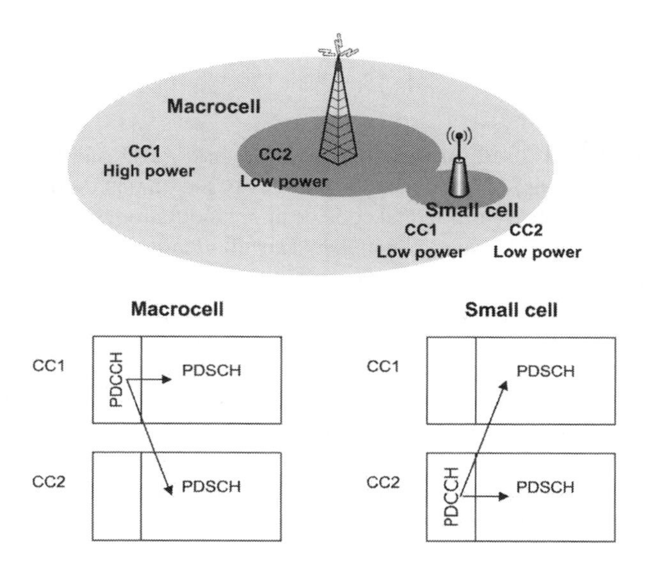

Figure 28.8: A typical heterogeneous network deployment with macrocells and small cells sharing two CCs.

28.3.1.2 PCFICH

Because of the potential for different loadings on different CCs, the number of OFDM[14] symbols used for the downlink control region, and hence the starting point of the PDSCH, can be set independently on each CC. However, in cases when cross-carrier scheduling is employed due to unreliable PDCCH reception on an SCell CC that is used for data transmission, the same inter-cell interference affecting the PDCCH would typically also affect the PCFICH, as they are both in the same control region of the CC. To address this issue without incurring a high signalling overhead, Release 10 provides a mechanism whereby the index of the first OFDM symbol of any cross-scheduled PDSCH can be signalled semi-statically for each CC. This obviates the need for cross-scheduled UEs to decode the PCFICH on the target CC. It should be noted, however, that this does not prevent the eNodeB from

[12]These issues are not problematic for homogeneous networks of macrocells transmitting at similar power levels, where the Release 8 design techniques aimed at maximizing frequency diversity and randomizing the inter-cell interference are entirely appropriate.

[13]For the data, the Release 8 ICIC mechanisms (e.g. based on Relative Narrowband Transmit Power (RNTP) indicators for the downlink – see Section 12.5) enable Fractional Frequency Reuse (FFR) to be configured between cells with a granularity of one RB; this X2-interface-based ICIC mechanism for the data channels is equally applicable in heterogeneous network carrier aggregation scenarios to share frequency resources on the different CCs between macrocells and neighbouring small cells.

[14]Orthogonal Frequency Division Multiplexing.

varying the control region size dynamically (i.e. from subframe to subframe) on each CC (although a relatively static control region size is likely to be suitable for ICIC purposes in many heterogeneous network scenarios). If a shorter control region than the semi-statically signalled one is used, the additional OFDM symbol(s) can still be used for data transmission in RBs assigned to non-cross-scheduled UEs; if a longer control region is used, it will cause some degradation to the PDSCH decoding for cross-scheduled UEs.

28.3.1.3 PHICH

The design of the PHICH (used for transmission of Hybrid ARQ ACK/NACKs in response to uplink data transmissions) for carrier aggregation is based on that defined in Release 8: the physical transmission aspects (orthogonal code design, modulation, scrambling sequence and mapping to resource elements – see Section 9.3.4) are all identical. The PHICH is transmitted on the downlink CC that was used to transmit the corresponding uplink resource grant; this is particularly beneficial for heterogeneous network deployments where some CCs may experience high inter-cell interference in the control channel region as explained above. If cross-carrier scheduling is used, one downlink CC may have to carry PHICH transmissions for multiple uplink CCs, and therefore there is an increased probability of PHICH collisions occurring (since the PHICH index is determined from the lowest PRB of the corresponding PUSCH transmission (see Section 9.3.4), which may be the same on multiple uplink CCs). To mitigate this, the PHICH index can be shifted by configuring different cyclic shifts of the PUSCH demodulation Reference Signals (RSs) among the uplink CCs whose PHICHs are transmitted in the same downlink CC control region (see Figure 9.9). In addition, the eNodeB scheduler can aim to avoid collisions by selecting different starting PRBs for the uplink resource allocations on the different CCs.

28.3.2 Uplink Control Signalling

For carrier aggregation, the uplink control signalling (HARQ ACK/NACK signalling, scheduling requests and Channel State Information (CSI) feedback) has to support up to five downlink CCs.

A UE may send a HARQ ACK/NACK for every downlink transport block – i.e. up to ten per subframe in the case of downlink spatial multiplexing with five downlink CCs. Since the Release 8 Physical Uplink Control Channel (PUCCH) (see Section 16.3.2) was not designed to carry such large numbers of ACK/NACK bits, new mechanisms are defined for carrier aggregation in Release 10.

Similarly, CSI feedback may be needed for up to five downlink CCs, although not necessarily all in the same subframe.

All PUCCH control signalling (corresponding to all configured CCs) is transmitted on the uplink PCC of the PCell (the uplink CC that is SIB2-linked to the configured downlink PCC). Thus PUCCH is never transmitted on more than one uplink CC.

Uplink control signalling may also be mapped to the PUSCH, as in Release 8 (see Section 16.4).

In addition to the Release 8 multiplexing modes, Release 10 supports simultaneous transmission of PUCCH for control information and PUSCH for data. Some potential benefits of simultaneous PUCCH and PUSCH transmission include:

- The IoT operating point can be set independently for control and data, which can improve efficiency as HARQ enables the PUSCH to operate at higher IoT levels than is possible for reliable control information reception on the PUCCH;
- Interference fluctuations on RBs used for PUCCH may be reduced.

28.3.2.1 HARQ Feedback

In order to provide HARQ feedback (ACK/NACK) for PDSCH transmissions on multiple CCs, new multibit ACK/NACK PUCCH formats are defined in Release 10 in support of carrier aggregation:

- PUCCH format 3;
- PUCCH format 1b with PUCCH 'channel selection', whereby some of the ACK/NACK information to be conveyed is indicated by selecting one of a number of possible PUCCH resources, in a similar way to Release 8 TDD operation – see Section 23.4.3.

For UEs that support no more than four ACK/NACK bits and are configured with up to two CCs, PUCCH format 1b with channel selection is utilized. For UEs that support more than four ACK/NACK bits, both PUCCH format 1b with channel selection and format 3 are supported, where PUCCH format 1b with channel selection can be used for up to four ACK/NACK bits and two configured CCs and format 3 for the full range of ACK/NACK bits; RRC signalling configures which PUCCH format is used in this case.

PUCCH Format 3

PUCCH format 3 is designed to convey large ACK/NACK payloads. Unlike the Release 8 PUCCH formats (namely PUCCH formats 1, 1a, 1b and 2 – see Section 16.3.2), PUCCH format 3 is not based on Zadoff-Chu sequences and is more similar to PUSCH transmissions. It has the following characteristics, illustrated in Figure 28.9:

- DFT-S-OFDM[15] waveform;
- The same demodulation RS structure as PUCCH format 2 (for both normal and extended cyclic prefix);
- Orthogonal cover sequence applied to the SC-FDMA symbols used for ACK/NACK data: these sequences are DFT sequences of length 5, allowing multiplexing of up to 5 format 3 transmissions in the same RB;
- No orthogonal cover sequence applied in the SC-FDMA symbols used for demodulation RSs; the RSs of multiple UEs are multiplexed by means of different cyclic shifts;
- A shortened format is defined in which the last SC-FDMA symbol is punctured and the orthogonal cover sequence is shortened to a Hadamard sequence of length 4.
- QPSK modulation.

The resulting PUCCH format 3 supports transmission of 48 coded bits. The actual number of bits of ACK/NACK feedback is determined from the number of configured CCs, the

[15]Discrete Fourier Transform Spread Orthogonal Frequency Division Multiplexing.

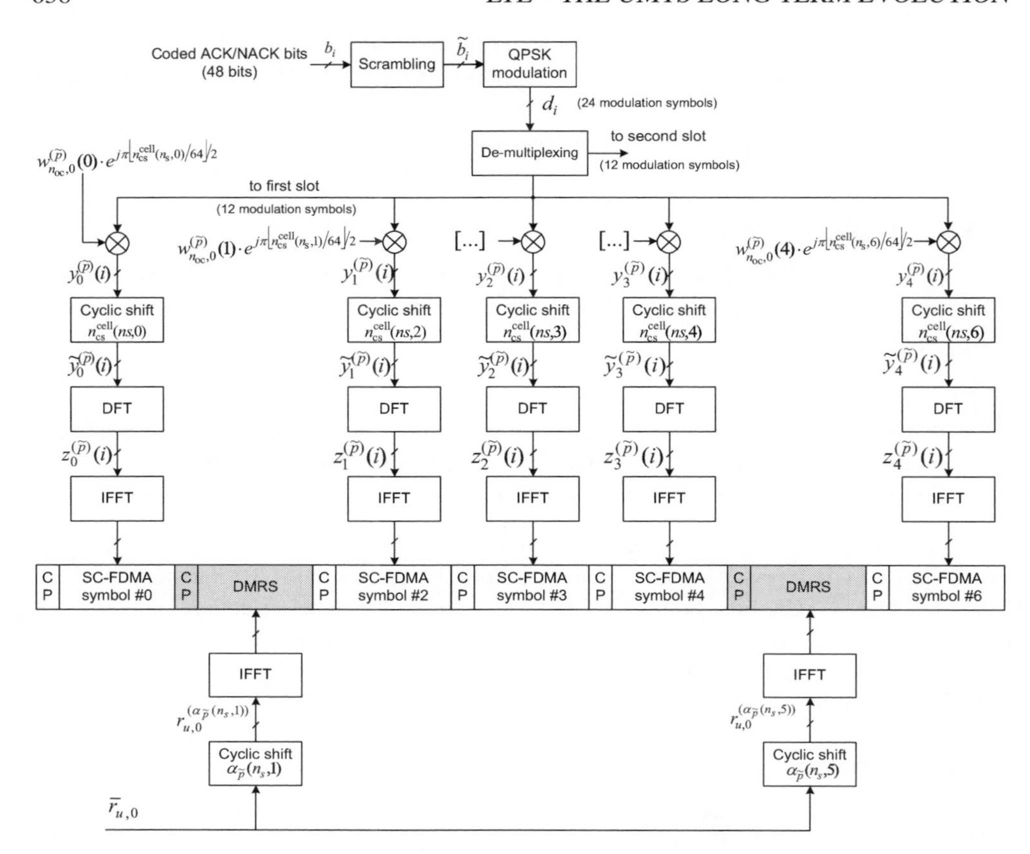

Figure 28.9: Structure of PUCCH format 3.

configured transmission modes on each of them, and, in TDD, the ACK/NACK bundling window size (the number of downlink subframes associated with a single uplink subframe – see Section 23.4.3). For FDD, a maximum payload of 10 ACK/NACK bits is supported, covering up to five CCs configured for MIMO transmission (i.e. two ACK/NACK bits per CC). For TDD, PUCCH format 3 supports an ACK/NACK payload size of up to 20 bits; if the number of ACK/NACK bits to be fed back for multiple downlink subframes associated with a single uplink subframe is greater than 20, 'spatial bundling' (i.e. a logical AND) of the ACK/NACK bits corresponding to the two codewords within a downlink subframe is performed for each of the serving cells. The maximum payload size carried by PUCCH format 3 in Release 10 is 21 bits (i.e. a code rate of 0.4375), corresponding to 20 bits of ACK/NACK information and one bit for a Scheduling Request (SR)[16] appended at the end of the ACK/NACK bits.

The ACK/NACK bits are concatenated in ascending order of the downlink CC index. For payload sizes less than or equal to 11 bits, channel coding uses the Reed-Muller (RM)

[16]See Section 4.4.2.2.

code from Release 8, with circular buffer rate matching (as explained in Section 10.3.2.4). When the payload is larger than 11 bits, alternate ACK/NACK bits are input to two separate RM encoders. Finally, in order to mitigate inter-cell interference, cell-specific scrambling per SC-FDMA symbol is introduced. This structure is shown in Figure 28.10.

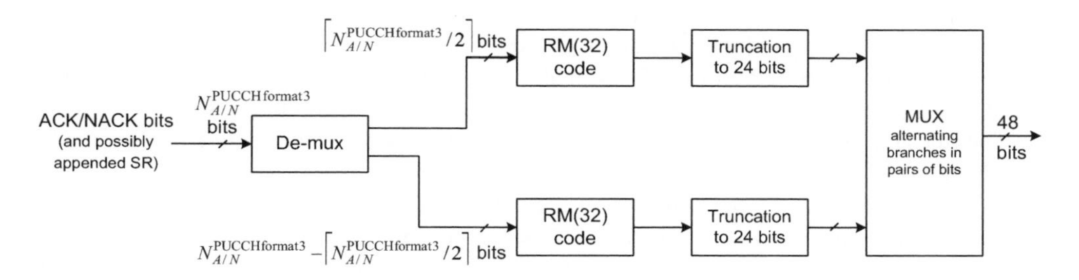

Figure 28.10: Coding and scrambling for PUCCH format 3.

The PUCCH resource to be used for format 3 is signalled explicitly to the UE. A set of four resources is configured by RRC signalling, of which one resource is then indicated dynamically for each ACK/NACK occasion using an indicator transmitted in the Transmitter Power Control (TPC) field of the PDCCH corresponding to PDSCH on the SCCs. All SCC PDCCH assignments in a given subframe indicate the same value.

If transmit diversity is used for PUCCH format 3 (see Section 29.4.2), the RRC signalling configures four pairs of PUCCH resources, and the PDCCH assigning resources for the SCC PDSCH indicates one of these pairs to be used by the two antenna ports.

If no PDCCH corresponding to PDSCH on an SCC is received in a given subframe and a single PDSCH is received on PCC, a UE that is configured for PUCCH format 3 would instead use the Release 8 format 1a or 1b.

PUCCH Format 1b with Channel Selection

PUCCH format 1b with channel selection involves configuring up to four PUCCH format 1b resources ('channels'); the selection of one of these resources indicates some of the ACK/NACK information to be conveyed.

For FDD, the use of PUCCH format 1b with channel selection to convey the ACK/NACK information for two CCs is straightforward. For TDD, it is necessary to use spatial bundling of ACK/NACK bits across the two codewords within a downlink subframe for each of the serving cells if the number of ACK/NACK bits to be fed back is greater than four. If the number of ACK/NACK bits after performing spatial bundling is still larger than four, time-domain bundling is employed in addition.

Mapping tables are specified for the cases of two, three or four ACK/NACK bits to define the mapping of ACK/NACK combinations to the configured PUCCH resources. These tables are designed to support fully implicit resource indication, fallback to Release 8 operation in the case of a single configured CC, and equalization of the performance of individual ACK/NACK bits. Separate mapping tables are defined depending on whether or not time-domain bundling of the ACK/NACK feedback is performed. A Release 10 UE configured with a single TDD serving cell can be configured to use either the mapping tables defined for TDD carrier aggregation without time-domain bundling or the tables defined for Release 8.

If a UE is configured for PUCCH format 1b with channel selection, implicit ACK/NACK resource allocation is used for dynamically scheduled PDSCH transmissions on the PCC and for cross-carrier scheduling of SCCs from the PCC, as well as for release of an SPS resource. In the case of PDSCH transmissions on SCCs, for non cross-carrier scheduling or for cross-carrier scheduling from an SCC, explicit PUCCH resource allocation is used: a set of resources is configured by RRC signalling, and the PDCCH(s) corresponding to the PDSCH on the SCC(s) indicate resources derived from this set using the TPC field.[17]

The rules for implicit PUCCH resource allocation are based on those defined in Release 8. For dynamic scheduling, a set of implicit ACK/NACK resources is derived from the indexes of the first CCEs used for transmission of the corresponding DCI assignments. When time domain bundling is used and the Downlink Assignment Index (DAI – see Section 23.4.3) is equal to either '1' or '2', the PUCCH resources are derived from the combination of the DAI value and the indexes of the first CCEs of the corresponding PDCCHs.

In the case of simultaneous ACK/NACK feedback and SR transmission, a UE that is configured to use PUCCH format 1b with channel selection transmits spatially bundled ACK/NACK feedback on a single SR resource. For TDD, the UE transmits two bits on the SR resource, representing the number of ACKs over both CCs within the bundling window.

ACK/NACK Repetition

ACK/NACK repetition (see Section 16.5) on PUCCH is not supported for carrier aggregation, because it can impact downlink performance and a UE configured for carrier aggregation is generally not assumed to suffer from a transmission power limitation for control signalling.

28.3.2.2 Channel State Information Feedback

Periodic Channel State Information (CSI) feedback (see Section 10.2.1.2) is independently configured for each downlink CC by RRC signalling. With carrier aggregation, periodic CSI is reported for only one downlink CC in any given subframe. Different offsets and periodicities should as far as possible be configured for each CC to aim to minimize collisions between CSI reports of different CCs in one subframe. If a collision of multiple CSI reports does occur, one report is selected for transmission according to defined prioritization rules (the others being dropped):

- First priority is given to the CSI reports that contain a Rank Indicator (RI) or a wideband 'first'[18] Precoding Matrix Indicator (PMI);
- Second priority is given to other wideband CQI and/or PMI reports;
- The lowest priority is assigned to the sub-band CQI/PMI reports.
- If there are multiple CCs with a report of the same priority, the CC of the serving cell with lowest cell index is prioritized.

[17] This indication on the PDCCH does not increase the DCI message size, since these TPC fields are not used for PUCCH power control (the PUCCH power being controlled by the PCC grant).

[18] 'First' refers to the first part of the dual-stage PMI codebook structure introduced in Release 10 for the case of 8 downlink transmit antennas – see Section 29.3.3.

If simultaneous PUSCH and PUCCH transmission is supported by the UE and is enabled, a collision between periodic CSI feedback and ACK/NACK can be resolved by transmitting the CSI feedback on the PUSCH and ACK/NACK on PUCCH. Also, if there is a PUSCH transmission and the uplink control signalling consists only of periodic CSI feedback, the periodic CSI feedback is in this case transmitted on the PUCCH (not the PUSCH as in Release 8).

28.3.2.3 Uplink Control Information (UCI) on PUSCH

If a UE is configured with multiple serving cells and simultaneous PUCCH and PUSCH is not enabled, and there is at least one PUSCH transmission, all Uplink Control Information (UCI) is multiplexed onto a PUSCH. Any periodic CSI feedback is normally[19] transmitted on the PCC PUSCH if it is transmitted. If the PCC PUSCH is not transmitted, the UCI is transmitted on an SCC PUSCH if one is transmitted; if more than one SCC PUSCH transmission occurs in the subframe, the periodic UCI is transmitted on the SCC PUSCH of the serving cell with the lowest cell index.

The ACK/NACK payload size for transmission on the PUSCH is determined from the number of configured downlink CCs, the configured transmission mode for each downlink CC, and, in the case of TDD, the bundling window size and the signalled DAI value in the UL grant.

Aperiodic CSI is reported on a PUSCH if it is triggered by a request in an uplink DCI message or a Random Access Response grant. When a UE is configured with multiple serving cells in carrier aggregation, a CSI request transmitted in the UE-specific PDCCH search space can trigger CSI reports for one or more downlink CCs, as shown in Table 28.1.

Table 28.1: Combinations of downlink CCs for which aperiodic CSI may be triggered.

Value of CSI request field	Meaning
00	No CSI report is triggered
01	CSI report is triggered for the cell on which the trigger is sent
10	CSI is triggered for a first set of serving cells configured by higher layers
11	CSI is triggered for a second set of serving cells configured by higher layers

If aperiodic CSI feedback is triggered using the common search space, the feedback is transmitted for an RRC-configured set of CCs.

In the case of aperiodic CSI triggering when multiple PUSCH transmissions are taking place on different uplink CCs in the same subframe, the UCI is always transmitted on the PUSCH on the CC indicated by the uplink grant containing the aperiodic CSI trigger. UEs are not expected to receive more than one positive aperiodic CSI trigger for a given subframe.

In the case of a collision between periodic CSI and aperiodic CSI for the same or different downlink CCs, the periodic CSI is dropped and only the aperiodic CSI feedback

[19]The exceptions relate to PUSCH transmissions that are part of the random access procedure.

is transmitted. This applies even if the periodic and aperiodic CSIs are for different downlink CCs.

28.3.3 Sounding Reference Signals

Sounding Reference Signals (SRSs) can be triggered on any serving cell either by higher-layer signalling or dynamically via the DCI messages for UL grants, as explained in Section 29.2.2. When carrier aggregation is configured, a UE may be configured with SRS parameters for both types of SRS triggering on each serving cell.

The following rules are defined for SRSs in relation to transmissions on multiple CCs:

- SRSs may be transmitted simultaneously on different CCs;
- If a PUCCH transmission on the PCC coincides with SRS on an SCC, the UE transmits the SRS on the SCC if the PUCCH uses a shortened format, and not otherwise;
- If a PUSCH transmission coincides with SRS on different CCs:
  - PUSCH transmitted in any cell-specifically configured SRS subframe is rate matched around the SRS resources on the same CC;
  - If PUSCH is transmitted in the same SC-FDMA symbol on a different CC from the same UE, the SRS is dropped.

28.3.4 Uplink Timing Advance

As mentioned in Section 28.1, the timing of the PCell and all SCells configured for a UE is expected to be synchronized. A single Timing Advance (TA) command (see Section 18.2) is therefore sufficient to control the UE's uplink transmission timing for all the uplink CCs together. This simplifies the UE implementation since a single time reference for the baseband processing can be used. In addition, for contiguous CCs, a single Inverse Fast Fourier Transform (IFFT) can be used for the generation of the signals for multiple uplink CCs.

It is, however, expected that Release 11 will introduce the possibility of independent TA per CC, which may be beneficial for some scenarios such as the use of frequency-selective repeaters on certain of the configured CCs.

28.3.5 Uplink Power Control

Uplink power control with carrier aggregation follows the same principles as for single carrier transmission in Release 8 (see Section 18.3). When multiple CCs are configured, uplink power control operates independently for each CC. This allows the different operating conditions of each CC (e.g. different frequency bands or different interference scenarios) to be taken into account. The parameters for open-loop power control (P_0 for both PUSCH and PUCCH, to set the operating point, and α, the fractional path-loss compensation factor) are therefore all CC-specific, as are the closed-loop TPC commands and any MCS-dependent offsets. TPC commands in uplink resource grants are applied to the PUSCH on the uplink CC for which the grant applies. The TPC commands in PCC downlink resource assignments are applied to the uplink PCC on which the corresponding HARQ ACK/NACK signalling is transmitted.[20]

Power control for groups of UEs using DCI Formats 3 and 3A is supported only for the same CC on which the TPC commands are transmitted. Cross-carrier scheduling of grouped power control commands is not supported, since SPS and periodic CSI reporting, which are some of the main uses for group power control, take place only on the PCC.

28.3.5.1 Path-Loss Estimation

Since the uplink transmission power is based on the path-loss estimated on a downlink CC, a reference downlink CC is defined for each uplink CC. The path-loss reference for an uplink CC can be either the SIB2-linked downlink CC or the downlink PCC, according to network configuration. The downlink CC used for path-loss estimation should always be in the same frequency band as the uplink CC.

Configurability of the path-loss reference allows appropriate operation for different deployment scenarios. For example, in heterogeneous network deployments it may happen that reliable uplink transmission on an SCC is possible, but the path-loss estimation on the SIB2-linked downlink SCC is not sufficiently reliable due to interference. The ability to configure the path-loss to be estimated on the downlink PCC for power control of the transmissions on such an uplink SCC can facilitate efficient resource usage and better load balancing among CCs.

A configured but deactivated CC can be used as the path-loss reference according to the configuration (SIB2 or PCC), in order to provide reasonably good path-loss estimation for power control upon SCell reactivation. The UE would measure a deactivated CC less frequently (similarly to the case of long DRX cycles) in order not to reduce too significantly the potential power savings from the CC deactivation.

28.3.5.2 Maximum Power Behaviour

As in Release 8, the UE's transmission power is limited by the power class to which it belongs (e.g. 23 dBm for class 3 – see Section 21.3.1.2). In addition, for carrier aggregation a maximum transmission power is set for each individual CC; this must be satisfied first, although this is also typically 23 dBm.

The possibility of simultaneous PUCCH and PUSCH transmission also affects the behaviour of the UE when it reaches its maximum allowed transmission power. Since the control information is essential for the correct reception of data, and, unlike PUSCH transmissions, the PUCCH cannot benefit from HARQ, any PUCCH transmission is always prioritized over concurrent PUSCH transmissions. Hence, the required power is first set for the PUCCH, and then any remaining power is used for PUSCH transmissions. Similarly, among the PUSCH transmissions, a PUSCH carrying uplink control information is prioritized over PUSCH transmissions without it. When a UE reaches its maximum power, it therefore first scales down the power of the PUSCHs without control information. The scaling factor is normally common for all serving cells, although for some serving cells it may be set to

[20]Note that TPC commands in SCC downlink resource assignments may be used for PUCCH resource assignment – see Section 28.3.2.1.

zero (e.g. if the power after scaling falls below a useful level). Full details can be found in [3, Section 5.1.1.1].

Other factors affecting the UE's maximum transmission power are discussed in Section 21.3.1.2.

28.3.5.3 Power Headroom Reporting

For a UE configured with multiple uplink CCs, the Power Headroom Reports (PHRs) are independent for each CC. PHRs are used to provide the serving eNodeB with information about the difference between the nominal UE maximum transmission power and the estimated power for PUSCH transmission, and the difference between the nominal UE maximum power and the estimated power for PUSCH and PUCCH transmission on the PCell.

To support simultaneous PUCCH and PUSCH transmission, two types of PHR (known as extended PHRs (ePHR)) are supported in Release 10:

- Type 1, which only takes into account the PUSCH transmission power;
- Type 2, which indicates the power headroom when both PUCCH and PUSCH are present.

For a UE configured with simultaneous PUCCH and PUSCH, a Type 2 PHR for the PCC is always reported when Type 1 PHR is reported. For subframes where PUCCH is not actually transmitted, a hypothetical PUCCH format 1a transmission is assumed. Similarly, the UE assumes a hypothetical reference format for the PUSCH when reporting a PHR if no PUSCH transmission is scheduled on the PCC. PHRs based on these hypothetical reference formats are known as virtual PHRs.

In order to provide sufficient information on the total available UE transmission power, each Release 10 ePHR additionally includes the value of the current maximum power for the CC, $P_{\text{CMAX,c}}$, after taking into account the maximum power reductions explained in Section 21.3.1.2. A single MAC control element contains both $P_{\text{CMAX,c}}$ and the PHR for a given CC; a bitmap is used to indicate for which SCC the information is reported, and a 'virtual PHR indication' indicates whether the PUSCH power is based on an assumed reference format or not. $P_{\text{CMAX,c}}$ is not reported for virtual PHRs.

The use of the new Release 10 ePHR is configured by RRC signalling. The support of ePHR is mandatory for UEs supporting uplink carrier aggregation and for UEs supporting simultaneous PUSCH and PUCCH transmission.

ePHR is calculated based on the power before any power scaling due to maximum power limitations and can therefore be positive or negative (similar to the single-carrier PHR of Release 8 – see Section 18.3.3). There is one prohibit timer per UE to control how often the PHR can be transmitted. The prohibit timer is started after a PHR transmission, and a new PHR cannot be transmitted until the timer expires.

In addition to the PHR triggers described in Section 18.3.3, the activation of an SCell with a configured uplink CC triggers a PHR. This is useful for the eNodeB scheduler, since the power state of a UE changes upon activation of a CC.

28.3.6 Uplink Multiple Access Scheme Enhancements

Although not directly related to carrier aggregation, some further enhancements to the uplink multiple access scheme are introduced in Release 10 for LTE-Advanced.

28.3.6.1 Analysis of Candidate Enhancements

As described in Chapter 14, the uplink multiple access scheme of LTE in Releases 8 and 9 is Single Carrier Frequency Division Multiple Access (SC-FDMA) (also known as DFT-Spread-OFDM (DFT-S-OFDM)). This has the desirable property of being 'single carrier' and therefore maintaining a low Cubic Metric (CM). The main motivations for modifying this scheme for LTE-Advanced in Release 10 were the possibilities of performance improvements, especially in conjunction with the introduction of uplink Single-User Multiple-Input Multiple-Output (SU-MIMO), and the opportunity to increase the flexibility of uplink resource allocation to maximize the utilization of the spectrum.

Figures 28.11 and 28.12 show block diagrams of two candidate schemes considered for LTE-Advanced, namely clustered DFT-S-OFDM and multiple SC-FDMA respectively.

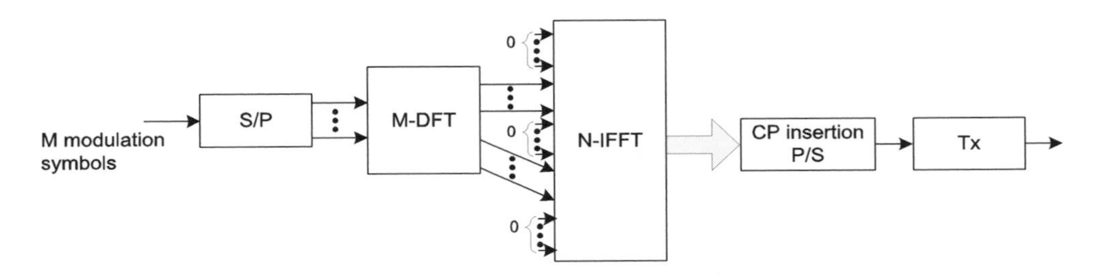

Figure 28.11: Block diagram of clustered DFT-S-OFDM.

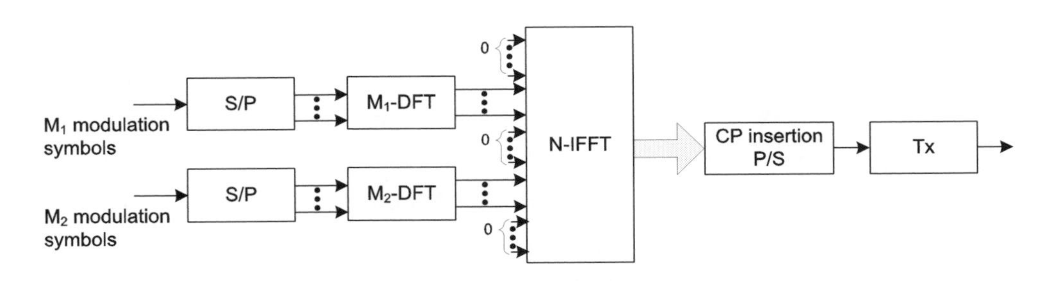

Figure 28.12: Block diagram of multiple SC-FDMA.

Clustered DFT-S-OFDM retains a single DFT operation but modifies the resource element mapping at the output of the DFT operation from a single cluster (as used for SC-FDMA) to multiple clusters which are multiplexed with $N - M$ zeros to form the input of the IFFT[21] operation over N-virtual subcarriers. The resulting waveform is no longer single-carrier but still has a low CM.

Multiple SC-FDMA consists of a number of DFT operations, where the xth DFT is of size M_x; the output symbols are then multiplexed with $N - \sum_{x=1}^{X} M_x$ zeros to fit the N-point IFFT. This waveform is no longer single carrier and experiences a worse CM than that of clustered DFT-S-OFDM for the same number of clusters.

[21] Inverse Fast Fourier Transform.

Figure 28.13 shows the CM of the schemes described above, compared to the SC-FDMA of Release 8 and to OFDMA, for different numbers of clusters and different modulation schemes. It can be seen that the CM for SC-FDMA and QPSK is the lowest, while the CM for OFDMA is the largest and invariant with respect to the number of clusters. The CM for clustered DFT-S-OFDM and multiple SC-FDMA increases with the number of clusters but never gets worse than that of OFDMA. The CM of multiple SC-FDMA is always higher than that of clustered DFT-S-OFDM for the same number of clusters.

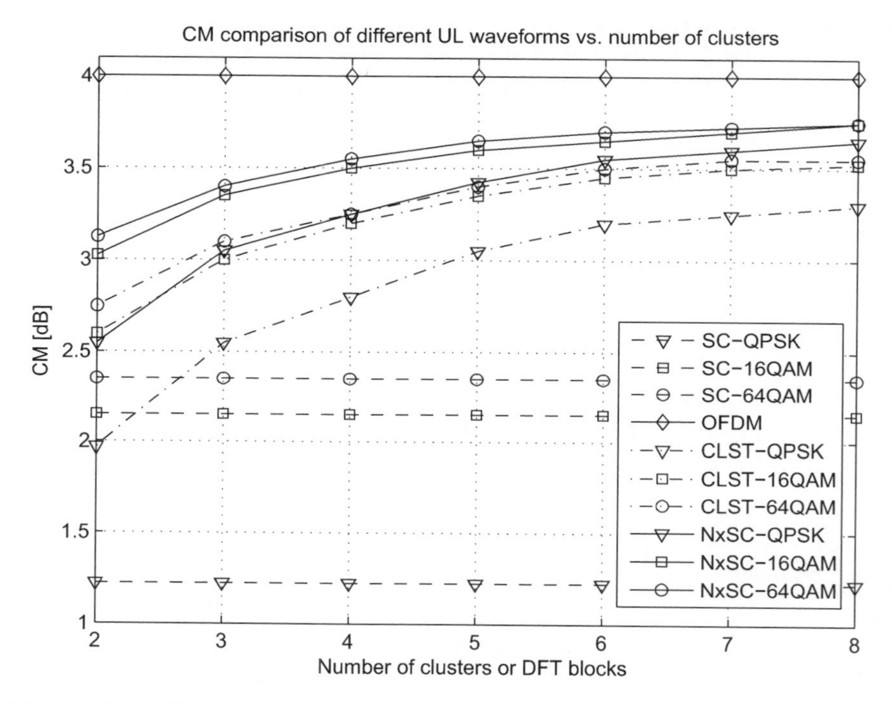

Figure 28.13: CM of SC-FDMA, clustered DFT-S-OFDM, multiple SC-FDMA and OFDMA.

Figure 28.14 shows the UE throughput Cumulative Distribution Function (CDF) for SC-FDMA, clustered DFT-S-FDMA and OFDMA. The simulation is for scenario D1 from [4] with 10 MHz system bandwidth, two RBs reserved for PUCCH transmission, a sub-band size of 6 RBs for scheduling and feedback reporting, Proportional Fair Scheduling (PFS), a maximum of 4 uplink grants per subframe, 10 UEs/cell each with a 1×2 antenna configuration, ideal channel estimation and Interference over Thermal (IoT) target of 7 dB. Different power backoffs are modelled in the simulation according to the difference in CM of the various schemes (see Figure 28.13). The figure shows that the performance of OFDMA and clustered DFT-S-OFDMA is practically identical. Both these schemes outperform SC-FDMA thanks to the additional flexibility in resource allocation which maximizes utilization of the spectrum.

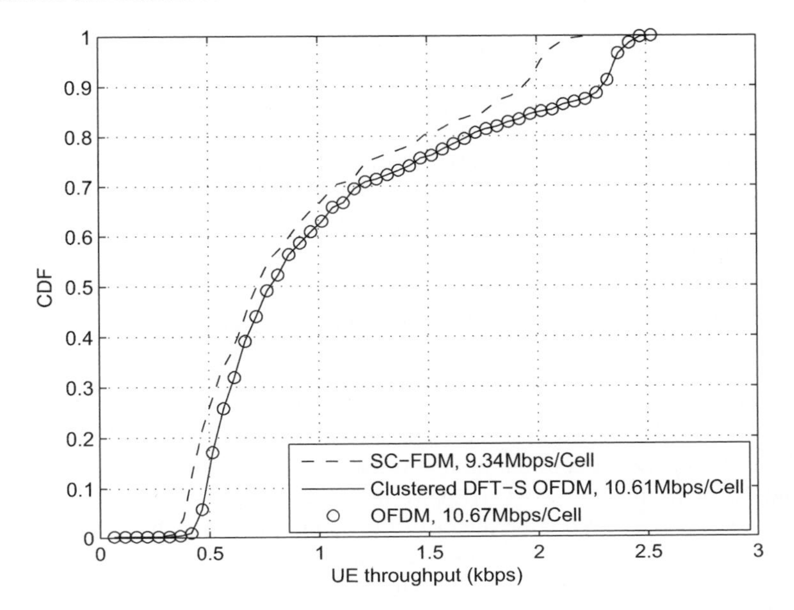

Figure 28.14: System-level performance comparison between SC-FDMA, clustered DFT-S-OFDMA and OFDMA.

28.3.6.2 Enhancements Included in Release 10

As a result of these considerations, DFT-S-OFDM continues to be the basis of the uplink multiple access scheme for the PUSCH in LTE-Advanced, but the possibility of frequency-non-contiguous resource allocation is introduced for an individual Release 10 UE, using clustered DFT-S-OFDM with a maximum of two clusters.

The uplink grant on the PDCCH (i.e. DCI formats 0 or 4 – see Section 9.3.5.1) indicates whether the PUSCH resource allocation is multiclustered or not, by means of a *resource allocation type* bit. If this bit is set, in the case of DCI format 0 the 'frequency hopping flag' is used as an extra bit for the multiclustered resource allocation signalling. In general, PUSCH frequency hopping is not supported in conjunction with multiclustered PUSCH transmission.

In order to signal the frequency-domain locations of the RB Groups (RBGs) of the non-contiguous resource allocations, the indexing scheme used for signalling the CQI sub-bands in Release 8 is re-used (see Section 10.2.1.1). The RBG size depends on the system bandwidth in the same way as for downlink resource allocation type 0 (see Section 9.3.5.4 and Table 9.5).

If the resource allocation is non-contiguous, the DeModulation Reference Signals (DM-RSs) transmitted in the PUSCH (see Section 15.5) are adapted to match the resource allocation: a single DM-RS base sequence is generated according to the total number of allocated RBs, and then split into sections for transmission in the PUSCH RBs.

In the case of carrier aggregation, one DFT is used per CC, thus yielding a multiple SC-FDMA scheme. If a different Power Amplifier (PA) is used for each CC, each PA will amplify a single-carrier waveform and hence benefit from the low CM (unless simultaneous PUCCH and PUSCH transmission is used).

28.4 UE Transmitter and Receiver Aspects

In LTE-Advanced Release 10 the spacing between the centre frequencies of contiguously aggregated CCs is a multiple of 300 kHz. The rationale behind this choice is to preserve backward compatibility with the 100 kHz frequency raster used in LTE Release 8 as well as preserving the orthogonality of the subcarriers with the 15 kHz spacing. Depending on the aggregation scenario, the actual spacing (a multiple of 300 kHz) may be facilitated by insertion of a number of unused subcarriers between contiguous CCs.

28.4.1 UE Transmitter Aspects of Carrier Aggregation

The output power dynamics are impacted by the UE architecture, which may be based on single or multiple PAs. Figure 28.15 [2] illustrates various options for PA architectures at the UE which can be used to support carrier aggregation.

When considering the PA configuration, it is necessary to take into account any additional back-off requirements that may exist. The CM, introduced as a predictor of required back-off in Section 21.3.3, is only a good predictor of the additional power back-off required if the third-order InterModulation (IM3) distortion product lands in the Adjacent Channel Leakage Ratio (ACLR)[22] band (as it does, for instance, for LTE Release 8 with full resource allocation, or for WCDMA-based system such as UMTS and HSPA – see Chapter 21).

The new multiple SC-FDMA and clustered DFT-S-OFDM waveforms supported in Release 10 (due to carrier aggregation and the concurrent transmission of PUSCH and PUCCH) impose more stringent linearity requirements on the PA than was the case for LTE Release 8.

The factors that determine the necessary UE PA back-off are compliance to the ACLR, Spectrum Emission Mask (SEM), spurious emissions and Error Vector Magnitude (EVM) requirements [2, 5].

Small resource assignments at the band edge behave as tones and hence produce highly concentrated InterModulation Distortion (IMD) products. Therefore, for the concurrent transmission of PUCCH and PUSCH, the SEM is expected to be the limiting requirement.

28.4.2 UE Receiver Aspects of Carrier Aggregation

For the baseband aspects of the UE receiver, the main impact of carrier aggregation is on the soft buffer allocation, where the total HARQ buffer has to be shared between the configured CCs.

For the RF aspects, two options were considered for the baseline UE receiver architecture as part of the carrier aggregation feasibility study:

- **Option A**: Single RF, and baseband processing with bandwidth $\geq$ 20 MHz;
- **Option B**: Multiple RF, and baseband processing with bandwidth $\leq$ 20MHz.

Clearly, Option A is only applicable for intra-band aggregation of contiguous CCs, but it has the advantage of keeping the UE receiver complexity low. Option B is applicable for intra-band and inter-band aggregations for contiguous or non-contiguous scenarios, but this flexibility comes at the expense of increased complexity.

[22] See Chapter 21.

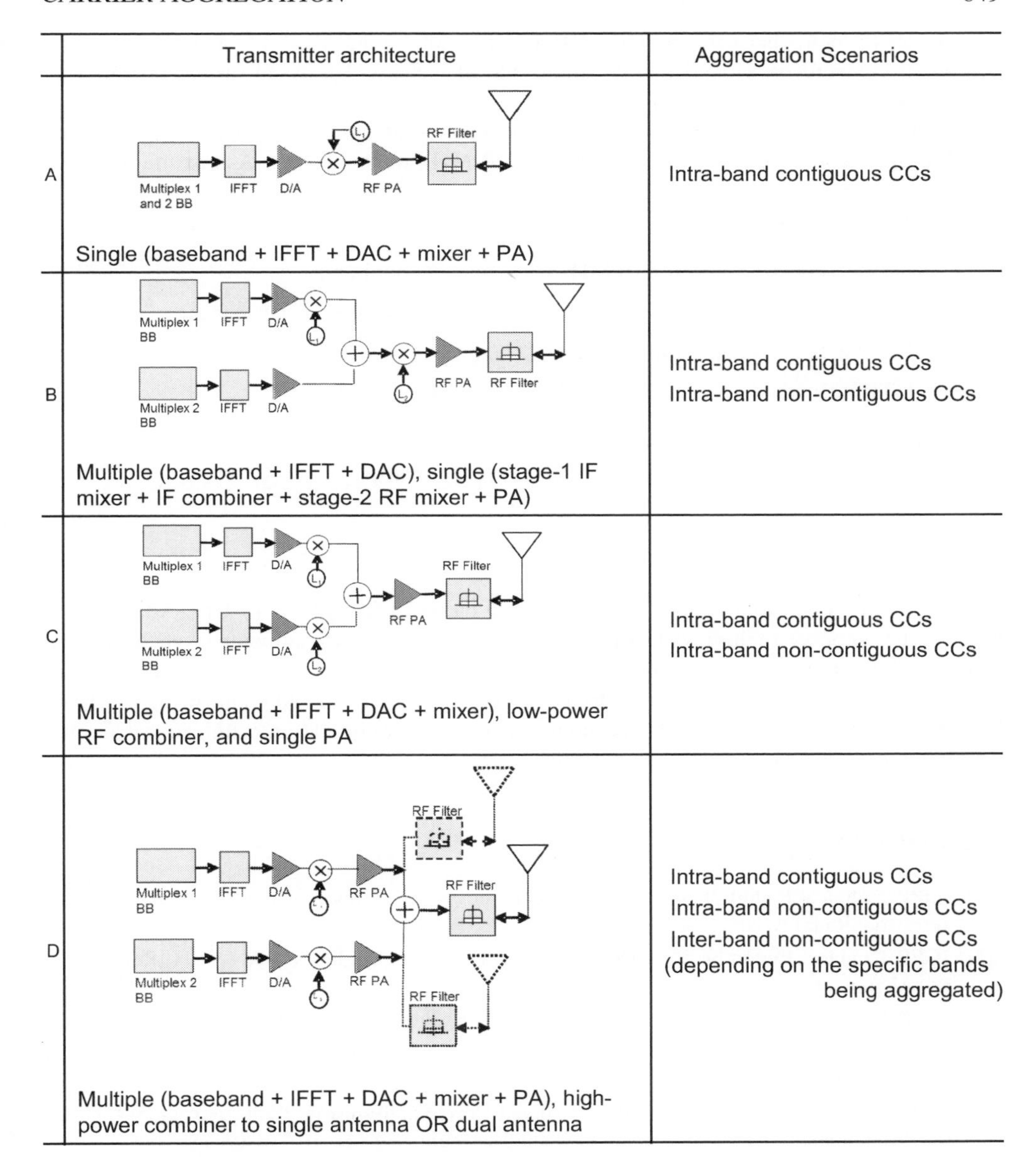

Figure 28.15: Some examples of PA configuration options for carrier aggregation. Reproduced by permission of © 3GPP.

28.4.3 Prioritized Carrier Aggregation Scenarios

Many carrier aggregation scenarios of relevance to different operators around the world were considered as part of the feasibility study for LTE-Advanced [2].

In order to focus the work to define RF requirements in 3GPP, some carrier aggregation combinations were prioritized for the timeframe of Release 10, based on the priorities of network operators.

For intra-band carrier aggregation, the first supported carrier bandwidths are 15 and 20 MHz in E-UTRA Band 1, and 10, 15 and 20 MHz in Band 40.[23] A maximum of two aggregated carriers are supported initially.

For inter-band carrier aggregation, the first defined scenarios are likely to include:

- E-UTRA Bands 1 and 5 for 10 MHz CCs, one CC per band;
- E-UTRA Bands 3 and 7 for 10, 15, and 20 MHz CCs, one CC per band;
- E-UTRA Bands 4 and 13 for 10 MHz CCs, one CC per band;
- E-UTRA Bands 4 and 17 for 10 MHz CCs, one CC per band.

Other combinations will be added in a release-independent manner, so that they can be implemented by UEs of any release from Release 10 onwards.

28.5 Summary

There are three main motivations for introducing carrier aggregation for LTE-Advanced in Release 10: support of high data rates, efficient utilization of fragmented spectrum, and support of heterogeneous network deployments by means of cross-carrier scheduling. In addition, enhancements to the uplink transmission scheme are included, allowing clustered DFT-S-OFDM transmission with non-contiguous resource allocation within one CC, and simultaneous PUCCH/PUSCH transmission.

References[24]

[1] 3GPP Technical Report 36.913, 'Requirements for further advancements for Evolved Universal Terrestrial Radio Access (E-UTRA) (LTE-Advanced)', www.3gpp.org.

[2] 3GPP Technical Report 36.815, 'LTE-Advanced Feasibility Studies in RAN WG4 (Release 9)', www.3gpp.org.

[3] 3GPP Technical Specification 36.213, 'Physical Layer Procedures (Release 10)', www.3gpp.org.

[4] 3GPP Technical Report 25.814, 'Physical Layer Aspects for Evolved Universal Terrestrial Radio Access (UTRA) (Release 7)', www.3gpp.org.

[5] 3GPP Technical Specification 36.101, 'Evolved Universal Terrestrial Radio Access (E-UTRA); User Equipment (UE) Radio Transmission and Reception (Release 10)', www.3gpp.org.

[23] See Section 21.2 for the E-UTRA band definitions.

[24] All web sites confirmed 1st March 2011.

29

Multiple Antenna Techniques for LTE-Advanced

Alex Gorokhov, Amir Farajidana, Kapil Bhattad, Xiliang Luo and Stefan Geirhofer

Multiple antenna techniques play a key role in LTE-Advanced. In the downlink, the goals are to support higher data rates than LTE Releases 8 and 9 through high-order Single-User Multiple-Input Multiple-Output (SU-MIMO), and higher spectral efficiency via enhanced Multi-User MIMO (MU-MIMO) techniques. To support these advances, new reference signals and enhanced UE feedback are introduced. In the uplink, SU-MIMO is introduced, and the control channel performance is enhanced using transmit diversity.

This chapter explains the techniques adopted for Release 10 and gives some insight into additional MIMO enhancement features that may be developed in future releases of LTE-Advanced.

29.1 Downlink Reference Signals

As discussed in detail in Section 8.2, LTE Release 8 provides cell-specific Reference Signals (RSs), also known as Common Reference Signals (CRSs) for up to 4 antenna ports. Cell-specific RSs are used by UEs both to perform channel estimation for demodulation of data and to derive feedback on the quality and spatial properties of the downlink radio channel. Together with signalling of the precoder used for data transmissions to the UE, the four cell-specific RS ports enable spatial multiplexing of up to four layers using codebook-based precoding, as explained in Section 11.2.2.2. LTE Release 9 additionally supports two-layer beamforming spatial multiplexing using precoded UE-specific RSs, which enable non-codebook-based precoding to be used, as explained in Section 11.2.2.3.

LTE – The UMTS Long Term Evolution: From Theory to Practice, Second Edition.
Stefania Sesia, Issam Toufik and Matthew Baker.

For LTE-Advanced, downlink SU-MIMO transmission is extended to support up to eight spatial layers, and for this purpose the precoded UE-specific RS approach is further developed for the data demodulation. This has the advantage that transmission of the new Release 10 UE-specific RSs can be limited to only those Resource Blocks (RBs) where they are needed for demodulation, thus avoiding incurring a large overhead across the whole system bandwidth (as would have been the case with an extension to the cell-specific RSs, which would also have adversely impacted 'legacy' pre-Release 10 UEs that are unaware of the new RS structure).

In addition, in order to enable the UE to estimate and feed back the Channel State Information (CSI) corresponding to up to eight antenna ports across a wide bandwidth, new RSs referred to as CSI-RSs are provided. Since CSI-RSs are used only for feedback purposes they can be sparse and incur only a small overhead.

These two new types of RS are explained in detail in the following subsections.

29.1.1 Downlink Reference Signals for Demodulation

In the same way as the UE-specific RSs of earlier releases (see Sections 8.2.3 and 8.2.3), the extended UE-specific RSs of LTE-Advanced are embedded in the RBs used for the Physical Downlink Shared CHannel (PDSCH) for a specific UE. The UE-specific RSs for each layer undergo the same precoding as the data symbols, and therefore there is no need for explicit signalling of precoding information to the UE. A variety of multi-antenna beamforming techniques can therefore be supported efficiently and transparently to the UE.

The pattern of Resource Elements (REs) designed for the new UE-specific RSs had to satisfy certain criteria. Firstly, it had to avoid overlapping with the cell-specific RSs and the control channels in order to ensure backward compatibility. Secondly, the UE-specific RSs of different layers should be orthogonally multiplexed to avoid inter-layer RS interference degrading the channel estimation accuracy; this could in principle be achieved by assigning different sets of REs to the RSs of different layers (by Frequency-Division and/or Time-Division Multiplexing FDM/TDM), and/or by Code Division Multiplexing (CDM). CDM has the advantages of more flexible power balancing across different layers and potentially better interference estimation accuracy in MU-MIMO scenarios.

For Release 10, the UE-specific RS pattern for up to 2 layers (referred to as 'rank-2' transmission) is identical to that defined in Release 9, in order to ensure backward compatibility. For up to 4 layers, the pattern is obtained by extending the Release 9 rank-2 UE-specific RS pattern in a hybrid CDM/FDM fashion, as shown in Figure 29.1(a) (for the case of the normal Cyclic Prefix (CP) length)[1]. The four precoded layers (antenna ports 7–10) are grouped into two groups of 2 REs, with the same length-2 Walsh–Hadamard Orthogonal Cover Codes (OCC) as in Release 9 (i.e. [1, 1] and [1, −1]) being used to multiplex the layers within each group. The UE-specific RSs in different groups are frequency-multiplexed on adjacent subcarriers. This pattern has been shown to provide reasonable performance for a variety of channel conditions and UE speeds [2, 3].

[1]By following the same principles, similar patterns are defined in Release 10 for the case of the extended CP and for the Downlink Pilot TimeSlot (DwPTS) field of the special mixed downlink-uplink subframe in TDD operation – see [1, Section 6.10.3]. Use of the extended CP with more than two UE-specific antenna ports is not supported in Release 10.

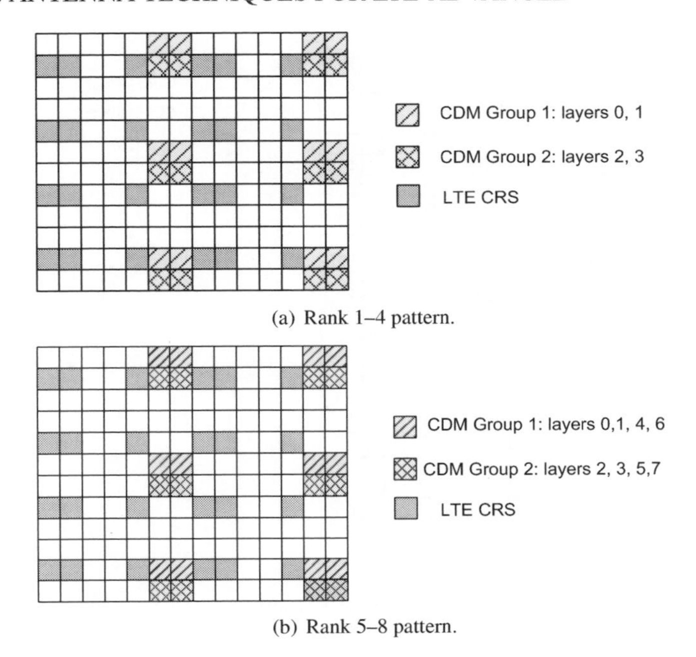

(a) Rank 1–4 pattern.

(b) Rank 5–8 pattern.

Figure 29.1: Release 10 UE-specific RS patterns.

For 8-layer transmission (antenna ports 7–14), the UE-specific RS structure of Figure 29.1(a) is further extended using hybrid CDM/FDM with two CDM groups each using a Walsh–Hadamard OCC of length 4 as shown in Figure 29.1(b). Exactly the same set of REs are used as for the rank-4 case, with the UE-specific RSs for each layer now being spread across four REs, all having the same frequency location but different time locations within the subframe. This approach maintains the power-balancing property of the rank-4 design and is optimized for pedestrian mobility – the most likely scenario for 8-layer SU-MIMO.

The length-2 and length-4 OCCs have a nested structure (the OCC of length two is nested into the OCC of length four) which ensures backward compatibility with the Release 9 design.

The mapping of the OCCs to REs is shown in Figure 29.2 for the case of normal CP. The mapping for odd and even RBs (in the frequency domain) is different in order to provide orthogonality in both time and frequency. Such orthogonality improves the performance of channel estimation in higher Doppler scenarios and reduces the inter-subcarrier interference across the two CDM groups in the presence of frequency offsets. This OCC mapping also enables power balancing across OFDM[2] symbols by peak power randomization [4].

The RS sequence for up to 8 layers is the same as in Release 9 (as described in Section 8.2.3), except that for antenna ports 9–14 only one sequence initialization is possible.

[2]Orthogonal Frequency Division Multiplexing.

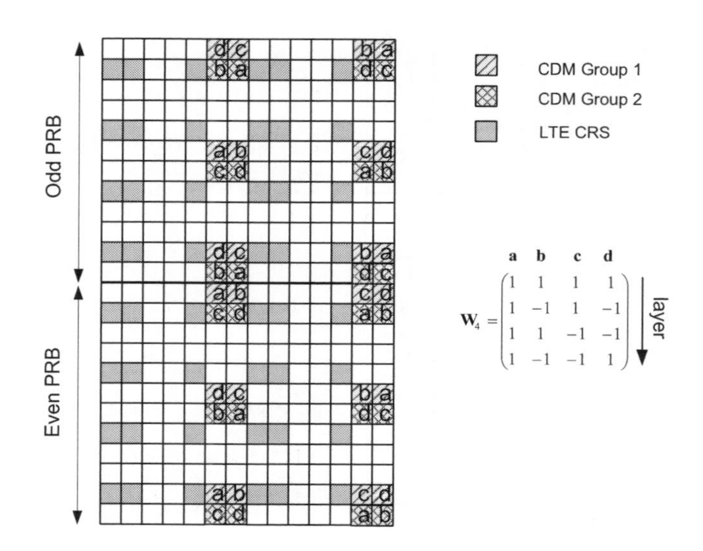

$$\mathbf{W}_4 = \begin{pmatrix} 1 & 1 & 1 & 1 \\ 1 & -1 & 1 & -1 \\ 1 & 1 & -1 & -1 \\ 1 & -1 & -1 & 1 \end{pmatrix}$$

Figure 29.2: Length-4 OCC mapping to REs.

29.1.2 Downlink Reference Signals for Estimation of Channel State Information (CSI-RS)

The main goal of CSI-RSs is to obtain channel state feedback for up to eight transmit antenna ports to assist the eNodeB in its precoding operations. Release 10 supports transmission of CSI-RS for 1, 2, 4 and 8 transmit antenna ports. CSI-RSs also enable the UE to estimate the CSI for multiple cells rather than just its serving cell, to support future multicell cooperative transmission schemes (see Section 29.5.1).

The following general design principles can be identified for the CSI-RS:

- In the frequency domain, uniform spacing of CSI-RS locations is highly desirable, as explained in Chapter 8.
- In the time domain, it is desirable to minimize the number of subframes containing CSI-RS, so that a UE can estimate the CSI for different antenna ports and even different cells with a minimal wake-up duty cycle when the UE is in Discontinuous Reception (DRX) mode, to preserve battery life.
- The overall CSI-RS overhead involves a trade-off between accurate CSI estimation for efficient operation and minimizing the impact on legacy pre-Release 10 UEs which are unaware of the presence of CSI-RS and whose data are punctured by the CSI-RS transmissions. Figure 29.3 shows that a CSI-RS density of one RE per RB per antenna port is a good choice, as the throughput degradation compared to ideal CSI estimation is negligible.
- CSI-RSs of different antenna ports within a cell, and, as far as possible, from different cells, should be orthogonally multiplexed to enable accurate CSI estimation.

- To ensure backward compatibility, CSI-RSs should avoid REs used for cell-specific RSs and control channels, as well as avoiding REs used for the Release 10 UE-specific RSs.

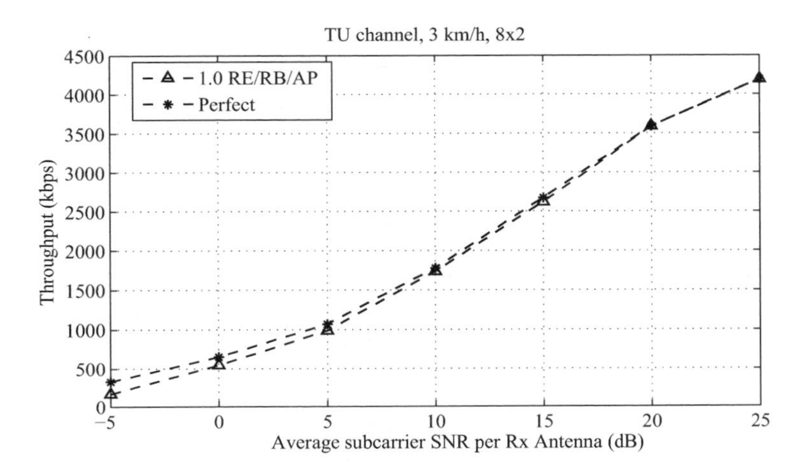

Figure 29.3: Throughput performance with a CSI-RS density of 1 RE per RB per antenna port: uncorrelated 8×2 SU-MIMO, 5 MHz, Typical Urban channel model, 3 km/h.

Taking these considerations into account, the CSI-RS patterns selected for Release 10 are shown in Figure 29.4. CDM codes of length 2 are used, so that CSI-RSs on two antenna ports share two REs on a given subcarrier.

The pattern shown in Figure 29.4(a) can be used in both frame structure 1 (FDD)[3] and frame structure 2 (TDD).[4] In Figure 29.4, the REs used for CSI-RSs are labelled using two letters, the first indicating the cell index and the second referring to the antenna ports of the CSI-RS transmitted on that RE. These patterns follow a 'nested' structure, meaning that the REs used in the case of two CSI-RS antenna ports are a subset of those used for four and eight antenna ports; this helps to simplify the implementation. The total number of supported antenna ports is 40, which can be used to give a frequency-reuse factor of 5 between cells with 8 antenna ports per cell, or a factor of 20 in the case of 2 antenna ports. It can be seen that collisions may occur with REs used for the UE-specific RS antenna ports defined in Release 8 for PDSCH transmission mode 7 (see Section 8.2.2); it may therefore be desirable to avoid scheduling UEs in transmission mode 7 in subframes containing CSI-RS.

The pattern shown in Figure 29.4(b) can only be used for frame structure 2 for TDD operation. This pattern is designed to avoid collisions with the Release 8 UE-specific RS antenna port, as this port is more suited to TDD operation where channel reciprocity can be more effectively exploited to support the beamforming. However, this pattern only offers frequency reuse factors of between 3 and 12 depending on the number of antenna ports per cell.

[3]Frequency Division Duplex.
[4]Time Division Duplex.

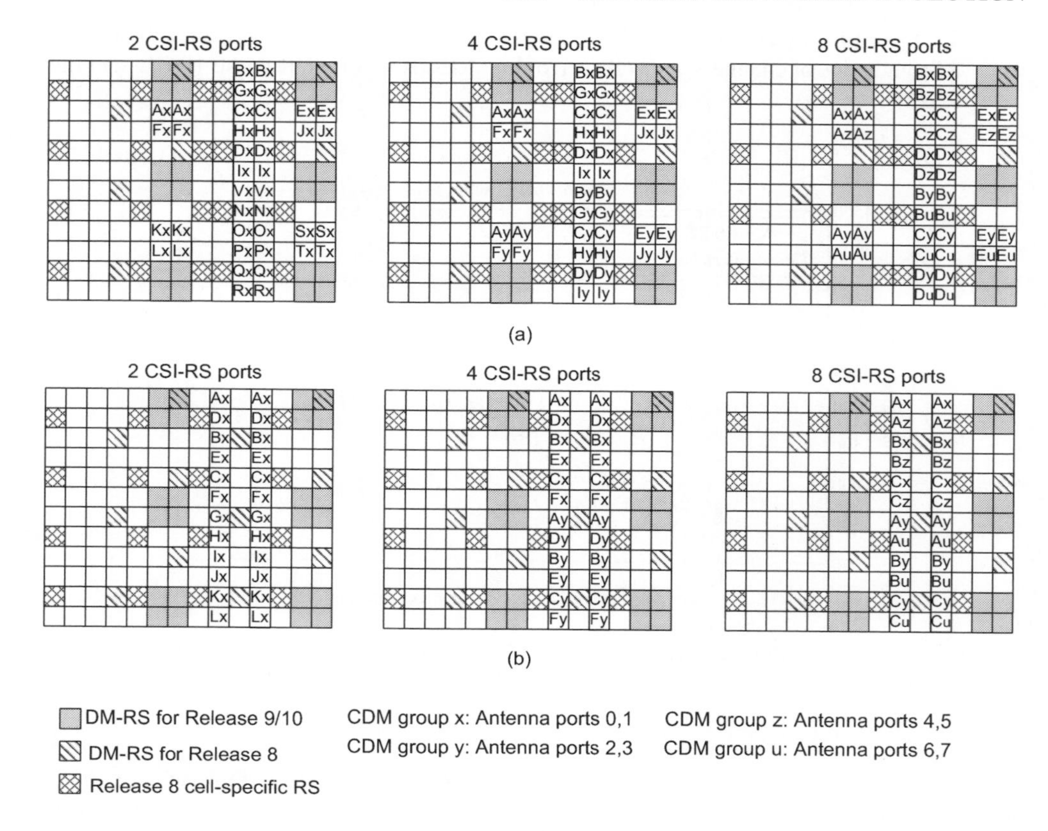

Figure 29.4: CSI-RS patterns for LTE-Advanced.

Similar CSI-RS patterns are provided for the case of the extended CP and the DwPTS field of the special mixed downlink-uplink subframe in frame structure 2 (see [1, Section 6.10.5]).

The CSI-RS configuration is UE-specific. When configured, CSI-RSs are present only in some specific subframes following a given duty cycle and subframe offset. The duty cycle and offset of the subframes containing CSI-RSs and the CSI-RS pattern used in those subframes are provided to a Release 10 UE through RRC signalling. The duty cycle and subframe offset are jointly coded, while the CSI-RS pattern is configured independently of these two parameters.

It should be noted that, in subframes containing CSI-RSs, rate-matching (see Section 10.3.2.4) for PDSCH transmissions to Release 10 UEs assumes that the CSI-RS REs are not available for PDSCH data, and the coded PDSCH data is only mapped to the surrounding REs. However, PDSCH transmissions to Release 8 and 9 UEs are punctured by CSI-RS transmissions. The density and periodicity of the CSI-RSs should therefore be chosen such that the impact of puncturing on the performance of these users is acceptable.

In the context of cooperative MIMO, it may be possible to improve the performance of channel estimation, and especially interference estimation, by coordinating CSI-RS transmissions across multiple cells. In Release 10 it is therefore possible to 'mute' a specific set of REs in data transmissions from a cell. The locations of these REs, known as the 'muting pattern', can be chosen to avoid colliding with CSI-RS transmissions from other cells and hence improve the inter-cell measurement quality. The muting pattern is indicated to the UEs by a 16-bit bitmap, where each bit corresponds to a 4-port CSI-RS configuration for frame structure 1 or 2 as shown in Figure 29.4(a). All the REs in an indicated 4-port CSI-RS configuration are muted, and the UE can assume zero transmission power on those REs unless they are used to transmit CSI-RSs.

In subframes in which muting is configured, data transmissions for Release 10 UEs are rate-matched around the muted REs in the same way as for CSI-RSs. Data transmitted to Release 8 and 9 UEs is punctured in the muted REs.[5]

29.2 Uplink Reference Signals

In Chapter 15, the LTE Release 8 uplink RS design is explained, including DeModulation RSs (DM-RSs) and Sounding RSs (SRSs). In Release 10, the uplink DM-RSs are extended to support uplink SU-MIMO transmission with up to four spatial layers, and the SRSs are enhanced to provide improved support for channel sounding.

29.2.1 Uplink DeModulation Reference Signals (DM-RS)

In order to support uplink SU-MIMO transmission with up to 4 spatial layers, DM-RSs for all spatial layers need to be multiplexed together to enable channel estimation of each layer at the eNodeB. Like the downlink UE-specific RSs, the uplink DM-RSs are precoded using the same precoding as the Physical Uplink Shared CHannel (PUSCH) data transmissions in the same RBs when multiple antenna transmission is used.

Release 10 uses different Cyclic Shifts (CSs) of the DM-RS base sequence to multiplex the DM-RSs for the different layers. For example, in a rank-4 SU-MIMO transmission, CS offsets 0, 6, 3 and 9 are used relative to the base sequence, for layers 1, 2, 3 and 4 respectively. In addition to separation of the antenna ports by selection of different CSs, two length-2 OCCs can be applied to the two DM-RS symbols in the two slots of one subframe to further separate the multiplexed DM-RSs. The codes used are [+1, +1] and [+1 − 1]; they enable better orthogonality to be achieved between DM-RSs that use close CSs. Figure 29.5 shows an example of the link-level throughput performance improvement that can be achieved thanks to the better orthogonality provided by the use of the OCCs in addition to the CS separation.

Different OCCs can also be assigned to different UEs for uplink MU-MIMO transmission. In this way, the DM-RSs from different UEs can be kept orthogonal to each other even if the PUSCH transmission bandwidths differ between the UEs (provided that there is no sequence group hopping (see Section 15.3.1) between the two slots within one subframe).

In order to reduce the control signalling overhead and make the benefits of the OCCs available to both SU-MIMO and MU-MIMO UEs, the identities of the OCCs applied to each spatial layer are implicitly derived from the 3-bit cyclic shift index signalled in the corresponding PDCCH uplink grants (see Section 9.3.5.1). The cyclic shift indices are divided into two groups: one group of 4 cyclic shift indices is targeted for SU-MIMO, and

[5] See [5, 6] for more background.

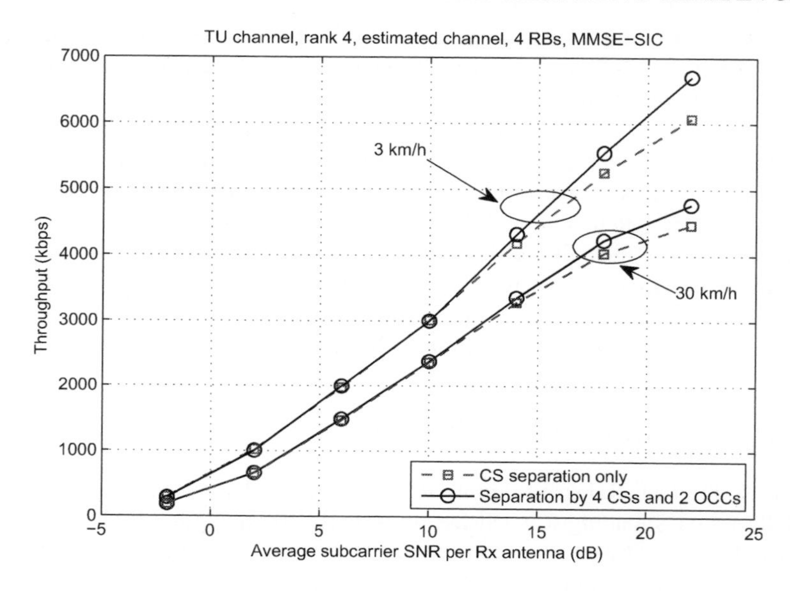

Figure 29.5: The benefit of the uplink DM-RS OCCs in terms of link throughput.

the layers with close CSs are assigned different OCCs; the other group is optimized for MU-MIMO, and the same OCC is applied to the different layers.

29.2.2 Sounding Reference Signals (SRSs)

The SRSs are important for uplink channel sounding to support dynamic uplink resource allocation, as well as for reciprocity-aided beamforming in the downlink, especially in TDD operation.

As explained in Section 15.6, the Release 8 SRSs allow for uplink sounding from a single transmit antenna only. With the introduction of uplink multiple-antenna transmission in Release 10, SRS transmission from all the uplink antenna ports is needed in order to enable all the spatial dimensions of the channel to be sounded. The same mechanisms as are available in Release 8 are used to separate the SRS transmissions from different antenna ports – i.e. different CSs and different transmission comb indices of the IFDMA[6] structure.

Nevertheless, the need for SRSs from up to four antenna ports increases the demand on SRS resources if all transmit antennas are to be sounded at a reasonable rate, and therefore semi-static configuration of the resources via higher layer (RRC) signalling is less appropriate than it was for Release 8. Therefore, Release 10 introduces the possibility of dynamically triggering aperiodic SRS transmissions via the Physical Downlink Control CHannel (PDCCH); these dynamic aperiodic SRS transmissions are known as 'type 1' SRSs, while the Release 8 RRC-configured SRSs described in Section 15.6 are known as 'type 0' in Release 10.

[6]Interleaved Frequency Division Multiple Access.

An indicator in an uplink resource grant on the PDCCH can be used to trigger a single type 1 SRS transmission. This facilitates rapid channel sounding to respond to changes in traffic or channel conditions, without tying up SRS resources for a long period. In DCI Format 0, one new bit can indicate activation of a type 1 SRS[7] according to a set of parameters that is configured beforehand by RRC signalling. In DCI Format 4, which is used for scheduling uplink SU-MIMO transmissions, two new bits allow one of three sets of RRC-configured type 1 SRS transmission parameters to be triggered (the remaining state indicates no type 1 SRS activation).

In the case of uplink multiple-antenna transmission (see Section 29.4), SRSs may be triggered on multiple antenna ports simultaneously.

Other enhancements to the SRSs, including coordination of SRS transmission resource configurations and frequency-hopping patterns, may also be considered for cooperative MIMO schemes and inter-cell interference coordination in the future (see, for example, [7] and references therein).

29.3 Downlink MIMO Enhancements

The main enhancements to the downlink multiple antenna transmission schemes in Release 10 are the extension of SU-MIMO to support 8-layer transmission, and improved support for MU-MIMO. These are introduced in a new PDSCH transmission mode, known as Transmission Mode 9.

29.3.1 Downlink 8-Antenna Transmission

The new RSs to support transmissions from 8 antenna ports in Release 10 are explained in Section 29.1.

For the data transmissions from 8 antenna ports, the Release 8/9 transmit diversity and spatial multiplexing modes are extended. For transmit diversity, no new schemes are specified in Release 10. Instead, if an eNodeB has more than four antennas and wishes to use transmit diversity, the Release 8 SFBC/FSTD[8] schemes are re-used via 'antenna virtualization'; this means that multiple antennas are treated as a single port by the application of suitable precoding, with a single Release 8 cell-specific RS being transmitted from all the antennas comprising the antenna port. The antenna virtualization is thus transparent to the UE.

For 8-layer spatial multiplexing, two aspects of the codeword-to-layer mapping needed to be considered, namely the number of codewords and the mapping of the codewords to layers. An increase in the number of codewords from the two supported in Release 8 could in theory have provided a throughput gain for some receivers such as those based on Minimum Mean Squared Error Serial Interference Cancellation (MMSE-SIC), but on the other hand a larger signalling overhead would have been incurred in both downlink and uplink, and the complexity of link adaptation would have been increased. Figure 29.6 shows the throughput performance of an 8×8 SU-MIMO transmission scheme with different numbers of codewords for a SIC receiver and a Linear MMSE (LMMSE) receiver. It can be seen that the performance gain from increasing the number of codewords beyond two is small even for the MMSE-SIC receiver.

[7]Only applicable if the DCI message is sent in the UE-specific search space – see Section 9.3.5.5.

[8]Space-Frequency Block Code / Frequency Switched Transmit Diversity (SFBC/FSTD).

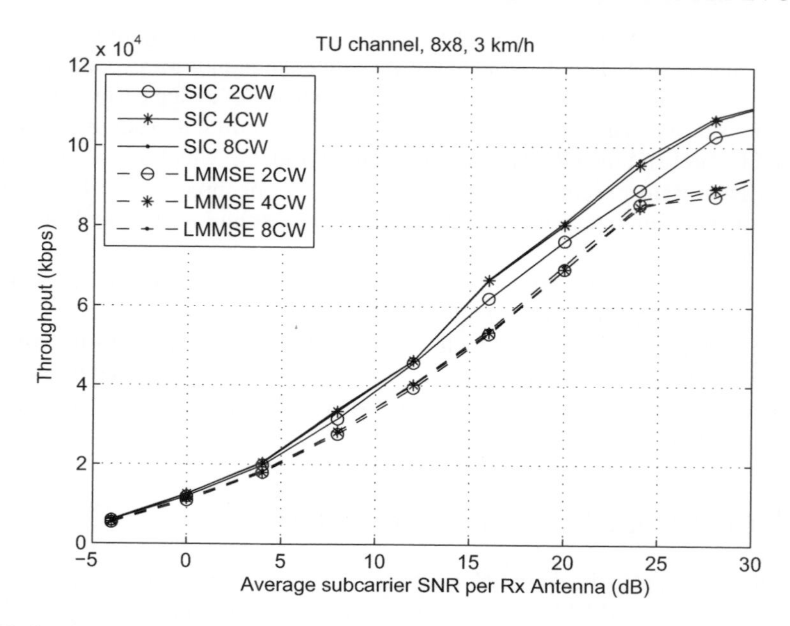

Figure 29.6: Throughput performance of LMMSE and MMSE-SIC receivers with different numbers of codewords.

Therefore Release 10 continues to use a maximum of two codewords even in the case of eight antenna ports. The codeword-to-layer mapping for 5 to 8 layers is illustrated in Figure 29.7. Note that, for up to 4 antenna ports, the mapping is the same as in Release 8.

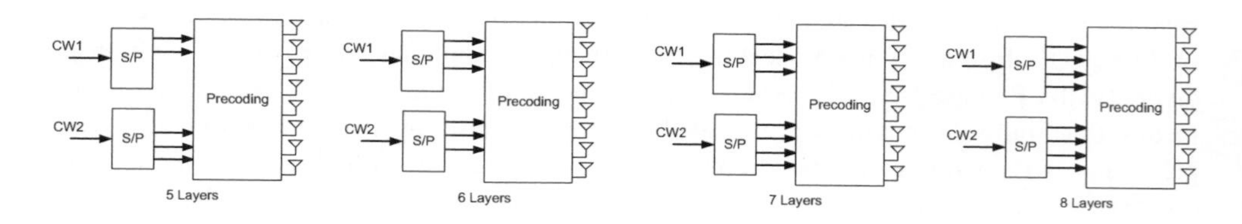

Figure 29.7: Codeword-to-layer mapping for up to 8 antenna ports.

As noted in Section 29.1, the Release 10 high-rank PDSCH transmissions use UE-specific RSs. This means that they are not constrained to being transmitted in subframes where the Release 8 cell-specific RSs are transmitted, and PDSCH transmission mode 9 in Release 10 therefore allows the PDSCH to be transmitted in the data region of 'MBSFN'[9] subframes, provided that those subframes are not actually used for MBMS[10] transmissions.

The use of non-codebook-based precoding for the PDSCH means that the eNodeB has a substantial degree of freedom in determining the actual precoding to apply for each UE

[9]Multimedia Broadcast Single Frequency Network.

[10]Multimedia Broadcast/Multicast Service.

and for each RB. However, in some cases the eNodeB may use the same precoding matrix across several RBs. For example, when a UE is configured to feed back sub-band Precoding Matrix Indicators (PMIs), the eNodeB is quite likely to use the same precoding matrix across the RBs in a sub-band, in the absence of other relevant information. In this case, the UE can improve its channel estimation quality if it has information about the RBs on which the same precoding matrix is applied. Therefore, in Release 10 it is specified that when the UE is configured to feed back PMI, it can assume that the precoder is the same across all the RBs within a *Precoding Resource block Group* (PRG). In resource allocation type 0 (see Section 9.3.5.4), the data allocation to the UE is performed in groups of contiguous RBs called Resource Block Groups (RBG); in this case, the PRG size is chosen such that it is a factor of the RBG size, in order to be consistent with the resource allocations to individual UEs. The sizes of the PMI sub-bands, RBGs and PRGs for different bandwidths are shown in Table 29.1.

Table 29.1: PRG sizes.

System bandwidth	Sub-band size	RBG size	PRG size
<10	4	1	1
11–26	4	2	2
27–63	6	3	3
64–110	8	4	2

29.3.2 Enhanced Downlink Multi-User MIMO

MU-MIMO was introduced in LTE Releases 8 and 9 as explained in Section 11.1.4. Further enhancements to MU-MIMO operation are included in PDSCH transmission mode 9 in Release 10.

Like PDSCH transmission mode 8 in Release 9, transmission mode 9 in Release 10 supports dynamic switching between SU-MIMO and MU-MIMO. The provision of UE-specific RS patterns with rank greater than one enables the scheduling of more than one spatial stream to UEs participating in MU-MIMO. Cell-specific scrambling of the UE-specific RS sequences, as well as orthogonal multiplexing of UE-specific RS ports assigned to co-scheduled UEs, enables accurate channel estimation and improved interference nulling at the UE.

Although Release 10 supports orthogonal multiplexing of up to 8 layers, the maximum intended multi-user multiplexing order is 4 in Release 10. The gains, if any, of supporting more than 4 layers with MU-MIMO are expected to be small since the channel estimation quality is expected to drastically decrease, and feedback inaccuracies make efficient selection of UEs for co-scheduling difficult.

Another salient aspect of MU-MIMO that was introduced in LTE Release 9 and is carried over into LTE-Advanced is the transparency of MU-MIMO operation from the point of view of the UE. Specifically, no signalling is provided to inform UEs of the presence of co-scheduled transmissions, nor of any parameters of such transmissions (such as rank, antenna port, RB allocation or power offset); in particular, the use of UE-specific RSs whose amplitude can be varied along with the data symbols means that the Release 8 signalling of a power offset to indicate that the eNodeB's transmit power in an RB is divided between multiple UEs is no longer necessary. Control signalling is limited to indicating the set of UE-specific RS ports allocated to the UE itself. As in Release 9, the 4-layer multiplexing is achieved by using two scrambling sequences and the rank-2 UE-specific RS pattern.

A new DCI format (Format 2C) is introduced in PDSCH transmission mode 9 to support dynamic switching between SU-MIMO transmission (up to rank 8) and MU-MIMO transmission (up to rank 4). DCI Format 2C is based on Format 2B with the addition of 2 new bits; also, the SCrambling IDentity (SCID) bit of Format 2B is reused to encode jointly the number of layers, UE-specific RS SCID and antennna port mapping. The encoding is shown in Table 29.2.[11]

Table 29.2: Encoding of antenna ports, layers and SCID in DCI Format 2C.

One codeword: Codeword 0 enabled, Codeword 1 disabled		Two codewords: Both codewords enabled	
State	Message	State	Message
0	1 layer, port 7, SCID=0	0	2 layers, ports 7–8, SCID=0
1	1 layer, port 7, SCID=1	1	2 layers, ports 7–8, SCID=1
2	1 layer, port 8, SCID=0	2	3 layers, ports 7–9
3	1 layer, port 8, SCID=1	3	4 layers, ports 7–10
4	2 layers, ports 7–8	4	5 layers, ports 7–11
5	3 layers, ports 7–9	5	6 layers, ports 7–12
6	4 layers, ports 7–10	6	7 layers, ports 7–13
7	Reserved	7	8 layers, ports 7–14

It is also well known that the system capacity of MU-MIMO is highly dependent on the accuracy of the CSI feedback from the UEs. Therefore, improved feedback is also a key factor for improved MU-MIMO performance; this is discussed in the next section.

29.3.3 Enhanced CSI Feedback

In Release 10, the same CQI/PMI/RI[12] feedback types are used as in Release 8. As explained in Section 11.2.3, this is often described as *implicit* feedback, as it provides an implicit representation of the channel consisting of an indication of the data rate that could be achieved if the eNodeB used a certain precoder.

[11] Note that transmissions with more than one layer but a single codeword only occur in the case of retransmissions of SU-MIMO transmissions. Hence, for transmission with one codeword and 2 layers, only one scrambling sequence is used.

[12] Channel Quality Indicator / Precoding Matrix Indicator / Rank Indicator.

The concept of implicit (i.e. recommended precoder) feedback can be compared to what is sometimes called *explicit feedback* (not supported in LTE or LTE-Advanced), whereby a UE would instead explicitly report a quantized representation of the physical CSI without making assumptions about the nature of the eNodeB precoder. Since the capacity achieved by MU-MIMO is very dependent on the accuracy of the CSI at the transmitter, explicit feedback mechanisms are often favoured in theoretical studies. However, the use of implicit feedback for MU-MIMO in LTE does have some practical advantages, including minimizing the implementation effort required to support both SU- and MU-MIMO. Furthermore, the accuracy of a UE's implicit feedback can very easily be tested, by transmitting data to the UE using the UE's recommended precoder and the Modulation and Coding Scheme (MCS) corresponding to the UE's CQI report.

Moreover, it can be argued that, under certain assumptions, implicit and explicit feedback schemes have similar results. As an example, in the case of rank-1 feedback, an explicit short-term CSI report would signal the direction of the highest channel gain, i.e. the strongest eigenvector of the channel matrix, after some form of quantization to reduce the overhead to a reasonable level. On the other hand, with implicit feedback, a UE would select the vector from the precoding codebook that yields the largest Signal-to-Interference-plus-Noise Ratio (SINR), which is necessarily an approximation of the strongest eigenvector, since the latter identifies the direction of maximum beamforming gain of the channel.

29.3.3.1 PMI Feedback Codebook Enhancements

Enhanced spatial feedback in PDSCH transmission mode 9 is available for the case when 8 CSI-RS ports are configured.

The principle of this enhanced feedback is based on an observation that large antenna arrays at the eNodeB typically comprise two cross-polarized subarrays, in which each subarray is a Uniform Linear Array (ULA). A close spacing of the antenna elements (of the order of half a wavelength) gives rise to fairly strong long-term correlation within a subarray, which is seen in particular in many practical scenarios in macrocellular networks where the angular spread at the eNodeB is relatively small. A high long-term correlation within each subarray can be exploited to reduce the rate of feedback required for each subarray, in time and/or frequency.

In general, the relatively persistent nature of the spatial signatures of the subarrays facilitates robust MU-MIMO scheduling with predictable multi-user interference. Conversely, the low correlation between the cross-polarized subarrays yields improved diversity for SU-MIMO scheduling. Following these principles, the PMI feedback codebooks for 8 CSI-RS antenna ports have been designed such that precoders within each 4-element ULA subarray are represented by a grid of beams, with the number of available beams from which to choose being 32 when 1- or 2-layer transmission is preferred, 16 for 3 or 4 layers, 4 for 5 to 7 layers and only one for 8 layers. Cross-polarized dimensions (corresponding to low correlation) are combined with *co-phasing factors* represented by quaternary symbols, following the same codebook design principle as in Release 8 for uncorrelated antennas.

An example of the resulting Release 10 codebook design for the case of 2-layer transmission being preferred is shown in Table 29.3.3.1. Full details of the PMI feedback codebooks for 8 CSI-RS antenna ports when other transmission ranks are preferred can be found in [1, Section 6.3.4.2.3].

Table 29.3: PMI feedback codebook design for the case of 2-layer feedback with 8 CSI-RS antenna ports.

i_2	0	1	2	3
Codebook	$W_{2i_1,2i_1,0}$	$W_{2i_1,2i_1,1}$	$W_{2i_1+1,2i_1+1,0}$	$W_{2i_1+1,2i_1+1,1}$

i_2	4	5	6	7
Codebook	$W_{2i_1+2,2i_1+2,0}$	$W_{2i_1+2,2i_1+2,1}$	$W_{2i_1+3,2i_1+3,0}$	$W_{2i_1+3,2i_1+3,1}$

i_2	8	9	10	11
Codebook	$W_{2i_1,2i_1+1,0}$	$W_{2i_1,2i_1+1,1}$	$W_{2i_1+1,2i_1+2,0}$	$W_{2i_1+1,2i_1+2,1}$

i_2	12	13	14	15
Codebook	$W_{2i_1,2i_1+3,0}$	$W_{2i_1,2i_1+3,1}$	$W_{2i_1+1,2i_1+3,0}$	$W_{2i_1+1,2i_1+3,1}$

$$W_{m,k,n} = \frac{1}{4}\begin{bmatrix} \mathbf{v}_m & \mathbf{v}_k \\ e^{j\pi n/2}\mathbf{v}_m & -e^{j\pi n/2}\mathbf{v}_k \end{bmatrix}, \quad \mathbf{v}_m = \begin{bmatrix} 1 & e^{j2\pi m/32} & e^{j4\pi m/32} & e^{j6\pi m/32} \end{bmatrix}^T.$$

The index $0 \leq i_1 \leq 15$ selects a cluster of beams from a set of partially overlapping clusters in a co-polarized subarray. The LSB of the index $0 \leq i_2 \leq 15$ selects one of the two co-phasing factors while the 3 MSBs of i_2 select one of 8 pairs of beams within a cluster.

This codebook structure can be viewed as a two-stage codebook, in which the precoding matrix $\mathbf{W}$ corresponding to the UE's CSI feedback can be decomposed into a product of two matrices, such that $\mathbf{W} = \mathbf{W}_1 \cdot \mathbf{W}_2$, where $\mathbf{W}_1$ corresponds to the ULA grid-of-beams component of the PMI and $\mathbf{W}_2$ corresponds to the frequency-selective co-phasing component of the PMI. The codebooks for PMI feedback when less than 8 CSI-RS antenna ports are configured, and indeed the Release 8/9 codebooks, can be considered to have the same structure with $\mathbf{W}_1$ being set to the identity matrix.

29.3.3.2 Periodic and Aperiodic CSI Reporting Enhancements

Aperiodic CSI feedback on the Physical Uplink Shared Channel (PUSCH) as described in Section 10.2.1.1 follows the same design in Release 10 as in Release 8, except for the use of the new PMI codebooks described above if 8 CSI-RS ports are configured. However, the periodic feedback on the PUCCH (Physical Uplink Control CHannel) described in Section 10.2.1.2 is extended in Release 10 to take advantage of the new two-stage PMI feedback codebook structure explained above.

As spatial signatures within a co-polarized subarray are not expected to change much in the frequency domain, sub-band PMI feedback for this component (referred to as $\mathbf{W}_1$ above) was not considered to be necessary. On the other hand, sub-band reporting can be useful for the spatial component that captures co-phasing across polarizations (referred to as $\mathbf{W}_2$).

These, together with the aim of keeping the uplink overhead as low as possible, led to the following PUCCH feedback modes being supported for the case of 8 CSI-RS ports:

Wideband periodic feedback. The RI/PMI/CQI information is carried in two separate types of PUCCH feedback report: the first contains the RI, while the second contains associated PMI and CQI information. Two alternative modes of operation exist in this case, known as submode 1 and submode 2, which are selected by means of an RRC signalling parameter 'PUCCH_format1-1_CSI_reporting_mode'. In submode 1, the wideband grid-of-beams component of the PMI is included in the first feedback report together with the RI. In submode 2, all of the PMI information (i.e. including both the grid-of-beams component and the sub-array co-phasing component) is contained in the second PUCCH report; in order to achieve this, the PMI feedback codebook has to be subsampled (i.e. only certain specified entries are available from which the UE may select) so that the second report can fit within the available number of bits on the PUCCH. Submode 2 minimizes the code rate for the RI, whereas submode 1 avoids the need for subsampling of the PMI codebook.

Sub-band periodic feedback. This mode allows the fact that the grid-of-beams component of the feedback may be relatively time-invariant [8] to be exploited, by reducing the feedback rate for that component and thereby making some feedback bits available for sub-band reporting of the co-phasing component. The RI/PMI/CQI information is carried in three separate types of PUCCH feedback report. A one-bit Precoder Type Indication (PTI) report is added to the RI reports so that the UE can indicate the contents of the second and third reports. If the PTI is set to '0', the second report contains the grid-of-beams PMI component and the third report contains the wideband co-phasing component of the PMI together with the associated CQI; on the other hand, if the UE estimates that the grid-of-beams component has not changed since the previous time it was reported, the PTI is set to '1', the second report contains wideband co-phasing PMI feedback together with associated CQI, and the third report contains sub-band co-phasing PMI and associated CQI; multiple such third reports are transmitted, one for each sub-band. An example of this sub-band periodic feedback using the PTI is illustrated in Figure 29.8, showing how the sub-band resolution of the co-phasing feedback is increased if the grid-of-beams state is static.

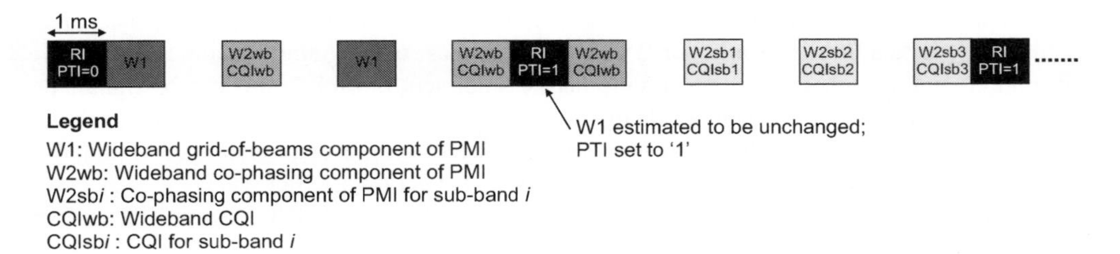

Figure 29.8: Example of a possible sequence of PUCCH CSI feedback reports when 8 CSI-RS ports and sub-band PMI feedback are configured.

29.4 Uplink Multiple Antenna Transmission

In Release 10, SU-MIMO transmission with up to four spatial layers is introduced to increase the data rate for the PUSCH, and the reliability of the control signalling on the PUCCH is increased using transmit diversity.

29.4.1 Uplink SU-MIMO for PUSCH

In order to support uplink SU-MIMO, the concept of transmission modes is introduced for the PUSCH in Release 10. The transmission modes are as follows:

PUSCH Transmission Mode 1: Transmission from a single antenna port;[13]

PUSCH Transmission Mode 2: Transmission from multiple antenna ports; within this mode, the UE can be configured to transmit from either 2 or 4 antenna ports.

If a UE is configured in PUSCH transmission mode 2, it can transmit up to two transport blocks per subframe.[14] The transport blocks are mapped onto one or more layers according to the same rules as are used in the downlink in Release 8, as shown in Table 11.2.

Each transport block is independently acknowledged using the Physical HARQ Indicator CHannel (PHICH). The PHICH index for the first codeword is the same as is used in Release 8 for single-codeword PUSCH transmission (i.e. associated with the lowest PRB index of the PUSCH resource allocation); the PHICH index for the second codeword is related to the first by an offset.

The PDCCH resource grant for uplink MIMO transmissions includes two independent New Data Indicators (NDIs) to indicate whether a retransmission is expected.

Closed-loop codebook-based precoding is used for the PUSCH in a very similar way to the Release 8 PUSCH in transmission mode 4 (see Section 11.2.2.2). The UE is instructed by the eNodeB as to which rank and precoding matrix to use, by means of a dynamic precoding matrix indicator transmitted in the uplink grant on the PDCCH.

In designing the codebooks for uplink SU-MIMO precoding, the primary considerations were preserving the single-carrier waveform at each of the transmit antennas and achieving as much precoding gain as possible. In order to comply with the former, there can be at most one non-zero entry in each row of the precoding matrix. For the rank-1 codebook, some vectors are introduced to turn off certain transmit antennas to save power.

The number of codewords in the codebook is a trade-off between performance and signalling overhead. As a result, for 2 transmit antennas, 6 precoding vectors are defined for rank-1, and a single identity precoding matrix is defined for rank-2; this can be signalled using 3 bits. For 4 transmit antennas, there are 24 precoding vectors for rank-1, 16 for rank-2, 12 for rank-3, and a single identity matrix for rank-4, requiring 6-bit signalling. Full details of the PUSCH precoding codebooks can be found in [1, Section 5.3.3A].

In view of the performance of closed-loop precoding for the PUSCH, transmit diversity is not considered useful. In Figure 29.9, the throughputs using Space-Time Block Coding (STBC) and long-term closed-loop precoding are compared at 3 km/h and 120 km/h under different transmit antenna correlations. Even with uncorrelated transmit antennas, STBC only

[13]Two configurations exist within this mode: one is the Release 8 PUSCH transmission scheme, while the other supports both contiguous and non-contiguous resource allocation (see Section 28.3.6.2).

[14]Per component carrier if carrier aggregation is employed – see Chapter 28.

shows better link throughput at high SINRs, corresponding to a BLock Error Rate (BLER) below 10% (which is an unlikely operating point). Furthermore, as the transmit antennas become more correlated, long-term precoding exhibits more beamforming gain.

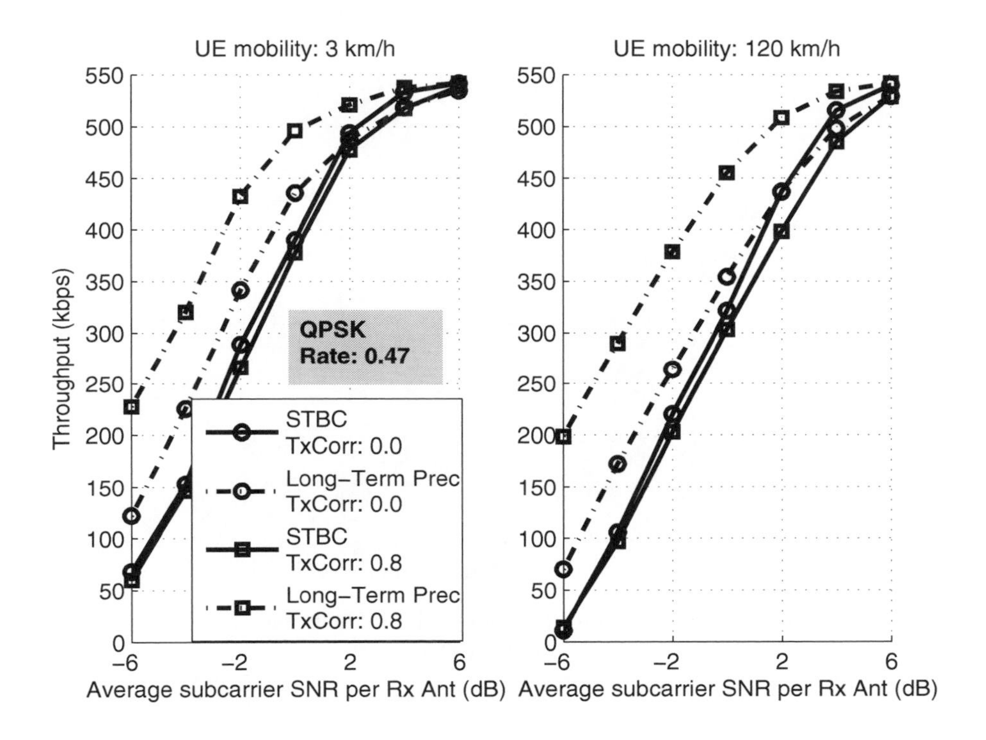

Figure 29.9: Link throughput performance of STBC vs long-term precoding [9].

If uplink control signalling is carried on the PUSCH[15] in a subframe when two codewords are transmitted, any Acknowledgement/Negative Acknowledgement (ACK/NACK) and RI are replicated across all layers of both codewords before channel coding, with the coded modulation symbols for ACK/NACK and RI being time-domain multiplexed with the data and time-aligned across all layers. CQI and PMI signalling is mapped only to the codeword with the highest MCS as indicated by the initial uplink grant. When the two codewords have the same MCS, codeword 0 is always selected. The same multiplexing and channel interleaving mechanisms are used as in Release 8, with the control information being mapped to the same REs on the two spatial layers if the codeword is transmitted on two layers. In Figure 29.10, the placement of the Uplink Control Information (UCI) is illustrated for the case when the uplink PUSCH transmission is of rank 2 and codeword 0 has the higher MCS.

[15]See Section 16.4.

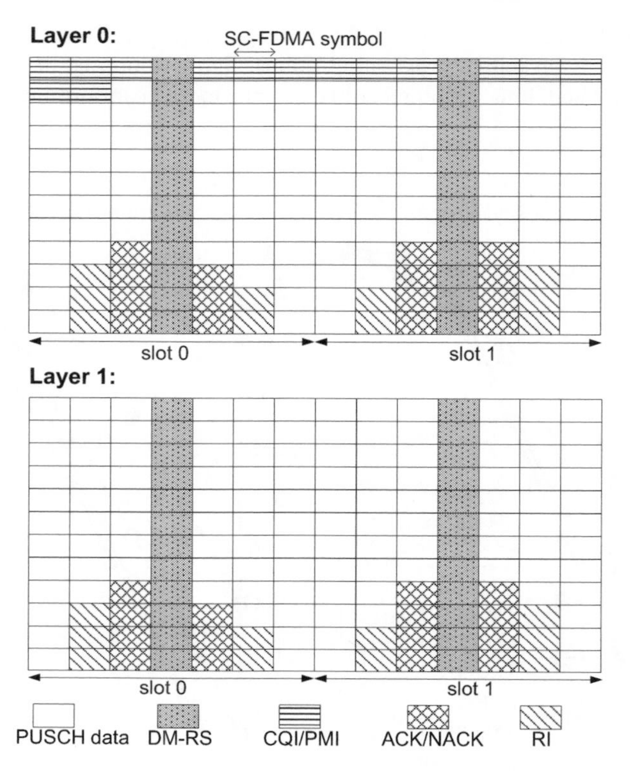

Figure 29.10: Multiplexing of control signalling with PUSCH data in the case of rank-2 SU-MIMO.

29.4.2 Uplink Transmit Diversity for PUCCH

The transmit diversity scheme introduced for the PUCCH in Release 10 is designed to ensure backward compatibility with the Release 8 PUCCH design (see Chapter 16). For PUCCH formats 1/1a/1b (see Section 16.3.2), Space Orthogonal-Resource Transmit Diversity (SORTD) [10] is used, whereby the UE transmits the same control information from different transmit antennas with different orthogonal resources, including cyclic shifts and OCCs. For PUCCH formats 1a/1b, the two orthogonal resources for SORTD transmission are derived from n_{CCE} and $n_{\text{CCE}} + 1$, where n_{CCE} is the number of the first Control Channel Element (CCE) used for the transmission of the corresponding uplink grant on the PDCCH. SORTD is also defined for PUCCH formats 2/2a/2b and 3. Transmit diversity is not supported in Release 10 for PUCCH format 1b with channel selection (see Section 28.3.2.1).

If the transmit antennas are uncorrelated, SORTD provides the best diversity performance, at the expense of consuming multiple orthogonal resources which potentially reduces the PUCCH multiplexing capability.

29.5 Coordinated MultiPoint (CoMP) Transmission and Reception

Coordinated MultiPoint (CoMP) transmission/reception, also known as Cooperative MIMO, has received significant attention in academic literature and is being studied as a technique to increase performance, especially at the cell edge, within the evolution of LTE-Advanced for Release 11 or beyond.

While the scope of CoMP includes both downlink and uplink cooperation, the downlink has received significantly more attention in the literature, primarily due to the more challenging nature of the transmission coordination problem. From the point of view of the air interface design, uplink CoMP basically consists of coordination of eNodeB scheduling and/or receiver processing, and hence the main standardization effort would lie in the definition of appropriate information exchange protocols between eNodeBs if multivendor operation is required. Indeed, uplink CoMP schemes can already be realized in Release 8 by proprietary mechanisms; even in the downlink, basic CoMP schemes can be realized in Release 8 between the cells controlled by a given eNodeB.

In addition to backhaul protocol support, downlink cooperation would require enhancements to the CSI feedback design. In the remainder of this section, the downlink is the primary focus.

29.5.1 Cooperative MIMO Schemes and Scenarios

Various forms of downlink cooperation are illustrated in Figure 29.11. Case (A) corresponds to the traditional uncoordinated single-cell transmission. Case (B) highlights a simple form of cooperation known as *coordinated scheduling* where Cell_B reduces the transmission power on a set of time-frequency resources for the benefit of UE_A and UE_C, served by Cell_A and Cell_C respectively. Note that Inter-Cell Interference Coordination (ICIC) via partial frequency reuse is a form of coordinated scheduling where the coordination is achieved by exchanging transmission power levels used on different time-frequency resources between eNodeBs via the X2 interface (see Section 12.5). Dynamic scheduling coordination is valuable in Home eNodeB (HeNB) Closed Subscriber Group (CSG) deployments where severe interference conditions can arise whenever UE moves into the coverage area of a different CSG, as explained in Chapter 24.

Case (C) illustrates an example of coordinated beamforming in the context of CSG deployments which adds a spatial dimension to ICIC and is applicable in deployments with multi-antenna eNodeBs. Unlike coordinated scheduling, coordinated beamforming allows an interfering HeNB_1 to transmit to UE_1 as long as transmit beams are steered away from UE_2 which is being served by HeNB_2 but located in the coverage area of HeNB_1. Note that opportunistic beamforming [11, 12] may be viewed as a simple form of coordinated beamforming whereby each cell performs beam-swapping according to predefined patterns while cells schedule UEs based on CSI feedback computed according to these patterns. This type of operation is well suited to the presence of a large number of UEs per cell and full buffer traffic. It is worth mentioning, however, that recent studies [13] have shown very limited spectral efficiency improvement due to coordinated beamforming forms of CoMP compared to single cell MU-MIMO in homogeneous macro-cellular deployments, even with full buffer traffic.

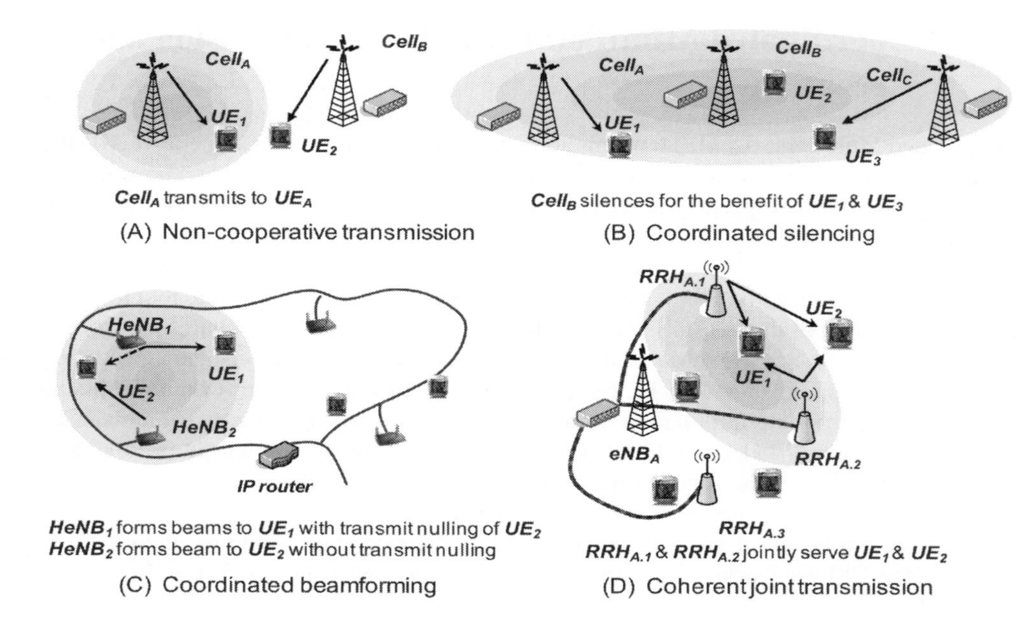

Figure 29.11: Downlink transmission schemes:
(A) non-cooperative transmission; (B) coordinated scheduling;
(C) coordinated beamforming; (D) coherent joint transmission.

Coherent Joint Transmission (JT) (also known as multicell MU-MIMO or 'network MIMO') involves simultaneous transmission of data packets to one or more UEs from multiple cells with cophasing. This form of coordinated transmission requires fast connectivity (low latency) between the transmission points to enable the exchange of control information, as well as high bandwidth backhaul to the eNodeBs involved in data sharing. While general applicability of JT in macrocellular scenarios needs further evolution of the network interfaces, practical scenarios of interest include eNodeBs equipped with multiple Remote Radio Heads (RRHs), as illustrated in case (D) of Figure 29.11, or heterogeneous networks of macro-eNodeBs and picocells. Multiple RRHs or picocells can be placed according to coverage and/or capacity requirements and connected to one or more host eNodeBs via fast broadband links. In such configurations, the processing functionality and resource management may be centralized in a host eNodeB. Deployment of RRHs or picocells can be an attractive way to enhance throughput and coverage, especially in hot-zone and indoor scenarios where JT can play a significant role. From the air interface point of view, RRHs can be treated as cells of the same eNodeB. A key enabler of JT in heterogeneous network scenarios would be multicell channel state feedback from the UE.

While coherent JT across multiple eNodeBs is the ultimate approach towards maximizing spectral efficiency for low-mobility UEs, the practically achievable gains remain to be understood. Most of the studies available to date have been focused on intra-eNodeB JT as well as intra-cluster JT with network-based fixed clustering of a limited number of cells [14]. This kind of limited technique, however, shows limited spectral efficiency gains [15]. On the other hand, theoretical analysis of JT transmission across a large number of eNodeBs suggests potential for substantial gains [16]. Besides the basic requirement of low latency and high throughput backhaul, major practical challenges yet to be addressed include dynamic and distributed multicell multi-user scheduling and transmit processing at the network side, and multicell channel quality prediction, and means for accurate measurement and delivery of the multicell channel state feedback from the UEs.

The greatest benefits in practice are likely to be seen in heterogeneous networks, where the proportion of UEs in interference-limited scenarios is relatively high.

29.6 Summary

As explained in this chapter, Release 10 incorporates some important enhancements to multi-antenna transmission in both uplink and downlink, aimed at substantially increasing peak data rates and system spectral efficiency. This is achieved by the introduction of 4×4 SU-MIMO for uplink data and transmit diversity for uplink control signalling, as well as in the downlink higher-order (8×8) SU-MIMO and enhancements for MU-MIMO. The downlink developments are supported by new reference signals for data demodulation and CSI estimation, and enhanced feedback for improved CSIT accuracy with low overhead.

Further evolution of the multi-antenna transmission schemes supported by LTE-Advanced is likely by means of CoMP, for which studies of potential performance improvements with practical signalling schemes are ongoing.

References[16]

[1] 3GPP Technical Specification 36.211, 'Evolved Universal Terrestrial Radio Access (E-UTRA); Physical Channels and Modulation (Release 10)', www.3gpp.org.

[2] Ericsson and ST-Ericsson, 'R1-093485: Downlink demodulation RS design for Rel-9 and beyond', www.3gpp.org, 3GPP TSG RAN WG1, meeting 58, Shenzhen, China, August 2009.

[3] Qualcomm Europe, 'R1-094211: UE-RS Patterns for Rank 3-4,' www.3gpp.org, 3GPP TSG RAN WG1, meeting 58bis, Miyazaki, Japan, October 2009.

[4] Qualcomm Europe, 'R1-104795: Details of OCC mapping for DL MIMO operation,' www.3gpp.org, 3GPP TSG RAN WG1, meeting 62bis, Xi'an, China, October 2010.

[5] Huawei, LG Electronics, Samsung, Panasonic, Intel, HiSilicon, New Postcom, CATR, Potevio and CMCC, 'R1-105132: Proposal for specification of PDSCH Muting', www.3gpp.org, 3GPP TSG RAN WG1, meeting 62bis, Xi'an, China, October 2010.

[6] Nokia Siemens Networks and Nokia, 'R1-094648: Inter-cell CSI-RS design and performance', www.3gpp.org, 3GPP TSG RAN WG1, meeting 59, Jeju, Korea, November 2009.

[7] Alcatel-Lucent and Alcatel-Lucent Shanghai Bell, 'R1-094609: UL SRS enhancements to support CoMP and Transmit Diversity,' www.3gpp.org, 3GPP TSG RAN WG1, meeting 59, Jeju, Korea, November 2009.

[8] Alcatel-Lucent Shanghai Bell and Alcatel-Lucent, 'R1-104088: Discussion of two-stage feedback proposals,' www.3gpp.org, 3GPP TSG RAN WG1, meeting 61bis, Dresden, Germany, June 2010.

[9] Qualcomm Europe, 'R1-092716: Rank-1 Precoding vs OL Tx Diversity for PUSCH', www.3gpp.org, 3GPP TSG RAN WG1, meeting 57bis, Los Angeles, USA, July 2009.

[16]All web sites confirmed 1st March 2011.

[10] LG Electronics, 'R1-090786: PUCCH TxD Schemes for LTE-A', www.3gpp.org, 3GPP TSG RAN WG1, meeting 56, Athens, Greece, February 2009.

[11] P. Vishwanath, D. Tse, and R. Laroia, 'Opportunistic Beamforming Using Dumb Antennas', *IEEE Transaction on Information Theory*, Vol. 48, pp. 1277–1294, June 2002.

[12] Huawei, 'R1-093037: Performance of DL coordinated beam switching with bursty traffic', www.3gpp.org, 3GPP TSG RAN WG1, meeting 58, Shenzhen, China, August 2009.

[13] NTT DOCOMO, 'R1-094953: TP for TR36.814 on Self-evaluation results', www.3gpp.org, 3GPP TSG RAN WG1, meeting 59, Jeju, Korea, November 2009.

[14] P. Marsch and G. Fettweis, 'A Decentralized Optimization Approach to Backhaul-Constrained Distributed Antenna Systems' in *Proceedings of the 16th IST Mobile & Wireless Communications Summit*, July 2007.

[15] NTT DOCOMO, 'R1-090314: Investigation on Coordinated Multipoint Transmission Schemes in LTE-Advanced Downlink', www.3gpp.org, 3GPP TSG RAN WG1, meeting 55bis, Ljubljana, Slovenia, January 2009.

[16] P. Marsch and G. Fettweis, 'A Framework for Determining Realistic Capacity Bounds for Distributed Antenna Systems', *Proceedings of the IEEE Information Theory Workshop*, October 2006.

30

Relaying

Eric Hardouin, J. Nicholas Laneman, Alexander Golitschek, Hidetoshi Suzuki, Osvaldo Gonsa

30.1 Introduction

30.1.1 What is Relaying?

Relays are a key new feature of LTE-Advanced, introduced in Release 10 of the LTE specifications. These additional network nodes are designed to complement a macro-cellular network of regular eNodeBs with reduced cost, by expanding coverage or increasing capacity.

Early relays, in the form of repeaters, are already present in legacy radio interface technologies such as UMTS and Release 8 of LTE [1,2]. Repeaters simply amplify the radio-frequency signal received from a macro base station. Compared to a base station, repeaters have a lower cost since they involve no baseband processing, backhaul network installation or subscription fee for access to the fixed public network. They are typically used to improve coverage in zones where the traffic is too light to justify the deployment of a base station, or where there is no easy backhaul network access, such as road segments in rural areas. Repeaters are also convenient for the provision of indoor coverage (e.g. in shopping malls). In the majority of cases, repeaters are deployed by network operators, but user-deployed low-power repeaters are also now available (e.g. to cover a shop or flat).

Although useful, repeaters exhibit two significant drawbacks:

- Since a repeater only amplifies the received signal, any received interference is also amplified. The signal quality in the zone covered by the repeater is thus degraded compared to a regular base station.
- Repeaters are operated independently of the radio access network. Separate Operation and Maintenance (O&M) functionality is therefore required for the repeater to signal

LTE – The UMTS Long Term Evolution: From Theory to Practice, Second Edition.
Stefania Sesia, Issam Toufik and Matthew Baker.

its status and any potential malfunction to the network, which results in additional OPerational EXpenditure (OPEX) for the network operator.

Relays can be seen as an evolution of repeaters to solve the above drawbacks. A Relay Node (RN) is a network node connected wirelessly to a source eNodeB, called the *donor eNodeB*. An important characteristic of RNs is that they are under the full control of the radio access network, which permits similar monitoring and remote control capabilities as for an eNodeB. In contrast to a repeater, an RN processes the received signal before forwarding it; this may involve Layer 1, Layer 2 or Layer 3 operations, such that an RN can, in principle, range from an enhanced repeater to a fully fledged eNodeB with a wireless backhaul connection, as discussed further in Section 30.3.

In contrast to the minimal delay introduced by the simple amplification process of a repeater, the processing performed by an RN requires at least two transmission occasions to deliver the signal from the donor eNodeB to a UE, as illustrated in Figure 30.1.

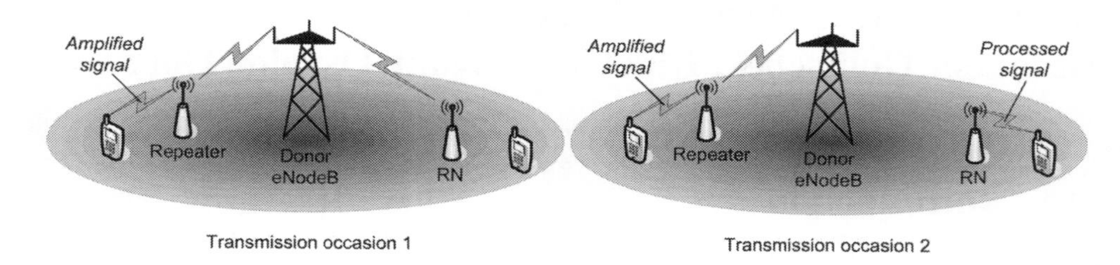

Figure 30.1: Reception and transmission phases of a repeater and a relay node.

In 3GPP, the following terminology is used, some of which is illustrated in Figure 30.2:

- **Donor eNodeB/cell**. The source eNodeB/cell from which the RN receives its signal.
- **Relay cell**. The coverage area of the RN.
- **Backhaul link**. The link between the donor eNodeB and the RN.
- **Access link**. The link between the RN and a UE.
- **Direct link**. The link between the donor eNodeB and a UE.
- **Inband/outband**. An inband RN uses the same carrier frequency for the backhaul link as for the access link; otherwise, the RN is said to be outband.
- **Half/full duplex**. A half-duplex RN cannot receive on the backhaul link at the same time as transmitting on the access link, and vice versa, whereas a full-duplex RN has sufficient antenna isolation to be able to operate without this restriction. This distinction applies to inband RNs only, since outband RNs are always full-duplex.
- **Donor and coverage antennas**. At the RN, the donor antenna(s) are used for the backhaul link, while the coverage antenna(s) are used for the access link. In some cases, the physical donor and coverage antennas may be the same.

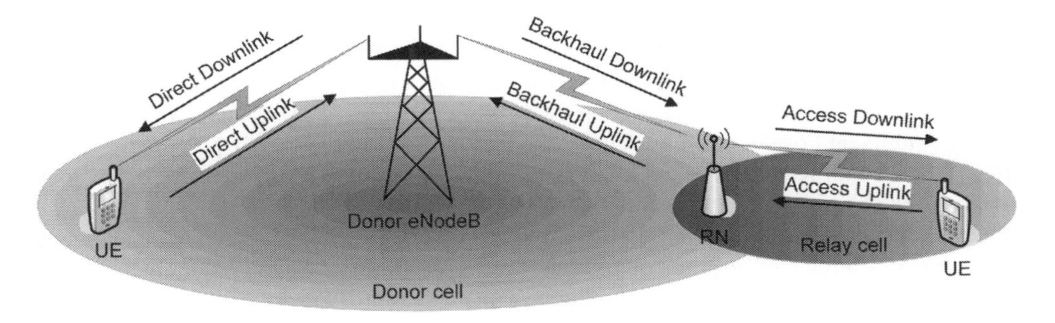

Figure 30.2: Terminology for relaying.

30.1.2 Characteristics of Relay Nodes

The characteristics of RNs have a significant impact on the network operation, giving rise to a trade-off between the cost and efficiency of RNs.

Inband RNs are expected to be the most common, in which case the backhaul link consumes radio resources from the donor cell, thus reducing the capacity of the latter. If the RN is also half-duplex, there are further complications for the system design, as discussed in Section 30.3. Full-duplex RNs provide better relay cell capacity and allow simpler system design than half-duplex RNs. On the other hand, full-duplex RNs need separate antennas for the access and backhaul links, with sufficient isolation between them, and are therefore expected to be more costly than half-duplex RNs. Outband relaying increases the relay cell capacity and simplifies the system design by allowing the backhaul link to operate on a separate carrier frequency, but this is expensive for the network operator, if indeed a second carrier frequency is available.

The transmit power and the type and number of antennas directly impact the complexity, cost, size and weight of an RN, which in turn affect the site costs (engineering and site rental).

The RN transmission power depends on the deployment scenario (see Section 30.1.4) and can range from 30 dBm or less to levels similar to a macro eNodeB (e.g. 46 dBm).

If the RN uses a single physical antenna for both the backhaul and the access links, an omni-directional antenna is typical. However, the backhaul link quality can be enhanced by the use of a directional antenna towards the donor eNodeB, which also reduces interference to other cells when the RN transmits in the uplink. In this case, a separate coverage antenna is needed, which may also be directional, away from the donor eNodeB. Multiple physical antennas at the RN are needed to support Multiple-Input Multiple-Output (MIMO) techniques, which can be especially beneficial for enhancing the backhaul link capacity – expected to be the bottleneck of relaying.

The required processing capabilities of an RN are also affected by the number of UEs the RN is designed to support, which may depend on whether it is deployed indoors or outdoors.

The main expected benefit of installing an RN in an LTE network compared to an additional eNodeB is a reduction of the infrastructure and operational costs, as well as allowing easier and quicker installation.

Unlike eNodeBs, RNs do not require any wired or microwave backhaul connection to the network. This avoids a significant part of the network CAPital EXpenditure (CAPEX) and OPEX, since wired backhaul is typically costly for both the initial connection to the site and the monthly subscription to a network provider, while microwave backhaul suffers from high initial equipment costs as well as licence fees for the microwave carrier frequency. For an RN, the backhaul-related costs are much reduced, since the same radio interface technology is used for the backhaul and access links, and, if the RN operates inband, no additional spectrum licence is needed. An RN also allows more flexibility in site selection than a wired node, by alleviating the constraint on fixed network availability. The main requirements for an RN site are a power supply and good radio link quality towards the donor eNodeB. This can also enable RNs to be deployed very quickly, so that RNs can provide a convenient solution for temporary or emergency deployments (see Section 30.1.4).

Nevertheless, RNs also have some drawbacks compared to eNodeBs. In addition to the increased delay imposed on the transmissions compared to normal eNodeB operation, the capacity offered by an RN will be reduced compared to an additional eNodeB, due to the radio resources consumed by the wireless backhaul. In terms of coverage extension, however, simulations show that the performance difference between outdoor inband RNs and pico eNodeBs (or outband RNs) is small (e.g. [3]).

30.1.3 Protocol Functionality of Relay Nodes

RNs can be categorized by the protocol layer functionality provided [4, 5]. Simplified diagrams of the protocol stacks for Layer 1, 2 and 3 RNs are shown in Figure 30.3.

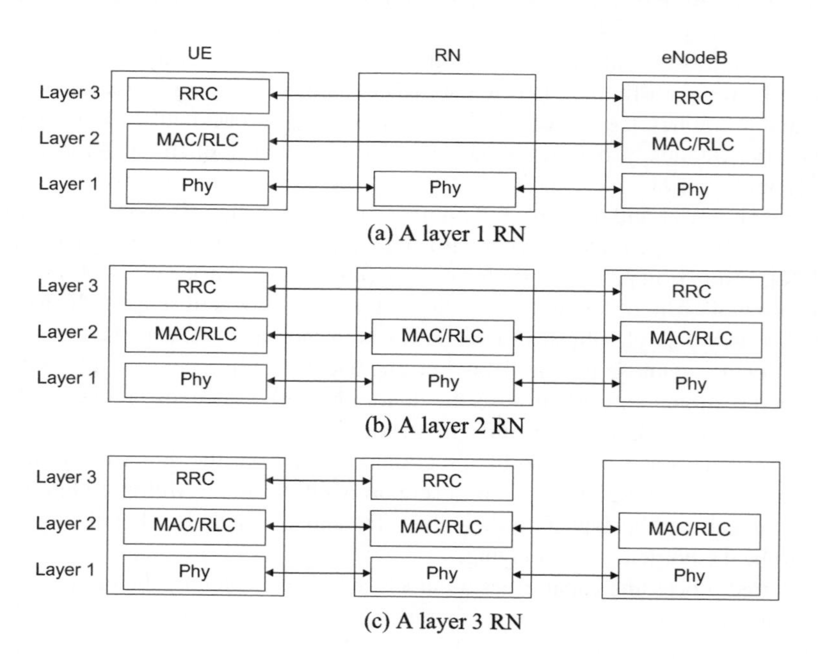

Figure 30.3: Protocol stack alternatives for an RN.

30.1.3.1 Layer 1 RN

In a Layer 1 RN, no Layer 2 processing is carried out. If the RN only performs Radio Frequency (RF) processing, it is a repeater (sometimes known as a 'Layer 0' RN) [2], as described in Section 30.1.1. Layer 1 RNs that are not simple repeaters can include baseband processing such as Forward Error Correction (FEC), but do not include scheduling functions.

30.1.3.2 Layer 2 RN

In a Layer 2 RN, Medium Access Control (MAC) functions such as scheduling (see Section 4.4.2.1) are carried out within the RN. In order to be identified by a UE as a network transmission/reception point, a Physical Cell-ID (PCI) signalled by the Primary and Secondary Synchronization Signals (PSS/SSS, see Section 7.2) is required. The presence of a PCI requires Layer 3 identification by the UE. On the other hand, some Layer 3 functions, such as Radio Resource Control (RRC, see Section 3.2), can be located within the eNodeB rather than the RN. Thus although the cell itself is realized within the RN, only Layer 2 functions are implemented there. Alternatively, a Layer 2 RN could be realized without the UE identifying it as a network transmission/reception point (no PCI is required in this case); in such a case, the relay might work with the eNodeB in a cooperative way, for example only for retransmission of packets. A Layer 2 RN may also support Radio Link Control (RLC) Automatic Retransmission reQuest (ARQ) and segmentation/concatenation functions (see Section 4.3.1).

30.1.3.3 Layer 3 RN

A Layer 3 RN has its own PCI signalled by the PSS/SSS. Mobility between RNs and eNodeBs is based on RRC in the Layer 3 control plane (see Chapter 3). In the user plane, a Layer 3 RN supports protocols at least up to PDCP (i.e. IP packet handling) (see Chapter 4). All Layer 1 and Layer 2 functions are supported within such an RN.

30.1.4 Relevant Deployment Scenarios

RNs can be used either to extend the cell coverage or to provide additional traffic capacity (hot spots). Coverage is defined by the geographical area of reception of the control channels – i.e. the *cell footprint*. Increased traffic capacity is provided by higher Signal-to-Interference-plus-Noise Ratios (SINRs), and possibly by cell splitting within the existing cell footprint.

The main practical deployment scenarios are illustrated in Figure 30.4 and discussed below.

30.1.4.1 Cell Coverage Extension

In rural scenarios, RNs can be used to provide cost-effective coverage of a large zone to extend the cell footprint provided by a macro eNodeB, as in Figure 30.4(a). Such an RN is typically mounted on a high mast, with a transmit power similar to that of a macro eNodeB, i.e. 46 dBm. RNs can also be useful for deployments in a higher frequency band than a 3G network, enabling the same coverage to be retained despite the worse radio propagation conditions, without increasing the number of base station sites.

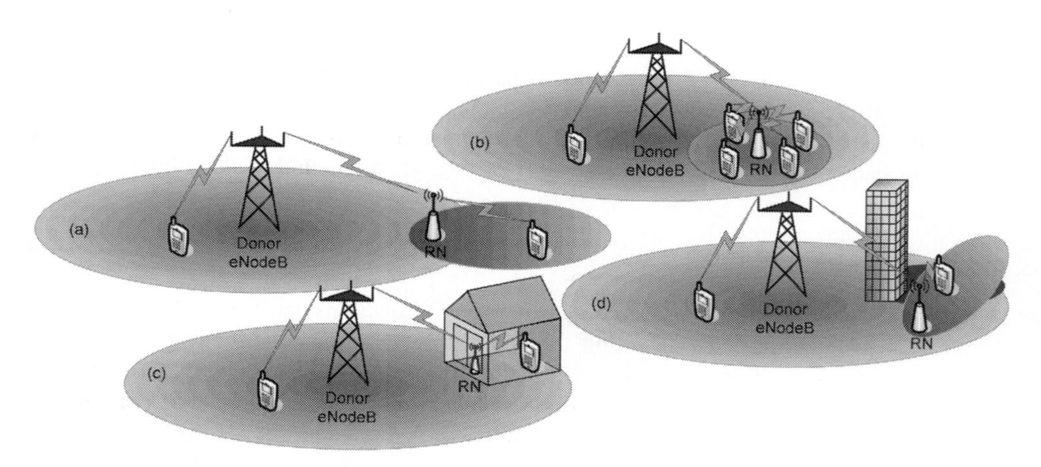

Figure 30.4: Typical use cases for relays: (a) cell coverage extension; (b) capacity boost; (c) indoor coverage enhancement; (d) dead spot mitigation.

30.1.4.2 Urban or Suburban Outdoor Capacity Boost

In this scenario, illustrated in Figure 30.4(b), the RNs are used to provide capacity boost in particular areas of the cell (hot spots), or at the cell edge to homogenize the throughput offered throughout the cell. Lamp posts often provide convenient outdoor sites for RNs in such environments. Transmit powers of 30 dBm or less are typically appropriate for urban environments due to the small cell size and inter-cell interference issues. Suburban deployments might use up to 37 dBm.

30.1.4.3 Indoor Coverage Enhancement

RNs can be placed inside a building to provide enhanced indoor coverage (Figure 30.4(c)); in such a case, they can be seen as femto eNodeBs (or Home eNodeBs – see Chapter 24) with a wireless backhaul. This can be useful in cases where the user has no ADSL link, or in countries where there is no wired network infrastructure to provide the backhaul access. Indoor RNs must have low transmission power to avoid interference problems.

30.1.4.4 Dead Spot Mitigation

RNs may be used for the mitigation of dead spots – i.e. filling coverage holes in the macro network caused by large obstacles such as high buildings, as in Figure 30.4(d). In such a scenario, the RN is typically placed in line of sight of the donor eNodeB and radiates toward the dead spot. If the extra capacity required by the dead spot compared to the donor cell is limited, an RN represents a more cost-effective solution than a conventional eNodeB.

30.1.4.5 Temporary Deployments

RNs provide a convenient solution for temporary network deployments such as special events (e.g. sports games, street fairs) or emergency deployments (e.g. if network equipment has been damaged by a natural disaster). Suitable RNs would be low power in order to be able to operate without a special eNodeB-type power supply.

30.1.4.6 Group Mobility

An RN can provide coverage to passengers in a moving vehicle such as a bus, train or ship. The RN is placed on the vehicle and connects to the most appropriate donor eNodeB as the vehicle moves. Unlike the other scenarios, such an RN is required to support mobility procedures (handover). In-vehicle RNs offer better coverage than macro eNodeBs and save the UEs' batteries since the UEs' uplink transmission power is reduced. A vehicle-mounted RN can have a higher transmit power and/or more efficient antennas than a UE and can therefore reduce the incidence of call drops during handovers.

30.2 Theoretical Analysis of Relaying

The capacity of RN-assisted cellular network deployments can be considered in the framework of the point-to-point *relay channel* [6], the *multi-access relay channel* [7], and the *broadcast relay channel* [8]. Fundamental performance limits for the channel capacity are generally unknown for these cases, but a number of communication strategies and upper/lower bounds on the fundamental limits have been obtained. For LTE-Advanced, the most relevant model for the cell coverage extension scenario of Figure 30.4(a) is a collection of relay channels, while the most relevant model for the capacity boost scenario of Figure 30.4(b) consists of multiple-access and broadcast relay channels for the uplink and downlink respectively.

This section summarizes some of the relevant theoretical developments and fundamental limits in the context of LTE-Advanced and illustrates how practical system constraints reduce the complexity of the models and of the architectural designs. A more detailed introduction to relaying and cooperative communications can be found in [9].

30.2.1 Relaying Strategies and Benefits

We begin with a summary of the general Gaussian relay channel [6, 10], on which much of the following analysis is derived.

A general Gaussian relay channel relevant to both the coverage extension and capacity boost cases of Figure 30.4 can be modelled as follows (also illustrated for the case of the downlink in Figure 30.5):

$$Y_2(t) = H_{2,1}X_1(t) + Z_2(t)$$

$$Y_3(t) = H_{3,1}X_1(t) + H_{3,2}X_2(t) + Z_3(t)$$

where $X_1(t)$ is the source transmission, $Y_3(t)$ is the destination received signal, and $X_2(t)$ and $Y_2(t)$ are the RN transmitted and received signal, respectively. The additive noises $Z_j(t)$ are modelled as white Gaussian, with one-sided power spectral densities N_j, $j = 2, 3$. The channel coefficients $H_{j,i}$, $i = 1, 2$, $j = 2, 3$ capture the effects of path-loss and shadowing; for simplicity, we assume here that these coefficients vary slowly relative to the signalling interval, are non-selective with frequency and are known to the source, the RN and the destination. The model can readily be extended to incorporate other channel effects, such as multiple antennas, resolvable multipath and limited knowledge of the channel coefficients at the transmitters and receivers.

For the cell coverage extension scenario of Figure 30.4(a), the destination is outside the coverage range of the control channels, and therefore $H_{3,1}$ is set to zero.

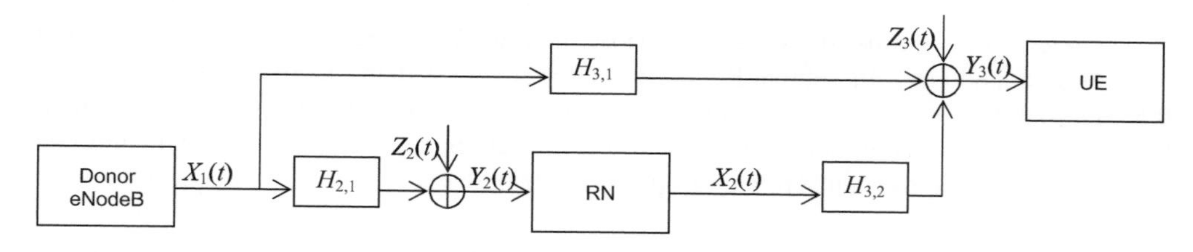

Figure 30.5: The general Gaussian relay channel model in the context of downlink transmission in an RN-assisted cellular network. The same model can be applied to the uplink case, with the UE and the donor eNodeB reversed.

The source and RN are assumed to have average power constraints P_1 and P_2 respectively. The waveforms are essentially band-limited to the same frequency band of width W.

The received Signal-to-Noise Ratio (SNR) between transmitter $i = 1, 2$ and receiver $j = 2, 3$ can then be written as:

$$\rho_{j,i} = \frac{|H_{j,i}|^2 P_i}{N_j W} \, . \tag{30.1}$$

We consider here distance-dependent path-loss models such that the average channel gain $\mathbb{E}[|H_{j,i}|^2] \propto d_{j,i}^{-\nu}$, where $d_{j,i}$ is the distance between transmitter i and receiver j, and $2 \leq \nu \leq 5$ is the path-loss exponent [11, 12].

For the capacity boost scenario of Figure 30.4(b), we first analyse the potential capacity increase compared to the case when the RN is not present. If $X_2(t) = 0$, then the channel model reduces to a point-to-point Gaussian channel with capacity given by [10, 13]

$$C_{\text{direct}} = W \log_2 (1 + \rho_{3,1}) \, . \tag{30.2}$$

By processing the received signal $Y_2(t)$, the RN can create its transmission $X_2(t)$ so as to enhance performance beyond Equation (30.2) which therefore represents a lower bound on the capacity of an RN-aided channel.

An upper bound on the capacity of the general Gaussian relay channel can be obtained by considering the total information flows across a set of 'cuts' in the RN-assisted network, as shown in Figure 30.6. It is therefore known as the *cut-set bound* [6, 10].

The cut-set upper bound is given by the minimum of the rate of the broadcast cut from the source to the RN and destination, and the rate of the multiple-access cut from the source and RN to the destination, maximized over the parameter α $(0 \leq \alpha \leq 1)$ which controls the amount of correlation between the signals $X_1(t)$ and $X_2(t)$: $C_{\text{relay}} \leq C_{\text{cutset}}$, where

$$\begin{aligned} C_{\text{cutset}} = W \max_{0\leq\alpha\leq1} \min \{ & \log_2 (1+(1-\alpha)(\rho_{2,1}+\rho_{3,1})) , \\ & \log_2 \left(1+\rho_{3,1}+\rho_{3,2}+ \sqrt{\alpha\rho_{3,1}\rho_{3,2}}\right)\} \end{aligned} \tag{30.3}$$

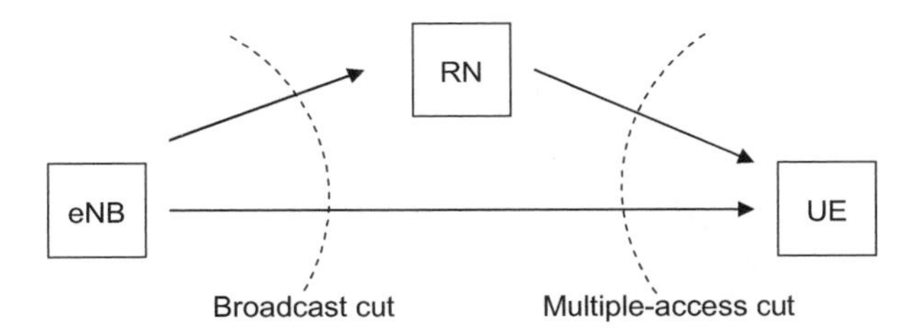

Figure 30.6: Broadcast and multiple-access cuts through a relay network.

As introduced in Section 30.1.1, a decode-and-forward RN decodes the source transmission $X_1(t)$ and re-encodes it to produce $X_2(t)$. This approach in theory allows for the design of a joint signalling scheme among $X_1(t)$ and $X_2(t)$, roughly approximating a system with two transmit antennas and one receive antenna. The resulting lower bound on channel capacity for this approach is $C_{\text{relay}} \geq C_{\text{decode}}$ [6, 10], where

$$C_{\text{decode}} = W \max_{0\leq\alpha\leq 1} \min \Big\{ \log_2 \left(1+(1-\alpha)\rho_{2,1}\right), \log_2 \left(1+\rho_{3,1}+\rho_{3,2}+\sqrt{\alpha\rho_{3,1}\rho_{3,2}}\right)\Big\} . \tag{30.4}$$

The first term in the minimum in Equation (30.4) corresponds to the capacity of the link between the source and RN assuming that the fraction $(1 - \alpha)$ of the source power P_1 is used to convey information to the RN. The second term in the minimum in Equation (30.4) corresponds to the multiple-access cut from the source and RN to the destination, and is identical to the second term in the minimum in Equation (30.3).

Examination of Equations (30.2)–(30.4) provides some theoretical insights about the potential benefits of relaying. To see these benefits, assume that the RN lies between the source and destination, so that we can expect both $\rho_{2,1} \geq \rho_{3,1}$ and $\rho_{3,2} \geq \rho_{3,1}$. If $\alpha = 0$ (i.e. no correlation between $X_1(t)$ and $X_2(t)$), then both terms in the minima of Equations (30.3) and (30.4) are larger than Equation (30.2). These observations, illustrated in Figure 30.7, suggest that relaying increases the capacity for a given amount of source power and bandwidth relative to direct transmission, extends coverage for the same target data rate, or some combination of these two benefits.

Additional observations result from carefully examining the structure of the gains that relaying can provide. These gains can be classified into the following three basic types:

- **Multihop Gain** results from the fact that, if the RN lies physically between the source and destination, then transmissions from the source to the RN and from the RN to the destination occur over shorter distances than direct transmission from the source to the destination. If we assume $\rho_{2,1} \gg \rho_{3,1}$ and $\rho_{3,2} \gg \rho_{3,1}$ in Equations (30.3) and (30.4), as is the case in the coverage extension scenario, then in the best case of a full-duplex relay

  $$C_{\text{relay}} \approx W \min \{\log_2 (1+\rho_{2,1}), \log_2 (1+\rho_{3,2})\} \gg C_{\text{direct}} .$$

  It needs to be borne in mind that this gain comes at the expense of power and bandwidth usage in the cell of the donor eNodeB. From a system perspective, there may be cases

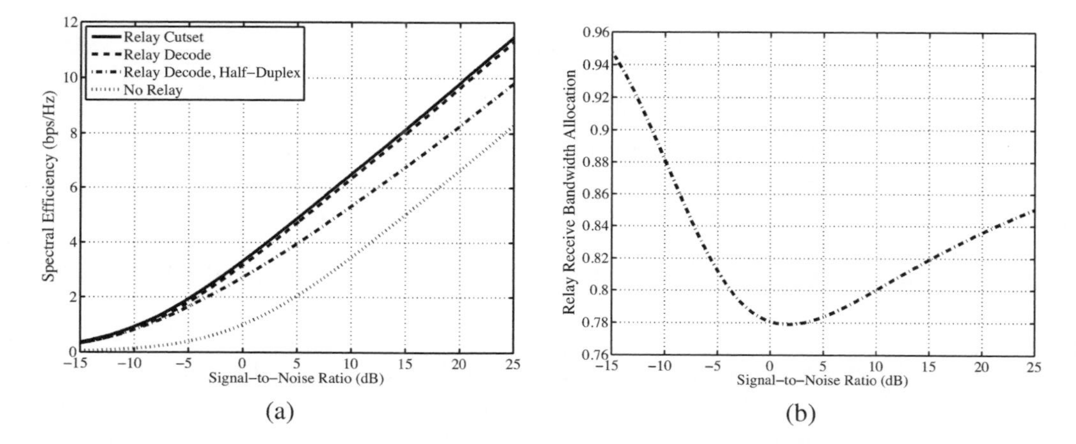

Figure 30.7: Example theoretical performance of various relaying schemes with the RN located halfway between the source and destination, path-loss exponent $\nu = 3$, and no correlation between the backhaul and access links:
(a) spectral efficiency C/W versus SNR between the source and RN,
(b) fractional bandwidth β allocated to RN reception for a half-duplex RN strategy.
(As explained in Section 30.2.2, the parameter β is the fraction of the channel dimensions during which the RN can receive when it is operating in half-duplex mode; the RN can transmit in the remaining $1 - \beta$ opportunities.)

in which total capacity is higher if these resources are instead devoted to UEs closer to the donor eNodeB. However, in general, the use of RNs will have the effect of increasing fairness, as the capacity to distant terminals is increased at the expense of capacity to terminals close to the donor eNodeB.

- **Power / Diversity Gain** results from the summing and averaging of SNRs in the various expressions. For example, the second term in the minimum in Equation (30.4) has effective SNR at least

$$\rho_{3,1} + \rho_{3,2} = \left(\frac{|H_{3,1}|^2 P_1 + |H_{3,2}|^2 P_2}{P_1 + P_2} \right) \cdot \left(\frac{P_1 + P_2}{N_3 W} \right). \tag{30.5}$$

The first term in Equation (30.5) corresponds to a weighted average of the channel gains, $|H_{3,1}|^2$ and $|H_{3,2}|^2$, that is reminiscent of transmit diversity [14, 15]. Relaying protocols that provide diversity benefits are analysed in [16, 17] and references therein. The second term in Equation (30.5) highlights the additional power that the relay can provide to the system.

- **Coherence Gain** results from the ability to create correlated signals $X_1(t)$ and $X_2(t)$ with the parameter α, leading to the additional SNR of $\sqrt{\alpha\rho_{3,1}\rho_{3,2}}$ in the second term in the minima in Equations (30.3) and (30.4). Note that coherence gain is only possible with accurate carrier and symbol timing synchronization, which is implicitly assumed here. Such tight synchronization may in theory be possible on the downlink but may not be realistic on the uplink. It is possible to show that, with phase variations known to the receivers but not the transmitters, the optimal choice is $\alpha = 0$, i.e. no correlation between $X_1(t)$ and $X_2(t)$ [8].

30.2.2 Duplex Constraints and Resource Allocation

Due to circuit isolation and radio cost issues, it is often necessary to preclude the relay from transmitting and receiving simultaneously in the same frequency band. This *half-duplex* constraint is often modelled by allowing the relay to receive in a fraction $0 \leq \beta \leq 1$ of the channel dimensions (time and frequency) and to transmit in the remaining fraction $\overline{\beta} = 1 - \beta$; the source utilizes a fraction $0 \leq \gamma \leq 1$ of its power while the relay receives and the remaining fraction $\overline{\gamma} = 1 - \gamma$ while the relay transmits. Time-sharing can then be performed between two general Gaussian relay channels of the form discussed in Section 30.2.1, one in which the relay only receives ($X_2(t) = 0$), and the other in which the relay only transmits ($Y_2(t) = 0$). As an example, under this half-duplex constraint, Equation (30.4) becomes:

$$\begin{aligned} C_{\text{decode,hd}} = W \max_{0 \leq \beta, \gamma \leq 1} \min \{ & \beta \log_2 (1 + \gamma \rho_{2,1} / \beta) , \\ & \beta \log_2 (1 + \gamma \rho_{3,1} / \beta) \\ & + \overline{\beta} \log_2 \left(1 + (\overline{\gamma} \rho_{3,1} + \rho_{3,2} + \sqrt{\overline{\gamma} \rho_{3,1} \rho_{3,2}}) / \overline{\beta} \right) \} . \end{aligned} \tag{30.6}$$

The rate expression in Equation (30.6) simplifies considerably for the special case where $\beta = 1/2$ and $\gamma = 1$. In this case, the half-duplex constraint affects the multihop gain in the sense that, if $\rho_{2,1} \gg \rho_{3,1}$ and $\rho_{3,2} \gg \rho_{3,1}$, then

$$C_{\text{decode,hd}} \approx W/2 \min \{ \log_2(1 + 2\rho_{2,1}), \log_2(1 + 2\rho_{3,2}) \} .$$

Here the SNR gain of multihop transmission can be eliminated by the bandwidth cost of half-duplex, so that $C_{\text{decode,hd}} < C_{\text{direct}}$, especially for high rates [18].

Because modern cellular systems such as LTE use basically orthogonal transmission schemes, the key challenges within a cell relate to resource allocation in time and frequency. This view is oversimplified relative to general information-theoretic treatments of multiple access and broadcast, but is generally adopted in practice to avoid excessive complexity. The result of transmission resources being orthogonal is that the general multiple access channel and broadcast channel can be converted into a collection of point-to-point channels with limited interference over which power and transmission resources can be allocated. Similarly, orthogonal transmissions convert the general multiple-access relay channel and broadcast relay channel into collections of simpler channels. For example, orthogonal transmissions on the uplink in the capacity-boost scenario of Figure 30.4(b) can result in a subset of point-to-point channels from the UEs to the eNodeB and subsets of channels between a set of UEs and each relay whose rates are aggregated into a backhaul channel between the relay and eNodeB. For the decode strategy, a relay must decode *all* signals of the UEs associated with it, and convey the sum rate over the backhaul channel. Careful resource allocation must be performed by the cell scheduler in order to ensure that the half-duplex constraint and the backhaul throughput do not become system bottlenecks.

Finally, it should be noted that, although not currently considered in the context of LTE-Advanced, network coding of data streams at the relays can lead to increased system capacity and more efficient use of resources [19–21].

30.3 Relay Nodes in LTE-Advanced

As outlined in Section 30.1, the term 'Relay Node' can encompass a wide variety of different implementations and operation modes. The RN functionality defined in Release 10 is the result of careful consideration of these different possibilities.

30.3.1 Types of RN

An important aspect of RN design is the extent to which the RN can be seen by a UE. RNs in Release 10 are designed to be backward compatible to legacy UEs (i.e. UEs conforming to Releases 8 and 9 of the LTE specifications). During the study for LTE-Advanced, 3GPP identified two major types of RNs – Type 1 and Type 2 – and some subtypes of Type 1 RNs.

30.3.1.1 Type 1, 1a and 1b RNs

These types of RN are Layer 3 RNs. A UE sees them as separate cells (i.e. normal eNodeBs) with their own PCIs. The RN transmits all control and data channels including the PSS/SSS. UEs receive scheduling information such as the Physical Downlink Control CHannel (PDCCH) from the RN, and send feedback information such as Channel State Information (CSI) and ACKnowledgement/Negative ACKnowledgements (ACK/NACKs) to the RN. The handover and cell reselection procedures are identical to Release 8 (see Chapter 3), and the scheduler is located in the RN in order to respond quickly to a UE's feedback.

Type 1 RNs are inband half-duplex RNs, while Type 1a RNs operate outband.

Type 1b RNs have sufficient isolation between the received and transmitted signals to enable full-duplex operation – i.e. the backhaul and access links can be active simultaneously without the need for time-division multiplexing. In one example of a Type 1b RN, the backhaul antenna may be located outside the building while the coverage antenna is located inside the building for the improved support of UEs inside the building. Antenna isolation may also be aided by mechanical or adaptive beam-forming.

LTE-Advanced supports Type 1 and Type 1a RNs in Release 10 of the specifications. No specific support is provided for Type 1b RNs, although such RNs may be able to be deployed by implementation-dependent means.

30.3.1.2 Type 2 RNs

Type 2 RNs, which were also studied for LTE-Advanced but not adopted, are Layer 2 RNs. They cannot be identified by a UE since these RNs would not transmit control channels or have their own physical cell-IDs. These RNs would only transmit the Physical Downlink Shared Channel (PDSCH), and the scheduler is located in the eNodeB.

Type 2 RNs would be operated in a non-cooperative way with the donor eNodeB [22] or in a cooperative way. In the latter case, the eNodeB and RN would jointly transmit and receive the signal to and from a UE, as illustrated in Figure 30.8. The initial transmission from the eNodeB would be received at both the RN and the UE at the same time. As the backhaul link should usually experience good radio conditions, the RN would be expected to receive the data correctly more often than the UE. For the retransmissions, both the RN and the eNodeB would transmit to the UE, and the UE would combine the two signals. The coverage area of a Type 2 RN would always overlap the coverage area of the donor eNodeB, as the UE would rely on control channel reception from the donor eNodeB.

Figure 30.8: Cooperative operation of Type 2 RNs.

30.3.2 Backhaul and Access Resource Sharing

It can be seen from Figure 30.2 that an RN competes for radio resources with UEs in the donor eNodeB's cell coverage areas. This can, in principle, be solved either by dynamic resource allocation, where the RN is scheduled like any other UE in the donor eNodeB's cells, or by reserving a specific band or part of a band for the backhaul. On the backhaul link, an RN has to receive and transmit signals from and to the donor eNodeB, while on the access link, an RN has to transmit and receive signals to and from the UEs. Some coordination or separation is necessary in order to avoid the transmitted signal of the RN causing interference to its own receiver. This separation can be in frequency, in time or, in the case of Type 1b relays as discussed above, in space. Alternatively a transmission signal canceller can be implemented in the receiver [23] to limit the self-interference, although this increases the RN complexity.

30.3.2.1 Separation in Frequency

Outband RNs (i.e. Type 1a RNs) use different frequencies for the backhaul and access links. Nevertheless, if the frequencies for the backhaul and access links are not sufficiently separated, interference can still occur due to out-of-band and spurious emissions.

Alternatively, frequency separation in the same band could in theory be achieved if the eNodeB scheduler were to allocate dynamically the Resource Blocks (RBs) for the backhaul and access links, provided that the RN's receiver could be sufficiently isolated against sidelobes from the transmitted signal. This would allow dynamic characteristics of the data traffic to be taken into account. However, the scheduler design for such an arrangement may be complex, and it would require additional signalling from the eNodeB to the RN to indicate the access link resource assignments. Therefore, the simpler realization is to assign the resources for the backhaul and access links relatively statically.

30.3.2.2 Separation in Time

Separation in time is realized by time-division multiplexing between the backhaul and access links within the same spectrum. The schedulers for the backhaul and access links are each responsible for not scheduling transmissions when the respective link is not available.

If arbitrary time instances were used for backhaul downlink reception at the RN, with no access link transmissions from the RN at these times, a UE could lose its connection to the RN due of the lack of Reference Signals (RSs), synchronization signals and control channels such as the Physical Control Format Indicator CHannel (PCFICH) and Physical Broadcast CHannel (PBCH). Therefore, to meet the requirement for backward compatibility, the time instances at which no transmission will occur from the RN have to be designed in such a way that even Release 8 UEs are not disturbed by them.

The Release 8 LTE specifications already provided a mechanism to inform UEs by RRC signalling [24] that certain subframes are designated as 'Multimedia Broadcast Single Frequency Network' (MBSFN) subframes. In such subframes, the UE expects control signals and RSs only in the first 1–2 OFDM symbols and ignores the rest of the subframe.[1] Originally this mechanism was introduced purely for MBSFN transmissions (see Section 13.4), but the signalling can be reused in Release 10 for other purposes, such as relaying, while avoiding disturbance to Release 8 and 9 UEs. Therefore these subframes in which RN backhaul downlink reception occurs are still referred to as 'MBSFN' subframes, even though no actual MBSFN transmission takes place there.

MBSFN subframes are therefore used for time-division multiplexing between the backhaul and access links for Type 1 RNs for LTE-Advanced.

As shown in Figure 30.9, the number of available OFDM symbols for an RN to receive the backhaul signal in an MBSFN subframe is less than the full subframe length because the RN has to transmit control signalling on the access link in the first 1–2 OFDM symbols for backward compatibility; additionally, some time is required for the RN to switch from transmission to reception and vice versa.

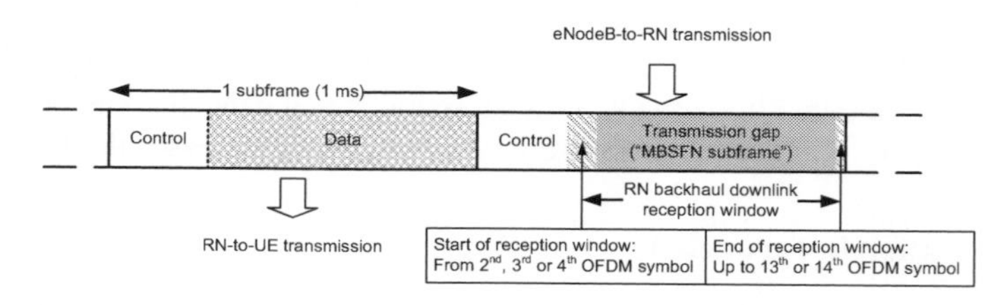

Figure 30.9: Transmission from an RN to a UE in a normal subframe, and from the donor eNodeB to the RN in an MBSFN subframe.

The starting symbol of the RN's backhaul downlink reception window is configured by RRC signalling as the start of the second, third or fourth OFDM symbol, and can therefore be changed during the operation of an RN by RRC reconfiguration. The last symbol of the reception window is a function of the eNodeB–RN synchronization, the backhaul signal

[1]UEs of Release 9 and later may be instructed to receive the Physical Multicast Channel (PMCH) for MBMS in MBSFN subframes (see Section 13.4.1).

propagation delay and the RN's transmission–reception switching time, all of which can be considered to be matters of RN implementation and deployment. Therefore the last symbol of the RN's backhaul downlink reception window does not need to be configurable by RRC, and the reception window continues up to the last or second-to-last OFDM symbol. Consequently, the resulting number of OFDM symbols within an RN's backhaul downlink reception window ranges from 10 to 13 (in the case of the normal cyclic prefix being configured). These reception restrictions need to be taken into account by the donor eNodeB when transmitting on the backhaul link, by suitable RRC configuration, resource assignment and link adaptation.

30.3.3 Relay Architecture

RNs in LTE-Advanced support the full eNodeB functionality, including termination of the radio protocols and the S1 and X2 interfaces (see Sections 2.5 and 2.6 respectively). The network interfaces relevant to RNs are shown in Figure 30.10. The backhaul interface is known as the Un interface, over which the RN is connected to the donor eNodeB via S1 and X2 interfaces.

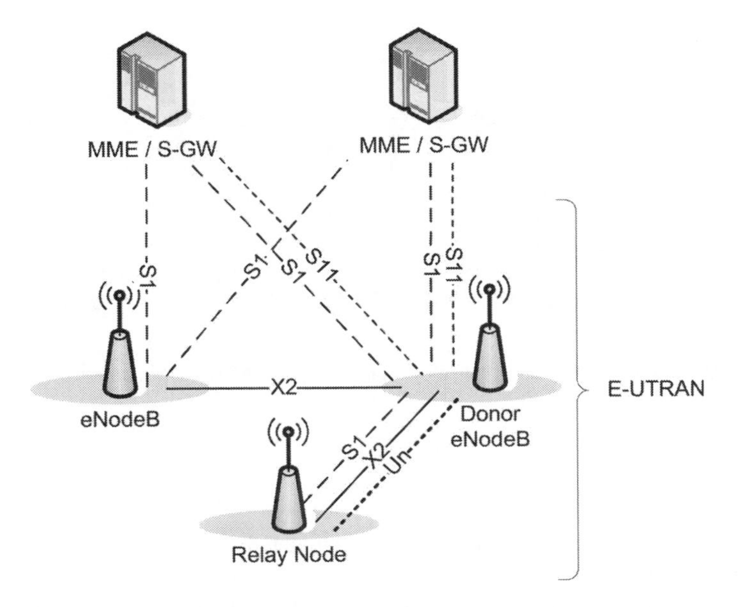

Figure 30.10: Network architecture and interfaces for RNs.
Reproduced by permission of © 3GPP.

The donor eNodeB provides S1 and X2 proxy functionality between the RN and other eNodeBs, Mobility Management Entities (MMEs) and Serving Gateways (S-GWs) (see Chapter 2); this includes passing UE-specific S1 and X2 signalling messages, as well as GPRS[2] Tunnelling Protocol (GTP) data packets, between the S1/X2 interfaces associated

[2]General Packet Radio Service.

with the RN and the S1/X2 interfaces associated with other network nodes. Due to the proxy-like functionality, the donor eNodeB appears to the RN as an MME (for S1), as an eNodeB (for X2) and as an S-GW.

30.3.3.1 Relay Node User Plane

The donor eNodeB provides the S-GW and Packet Data Network (PDN) Gateway (P-GW) functionality for the RN (see Figure 30.11). This includes creating a session for the RN and managing Evolved Packet System (EPS) bearers (see Section 2.4) for the RN. The RN (in the uplink) and donor eNodeB (in the downlink) also perform mapping of signalling and data packets onto EPS bearers that are set up for the RN. The mapping is based on existing Quality of Service (QoS) mechanisms defined for the UE and the P-GW such as Quality Class Identifier (QCI) values (see Section 2.4). On the backhaul link, there can be up to eight Data Radio Bearers (DRBs) per RN. Data on these RN bearers is mapped from the EPS bearers of the UEs connected to the RN according to the QCI values of the EPS bearers. The mapping is configured by O&M and supports many-to-one mapping. The set up, timing and modification of Un bearers is left to the donor eNodeB implementation.

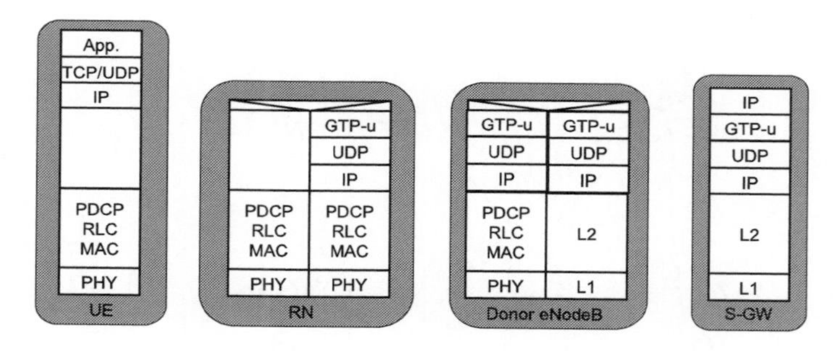

Figure 30.11: RN user plane protocol stack. Reproduced by permission of © 3GPP.

30.3.3.2 Relay Node Control Plane

There is one S1 interface relation between an RN and the donor eNodeB, and one S1 interface between the donor eNodeB and each MME in the 'MME pool' (see Section 2.2.2). The donor eNodeB processes and forwards all S1 messages between the RN and the MMEs for all UE-dedicated procedures (see Figure 30.12). The processing of S1 Application Protocol (S1-AP) messages includes modifying S1-AP UE identities and GTP Tunnelling End IDs (TEIDs – see Section 2.5) but leaves other parts of the messages unchanged. All non-UE-dedicated procedures are handled locally, between the RN and donor eNodeB, and between the donor eNodeB and MMEs.

There is one X2 interface relation between an RN and the donor eNodeB, and one X2 interface relation between the donor eNodeB and each other eNodeB with which the donor eNodeB has an X2 interface relationship. X2 messages for mobility are processed in a similar way to the S1 messages.

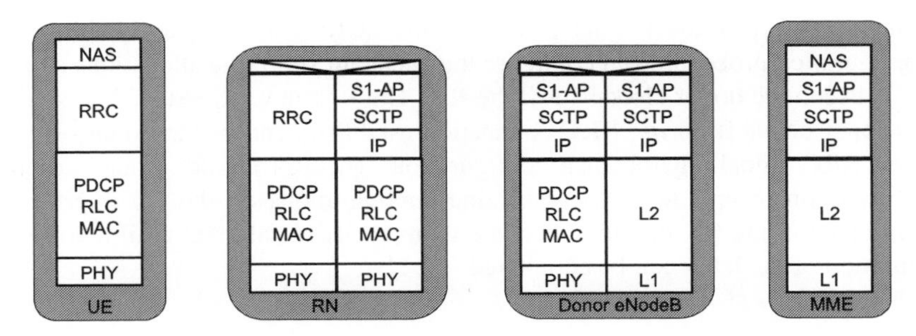

Figure 30.12: RN control plane protocol stack. Reproduced by permission of © 3GPP.

30.3.4 RN Initialization and Configuration

RN initialization is similar to the process of a UE performing 'initial attach' (see Sections 2.2.1.1 and 3.2.3.2). The only differences are that the S-GW/P-GW functionality is performed by the donor eNodeB and that during the RRC connection establishment, an RN signals an RN Indicator to the donor eNodeB. Based on this indicator, the donor eNodeB selects an MME that supports relay functionality.

After initial attachment, performed as if it was a UE, an RN has to enable and configure its RN functionality; this is done in two phases:

Phase I. The RN retrieves initial configuration parameters, including a list of donor eNodeBs to which it is allowed to attach, from an RN O&M server. The RN then detaches from the network as a UE and triggers Phase II.

Phase II. The RN connects to a donor eNodeB selected from the list acquired during Phase I. In order for the donor eNodeB to be authorized to pass S1-AP traffic to the RN, the donor eNB needs to know that the RN has an RN subscription rather than a UE subscription. The preferred approach for the donor eNodeB to obtain this information is via an S1-AP message from the MME, which can in turn obtain the relevant information from the HSS by means of an indication from the subscription profile in the HSS related to the Universal Subscriber Identity Module(s) (USIM(s)) in the RN.[3] During Phase II, the donor eNodeB also embeds and provides the S-GW/P-GW-like functions needed to support the RN operation. This includes creating a session for the RN and managing EPS bearers for the RN, as well as terminating the S11 interface towards the MME serving the RN.

After the donor eNodeB has set up the bearers for the S1/X2 interfaces, the RN initiates the setup of S1 and X2 associations with the donor eNodeB and begins relay operations.

The RRC layer of the Un interface running on the backhaul link can send updated System Information (SI) in a dedicated message to RNs. When an RN receives such signalling, it applies it immediately.

In addition, RN-specific RRC functionality over the backhaul link is provided through a new Un reconfiguration procedure that applies after an RN has attached and commenced

[3]This indication may, for example, be based on subscription type, or on particular Access Point Names (APNs) which are used only by RNs.

relay behaviour; this procedure is independent of the Uu RRC reconfiguration procedure. The RN reconfiguration procedure can configure the backhaul subframe allocations and physical channels between the donor eNodeB and the RN. The RN may request such a configuration from the donor eNodeB during RRC connection establishment, or the donor eNodeB may initiate the RRC signalling for such configuration. The RN applies a new configuration immediately upon reception, so the subframe configurations on the Un interface and on the Uu interface in the RN cell may become temporarily misaligned until a corresponding reconfiguration of the latter can be completed.

30.3.5 Random Access on the Backhaul Link

RNs support both contention-based and non-contention-based random access (see Section 17.3 over the backhaul link in exactly the same way as a UE. When an RN performs the random access procedure, it suspends any current backhaul subframe configuration, meaning that it temporarily disregards any Un-interface-specific configuration. The backhaul subframe configuration is resumed after the random access procedure is successfully completed.

30.3.6 Radio Link Failure on the Backhaul Link

RNs detect radio link failure on the backhaul link using the same parameters as UEs (see Section 22.7). If an attempt to re-establish an RRC connection fails, then the RN goes to RRC_IDLE state and tries to recover. The details of RN behaviour from RRC_IDLE are left to RN implementation; it is assumed that an RN would always try to recover the RRC connection as quickly as possible and not remain in RRC_IDLE for longer than necessary.

If an RN experiences a Scheduling Request (SR)[4] failure (e.g. a preconfigured number of SR attempts have been made), it follows the same procedure as a UE.

30.3.7 RN Security

RNs have different security considerations from an eNodeB or a UE. This arises from the fact that the behaviour of RNs comprises aspects of both eNodeB and UE behaviour, and because the connection of an RN to the network is wireless. The latter issue imposes additional security considerations and is especially important when the RN has to send RRC, S1 or X2 signalling messages over the backhaul radio bearers. Some such security threats are as follows (further details can be found in [25, Section 5.3]):

- Impersonation of an RN to attack the user(s) attached to the RN;
- Attacks on the backhaul link between the RN and donor eNodeB;
- Inserting a 'Man in the Middle' (MitM);
- Attacking the user data traffic;
- Impersonation of an RN to attack the network;
- Attacks on the RN itself;
- Denial of Service (DoS) attacks.

[4] See Section 4.4.2.2.

RNs support the Transport Network Layer (TNL) of the S1-MME and S1-U interfaces (see Figure 2.1), and hence a function to ensure secure transport over the backhaul link is needed. The solution for RN security (see [26, Annex D]) is to realize a one-to-one binding of an RN and a USIM called USIM-RN, which is not realized by eNodeBs. This one-to-one binding uses symmetric pre-shared keys or certificates. The donor eNodeBs store the IP security credentials in the secure part of the eNodeB platform, thus providing a secure anchor for Network Domain Security (NDS)/IP.

Over the Un interface, security is provided by Un PDCP between the RN and the donor eNodeB, and by NDS/IP between the donor eNodeB and the MME. The native SEcurity Gateway (SEG) can be reused for NDS/IP traffic between the donor eNodeB and the MME. Un security is modified compared to Uu security such that integrity protection is provided in the Un user plane at least for PDCP PDUs carrying S1 signalling.

The security architecture for RNs is shown in Figure 30.13.

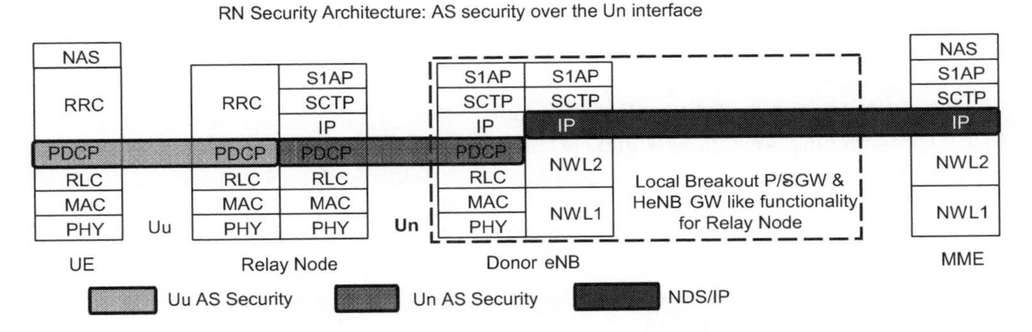

Figure 30.13: RN security architecture. Reproduced by permission of © 3GPP.

More information about the security considerations for the backhaul link and RNs in general can be found in [25] and in [26].

Security over the access link is provided by the RN in the same way as a normal eNodeB (see Sections 3.2.3.1 and 4.2.3).

30.3.8 Backhaul Physical Channels

The nature of the relay backhaul link results in a number of differences between the physical channels or signals defined for the backhaul link and those defined for the access link. These are reviewed below; full details can be found in [27].

30.3.8.1 Backhaul Reference Signals

Most of the Reference Signals (RSs) defined for the access link are also usable on the backhaul link, including the cell-specific RSs (see Section 8.2.1) and the new UE-specific RSs defined in Release 10 (see Section 29.1.1).

One exception may occur if the last OFDM symbol of a subframe is not within the RN's backhaul downlink reception window, in which case the second slot of subframes addressed to such an RN does not carry the RSs of the UE-specific DeModulation RS (DM-RS) antenna ports 7 to 10. In this case, only the DM-RSs from the first slot are available for channel estimation and demodulation.

A further restriction is that the backhaul link in Release 10 is limited to a maximum of four spatial layers, and DM-RS antenna ports 11 to 14 are never transmitted in Physical Resource Blocks (PRBs)[5] addressed to an RN.

30.3.8.2 Backhaul Control Channels

As explained in Section 30.3.2, the time-division separation of transmission and reception for the Type 1 RN in LTE-Advanced utilizes MBSFN subframes.

Due to the timing synchronization between the access and backhaul links, and the need to transmit one or two OFDM symbols for the PDCCH on the access link at the start of each MBSFN subframe, the RN is not able to receive the PDCCH from the donor eNodeB. To cater for this, a new backhaul control channel known as the Relay PDCCH (R-PDCCH) is provided. The R-PDCCH has to be multiplexed into the backhaul subframes within the RN's reception window, and it therefore occupies Resource Elements (REs) that otherwise would be used for the PDSCH. In the first slot of a subframe, the R-PDCCH starts in the fourth OFDM symbol (even though the RN's reception window may start earlier), and only R-PDCCHs containing downlink assignments are transmitted, enabling the RN to try to decode the downlink assignments immediately after the first slot. In the second slot, R-PDCCHs containing uplink grants are transmitted (the last OFDM symbol being determined by the length of the reception window (see Section 30.3.2.2)). This is illustrated in Figure 30.14.

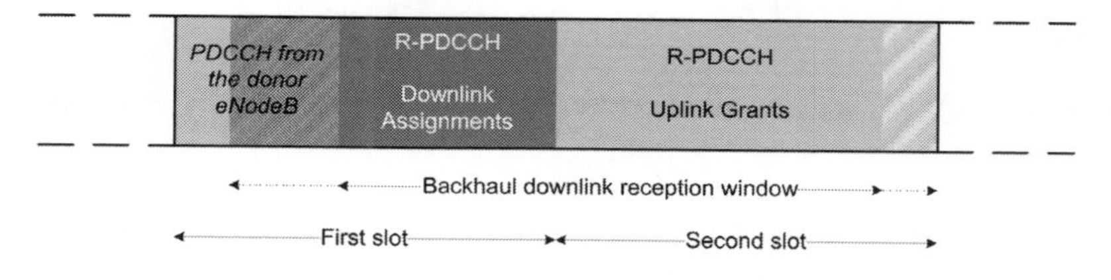

Figure 30.14: Placement of the R-PDCCH.

Since the backhaul link conditions are likely to be relatively static and predictable, the R-PDCCH is mapped to certain PRBs instead of being distributed in frequency over the system bandwidth like the PDCCH. The PRBs used for the R-PDCCH are configurable for each RN. Within these PRBs, two alternative R-PDCCH formats are supported: a 'cross-interleaved' R-PDCCH and a 'non-cross-interleaved' R-PDCCH. These are illustrated in Figure 30.15 and described below.

Cross-interleaved R-PDCCH

The cross-interleaved R-PDCCH is only supported if the RN is configured to use the donor eNodeB's cell-specific RSs for demodulation of the R-PDCCH.

[5]See Section 6.2.

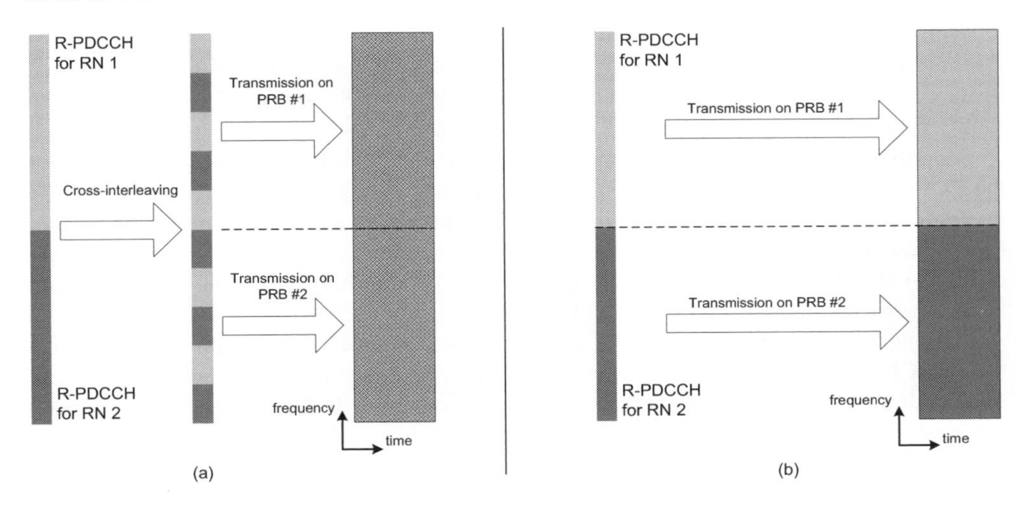

Figure 30.15: R-PDCCH formats: (a) cross-interleaved; (b) non-cross-interleaved.

The cross-interleaved R-PDCCH is designed to re-use as far as possible the design of the PDCCH. It is constructed in the same way as a regular Uu PDCCH (see Section 9.3.5) in respect of multiplexing, scrambling, modulation and mapping to resource elements and antenna ports, with the following exceptions:

- A Resource Element Group (REG) is composed of 4 consecutively available REs in one OFDM symbol in a PRB, counted in ascending order of subcarriers, where an RE is assumed not to be available for the R-PDCCH in the following cases:
  - if the RE is used for cell-specific RSs;[6]
  - if any CSI-RSs (see Section 29.1.2) are present in a subframe (even if they are muted), all REs for transmission of CSI-RSs on antenna ports 15 to 22 are unavailable for the R-PDCCH in that subframe.
- UE-specific RSs are not mapped onto PRB pairs used for the transmission of cross-interleaved R-PDCCH.

These exceptions are designed to simplify the RN receiver design and to reduce the need for excessive configuration parameters, even though it means that a few REs are not able to be used.

Non-cross-interleaved R-PDCCH

The non-cross-interleaved R-PDCCH can be supported regardless of whether the RN is configured to use the donor eNodeB's cell-specific RSs or the UE-specific DM-RSs for demodulation of the R-PDCCH.

[6]If the cell-specific RSs are configured to be transmitted only on antenna port 0, the REs that would be used for transmission of cell-specific RSs on antenna port 1 are also deemed to be unavailable for the R-PDCCH.

With this format, a single R-PDCCH occupies either 1, 2, 4, or 8 contiguous virtual resource blocks, depending on the aggregation level used for the R-PDCCH. The virtual resource blocks are configured for each RN individually by RRC signalling, using any of the PDSCH resource allocation types 0, 1 or 2 (see Section 9.3.5.4).

Scrambling, modulation and mapping to antenna ports follow the same methods as for the PDCCH. However, in contrast to the PDCCH and the cross-interleaved R-PDCCH, the mapping to REs fills REs across the aggregated VRBs before using the next OFDM symbol (not including REs that carry RS). This helps the RN to detect correctly the aggregation level of the R-PDCCH, which can be important for correctly determining whether R-PDCCH and PDSCH are multiplexed into the same PRB pair.

If the non-cross-interleaved R-PDCCH is transmitted to an RN that is configured to receive the PDSCH on the backhaul link in transmission mode 9 (see Section 9.2.2.1), the control information on the R-PDCCH indicates which and how many UE-specific RS antenna ports are transmitted. However, this information affects the R-PDCCH decoding due to the fact that REs used for UE-specific RSs are unavailable for the R-PDCCH itself. Therefore, in order to avoid excessive receiver complexity, REs should not be used for R-PDCCH transmission in the first slot if they could potentially be used for UE-specific RSs according to the maximum possible rank for PDSCH transmission according to the RRC configuration (the parameter 'codebookSubsetRestriction-r10').

R-PDCCH Search Space

For both R-PDCCH formats, similar blind decoding and search space functionality is defined as for the Uu link (see Section 9.3.5.5). However, for the R-PDCCH, the search space is defined separately in the first and second slots. The number of blind decoding trials per slot for the RN is 6 for aggregation levels 1 and 2, and 2 for aggregation levels 4 and 8.

Control Format Indicator

A backhaul version of the PCFICH is not needed, because the size of the control signalling region of the backhaul subframes can be configured semi-statically by RRC signalling. REs that are semi-statically assigned for RN control signalling but are not actually needed for that purpose can still be used by the donor eNodeB for normal data transmissions to UEs in its cells; in such cases, the RN's blind decoding attempt for those REs would simply fail.

ACK/NACK feedback for backhaul uplink transmissions

On the Uu interface, the Physical HARQ Indicator CHannel (PHICH – see Section 9.3.4) carries the HARQ ACK/NACK information in response to uplink data transmissions on the Physical Uplink Shared CHannel (PUSCH). In a synchronous non-adaptive HARQ protocol (see Section 4.4.1.1) the PHICH can be seen as a resource-efficient retransmission request, since it is tied to a given PUSCH transmission. However, for the backhaul uplink, a synchronous non-adaptive HARQ operation is not always straightforward, and therefore an RN-specific PHICH for the backhaul was not considered to be sufficiently beneficial to be worth specifying such a channel. Consequently, all backhaul uplink retransmissions have to be triggered by an R-PDCCH, and are therefore adaptive. If an RN does not detect an R-PDCCH to trigger a PUSCH transmission, the RN simply does not make a PUSCH transmission.

Backhaul uplink control signalling

The uplink control signalling is transmitted by the RN in the same way as specified for the Uu link (see Section 16.3). However, the transmission resources for the backhaul PUCCH are determined by higher layer signalling and can therefore be considered as reserved. Consequently, in contrast to the PDCCH, there is no relationship between the resources used for R-PDCCH transmission and the resources used for PUCCH transmission.

30.3.8.3 Backhaul Data Transmission

Data transmission on the backhaul link uses the same physical channels (PDSCH and PUSCH) as are defined for the access link. For the multiplexing of the PDSCH and R-PDCCH, three possibilities exist in terms of occupation by R-PDCCHs within a pair of backhaul PRBs (illustrated in Figure 30.16):

- a downlink assignment in the first slot and an uplink grant in the second slot;
- only a downlink assignment in the first slot;
- only an uplink assignment in the second slot.

In the second case, it is desirable to use the second slot of such a PRB pair for PDSCH data transmission to the RN to which the R-PDCCH in the first slot is addressed, instead of wasting the PRB by not using it. Therefore, a downlink assignment may include the PRB pair containing the downlink assignment itself. In this case, the RN is aware that the first slot was used to transmit the R-PDCCH, and can therefore correctly assume that only the second slot of that pair of PRBs carries PDSCH data. Although this is in principle applicable to both the cross-interleaved and the non-cross-interleaved R-PDCCH formats, the non-cross-interleaved format benefits more because a second slot is either completely used or completely unused for an R-PDCCH transmission; by contrast, for the cross-interleaved R-PDCCH, a partial occupation of the second slot by an R-PDCCH is likely.

On the other hand, if a donor eNodeB uses at least part of the second slot of a pair of PRBs for any kind of R-PDCCH transmission, that pair of PRBs should not be assigned to any RN;

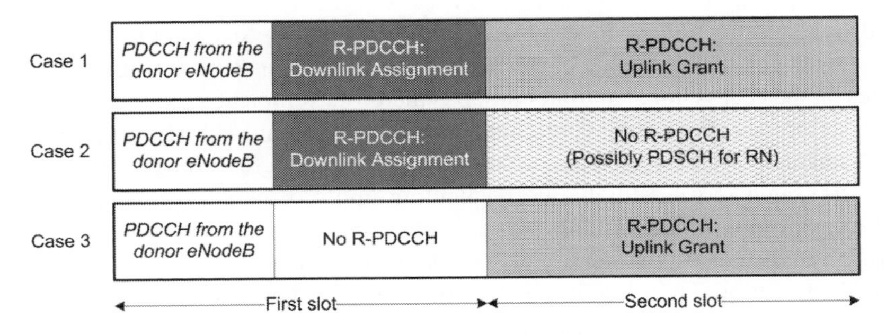

Figure 30.16: Supported R-PDCCH/PDSCH multiplexing possibilities.

otherwise the RN would have to perform additional blind decoding to establish whether the second slot contained data or an uplink grant.

In the third case, it could, in theory, be possible to use the resources of the first slot of such a pair of PRBs for a PDSCH transmission to the same or another RN. However, the practical benefit is smaller compared to the second case, because the available number of REs that could be used for a PDSCH in the first slot is substantially smaller than in the second slot due to the constraint of the RN's backhaul downlink reception window. Moreover, it is simpler to transmit a downlink assignment in the first PRB of the pair (as in the first case) than to use it for a PDSCH. Therefore, the transmission of PDSCH in the first PRB in the third case is not supported.

30.3.9 Backhaul Scheduling

In the case of an outband RN, there are no special restrictions for the backhaul scheduler located at the donor eNodeB, because no adverse effects from or for the access link need to be taken into account. The donor eNodeB can schedule an outband RN for backhaul uplink and downlink just like any UE in the donor eNodeB's cells.

On the other hand, in the case of an inband Type 1 RN, the donor eNodeB needs to consider at least the following constraints:

- which subframes are available for backhaul data transmission or reception, as dictated typically by the MBSFN subframe configuration advertised by the RN;
- the availability of transmission resources for feedback to or from the RN with suitable timing for the corresponding data transmissions.

Regardless of such constraints, the scheduler in the donor eNodeB should also try to take into account the added delay arising from the additional network hop from the donor eNodeB to the RN. Depending on the overall delay requirements, the donor eNodeB should limit the required number of backhaul retransmissions, as each backhaul retransmission adds at least 8 ms to the total delay experienced (see Section 10.3.2.5). A donor eNodeB may, for example, use more conservative link adaptation for an RN than it would for a regular UE.

For FDD operation (i.e. using Frame Structure Type 1), the subframes configured for backhaul transmission follow a periodicity of 8 ms, with the exception of subframes that cannot be used for the downlink backhaul because they cannot be configured as MBSFN subframes on the access link. An 8-bit bitmap is therefore sufficient to configure the downlink backhaul subframes. For every configured downlink backhaul subframe, a corresponding uplink backhaul subframe 4 ms later is configured. An uplink grant transmitted in a downlink backhaul subframe is valid for the corresponding uplink backhaul subframe. Similarly, a PDSCH transmission in a downlink backhaul subframe is associated with an ACK/NACK transmission in the corresponding uplink backhaul subframe.

For TDD operation (i.e. using Frame Structure Type 2), the backhaul and access subframe configurations must be mutually compatible. While Frame Structure Type 2 supports a total of seven uplink-downlink configurations (0 to 6 – see Section 6.2), an RN only needs to support five of these configurations for the access link. For most of these access link configurations, several backhaul configurations exist, allowing a variety of numbers and/or positions of the backhaul downlink and uplink subframes, as shown in Figure 30.17.[7]

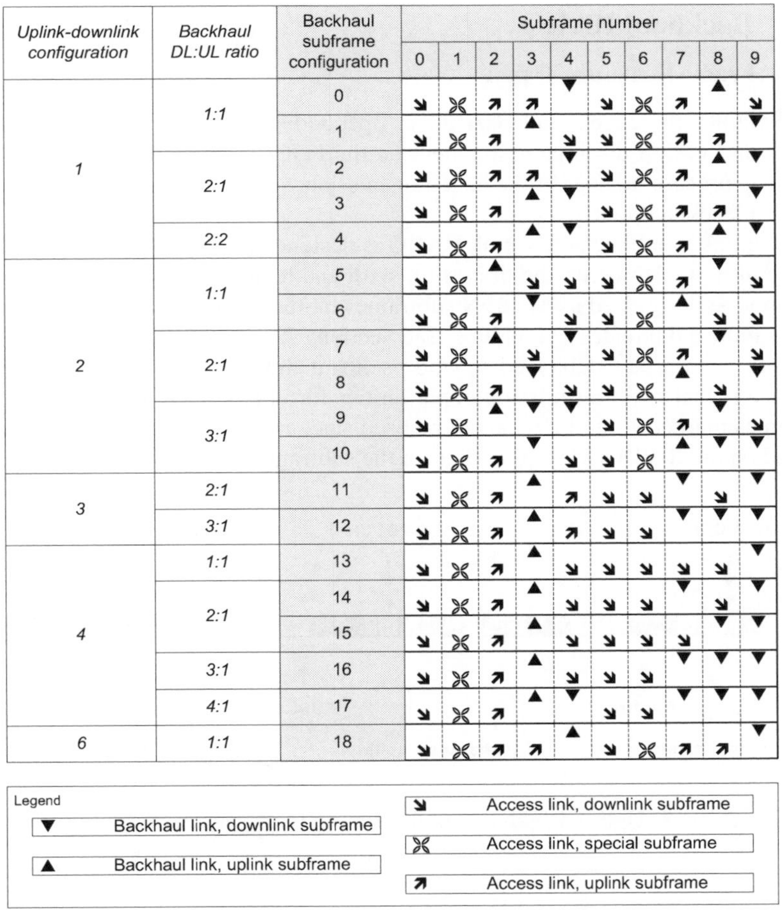

Uplink-downlink configuration	*Backhaul DL:UL ratio*	Backhaul subframe configuration	Subframe number 0	1	2	3	4	5	6	7	8	9
1	*1:1*	0	↘	⌘	↗	↗	▼	↘	⌘	↗	▲	↘
		1	↘	⌘	↗	▲	↘	↘	⌘	↗	↗	▼
	2:1	2	↘	⌘	↗	↗	▼	↘	⌘	↗	▲	▼
		3	↘	⌘	↗	▲	▼	↘	⌘	↗	↗	▼
	2:2	4	↘	⌘	↗	▲	▼	↘	⌘	↗	▲	▼
2	*1:1*	5	↘	⌘	▲	↘	↘	↘	⌘	↗	▼	↘
		6	↘	⌘	↗	▼	↘	↘	⌘	▲	↘	↘
	2:1	7	↘	⌘	▲	↘	▼	↘	⌘	↗	▼	↘
		8	↘	⌘	↗	▼	↘	↘	⌘	▲	↘	▼
	3:1	9	↘	⌘	▲	▼	▼	↘	⌘	↗	▼	↘
		10	↘	⌘	↗	▼	↘	↘	⌘	▲	▼	▼
3	*2:1*	11	↘	⌘	↗	▲	↗	↘	↘	▼	↘	▼
	3:1	12	↘	⌘	↗	▲	↗	↘	↘	▼	▼	▼
4	*1:1*	13	↘	⌘	↗	▲	↘	↘	↘	↘	↘	▼
	2:1	14	↘	⌘	↗	▲	↘	↘	↘	▼	↘	▼
		15	↘	⌘	↗	▲	↘	↘	↘	↘	▼	▼
	3:1	16	↘	⌘	↗	▲	↘	↘	↘	▼	▼	▼
	4:1	17	↘	⌘	↗	▲	▼	↘	↘	▼	▼	▼
6	*1:1*	18	↘	⌘	↗	↗	▲	↘	⌘	↗	↗	▼

Legend
▼ Backhaul link, downlink subframe
▲ Backhaul link, uplink subframe
↘ Access link, downlink subframe
⌘ Access link, special subframe
↗ Access link, uplink subframe

Figure 30.17: TDD subframe configurations.

For a backhaul uplink grant transmitted in backhaul downlink subframe n by the donor eNodeB, the corresponding backhaul PUSCH transmission by the RN occurs in backhaul uplink subframe $n + k_1$, where k_1 is determined in the same way as for the Uu interface.

For a backhaul downlink PDSCH transmission by the donor eNodeB in subframe n, the corresponding HARQ ACK/NACK feedback is transmitted by the RN in backhaul

[7]The RN is not required to use the same uplink-downlink configuration for the access link as for the backhaul link; however, it is likely that this can be assumed in most, if not all, practical deployments. Figure 30.17 reflects this assumption.

uplink subframe $n + k_2$, where k_2 is the smallest integer greater than or equal to 4. If multiple ACK/NACKs need to be transmitted in a single subframe, the same 'bundling' and 'multiplexing' methods are available as for the Uu interface (see Section 23.4.3).

30.3.10 Backhaul HARQ

30.3.10.1 Frame Structure Type 1

In the downlink, the access link HARQ is operated in an asynchronous manner. Each downlink transmission resource assignment on the PDCCH includes an explicit indication of the HARQ process number of the data transmission, so the same design can be applied to the downlink backhaul link.

For the uplink, the FDD access link HARQ is designed as a synchronous protocol with a periodicity of 8 ms. However, the first, fifth, sixth and tenth subframes in each 10 ms radio frame cannot be designated as MBSFN subframes on the access link due to the presence of synchronization and broadcast signalling (see Sections 7.2 and 9.2.1), and therefore a regular 8 ms pattern cannot be established for the backhaul link. As shown in Figure 30.18, the consequence is that two transmission opportunities for uplink data on the backhaul link (in the subframes marked 13 and 29) cannot be used since no corresponding resource grant can be received on the R-PDCCH 4 ms earlier (in the subframes marked 9 and 25).

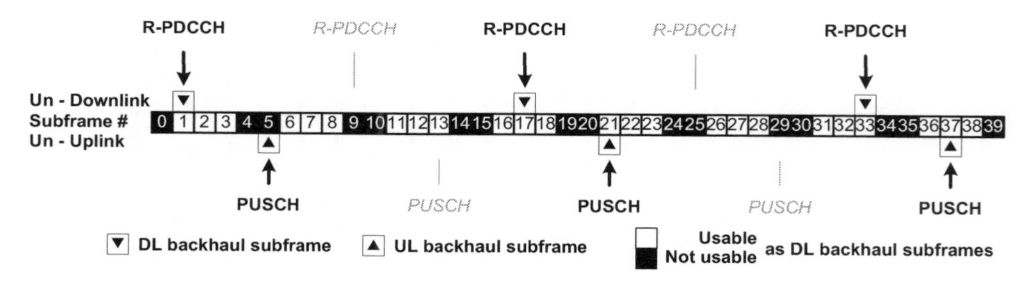

Figure 30.18: Uplink backhaul HARQ timing with 8 ms periodicity (FDD).

The backhaul uplink therefore uses a modification of the access link protocol. The synchronous aspect of the identification of HARQ processes is retained in such a way that the HARQ processes are mapped onto backhaul uplink subframes sequentially; however, the time aspect of the synchronous operation is modified. This is similar to the way in which the uplink HARQ works in the TDD access link, where the availability of subframes for uplink transmission is configured by an RRC parameter. In order to simplify the backhaul uplink subframe configuration and to disrupt as few HARQ processes as possible on the access uplink, only configurations with a basic periodicity of 8 ms are supported. Since an uplink transmission needs to be assigned by an uplink grant 4 ms earlier by means of an R-PDCCH, this implies that if a downlink subframe cannot be used for backhaul transmission due to an MBSFN subframe restriction on the access link, the corresponding uplink subframe cannot be configured as a backhaul uplink subframe, even if it were necessary to do so to follow the 8 ms periodicity.

A minimum Round Trip Time (RTT) of 8 ms applies for the backhaul HARQ (just as for the access link HARQ). In order to minimize the delay, the number of backhaul uplink HARQ processes is minimized as a direct function of the backhaul uplink subframe configuration. This results in the number of HARQ processes on the backhaul uplink being between 1 and 6.

30.3.10.2 Frame Structure Type 2

In the downlink, the access link HARQ is operated in an asynchronous manner. Each downlink transmission resource assignment on the PDCCH includes an explicit indication of the HARQ process number of the PDSCH transmission, so the same design can be applied to the backhaul downlink. In the uplink, a synchronous protocol is employed with a fixed RTT of 10 ms, regardless of the uplink-downlink configuration. For backhaul subframe configuration 4 (see Figure 30.17), two uplink HARQ processes are employed, while for all other backhaul subframe configurations, only a single uplink HARQ process is used.

30.4 Summary

The introduction of relaying represents a significant new step in the LTE radio access network as it is evolved for LTE-Advanced. The incorporation of Relay Nodes (RNs) in an LTE network can offer attractive cost advantages for network operators compared to a homogeneous deployment of macro eNodeBs. Key deployment scenarios for RNs include coverage extension, to extend the range of macro-cells, and capacity enhancement, to boost the supported data rates within the coverage area of a macro-cell.

In LTE-Advanced, RNs operate as a type of eNodeB, with a wireless backhaul link that is specially designed to reuse as much of the LTE access link technology as possible to connect to a donor eNodeB. This wireless backhaul link may operate on the same or a different frequency as the access link from the RN to the RN-assisted UEs (inband and outband operation respectively), depending on the available spectrum of the network operator.

References[8]

[1] 3GPP Technical Specification 25.106, 'UTRA repeater radio transmission and reception', www.3gpp.org.

[2] 3GPP Technical Specification 36.106, 'Evolved Universal Terrestrial Radio Access (E-UTRA); FDD Repeater radio transmission and reception', www.3gpp.org.

[3] Nokia Siemens Networks and Nokia, 'R1-100353: Comparing In-band vs. Out-band Relays in coverage limited scenario', www.3gpp.org, 3GPP TSG RAN WG1, meeting 59bis, Valencia, Spain, January 2010.

[4] Panasonic, 'R1-082397: Discussion on the various types of relays', www.3gpp.org, 3GPP TSG RAN WG1, meeting 54, Warsaw, Poland, June 2008.

[5] Ericsson, 'R1-082024: A discussion on some technology components for LTE-Advanced', www.3gpp.org, 3GPP TSG RAN WG1, meeting 53, Kansas City, USA, May 2008.

[6] T. M. Cover and A. A. El Gamal, 'Capacity Theorems for the Relay Channel', *IEEE Trans. Inf. Theory*, Vol. 25, pp. 572–584, September 1979.

[8]All web sites confirmed 1st March 2011.

[7] G. Kramer and A. J. van Wijngaarden, 'On the White Gaussian Multiple-Access Relay Channel' in *Proc. IEEE Int. Symp. Information Theory (ISIT)*, (Sorrento, Italy), p. 40, June 2000.

[8] G. Kramer, M. Gastpar, and P. Gupta, 'Cooperative Strategies and Capacity Theorems for Relay Networks', *IEEE Trans. Inf. Theory*, Vol. 51, pp. 3037–3063, September 2005.

[9] G. Kramer, I. Marić, and R. D. Yates, 'Cooperative Communications', Vol. 1 of *Foundations and Trends in Networking*. Hanover, MA: now Publishers, Inc., 2007.

[10] T. M. Cover and J. A. Thomas, *Elements of Information Theory*. New York: John Wiley & Sons, Inc., 1991.

[11] J. G. Proakis, *Digital Communications*, Fourth Edition. New York: McGraw-Hill, Inc., fourth ed., 2001.

[12] T. S. Rappaport, *Wireless Communications: Principles and Practice*, Second Edition. Upper Saddle River, New Jersey: Prentice-Hall, Inc., 2002.

[13] C. E. Shannon, 'A Mathematical Theory of Communication', *Bell Syst. Tech. J.*, Vol. 27, pp. 379–423, 623–656, 1948.

[14] I. E. Telatar, 'Capacity of Multi-Antenna Gaussian Channels', *European Trans. on Telecomm.*, Vol. 10, pp. 585–596, November-December 1999.

[15] G. J. Foschini and M. J. Gans, 'On Limits of Wireless Communications in a Fading Environment when Using Multiple Antennas', *Wireless Personal Communications*, Vol. 6, pp. 311–335, March 1998.

[16] J. N. Laneman, D. N. C. Tse, and G. W. Wornell, 'Cooperative Diversity in Wireless Networks: Efficient Protocols and Outage Behavior', *IEEE Trans. Inf. Theory*, Vol. 50, pp. 3062–3080, December 2004.

[17] K. Azarian, H. El Gamal, and P. Schniter, 'On the Achievable Diversity-Multiplexing Tradeoff in Half-Duplex Cooperative Channels', *IEEE Trans. Inf. Theory*, Vol. 51, pp. 4152–4172, July 2005.

[18] Sikora M., J. N. Laneman, M. Haenggi, D. J. Costello, and T. E. Fuja, 'Bandwidth- and Power-Efficient Routing in Linear Wireless Networks', *IEEE Trans. Inf. Theory*, Vol. 52, pp. 2624–2633, June 2006.

[19] R. Koetter and M. Medard, 'An Algebraic Approach to Network Coding', *IEEE/ACM Trans. Netw.*, Vol. 11, pp. 782–795, October 2003.

[20] S. Katti, S. Gollakota, and D. Katabi, 'Embracing wireless interference: analog network coding' in *Proc. Conf. Applications, Technologies, Architectures, and Protocols for Computer Communications (SIGCOMM)*, pp. 397–408, 2007.

[21] S. Katti, H. Rahul, W. Hu, D. Katabi, M. Médard, and J. Crowcroft, 'XORs in the air: practical wireless network coding', *IEEE/ACM Trans. Netw.*, Vol. 16, pp. 497–510, may 2008.

[22] Alcatel-Lucent and CHTTL, 'R1-092321: System Design Frameworks to Support Type II Relay Operation in LTE-A', www.3gpp.org, 3GPP TSG RAN WG1, meeting 57bis, Los Angeles, USA, June 2009.

[23] Samsung, 'R1-100139: Full duplex configuration of Un and Uu subframes for Type I relay', www.3gpp.org, 3GPP TSG RAN WG1 meeting 59bis, Valencia, Spain, January 2010.

[24] 3GPP Technical Specification 36.331, 'Evolved Universal Terrestrial Radio Access (E-UTRA) Radio Resource Control (RRC); Protocol specification', www.3gpp.org.

[25] 3GPP Technical Report 33.816, 'Feasibility Study on LTE Relay Node Security (Release 10)', www.3gpp.org.

[26] 3GPP Technical Specification 33.401, '3GPP System Architecture Evolution (SAE); Security architecture', www.3gpp.org.

[27] 3GPP Technical Specification 36.216, 'Evolved Universal Terrestrial Radio Access (E-UTRA); Physical Layer for Relaying Operation (Release 10)', www.3gpp.org.

31

Additional Features of LTE Release 10

Teck Hu, Philippe Godin and Sudeep Palat

31.1 Introduction

In addition to the features described in Chapters 28 to 30, a number of additional enhancement features are included in Release 10 of the LTE specifications. These are explained in the following sections.

31.2 Enhanced Inter-Cell Interference Coordination

An important feature of LTE-Advanced is the support of heterogeneous network deployments consisting of small cells overlaid within the coverage area of a macrocellular network. The small cells may comprise picocells or femtocells (see Chapter 24) or Relay Nodes (RNs) (see Chapter 30). Overlaying small cells in this way enables higher spectral reuse due to cell-splitting, which delivers an increase in capacity and can support localized high traffic-densities ('hot-spots'). Overlaying small cells is particularly attractive when the availability of suitable sites for macrocells is limited. An alternative would be to use separate spectrum (see Section 28.3.1.1), but this depends on the availability of such additional spectrum.

However, the improved system capacity from heterogeneous deployments sharing the same spectrum present interference scenarios that are different from a homogeneous macrocellular networks. Sections 31.2.1 to 31.2.6 explore these scenarios in more detail and explain the enhanced Inter-Cell Interference Coordination (eICIC) techniques introduced in Release 10 to address them.

LTE – The UMTS Long Term Evolution: From Theory to Practice, Second Edition.
Stefania Sesia, Issam Toufik and Matthew Baker.

The enhanced ICIC techniques complement those already available since Release 8 (see Sections 12.5.1 and 12.5.2) and are designed for the case of single-carrier operation (i.e. without carrier aggregation).

Co-channel heterogeneous network deployments can in general be categorized into two scenarios: macro-pico and macro-femto. In the macro-pico case, the small cells are Open Subscriber Group (OSG) cells, open to all users of the macrocellular network. In the macro-femto case, the small cells are Closed Subscriber Group (CSG) cells accessible only to a limited group of users. Although both scenarios are considered in Release 10, the main focus of the eICIC techniques, and hence of this section, is the OSG macro-pico scenario.[1] The differences in the applicable eICIC solutions between the two scenarios are mentioned where relevant. Figure 31.1 illustrates the interference scenarios for the macro-pico scenario in both downlink and uplink.

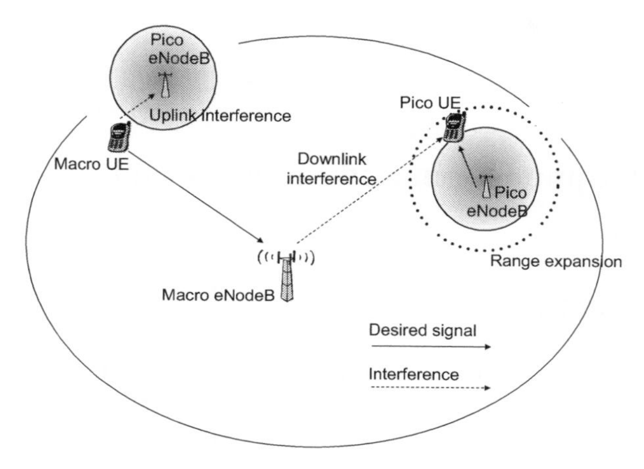

Figure 31.1: Heterogeneous network interference scenarios in downlink and uplink.

In a homogeneous macrocellular network, a UE is served by the strongest cell, and the number of UEs served by the picocells would be limited due to the lower power of the picocells compared to the macrocells. To derive the full gain of cell splitting, the serving eNodeB can intentionally 'bias' the handover offset values of some UEs in RRC_CONNECTED mode to transfer them to the picocells. This has the effect of expanding the coverage of the picocells and is therefore sometimes known as *Cell Range Expansion* (CRE). If a UE is transferred in this way, the signal from the pico eNodeB would be low due to its lower transmission power, while the interference received from the macro eNodeB would be significantly higher, as highlighted in Figure 31.1.

To illustrate the level of interference, Figure 31.2 plots the Cumulative Distribution Function (CDF) of the received Signal-to-Interference-plus-Noise Ratio (SINR) with different bias values used for the CRE. The locations of the picocells and the UE distributions are according to configuration 4b in [1, Section A.2.1.1.2], with four picocells per macrocell.[2] The figure clearly shows how the SINR degrades as the bias is increased.

[1]The corresponding interference scenarios for the macro-femto case are discussed in Section 24.3.1.

[2]Additional simulation results can be found in [2].

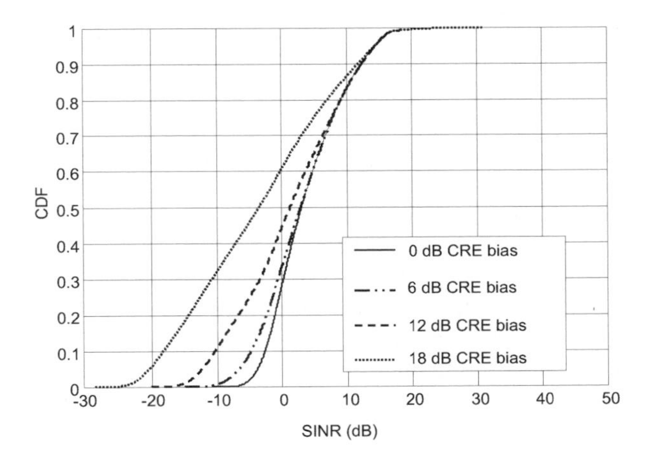

Figure 31.2: SINR for macro-pico scenario with various degrees of range expansion.

31.2.1 LTE Interference Management

The interference mitigation and ICIC techniques available in Release 8 can be summarized as frequency-domain scheduling (including frequency-selective scheduling and frequency hopping), power setting, increasing robustness (for example by beamforming or interference cancellation for the data channels, or by increasing the number of Control Channel Elements (CCEs) used for each message on the PDCCH – see Section 9.3.5) and X2-interface-based data channel interference coordination (see Sections 12.5.1 and 12.5.2).

Interference avoidance based on frequency-domain partitioning between different cells are of limited benefit for the synchronization signals, Physical Broadcast Channel (PBCH), cell-specific Reference Signals (RSs) or control channels (PDCCH, PCFICH, and PHICH). These are needed for initial access to the network and/or thereafter for maintaining the radio link. Therefore, their time-frequency locations are fixed (with the exception of the cell-specific RSs, which can use a frequency reuse factor of 3 or 6 depending on the number of antenna ports configured) and frequency partitioning of these channels and signals would not be backward-compatible with Release 8/9 UEs. However, the interference experienced by the picocell UEs in a co-channel macro-pico deployment affects these channels also, and, if very large range expansion is employed, the control channel reception at the picocell UEs may fail, resulting in outage.

The need for interference mitigation of the control channels was the motivation for the specification of time-domain-based ICIC in Release 10. The overall objective of this eICIC is to mute certain subframes of one layer of cells in order to reduce the interference to the other layer. These muted subframes are called *Almost Blank Subframes* (ABSs) [3].

31.2.2 Almost Blank Subframes

ABSs are defined as subframes with reduced downlink transmission power and/or activity. Ideally, ABSs would be totally blank when configured by a macrocell in a macro-pico scenario, in order to remove as much as possible of the interference towards the UEs served by the picocells. In practice, the transmission and exact content of ABSs would be an implementation choice, taking into account the gains from interference reduction versus the loss of transmission resources from being unable to transmit PDSCH data in the ABSs. As one example, uplink PUSCH data traffic could continue to be scheduled by means of uplink grants on the PDCCH, and corresponding HARQ[3] ACK/NACK feedback could be transmitted, while downlink assignments might be stopped.

The important consideration of backward-compatibility means that cells must remain accessible and measurable for Release 8/9 UEs. As a result, the cell-specific RSs (and CSI-RS if configured), PSS, SSS, Paging Channel (PCH) and PBCH must also be transmitted in ABSs. Nevertheless, even with these transmissions, the ABSs can contain much less energy than normal subframes and thus reduce the interference towards both the data and control signalling of the corresponding subframes of the co-channel victim cell.

A subframe designated as an ABS can be either a normal subframe or an MBSFN subframe,[4] as illustrated in Figure 31.3.

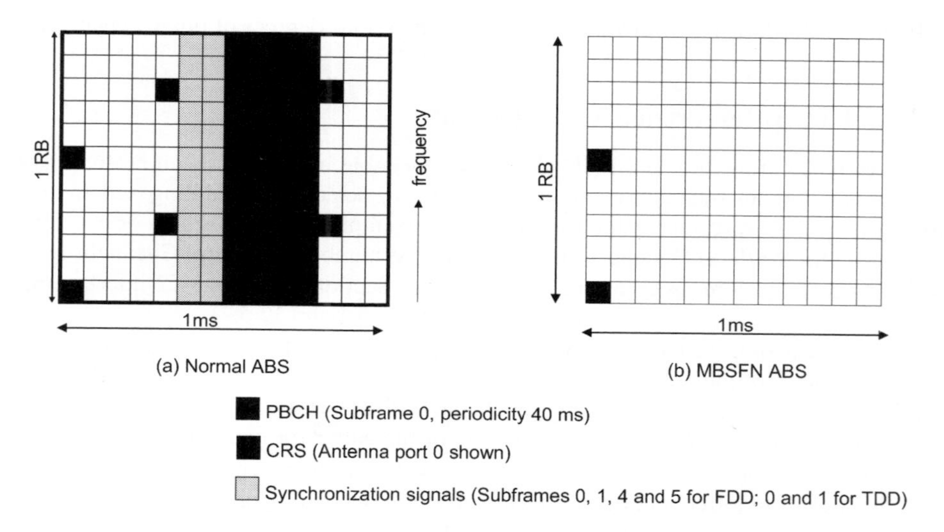

Figure 31.3: Normal and MBSFN Almost Blank Subframes.

ABSs which are also MBSFN subframes can reduce interference to the victim cell even more than ABSs which are not MBSFN subframes, since the former do not have cell-specific RSs in the PDSCH region and cannot contain PBCH or PCH transmissions. However, only subframes 1, 2, 3, 6, 7 and 8[5] can be configured as MBSFN subframes, and, as a result, MBSFN subframes alone cannot fully support the 8 ms periodicity of the synchronous uplink HARQ processes. Therefore, if uplink scheduling is required during downlink ABSs, some of the downlink ABSs would need to occur in subframes which cannot be configured as MBSFN subframes.

[3]Hybrid Automatic Repeat reQuest.

[4]Multimedia Broadcast Single Frequency Network subframe – see Section 13.4.1.

[5]For FDD; indexing starting from zero.

31.2.3 X2 Interface Enhancements for Time-Domain ICIC

The characteristics of ABSs allow a high degree of flexibility in their implementation. To ensure their effectiveness when configured by one eNodeB, coordination is required with the neighbouring eNodeBs they are intended to benefit. For this purpose, time-domain inter-cell coordination signalling via the X2 interface is defined in Release 10, whereby the pattern of ABSs configured by one eNodeB can be indicated to its neighbours.

This signalling comprises 'ABS bitmaps', of which two can be signalled over the X2 interface from one cell (typically a macrocell) to another (typically a picocell).[6]

The ABS patterns are updated 'semi-statically', i.e. not faster than the Release 8/9 frequency-domain RNTP[7] indications.

The first ABS bitmap, known as the ABS Pattern, indicates the complete set of ABS subframes to assist the receiving eNodeB with its scheduling operations. Like the RNTP indications, each bit of an ABS bitmap informs the receiving eNodeB of the sending cell's intention regarding its transmit power – but on a per-subframe basis rather than a per-RB basis. This information can be used by the receiving eNodeB to arrange its scheduling operations to avoid transmitting in non-ABS subframes to UEs believed to be in high-interference areas such as the cell edge. Instead, the receiving eNodeB can schedule such UEs in the ABS subframes where they should experience less interference. As with frequency-domain ICIC, the exact operation of ABS-based time-selective scheduling is left up to the eNodeB implementation.

The second ABS bitmap, known as the 'Measurement Subset', is a subset of the first and is expected to be less variable; it is intended to recommend to the receiving eNodeB a suitable set of subframes that can be used for measurements by the UEs that would potentially be suffering interference in the subframes that are not configured as ABSs.

In order to align with the uplink HARQ round trip times, the ABS bitmaps have the following periodicities:

- 40 ms for FDD;
- 20 ms for TDD configurations 1–5;
- 70 ms for TDD configuration 0
- 60 ms for TDD configuration 6.

The ABS bitmap signalling can be transferred as part of the 'Load Indication' and 'Resource Status Reporting Initiation' procedures over the X2 interface. In addition, a message known as an 'Invoke Indication' allows one eNodeB to request ABS configuration at another. Such a request would be based on the requesting eNodeB's assessment of the interference situation in its cell(s). The request initiates the Load Indication procedure, and the resulting 'Load Information' message from the eNodeB receiving the request contains the ABS bitmaps.

An eNodeB which has been informed about configured ABSs in another cell can return an 'ABS Status' message to the eNodeB that configured the ABSs. This message assists the latter eNodeB in determining whether the number of configured ABSs might need to be increased

[6]ABS bitmaps make no distinction as to the cell type; ABSs can be configured by macro, pico or femto cells, but, in Release 10, ABS bitmap signalling over the X2 interface is not supported from a macro eNodeB to a CSG Home eNodeB. ABS configuration in CSG cases is performed through Operation and Maintenance (O&M).

[7]Relative Narrowband Transmit Power – see Section 12.5.1.

or decreased, by indicating the percentage resource blocks in the currently configured ABSs that are allocated for UEs served by the victim eNodeB.

In scenarios where the UEs served by a pico eNodeB suffer interference from more than one strong interfering macro eNodeB, each with potentially different ABS patterns, the set of usable ABSs may be different from what any one individual macro eNodeB has indicated to the pico eNB. For the macro eNodeB to interpret correctly the ABS Status information from the pico eNodeB, the pico eNodeB should indicate the actual usable ABS pattern (in the form of a bitmap in the information element 'Usable ABS Pattern Info') to the macro eNodeB together with the ABS status information.

Figure 31.4 shows a typical message exchange over the X2 interface for ABS coordination between a macro eNodeB and a pico eNodeB. Box A in Figure 31.4 shows the invoke function being used to request the the macro eNodeB to configure ABSs. The macro eNodeB may take this request into consideration when configuring ABSs; the exact relationship between a request and any subsequent decision to configure ABSs is not standardized.

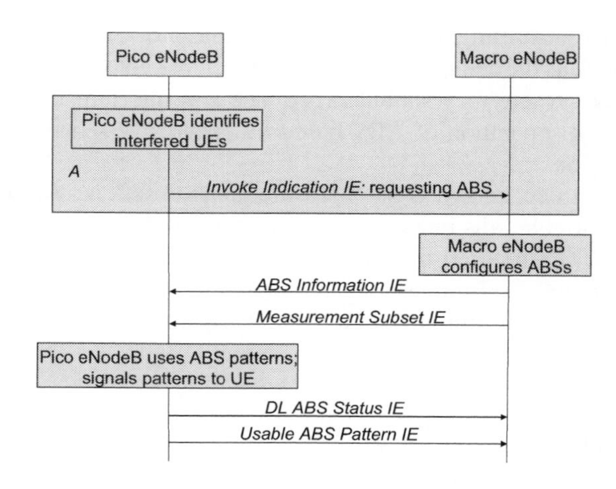

Figure 31.4: A typical exchange of ABS information over the X2 interface.

31.2.4 UE Measurements in Time-Domain ICIC Scenarios

For UEs which are suffering inter-cell interference, the measurement results will depend substantially on whether the measurements are restricted to the ABSs (which are protected from interference) or are made in all subframes.

Figure 31.5 illustrates the various UE measurement scenarios which are affected by the existence of protected subframes. The most common scenario for each of these measurements is when UEs served by a picocell suffer interference from a macrocell. UEs served by a macrocell do not need restrictions for their serving-cell measurements; however, they should use the configured ABSs to measure any picocells which might be handover targets, since during the ABS subframes the interference experienced at the picocell is lower than in other subframes; this allows interfered picocells to be detected by the macro UEs for possible handover.

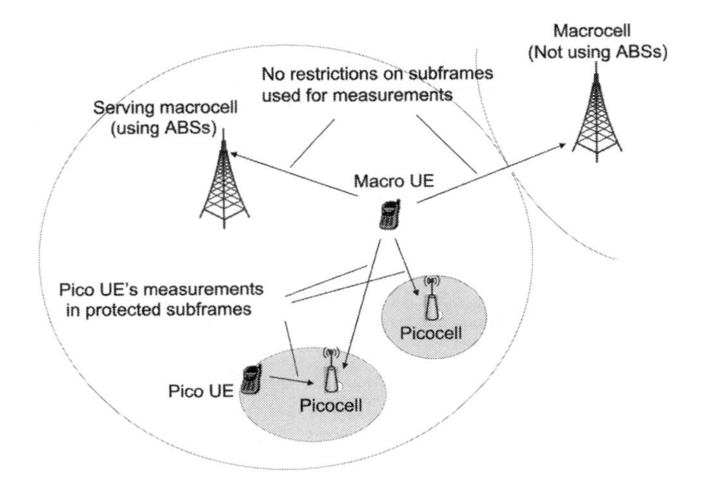

Figure 31.5: UE measurements of serving and neighbour cells.

The affected measurements can be classified as Radio Link Monitoring (RLM) measurements, Radio Resource Management (RRM) measurements and Channel State Information (CSI) measurements.

31.2.4.1 RLM Measurements

The RLM procedure is used for a UE to monitor the downlink radio link quality of the serving cell for the purpose of determining whether synchronization has been lost and Radio Link Failure (RLF) has occurred (see Section 22.7). In order to avoid unnecessary loss of synchronization being reported by a UE, when ABSs are configured the RLM measurements should be made in those subframes only.

31.2.4.2 RRM Measurements

For RRM-related measurements such as intra-frequency cell reselection, the UE needs to perform RSRQ and RSRP measurements.[8]

RSRQ. Since the measured RSSI[9] value used in deriving RSRQ will depend on whether the measurement is made in ABSs only or in all subframes, the RSRQ measurements need to be restricted to a limited set of subframes, which are indicated to the UE by RRC signalling. If the network restricts a UE's RSRQ measurements in this way, the RSSI definition is also modified so that it is measured over all OFDM symbols in the indicated subframes (rather than only the OFDM symbols containing cell-specific RSs). This modification reduces the impact of interference from the cell-specific RSs in the ABSs, especially in the case of ABSs which are MBSFN subframes; in MBSFN ABSs, cell-specific RSs are present only in the first

[8]Reference Signal Received Quality and Reference Signal Received Power – see Sections 22.3.1.2 and 22.3.1.1 respectively.

[9]Received Signal Strength Indicator.

symbol, and taking measurements only on this symbol would underestimate the RSRQ due to the strong neighbour-cell cell-specific RS interference. Figure 31.6 shows the effect of this modification on the RSRQ accuracy by comparing the CDFs of the RSRQ measurements with the Release 8 and Release 10 RSSI definitions.[10] The case when the neighbour-cell cell-specific RS interference is perfectly cancelled by the UE is also shown for comparison.

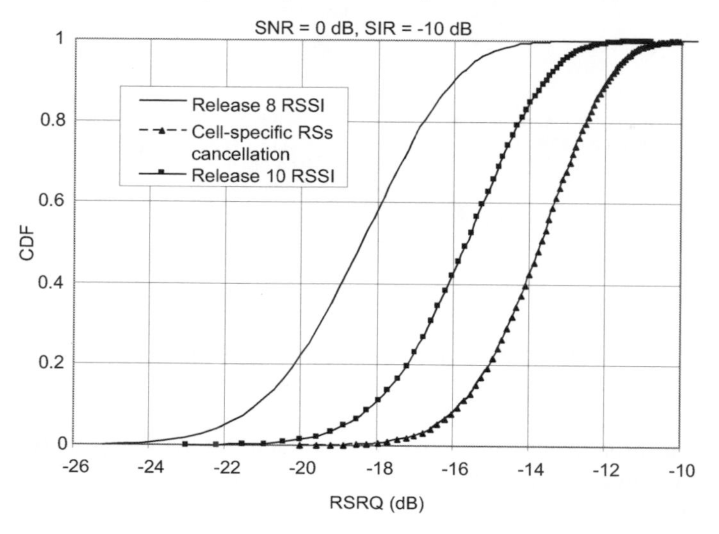

Figure 31.6: RSSI measurement modification.

RSRP. Since RSRP is a measurement of cell-specific RSs, RSRP measurement values are not affected as strongly by whether they are taken in ABSs or non-ABSs, and therefore the RSRP definition is unchanged in Release 10. However, the RSRP measurement accuracy is likely to be strongly impacted when there is collision between the cell-specific RSs of an interfering cell and those of the cell being measured.

31.2.4.3 CSI Measurements

CSI measurements, especially the CQI reports, provide the serving eNodeB with information related to the SINR, which is useful for scheduling and selecting an appropriate Modulation and Coding Scheme (MCS) to use for each UE. When ABSs are configured in an interfering cell, the SINR experienced by the UEs in a victim cell will fluctuate dramatically from subframe to subframe. In order to exploit the potential performance gain of ABSs for scheduling and link adaptation, the CSI measurements made by these UEs should be performed on restricted subsets of the subframes.

31.2.5 RRC Signalling for Restricted Measurements

When an eNodeB receives the first and second bitmaps of the ABS information over the X2 interface from another eNodeB, it has to derive suitable subframe patterns to signal to the UEs it is serving, to restrict their measurements in the time-domain. The signalled patterns should

[10]The simulation assumptions used here can be found in [4].

be a subset of the ABS bitmaps received over the X2 interface, although the exact relationship between the patterns signalled over the X2 interface and the sets of subframes signalled to the UEs for restricted measurements is not standardized; it is entirely an implementation matter and can take into account other factors related to the scheduling algorithm.

Dedicated (i.e. UE-specific) RRC signalling is used to indicate to the UEs in connected mode[11] the sets of subframes to which their measurements should be restricted. The use of dedicated signalling (as opposed to broadcast) enables different UEs to be informed of different measurement subframe patterns, depending, for example, on their proximity to interfering neighbour cells.

The RRC signalling can configure three independent measurement-subframe patterns for a UE that supports time-domain measurement restrictions:

- **Pattern 1** applies to RLM and RRM measurements on the serving cell.
- **Pattern 2** applies to RRM measurements on certain intra-frequency neighbour cells with specifically indicated Physical Cell Identities (PCIs).[12] For neighbour cells whose PCIs are not listed, no RRM measurement restrictions apply; this is the default for neighbour cells which have not configured ABSs.
- **Pattern 3** applies to CSI measurements (including Channel Quality Indicator (CQI), Precoding Matrix Indicator (PMI) and Rank Indicator (RI)). This pattern enables the eNodeB to receive accurate CSI for the subframes in which the UE is expected to be scheduled. Pattern 3 comprises two independent sets of subframes per UE. This enables the serving eNodeB to compare a UE's CQI reports for different sets of subframes, for example for ABSs and non-ABSs, or for ABSs configured by two different neighbour (e.g. macro) eNodeBs which have both signalled their ABSs to the serving (e.g. pico) eNodeB. In the latter example, the two signalled sets of subframes could comprise one set that is common to both the interfering eNodeBs and another that contains only subframes configured as ABSs by one of the interfering eNodeBs; it should be noted that the two sets are not permitted to overlap.

  For periodic CSI reporting (see Section 10.2.1.2), two sets of reporting periodicities and offsets are configured for the UE; each set is associated with one of the sets of subframes. For aperiodic CSI reporting (see Section 10.2.1.1), the UE reports CSI for whichever set of subframes contains the 'CQI reference resource' corresponding to the aperiodic CSI trigger (see [5, Section 7.2.3]).

Table 31.1 summarizes typical measurement restrictions that could be applied.

31.2.6 ABS Deployment Considerations

The decision as to whether and when to configure ABSs in a network involves a trade-off between capacity loss due to inability to schedule UEs in the ABSs of the interfering cells and capacity gain due to range expansion of the small cells. A number of factors should therefore be taken into account when reaching this decision for a practical deployment:

- Caution should be exercised if attempting to use CQI reports to identify the need for ABSs, since CQI reports are designed to indicate a suitable MCS for the PDSCH, for

[11]Restriction of measurements to specific sets of subframes is not supported for UEs in idle mode in Release 10.
[12]Restricted sets of measurement subframes for inter-frequency measurements are not supported in Release 10.

which frequency-domain ICIC techniques have been shown to be very effective. CQI reports may not adequately predict the likelihood of PDCCH decoding failure.

- At low system load, the probability of users experiencing co-channel interference is expected to be low, as is the need for ABSs. At higher loadings, careful planning of the small cell coverage areas may result in sufficient UEs being served by the small cells without needing significant range expansion and ABSs to mitigate the outage probability for cell edge users.
- If ABSs are configured, the number of UEs handed over into the small cells due to range expansion will affect the number of ABSs required.
- The pattern of ABSs configured by an eNodeB needs to ensure that any active guaranteed bit-rate services can be supported in downlink and uplink. The uplink HARQ timeline must also be taken into account.

Some example ABS patterns are shown in Figure 31.7 for both normal ABSs and MBSFN ABSs. Normal ABS densities of 12.5% and 25% are shown, corresponding to 1 ABS and 2 ABSs respectively per eight subframes. An MBSFN ABS density of 15% is also shown. In all the ABS patterns shown, the 40 ms ABS pattern periodicity for FDD is maintained.

Table 31.1: Typical UE measurement restrictions.

	Time-domain measurement restrictions		
	RLM and serving-cell RRM measurements (Pattern 1)	Neighbour-cell measurements (Pattern 2)	CSI measurements (Pattern 3)
UE with severe interference	ABSs of interfering cell	ABSs of interfering cell (if in PCI list)	ABSs of interfering cell
UE without severe interference	No restriction	ABSs of interfering cell (if in PCI list)	No restriction

31.3 Minimization of Drive Tests

Drive tests, whereby UEs are driven through a deployed access network, are widely used to detect possible coverage holes or inefficiencies in the radio network planning. The Minimization of Drive Tests (MDT) feature aims to enable some of these drive test campaigns to be replaced by the automatic collection of UE measurements triggered by the eNodeBs, in order to reduce the cost of network optimization and carbon dioxide emissions.

Two modes of MDT are specified in Release 10 [6, 7]:

- **Logged MDT**, where the UE logs measurements in idle mode, together with time-stamps and, optionally, accurate location information;
- **Immediate MDT**, where the UE makes measurements in connected mode.

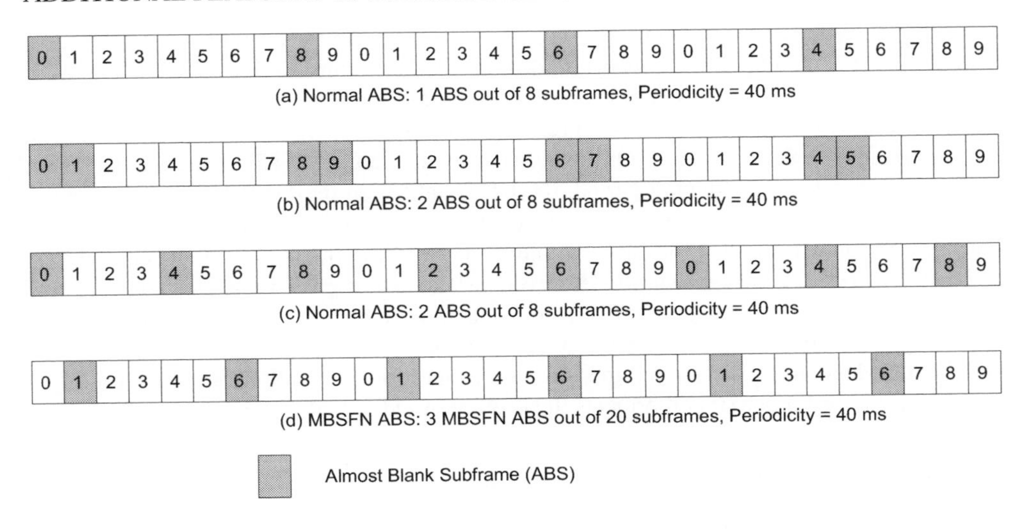

Figure 31.7: Examples of ABS patterns.

The two modes are configured by Operation and Maintenance (O&M) to give the operator more flexibility, even though the configuration and the measurements collected for the two modes may be different.

Both MDT modes can be requested as either a ‘Signalling-Based Trace’ or a ‘Management-Based Trace’ (see Section 2.5.8). Signalling-based MDT would typically be triggered when one specific UE is targeted, whereas with management-based MDT the eNodeB can pick any arbitrary UE that satisfies an MDT requirement.

The two MDT modes are specified for the control plane, but they could also potentially be used as diagnostic monitoring functions [8] in the user plane.

31.3.1 Logged MDT

To perform logged MDT, a UE is configured with the area in which it should log measurements (defined in terms of cells, tracking areas or a whole PLMN[13]), a logging interval, the logging duration and an absolute time reference which it has to feed back in the measurement log. Whenever the UE enters the configured area, it will start logging radio measurements (RSRP and RSRQ) together with the relative time compared to the configured time reference; it may additionally store the location (in terms of cell identity, GNSS[14] coordinates or RF ‘fingerprints’ of neighbour cells (i.e. signal strengths of a number of intra-frequency neighbour cells)). The MDT log is suspended whenever the UE transitions to idle mode, changes RAT or exits its Registered PLMN, and it is resumed when the UE re-enters connected mode in its registered PLMN under the RAT in which it received its configuration.

The reporting of the log to the eNodeB can be done when the UE is in connected mode even if the logging period has not expired, provided that the UE is in a cell of the RAT in

[13]Public Land Mobile Network.

[14]Global Navigation Satellite System.

which it received the configuration. The UE first provides only an indication that MDT data is available. It is then up to the network to decide whether to retrieve the log at this point in time or later. The network thus keeps control of the MDT log reporting.

Dedicated RRC signalling is used to retrieve MDT logs, using both integrity protection and ciphering (see Section 4.2.3). Once successively delivered, the UE clears the data from memory. At the end of a logging period, the UE should keep the MDT log for a minimum period of 48 hours so that the network can retrieve it. Once retrieved, the MDT log is transferred to an MDT collection entity within the network where, in general, it is possible to associate the log with the UE identity. However, special care is given to privacy and data protection in Release 10, such that MDT trace information can only be collected with prior user consent, which users can give or revoke at any time. In addition, processing of management-based MDT data has to be handled in such a way that it does not allow identification of an individual user or UE and that it protects the privacy of individual user locations.

31.3.2 Immediate MDT

To perform immediate MDT, the eNodeB selects UEs that are in connected mode within the area where the MDT log is required and that have the necessary capabilities. Immediate MDT reuses the existing Rel-8/9 measurement configuration and reporting procedures for Radio Resource Management (RRM) purposes.

The UE reports the measurements to the network every time a configured reporting condition is met. The measurements include:

- RSRP and RSRQ measurements, either provided periodically according to a configured reporting interval and number of reports, or provided with Event A2 (see Section 3.2.5.2) or on Radio Link Failure (RLF) (see Section 22.7);
- The UE's power headroom (see Sections 18.3.3 and 28.3.5.3);
- The uplink signal strength measured by the eNodeB.

In addition to these measurements, the overall report may additionally include location information similar to that for logged MDT, but it does not contain time information. RLF reports are stored by the UE even if it transitions to idle mode.

31.4 Machine-Type Communications

Machine-Type Communication (MTC) is a form of data communication which involves one or more entities that do not necessarily need human interaction. MTC devices are UEs equipped for MTC, which communicate through a Public Land Mobile Network (PLMN) with MTC servers and/or other MTC devices.

Applications for MTC are many and diverse. The following are some examples:

- MTC devices can be used by utility companies to read smart meters remotely at short but regular intervals.
- Automatic earthquake monitoring can provide vital public warnings.
- In the transport sector, MTC can be used for toll collection and road-pricing schemes, for monitoring the location of vehicles and for making automatic emergency calls in case of accidents.

- In the future, many consumer electronic devices are likely to incorporate MTC capability, allowing automatic remote maintenance as well as monitoring of heating, alarms and other applications.

An MTC device and its communication requirements are often different from normal human communications. Their traffic can typically be characterized as small, delay-tolerant data packets which are sent infrequently. Other considerations such as the fact that the traffic may originate predominantly from the device itself, low mobility, and requirements for low cost and low battery consumption may also be relevant depending on the type of application. The data packets are sometimes timer-controlled or they may be triggered by certain events such as earthquakes. There could potentially be a large number of MTC devices in a cell.

These characteristics of MTC devices motivate some special handling compared to human-generated traffic. One of the primary considerations addressed first in 3GPP is the potential for overload that could be caused by a large number of MTC devices. Since the MTC traffic can be timer-controlled or event-triggered, there is a possibility of a sudden surge in MTC traffic in a cell in which there is a large number of MTC devices. It is important to ensure that MTC traffic, which is delay-tolerant, does not disrupt (or at least causes only minimal additional delay for) human-generated traffic.

Release 10 support for MTC therefore focuses on overload protection for the Core Network (CN). This is achieved by exploiting the delay-tolerant nature of the MTC traffic by rejecting connection attempts from MTC devices as necessary. Specific CN nodes can provide overload indications to eNodeBs to request eNodeBs to prevent MTC access to these CN nodes. When an MTC device sends an RRC Connection Establishment request including a 'delay tolerant' cause code towards such a CN node, the eNodeB rejects the RRC connection request. In addition, the eNodeB provides an 'extended wait timer' in the RRC 'Connection Reject' message which prevents the MTC device from accessing the network again for delay-tolerant reasons for the duration of the timer. An example message flow for this type of connection rejection is shown in box A in Figure 31.8.

Depending on the type of access, identification of the CN nodes serving the MTC device is not always possible from the RRC 'Connection Request' message. For such cases, the CN node ID is provided in the RRC Connection Setup Complete message. Since it is then not possible to reject the connection request, the extended wait timer is also included in the RRC Connection Release message. This flow is shown in box B in Figure 31.8.

Additional overload-prevention mechanisms such as a longer periodic tracking area update timer are defined in the CN specifications [9].

While Release 10 does not support dynamic change of access type from MTC devices, some exceptions such as emergency calls are still allowed with higher priority.

The Release 10 support for MTC is expected to further optimized in Release 11 and beyond. Several other areas could be addressed. Mechanisms for handling RAN overload are expected to be included in Release 11 and are likely to include the possibility of barring MTC devices from even performing the RACH access for connection establishment during periods of overload, in order to further minimize impact on normal-priority traffic. Optimizing signalling overhead will also be a priority for Release 11, as the normal signalling connection establishment procedure that is required before a device can send any data causes significant overhead in the case of the small and infrequent packets sent by MTC devices. Other prioritized work areas will include handling device addressing for large numbers of

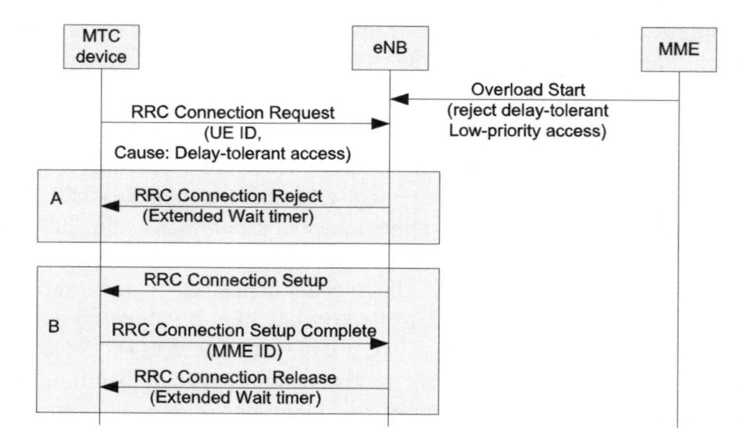

Figure 31.8: Mechanisms for handling CN overload.

devices without a Mobile Station International Subscriber Directory Number (MSISDN), power saving and optimizations for efficient handling of groups of similar MTC devices.

References[15]

[1] 3GPP Technical Report 36.814, 'Evolved Universal Terrestrial Radio Access (E-UTRA); Further advancements for E-UTRA Physical Layer Aspects (Release 9)', www.3gpp.org.

[2] Alcatel-Lucent and Alcatel-Lucent Shanghai Bell, 'Co-Channel Control Channel Performance for Hetnet', www.3gpp.org, 3GPP TSG RAN WG1, meeting 60bis, Beijing, China, April 2010.

[3] Alcatel-Lucent, 'R1-093340: Blank Subframes for LTE', www.3gpp.org, 3GPP TSG RAN WG1, meeting 58, Shenzhen, China, August 2009.

[4] NTT DOCOMO, 'R4-110224: Simulation results for e-ICIC RSRP/RSRQ measurements', www.3gpp.org, 3GPP TSG RAN WG4, meeting 57AH, Austin, USA, January 2011.

[5] 3GPP Technical Specification 36.213, 'Evolved Universal Terrestrial Radio Access (E-UTRA); Physical Layer Procedures (Release 10)', www.3gpp.org.

[6] 3GPP Technical Specification 32.422, 'Telecommunication management; Subscriber and equipment trace; Trace control and configuration management (Release 10)', www.3gpp.org.

[7] 3GPP Technical Specification 37.320, 'Universal Terrestrial Radio Access (UTRA) and Evolved Universal Terrestrial Radio Access (E-UTRA); Radio measurement collection for Minimization of Drive Tests (MDT); Overall description; Stage 2 (Release 10)', www.3gpp.org.

[8] Open Mobile Alliance, 'DiagMon Functions Supplemental Specification OMA-TS-DiagMon_Functions-V1', www.openmobilealliance.org.

[9] 3GPP Technical Specification 24.301, 'Non-Access Stratum (NAS) Protocol for Evolved Packet System (EPS) (Release 10)', www.3gpp.org.

[15] All web sites confirmed 1st March 2011.

32

LTE-Advanced Performance and Future Developments

Takehiro Nakamura and Tetsushi Abe

LTE Release 8 incorporates the latest advances in radio technology and has attracted global industry support. Intensive field trials conducted by equipment vendors and network operators have confirmed the high performance of LTE, and commercial services using LTE were first launched in December 2009. Many leading operators worldwide have committed to deploying LTE, and the roll-out of LTE is expanding and accelerating.

LTE Release 9 brings enhancements and performance improvements in a number of aspects, including Home eNodeBs, MBMS, location services and MIMO. For Release 10 and beyond (LTE-Advanced), the standardization activities in 3GPP were aligned with the IMT-Advanced standardization process in ITU-R[1] to meet future demands for continuous improvement of system performance and satisfy the constantly increasing expectations and demands of the consumers. It was confirmed that LTE-Advanced fully satisfies the requirements set by ITU-R, and LTE-Advanced was therefore accepted as an IMT-Advanced technology. This chapter provides a summary of the system performance achievable with LTE-Advanced using the Release 10 version of the specifications, followed by a discussion of future developments beyond Release 10.

32.1 LTE-Advanced System Performance

As listed in Table 27.1, ITU-R specified eight items as the minimum requirements for IMT-Advanced:

[1] International Telecommunication Union, Radio Communication Sector.

LTE – The UMTS Long Term Evolution: From Theory to Practice, Second Edition.
Stefania Sesia, Issam Toufik and Matthew Baker.

1. Peak spectral efficiency;
2. Cell spectral efficiency;
3. Cell-edge user spectral efficiency;
4. Bandwidth;
5. Latency;
6. Mobility;
7. Handover interruption time;
8. VoIP capacity.

Some of the results of the assessment of LTE-Advanced conducted by 3GPP against these criteria are presented below.

The peak spectral efficiencies for the uplink and downlink are shown in Table 32.1. Details of the assumptions, including overheads, can be found in [1]. The table shows that the ITU-R requirements can be satisfied by using 4-layer spatial multiplexing (based on LTE Release 8 Single-User MIMO (SU-MIMO)) for the downlink and 2-layer spatial multiplexing (based on Release 10 (LTE-Advanced) SU-MIMO) for the uplink.

The more stringent peak spectral efficiency requirements set by 3GPP for LTE-Advanced (see Table 27.2) can be satisfied using up to 8-layer spatial multiplexing in the downlink and up to 4-layer spatial multiplexing in the uplink, according to the Release 10 specifications for SU-MIMO.

Table 32.1: Peak spectral efficiency for LTE-Advanced Release 10.

	Downlink	Uplink
ITU-R requirements (bps/Hz)	15	6.75
LTE-Advanced peak spectral efficiency (bps/Hz)	16.3 (4 MIMO layers) 30.6 (8 MIMO layers)	8.4 (2 MIMO layers) 16.8 (4 MIMO layers)

The average cell spectral efficiencies and cell-edge user spectral efficiencies for the downlink and uplink in the ITU-R test environments are shown in Figures 32.1 and 32.2 respectively. The downlink performance was evaluated using LTE-Advanced Multi-User MIMO (MU-MIMO) and LTE Release 8 SU-MIMO, assuming four transmit and two receive antennas (4×2).

Figure 32.1 shows that in the downlink, LTE Release 8 SU-MIMO can satisfy the ITU-R requirements in the Indoor Hotspot (InH) and Rural Macrocell (RMa) deployment scenarios, but not in the Urban Microcell (UMi) and Urban Macrocell (UMa) scenarios.[2] On the other hand, LTE-Advanced MU-MIMO can satisfy the requirements in all the test environments and provides substantial gain over LTE Release 8 SU-MIMO.

The uplink performance was evaluated using LTE-Advanced SU-MIMO with two transmit and four receive antennas (2×4) and LTE Release 8 single-antenna transmission with four receive antennas (1×4).

[2]See Section 20.3.6.1 for details of the ITU-R test environments.

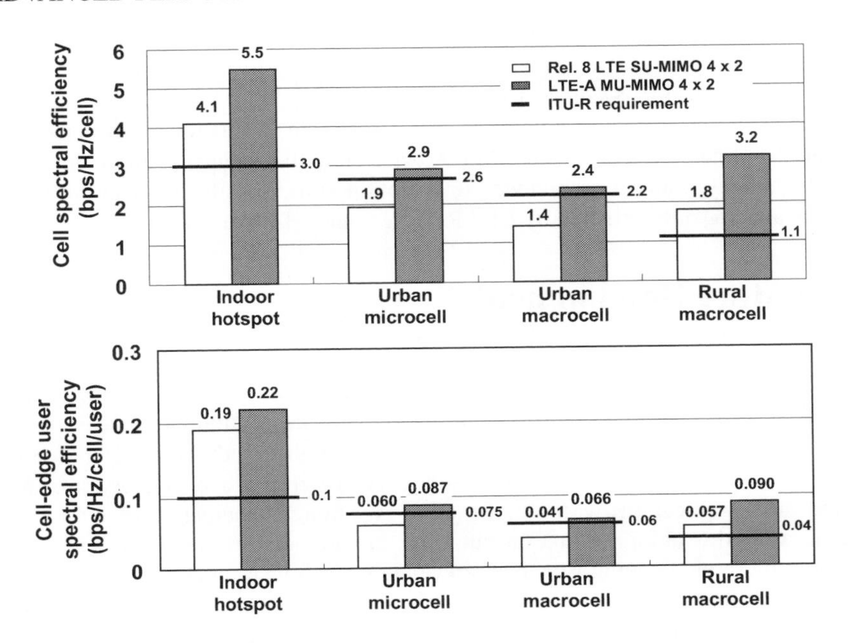

Figure 32.1: Downlink performance of LTE-Advanced in ITU-R deployment scenarios.

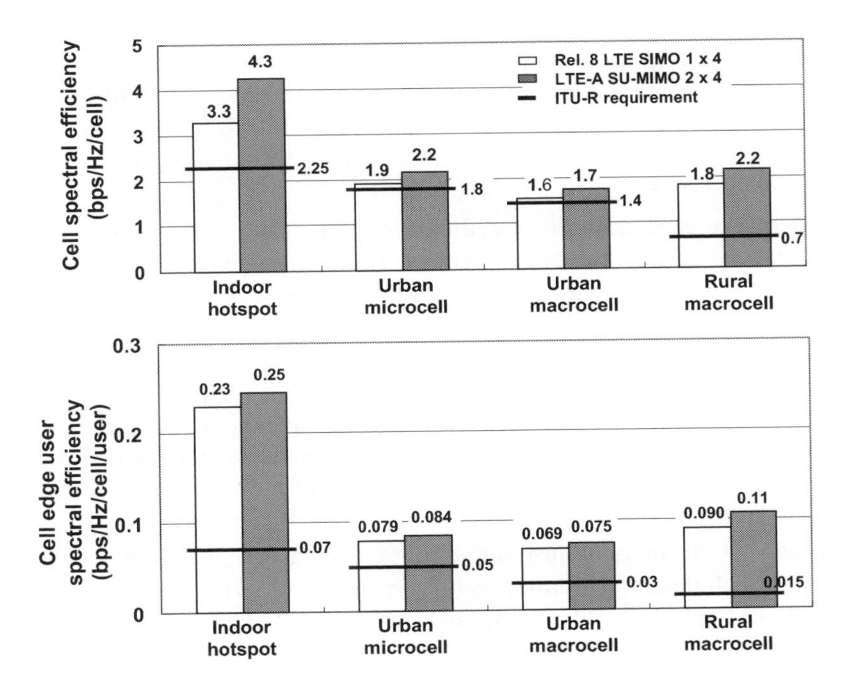

Figure 32.2: Uplink performance of LTE-Advanced in ITU-R deployment scenarios.

Figure 32.2 shows that in the uplink, LTE Release 8 single-antenna transmission satisfies the requirements in all of the environments. Additionally, LTE-Advanced SU-MIMO provides even further improved spectral efficiency.

In addition to these spectral efficiency evaluations, the evaluations conducted by 3GPP show that the other aspects of the IMT-Advanced requirements, namely in relation to bandwidth, latency, mobility, handover interruption time and the number of VoIP users accommodated, are fully satisfied by LTE Release 8 and LTE-Advanced [1].

32.2 Future Developments

Compared to previous mobile radio systems such as those based on HSPA, the achievable system performance is greatly improved by LTE Releases 8 and 9 followed by LTE-Advanced. Nevertheless, expectations from the market side continue to increase inexorably. In recent years, unprecedented market trends have been observed: the proliferation of high-specification terminals, especially smartphones, is bringing video delivery/streaming and other innovative applications within the reach of ever more consumers. In the future, mobile data traffic will grow at a pace that far outstrips that previously experienced. Meanwhile, it is becoming more challenging to achieve revenue growth as a result of the introduction of flat-rate data tariffs by many network operators around the world. Further reduction of the cost per bit will be a necessity in the future more than ever before in the history of mobile communication technology.

From another perspective, ecological concerns are steadily attracting more attention. Mobile communications needs innovative solutions to decrease the power consumed.

A further global concern is the so-called ‘digital divide’ – the regional differences in service availability, for example between rural and urban areas. In the future, comparable mobile services need to be provided for all areas in a cost-effective way.

Following these global trends, the main requirements for future releases of LTE can be broadly identified as follows:

1. **Increased capacity and spectral efficiency.** It is necessary to improve system capacity in order to support the tremendous growth of data traffic. In particular, further improvement of spectrum utilization efficiency is very important, since the available spectrum bands that are suitable for mobile communications are limited.
2. **Improvement of throughput experienced by the user.** For video delivery/streaming and new content-rich applications, the throughput actually experienced by the individual user is paramount and needs further improvement, not just the system capacity. In indoor scenarios, non-voice traffic is becoming dominant, and this will become even more obvious in the future. Further optimization of the system specifications for the indoor environment is therefore important.
3. **Fairness of throughput provision.** HSPA system design is primarily best-effort-based, which results in a throughput gap between cell-centre and cell-edge users. In both LTE and LTE-Advanced, potential solutions to reduce the throughput gap have been studied. To further improve fairness among users in future systems, improvements of cell-edge performance are required without sacrificing cell-centre performance. Fairness between cells and different geographical regions is also important from the perspective of the digital divide.

4. **Reduction of cost per bit.** Growth in data traffic is out-pacing growth in revenues. Besides, further network expansion and performance improvement in rural areas are key to addressing the digital divide.
5. **Energy saving.** Solutions to reduce energy consumption need to be considered at both system and device levels, in order to address both environmental and operational costs.

These requirements will need to be considered by future releases of the LTE specifications. Prime candidates for near-term system improvements of LTE in Release 11 include features that were partially studied but not specified within the timeframe of Release 10. In addition, further enhancements of features that were specified in Release 10 are potential aspects for consideration. Some of the main topics to be addressed in the short-term are therefore as follows:

- **Coordinated Multi-Point Transmission/Reception (CoMP).** As discussed in Section 29.5, some initial studies were performed into coordinated scheduling, coordinated beamforming and joint transmission techniques during the feasibility study for LTE-Advanced. Intra-eNodeB CoMP is already possible in LTE in a manner independent of standardization, but standardized inter-eNodeB coordination is not specified. Further study of the feasibility and performance gains of CoMP during the Release 11 timeframe is expected to result in a clear understanding of these aspects and of the potential use cases and deployment scenarios.
- **Downlink and uplink MIMO enhancement.** As MIMO deployments become more common, additional features for codebook and feedback enhancements, and higher-order MIMO are likely areas for improved performance to be achieved. MIMO performance in new deployment scenarios, including small cells, will also be studied with a view to identifying whether enhancements can deliver useful performance improvements in such conditions.
- **Carrier aggregation enhancements.** As spectrum availability changes for different network operators around the world, new combinations of adjacent and non-adjacent carriers and bands will continue to become relevant. RF requirements will be developed for such combinations and typically introduced in a release-independent manner so that user equipment of any release may support them; this will impose ever-increasing demands on equipment hardware capabilities and complexity. In addition, multiple timing advance controls are likely to be introduced for carrier aggregation cases where the end-to-end propagation delay on one aggregated carrier is significantly different from that on another, for example where one carrier makes use of repeaters. Control signalling improvements may also be considered to reduce overhead, and enhanced transmit diversity for the uplink HARQ acknowledgement feedback is expected to be introduced. Additional types of component carrier that are not backward-compatible (e.g. without cell-specific reference signals or synchronization signals) may also be considered.
- **Inter-Cell Interference Coordination (ICIC) enhancement.** ICIC has been supported since the beginning of LTE and was enhanced in Release 10 (see Section 31.2). In future releases, further improvements are expected. In particular, new features for the X2 interface may be further investigated. Current market conditions set high

expectations for low-cost dense networks and heterogeneous networks, including deployment of both femtocells and picocells. Such network deployments have the potential to provide attractive solutions to enhance system performance with little impact on the radio interface and terminals, and enhanced ICIC will play an important role in achieving this.

- **Self-Optimizing Networks (SON) and Minimzation of Drive Tests (MDT).** SON was initially specified in Release 8. It enables the cost of network deployment and maintenance to be reduced, especially for dense networks. Further enhancements are to be investigated in future releases as reduction of the cost of network optimization is an increasingly important part of minimizing the cost per bit. Reducing the number of drive tests that have to be conducted is another effective way to enable operators to reduce the network cost; some MDT features are included in the Release 10 specifications, and more will be investigated for further practical use cases, with consideration being paid to the tradeoffs between benefits and network complexity.
- **Energy saving.** Energy saving is important for base stations in particular. Reduction of power consumption of equipment is essentially an implementation matter, but the possibility of designing efficient implementations should be taken into account when specifying any new features for the radio interface.
- **Machine-Type Communications.** As discussed in Section 31.4, Machine-Type Communications (MTC) are proliferating in cellular networks for numerous applications. Such communications have certain particular characteristics, typically including low mobility, low frequency of use, a need for long battery life, and potentially a very large number of terminals. Future releases of LTE may contain specific radio access network features to provide optimized support for communications with these characteristics, if such features are identified.

Besides the above aspects, further improvement of features already included in Releases 8, 9 and 10, such as MBMS, UE location technologies and relaying (for example relay backhaul enhancements or improved support for mobile relays on high-speed trains) are likely to be considered for future releases of LTE.

Market requirements will continue to evolve; thus mid- and long-term system improvements of LTE will need to keep pace with market growth. For long-term improvements, the most useful developments will be brand-new features that deliver very substantial benefits. Potential candidates may include features related to new multiple access schemes and advanced multi-antenna solutions for the physical layer, features for new architectures and network deployment scenarios, and new performance specifications for the mobile terminal with more advanced receivers.

References[3]

[1] 3GPP Technical Report 36.912, ‘Feasibility Study for Further Advancements for E-UTRA (LTE-Advanced) (Release 9)’, www.3gpp.org.

[3]All web sites confirmed 1st March 2011.

Index

LTE – The UMTS Long Term Evolution: From Theory to Practice, Second Edition.
Stefania Sesia, Issam Toufik and Matthew Baker.